1624
Drebble, Incubator

1728
Watt, Flyball governor

1868
Maxwell, Flyball stability analysis

1877
Routh, Stability

1890
Liapunov, Nonlinear stability

1910
Sperry, Gyroscope and autopilot

1927
Black, Feedback electronic amplifier: Bush, Differential analyzer

1932
Nyquist, Nyquist stability criterion

1938
Bode, Frequency response methods

1942
Wiener, Optimal filter design
Ziegler-Nichols PID tuning

1947
Hurewicz, Sampled data systems; Nichols, Nichols chart

1948
Evans, Root locus

1950
Kochenberger, Nonlinear analysis

1956
Pontryagin, Maximum principle

1957
Bellman, Dynamic programming

1960
Draper, Inertial navigation; Kalman, Optimal estimation

1969
Hoff, Microprocessor

Feedback Control of Dynamic Systems

Fourth Edition

Gene F. Franklin
Stanford University

J. David Powell
Stanford University

Abbas Emami-Naeini
SC Solutions, Inc.

Prentice Hall
Upper Saddle River, New Jersey 07458

Library of Congress Cataloging-in-Publication Data

Franklin, Gene F.
    Feedback control of dynamic systems / Gene F. Franklin,
      J. David Powell, Abbas Emami-Naeini.—4th ed.
    p.   cm.
    Includes index.
    ISBN 0-13-032393-4
1. Feedback control systems.   I. Powell, J. David.   II. Emami-Naeini, Abbas.   III. Title.
CIP Data available.

Vice President and Editorial Director, ECS: *Marcia J. Horton*
Acquisitions Editor: *Eric Frank*
Editorial Assistant: *Jessica Romeo*
Vice President and Director of Production and Manufacturing, ESM: *David W. Riccardi*
Executive Managing Editor: *Vince O'Brien*
Managing Editor: *David A. George*
Production Editor: *Irwin Zucker*
Composition: *PreTEX, Inc.*
Director of Creative Services: *Paul Belfanti*
Creative Director: *Carole Anson*
Art Director: *Jonathan Boylan*
Assistant to the Art Director: *John Christiana*
Art Editor: *Xiaohong Zhu*
Interior Designer: *Maria Guglielmo*
Cover Designer: *Stacey Abraham*
Manufacturing Manager: *Trudy Pisciotti*
Manufacturing Buyer: *Lisa McDowell*
Marketing Manager: *Holly Stark*

About the Cover: *Photograph of computer hard drive courtesy of Corbis Images. Photograph of satellite orbiting the earth courtesy of NASA. Photograph of Boeing 747 jet aircraft courtesy of Tony Stone Images.*

© 2002, 1994, 1991, 1986 by Prentice Hall
Prentice-Hall, Inc.
Upper Saddle River, New Jersey 07458

The author and publisher of this book have used their best efforts in preparing this book. These efforts include the development, research, and testing of the theories and programs to determine their effectiveness. The author and publisher make no warranty of any kind, expressed or implied, with regard to these programs or the documentation contained in this book. The author and publisher shall not be liable in any event for incidental or consequential damages in connection with, or arising out of, the furnishing, performance, or use of these programs.

MATLAB and Simulink are registered trademarks of The MathWorks, Inc., 3 Apple Hill Drive, Natick, MA, 01760-2098.

Printed in the United States of America

10 9 8 7 6 5 4 3 2 1

ISBN 0-13-032393-4

Pearson Education Ltd., *London*
Pearson Education Australia Pty. Limited, *Sydney*
Pearson Education Singapore, Pte. Ltd.
Pearson Education North Asia Ltd., *Hong Kong*
Pearson Education Canada Inc., *Toronto*
Pearson Educatíon de Mexico, S.A. de C.V.
Pearson Education—Japan, *Tokyo*
Pearson Education Malaysia, Pte. Ltd.
Pearson Education, *Upper Saddle River, New Jersey*

To Gertrude, David, Carole
Valerie, Daisy, Annika, Davenport
Malahat, Sheila, Nima

Contents

Preface

In this fourth edition we again had the objectives of retaining the best of the previous editions, to rewrite key sections where we felt it was possible to improve the presentations and enhance the book's pedagogical effectiveness, and to take better advantage of the wide use of computers in control design, especially the toolboxes of MATLAB and Simulink, from The Mathworks, Inc.

The basic structure of the book is unchanged and we continue to combine analysis with design using the three approaches of the root locus, frequency response, and state variable equations. The text continues to include carefully worked out examples, many of them new to this edition, to illustrate the material. As a new feature, to assist the students in verifying that they have learned the material, we provide a set of review questions at the end of each chapter with answers in the back of the book. While modest changes were made throughout the entire book, special attention was given to the introduction of transforms in Chapter 3, to the introduction to feedback in Chapter 4, and to the organization and statements of the problems appearing at the end of each chapter.

In the three central chapters on the design methods, we continue to expect the students to learn how to perform the basic calculations by hand in order to be able to guide a design by understanding (and frequently by a quick sketch) rather than by computer rote. However, more than in previous editions, we de-emphasize the manual work and introduce computer tools early on in recognition of the universal use of these tools in control analysis and design. For example, we no longer mark certain problems as requiring a computer but, rather, expect that the student has access to a computer in every case, as needed.

Furthermore, in recognition of the fact that, increasingly, controllers are implemented in embedded computers, we introduce digital control in Chapter 4 and in a number of cases compare the responses of feedback systems using analog controllers with those having a digital "equivalent" controller. As before, we have prepared a collection of all the MATLAB ".m" files used to produce the figures in the book and these are available at the companion web site for this title:

http://www.prenhall.com/franklin

or at the homepage for SC Solutions, Inc.:

http://scsolutions.com/scsolutions.control.html

As representative applications of control, we again present extensive case studies in Chapter 9. In this edition we have added new studies of the control of the read-write head assembly of a computer hard disk and the temperature control of a silicon wafer in a Rapid Thermal Processor used in the fabrication of integrated circuits.

We feel that this fourth edition presents the material with good pedagogical support, provides strong motivation for the study of control, and represents a solid foundation for meeting the educational challenges of a study of feedback control.

Addressing the Educational Challenges

Some of the educational challenges facing students of feedback control are long-standing; others have emerged in recent years. Some of the challenges remain for students across their entire engineering education; others are unique to this relatively sophisticated course. Whether they are old or new, general or particular, the educational challenges we perceived were critical to the evolution of this text. Here we will state several educational challenges and describe our approaches to each of them.

- CHALLENGE: *Students must master design as well as analysis techniques.*

Design is central to all of engineering and especially to control systems. Students find that design issues, with their corresponding opportunities to tackle practical applications, particularly motivating. But students also find design problems difficult because design problem statements are usually poorly posed and lack unique solutions. Because of both its inherent importance for and its motivational effect on students, design is emphasized throughout this text so that confidence in solving design problems is developed from the start.

The emphasis on design begins in Chapter 4, following the development of modeling and dynamic response. The basic idea of feedback is introduced first, showing its influence on disturbance rejection, tracking accuracy, and robustness to parameter changes. The design orientation continues with uniform treatments of the root locus, frequency response, and state variable feedback techniques. All of the treatments are aimed at providing the knowledge necessary to find a good feedback control design with no more complex mathematical development than is essential to clear understanding.

Throughout the text, examples are used to compare and contrast the design techniques afforded by the different design methods and, in the capstone case studies of Chapter 9, complex real-world design problems are tackled using all of the methods in a unified way.

- CHALLENGE: *New ideas continue to be introduced into control.*

Control is an active field of research and hence there is a steady influx of new concepts, ideas, and techniques. In time, some of these elements develop to the point where they join the list of things every control engineer must know. This text is devoted to supporting students equally in their need to grasp both traditional and more modern topics.

In each of our previous editions we have tried to give equal time to root locus, frequency response, and state variable methods for design. In this edition we have shifted the emphasis from manual design methods augmented with computer tools to an emphasis on computer-aided methods augmented with a solid mastery of the underlying techniques. Included in this re-emphasis is the early introduction of sampling, which enables one to design digital controllers. While this material can be skipped to save time without disruption of the flow of the text, we feel that it is very important for students to recognize that digital control is being used increasingly and that the most basic techniques of digital control are easily mastered.

With regret we acknowledge that we are not able at this time to introduce the important topics of hybrid control or designs based on various optimization methods.

- CHALLENGE: *Students need to manage a great deal of information.*

The vast array of systems to which feedback control is applied and the growing variety of techniques available for the solution of control problems means that today's student of feedback control must learn many new ideas. How do students keep their perspective as they plow through lengthy and complex textual passages? How do they identify highlights and draw conclusions? How do they review for exams? Helping students with these tasks was a criterion for the fourth edition. We outline these features in the accompanying table on page xiv.

- CHALLENGE: *Students of feedback control come from a wide range of disciplines.*

Feedback control is an interdisciplinary field in that control is applied to systems in every conceivable area of engineering. Consequently, some schools have separate introductory courses for control within the standard disciplines and some, such as Stanford University, have a single set of courses taken by students from many disciplines. However, to restrict the examples to one field is to miss much of the range and power of feedback; but to cover the whole range of applications is overwhelming. In this book we develop the interdisciplinary nature of the field and provide review material for several of the most common technologies so that students from many disciplines will be comfortable with the presentation. For electrical engineering students who typically have a good background in transform analysis, we include an introduction to writing equations of motion for mechanical mechanisms in Chapter 2. For mechanical engineers, we include in Chapter 3 a review of the Laplace Transform and dynamic response as needed in control. In addition, we introduce other technologies briefly and, from time to time, we present the equations of motion of a physical system without derivation but with enough physical description to

FEATURE	REFERENCE EXAMPLE
Chapter openers offer perspective and overview. They place the specific chapter topic in the context of the discipline as a whole and they briefly overview the chapter sections.	Chapter 3 opener, pp. 94–95
Margin notes help students scan for chapter highlights. They point to important definitions, equations, and concepts.	pp. 49–50
Boxed highlights identify key concepts within the running text. They also function to summarize important design procedures.	Advantage of feedback, p. 206; compensation design, p. 440
Bulleted chapter summaries help with student review and prioritization. These summaries briefly reiterate the key concepts and conclusions of the chapter.	Chapter 2 summary, pp. 77–78
Synopsis of design aids. Relationships used in design and throughout the book are collected in one place for easy reference.	Inside back cover
The *color blue* is used (1) to highlight useful; pedagogical features; (2) to highlight components under particular scrutiny within block diagrams; (3) to distinguish curves on graphs; and (4) to lend a more realistic look to figures of physical systems.	Fig. 5.43, p. 330 Fig. 2.9, p. 32
Review questions at the end of each chapter with solutions in the back guide the student in self-study.	Chapter 2, p. 78

be understood from a response point of view. Examples of some of the physical systems represented in the text include the read-write head for a computer disk drive, a satellite tracking system, the fuel-air ratio in an automobile engine, and an airplane autopilot system.

Outline of the Book

The contents of the book is organized into nine chapters and seven appendixes. The chapters include some sections of advanced or enrichment material marked with a triangular blue icon that can be omitted without interfering with the flow of the material. Examples and problems based on this material are also marked with these icons. The appendixes include background and reference material such as Laplace transform tables, a review of complex variables, a review of matrix theory, and answers to the end-of-chapter review questions.

In Chapter 1, the essential ideas of feedback and some of the key design issues are introduced. The chapter also contains a brief history of control, from the ancient beginnings of process control to the contributions of flight control and electronic feedback amplifiers. It is hoped that this brief history will give a context for the field, introduce some of the key figures who contributed to its development, and provide motivation to the student for the studies to come.

Chapter 2 is a short presentation of dynamic modeling and includes mechanical, electrical, electro-mechanical, fluid, and thermodynamic devices. It also discusses the state variable formulation of differential equations. This material can be omitted, used as the basis for review homework to smooth out the usual non-uniform preparation of students, or covered in depth.

Chapter 3 covers dynamic response as used in control. Again, much of this material may have been covered previously, especially by electrical engineering students. For many students, the correlation between pole locations and transient response and the effects of extra zeros and poles on dynamic response is new material, as is the notion of stability of a closed-loop system. This material needs to be covered carefully.

Chapter 4 introduces feedback in the most elementary context, permitting concentration on the essential effects of feedback on tracking accuracy, disturbance rejection, and sensitivity to model errors. The basic equation and transfer functions of feedback are introduced along with the definitions of the sensitivity and complementary sensitivity functions. In the context of a first-order model for speed control, the concepts of proportional, integral, and derivative (PID) control are introduced. In this way, the student gets the idea of what control is all about before the tedious rules of root locus or the Nyquist Stability Criterion are developed. Finally, in this chapter the basic issues of digital control are introduced, along with the idea of a digital equivalent controller. In this approach, the central issues of control design are brought forward and can remain in the foreground during the development of the necessary analysis that goes with construction of sophisticated design tools. The concepts of steady-state tracking error and system type are also treated here.

Following the overview of feedback, the core of the book presents the design methods based on root locus, frequency response, and state variable feedback in Chapters 5, 6, and 7, respectively.

Chapter 8 develops in more detail the tools needed to design feedback control for implementation in a digital computer. However, for a complete treatment of feedback control using digital computers, the reader is referred to the companion text, *Digital Control of Dynamic Systems*, by Franklin, Powell, and Workman (Prentice Hall, 1998).

In Chapter 9, the three primary approaches are integrated in several case studies and a framework for design is described that includes a touch of the real-world context of practical control design.

Course Configurations

The material in this text can be covered flexibly. Most first-course students in controls will have some background in dynamics and Laplace transforms.

Therefore, Chapter 2 and most of Chapter 3 would be a review for those students. In a 10-week quarter, it is possible to review Chapter 3, and cover all of Chapters 1, 4, 5, and 6. Most optional sections noted with a blue triangle should be omitted. In the second quarter, Chapters 7 and 9 can be covered comfortably including these optional sections. Alternatively, some optional sections could be omitted and selected portions of Chapter 8 included. A semester course should comfortably accommodate Chapters 1–7, including the review material of Chapters 2 and 3, if needed. If time remains after this core coverage, selected case studies from Chapter 9 or some introduction of digital control from Chapter 8 may be added.

The entire book can also be used for a three-quarter sequence of courses consisting of modeling and dynamic response (Chapters 2 and 3), classical control (Chapters 4–6), and modern control (Chapters 7–9).

Two basic 10-week courses are offered at Stanford and are taken by seniors and first-year graduate students who have not had a course in control, mostly in the Departments of Aeronautics and Astronautics, Mechanical Engineering, and Electrical Engineering. The first course reviews Chapters 2 and 3 and covers Chapters 4–6. The more advanced course is intended for graduate students and reviews Chapters 4–6 and covers Chapters 7–9. This sequence complements a graduate course in linear systems and is the prerequisite to courses in digital control, optimal control, flight control, and smart product design. Several of the subsequent courses include extensive laboratory experiments. Prerequisites for the course sequence include dynamics or circuit analysis and Laplace transforms.

Prerequisites to this Feedback Control Course

This book is for a first course at the senior level for all engineering majors. For the core topics in Chapters 4–7, prerequisite understanding of modeling and dynamic response is necessary. Many students will come into the course with sufficient background in those concepts from previous courses in physics, circuits, and dynamic response. For those needing review, Chapters 2 and 3 should fill in the gaps.

An elementary understanding of matrix algebra is necessary to understand the state-space material. While all students will have much of this in prerequisite math courses, a review of the basic relations is given in Appendix C and a brief treatment of particular material needed in control is given at the start of Chapter 7. The emphasis is on the relations between linear dynamic systems and linear algebra.

Supplements

An Instructor's Manual with complete solutions to homework problems is available to faculty who adopt the fourth edition. The web sites mentioned above include the .m files used to generate all of the MATLAB figures in the book.

Acknowledgments

Finally, we wish to acknowledge our great debt to all those who have contributed to the development of feedback control into the exciting field it is today and specifically to the considerable help and education we have received from our students and our colleagues. In particular, we have benefited in this effort by many discussions with the following individuals, who have taught introductory control at Stanford: A. E. Bryson, Jr., R. H. Cannon, Jr., D. B. DeBra, S. Rock, C. Tomlin, S. Boyd, and P. Enge. Special thanks go to Prof. Dan DeBra for his careful reading of the manuscript and many useful comments. In addition, our colleagues M. Spong, L. Pao, D. Meyer, K. Pasino, P. Dorato, and M. Saif have provided valuable feedback. We also appreciate the help of D. de Roover, G. van der Linden, J. Ebert, R. Kosut, M. Tao, and A. Rahimi. Furthermore, L. Kobayashi, H-T. Lee, E. Thuriyasena, and M. Matsuoka provided valuable help in proofreading the chapters as well as in the preparation of the solutions manual.

We would also like to thank the following reviewers: Hy D. Tran at the University of New Mexico; Paul I. Ro at North Carolina State University, Lucy Y. Pao at University of Colorado, Arnold Lee Swindlehurst at Brigham Young University, and E. Harry Law at Clemson University.

GENE F. FRANKLIN
J. DAVID POWELL
ABBAS EMAMI-NAEINI
Stanford, California

1 An Overview and Brief History of Feedback Control

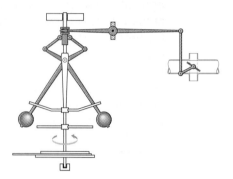

Chapter Overview

In this chapter we begin our exploration of feedback control using a simple familiar example: a household furnace controlled by a thermostat. The generic components of a control system are identified within the context of this example. In another example—an automobile cruise control—we develop the elementary static equations and assign numerical values to elements of the system model in order to compare the performance of open-loop control to that of feedback control when dynamics are ignored. In order to provide a context for our studies and to give you a glimpse of how the field has evolved, Section 1.3 provides a brief history of control theory and design. Finally, Section 1.4 provides a brief overview of the contents and organization of the entire book.

A Perspective on Feedback Control

Control of dynamic systems is a very common concept with many characteristics. A system that involves a person controlling a machine, as in driving an automobile, is called **manual control**. A system that involves machines only, as when room temperature can be set by a thermostat, is called **automatic control**. Systems designed to hold an output steady against unknown disturbances are called **regulators**, while systems designed to track a reference signal are called **tracking** or **servo** systems. Control systems are also classified according to the information used to compute the controlling action. If the controller does *not* use a measure of the system output being controlled in computing the control action to take, the system is called **open-loop control**. If the controlled output signal *is* measured and fed back for use in the control computation, the system is called closed-loop or **feedback control**. There are many other important properties of control systems in addition to these most basic characteristics. For example, in this book we will mainly be concerned with controlling processes that can be adequately described by **linear, time invariant** equations, whereas all physical processes are nonlinear if the signals are large and their characteristics vary with time if observed for a long time. We will also mainly consider feedback of the present output only, but a very familiar example illustrates the limitation imposed by that assumption. When driving a car, the use of simple feedback corresponds to driving in a thick fog where one can *only see the road immediately at the front of the car* and is unable to see the future required position! Looking at the road ahead is a form of predictive control. This information, which has obvious advantages, would always be used where it is available; but in most automatic control situations studied in this book, observation of the future track or disturbance is not possible. In any case the control designer should study the process to see if any sensor could anticipate either a track to be followed or a disturbance to be rejected. If such a possibility is feasible, the control designer should use it to **feed forward** an early warning to the control system. An example of this is in the control of steam pressure in the boiler of an electric power generation plant. A measure of the *electric* power demand at the output of the plant can be fed forward to the boiler controller in anticipation of a soon-to-be-demanded increase in steam flow.

Open-loop control

Feedback control

The evolution of cheap and powerful digital computers has had a major impact on control design and control implementation. Software such as MATLAB is a great aid to solving the equations and realizing the graphics of control design methods. For analyzing system response, students of control have tools such as Simulink which can easily compute the response of linear as well as nonlinear models of processes and controls. The controllers which compute the signals necessary to effect control are mainly electronic units because of the flexibility and cost-effectiveness of these devices. While analog units are typically faster and cheaper to implement simple equations than digital logic, the greater programming flexibility and increasing cost-effectiveness of embedded digital processors is causing them to become ever more common in controller implementation. The influence of these trends on our introduction to the stimulating field of control is evident throughout the text.

The applications of feedback control have never been more exciting than they are today. Landing and collision avoidance systems using the Global Positioning System (GPS) are now under development and promise a revolution in our ability to navigate in an ever more crowded airspace. In the magnetic data storage devices for computers known as hard disks, control of the read/write head assembly is often designed to have tracking errors on the order of microns and to move at speeds of a fraction of a millisecond. Control is essential to the operation of systems from cell phones to jumbo jets and from washing machines to oil refineries as large as a small city. The list goes on and on. In fact, many engineers refer to control as a *hidden technology* because of its essential importance to so many devices and systems while being mainly out of sight. The future will no doubt see engineers create even more imaginative applications of feedback control. Study of control problems over the past 200 years has led to an extensive body of knowledge common to both manual and automatic control which has evolved into the discipline of control systems design, the subject of this book.

1.1 A Simple Feedback System

In feedback systems the variable being controlled—such as temperature or speed—is measured by a sensor, and the measured information is fed back to the controller to influence the controlled variable. The principle is readily illustrated by a very common system, the household furnace controlled by a thermostat. The components of this system and their interconnections are shown in Fig. 1.1(a). Such a picture identifies the major parts of the system and shows the directions of information flow from one component to another.

We can easily analyze the operation of this system qualitatively from the graph. Suppose both the temperature in the room where the thermostat is located and the outside temperature are significantly below the reference temperature (also called the set point) when power is applied. The thermostat will be *on*, and the control logic will open the furnace gas valve and light the fire

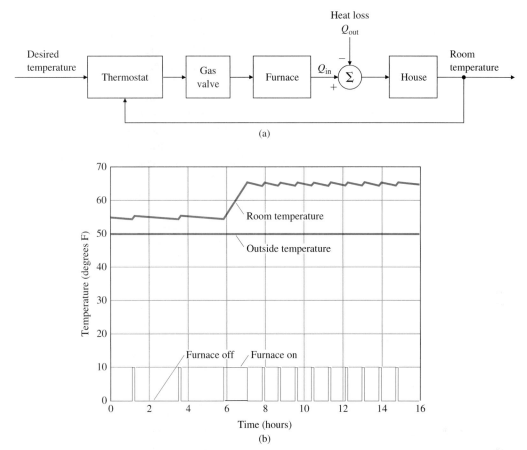

Figure 1.1 (a) Component block diagram of a room temperature control system (b) Plot of room temperature and furnace action

box. This will cause heat Q_{in} to be supplied to the house at a rate that will be significantly larger than the heat loss Q_{out}. As a result, the room temperature will rise until it exceeds the thermostat reference setting by a small amount. At this time the furnace will be turned off and the room temperature will start to fall toward the outside value. When it falls a small amount below the set point, the thermostat will come on again and the cycle will repeat. Typical plots of room temperature along with the furnace cycles of on and off are shown in Fig. 1.1. The outside temperature is held at 50°F and the thermostat is initially set at 55°F. At 6 a.m., the thermostat is stepped to 65°F and the furnace brings it to that level and cycles the temperature around that figure thereafter.[1] Notice that the house is well-insulated so that the fall of temperature with the furnace

---

[1] Notice that the furnace had come on a few minutes before 6 a.m. on its regular nighttime schedule.

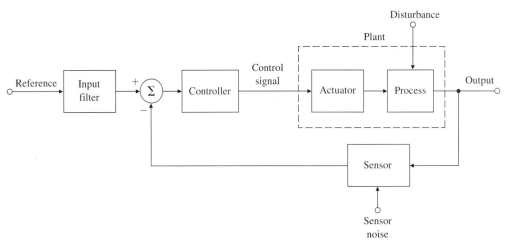

Figure 1.2 Component block diagram of an elementary feedback control

off is significantly slower than the rise with the furnace on. From this example we can identify the generic components of the elementary feedback control system as shown in Fig. 1.2.

The central component of this feedback system is the **process** whose output is to be controlled. In our example the process would be the house whose output is the room temperature, and the **disturbance** to the process is the flow of heat from the house due to conduction through the walls and roof to the lower outside temperature. (The outward flow of heat also depends on other factors such as wind, open doors, etc.) The design of the process can obviously have a major impact on the effectiveness of the controls. The temperature of a well-insulated house with thermopane windows is clearly easier to control than otherwise. Similarly, the design of aircraft with control in mind makes a world of difference to the final performance. In every case, the earlier the issues of control are introduced into the process design, the better. The **actuator** is the device that can influence the controlled variable of the process, and in our case the actuator is a gas furnace. Actually, the furnace usually has a pilot light or striking mechanism, a gas valve, and a blower fan which turns on or off depending on the air temperature in the furnace. These details illustrate the fact that many feedback systems contain components that themselves form other feedback systems.[2] The central issue with the actuator is its ability to move the process output with adequate speed and range. The furnace must produce more heat than the house loses on the worst day and must distribute it quickly if the house temperature is to be kept in a narrow range. Power, speed, and reliability are usually more important than accuracy. Generally, the process

---

[2] Jonathan Swift (1733) said it this way: "So, Naturalists observe, a flea Hath smaller fleas that on him prey; And these have smaller still to bite 'em; And so proceed, *ad infinitum*."

and the actuator are intimately connected and the control design centers on finding a suitable input or control signal to send to the actuator. The combination of process and actuator is called the **plant**, and the component that actually computes the desired control signal is the **controller**. Because of the flexibility of electrical signal processing, the controller typically works on electrical signals, although the use of pneumatic controllers based on compressed air has a long and important place in process control. With the development of digital technology, cost effectiveness and flexibility have led to the use of digital signal processors as the controller in an increasing number of cases. The component labeled **thermostat** in Fig. 1.1 measures the room temperature and is called the **sensor** in Fig. 1.2, a device whose output inevitably contains sensor noise. Sensor selection and placement are very important in control design because it is sometimes not possible for the true controlled variable and the sensed variable to be the same. For example, although we may really wish to control the house temperature as a whole, the thermostat is in one particular room, which may or may not be at the same temperature as the rest of the house. For instance, if the thermostat is set to 68°F but is placed in the living room near a roaring fireplace, a person working in the study could still feel uncomfortably cold.[3,4]

As we will see in addition to placement, important properties of a sensor are the accuracy of the measurements as well as low noise, reliability, and linearity. The sensor will typically convert the physical variable into an electrical signal for use by the controller. Our general system also includes an **input shaping filter** whose role is to convert the reference signal to electrical form for later manipulation by the controller. In some cases the input shaping filter can modify or shape the reference command input in ways that improve the system response. Finally, there is a **comparator** to compute the difference between the reference signal and the sensor output to give the controller a measure of the system error.

This text will present methods for analyzing feedback control systems and their components and will describe the most important design techniques engineers can use with confidence in applying feedback to solve control problems. We will also study the specific advantages of feedback that compensate for the additional complexity it demands. However, although the temperature control system is easy to understand, it is nonlinear as seen by the fact that the furnace is either on or off and to introduce linear controls we need another example.

---

[3] In the renovations of the kitchen in the house of one of the authors the new ovens were placed against the wall where the thermostat was mounted on the other side. Now when dinner is baked in the kitchen on a cold day, the author freezes in his study unless the thermostat is reset.

[4] The story is told of the new employee at the nitroglycerin factory who was to control the temperature of a critical part of the process manually. He was told to "keep that reading below 300°." On a routine inspection tour, the supervisor realized that the batch was dangerously hot and found the worker holding the thermometer under the cold water tap to bring it down to 300°. They got out just before the explosion. Moral: Sometimes automatic control is better than manual.

Figure 1.3
Component block diagram
of automobile cruise control

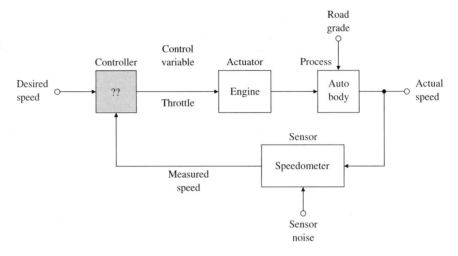

1.2   A First Analysis of Feedback

The value of feedback can be readily demonstrated by quantitative analysis of a simplified model of a familiar system, the cruise control of an automobile (Fig. 1.3). To study this situation analytically, we need a mathematical **model** of our system in the form of a set of quantitative relationships among the variables. For this example we ignore the dynamic response of the car and consider only the steady behavior. (Dynamics will of course play a major role in later chapters.) Furthermore, we assume that for the range of speeds to be used by the system, we can approximate the relations as linear. After measuring the speed of the vehicle on a level road at 65 mph, we find that a 1° change in the throttle angle (our control variable) causes a 10-mph change in speed. From observations while driving up and down hills it is found that when the grade changes by 1%, we measure a speed change of 5 mph. The speedometer is found to be accurate to a fraction of 1 mph and will be considered exact. With these relations, we can draw the **block diagram** of the plant (Fig. 1.4), which shows these mathematical relationships in graphical form. In this diagram the connecting lines carry signals and a block is like an ideal amplifier which

Figure 1.4
Block diagram of the cruise
control plant

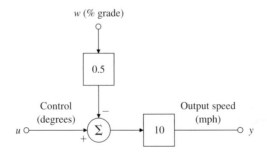

multiplies the signal at its input by the value marked in the block to give the output signal. To sum two or more signals, we show lines for the signals coming into a summer, a circle with the summation sign Σ inside. An algebraic sign (plus or minus) beside each arrow head indicates whether the input adds to or subtracts from the total output of the summer. For this analysis, we wish to compare the effects of a 1% grade on the output speed when the reference speed is set for 65 with and without feedback to the controller.

In the first case shown in Fig. 1.5 the controller does not use the speedometer reading but sets $u = r/10$. This is an example of an **open-loop control system**. The term *open-loop* refers to the fact that there is no closed path or loop around which the signals go in the block diagram. In our simple example the open-loop output speed, y_{ol}, is given by the equations

Open-loop control

$$y_{ol} = 10(u - 0.5w)$$

$$= 10\left(\frac{r}{10} - 0.5w\right)$$

$$= r - 5w.$$

The error in output speed is

$$e_{ol} = r - y_{ol} \tag{1.1}$$

$$= 5w, \tag{1.2}$$

and the percent error is

$$\% \text{ error} = 500\frac{w}{r}. \tag{1.3}$$

If $r = 65$ and the road is level, then $w = 0$ and the speed will be 65 with no error. However, if $w = 1$ corresponding to a 1% grade, then the speed will be 60 and we have a 5-mph error, which is a 7.69% error in the speed. For a grade of 2%, the speed error would be 10 mph, which is an error of 15.38%, and so on. The example shows that there would be no error when $w = 0$, but this result depends on the controller gain being the exact inverse of the plant gain of 10. In practice, the plant gain is subject to change; and if it does change, errors are introduced by this means also. If there is an error in the plant gain in open-loop control, the percent speed error would be the same as the percent plant-gain error.

Figure 1.5
Open-loop cruise control

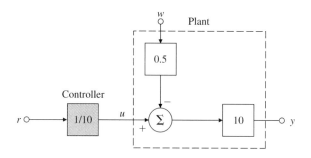

The block diagram of a feedback scheme is shown in Fig. 1.6, where the controller gain has been set to 10. Recall that in this simple example, we have assumed that we have an ideal sensor whose block is not shown. In this case the equations are

$$y_{cl} = 10u - 5w,$$

$$u = 10(r - y_{cl}).$$

Combining them yields

$$y_{cl} = 100r - 100y_{cl} - 5w,$$

$$101y_{cl} = 100r - 5w,$$

$$y_{cl} = \frac{100}{101}r - \frac{5}{101}w,$$

$$e_{cl} = \frac{r}{101} + \frac{5w}{101}.$$

Thus the feedback has reduced the sensitivity of the speed error to the grade by a factor of 101 when compared with the open-loop system. Note, however, that there is now a small speed error on level ground, because even when $w = 0$,

$$y_{cl} = \frac{100}{101}r = 0.99r \text{ mph}.$$

This error will be small as long as the loop gain (product of plant and controller gains) is large.[5] If we again consider a reference speed of 65 mph and compare

Figure 1.6
Closed-loop cruise control

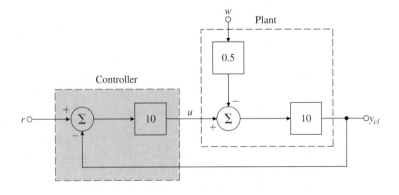

---

[5] In case the error is too large, it is common practice to *reset* the reference—in this case to $\frac{101}{100}r$—so the output reaches the true desired value.

speeds with a 1% grade, the percent error in the output speed is

$$\%error = 100 \frac{\dfrac{65 \times 100}{101} - \left(\dfrac{65 \times 100}{101} - \dfrac{5}{101} \right)}{\dfrac{65 \times 100}{101}} \tag{1.4}$$

$$= 100 \frac{5 \times 101}{101 \times 65 \times 100} \tag{1.5}$$

$$= 0.0769\%. \tag{1.6}$$

The reduction of the speed sensitivity to grade disturbances and plant gain in our example is due to the loop gain of 100 in the feedback case. Unfortunately, there are limits to how high this gain can be made; when dynamics are introduced, the feedback can make the response worse than before, or even cause the system to become unstable. The dilemma is illustrated by another familiar situation where it is easy to change a feedback gain. If one tries to raise the gain of a public-address amplifier too much, the sound system will squeal in a most unpleasant way. This is a situation where the gain in the feedback loop—from the speakers to the microphone through the amplifier back to the speakers—is too much. The issue of how to get the gain as large as possible to reduce the errors without making the system become unstable and squeal is what much of feedback control design is all about.

The design tradeoff

1.3   A Brief History

An interesting history of early work on feedback control has been written by O. Mayr (1970), who traces the control of mechanisms to antiquity. Two of the earliest examples are the control of flow rate to regulate a water clock and the control of liquid level in a wine vessel, which is thereby kept full regardless of how many cups are dipped from it. The control of fluid flow rate is reduced to the control of fluid level, since a small orifice will produce constant flow if the pressure is constant, which is the case if the level of the liquid above the orifice is constant. The mechanism of the liquid-level control invented in antiquity and still used today (for example, in the water tank of the ordinary flush toilet) is the **float valve**. As the liquid level falls, so does the float, allowing the flow into the tank to increase; as the level rises, the flow is reduced and, if necessary, cut off. Figure 1.7 shows how a float valve operates. Notice here that sensor and actuator are not separate devices but are, instead, contained in the carefully shaped float-and-supply-tube combination.

Liquid-level control

A more recent invention described by Mayr (1970) is a system, designed by Cornelis Drebbel in about 1620, to control the temperature of a furnace

Drebbel's incubator

Figure 1.7
Early historical control of
liquid level and flow

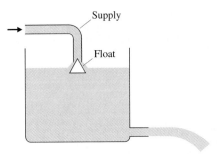

used to heat an incubator[6] (Fig. 1.8). The furnace consists of a box to contain
the fire, with a flue at the top fitted with a damper. Inside the fire box is the
double-walled incubator box, the hollow walls of which are filled with water
to transfer the heat evenly to the incubator. The temperature sensor is a glass
vessel filled with alcohol and mercury and placed in the water jacket around
the incubator box. As the fire heats the box and water, the alcohol expands
and the riser floats up, lowering the damper on the flue. If the box is too cold,
the alcohol contracts, the damper is opened, and the fire burns hotter. The
desired temperature is set by the length of the riser, which sets the opening of
the damper for a given expansion of the alcohol.

A famous problem in the chronicles of control systems was the search for
a means to control the rotation speed of a shaft. Much early work (Fuller,
1976) seems to have been motivated by the desire to automatically control
the speed of the grinding stone in a wind-driven flour mill. Of various meth-

Figure 1.8
Drebbel's incubator for
hatching chicken eggs.
(Adapted from Mayr, 1970)

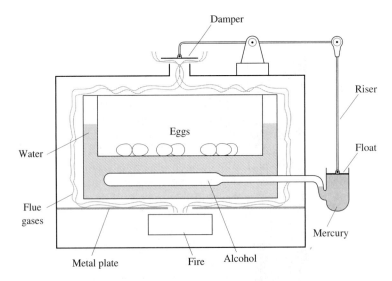

[6] French doctors introduced incubators into the care of premature babies over 100 years ago.

Figure 1.9
A steam engine from the shop of James Watt. (*British Crown Copyright, Science Museum, London*)

Fly-ball governor

ods attempted, the one with the most promise used a conical pendulum, or **fly-ball governor**, to measure the speed of the mill. The sails of the driving windmill were rolled up or let out with ropes and pulleys, much like a window shade, to maintain fixed speed. However, it was adaptation of these principles to the steam engine in the laboratories of James Watt around 1788 that made the fly-ball governor famous. An early version is shown in Fig. 1.9, while Figs. 1.10 and 1.11 show a close-up of a fly-ball governor and a sketch of its components.

The action of the fly-ball governor (also called a centrifugal governor) is simple to describe. Suppose the engine is operating in equilibrium. Two

Figure 1.10
Watt's steam engine (1789–1800) with fly-ball governor. (*British Crown Copyright, Science Museum, London*)

Figure 1.11
Operating parts of a fly-ball governor

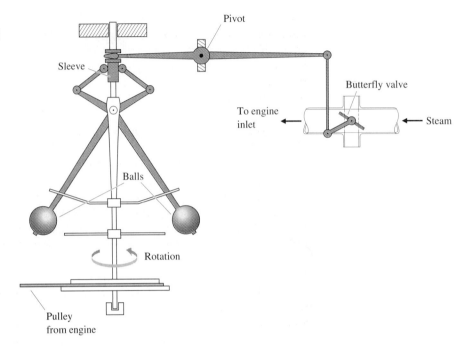

weighted balls spinning around a central shaft can be seen to describe a cone of a given angle with the shaft. When a load is suddenly applied to the engine, its speed will slow, and the balls of the governor will drop to a smaller cone. Thus the ball angle is used to sense the output speed. This action, through the levers, will open the main valve to the steam chest (which is the actuator) and admit more steam to the engine, restoring most of the lost speed. To hold the steam valve at a new position it is necessary for the fly balls to rotate at a different angle, implying that the speed under load is not exactly the same as before. We saw this effect earlier with cruise control, where feedback control gave a very small error. To recover the exact same speed in the system, it would require resetting the desired speed setting by changing the length of the rod from the lever to the valve. Subsequent inventors introduced mechanisms that integrated the speed error to provide automatic reset. In Chapter 4 we will analyze these systems to show that such integration can result in feedback systems with zero steady-state error to constant disturbances.

Because Watt was a practical man, like the millwrights before him, he did not engage in theoretical analysis of the governor. Fuller (1976) has traced the early development of control theory to a period of studies from Christian Huygens in 1673 to James Clerk Maxwell in 1868. Fuller gives particular credit to the contributions of G. B. Airy, professor of mathematics and astronomy at Cambridge University from 1826 to 1835 and Astronomer Royal at Greenwich Observatory from 1835 to 1881. Airy was concerned with speed control; if his telescopes could be rotated counter to the rotation of the Earth, a fixed

Beginnings of Control Theory

star could be observed for extended periods. Using the centrifugal-pendulum governor he discovered that it was capable of unstable motion: "and the machine (if I may so express myself) became perfectly wild" (Airy, 1840; quoted in Fuller, 1976). According to Fuller, Airy was the first worker to discuss instability in a feedback control system and the first to analyze such a system using differential equations. These attributes signal the beginnings of the study of feedback control dynamics.

Stability analysis

The first systematic study of the stability of feedback control was apparently given in the paper "On Governors" by J. C. Maxwell (1868).[7] In this paper, Maxwell developed the differential equations of the governor, linearized them about equilibrium, and stated that stability depends on the roots of a certain (characteristic) equation having negative real parts. Maxwell attempted to derive conditions on the coefficients of a polynomial that would hold if all the roots had negative real parts. He was successful only for second- and third-order cases. Determining criteria for stability was the problem for the Adams Prize of 1877, which was won by E. J. Routh.[8] His criterion, developed in his essay, remains of sufficient interest that control engineers are still learning how to apply his simple technique. Analysis of the characteristic equation remained the foundation of control theory until the invention of the electronic feedback amplifier by H. S. Black in 1927 at Bell Telephone Laboratories.

Shortly after publication of Routh's work, the Russian mathematician A. M. Lyapunov (1893) began studying the question of stability of motion. His studies were based on the nonlinear differential equations of motion and also included results for linear equations that are equivalent to Routh's criterion. His work was fundamental to what is now called the state-variable approach to control theory but was not introduced into the control literature until about 1958.

Frequency response

The development of the feedback amplifier is briefly described in an interesting article based on a talk by H. W. Bode (1960) reproduced in Bellman and Kalaba (1964). With the introduction of electronic amplifiers, long-distance telephoning became possible in the decades following World War I. However, as distances increased, so did the loss of electrical energy; in spite of using larger diameter wire, increasing numbers of amplifiers were needed to replace the lost energy. Unfortunately, large numbers of amplifiers resulted in much distortion because the small nonlinearity of the vacuum tubes then used in electronic amplifiers were multiplied many times. To solve the problem of reducing distortion, Black proposed the feedback amplifier. As mentioned earlier in connection with the automobile cruise control, the more we wish to reduce errors (or distortion), the more feedback we need to apply. The loop gain from actuator to plant to sensor to actuator must be made very large.

---

[7]   An exposition of Maxwell's contribution is given in Fuller (1976).

[8]   E. J. Routh was first academically in his class at Cambridge University in 1854, while J. C. Maxwell was second. In 1877 Maxwell was on the Adams Prize Committee that chose the problem of stability as the topic for the year.

With high gain the feedback loop begins to squeal and is unstable. Here was Maxwell's and Routh's stability problem again, except that in this technology the dynamics were so complex (with differential equations of order 50 being common) that Routh's criterion was not very helpful. So the communications engineers at Bell Telephone Laboratories, familiar with the concept of frequency response and the mathematics of complex variables, turned to complex analysis. In 1932 H. Nyquist published a paper describing how to determine stability from a graphical plot of the loop frequency response. From this theory there developed an extensive methodology of feedback-amplifier design described by Bode (1945) and extensively used still in the design of feedback controls. Nyquist and Bode plots are discussed in more detail in Chapter 6.

PID control

Simultaneous with the development of the feedback amplifier, feedback control of industrial processes was becoming standard. This field, characterized by processes that are not only highly complex but also nonlinear and subject to relatively long time delays between actuator and sensor, developed **proportional integral-derivative (PID) control**. The PID controller was first described by Callender et al. (1936). This technology was based on extensive experimental work and simple linearized approximations to the system dynamics. It led to standard experiments suitable to application in the field and eventually to satisfactory "tuning" of the coefficients of the PID controller. (PID controllers are covered in Chapter 4.) Also under development at this time were devices for guiding and controlling aircraft; especially important was the development of sensors for measuring aircraft altitude and speed. An interesting account of this branch of control theory is given in McRuer (1973).

An enormous impulse was given to the field of feedback control during World War II. In the United States, engineers and mathematicians at the MIT Radiation Laboratory combined their knowledge to bring together not only Bode's feedback amplifier theory and the PID control of processes but also the theory of stochastic processes developed by N. Wiener (1930). The result was the development of a comprehensive set of techniques for the design of **servomechanisms**, as control mechanisms came to be called. Much of this work was collected and published in the records of the Radiation Laboratory by James et al. (1947).

Root locus

Another approach to control systems design was introduced in 1948 by W. R. Evans, who was working in the field of guidance and control of aircraft. Many of his problems involved unstable or neutrally stable dynamics, which made the frequency methods difficult, so he suggested returning to the study of the characteristic equation that had been the basis of the work of Maxwell and Routh nearly 70 years earlier. However, Evans developed techniques and rules allowing one to follow graphically the paths of the roots of the characteristic equation as a parameter was changed. His method, the **root locus**, is suitable for design as well as for stability analysis and remains an important technique today. The root-locus method developed by Evans is covered in Chapter 5.

State-variable design

During the 1950s several authors, including R. Bellman and R. E. Kalman in the United States and L. S. Pontryagin in the U.S.S.R., began again to consider the ordinary differential equation (ODE) as a model for control systems. Much of this work was stimulated by the new field of control of artificial earth satellites, in which the ODE is a natural form for writing the model. Supporting this endeavor were digital computers, which could be used to carry out calculations unthinkable 10 years before. (Now, of course, these calculations can be done by any engineering student with a desktop computer.) The work of Lyapunov was translated into the language of control at about this time, and the study of optimal controls, begun by Wiener and Phillips during World War II, was extended to optimizing trajectories of nonlinear systems based on the calculus of variations. Much of this work was presented at the first conference of the newly formed International Federation of Automatic Control held in Moscow in 1960.[9] This work did not use the frequency response or the characteristic equation but worked directly with the ODE in "normal" or "state" form and typically called for extensive use of computers. Even though the foundations of the study of ODEs were laid in the late 19th century, this approach is now often called **modern control** to distinguish it from **classical control**, which uses the complex variable methods of Bode and others. In the period from the 1970s continuing through the present, we find a growing body of work that seeks to use the best features of each technique.

Modern control
Classical control

Thus we come to the current state of affairs where the principles of control are applied in a wide range of disciplines, including every branch of engineering. The well-prepared control engineer needs to understand the basic mathematical theory that underlies the field and must be able to select the best design technique suited to the problem at hand. With the ubiquitous use of computers it is especially important that the engineer is able to use his or her knowledge to guide and verify calculations done on the computer.[10]

1.4 An Overview of the Book

The central purpose of this book is to introduce the most important techniques for single-input–single-output control systems design. **Chapter 2** will review the techniques necessary to obtain models of the dynamic systems that we wish to control. These include model making for mechanical, electric, electromechanical, and a few other physical systems. Also described in Chapter 2 is the linearization of nonlinear models.

In **Chapter 3** and **Appendix A** we will discuss the analysis of dynamic response using Laplace transforms along with the relationship between time response and the poles and zeros of a transfer function. The chapter also

---

[9] Optimal control gained a large boost when Bryson and Denham (1962) showed that the path of a supersonic aircraft should actually dive at one point in order to reach a given altitude in minimum time. This nonintuitive result was later demonstrated to skeptical fighter pilots in flight tests.

[10] For more background on the history of control, see the survey papers appearing in the *IEEE Control Systems Magazine* of November 1984 and June 1996.

includes a discussion of the critical issue of system stability including the Routh test. The chapter includes a brief discussion of numerical simulation and how to construct models from experimental data.

In **Chapter 4** we will cover the basic equations and features of feedback. An analysis of the effects of feedback on disturbance rejection, tracking accuracy, sensitivity to parameter changes, and on dynamic response will be given. The idea of elementary proportional-integral-derivative (PID) control is discussed. Also in this chapter a brief introduction is given to the digital implementation of transfer functions and thus of linear time-invariant controllers so that the effects of digital control can be compared with analog controllers as these are designed.

In **Chapters 5, 6, and 7** we introduce the techniques for realizing the control objectives first identified in Chapter 4 in more complex dynamic systems. These methods include the root locus, frequency response, and state-variable techniques. These are alternative means to the same end and have different advantages and disadvantages as guides to design of controls. The methods are fundamentally complementary, and each needs to be understood to achieve the most effective control systems design.

In **Chapter 8** we develop further the ideas of implementing controllers in a digital computer that were introduced in Chapter 4. The chapter addresses how one "digitizes" the control equations developed in Chapters 5 through 7, how the analysis of sampled systems requires another analysis tool—the z-transform—and how the sampling introduces a delay that tends to destabilize the system.

Application of all the techniques to problems of substantial complexity are discussed in **Chapter 9**. There all the design methods are brought to bear simultaneously on specific case studies.

Computer aids

Control designers today make extensive use of computer-aided control systems design software that is commercially available. Furthermore, most instructional programs in control systems design make software tools available to the students. The most widely used software for the purpose is MATLAB and Simulink from The Mathworks. MATLAB routines have been included throughout the text to help illustrate this method of solution and many problems require computer aids for solution. Many of the figures in the book were created using MATLAB and the files for their creation are available free of charge on the web at the sites

http://www.prenhall.com/franklin

or

http://www.scsolutions.com/scsolutions.control.html.

Students and instructors are invited to use these files because it is believed that they should be helpful in learning how to use computer methods to solve control problems.

Needless to say, many topics are not treated in the book. We do not extend the methods to multivariable controls (which are systems with more than one input and/or output). Nor is optimal control treated in more than a very introductory manner in Chapter 7. Despite the fact that essentially all real design problems are for nonlinear plants, we have omitted any real consideration of nonlinear control, although we include brief sections in Chapters 5, 6, and 7 to illustrate the first steps in extending the several techniques to nonlinear systems. For example, the issue of the anti-windup controller is presented in Chapter 4, a brief treatment of saturation is given in Chapter 5, a discussion of the describing function is in Chapter 6, and the Lyapunov stability theory is introduced in Chapter 7.

Also beyond the scope of this text is a detailed treatment of the experimental testing and modeling of real hardware, which is the ultimate test of whether any design really works. The book concentrates on analysis and design of linear controllers for linear plant models—not because we think that is the final test of a design, but because that is the best way to grasp the basic ideas of feedback and is usually the first step in arriving at a satisfactory design. We believe that mastery of the material here will provide a foundation of understanding on which to build knowledge of these more advanced and realistic topics—a foundation strong enough to allow one to build a personal design method in the tradition of all those who worked to give us the knowledge we present here.

SUMMARY

- **Control** is the process of making a system variable adhere to a particular value, called the **reference value.** A system designed to follow a changing reference is called **tracking control** or a **servo.** A system designed to maintain an output fixed regardless of the disturbances present is called a **regulating control** or a **regulator.**

- Two kinds of control were defined and illustrated based on the information used in control and named by the resulting structure. In **open-loop control** the system does *not* measure the output and there is no correction of the actuating signal to make that output conform to the reference signal. In **closed-loop control** the system includes a sensor to measure the output and uses **feedback** of the sensed value to influence the control variable.

- A simple feedback system consists of the **process** whose output is to be controlled, the **actuator** whose output causes the process output to change, **reference** and **output sensors** that measure these signals, and the **controller**, which implements the logic by which the control signal that commands the actuator is calculated.

- **Block diagrams** are helpful for visualizing system structure and the flow of information in control systems. The most common block diagrams represent the mathematical relationships among the signals in a control system

- The theory and design techniques of control have come to be divided into two categories: **Classical control** methods use the Laplace or Fourier transforms and were the dominant methods for control design until about 1960, whereas **Modern control** methods are based on ordinary differential equations (ODEs) in state form and were introduced into the field starting in the 1960s. Many connections have been discovered between the two categories, and well-prepared engineers must be familiar with both techniques.

Review Questions

1. What are the main components of a feedback control system?

2. What is the purpose of the sensor?

3. Give three important properties of a good sensor.

4. What is the purpose of the actuator?

5. Give three important properties of a good actuator.

6. What is the purpose of the compensator? Give the input(s) and output(s) of the compensator.

7. What physical variable(s) of a process can be directly measured by a Hall effect sensor?

8. What physical variable is measured by a tachometer?

9. Describe three different techniques for measuring temperature.

10. Why do most sensors have an electrical output, regardless of the physical nature of the variable being measured?

Problems

1.1. Draw a component block diagram for each of the following feedback control systems.

   (a) The manual steering system of an automobile

   (b) Drebbel's incubator

   (c) The water level controlled by a float and valve

   (d) Watt's steam engine with fly-ball governor

In each case, indicate the location of the elements listed below and give the units associated with each signal.

   - The process
   - The process desired output signal
   - The sensor
   - The actuator
   - The actuator output signal
   - The controller
   - The controller output signal
   - The reference signal
   - The error signal

Notice that in a number of cases the same physical device may perform more than one of these functions.

1.2. Identify the physical principles and describe the operation of the thermostat in your home or office.

1.3. A machine for making paper is diagrammed in Fig. 1.12. There are two main parameters under feedback control: the density of fibers as controlled by the consistency of the thick stock that flows from the headbox onto the wire, and the moisture content of the final product that comes out of the dryers. Stock from the machine chest is diluted by white water returning from under the wire as controlled by a control valve (CV). A meter supplies a reading of the consistency. At the "dry end" of the machine, there is a moisture sensor. Draw a block diagram and identify the seven components listed in Problem 1.1 for

   (a) Control of consistency

   (b) Control of moisture

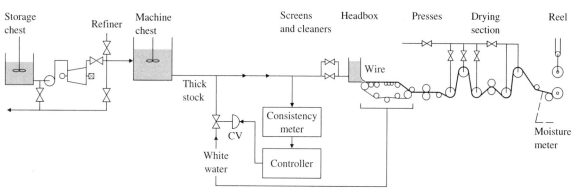

Figure 1.12 A papermaking machine (*From Åström, 1970, p. 192. Reprinted with permission*)

1.4. Many variables in the human body are under feedback control. For each of the following controlled variables, draw a block diagram showing the process being controlled, the sensor that measures the variable, the actuator that causes it to increase and/or decrease, the information path that completes the feedback path, and the disturbances that upset the variable. You may need to consult an encyclopedia or textbook on human physiology for information on this problem.

   (a) Blood pressure

   (b) Blood sugar concentration

   (c) Heart rate

   (d) Eye-pointing angle

   (e) Eye-pupil diameter

1.5. Draw a block diagram of the components for temperature control in a refrigerator or automobile air-conditioning system.

1.6. Draw a block diagram of the components for an elevator-position control. Indicate how you would measure the position of the elevator car. Consider a combined coarse and fine measurement system. What accuracies do you suggest for each sensor? Your system should be able to correct for the fact that in elevators for tall buildings there is significant cable stretch as a function of cab load.

1.7. Feedback control requires being able to sense the variable being controlled. Because electrical signals can be transmitted, amplified, and processed easily, often we want to have a sensor whose output is a voltage or current proportional to the variable being measured. Describe a sensor that would give an electrical output proportional to:

(a) Temperature

(b) Pressure

(c) Liquid level

(d) Flow of liquid along a pipe (or blood along an artery) force

(e) Linear position

(f) Rotational position

(g) Linear velocity

(h) Rotational speed

(i) Translational acceleration

(j) Torque

1.8. Each of the variables listed in Problem 1.7 can be brought under feedback control. Describe an actuator that could accept an electrical input and be used to control the variables listed. Give the units of the actuator output signal.

2 Dynamic Models

Chapter Overview

The fundamental step in building a dynamic model is writing the equations of motion for the system. Through discussion and a variety of examples, Section 2.1 demonstrates how to write the equations of motion for a variety of mechanical systems. Once such equations are written, they can be expressed in the state-variable form, as defined and shown in Section 2.2. Models in state-variable form enhance our ability to apply the computational efficiency of analysis tools such as MATLAB; an example using this tool is shown in Section 2.2.

In addition to the mechanical systems discussed, electric circuits, electromechanical systems, and heat and fluid-flow systems are modeled in Sections 2.3 to 2.5, respectively. Finally, Section 2.6 concludes with a discussion of linearization and scaling. The differential equations developed in modeling are often nonlinear. Because they are significantly more challenging to solve than linear ones and because linear models are usually adequate, this concluding explanation of how to linearize nonlinear equations is intended to make possible more efficient analysis and design.

In order to focus on the important first step of developing mathematical models, we will defer explanation of the computational methods used to solve the equations of motion developed in this chapter until Chapter 3.

A Perspective on Dynamic Models

The overall goal of feedback control is to use the principle of feedback to cause the output variable of a dynamic process to follow a desired reference variable accurately regardless of the reference variable's path and of any external disturbances or any changes in the dynamics of the process. This complex goal is met as the result of a number of simple, distinct steps. The first of these is to develop a mathematical description (called a **dynamic model**) of the process to be controlled. The term **model**, as it is used and understood by control engineers, means a set of differential equations that describe the dynamic behavior of the process.

In many cases the modeling of complex processes is difficult and expensive, especially when the important steps of building and testing prototypes are included. However, in this introductory text, we will focus on the most basic principles of modeling for the most common physical systems. More comprehensive sources and specialized texts will be referenced throughout the text where appropriate for those wishing more detail.

In later chapters we will explore a variety of analysis methods for dealing with the equations of motion, including the state-variable method, which is introduced briefly in this chapter. Because it lends itself to computer analysis, this method is of particular interest to contemporary students of control.

2.1 Dynamics of Mechanical Systems

Newton's law for translational motion

The cornerstone for obtaining a mathematical model, or the **equations of motion**, for any mechanical system is Newton's law,

$$\mathbf{F} = m\mathbf{a}, \tag{2.1}$$

where

> $\mathbf{F} = $ the vector sum of all forces applied to each body in a system, newtons (N) or pounds (lb),
>
> $\mathbf{a} = $ the vector acceleration of each body with respect to an inertial reference frame (that is, one that is neither accelerating nor rotating with respect to the stars); often called **inertial acceleration**, m/sec^2 or ft/sec^2,
>
> $m = $ mass of the body, kg or slug.

Note that here in Eq. (2.1), as throughout the text, we use the convention of boldfacing the type to indicate that the quantity is a matrix or vector, possibly a vector function.

In SI units a force of one newton will impart an acceleration of 1 m/sec^2 to a mass of 1 kilogram. In English units a force of 1 lb will impart an acceleration of 1 ft/sec^2 to a mass of 1 slug. The "weight" of an object is mg, where g is the acceleration of gravity ($= 9.81$ m/sec$^2 = 32.2$ ft/sec^2). In English units it is common usage to refer to the mass of an object in terms of its weight in pounds, which is the quantity measured on scales. To obtain the mass in slugs for use in Newton's law, divide the weight by g. Therefore, an object weighing 1 lb has a mass of 1/32.2 slugs. A slug has units lb $\cdot$ sec^2/ft. In metric units, scales are typically calibrated in kilograms, which is a direct measure of mass.

Use of free-body diagram in applying Newton's law

Application of this law typically involves defining convenient coordinates to account for the body's motion (position, velocity, and acceleration), determining the forces on the body using a free-body diagram, and then writing the equations of motion from Eq. (2.1). The procedure is simplest when the coordinates chosen express the position with respect to an inertial frame because in this case the accelerations needed for Newton's law are simply the second derivatives of the position coordinates.

EXAMPLE 2.1

A Simple System; Cruise-Control Model

Write the equations of motion for the speed and forward motion of the car shown in Fig. 2.1 assuming that the engine imparts a force u as shown.

Figure 2.1
Cruise-control model

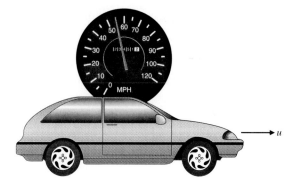

Figure 2.1
Cruise-control model

Solution.   For simplicity we assume that the rotational inertia of the wheels is negligible and that there is friction retarding the motion of the car that is proportional to the car's speed.[1] The car can then be approximated for modeling purposes using the free-body diagram seen in Fig. 2.2, which defines coordinates, shows all forces acting on the body (heavy lines), and indicates the acceleration (dashed lines). The coordinate of the car's position, x, is the distance from the reference line shown and chosen so that positive is to the right. Note that in this case the inertial acceleration is simply the second derivative of x (that is, $\mathbf{a} = \ddot{x}$) because the car position is measured with respect to an inertial reference. The equation of motion is found using Eq. (2.1). The friction force acts opposite to the direction of motion; therefore it is drawn opposite the direction of positive motion and entered as a negative force in Eq. (2.1). The result is

$$u - b\dot{x} = m\ddot{x}, \tag{2.2}$$

or

$$\ddot{x} + \frac{b}{m}\dot{x} = \frac{u}{m}. \tag{2.3}$$

For the case of the automotive cruise control where the variable of interest is the speed, $v(= \dot{x})$, the equation of motion becomes

$$\dot{v} + \frac{b}{m}v = \frac{u}{m}. \tag{2.4}$$

The solution of such an equation will be covered in detail in Chapter 3; however, the essence is that you assume a solution of the form $v = V_o e^{st}$ given an input of the

Figure 2.2
Free-body diagram for
cruise control

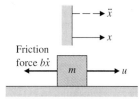

---

[1] If the speed is v, the aerodynamic friction force is proportional to v^2. In this simple model we have taken a linear approximation.

form $u = U_o e^{st}$. Then because $\dot{v} = s V_o e^{st}$, the differential equation can be written as

$$\left(s + \frac{b}{m} \right) V_o e^{st} = \frac{1}{m} U_o e^{st}. \tag{2.5}$$

The e^{st} term cancels out, and we find that

$$\frac{V_o}{U_o} = \frac{\frac{1}{m}}{s + \frac{b}{m}}. \tag{2.6}$$

For reasons that will become clear in Chapter 3, this is usually written as

$$\frac{V(s)}{U(s)} = \frac{\frac{1}{m}}{s + \frac{b}{m}}. \tag{2.7}$$

Transfer function

This expression of the differential equation in (2.4) is called the **transfer function** and will be used extensively in later chapters. Note that, in essence, we have substituted s for d/dt in Eq. (2.4).[2]

The solution of $v(t)$ is an exponential; therefore, the value of b could be obtained at a particular speed by providing the car with a step throttle change and observing the speed. The best-fit exponential to the experimental $v(t)$ would then yield the value of b/m in the model. Weighing the car will yield m.

For a rough determination of the parameters, refer to Fig. 3.18(a). Here we see that the time constant of the exponential decay can be determined experimentally by measuring the slope of the initial speed decay. When driving at a steady 60 mph, then releasing the throttle, we observed that the speed decayed to 55 mph in 5 sec. Extrapolating this curve yields a time constant of approximately 60 sec. Thus, $b/m \approx 1/60 \ \text{sec}^{-1}$ and, because the car's mass is 1500 kg, we have $m = 1580$ kg and $b \approx 26$ N·sec/m.

Newton's law also can be applied to systems with more than one mass. In this case it is particularly important to draw the free-body diagram of each mass showing the applied external forces as well as the equal and opposite internal forces that act from each mass on the other.

EXAMPLE 2.2   *A Two-mass System: Suspension Model*

Figure 2.3 shows an automobile suspension system. Write the equations of motion for the automobile and wheel motion assuming one-dimensional vertical motion of one-quarter of the car mass above one wheel. A system comprised of one of the four-wheel suspensions is usually referred to as a quarter-car model. Assume the model is for a car with a mass of 1580 kg including the four wheels, which have a mass of 20 kg each. By placing a known weight (an author) directly over a wheel and measuring the car's deflection, we find that $k_s = 130{,}000$ N/m. Measuring the wheel's deflection for the same applied weight, we find that $k_w \simeq 1{,}000{,}000$ N/m. By using the results in Section 3.3, Fig. 3.23(b), and qualitatively observing that the car's response as the author jumps off matches the $\zeta = 0.7$ curve, we conclude that $b = 9800$ N·sec/m.

---

[2] The use of an operator for differentiation was developed by Cauchy about 1820 based on the Laplace transform, which was developed about 1780. In Chapter 3 we will show how to derive transfer functions using the Laplace Transform. Reference: Gardner and Barnes (1942).

Figure 2.3
Automobile suspension

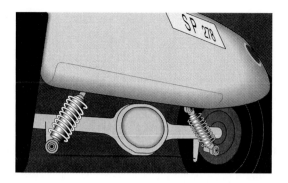

Solution.   The system can be approximated by the simplified system shown in Fig. 2.4. The coordinates of the two masses, x and y, with the reference directions as shown, are the displacements of the masses from their equilibrium conditions. The equilibrium positions are offset from the springs' unstretched positions because of the force of gravity. The shock absorber is represented in the schematic diagram by a dashpot symbol with friction constant b. The magnitude of the force from the shock absorber is assumed to be proportional to the rate of change of the relative displacement of the two masses—that is, to $b(\dot{y} - \dot{x})$. The force of gravity could be included in the free-body diagram; however, its effect is to produce a constant offset of x and y. By defining x and y to be the distance from the equilibrium position, the need to include the gravity forces is eliminated.

The force from the car suspension acts on both masses in proportion to their relative displacement with spring constant k_s. Figure 2.5 shows the free-body diagram of each mass. Note that the forces from the spring on the two masses are equal in magnitude but act in opposite directions, which is also the case for the damper. A positive displacement y of mass m_2 will result in a force from the spring on m_2 in the direction shown and a force from the spring on m_1 in the direction shown. However, a positive displacement x of mass m_1 will result in a force from the spring k_s on m_1 in the opposite direction to that drawn in Fig. 2.5 as indicated by the *minus x*-term for the spring force.

The lower spring k_w represents the tire compressibility, for which there is insufficient damping (velocity-dependent force) to warrant including a dashpot in the model.

Figure 2.4
The quarter-car model

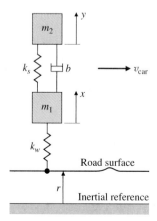

Figure 2.5
Free-body diagrams for
suspension system

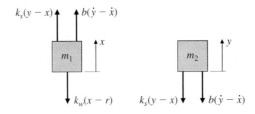

The force from this spring is proportional to the distance the tire is compressed and the nominal equilibrium force would be that required to support m_1 and m_2 against gravity. By defining x to be the distance from equilibrium, a force will result if either the road surface has a bump (r changes from its equilibrium value of zero) or the wheel bounces (x changes). The motion of the simplified car over a bumpy road will result in a value of $r(t)$ that is not constant.

As noted above, there is a constant force of gravity acting on each mass; however, this force has been omitted, as have the equal and opposite forces from the springs. Gravitational forces can always be omitted from vertical-spring mass systems (1) if the position coordinates are defined from the equilibrium position that results when gravity is acting and (2) if the spring forces used in the analysis are actually the perturbation in spring forces from those forces acting at equilibrium.

Applying Eq. (2.1) to each mass and noting that some forces on each mass are in the negative (down) direction yields the system of equations

$$b(\dot{y} - \dot{x}) + k_s(y - x) - k_w(x - r) = m_1\ddot{x}, \tag{2.8}$$

$$-k_s(y - x) - b(\dot{y} - \dot{x}) = m_2\ddot{y}. \tag{2.9}$$

Some rearranging results in

$$\ddot{x} + \frac{b}{m_1}(\dot{x} - \dot{y}) + \frac{k_s}{m_1}(x - y) + \frac{k_w}{m_1}x = \frac{k_w}{m_1}r, \tag{2.10}$$

$$\ddot{y} + \frac{b}{m_2}(\dot{y} - \dot{x}) + \frac{k_s}{m_2}(y - x) = 0. \tag{2.11}$$

Check for sign errors

The most common source of error in writing equations for systems like these are sign errors. The method for keeping the signs straight in the development above entailed mentally picturing the displacement of the masses and drawing the resulting force in the direction that the displacement would produce. Once you have obtained the equations for a system, a check on the signs for systems that are obviously stable from physical reasoning can be quickly carried out. As we will see when we study stability in Section 3.6, a stable system always has the same signs on similiar variables. For this system, Eq. (2.10) shows that the signs on the $\ddot{x}$-, $\dot{x}$-, and x-terms are all positive, as they must be for stability. Likewise, the signs on the $\ddot{y}$-, $\dot{y}$-, and y-terms are all positive in Eq. (2.11).

The transfer function is obtained in a similar manner as before. Substituting s for d/dt in the differential equations yields

$$s^2X(s) + s\frac{b}{m_1}(X(s) - Y(s)) + \frac{k_s}{m_1}(X(s) - Y(s)) + \frac{k_w}{m_1}X(s) = \frac{k_w}{m_1}R(s),$$

$$s^2Y(s) + s\frac{b}{m_2}(Y(s) - X(s)) + \frac{k_s}{m_2}(Y(s) - X(s)) = 0$$

and after some algebra and rearranging yields the transfer function

$$\frac{Y(s)}{R(s)} = \frac{\dfrac{k_w b}{m_1 m_2}\left(s + \dfrac{k_s}{b}\right)}{s^4 + \left(\dfrac{b}{m_1} + \dfrac{b}{m_2}\right)s^3 + \left(\dfrac{k_s}{m_1} + \dfrac{k_s}{m_2} + \dfrac{k_w}{m_1}\right)s^2 + \left(\dfrac{k_w b}{m_1 m_2}\right)s + \dfrac{k_w k_s}{m_1 m_2}}. \tag{2.12}$$

To determine numerical values, we subtract the mass of the four wheels from the total car mass of 1580 kg and divide by 4 to find that $m_2 = 375$ kg. The wheel mass was measured directly to be $m_1 = 20$ kg. Therefore, the transfer function with the numerical values is

$$\frac{Y(s)}{R(s)} = \frac{1.31e06(s + 13.3)}{s^4 + (516.1)s^3 + (5.685e04)s^2 + (1.307e06)s + 1.733e07}. \tag{2.13}$$

Newton's law for rotational motion

Application of Newton's law to one-dimensional rotational systems requires that Eq. (2.1) be modified to

$$M = I\alpha, \tag{2.14}$$

where

M = the sum of all external moments about the center of mass of a body, N · m or lb · ft,

I = the body's mass moment of inertia about its center of mass, kg · m^2 or slug · ft^2,

α = the angular acceleration of the body, rad/sec^2.

EXAMPLE 2.3   *Rotational Motion: Satellite Attitude Control Model*

Satellites, as shown in Fig. 2.6, usually require attitude control so that antennas, sensors, and solar panels are properly oriented. Antennas are usually pointed toward a particular location on earth, while solar panels need to be oriented toward the sun for maximum power generation. To gain insight into the full three-axis attitude-control system, it is helpful to consider one axis at a time. Write the equations of motion for one axis of this system.

Solution.   Figure 2.7 depicts this case, where motion is only allowed about the axis perpendicular to the page. The angle θ that describes the satellite orientation must be measured with respect to an inertial reference—that is, a reference that has no angular acceleration. The control force comes from reaction jets that produce a moment of $F_c d$ about the mass center. There may also be small disturbance moments on the satellite,

Figure 2.6
Communications
satellite *(Courtesy Space
Systems/Loral)*

Figure 2.7
Satellite control schematic

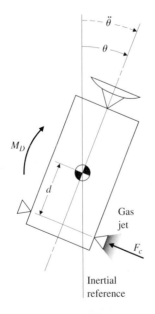

M_D, which arise primarily from solar pressure acting on any asymmetry in the solar panels. Applying Eq. (2.14) yields the equation of motion

$$F_c d + M_D = I\ddot{\theta}. \tag{2.15}$$

Double-integrator plant

The output of this system, θ, results from integrating the sum of the input torques twice; hence this type of system is often referred to as the **double-integrator plant**. The transfer function can be obtained as described for Eq. (2.7),

$$\frac{\Theta(s)}{U(s)} = \frac{1}{I}\frac{1}{s^2}, \tag{2.16}$$

$\frac{1}{s^2}$ **plant**

where $u = F_c d + M_D$. In this form, the system is often referred to as the $\frac{1}{s^2}$ **plant** .

In many cases a system, such as the disk-drive read/write head shown in Fig. 2.8, in reality has some flexibility, which can cause problems in the design

Figure 2.8
Disk read/write mechanism
(*Photo courtesy of
Hewlett-Packard Company*)

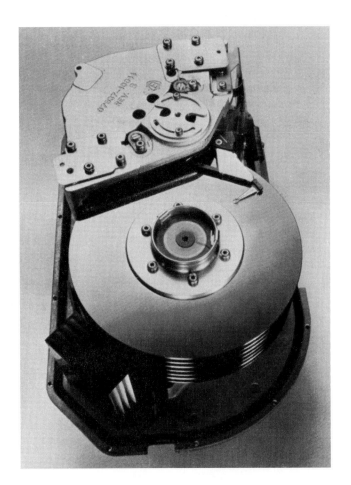

of a control system. Particular difficulty arises when there is flexibility, as in this case, between the sensor and actuator locations. Therefore, it is often important to include this flexibility in the model even when the system seems to be quite rigid.

EXAMPLE 2.4   *Flexibility: Flexible Read/Write for a Disk Drive*

Assume there is some flexibility between the read head and the drive motor in Fig. 2.8. Find the equations of motion relating the motion of the read head to a torque applied to the base.

Solution.   Schematically the dynamic model for this situation is as shown in Fig. 2.9. This model is dynamically similar to the resonant system shown in Fig. 2.4 and results in equations of motion that are similar in form to Eqs. (2.10) and (2.11). The moments on each body are shown in the free-body diagrams in Fig. 2.10. The discussion of the moments on each body is essentially the same as the discussion for Example 2.2, except the springs and damper in that case produced forces, instead of moments that act on each inertia, as in this case. When the moments are summed, equated to the accelerations according to Eq. (2.14), and rearranged, the result is

$$I_1\ddot{\theta}_1 + b(\dot{\theta}_1 - \dot{\theta}_2) + k(\theta_1 - \theta_2) = M_c + M_D \tag{2.17}$$

$$I_2\ddot{\theta}_2 + b(\dot{\theta}_2 - \dot{\theta}_1) + k(\theta_2 - \theta_1) = 0. \tag{2.18}$$

Ignoring the disturbance torque M_D and the damping b for simplicity, we find the transfer function from the applied torque, M_c, to the read head motion to be

$$\frac{\Theta_2(s)}{M_c(s)} = \frac{k}{I_1 I_2 s^2 \left(s^2 + \dfrac{k}{I_1} + \dfrac{k}{I_2}\right)}. \tag{2.19}$$

It might also be possible to sense the motion of the inertia where the torque is applied, θ_1, in which case the transfer function with the same simplifications would be

$$\frac{\Theta_1(s)}{M_c(s)} = \frac{I_2 s^2 + k}{I_1 I_2 s^2 \left(s^2 + \dfrac{k}{I_1} + \dfrac{k}{I_2}\right)}. \tag{2.20}$$

Figure 2.9
Disk read/write head schematic for modeling

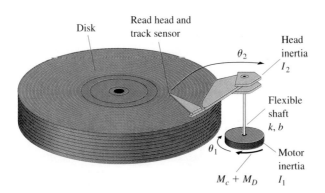

Disk

Read head and track sensor

θ_2

Head inertia I_2

Flexible shaft k, b

Motor inertia I_1

θ_1

$M_c + M_D$

Figure 2.10
Free-body diagrams of the
disk read/write head

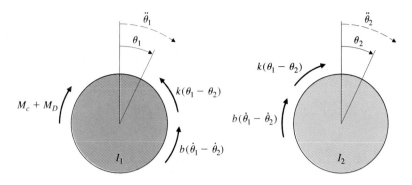

Collocated sensor and
actuator

These two cases are typical of many situations where the sensor and actuator may or may not be placed in the same location in a flexible body. We refer to the situation between sensor and actuator in Eq. (2.19) as the "noncollocated" case whereas Eq. (2.20) describes the "collocated" case. You will see in Chapter 5 that it is far more difficult to control a system when there is flexibility between the sensor and actuator (non-collocated case) than when the sensor and actuator are rigidly attached to one another (the collocated case).

In the special case where a point in a rotating body is fixed with respect to an inertial reference, as is the case with a pendulum, Eq. (2.14) can be applied such that M is the sum of all moments about the *fixed* point and I is the moment of inertia about the fixed point.

EXAMPLE 2.5

Rotational Motion: Pendulum

(a) Write the equations of motion for the simple pendulum shown in Fig. 2.11, where all the mass is concentrated at the endpoint.

(b) Use MATLAB to determine the time history of θ to a step input in T_c of $1 \text{ N} \cdot \text{m}$. Assume $l = 1$ m, $m = 0.5$ kg, and $g = 9.81 \text{ m/sec}^2$.

Figure 2.11
Pendulum

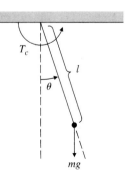

Solution.

(a) **Equations of Motion:** The moment of inertia about the pivot point is $I = ml^2$. The sum of moments about the pivot point contains a term from gravity as well as the applied torque T_c. The equation of motion, obtained from Eq. (2.14), is

$$T_c - mgl \sin \theta = I\ddot{\theta}, \tag{2.21}$$

which is usually written in the form

$$\ddot{\theta} + \frac{g}{l} \sin \theta = \frac{T_c}{ml^2}. \tag{2.22}$$

This equation is nonlinear due to the $\sin \theta$ term. A general discussion of nonlinear equations is contained in Section 2.6.1; however, we can proceed with a linearization of this case by assuming the motion is small enough that $\sin \theta \cong \theta$. Then Eq. (2.22) becomes the linear equation

$$\ddot{\theta} + \frac{g}{l}\theta = \frac{T_c}{ml^2}. \tag{2.23}$$

If there is no applied torque, this represents the motion of a harmonic oscillator with a natural frequency of [3]

$$\omega_n = \sqrt{\frac{g}{l}}. \tag{2.24}$$

The transfer function can be obtained as described for Eq. (2.7) yielding

$$\frac{\Theta(s)}{T_c(s)} = \frac{\dfrac{1}{ml^2}}{s^2 + \dfrac{g}{l}}. \tag{2.25}$$

(b) **Time History:** The dynamics of a system can be prescribed to MATLAB in terms of row vectors containing the coefficients of the polynomials describing the numerator and denominator of its transfer function. In this case, the numerator (called num) is simply one number because there are no powers of s, so that

$$\text{num} = \frac{1}{ml^2} = \frac{1}{(0.5)(1)^2} = [2]$$

and the denominator (called den) contains the powers of s in $s^2 + \frac{g}{l}$, which is

$$\text{den} = \begin{bmatrix} 1 & 0 & \frac{g}{l} \end{bmatrix} = \begin{bmatrix} 1 & 0 & 9.81 \end{bmatrix}$$

The desired response of the system can be obtained by using the MATLAB step response function, called step. The MATLAB statements

```
t = 0:0.02:10;          % vector of times for output
num = 2;
den = [1  0  9.81];
sys = tf(num, den);     % defines the system by its numerator and denominator
y = step(sys, t);       % computes step response
plot(t,y)               % plots step response
```

will produce the desired time history similar to that shown in Fig. 2.12. (t is a vector of the times desired in the solution output.)

---

[3] In a grandfather clock it is desired to have a pendulum period of exactly 2 sec. How long should the pendulum be?

Figure 2.12
Response of the pendulum
to a step input in the
applied torque

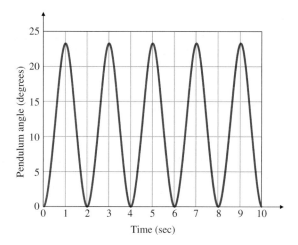

In some cases, mechanical systems contain both translational and rotational portions. The procedure is the same; sketch the free-body diagrams, define coordinates and positive directions, determine all forces and moments acting, and apply Eqs. (2.1) and/or (2.14).

As we saw in the last example, the resulting equations of motion are often nonlinear. Such equations are much more difficult to solve than linear ones, and the kinds of possible motions resulting from a nonlinear model are much more difficult to categorize than those resulting from a linear model. It is therefore useful to linearize models in order to gain access to linear analysis methods. It may be that the linear models and linear analysis are used only for the design of the control system (whose function may be to maintain the system in the linear region). Once a control system is synthesized and shown to have desirable performance based on linear analysis, it is then prudent to carry out an accurate numerical simulation of the system with all the nonlinearities in order to validate that performance. Section 2.6.1 expands on these ideas.

EXAMPLE 2.6   *Rotational and Translational Motion: Hanging Crane*

Write the equations of motion for the hanging crane pictured in Fig. 2.13 and shown schematically in Fig. 2.14. Linearize the equations about $\theta = 0$, which would typically be valid for the hanging crane. Also linearize the equations for $\theta = \pi$, which represents the situation for the inverted pendulum shown in Fig. 2.15.

Solution.   A schematic diagram of the hanging crane is shown in Fig. 2.14, while the free-body diagrams are shown in Fig. 2.16. In the case of the pendulum, the forces are shown with bold lines, while the components of the inertial acceleration of its center of mass are shown with dashed lines. Because the pivot point of the pendulum is *not* fixed with respect to an inertial reference, the rotation of the pendulum and the motion of

Figure 2.13
Crane with a hanging
load (*Photo courtesy of
Harnischfeger Corporation,
Milwaukee, Wisconsin*)

Figure 2.14
Schematic of the crane with
hanging load

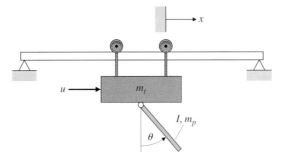

Figure 2.15
Inverted pendulum

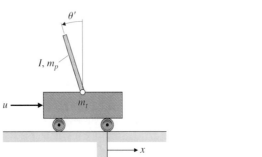

Figure 2.16
Hanging crane:
(a) free-body diagram of
the trolley; (b) free-body
diagram of the pendulum;
(c) position vector of the
pendulum

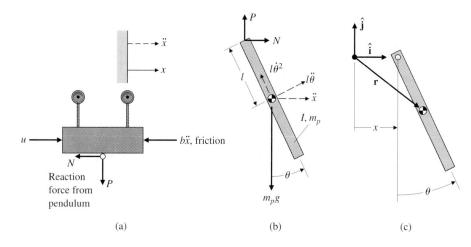

(a)   (b)   (c)

its mass center must be considered. The inertial acceleration needs to be determined because the vector **a** in Eq. (2.1) is given with respect to an inertial reference. The inertial acceleration of the pendulum's mass center is the vector sum of the three dashed arrows shown in Fig. 2.16(b). The derivation of the components of an object's acceleration is called **kinematics** and is usually studied as a prelude to the application of Newton's laws. The results of a kinematic study are shown in Fig. 2.16(b). The component of acceleration along the pendulum is $l\dot{\theta}^2$ and is called the centripetal acceleration. It is present for any object whose velocity is changing direction. The $\ddot{x}$-component of acceleration is a consequence of the pendulum pivot point accelerating at the trolley's acceleration and will always have the same direction and magnitude as those of the trolley's. The $l\ddot{\theta}$ component is a result of angular acceleration of the pendulum and is always perpendicular to the pendulum.

These results can be confirmed by expressing the center of mass of the pendulum as a vector from an inertial reference and then differentiating that vector twice to obtain an inertial acceleration. Figure. 2.16 (c) shows $\hat{\mathbf{i}}$ and $\hat{\mathbf{j}}$ axes which are inertially fixed and a vector **r** describing the position of the pendulum center of mass. The vector can be expressed as

$$\mathbf{r} = x\hat{\mathbf{i}} + l(\hat{\mathbf{i}}\sin\theta - \hat{\mathbf{j}}\cos\theta).$$

The first derivative of **r** is

$$\dot{\mathbf{r}} = \dot{x}\hat{\mathbf{i}} + l\dot{\theta}(\hat{\mathbf{i}}\cos\theta + \hat{\mathbf{j}}\sin\theta).$$

Likewise, the second derivative of **r** is

$$\ddot{\mathbf{r}} = \ddot{x}\hat{\mathbf{i}} + l\ddot{\theta}(\hat{\mathbf{i}}\cos\theta + \hat{\mathbf{j}}\sin\theta) - l\dot{\theta}^2(\hat{\mathbf{i}}\sin\theta - \hat{\mathbf{j}}\cos\theta).$$

Note that the equation for $\ddot{\mathbf{r}}$ confirms the acceleration components shown in Fig. 2.16(b). The $l\dot{\theta}^2$ term is aligned along the pendulum pointing toward the axis of rotation, and the $l\ddot{\theta}$ term is aligned perpendicular to the pendulum pointing in the direction of a positive rotation.

Having all the forces and accelerations for the two bodies, we now proceed to apply Eq. (2.1). In the case of the trolley, Fig. 2.16 (a), we see that it is constrained by the tracks to move only in the x-direction; therefore, application of Eq. (2.1) in this direction yields

$$m_t\ddot{x} + b\dot{x} = u - N, \tag{2.26}$$

where N is an unknown reaction force applied by the pendulum. Conceptually, Eq. (2.1) can be applied to the pendulum of Fig. 2.16(b) in the vertical and horizontal directions, and Eq. (2.14) can be applied for rotational motion to yield three equations in the three unknowns: N, P, and θ. These three equations can then be manipulated to eliminate the reaction forces N and P so that a single equation results describing the motion of the pendulum—that is, a single equation in θ. For example, application of Eq. (2.1) for pendulum motion in the x-direction yields

$$N = m_p\ddot{x} + m_pl\ddot{\theta}\cos\theta - m_pl\dot{\theta}^2\sin\theta. \tag{2.27}$$

However, considerable algebra will be avoided if Eq. (2.1) is applied perpendicular to the pendulum to yield

$$P\sin\theta + N\cos\theta - m_pg\sin\theta = m_pl\ddot{\theta} + m_p\ddot{x}\cos\theta. \tag{2.28}$$

Application of Eq. (2.14) for the rotational pendulum motion where the moments are summed about the center of mass yields

$$-Pl\sin\theta - Nl\cos\theta = I\ddot{\theta}, \tag{2.29}$$

where I is the moment of inertia about the pendulum's mass center. The reaction forces N and P can now be eliminated by combining Eqs. (2.28) and (2.29). This yields the equation

$$(I + m_pl^2)\ddot{\theta} + m_pgl\sin\theta = -m_pl\ddot{x}\cos\theta. \tag{2.30}$$

It is identical to a pendulum equation of motion, except that it contains a forcing function that is proportional to the trolley's acceleration.

An equation describing the trolley motion was found in Eq. (2.26), but it contains the unknown reaction force N. By combining Eqs. (2.27) and (2.26), N can be eliminated to yield

$$(m_t + m_p)\ddot{x} + b\dot{x} + m_pl\ddot{\theta}\cos\theta - m_pl\dot{\theta}^2\sin\theta = u. \tag{2.31}$$

Equations (2.30) and (2.31) are the nonlinear differential equations that describe the motion of the crane with its hanging load. For an accurate calculation of the motion of the system, these nonlinear equations need to be solved.

To linearize the equations for small motions about $\theta = 0$, let $\cos\theta \cong 1$, $\sin\theta \cong \theta$, and $\dot{\theta}^2 \cong 0$; thus the equations are approximated by

$$\begin{aligned} (I + m_pl^2)\ddot{\theta} + m_pgl\theta &= -m_pl\ddot{x}, \\ (m_t + m_p)\ddot{x} + b\dot{x} + m_pl\ddot{\theta} &= u. \end{aligned} \tag{2.32}$$

Neglecting the friction term, b, this leads to the transfer function from the control input, u, to hanging crane angle, θ,

$$\frac{\Theta(s)}{U(s)} = \frac{-m_pl}{((I + m_pl^2)(m_t + m_p) - m_p^2l^2)s^2 + m_pgl(m_t + m_p)} \tag{2.33}$$

Inverted pendulum equations

For the inverted pendulum in Fig. 2.15 where $\theta \cong \pi$, assume $\theta = \pi + \theta'$, where θ' represents motion from the vertical *upward* direction. In this case, $\cos \theta \cong -1$, $\sin \theta \cong -\theta'$ in Eqs. (2.30) and (2.31), and Eqs. (2.32) become[4]

$$(I + m_p l^2)\ddot{\theta}' - m_p g l \theta' = m_p l \ddot{x}$$

$$(m_t + m_p)\ddot{x} + b\dot{x} - m_p l \ddot{\theta}' = u. \tag{2.34}$$

As noted in Example 2.2, a stable system will always have the same signs on each variable which is the case for the stable hanging crane modeled by Eqs. (2.32). However, the signs on θ' and $\ddot{\theta}'$ in the first equation of Eqs. (2.34) are opposite; this indicates instability, which is the characteristic of the inverted pendulum.

The transfer function, again without friction, is

$$\frac{\Theta'(s)}{U(s)} = \frac{m_p l}{((I + m_p l^2) - m_p^2 l^2)s^2 - m_p g l (m_t + m_p)} \tag{2.35}$$

Summary: Developing Equations of Motion for Rigid Bodies The physics necessary to write the equations of motion of a rigid body is entirely given by Newton's laws of motion. The method is as follows:

1. Assign variables such as x and θ that are both necessary and sufficient to describe an *arbitrary* position of the object.

2. Draw a free-body diagram of each component. Indicate *all* forces acting on each body and their reference directions. Also indicate the accelerations of the center of mass with respect to an inertial reference for each body.

3. Apply Newton's law in translation [Eq. (2.1)] and/or rotation [Eq. (2.14)] form.

4. Combine the equations to eliminate internal forces.

5. The number of independent equations should equal the number of unknowns.

The combination of the equations to eliminate internal forces is sometimes expedited by an intelligent choice of directions for the application of Eq. (2.1), as illustrated by Example 2.6. It is often useful to try alternate directions and then evaluate the algebra required to eliminate the internal reaction forces. Applying Newton's laws in different directions sometimes yields what at first appear to be quite different sets of equations; however, after manipulation, you should always be able to show the equations are equivalent.

---

[4] The inverted pendulum is often described with the angle of the pendulum being positive for *clockwise* motion. If defined that way, then reverse the sign on all terms in Eqs. (2.34) in θ' or $\ddot{\theta}'$.

Distributed Parameter Systems

All the preceding examples contained one or more rigid bodies, although some were connected to others by springs. Actual structures—for example, satellite solar panels, airplane wings, or robot arms—usually bend, as shown by the flexible beam in Fig. 2.17(a). The equation describing its motion is a fourth-order *partial* differential equation that arises because the mass elements are continuously distributed along the beam with a small amount of flexibility between each element. This type of system is called a **distributed parameter system**. The dynamic analysis methods presented in this section are not sufficient to analyze this case; however, more advanced texts (Thomson, 1998) show that the result is

$$EI \frac{\partial^4 w}{\partial x^4} + \rho \frac{\partial^2 w}{\partial t^2} = 0, \tag{2.36}$$

where

$$E = \text{Young's modulus,}$$

$$I = \text{beam area moment of inertia,}$$

$$\rho = \text{beam density,}$$

$$w = \text{beam deflection at length } x \text{ along the beam.}$$

The exact solution to Eq. (2.36) is too cumbersome to use in designing control systems, but it is often important to account for the gross effects of bending in control systems design.

The continuous beam in Fig. 2.17(b) has an infinite number of vibration-mode shapes, all with different frequencies. Typically, the lowest-frequency modes have the largest amplitude and are the most important to approximate

Figure 2.17
(a) Flexible robot arm used for research at Stanford University; (b) model for a continuous flexible beam; (c) simplified model for the first bending mode; (d) model for the first and second bending modes *(Photo courtesy of E. Schmitz)*

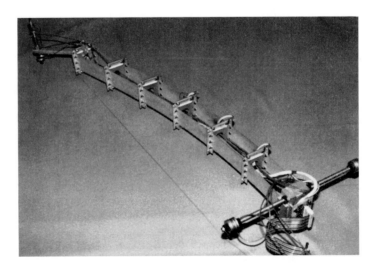

well. The simplified model in Fig. 2.17(c) can be made to duplicate the essential behavior of the first bending mode shape and frequency and would usually be adequate for controller design. If frequencies higher than the first bending mode are anticipated in the control system operation, it may be necessary to model the beam as shown in Fig. 2.17(d), which can be made to approximate the first two bending modes and frequencies. Likewise, higher-order models can be used if such accuracy and complexity are deemed necessary (Thomson, 1998; Schmitz, 1985). When a continuously bending object is approximated as two or more rigid bodies connected by springs, the resulting model is sometimes referred to as a **lumped parameter model**.

A flexible structure can be approximated by a lumped parameter model.

2.2   Differential Equations in State-Variable Form

Use of Newton's law and the free-body diagram in Section 2.1 typically lead to second-order differential equations, that is, equations that contain the second derivative such as $\ddot{x}$ in Eq. (2.3) or $\ddot{\theta}$ in Eq. (2.15). Differential equations also can be expressed as a set of simultaneous first-order differential equations. For example, Eq. (2.15) can be expressed as

$$\dot{x}_1 = x_2, \tag{2.37}$$

$$\dot{x}_2 = \frac{u}{I}, \tag{2.38}$$

where

$$u = F_c d + M_D,$$
$$x_1 = \theta,$$
$$x_2 = \dot{\theta},$$
$$\dot{x}_2 = \ddot{\theta}.$$

The output of this system is θ, the satellite attitude.

Standard form of linear differential equations

These same equations can be represented in the **state-variable form** as the vector equation

$$\dot{\mathbf{x}} = \mathbf{F}\mathbf{x} + \mathbf{G}u, \tag{2.39}$$

where the input is u and the output is

$$y = \mathbf{H}\mathbf{x} + Ju. \tag{2.40}$$

The column vector $\mathbf{x}$ is called the **state of the system** and contains n elements for an nth-order system. For mechanical systems, the state vector elements usually consist of the positions and velocities of the separate bodies, as is the case for the example in Eqs. (2.37) and (2.38). The quantity $\mathbf{F}$ is an $n \times n$ **system matrix**, $\mathbf{G}$ is an $n \times 1$ **input matrix**, $\mathbf{H}$ is a $1 \times n$ row matrix referred to as the **output**

matrix, and J is a scalar called the **direct transmission term**.[5] To save space we will sometimes refer to a state vector by its **transpose**, $\mathbf{x} = [\, x_1 \quad x_2 \, .. \,]^T$, which is equivalent to

$$\mathbf{x} = \begin{bmatrix} x_1 \\ x_2 \\ : \end{bmatrix}.$$

In Chapter 7 we will consider control systems design using the state-variable form. Some aspects of matrix theory that aid in the analysis of systems described in this manner are contained in Section 7.2 and Appendix C. For the case where the relationships are nonlinear [such as the case in Eqs. (2.22), (2.30), and (2.31)], the linear form cannot be used directly. One must linearize the equations to fit the form or use some of the techniques discussed in Section 2.6.1.

The state-variable method of specifying differential equations is used by computer-aided control systems design software packages (e.g., MATLAB). Therefore, in order to specify linear differential equations to the computer, you need to know the values of the matrices $\mathbf{F}$, $\mathbf{G}$, and $\mathbf{H}$ and the constant J. Another option available for specifying a dynamic system to software is to describe the system in terms of its **transfer function**, a concept introduced in Section 2.1 and that will be explained further in Chapter 3.

EXAMPLE 2.7   *Satellite Attitude Control Model in State-Variable Form*

Determine the $\mathbf{F}, \mathbf{G}, \mathbf{H}, J$ matrices in the state-variable form for the satellite attitude control model in Example 2.3 with $M_D = 0$.

Solution.   Define the attitude and the angular velocity of the satellite as the state-variables so that $\mathbf{x} \triangleq [\theta \quad \omega]^T$.[6] The single second-order equation (2.15) can then be written in an equivalent way as two first-order equations:

$$\dot{\theta} = \omega,$$

$$\dot{\omega} = \frac{d}{I} F_c.$$

These equations are expressed using Eq. (2.39), $\dot{\mathbf{x}} = \mathbf{F}\mathbf{x} + \mathbf{G}u$, as

$$\begin{bmatrix} \dot{\theta} \\ \dot{\omega} \end{bmatrix} = \begin{bmatrix} 0 & 1 \\ 0 & 0 \end{bmatrix} \begin{bmatrix} \theta \\ \omega \end{bmatrix} + \begin{bmatrix} 0 \\ d/I \end{bmatrix} F_c.$$

The output of the system is the satellite attitude, $y = \theta$. Using Eq. (2.40), $y = \mathbf{H}\mathbf{x} + Ju$, this relation is expressed as

$$y = \begin{bmatrix} 1 & 0 \end{bmatrix} \begin{bmatrix} \theta \\ \omega \end{bmatrix}.$$

---

[5] It is also common to use the notation $\mathbf{A}$, $\mathbf{B}$, $\mathbf{C}$, and D in place of $\mathbf{F}$, $\mathbf{G}$, $\mathbf{H}$, and J. We will typically use $\mathbf{F}$, $\mathbf{G}$ to represent plant dynamics and $\mathbf{A}$, $\mathbf{B}$ to represent a general linear system.

[6] The symbol $\triangleq$ means "is defined to be."

Therefore, the matrices for the state-variable form are

$$\mathbf{F} = \begin{bmatrix} 0 & 1 \\ 0 & 0 \end{bmatrix}, \qquad \mathbf{G} = \begin{bmatrix} 0 \\ d/I \end{bmatrix}, \qquad \mathbf{H} = [1 \quad 0], \qquad J = 0,$$

and the input $u \overset{\triangle}{=} F_c$.

For this very simple example the state-variable form is a more cumbersome way of writing the differential equation than the second-order version in Eq. (2.15). However, the method is not more cumbersome for most systems, and the advantages of having a standard form for use in computer-aided design have led to widespread use of the state-variable form.

The following example has more complexity and shows how to use MATLAB to find the solution of linear differential equations. A common solution method for evaluating the performance of a control system is to assume that the initial conditions are zero and the input $u(t)$ is a step with a magnitude of 1; that is, $u(t) = 0$ for $t < 0$ and $u(t) = 1$ for $t \geq 0$. This is called the **unit step response**. For steps in u of some magnitude other than 1, the solution is called simply the **step response**. The methods used in MATLAB to calculate the response will be discussed in Section 3.6 so, for now, you are not expected to understand how this calculation is made.

Step Response

EXAMPLE 2.8

Cruise-Control Step Response

(a) Rewrite the equation of motion from Example 2.1 in state-variable form where the output is the car position x.

(b) Use MATLAB to find the response of the velocity of the car for the case where the input jumps from being $u = 0$ at time $t = 0$ to a constant $u = 500$ N thereafter. Assume that the car mass m is 1000 kg and $b = 50$ N · sec/m.

Solution.

(a) **Equations of Motion:** First we need to express the differential equation describing the plant, Eq. (2.3), as a set of simultaneous first-order equations. To do so, we define the position and the velocity of the car as the state variables x and v, so that $\mathbf{x} = [x \ v]^T$. The single second-order equation, Eq. (2.3), can then be rewritten as a set of two first-order equations:

$$\dot{x} = v,$$

$$\dot{v} = -\frac{b}{m}v + \frac{1}{m}u.$$

Next, we use the standard form of Eq. (2.39), $\dot{\mathbf{x}} = \mathbf{F}\mathbf{x} + \mathbf{G}u$, to express these equations:

$$\begin{bmatrix} \dot{x} \\ \dot{v} \end{bmatrix} = \begin{bmatrix} 0 & 1 \\ 0 & -b/m \end{bmatrix} \begin{bmatrix} x \\ v \end{bmatrix} + \begin{bmatrix} 0 \\ 1/m \end{bmatrix} u. \tag{2.41}$$

The output of the system is the car position $y = x_1 = x$, which is expressed in matrix form as

$$y = \begin{bmatrix} 1 & 0 \end{bmatrix} \begin{bmatrix} x \\ v \end{bmatrix},$$

or

$$y = \mathbf{Hx}.$$

So the state-variable-form matrices defining this example are

$$\mathbf{F} = \begin{bmatrix} 0 & 1 \\ 0 & -b/m \end{bmatrix}, \qquad \mathbf{G} = \begin{bmatrix} 0 \\ 1/m \end{bmatrix}, \qquad \mathbf{H} = \begin{bmatrix} 1 & 0 \end{bmatrix}, \qquad J = 0.$$

(b) **Time Response:** The equations of motion are those given in part (a) except that now the output is $v = x_2$. Therefore, the output matrix is

$$\mathbf{H} = \begin{bmatrix} 0 & 1 \end{bmatrix}.$$

The coefficients required are $b/m = 0.05$ and $1/m = 0.001$. The numerical values of the matrices defining the system are thus

$$\mathbf{F} = \begin{bmatrix} 0 & 1 \\ 0 & -0.05 \end{bmatrix}, \qquad \mathbf{G} = \begin{bmatrix} 0 \\ 0.001 \end{bmatrix}, \qquad \mathbf{H} = \begin{bmatrix} 0 & 1 \end{bmatrix}, \qquad J = 0.$$

The **step** function in MATLAB calculates the time response of a linear system to a unit step input. Because the system is linear, the output for this case can be multiplied by the magnitude of the input step to derive a step response of any amplitude. Equivalently, the $\mathbf{G}$ matrix can be multiplied by the magnitude of the input step.

Step response with MATLAB

The statements

```
F = [0   1; 0   − 0.05];
G = [0; 0.001];
H = [0   1];
J = 0;
sys = ss(F, 500·G, H, J);     % step gives unit step response, so 500·G gives
                              u = 500 N.
step(sys);                    % plots the step response
```

calculate and plot the time response for an input step with a 500-N magnitude. The step response is shown in Fig. 2.18.

Figure 2.18
Response of the car velocity to a step in u

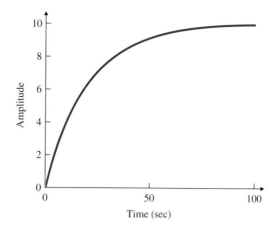

The state-variable form can be applied to a system of any order. Example 2.9 illustrates the method for a fourth-order system.

EXAMPLE 2.9

Flexible Disk-Drive in State-Variable Form

Find the state-variable form of the differential equations for Example 2.4, where the output is θ_2.

Solution.   Define the state vector to be

$$\mathbf{x} = [\,\theta_1 \quad \dot{\theta}_1 \quad \theta_2 \quad \dot{\theta}_2\,]^T .$$

Then solve Eqs. (2.17) and (2.18) for $\ddot{\theta}_1$ and $\ddot{\theta}_2$ so that the state-variable form is more apparent. The resulting matrices are

$$\mathbf{F} = \begin{bmatrix} 0 & 1 & 0 & 0 \\ -\frac{k}{I_1} & -\frac{b}{I_1} & \frac{k}{I_1} & \frac{b}{I_1} \\ 0 & 0 & 0 & 1 \\ \frac{k}{I_2} & \frac{b}{I_2} & -\frac{k}{I_2} & -\frac{b}{I_2} \end{bmatrix}, \qquad \mathbf{G} = \begin{bmatrix} 0 \\ \frac{1}{I_1} \\ 0 \\ 0 \end{bmatrix}, \qquad \mathbf{H} = [0 \ \ 0 \ \ 1 \ \ 0], \qquad J = 0.$$

In some cases, such as in Eqs. (2.32), more than one second derivative will appear in a differential equation. This makes transformation to state-variable form more difficult. Problem 2.9 illustrates both the difficulty and the solution. Another difficulty arises if the differential equation contains derivatives of the input u. Techniques to handle this situation will be discussed in Section 7.2.

2.3   Models of Electric Circuits

Electric circuits are frequently used in control systems largely because of the ease of manipulation and processing of electric signals. Although controllers are increasingly implemented with digital logic, many functions are still performed with analog circuits. Analog circuits are faster than digital and, for very simple controllers, an analog circuit would be less expensive than a digital implementation. Furthermore, the power amplifier for electromechanical control and the anti-alias prefilters for digital control must be analog circuits.

Electric circuits consist of interconnections of sources of electric voltage and current, and other electronic elements such as resistors, capacitors, and transistors. An important building block for circuits is an operational amplifier (or op-amp)[7] which is also an example of a complex feedback system. Some of the most important methods of feedback system design were developed by the designers of high-gain, wide-bandwidth feedback amplifiers, mainly at the Bell Telephone Laboratories between 1925 and 1940. Electric and electronic components also play a central role in electromechanical energy conversion devices such as electric motors, generators and electrical sensors. In this brief survey we cannot derive the physics of electricity or give a comprehensive review of all the important analysis techniques. We will define the variables, describe the relations imposed on them by typical elements and circuits, and describe a few of the most effective methods available for solving the resulting equations.

Symbols for some linear circuit elements and their current–voltage relations are given in Fig. 2.19. Passive circuits consist of interconnections of resistors, capacitors, and inductors. With electronics we increase the set of electrical elements by adding active devices, including diodes, transistors, and amplifiers.

Kirchhoff's laws

The basic equations of electric circuits are called Kirchhoff's laws, which are:

Kirchhoff's current law (KCL): *The algebraic sum of currents leaving a junction or node equals the algebraic sum of currents entering that node.*

Kirchhoff's voltage law (KVL): *The algebraic sum of all voltages taken around a closed path in a circuit is zero.*

With complex circuits of many elements it is essential to write the equations in a careful, well organized way. Of the numerous methods for doing this we choose for description and illustration the popular and powerful scheme known as **node analysis**. One node is selected as a reference and we assume the voltages of all other nodes to be unknowns. The choice of reference is arbitrary in theory, but in actual electronic circuits the common, or ground,

---

[7] Oliver Heaviside introduced the mathematical operation p to signify differentiation so that $pv = dv/dt$. The Laplace transform incorporates this idea, using the complex variable s. Ragazzini *et al.* (1947) demonstrated that an ideal, high-gain electronic amplifier permitted one to realize arbitrary "operations" in the Laplace transform variable s, so they named it the operational amplifier, commonly abbreviated to op-amp.

Figure 2.19
Elements of electric circuits

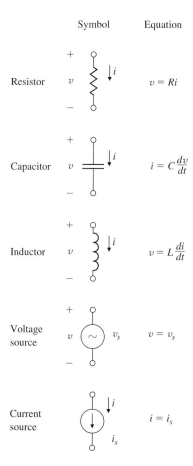

Symbol   Equation

Resistor $v = Ri$

Capacitor $i = C\dfrac{dv}{dt}$

Inductor $v = L\dfrac{di}{dt}$

Voltage source $v = v_s$

Current source $i = i_s$

terminal is the obvious and standard choice. Next, we write equations for the selected unknowns using the current law (KCL) at each node. We express these currents in terms of the selected unknowns by using the element equations in Fig. 2.19. If the circuit contains voltage sources, we must substitute a voltage law (KVL) for such sources. Example 2.10 illustrates how node analysis works.

EXAMPLE 2.10   *Equations for the Bridged Tee Circuit*

Determine the differential equations for the circuit shown in Fig. 2.20.

Solution.   We select node 4 as the reference and the voltages v_1, v_2, and v_3 at nodes 1, 2, and 3, as the unknowns. We start with a degenerate KVL relationship

$$v_1 = v_i. \tag{2.42}$$

At node 2 the KCL is

$$-\frac{v_1 - v_2}{R_1} + \frac{v_2 - v_3}{R_2} + C_1\frac{dv_2}{dt} = 0, \tag{2.43}$$

Figure 2.20
Bridged tee circuit

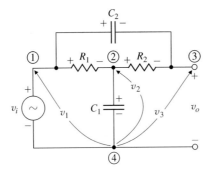

And at node 3 the KCL is

$$\frac{v_3 - v_2}{R_2} + C_2 \frac{d(v_3 - v_1)}{dt} = 0. \tag{2.44}$$

These three equations describe the circuit.

In order to write the equations in the state-variable form (i.e., a set of simultaneous first-order differential equations), we select the capacitor voltages v_{C_1} and v_{C_2} as the state elements (i.e., $\mathbf{x} = [v_{C_1} \; v_{C_2}]^T$) and select v_i as the input (i.e., $u = v_i$). Here $v_{C_1} = v_2$, $v_{C_2} = v_1 - v_3$, and still $v_1 = v_i$. Thus $v_1 = v_i$, $v_2 = v_{C_1}$, and $v_3 = v_i - v_{C_2}$. In terms of v_{C_1} and v_{C_2}, Eq. (2.43) is

$$\frac{v_{C_1} - v_i}{R_1} + \frac{v_{C_1} - (v_i - v_{C_2})}{R_2} + C_1 \frac{dv_{C_1}}{dt} = 0.$$

Rearranging this equation into standard form, we obtain

$$\frac{dv_{C_1}}{dt} = -\frac{1}{C_1}\left(\frac{1}{R_1} + \frac{1}{R_2}\right)v_{C_1} - \frac{1}{C_1}\left(\frac{1}{R_2}\right)v_{C_2} + \frac{1}{C_1}\left(\frac{1}{R_1} + \frac{1}{R_2}\right)v_i. \tag{2.45}$$

In terms of v_{C_1} and v_{C_2}, Eq. (2.44) is

$$\frac{v_i - v_{C_2} - v_{C_1}}{R_2} + C_2 \frac{d}{dt}(v_i - v_{C_2} - v_i) = 0.$$

In standard form, the equation is

$$\frac{dv_{C_2}}{dt} = -\frac{v_{C_1}}{C_2 R_2} - \frac{v_{C_2}}{C_2 R_2} + \frac{v_i}{C_2 R_2}. \tag{2.46}$$

Equations (2.42)–(2.44) are entirely equivalent to the state-variable form, Eqs. (2.45) and (2.46), in describing the circuit. The standard matrix definitions are

$$\mathbf{F} = \begin{bmatrix} -\frac{1}{C_1}\left(\frac{1}{R_1} + \frac{1}{R_2}\right) & -\frac{1}{C_1}\left(\frac{1}{R_2}\right) \\ -\frac{1}{C_2 R_2} & -\frac{1}{C_2 R_2} \end{bmatrix}, \qquad \mathbf{G} = \begin{bmatrix} \frac{1}{C_1}\left(\frac{1}{R_1} + \frac{1}{R_2}\right) \\ \frac{1}{C_2 R_2} \end{bmatrix}.$$

Figure 2.21
(a) Op-amp simplified circuit, (b) op-amp schematic symbol, (c) Reduced symbol for $v_+ = v_- = 0$

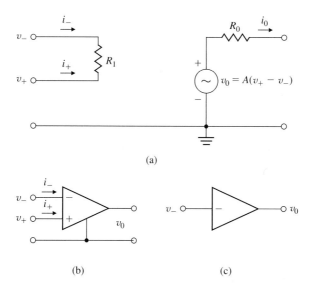

(a)

(b)   (c)

Operational amplifier

Kirchhoff's laws can also be applied to circuits that contain an **operational amplifier**. The simplified circuit of the op-amp is shown in Figure 2.21(a) and the schematic symbol is drawn in Figure 2.21(b). If the positive terminal is not shown, it is assumed to be connected to ground, $v_+ = 0$, and the reduced symbol of Fig. 2.21 (c) is used. For use in control circuits, it is usually assumed that the op-amp is *ideal* with the values $R_1 = \infty$, $R_o = 0$, and $A = \infty$. The equations of the ideal op-amp are extremely simple, being

$$i_+ = i_- = 0, \tag{2.47}$$

$$v_+ - v_- = 0. \tag{2.48}$$

The gain of the amplifier is assumed to be so high that the output voltage becomes v_{out} = *whatever it takes* to satisfy these equations. Of course a real amplifier only approximates these equations; but unless specifically described, we will assume all op-amps are ideal. More realistic models are the subject of several problems given at the end of the chapter.

EXAMPLE 2.11

Op-amp Summer

Find the equations and transfer functions of the circuit shown in Fig. 2.22.

Solution.   Equation (2.48) requires that $v_- = 0$ and thus the currents are $i_1 = v_1/R_1$, $i_2 = v_2/R_2$, and $i_{out} = v_{out}/R_f$. To satisfy Eq. (2.47), $i_1 + i_2 + i_{out} = 0$, from which $v_1/R_1 + v_2/R_2 + v_{out}/R_f = 0$ and we have

$$v_{out} = -\left[\frac{R_f}{R_1} v_1 + \frac{R_f}{R_2} v_2 \right]. \tag{2.49}$$

Figure 2.22
The op-amp summer

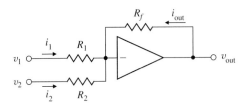

From this equation we see that the circuit output is a weighted sum of the input voltages with a sign change. The circuit is called a **summer**.

The op-amp summer

A second important example for control is given by the op-amp integrator.

EXAMPLE 2.12

Integrator

OP-amp as integrator

Find the transfer function for the circuit shown in Fig. 2.23.

Solution.   In this case the equations are differential and Eqs. (2.47) and (2.48) require

$$i_{in} + i_{out} = 0 \tag{2.50}$$

so that

$$\frac{v_{in}}{R_{in}} + C\frac{dv_{out}}{dt} = 0. \tag{2.51}$$

Equation (2.51) can be written in integral form as

$$v_{out} = -\frac{1}{R_{in}C} \int_0^t v_{in}(\tau)d\tau + v_{out}(0). \tag{2.52}$$

Using the operational notation that $d/dt = s$ in Eq.(2.51) the transfer function (which assumes zero initial conditions) can be written as

$$V_{out}(s) = -\frac{1}{s}\frac{V_{in}(s)}{R_{in}C}. \tag{2.53}$$

Figure 2.23
The op-amp integrator

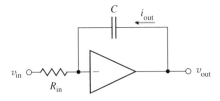

Thus the ideal op-amp in this circuit performs the operation of integration and the circuit is simply referred to as an **integrator**.

2.4   Models of Electromechanical Systems

Electric current and magnetic fields interact in two ways that are particularly important to an understanding of the operation of most electromechanical actuators and sensors. If a current of i amperes in a conductor of length l meters is arranged at right angles in a magnetic field of B tesla, then there is a force on the conductor at right angles to the plane of i and B, with magnitude

Law of motors

$$F = Bli \text{ newtons.} \tag{2.54}$$

This equation is the basis of conversion of electric energy to mechanical work and is called the **law of motors**.

EXAMPLE 2.13   *Modeling a Loudspeaker*

A typical geometry for a loudspeaker for producing sound is sketched in Fig. 2.24. The permanent magnet establishes a radial field in the cylindrical gap between the poles of the magnet. The force on the conductor wound on the bobbin causes the voice coil to move, producing sound.[8] The effects of the air can be modeled as if the cone had equivalent mass M and viscous friction coefficient b. Assume that the magnet establishes a uniform field B of 0.5 tesla and the bobbin has 20 turns at a 2-cm diameter. Write the equations of motion of the device.

Figure 2.24
Geometry of a loudspeaker:
(a) overall configuration,
(b) the electromagnet and
voice coil

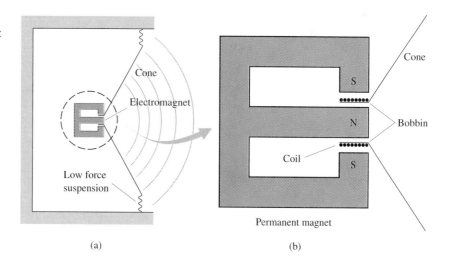

(a)

(b)

---
[8] Similar voice-coil motors are commonly used as the actuator for the read/write head assembly of computer hard-disk data access devices.

Solution.   The current is at right angles to the field, and the force of interest is at right angles to the plane of i and B, so Eq. (2.54) applies. In this case the field strength is $B = 0.5$ tesla and the conductor length is

$$l = 20 \times \frac{2\pi}{100} \text{ m} = 1.26 \text{ m}.$$

Thus the force is

$$F = 0.5 \times 1.26 \times i = 0.63i \text{ newtons.}$$

The mechanical equation follows from Newton's laws and for a mass M and friction coefficient b the equation is

$$M\ddot{x} + b\dot{x} = 0.63i. \tag{2.55}$$

This second-order differential equation describes the motion of the loudspeaker cone as a function of the input current i driving the system. To put the model in state-variable form, a logical state vector for this system would be $\mathbf{x} \triangleq [x \quad \dot{x}]^T$, which leads to the standard matrices

$$\mathbf{F} = \begin{bmatrix} 0 & 1 \\ 0 & -b/M \end{bmatrix} \quad \text{and} \quad \mathbf{G} = \begin{bmatrix} 0 \\ 0.63/M \end{bmatrix}.$$

Substituting s for d/dt in Eq. (2.55) as before, the transfer function is easily found to be

$$\frac{X(s)}{I(s)} = \frac{0.63/M}{s(s + b/M)}. \tag{2.56}$$

The second important electromechanical relationship is the effect of mechanical motion on electric voltage. If a conductor of length l meters is moving in a magnetic field of B teslas at a velocity v meters per second at mutually right angles, an electric voltage is established across the conductor with magnitude

$$e(t) = Blv \text{ volts.} \tag{2.57}$$

Law of the generator

This expression is called the **law of generators**.

EXAMPLE 2.14   *Loudspeaker with Circuit*

For the loudspeaker in Fig. 2.24 and the circuit driving it in Fig. 2.25, find the differential equations relating the input voltage v_a to the output cone displacement x. Assume the effective circuit resistance is R and the inductance is L.

Figure 2.25
A loudspeaker showing the electric circuit

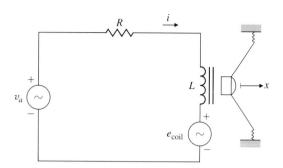

Solution.   The loudspeaker motion satisfies Eq. (2.55), and the motion results in a voltage across the coil as given by Eq. (2.57), with the velocity $\dot{x}$. The resulting voltage is

$$e_{\text{coil}} = Bl\dot{x} = 0.63\dot{x}. \tag{2.58}$$

This induced voltage effect needs to be added to the analysis of the circuit. The equation of motion for the electric circuit is

$$L\frac{di}{dt} + Ri = v_a - 0.63\dot{x}. \tag{2.59}$$

These two coupled equations, (2.55) and (2.59), comprise the dynamic model for the loudspeaker. A logical state vector for this third-order system would be $\mathbf{x} \overset{\triangle}{=} [x\,\dot{x}\,i]^T$, which leads to the standard matrices

$$\mathbf{F} = \begin{bmatrix} 0 & 1 & 0 \\ 0 & -b/M & 0.63/M \\ 0 & -0.63/L & -R/L \end{bmatrix} \quad \text{and} \quad \mathbf{G} = \begin{bmatrix} 0 \\ 0 \\ 1/L \end{bmatrix},$$

where now the input $u \overset{\triangle}{=} v_a$.

Again substituting s for d/dt in these equations, the transfer function between the applied voltage and the loudspeaker displacement is found to be

$$\frac{X(s)}{V_a(s)} = \frac{0.63}{s\,[(Ms+b)(Ls+R) + (0.63)^2]}. \tag{2.60}$$

DC motor actuators

     A common actuator based on these principles and used in control systems is the DC motor to provide rotary motion. A sketch of the basic components of a DC motor is given in Fig. 2.26. In addition to housing and bearings, the nonturning part (stator) has magnets which establish a field across the rotor. The magnets may be electromagnets or, for small motors, permanent magnets. The brushes contact the rotating commutator which causes the current always to be in the proper conductor windings so as to produce maximum torque. If the direction of the current is reversed, the direction of the torque is reversed.

Figure 2.26
Sketch of a DC motor

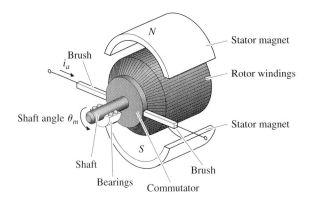

The motor equations give the torque, T, on the rotor in terms of the armature current, i_a, and express the back emf voltage in terms of the shaft's rotational velocity $\dot\theta_m$.[9] Thus

Back EMF

$$T = K_t\, i_a, \tag{2.61}$$

$$e = K_e\, \dot\theta_m. \tag{2.62}$$

Torque

In consistent units the torque constant K_t equals the electric constant K_e, but in some cases the torque constant will be given in other units such as ounce-inches per ampere and the electric constant may be expressed in units of volts per 1000 rpm. In such cases the engineer must make the necessary translations to be certain the equations are correct.

EXAMPLE 2.15

Modeling a DC Motor

Find the equations for a DC motor with the equivalent electric circuit shown in Fig. 2.27(a). Assume the rotor has inertia J_m and viscous friction coefficient b.

Solution.   The free-body diagram for the rotor, shown in Fig. 2.27(b), defines the positive direction and shows the two applied torques, T and $b\dot\theta_m$. Application of Newton's laws yields

$$J_m\ddot\theta_m + b\dot\theta_m = K_t i_a. \tag{2.63}$$

Analysis of the electric circuit including the back emf voltage shows the electrical equation to be

$$L_a\frac{di_a}{dt} + R_a i_a = v_a - K_e\dot\theta_m. \tag{2.64}$$

A state vector for this system is $\mathbf{x} \overset{\triangle}{=} [\theta_m \quad \dot\theta_m \quad i_a]^T$, which leads to the standard matrices

$$\mathbf{F} = \begin{bmatrix} 0 & 1 & 0 \\ 0 & -\frac{b}{J_m} & \frac{K_t}{J_m} \\ 0 & -\frac{K_e}{L_a} & -\frac{R_a}{L_a} \end{bmatrix} \quad \text{and} \quad \mathbf{G} = \begin{bmatrix} 0 \\ 0 \\ \frac{1}{L_a} \end{bmatrix},$$

[9] Because the generated electromotive force (emf) works against the applied armature voltage, it is called the **back emf**.

Figure 2.27
DC motor: (a) electric circuit of the armature, (b) free-body diagram of the rotor

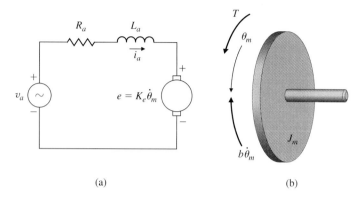

(a)   (b)

where the input $u \triangleq v_a$. When we substitute s for d/dt in Eq. (2.63) and Eq. (2.64), the transfer function for the motor is readily found to be

$$\frac{\Theta_m(s)}{V_a(s)} = \frac{K_t}{s[(J_m s + b)(L_a s + R_a) + K_t K_e]}. \tag{2.65}$$

In many cases the relative effect of the inductance is negligible compared to the mechanical motion and can be neglected in Eq. (2.64). In this case we can combine Eqs. (2.63) and (2.64) into one equation to get

$$J_m \ddot{\theta}_m + \left(b + \frac{K_t K_e}{R_a}\right)\dot{\theta}_m = \frac{K_t}{R_a} v_a. \tag{2.66}$$

From Eq. (2.66) it is clear that in this case the effect of the back emf is indistinguishable from the friction and the transfer function is

$$\frac{\Theta_m(s)}{V_a(s)} = \frac{\dfrac{K_t}{R_a}}{J_m s^2 + \left(b + \dfrac{K_t K_e}{R_a}\right) s} \tag{2.67}$$

$$= \frac{K}{s(\tau s + 1)}, \tag{2.68}$$

where

$$K = \frac{K_t}{b R_a + K_t K_e}, \tag{2.69}$$

$$\tau = \frac{R_a J_m}{b R_a + K_t K_e}. \tag{2.70}$$

In many cases, a transfer function between the motor input and the output speed ($\omega = \dot{\theta}_m$) is required. In that case, the transfer function would be

$$\frac{\Omega(s)}{V_a(s)} = s \frac{\Theta_m(s)}{V_a(s)} = \frac{K}{\tau s + 1}. \tag{2.71}$$

AC motor actuators

Another device used for electromechanical energy conversion is the alternating-current (AC) induction motor invented by N. Tesla. Elementary analysis of the AC motor is more complex than that of the DC motor. A typical experimental set of curves of speed versus torque for fixed frequency and varying amplitude of applied (sinusoidal) voltage is given in Fig. 2.28. Although the data in the figure are for a constant engine speed, they can be used to extract the motor constants that will provide a dynamic model for the motor. For analysis of a control problem involving an AC motor such as that described by Fig. 2.28, we make a linear approximation to the curves for speed near zero and at a midrange voltage to obtain the expression

$$T = K_1 v_a - K_2 \dot{\theta}_m. \tag{2.72}$$

The constant K_1 represents the ratio of a change in torque to a change in voltage at zero speed and is proportional to the distance between the curves at zero speed. The constant K_2 represents the ratio of a change in torque to a change in speed at zero speed and a midrange voltage; therefore, it is the slope of a curve at zero speed as shown by the line at V_2. For the electrical portion, values for the armature resistance R_a and inductance L_a are also determined by experiment. Once we have values for K_1, K_2, R_a, and L_a, the analysis proceeds as the analysis in Example 2.15 for the DC motor. For the case where the inductor can be neglected, we can substitute K_1 and K_2 into Eq. (2.66) in place of K_t/R_a and $K_t K_e/R_a$, respectively.

In addition to the DC and AC motors mentioned here, control systems use brushless DC motors (Reliance Motion Control Corp., 1980) and stepping motors (Kuo, 1972). Models for these machines, developed in the works just cited, do not differ in principle from the motors considered in this section. In general, the analysis, supported by experiment, develops the torque as a

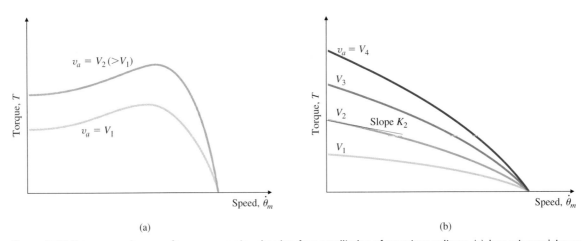

(a)   (b)

Figure 2.28 Torque-speed curves for a servo motor showing four amplitudes of armature voltage; (a) low-rotor-resistance machine; (b) high-rotor-resistance machine showing four values of armature voltage, v_a

function of voltage and speed similar to the AC-motor torque–speed curves given in Fig. 2.28. From such curves one can obtain (a) a linearized formula such as Eq. (2.72) to use in the mechanical part of the system and (b) an equivalent circuit consisting of a resistance and an inductance to use in the electrical part.

2.5   Heat- and Fluid-Flow Models

Thermodynamics, heat transfer, and fluid dynamics are each the subject of complete textbooks. For purposes of generating dynamic models for use in control systems, the most important aspect of the physics is to represent the dynamic interaction between the variables. Experiments are usually required to determine the actual values of the parameters and thus to complete the dynamic model for purposes of control systems design.

2.5.1   Heat Flow

Some control systems involve regulation of temperature for portions of the system. The dynamic models of temperature control systems involve the flow and storage of heat energy. Heat energy flows through substances at a rate proportional to the temperature difference across the substance; that is,

$$q = \frac{1}{R}(T_1 - T_2), \tag{2.73}$$

where

$q =$ heat energy flow, joules per second (J/sec) or British thermal unit/sec (BTU/sec),

$R =$ thermal resistance, $°C/J \cdot \sec$ or $°F/BTU$,

$T =$ temperature, $°C$ or $°F$.

The net heat-energy flow into a substance affects the temperature of the substance according to the relation

$$\dot{T} = \frac{1}{C}q, \tag{2.74}$$

where C is the thermal capacity. Typically, there are several paths for heat to flow into or out of a substance, and q in Eq. (2.74) is the sum of heat flows obeying Eq. (2.73).

EXAMPLE 2.16   *Equations for Heat Flow*

A room with all but two sides insulated ($1/R = 0$) is shown in Fig. 2.29. Find the differential equations that determine the temperature in the room.

Figure 2.29
Dynamic model for room
temperature

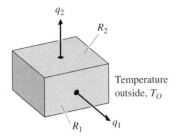

Solution.   Application of Eqs. (2.73) and (2.74) yields

$$\dot{T}_I = \frac{1}{C_I} \left(\frac{1}{R_1} + \frac{1}{R_2} \right) (T_O - T_I),$$

where

C_I = thermal capacity of air within the room,

T_O = temperature outside,

T_I = temperature inside,

R_2 = thermal resistance of the room ceiling,

R_1 = thermal resistance of the room wall.

Normally the material properties are given in tables as follows:

Specific heat

1. The specific heat at constant volume c_v, which is converted to heat capacity
by

$$C = mc_v, \tag{2.75}$$

where m is the mass of the substance;

Thermal conductivity

2. The thermal conductivity[10] k, which is related to thermal resistance R by

$$\frac{1}{R} = \frac{kA}{l},$$

where A is the cross-sectional area and l is the length of the heat-flow path.

---

[10] In the case of insulation for houses, resistance is quoted as R-values; for example, R–11 refers
to a substance that has a resistance to heat flow equivalent to that given by 11 in. of solid wood.

In addition to flow due to transfer as expressed by Eq. (2.73), heat can also flow when a warmer mass flows into a cooler mass, or vice versa. In this case,

$$q = wc_v(T_1 - T_2), \tag{2.76}$$

where w is the mass flow rate of the fluid at T_1 flowing into the reservoir at T_2. For a more complete discussion of dynamic models for temperature control systems, see Cannon (1967) or textbooks on heat transfer.

EXAMPLE 2.17

Equations for Modeling a Heat Exchanger

A heat exchanger is shown in Fig. 2.30. Steam enters the chamber through the controllable valve at the top, and cooler steam leaves at the bottom. There is a constant flow of water through the pipe that winds through the middle of the chamber so that it picks up heat from the steam. Find the differential equations that describe the dynamics of the measured water outflow temperature as a function of the area, A_s, of the steam-inlet control valve when open. The sensor that measures the water outflow temperature, being downstream from the exit temperature in the pipe, lags the temperature by t_d seconds.

Solution.   The temperature of the water in the pipe will vary continuously along the pipe as the heat flows from the steam to the water. The temperature of the steam will also reduce in the chamber as it passes over the maze of pipes. An accurate thermal model of this process is therefore quite involved because the actual heat transfer from the steam to the water will be proportional to the local temperatures of each fluid. For many control applications it is not necessary to have great accuracy because the feedback will correct for a considerable amount of error in the model. Therefore, it makes sense to combine the spatially varying temperatures into the single temperatures T_s and T_w for the outflow steam and water temperatures, respectively. We then assume that the heat transfer from steam to water is proportional to the difference in these temperatures as

Figure 2.30
Heat exchanger

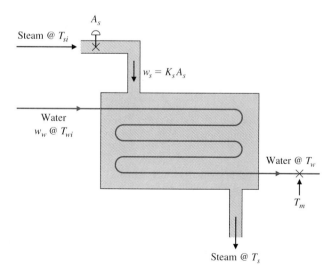

given by Eq. (2.73). There is also a flow of heat into the chamber from the inlet steam that depends on the steam flow rate and its temperature according to Eq. (2.76):

$$q_{\text{in}} = w_s c_{vs} (T_{si} - T_s),$$

where

$$w_s = K_s A_s, \text{ mass flow rate of the steam,}$$

$$A_s = \text{area of the steam inlet valve,}$$

$$K_s = \text{flow coefficient of the inlet valve,}$$

$$c_{vs} = \text{specific heat of the steam,}$$

$$T_{si} = \text{temperature of the inflow steam,}$$

$$T_s = \text{temperature of the outflow steam.}$$

The net heat flow into the chamber is the difference between the heat from the hot incoming steam and the heat flowing out to the water. This net flow determines the rate of temperature change of the steam according to Eq. (2.74):

$$C_s \dot{T}_s = A_s K_s c_{vs} (T_{si} - T_s) - \frac{1}{R}(T_s - T_w), \qquad (2.77)$$

where

$C_s = m_s c_{vs}$ is the thermal capacity of the steam in the chamber with mass m_s,

$R = $ the thermal resistance of the heat flow averaged over the entire exchanger.

Likewise, the differential equation describing the water temperature is

$$C_w \dot{T}_w = w_w C_{vw} (T_{wi} - T_w) + \frac{1}{R}(T_s - T_w), \qquad (2.78)$$

where

$$w_w = \text{mass flow rate of the water,}$$

$$C_{vw} = \text{specific heat of the water,}$$

$$T_{wi} = \text{temperature of the incoming water,}$$

$$T_w = \text{temperature of the outflowing water.}$$

To complete the dynamics, the time delay between the measurement and the exit flow is described by the relation
$$T_m(t) = T_w(t - t_d),$$

where T_m is the measured downstream temperature of the water and t_d is the time delay. There may also be a delay in the measurement of the steam temperature, T_s, which would be modeled in the same manner.

Equation (2.77) is nonlinear because the state variable T_s is multiplied by the control input A_s. The equation can be linearized about T_{so} (a specific value of T_s) so that $T_{si} - T_s$ is assumed constant for purposes of approximating the nonlinear term, which we will define as ΔT_s. In order to eliminate the T_{wi} term in Eq. (2.78), it is convenient to measure all temperatures in terms of deviation in degrees from T_{wi}. The resulting equations are then

$$C_s \dot{T}_s = -\frac{1}{R} T_s + \frac{1}{R} T_w + K_s c_{vs} \Delta T_s A_s,$$

$$C_w \dot{T}_w = -\left(\frac{1}{R} + w_w c_{vw} \right) T_w + \frac{1}{R} T_s,$$

$$T_m = T_w(t - t_d).$$

Although the time delay is not a nonlinearity, we will see in Chapter 3 that operationally, $T_m = e^{-t_d s} T_w$. Therefore, the transfer function of the heat exchanger has the form

$$\frac{T_m(s)}{A_s(s)} = \frac{K e^{-t_d s}}{(\tau_1 s + 1)(\tau_2 + 1)}. \tag{2.79}$$

For purposes of illustrating the standard state-variable form, we relate T_m to T_w and define the state vector to be $\mathbf{x} \triangleq [T_s \ T_w]^T$ with the control input, $u = A_s$ so that

$$\mathbf{F} = \begin{bmatrix} -\frac{1}{C_s R} & \frac{1}{C_s R} \\ \frac{1}{C_w R} & -\frac{1}{C_w}\left(\frac{1}{R} + w_w c_{vw}\right) \end{bmatrix} \quad \text{and} \quad \mathbf{G} = \begin{bmatrix} K_s c_{vs} \Delta T_s \\ 0 \end{bmatrix}. \tag{2.80}$$

2.5.2   Incompressible Fluid Flow

Fluid flows are common in many control systems components. One example is the hydraulic actuator, which is used extensively in control systems because it can supply a large force with low inertia and low weight. They are often used to move the aerodynamic control surfaces of airplanes, to gimbal rocket nozzles, to move the linkages in earth-moving equipment, farm tractor implements, and snow grooming machines, and to move robot arms.

The physical relations governing fluid flow are continuity, force equilibrium, and resistance. **The continuity relation** is simply a statement of the conservation of matter:

$$\dot{m} = w_{\text{in}} - w_{\text{out}}, \tag{2.81}$$

where

$$m = \text{fluid mass within a prescribed portion of the system,}$$

$$w_{\text{in}} = \text{mass flow rate into the prescribed portion of the system,}$$

$$w_{\text{out}} = \text{mass flow rate out of the prescribed portion of the system.}$$

EXAMPLE 2.18   **_Equations for Describing Water Tank Height_**

Determine the differential equation describing the height of the water in the tank in Fig. 2.31.

Solution.   Application of Eq. (2.81) yields

$$\dot{h} = \frac{1}{A\rho}(w_{in} - w_{out}),\qquad(2.82)$$

where

$$A = \text{area of the tank,}$$

$$\rho = \text{density of water,}$$

$$h = m/A\rho = \text{height of water,}$$

$$m = \text{mass of water in the tank.}$$

Figure 2.31
Water-tank example

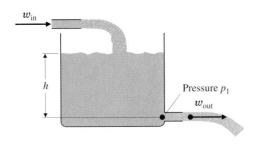

Force equilibrium must apply exactly as described by Eq. (2.1) for mechanical systems. Sometimes in fluid flow systems some forces result from fluid pressure acting on a piston. In this case the force from the fluid is

$$f = pA,\qquad(2.83)$$

where

$$f = \text{force,}$$

$$p = \text{pressure in the fluid,}$$

$$A = \text{area on which the fluid acts.}$$

EXAMPLE 2.19   **_Modeling a Hydraulic Piston_**

Determine the differential equation describing the motion of the piston actuator shown in Fig. 2.32 given that there is a force F_D acting on it and a pressure p in the chamber.

Figure 2.32
Hydraulic piston actuator

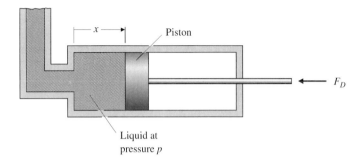

Piston

Liquid at
pressure p

Solution.   Equations (2.1) and (2.83) apply directly, where the forces include the fluid pressure as well as the applied force. The result is

$$M\ddot{x} = Ap - F_D,$$

where

$$A = \text{area of the piston,}$$

$$p = \text{pressure in the chamber,}$$

$$M = \text{mass of the piston,}$$

$$x = \text{position of the piston.}$$

In many cases of fluid-flow problems the flow is resisted either by a constriction in the path or by friction. The general form of the effect of resistance is given by

$$w = \frac{1}{R}(p_1 - p_2)^{1/\alpha}, \tag{2.84}$$

where

$$w = \text{mass flow rate,}$$

$$p_1, p_2 = \text{pressures at ends of the path through which flow is occurring,}$$

$$R, \alpha = \text{constants whose values depend on the type of restriction,}$$

or, as is more commonly used in hydraulics,

$$Q = \frac{1}{\rho R}(p_1 - p_2)^{1/\alpha}, \tag{2.85}$$

where

$$Q = \text{volume flow rate, where } Q = \frac{w}{\rho},$$

$$\rho = \text{fluid density.}$$

The constant α takes on values between 1 and 2. The most common value is approximately 2 for high flow rates (those having a Reynolds number Re $> 10^5$) through pipes or through short constrictions or nozzles. For very slow flows through long pipes or porous plugs where the flow remains laminar (Re $\lesssim 1000$), $\alpha = 1$. Flow rates in between these extremes can yield intermediate values of α. The Reynolds number indicates the relative importance of inertial forces and viscous forces in the flow. It is proportional to a material's velocity and density and to the size of the restriction, and it is inversely proportional to the viscosity. When Re is small, the viscous forces predominate and the flow is laminar. When Re is large, the inertial forces predominate and the flow is turbulent.

Note that a value of $\alpha = 2$ indicates that the flow is proportional to the square root of the pressure difference and therefore will produce a nonlinear differential equation. For the initial stages of control systems analysis and design, it is typically very useful to linearize these equations so that the design techniques described in this book can be applied. Linearization involves selecting an operating point and expanding the nonlinear term to be a small perturbation from that point.

EXAMPLE 2.20   *Linearization of Water Tank Height and Outflow*

Find the nonlinear differential equation describing the height of the water in the tank in Fig. 2.31. Assume there is a relatively short restriction at the outlet and that $\alpha = 2$. Also linearize your equation about the operating point h_o.

Solution.   Applying Eq. (2.84) yields the flow out of the tank as a function of the height of the water in the tank:

$$w_{\text{out}} = \frac{1}{R}(p_1 - p_a)^{1/2}, \qquad (2.86)$$

where

$$p_1 = \rho g h + p_a, \text{ the hydrostatic pressure,}$$

$$p_a = \text{ambient pressure outside the restriction.}$$

Substituting Eq. (2.86) into Eq. (2.82) yields the nonlinear differential equation for the height:

$$\dot{h} = \frac{1}{A\rho}\left(w_{\text{in}} - \frac{1}{R}\sqrt{p_1 - p_a}\right). \qquad (2.87)$$

Linearization involves selecting the operating point $p_o = \rho g h_o + p_a$ and substituting $p_1 = p_o + \Delta p$ into Eq. (2.86). Then we expand the nonlinear term according to the relation

$$(1 + \varepsilon)^\beta \cong 1 + \beta\varepsilon, \qquad (2.88)$$

where $\varepsilon \ll 1$. Equation (2.86) can thus be written as

$$w_{\text{out}} = \frac{\sqrt{p_o - p_a}}{R}\left(1 + \frac{\Delta p}{p_o - p_a}\right)^{1/2}$$

$$\cong \frac{\sqrt{p_o - p_a}}{R}\left(1 + \frac{1}{2}\frac{\Delta p}{p_o - p_a}\right). \tag{2.89}$$

The linearizing approximation made in Eq. (2.89) is valid as long as $\Delta p \ll p_o - p_a$; that is, as long as the deviations of the system pressure from the chosen operating point are relatively small.

Combining Eqs. (2.82) and (2.89) yields the following linearized equation of motion for the water tank level:

$$\Delta \dot{h} = \frac{1}{A\rho}\left[w_{\text{in}} - \frac{\sqrt{p_o - p_a}}{R}\left(1 + \frac{1}{2}\frac{\Delta p}{p_o - p_a}\right)\right].$$

Because $\Delta p = \rho g \,\Delta h$, this equation reduces to

$$\Delta \dot{h} = -\frac{g}{2AR\sqrt{p_o - p_a}}\Delta h + \frac{w_{\text{in}}}{A\rho} - \frac{\sqrt{p_o - p_a}}{\rho AR}, \tag{2.90}$$

which is a linear differential equation for $\Delta \dot{h}$. The operating point is not an equilibrium point because some control input is required to maintain it. In other words, when the system is at the operating point ($\Delta h = 0$) with no input ($w_{\text{in}} = 0$), it will move from that point because $\Delta \dot{h} \neq 0$. Therefore, the linearized equation does not fit the standard form of Eqs. (2.39) due to the extra constant on the right. To define an operating point that is also an equilibrium point so Eqs. (2.39) can be used, we need to require that there be a nominal flow rate,

$$\frac{w_{\text{in}_o}}{A\rho} = \frac{\sqrt{p_o - p_a}}{\rho AR}$$

and define the linearized input flow to be a perturbation from that value.

Hydraulic actuators obey the same fundamental relationships we saw in the water tank: continuity [Eq. (2.81)], force balance [Eq. (2.83)], and flow resistance [Eq. (2.84)]. Although the development here assumes the fluid is perfectly incompressible, in fact, hydraulice fluid has some compressibility due primarily to entrained air. This feature causes hydraulic actuators to have some resonance because the compressibility of the fluid acts like a stiff spring. This resonance limits their speed of response.

EXAMPLE 2.21   *Modeling a Hydraulic Actuator*

(a) Find the nonlinear differential equations relating the movement of the control surface, θ, to the input displacement x of the valve for the hydraulic actuator shown in Fig. 2.33.

(b) Find the linear approximation to the equations of motion when $\dot{y} = \text{constant}$, with and without an applied load—that is, when $F \neq 0$ and when $F = 0$. Assume that θ motion is small.

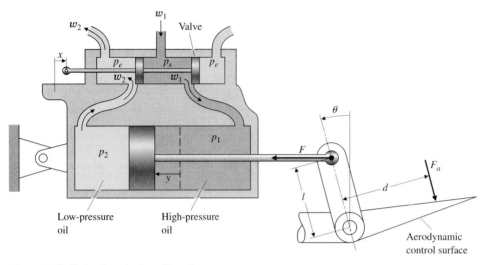

Figure 2.33 Hydraulic actuator with valve

Solution.

(a) **Equations of Motion:** When the valve is at $x = 0$, both passages are closed and no motion results. When $x > 0$, as shown in Fig. 2.33, the oil flows clockwise as shown and the piston is forced to the left. When $x < 0$, the fluid flows counterclockwise: The oil supply at high pressure p_s enters the *left* side of the large piston chamber, forcing the piston to the right. This causes the oil to flow out of the valve chamber from the rightmost channel instead of the leftmost one.

We assume the flow through the orifice formed by the valve is proportional to x; that is,

$$\frac{w_1}{\rho} = Q_1 = \frac{1}{\rho R_1}(p_s - p_1)^{1/2}x. \tag{2.91}$$

Similarly,

$$\frac{w_2}{\rho} = Q_2 = \frac{1}{\rho R_2}(p_2 - p_e)^{1/2}x. \tag{2.92}$$

The continuity relation yields

$$A\dot{y} = Q_1 = Q_2, \tag{2.93}$$

where

$$A = \text{piston area.}$$

The force balance on the piston yields

$$A(p_1 - p_2) - F = m\ddot{y}, \tag{2.94}$$

where

$m = $ mass of the piston and the attached rod,

$F = $ force applied by the piston rod to the control surface attachment point.

Furthermore, the moment balance of the control surface using Eq. (2.14) yields

$$I\ddot{\theta} = Fl\cos\theta - F_a d, \tag{2.95}$$

where

I = moment of inertia of the control surface and attachment about the hinge,

F_a = applied aerodynamic load.

To solve this set of five equations, we require the additional kinematic relationship between θ and y,

$$y = l\sin\theta. \tag{2.96}$$

The actuator is usually constructed so that the valve exposes the two passages equally; therefore, $R_1 = R_2$, and infer from Eqs. (2.91) to (2.93) that

$$p_s - p_1 = p_2 - p_e. \tag{2.97}$$

These relations complete the nonlinear differential equations of motion; they are formidable and difficult to solve.

(b) **Linearization and Simplification**: For the case where $\dot{y}$ = a constant ($\ddot{y} = 0$) and there is no applied load ($F = 0$), Eqs. (2.94) and (2.97) indicate that

$$p_1 = p_2 = \frac{p_s + p_e}{2}. \tag{2.98}$$

Therefore, using Eq. (2.93) and letting $\sin\theta = \theta$ (because θ is assumed to be small), we get

$$\dot{\theta} = \frac{\sqrt{p_s - p_e}}{\sqrt{2}A\rho Rl}x. \tag{2.99}$$

This represents a single integration between the input x and the output θ where the proportionality constant is a function only of the supply pressure and the fixed parameters of the actuator. Therefore, the parameters for the state-variable form of the equations are

$$\mathbf{x} \stackrel{\triangle}{=} \theta, \quad \mathbf{F} = 0, \quad \mathbf{G} = \frac{\sqrt{p_s - p_e}}{\sqrt{2}A\rho Rl}, \quad \text{and} \quad u \stackrel{\triangle}{=} x.$$

For the case $\dot{y}$ = constant but $F \neq 0$, Eqs. (2.94) and (2.97) indicate that

$$p_1 = \frac{p_s + p_e + F/A}{2}$$

and

$$\dot{\theta} = \frac{\sqrt{p_s - p_e - F/A}}{\sqrt{2}A_\rho Rl}x. \tag{2.100}$$

This result is also a single integration between the input x and the output θ, but the proportionality constant now depends on the applied load F.

As long as the commanded values of x produce θ motion that has a sufficiently small value of $\ddot{\theta}$, the approximation given by Eqs. (2.99) or (2.100) is valid and no other linearized dynamic relationships are necessary. However, as soon as the commanded values of x produce accelerations where the inertial forces ($m\ddot{y}$ and the reaction to $I\ddot{\theta}$) are a significant fraction of $p_s - p_e$, the approximations are no longer valid. We must then incorporate these forces into the equations, thus obtaining a dynamic relationship between x and θ that is much more involved than the pure integration implied by Eqs. (2.99) or (2.100). Typically, for initial control system designs, hydraulic actuators are assumed to obey the simple relationship of Eqs. (2.99) or (2.100).

2.6   Linearization and Scaling

The differential equations of motion for almost all processes selected for control are nonlinear. On the other hand, as will be evident in the next chapter, both analysis and control design are far easier for linear than for nonlinear models. **Linearization** is the process of finding a linear model that approximates a nonlinear one. Fortunately, as Lyapunov proved over 100 years ago, if a small-signal linear model is valid near an equilibrium and is stable, then there is a region (which may be small, of course) containing the equilibrium within which the nonlinear system is stable.[11] So we can safely make a linear model and design a linear control for it such that, at least in the neighborhood of the equilibrium, our design will be stable. Because a very important role of feedback control is to maintain the process variables near equilibrium, such small-signal linear models are a frequent starting point for control models. Small-signal linearization is discussed in Section 2.6.1.

An alternative approach to obtain a linear model for use as the basis of control system design is to use part of the control effort to cancel the nonlinear terms and to design the remainder of the control based on linear theory. This approach—linearization by feedback—is popular in the field of robotics, where it is called the **method of computed torque**. It is also a research topic for control of aircraft. Section 2.6.2 takes a brief look at this method. Finally, some nonlinear functions are such that an **inverse nonlinearity** can be found to be placed in series with it so the combination is linear. This method is often used to correct mild nonlinear characteristics of a sensor or actuator that have small variations in use.

The magnitude of the values of the variables in a problem is often very different, sometimes so much so that numerical difficulties arise. This was a serious problem years ago when equations were solved using analog computers,

---

[11] In 1949 the Russian scientist Aizerman conjectured that if a certain class of systems were stable with any linear gain between two limits, then the nonlinear system with a nonlinear gain characteristic that was kept between the same limits would also be stable. Unfortunately, this conjecture is not true.

and it was routine to *scale* the variables so that all had similar magnitudes. Today's widespread use of digital computers for solving differential equations has largely eliminated the need to scale a problem unless the number of variables is very large because computers are now capable of accurately handling numbers with wide variations in magnitude. Nevertheless, it is wise to understand the principle of scaling for the few cases where extreme variations in magnitude exist and scaling is necessary or the computer wordsize is limited. Sections 2.6.3 and 2.6.4 discuss two kinds of scaling.

2.6.1   Small-Signal Linearization

A nonlinear differential equation is one where the derivatives of the state have a nonlinear relationship to the state itself and/or the control. In other words, the differential equations *cannot* be written in the form

$$\dot{\mathbf{x}} = \mathbf{F}\mathbf{x} + \mathbf{G}u$$

but must be left in the form [12]

$$\dot{\mathbf{x}} = \mathbf{f}(\mathbf{x}, u).$$

For small-signal linearization we first determine equilibrium values of $\mathbf{x}_o$, $\mathbf{u}_o$, that is, values where $\dot{\mathbf{x}}_o = \mathbf{0} = \mathbf{f}(\mathbf{x}_o, u_o)$. We then expand the nonlinear equation in terms of perturbations from these equilibrium values; that is, we let $\mathbf{x} = \mathbf{x}_o + \delta\mathbf{x}$ and $u = u_o + \delta u$, so that

$$\dot{\mathbf{x}}_o + \delta\dot{\mathbf{x}} \cong \mathbf{f}(\mathbf{x}_o, u_o) + \mathbf{F}\delta\mathbf{x} + \mathbf{G}\delta u,$$

where $\mathbf{F}$ and $\mathbf{G}$ are the best linear fits to the nonlinear function $\mathbf{f}(\mathbf{x}, u)$ at $\mathbf{x}_o$ and u_o, that is,

$$\mathbf{F} = \left[\frac{\partial \mathbf{f}}{\partial \mathbf{x}}\right]_{x_o, u_o} \quad \text{and} \quad \mathbf{G} = \left[\frac{\partial \mathbf{f}}{\partial u}\right]_{x_o, u_o}. \tag{2.101}$$

Substracting out the equilibrium solution, this reduces to

$$\delta\dot{\mathbf{x}} = \mathbf{F}\delta\mathbf{x} + \mathbf{G}\delta u, \tag{2.102}$$

which is a linear differential equation approximating the dynamics of the motion *about* the equilibrium point. Normally, the δ notation is dropped and it is understood that x and u refer to the deviation from the equilibrium.

In developing the models discussed so far in this chapter, we have encountered nonlinear equations on several occasions: the pendulum in Example 2.5, the hanging crane in Example 2.6, the AC induction motor in Section 2.4, the

---

[12] This equation assumes the system is time-invariant. A more general expression would be $\dot{\mathbf{x}} = \mathbf{f}(\mathbf{x}, u, t)$.

tank flow in Example 2.20, and the hydraulic actuator in Example 2.21.   In each case, we assumed either that the motion was small or that motion from some operating point was small, so that nonlinear functions were approximated by linear functions. The steps followed in those examples essentially involved finding **F** and **G** in order to linearize the differential equations to the form of Eq. (2.102) as illustrated in the following example.

EXAMPLE 2.22   *Linearization of Motion in a Ball Levitator*

Figure 2.34 shows a magnetic bearing used in large turbo machinery. The magnetics are energized using feedback control methods so that the axle is always in the center and never touches the magnets, thus keeping friction to an almost nonexistent level. A simplified version of a magnetic bearing that can be built in a laboratory is shown in Fig. 2.35, where one electromagnet is used to levitate a metal ball. The physical arrangement of the levitator is depicted in Fig. 2.36. The equation of motion of the ball, derived from Newton's law, Eq. (2.1), is

$$m\ddot{x} = f_m(x, i) - mg, \qquad (2.103)$$

where the force $f_m(x, i)$ is caused by the field of the electromagnet. Theoretically, the force from an electromagnet falls off with an inverse square relationship to the distance from the magnet, but the exact relationship for the laboratory levitator is difficult to derive from physical principles because its magnetic field is so complex. However, the forces can be measured with a scale. Figure 2.37 shows the experimental curves for a ball with a 1-cm diameter and a mass of 8.4×10^{-3} kg. At the value for the current of $i_2 = 600$ mA and the displacement x_1 shown in the figure, the magnetic force f_m just cancels the gravity force $mg = 82 \times 10^{-3}$ N. (The mass of the ball is

Figure 2.34
A magnetic bearing *(Photo courtesy of Magnetic Bearings, Inc.)*

Figure 2.35
Magnetic ball levitator used
in the laboratory

Figure 2.36
Model for ball levitation

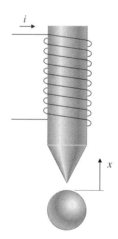

Figure 2.37
Experimentally determined
force curves

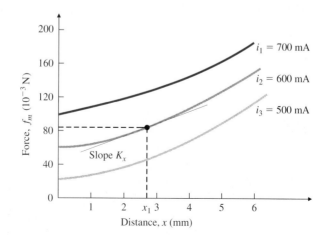

8.4×10^{-3} kg, and the acceleration of gravity is 9.8 m/sec².) Therefore the point (x_1, i_2) represents an equilibrium. Using the data, find the linearized equations of motion about the equilibrium point.

Solution.   First we write in expansion form the force in terms of deviations from the equilibrium values x_1 and i_2:

$$f_m(x_1 + \delta x, i_2 + \delta i) \cong f_m(x_1, i_2) + K_x \delta x + K_i \delta i. \qquad (2.104)$$

The linear gains are found as follows: K_x is the slope of the force versus x along the curve $i = i_2$, as shown in Fig. 2.37, and is found to be about 14 N/m. K_i is the change of force with current for the value of fixed $x = x_1$. We find that for $i = i_1 = 700$ mA at $x = x_1$, the force is about 122×10^{-3} N, and at $i = i_3 = 500$ mA at $x = x_1$, it is about 42×10^{-3} N. Thus

$$K_i \cong \frac{122 \times 10^{-3} - 42 \times 10^{-3}}{700 - 500} = \frac{80 \times 10^{-3} \text{ N}}{200 \text{ mA}}$$

$$\cong 400 \times 10^{-3} \text{ N/A}$$

$$\cong 0.4 \text{ N/A}.$$

Substituting these values into Eq. (2.104) leads to the following linear approximation for the force in the neighborhood of equilibrium:

$$f_m \cong 82 \times 10^{-3} + 14\delta x + 0.4\delta i.$$

Substituting this expression into Eq. (2.103) and using the numerical values for mass and gravity force, we get for the linearized model

$$(8.4 \times 10^{-3})\ddot{x} = 82 \times 10^{-3} + 14\delta x + 0.4\delta i - 82 \times 10^{-3}.$$

Because $x = x_1 + \delta x$, then $\ddot{x} = \delta\ddot{x}$. The equation in terms of δx is thus

$$(8.4 \times 10^{-3})\delta\ddot{x} = 14\delta x + 0.4\delta i,$$

$$\delta\ddot{x} = 1667\delta x + 47.6\delta i, \qquad (2.105)$$

which is the desired linearized equation of motion about the equilibrium point. A logical state vector is $\mathbf{x} = [\delta x \ \delta\dot{x}]^T$, which leads to the standard matrices

$$\mathbf{F} = \begin{bmatrix} 0 & 1 \\ 1667 & 0 \end{bmatrix} \quad \text{and} \quad \mathbf{G} = \begin{bmatrix} 0 \\ 47.6 \end{bmatrix}$$

and the control $u = \delta i$.

EXAMPLE 2.23   *Alternate Linearization of the Water Tank*

Repeat the linearization of Example 2.20 using the concepts presented in this section.

Solution.   Equation (2.87) may be written as

$$\dot{x} = f(x, u) \tag{2.106}$$

where $x \triangleq h$, $u \triangleq w_{in}$, and $f = -\frac{1}{RA\rho}\sqrt{p_1 - p_a} + \frac{1}{A\rho}w_{in} = -\frac{1}{RA\rho}\sqrt{\rho gh - p_a} + \frac{1}{A\rho}w_{in}$.
The linearized equations are of the form

$$\delta\dot{x} = F\delta x + G\delta u, \tag{2.107}$$

where

$$[F]_{x_o,u_o} = \frac{\partial f}{\partial x} = \left[\frac{\partial f}{\partial h}\right]_{h_o,u_o} = \frac{\partial}{\partial h}\left[-\frac{1}{RA\rho}\sqrt{\rho gh - p_a}\right]_{h_o,u_o} \tag{2.108}$$

$$= -\frac{g}{2AR}\frac{1}{\sqrt{\rho gh_o - p_a}} = -\frac{g}{2AR}\frac{1}{\sqrt{p_o - p_a}} \tag{2.109}$$

and

$$[G]_{x_o,u_o} = \frac{\partial f}{\partial u} = \frac{\partial f}{\partial w_{in}} = \frac{1}{A\rho}. \tag{2.110}$$

However, note that some flow is required to maintain the system in equilibrium so that Eq. (2.107) is valid; specifically, we see from Eq. (2.87) that

$$u_o = w_{in_o} = \frac{1}{R}\sqrt{p_o - p_a} \qquad \text{for } \dot{h} = 0, \tag{2.111}$$

and the δu in Eq. (2.107) is δw_{in}, where $w_{in} = w_{in_o} + \delta w_{in}$. Therefore, Eq. (2.107) becomes

$$\delta\dot{h} = F\delta h + G\delta w_{in} = F\delta h + Gw_{in} - G\frac{1}{R}\sqrt{p_o - p_a} \tag{2.112}$$

and matches Eq. (2.90) precisely.

2.6.2 Linearization by Feedback

Linearization by feedback is accomplished by substracting the nonlinear terms out of the equations of motion and adding them to the control. The result is a linear system, provided that the computer implementing the control has enough capability to compute the nonlinear terms fast enough. A more detailed understanding of the method is best achieved through example.

To illustrate linearization by feedback, we consider the equation of a simple pendulum developed in Example 2.5 [Eq. (2.21)]:

$$ml^2\ddot{\theta} + mgl\sin\theta = T_c. \tag{2.113}$$

If we compute the torque T_c to be

$$T_c = mgl \sin \theta + u,$$ (2.114)

then the motion is described by

$$ml^2 \ddot{\theta} = u.$$ (2.115)

Equation (2.115) is a linear equation *no matter how large the angle* θ *becomes.* We use it as the model for purposes of control design because it enables us to use linear analysis techniques. The resulting linear control will provide the value of u based on measurements of θ; however, the value of the torque actually sent to the equipment would derive from Eq. (2.114). For robots with two or three rigid links, this computed-torque approach has led to effective control. It is also being researched for the control of aircraft, where the linear models change considerably in character with the flight regime.

2.6.3   Amplitude Scaling

There are two types of scaling that are sometimes carried out: amplitude scaling and time scaling. **Amplitude scaling** is usually performed unwittingly by simply picking units that make sense for the problem at hand. For the ball levitator, expressing the motion in millimeters and the current in milliamps would keep the numbers within a range that is easy to work with. Equation (2.105) was developed in the standard SI units of meters, kilograms, and amperes, but in computing the motion of a rocket going into orbit, using kilometers makes more sense. The equations of motion are usually solved using computer-aided design software, which is often capable of working in any units. For higher-order systems it becomes important to scale the problem so that the elements of the state vector have similar numerical variations. A method for accomplishing the best scaling for a complex system is first to estimate the maximum values for each state element and then to scale the system so that each element varies between -1 and 1.

In general, we can perform amplitude scaling by defining the scaled variables for each state element: If

$$x' = S_x x,$$ (2.116)

then

$$\dot{x}' = S_x \dot{x} \quad \text{and} \quad \ddot{x}' = S_x \ddot{x}.$$ (2.117)

We then pick S_x to result in the appropriate scale change, substitute Eqs. (2.116) and (2.117) into the equations of motion, and recompute the coefficients.

EXAMPLE 2.24   *Scaling for the Ball Levitator*

Scale the variables for the ball levitator in Example 2.22 to result in units of millimeters and milliamps instead of meters and amps.

Solution.   Referring to Eq. (2.116), we define

$$\delta x' = S_x \delta x \quad \text{and} \quad \delta i' = S_i \delta i$$

such that both S_x and S_i have a value of 1000 in order to convert δx and δi in meters and amps to $\delta x'$ and $\delta i'$ in millimeters and milliamps. Substituting these relations into Eq. (2.105) and taking note of Eq. (2.117) yields

$$\delta \ddot{x}' = 1667 \delta x' + 47.6 \frac{S_x}{S_i} \delta i'.$$

In this case $S_x = S_i$, so Eq. (2.105) remains unchanged. Had we scaled the two quantities by different amounts, there would have been a change in the last coefficient in the equation.

2.6.4   Time Scaling

The unit of time when using SI units or English units is seconds. Computer-aided design software is *usually* able to compute results accurately no matter how fast or slow the particular problem at hand. However, if a dynamic system responds in a few microseconds or if there are characteristic frequencies in the system on the order of several MHz, the problem may become ill-conditioned, so that the numerical routines produce errors. This can be particularly troublesome for high-order systems. The same holds true for an extremely slow system. It is therefore useful to know how to change the units of time should you encounter an ill-conditioned problem.

We define the new scaled time to be

$$\tau = \omega_o t \tag{2.118}$$

such that, if t is measured in seconds and $\omega_o = 1000$, then τ will be measured in milliseconds. The effect of the time scaling is to change the differentiation so that

$$\dot{x} = \frac{dx}{dt} = \frac{dx}{d(\tau/\omega_o)} = \omega_o \frac{dx}{d\tau}, \tag{2.119}$$

and

$$\ddot{x} = \frac{d^2 x}{dt^2} = \omega_o^2 \frac{d^2 x}{d\tau^2}. \tag{2.120}$$

Putting the equation into state-variable form allows a more concise way of stating time scaling. For the system described by

$$\dot{\mathbf{x}} = \mathbf{F}\mathbf{x} + \mathbf{G}u, \tag{2.121}$$

we say it is time-scaled, using $\tau = \omega_o t$, by the equations

$$\frac{d\mathbf{x}}{d\tau} = \frac{1}{\omega_o} \mathbf{F}\mathbf{x} + \frac{1}{\omega_o} \mathbf{G}u. \tag{2.122}$$

EXAMPLE 2.25   *Time Scaling an Oscillator*

The equation for an oscillator was derived in Example 2.5. For a case with a very fast natural frequency $\omega_n = 15,000$ rad/sec (about 2 kHz), Eq. (2.23) can be rewritten as

$$\ddot{\theta} + 15,000^2 \cdot \theta = 10^6 \cdot T_c.$$

Determine the time-scaled equation so that the unit of time is milliseconds.

Solution.   The value of ω_o in Eq. (2.118) is 1000. Equation (2.80) shows that

$$\frac{d^2\theta}{d\tau^2} = 10^{-6} \cdot \ddot{\theta},$$

and the time-scaled equation becomes

$$\frac{d^2\theta}{d\tau^2} + 15^2 \cdot \theta = T_c$$

In practice, we would then solve the equation

$$\ddot{\theta} + 15^2 \cdot \theta = T_c \tag{2.123}$$

and label the plots in milliseconds instead of seconds.
    In state-variable form with a state vector $\mathbf{x} = [\theta \, \dot{\theta}]^T$, the unscaled matrices are

$$\mathbf{F} = \begin{bmatrix} 0 & 1 \\ -15,000^2 & 0 \end{bmatrix} \quad \text{and} \quad \mathbf{G} = \begin{bmatrix} 0 \\ 10^6 \end{bmatrix}.$$

Applying Eq. (2.122) results in

$$\mathbf{F}' = \begin{bmatrix} 0 & \frac{1}{1000} \\ -\frac{15,000^2}{1000} & 0 \end{bmatrix} \quad \text{and} \quad \mathbf{G}' = \begin{bmatrix} 0 \\ 10^3 \end{bmatrix},$$

which yields state-variable equations which are equivalent to the scaled system in Eq. (2.123).

SUMMARY

- Mathematical modeling of the system to be controlled is the first step in analyzing and designing the required system controls. In this chapter we developed models for representative systems. Important equations for each category of system are summarized in Table 2.1.

- An alternative way of expressing the differential equations that character-ize the model of a linear system is the **state-variable form**,

$$\dot{\mathbf{x}} = \mathbf{Fx} + \mathbf{G}u,$$

$$y = \mathbf{Hx} + Ju.$$

TABLE 2.1   **Key Equations for Dynamic Models**

System	Important Laws or Relationships	Associated Equations	Equation Number
Mechanical	Translation motion (Newton's law)	$\mathbf{F} = ma$	(2.1)
	Rotational motion	$M = I\alpha$	(2.14)
	Motion of nonrigid bodies	$m\ddot{x} + b\dot{x} + kx = F$	(2.10)
State-variable form	Linear system	$\dot{\mathbf{x}} = \mathbf{Fx} + \mathbf{G}u$	(2.39)
		$y = \mathbf{Hx} + Ju$	(2.40)
Electrical	Operational amplifier		(2.47), (2.48)
Electromechanical	Law of motors	$F = Bli$	(2.54)
	Law of the generator	$e(t) = Blv$	(2.57)
	Torque developed in a rotor	$T = K_t i_a$	(2.61)
Back emf	Voltage generated as a result of rotation of a rotor	$e = K_e \dot{\theta}_m$	(2.62)
Heat flow	Heat-energy flow	$q = \frac{1}{R}(T_1 - T_2)$	(2.73)
	Temperature as a function of heat-energy flow	$\dot{T} = \frac{1}{C}q$	(2.74)
	Specific heat	$C = mc_v$	(2.75)
Fluid flow	Continuity relation (conservation of matter)	$\dot{m} = w_{in} - w_{out}$	(2.81)
	Force of a fluid acting on a piston	$f = pA$	(2.83)
	Effect of resistance to fluid flow	$w = \frac{1}{R}(p_1 - p_2)^{1/\alpha}$	(2.84)

Equations in state-variable form are conducive to solution by computer packages that were developed especially for matrix equations (e.g., MATLAB). In Section 2.2 we briefly introduced the state-variable form; it will be explored in more depth in Chapter 7.

- **Linearization** and **scaling** (Section 2.6) are methods by which certain complications of dealing with differential equations can be minimized. In linearization, nonlinear differential equations are approximated by linear ones by either (1) considering a small-signal linear model that is acurate near an equilibrium, or (2) linearization by feedback, or (3) introducing an inverse nonlinearity. Scaling of variables results in numerical values that fall within a narrow-enough range of magnitude to minimize errors and allow for ease of computation.

Review Questions

1. What is a "free-body" diagram?

2. What are the two forms of Newton's law?

3. Why is it convenient to write equations of motion in the state-variable form?

4. For a structural process to be controlled such as a robot arm, what is the meaning of "collocated control"? "Noncollocated control"?

5. When, why, and by whom was the device named an "operational amplifier"?

6. What is the major benefit of having zero input current to an operational amplifier?

7. State Kirchoff's Current Law.

8. State Kirchoff's Voltage Law.

9. Why is it important to have a small value for the armature resistance, R_a, of an electric motor?

10. What are the definition and units of the electric constant of a motor?

11. What are the definition and units of the torque constant of an electric motor?

12. Give the relationships for (a) heat flow across a substance and (b) heat storage in a substance.

13. Name and give the equations for the three relationships governing fluid flow.

14. Why do we approximate a physical model of the plant (which is *always* nonlinear) with a linear model?

Problems

Problems for Section 2.1

2.1. Write the differential equations for the mechanical systems shown in Fig. 2.38.

2.2. Write the equations of motion of a pendulum consisting of a thin, 2-kg stick of length l suspended from a pivot. How long should the rod be in order for the period to be exactly 2 sec? (The inertia I of a thin stick about an endpoint is $\frac{1}{3}ml^2$. Assume θ is small enough that $\sin\theta \cong \theta$.)

Figure 2.38
Mechanical systems

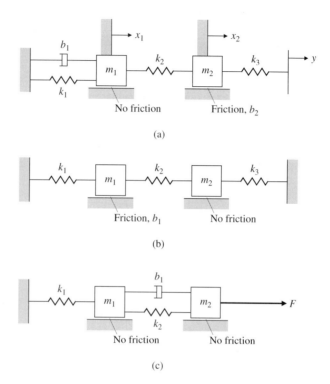

(a)

(b)

(c)

2.3. Write the equations of motion for the double-pendulum system shown in Fig. 2.39. Assume the displacement angles of the pendulums are small enough to ensure that the spring is always horizontal. The pendulum rods are taken to be massless, of length l, and the springs are attached 3/4 of the way down.

Figure 2.39
Double pendulum

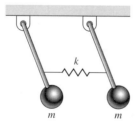

2.4. Write the equations of motion for a body of mass M suspended from a fixed point by a spring with a constant k. Carefully define where the body's displacement is zero.

2.5. For the car suspension discussed in Example 2.2,

   (a) Write the equations of motion [Eqs. (2.10) and (2.11)] in state-variable form. Use the state vector $\mathbf{x} = [x \quad \dot{x} \quad y \quad \dot{y}]^T$.

(b) Plot the position of the car and the wheel after the car hits a "unit bump" (i.e., r is a unit step) using MATLAB. Assume that $m_1 = 10$ kg, $m_2 = 350$ kg, $k_w = 500,000$ N/m, $k_s = 10,000$ N/m. Find the value of b that you would prefer if you were a passenger in the car.

2.6.   Automobile manufacturers are contemplating building active suspension systems. The simplest change is to make shock absorbers with a changeable damping, $b(u_1)$. It is also possible to make a device to be placed in parallel with the springs that has the ability to supply an equal force, u_2, in opposite directions on the wheel axle and the car body.

(a) Modify the equations of motion in Example 2.2 to include such control inputs.

(b) Is the resulting system linear?

(c) Is it possible to use the forcer, u_2, to completely replace the springs and shock absorber? Is this a good idea?

2.7.   Modify the equation of motion for the cruise control in Example 2.1, Eq. (2.4), so that it has a control law; that is, let

$$u = K(v_r - v) \tag{2.124}$$

where

$$v_r = \text{reference speed}, \tag{2.125}$$

$$K = \text{constant}. \tag{2.126}$$

This is a "proportional" control law where the difference between v_r and the actual speed is used as a signal to speed the engine up or slow it down. Put the equations in the standard state-variable form with v_r as the input and v as the state. Assume that $m = 1000$ kg and $b = 50$ N·sec/m, and find the response for a unit step in v_r using MATLAB. Using trial and error, find a value of K that you think would result in a control system in which the actual speed converges as quickly as possible to the reference speed with no objectionable behavior.

Problems for Section 2.2

2.8.   In many mechanical positioning systems there is flexibility between one part of the system and another. An example is shown in Fig. 2.6 where there is flexibility of the solar panels. Figure 2.40 depicts such a situation, where a force u is applied to the mass M and another mass m is connected to it. The coupling between the objects is often modeled by a spring constant k with a damping coefficient b, although the actual situation is usually much more complicated than this.

(a) Write the equations of motion governing this system, identify appropriate state variables, and express these equations in state-variable form.

(b) Find the transfer function between the control input, u, and the output, y.

Figure 2.40
Schematic of a system with
flexibility

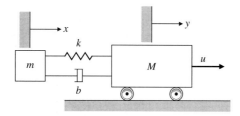

2.9. For the inverted pendulum, Eqs. (2.34),

(a) Try to put the equations of motion into state-variable form using the state vector $\mathbf{x} = [\theta \quad \dot\theta \quad x \quad \dot x]^T$. Why is it not possible?

(b) Write the equations in the "descriptor" form

$$\mathbf{E}\dot{\mathbf{x}} = \mathbf{F}'\mathbf{x} + \mathbf{G}'u,$$

and define values for $\mathbf{E}$, $\mathbf{F}'$, and $\mathbf{G}'$ (note that $\mathbf{E}$ is a 4×4 matrix). Then show how you would compute $\mathbf{F}$ and $\mathbf{G}$ for the standard state-variable description of the equations of motion.

2.10. The longitudinal linearized equations of motion of a Boeing 747 are given in Eq. (9.28). Using MATLAB or other computer aid:

(a) Determine the response of the altitude h for a 2-sec pulse of the elevator with a magnitude of $2°$. Note that, because Eq. (9.28) represents a set of linearized equations, the state variables actually represent the deviation of the state from the nominal operating point. For example, h represents the amount the altitude of the aircraft differs from 20,000 ft.

(b) Consider using the feedback law

$$\delta_e = K_h h + \delta_{e,ext}, \tag{2.127}$$

where the elevator input angle is the sum of a term proportional to the error in altitude h plus an external input (a disturbance or command input). Note from part (a) that a positive change in elevator causes a negative change in altitude, so that the proposed proportional feedback law has the logical sign to anticipate a stable system provided $K_h > 0$. By trial and error, try to find a value for the feedback gain K_h such that a $2°$ pulse of 2 sec on $\delta_{e,ext}$ yields a more stable altitude response.

(c) If you have trouble finding a value of K_h that produces a stable response, try modifying the feedback law to include information on pitch rate q :

$$\delta_e = K_h h + K_q q + \delta_{e,ext} \tag{2.128}$$

Use trial and error to pick appropriate values for both K_h and K_q. Assume the same type of pulse input for $\delta_{e,ext}$ as in part (b).

(d) Show that the further introduction of pitch-angle feedback, θ, such that

$$\delta_e = K_h h + K_q q + K_\theta \theta + \delta_{e,ext} \tag{2.129}$$

allows you to decrease the time it takes for the altitude to settle back to its nominal value, as well as to decrease the value of K_h required for a stable response. Note that, although $K_h = 0$ produces stable altitude behavior, we require $K_h > 0$ in order to guarantee that $h \to 0$ (so there will be no steady-state error).

2.11. A first step toward a realistic model of an op-amp is given by the equations below and shown in Fig. 2.41.

$$V_{out} = \frac{10^7}{s+1}[V_+ - V_-],$$

$$i_+ = i_- = 0.$$

Find the transfer function of the simple amplification circuit shown using this model.

Figure 2.41
Circuit for Problem 2.11

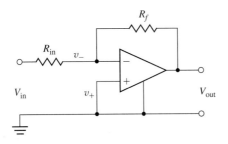

2.12. Show that the op-amp connection shown in Fig. 2.42 results in $V_o = V_{in}$ if the op-amp is ideal. Give the transfer function if the op-amp has the nonideal transfer function of Problem 2.11.

Figure 2.42
Circuit for Problem 2.12

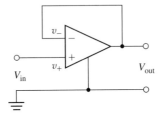

2.13. Show that, with the nonideal transfer function of Problem 2.11, the op-amp connection shown in Fig. 2.43 is *unstable*.

Figure 2.43
Circuit for Problem 2.13

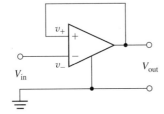

2.14. A common connection for a motor power amplifier is shown in Fig. 2.44. The idea is to have the motor current follow the input voltage and the connection is called a current amplifier. Assume that the sense resistor, R_s is very small compared with the feedback resistor, R and find the transfer function from V_{in} to I_a.

Figure 2.44
Op-Amp circuit for
Problem 2.14

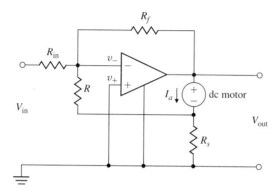

2.15. An op-amp connection with feedback to both the negative and the positive terminals is shown in Fig 2.45. If the op-amp has the nonideal transfer function given in Problem 2.11, give the maximum value possible for the positive feedback ratio, $P = r/(r + R)$ in terms of the negative feedback ratio, $N = R_{in}/(R_{in} + R_f)$ for the circuit to remain stable.

Figure 2.45
Op-Amp circuit for
Problem 2.15

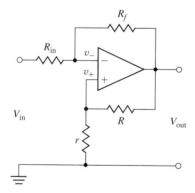

2.16. Write the dynamic equations and find the transfer functions for the circuits shown in Fig. 2.46.

    (a) Lead circuit,

    (b) Lag circuit,

    (c) Notch circuit.

Figure 2.46

Lead (a), lag (b), notch (c) circuits

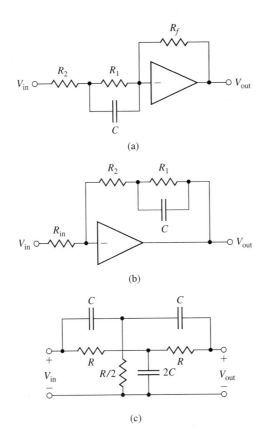

(a)

(b)

(c)

2.17. The very flexible circuit shown in Fig. 2.47 is called a biquad because its transfer function can be made to be the ratio of two second-order or quadratic polynomials. By selecting different values for R_a, R_b, R_c, and R_d the circuit can realize a low-pass, band-pass, high-pass, or band-reject (notch) filter.

   (a) Show that if $R_a = R$ and $R_b = R_c = R_d = \infty$, the transfer function from V_{in} to V_{out} can be written as the low-pass filter:

$$\frac{V_{out}}{V_{in}} = \frac{A}{\dfrac{s^2}{\omega_n^2} + 2\zeta \dfrac{s}{\omega_n} + 1}, \qquad (2.130)$$

where

$$A = \frac{R}{R_1},$$
$$\omega_n = \frac{1}{RC},$$
$$\zeta = \frac{R}{2R_2}.$$

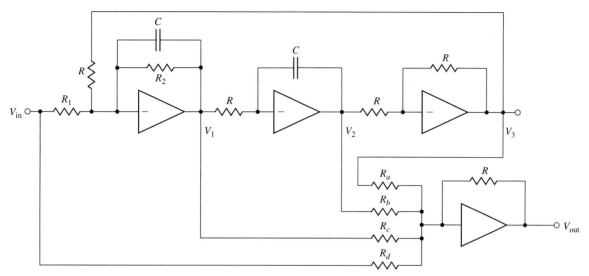

Figure 2.47 Op-amp

(b) Using the MATLAB comand step compute and plot on the same graph the step responses for the biquad of Fig. 2.47 for $A = 1$, $\omega_n = 1$, and $\zeta = 0.1$,   0.5, and 1.0.

2.18. Find the equations and transfer function for the biquad circuit of Fig. 2.47 if $R_a = R$, $R_d = R_1$, and $R_b = R_c = \infty$.

Problems for Section 2.4

2.19. The torque constant of a motor is the ratio of torque to current and is often given in ounce-inches per ampere (ounce-inches have dimension force–distance where an ounce is 1/16 of a pound). The electric constant of a motor is the ratio of back emf to speed and is often given in volts per 1000 rpm. In consistent units the two constants are the same for a given motor.

(a) Show that the units ounce-inches per ampere are proportional to volts per 1000 rpm by reducing both to MKS (SI) units.

(b) A certain motor has a back emf of 25 V at 1000 rpm. What is its torque constant in ounce-inches per ampere?

(c) What is the torque constant of the motor of part (b) in newton-meters per ampere?

2.20. A simplified sketch of a computer tape drive is given in Fig. 2.48.

(a) Write the equations of motion in terms of the parameters listed below. K and B represent the spring constant and the damping of tape stretch, respectively, and ω_1 and ω_2 are angular velocities. A positive current applied to the DC motor will provide a torque on the capstan in the clockwise direction as shown by the arrow. Assume that positive angular velocities of the two wheels are

Figure 2.48
Tape drive schematic

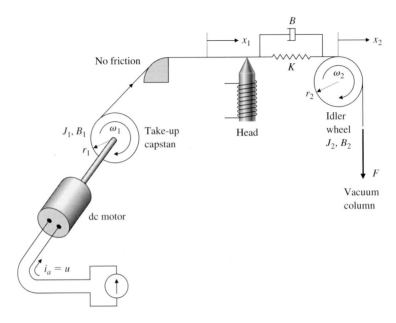

in the directions shown by the arrows.

$$J_1 = 5 \times 10^{-5} \text{ kg} \cdot \text{m}^2 \text{ (motor and capstan inertia)},$$

$$B_1 = 1 \times 10^{-2} \text{ N} \cdot \text{m} \cdot \text{sec (motor damping)},$$

$$r_1 = 2 \times 10^{-2} \text{ m},$$

$$K_t = 3 \times 10^{-2} \text{ N} \cdot \text{m/A (motor–torque constant)},$$

$$K = 2 \times 10^4 \text{ N/m},$$

$$B = 20 \text{ N/m} \cdot \text{sec},$$

$$r_2 = 2 \times 10^{-2} \text{ m},$$

$$J_2 = 2 \times 10^{-5} \text{ kg} \cdot \text{m}^2,$$

$$B_2 = 2 \times 10^{-2} \text{ N} \cdot \text{m} \cdot \text{sec (viscous damping, idler)},$$

$$F = 6 \text{ N, constant force,}$$

$$\dot{x}_1 = \text{tape velocity N/sec (variable to be controlled).}$$

(b) Write the equations in state-variable form as a set of first-order differential equations. Use the variables $(x_1, \omega_1, x_2, \omega_2, i_a)$.

(c) Use the values in part (a) and use MATLAB to find the response of x_1 to a step input in i_a.

2.21. Assume the driving force on the hanging crane of Fig. 2.14 is provided by a motor mounted on the cab with one of the support wheels connected directly to the motor's armature shaft. The motor constants are K_e and K_t, and the circuit driving the motor has a resistance R_a and negligible inductance. The wheel has a radius r. Write the equations of motion relating the applied motor voltage to the cab position and load angle.

2.22. The electromechanical system shown in Fig. 2.49 represents a simplified model of a capacitor microphone. The system consists in part of a parallel plate capacitor connected into an electric circuit. Capacitor plate a is rigidly fastened to the microphone frame. Sound waves pass through the mouthpiece and exert a force $f_s(t)$ on plate b, which has mass M and is connected to the frame by a set of springs and dampers. The capacitance C is a function of the distance x between the plates, as follows:

$$C(x) = \frac{\varepsilon A}{x},$$

where

ε = dielectric constant of the material between the plates,

A = surface area of the plates.

The charge q and the voltage e across the plates are related by

$$q = C(x)e.$$

The electric field in turn produces the following force f_e on the movable plate that opposes its motion:

$$f_e = \frac{q^2}{2\varepsilon A}.$$

(a) Write differential equations that describe the operation of this system. (It is acceptable to leave in nonlinear form.)

(b) Can one get a linear model?

(c) What is the output of the system?

Figure 2.49
Simplified model for capacitor microphone

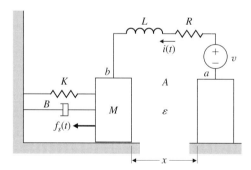

2.23. A very typical problem of electromechanical position control is an electric motor driving a load that has one dominant vibration mode. The problem arises in computer-disk-head control, reel-to-reel tape drives, and many other applications. A schematic diagram is sketched in Fig. 2.50. The motor has an electrical constant K_e, a torque constant K_t, an armature inductance L_a, and a resistance R_a. The rotor has an inertia J_1 and a viscous friction B. The load has an inertia J_2. The two inertias are connected by a shaft with a spring constant k and an equivalent viscous damping b.

Figure 2.50
Motor with a flexible load

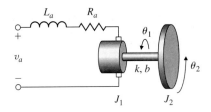

(a) Write the equations of motion.

(b) Write the equations as a set of simultaneous first-order equations in state-variable form. Use the state vector $\mathbf{x} = [\theta_2 \; \dot{\theta}_2 \; \theta_1 \; \dot{\theta}_1 \; i_a]^T$.

Problems for Section 2.5

2.24. A precision-table leveling scheme shown in Fig. 2.51 relies on thermal expansion of actuators under two corners to level the table by raising or lowering their respective corners. The parameters are:

$$T_{act} = \text{actuator temperature,}$$

$$T_{amb} = \text{ambient air temperature,}$$

$$R_f = \text{heat-flow coefficient between the actuator and the air,}$$

$$C = \text{thermal capacity of the actuator,}$$

$$R = \text{resistance of the heater.}$$

Assume that (1) the actuator acts as a pure electric resistance, (2) the heat flow into the actuator is proportional to the electric power input, and (3) the motion d is proportional to the difference between T_{act} and T_{amb} due to thermal expansion. Find the differential equations relating the height of the actuator d versus the applied voltage v_i.

Figure 2.51
(a) Precision table kept level
by actuators; (b) side view
of one actuator

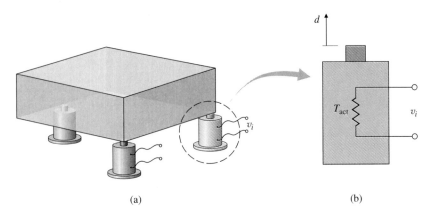

(a)   (b)

2.25. An air conditioner supplies cold air at the same temperature to each room on the fourth floor of the high-rise building shown in Fig. 2.52(a). The floor plan is shown in Fig. 2.52(b). The cold air flow produces an equal amount of heat flow q out of each room. Write a set of differential equations governing the temperature in each room, where

$$T_o = \text{temperature outside the building,}$$

$$R_o = \text{resistance to heat flow through the outer walls,}$$

$$R_i = \text{resistance to heat flow through the inner walls.}$$

Assume that (1) all rooms are perfect squares, (2) there is no heat flow through the floors or ceilings, and (3) the temperature in each room is uniform throughout the room. Take advantage of symmetry to reduce the number of differential equations to three.

Figure 2.52
Building air conditioning:
(a) high-rise building;
(b) floor plan of the fourth floor

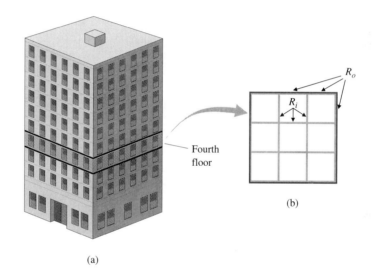

2.26. For the two-tank fluid-flow system shown in Fig. 2.53, find the differential equations relating the flow into the first tank to the flow out of the second tank.

Figure 2.53
Two-tank fluid-flow system
for Problem 2.26

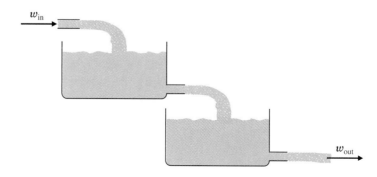

2.27. A laboratory experiment in the flow of water through two tanks is sketched in Fig. 2.54. Assume that Eq. (2.86) describes flow through the equal-sized holes at points A, B, or C.

(a) With holes at A and C but none at B, write the equations of motion for this system in terms of h_1 and h_2. Assume that $h_3 = 20$ cm, $h_1 > 20$ cm, and $h_2 < 20$ cm. When $h_2 = 10$ cm, the outflow is 200 g/min.

(b) At $h_1 = 30$ cm and $h_2 = 10$ cm, compute a linearized model and the transfer function from pump flow (in cubic centimeters per minute) to h_2.

(c) Repeat parts (a) and (b) assuming hole A is closed and hole B is open.

Figure 2.54

Two-tank fluid-flow system for Problem 2.27

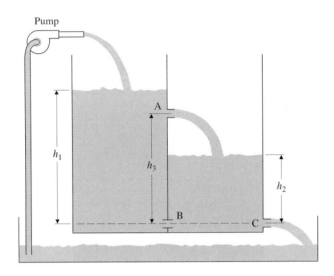

2.28. The equations for heating a house are given by Eqs. (2.73) and (2.74) and, in a particular case, can be written with time in hours as

$$C\frac{dT_h}{dt} = Ku - \frac{T_h - T_o}{R},$$

where

(a) C is the Thermal capacity of the house (BTU/°F),

(b) T_h is the temperature in the house (°F),

(c) T_o is the temperature outside the house (°F).

(d) K is the heat rating of the furnace ($= 90{,}000$ BTU/hr),

(e) R is the thermal resistance (°F per BTU/hr),

(f) u is the furnace switch ($= 1$ if the furnace is on and $= 0$ if the furnace is off).

It is measured that, with the outside temperature at 32°F and the house at 60°F, the furnace raises the temperature 2°F in 6 min (0.1 hr). With the furnace off, the house temperature falls 2°F in 40 min. What are the values of C and R for the house?

Problems for Section 2.6

2.29. Figure 2.55 shows a simple pendulum system in which a cord is wrapped around a fixed cylinder. The motion of the system that results is described by the differential equation

$$(l + R\theta)\ddot{\theta} + g\sin\theta + R\dot{\theta}^2 = 0,$$

where

$$l = \text{length of the cord in the vertical (down) position},$$

$$R = \text{radius of the cylinder}.$$

(a) Write the state-variable equations for this system.

(b) Linearize the equation around the point $\theta = 0$, and show that for small values of θ the system equation reduces to an equation for a simple pendulum, that is,

$$\ddot{\theta} + (g/l)\theta = 0.$$

Figure 2.55
Motion of cord wrapped around a fixed cylinder

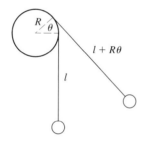

2.30. A schematic for the satellite and scientific probe for the Gravity Probe-B (GP-B) experiment is sketched in Fig. 2.56. Assume that the mass of the spacecraft plus helium tank, m_1, is 2000 kg and that the mass of the probe, m_2, is 1000 kg. A rotor will float inside of the probe and will be forced to follow the probe with a capacitive forcing mechanism; however, this will have no effect on m_2. The spring constant of the coupling, k, is 3.2×10^6. The viscous damping, b, is 4.6×10^3.

(a) Write the equations of motion for the system consisting of masses m_1 and m_2 using the inertial position variables, y_1 and y_2.

(b) The actual disturbance, u, is a micrometeorite and the resulting motion is very small. Therefore, rewrite your equations with the scaled variables $z_1 = \frac{y_1}{10^6}$, $z_2 = \frac{y_2}{10^6}$, and $v = 1000u$.

(c) Put the equations in state-variable form using the state $\mathbf{x} = [z_1 \quad \dot{z}_1 \quad z_2 \quad \dot{z}_2]^T$, the output $y = z_2$, and the input an impulse, $u = 10^{-3}\delta(t)$ N·sec on mass m_1.

(d) Using the numerical values, enter the equations of motion into MATLAB in the form

$$\dot{\mathbf{x}} = \mathbf{Fx} + \mathbf{G}u, \tag{2.131}$$

$$y = \mathbf{Hx} + Ju \tag{2.132}$$

Figure 2.56

Schematic diagram of the
GP-B satellite and probe

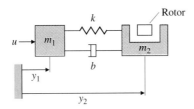

and define the MATLAB system: sysGPB = ss(F,G,H,J). Plot the response of
y caused by the impulse with the MATLAB command impulse(sysGPB). This
is the signal the rotor must follow.

2.31. The circuit shown in Fig. 2.57 has a nonlinear conductance G such that $i_G =$
$g(v_G) = v_G(v_G - 1)(v_G - 4)$. The state differential equations are

$$\frac{di}{dt} = -i + v,$$

$$\frac{dv}{dt} = -i + g(u - v),$$

where i and v are the states and u is the input.

(a) One equilibrium state occurs when $u = 1$ yielding $i_1 = v_1 = 0$. Find the
other two pairs of v and i that will produce equilibrium.

(b) Find the linearized model of the system about the equilibrium point $u = 1$,
$i = v_1 = 0$.

(c) Find the linearized models about the other two equilibrium points.

Figure 2.57

Nonlinear circuit for
Problem 2.31

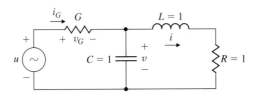

2.32. Consider the circuit shown in Fig. 2.58; u_1 and u_2 are voltage and current
sources, respectively, and R_1 and R_2 are nonlinear resistors with the following
characteristics:

Resistor 1 : $i_1 = G(v_1) = v_1^3$,
Resistor 2 : $v_2 = r(i_2)$,

where the function r is defined in Fig. 2.59.

(a) Show that the circuit equations can be written as

$$\dot{x}_1 = G(u_1 - x_1) + u_2 - x_3$$
$$\dot{x}_2 = x_3$$
$$\dot{x}_3 = x_1 - x_2 - r(x_3).$$

Figure 2.58
A non-linear circuit

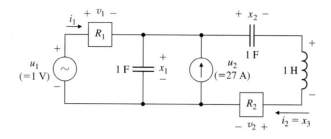

Figure 2.59
Non-linear resistance

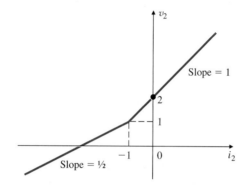

Suppose we have a constant voltage source of 1 V at u_1 and a constant current source of 27 A; that is, $u_1^0 = 1$ and $u_2^0 = 27$. Find the *equilibrium state* $\mathbf{x}^0 = (x_1^0, x_2^0, x_3^0)^T$ for the circuit. For a particular input $\mathbf{u}^0$, an equilibrium state of the system is defined to be any constant state vector whose elements satisfy the relation

$$\dot{x}_1 = \dot{x}_2 = \dot{x}_3 = 0.$$

Consequently, any system started in one of its equilibrium states will remain there indefinitely until a different input is applied.

(b) Due to disturbances, the initial state (capacitance, voltages, and inductor current) is slightly different from the equilibrium and so are the independent sources; that is,

$$u(t) = u^0 + \delta u(t),$$
$$x(t_0) = x^0(t_0) + \delta x(t_0).$$

Do a small-signal analysis of the network about the equilibrium found in (a), displaying the equations in the form

$$\delta \dot{x}_1 = f_{11}\delta x_1 + f_{12}\delta x_2 + f_{13}\delta x_3 + g_1\delta u_1 + g_2\delta u_2.$$

(c) Draw the circuit diagram that corresponds to the linearized model. Give the values of the elements.

3

Dynamic Response

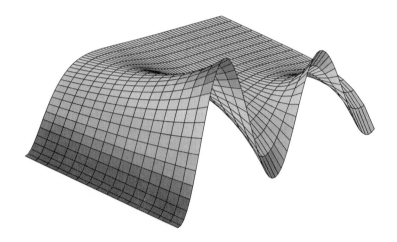

Chapter Overview

The Laplace transform, reviewed in Section 3.1 (and Appendix A), is the mathematical tool for transforming differential equations into an easier-to-manipulate algebraic form. In addition to the mathematical tools at our disposal, there are two graphical tools that can help us to visualize the model of a system and evaluate the pertinent mathematical relationships between elements of the system. One is the block diagram, which was introduced in Chapter 1; the other is the signal-flow graph. Both of these are discussed in Section 3.2.

Once the transfer function has been determined, we can identify its poles and zeros, which tell us a great deal about system characteristics including its frequency response introduced in Section 3.1. Sections 3.3 to 3.5 focus on poles and zeros and some of the ways for manipulating them to steer system characteristics in a desired way. When feedback is introduced, the possibility that the system may become *unstable* is introduced. To study this effect, in Section 3.6 we consider the definition of stability and Routh's test, which can determine stability by examining the coefficients of the system's characteristic equation. Instead of analyzing system dynamics through the transfer function, we may in certain circumstances prefer to use numerical solution techniques, as discussed in Section 3.7. Finally, a method for developing a model based on experimental time-response data is discussed in Section 3.8.

A Perspective on System Response

We saw in Chapter 2 how to obtain the dynamic model of a system. In designing a control system it is important to see how well a trial design matches the desired performance. We do this by solving the equations of the system model.

There are two ways to approach solving the dynamic equations. For a quick, *approximate* analysis we use linear analysis techniques. The resulting approximations of system response provide insight into why the solution has certain features and how the system might be changed to modify the response in a desired direction. In contrast, a *precise* picture of the system response typically calls for numerical techniques to solve the system equations. Although this chapter focuses primarily on linear analysis, we will also touch briefly on the numerical methods and computer tools that can be used to solve for the time response of both nonlinear and linear systems.

There are three domains within which to study dynamic response: the **s-plane**, the **frequency response**, and the **state space** (analysis using the state-variable description). The well-prepared control engineer needs to be fluent in all of them, so they will be treated in depth in Chapters 5 to 7, respectively. The purpose of this chapter is to discuss some of the fundamental mathematical tools needed before studying analysis in the s-plane and frequency response.

3.1 Review of Laplace Transforms

Two attributes of linear time-invariant systems form the basis for almost all analytical techniques applied to these systems:

1. A linear system response obeys the principle of superposition.
2. The response of a linear *time-invariant* system can be expressed as the convolution of the input with the unit impulse response of the system.

The concepts of superposition, convolution, and impulse response will be defined shortly.

From the second property (as we will show), it follows immediately that the response of a linear time-invariant system to an exponential input is also exponential. This result is the principal reason for the usefulness of Fourier and Laplace transforms in the study of linear time-invariant systems.

3.1.1 Response by Convolution

Superposition

The **principle of superposition** states that if the system has an input that can be expressed as a sum of signals, then the response of the system can be expressed as the sum of the individual responses to the respective signals. We can express superposition mathematically: Consider the system to have input u and output y; suppose further that, with the system at rest, we apply the input $u_1(t)$ and observe the output $y_1(t)$. After restoring the system to rest, we apply a second input $u_2(t)$ and again observe the output, which we call $y_2(t)$. Then, we form the composite input $u(t) = \alpha_1 u_1(t) + \alpha_2 u_2(t)$. Finally, if superposition applies, then the response will be $y(t) = \alpha_1 y_1(t) + \alpha_2 y_2(t)$. Superposition will apply if and only if the system is linear.

EXAMPLE 3.1

Superposition

Show that superposition holds for the system modeled by the first-order linear differential equation

$$\dot{y} + ky = u.$$

Solution.   We let $u = \alpha_1 u_1 + \alpha_2 u_2$ and assume that $y = \alpha_1 y_1 + \alpha_2 y_2$. Then $\dot{y} = \alpha_1 \dot{y}_1 + \alpha_2 \dot{y}_2$. If we substitute these expressions into the system equation, we get

$$\alpha_1 \dot{y}_1 + \alpha_2 \dot{y}_2 + k(\alpha_1 y_1 + \alpha_2 y_2) = \alpha_1 u_1 + \alpha_2 u_2.$$

From this it follows that

$$\alpha_1 (\dot{y}_1 + ky_1 - u_1) + \alpha_2 (\dot{y}_2 + ky_2 - u_2) = 0. \tag{3.1}$$

If y_1 is the solution with input u_1 and y_2 is the solution with input u_2, then Eq. (3.1) is satisfied, the response is the sum of the individual responses, and superposition holds.

Using the principle of superposition, we can solve for the system responses to a set of elementary signals. We are then able to solve for the response to a general signal simply by decomposing the given signal into a sum of the elementary components and, by superposition, concluding that the response to the general signal is the sum of the responses to the elementary signals. In order for this process to work, the elementary signals need to be sufficiently "rich" that any reasonable signal can be expressed as a sum of them, and their responses have to be easy to find. The most common candidates for elementary signals for use in linear systems are the impulse and the exponential.

The idea for the impulse comes from dynamics. Suppose we wish to study the motion of a baseball hit by a bat. The details of the collision between the bat and ball can be very complex as the ball deforms and the bat bends; however, for purposes of computing the path of the ball, we can summarize the effect of the collision as the net velocity change of the ball over a very short time period.

Impulse response

We assume that the ball is subjected to an **impulse**, a very intense force for a very short time. The physicist Paul Dirac suggested that such forces could be represented by the mathematical concept of an impulse, $\delta(t)$, which has the property that if $f(t)$ is continuous at $t = \tau$, then

$$\int_{-\infty}^{\infty} f(\tau)\delta(t - \tau)\, d\tau = f(t). \tag{3.2}$$

In other words, the impulse is so short and so intense that no value of f matters except over the short range where the δ occurs. Since integration is a limit of a summation process, Eq. (3.2) can be viewed as representing the function f as a sum of impulses. If we replace f by u, then Eq. (3.2) represents an input $u(t)$ as a sum of impulses of intensity $u(t - \tau)$. To find the response to an arbitrary input, the principle of superposition tells us we need only to find the response to a unit impulse. For a general linear system we can express the impulse response as $h(t, \tau)$, the response at t to an impulse applied at τ. The total response is then the sum (integral) of these with intensity u:

$$y(t) = \int_{-\infty}^{\infty} u(\tau)h(t, \tau)\, d\tau.$$

The superposition integral

This is the **superposition integral**. If the system is not only linear but also time-invariant, then the impulse response is given by $h(t - \tau)$ because the response at t to an input applied at τ depends only on the difference between the time the impulse is applied and the time we are observing the response. Time-invariant systems are called shift-invariant for this reason. For time-invariant systems the superposition integral then takes the special form

$$y(t) = \int_{-\infty}^{\infty} u(\tau)h(t - \tau)\, d\tau,$$

or

$$y(t) = \int_{-\infty}^{\infty} u(t - \tau)h(\tau)\, d\tau. \tag{3.3}$$

The convolution integral is a special form of the superposition integral.

Equation (3.3) is called the **convolution integral**. The integral occurs so often that the notation $y = u * h$ is often used for Eq. (3.3).

EXAMPLE 3.2   *Convolution*

We can illustrate convolution with a simple system. Consider the impulse response for the system described by the differential equation

$$\dot{y} + ky = u = \delta(t),$$

with an initial condition of $y(0) = 0$ before the impulse. Because $\delta(t)$ has an effect only near $t = 0$, we can integrate this equation from just before zero to just after zero with the result

$$\int_{0^-}^{0^+} \dot{y}\,dt + k \int_{0^-}^{0^+} y\,dt = \int_{0^-}^{0^+} \delta(t)\,dt.$$

The integral of $\dot{y}$ is simply y, the integral of y over so small a range is zero, and the integral of the impulse over the same range is unity. Therefore,

$$y(0^+) - y(0^-) = 1.$$

Because the system was at rest before application of the impulse, $y(0^-) = 0$. Thus the effect of the impulse is that $y(0^+) = 1$. For positive time we have the differential equation

$$\dot{y} + ky = 0, \qquad y(0^+) = 1.$$

If we assume a solution $y = Ae^{st}$, then $\dot{y} = Ase^{st}$. The above equation then becomes

$$Ase^{st} + kAe^{st} = 0,$$

$$s + k = 0,$$

$$s = -k.$$

Because $y(0^+) = 1$, it is necessary that $A = 1$. Thus the solution for the impulse response is $y(t) = h(t) = e^{-kt}$ for $t > 0$. To take care of the fact that $h(t) = 0$ for negative time, we define the **unit step function** as

$$1(t) = \begin{cases} 0, & t < 0, \\ 1, & t \geq 0 \end{cases}$$

With this definition the impulse response of the first-order system becomes

$$h(t) = e^{-kt} 1(t).$$

The response to a general input is given by the convolution of this impulse response and the input:

$$y(t) = \int_{-\infty}^{\infty} h(\tau) u(t - \tau)\,d\tau$$

$$= \int_{-\infty}^{\infty} e^{-k\tau} 1(\tau) u(t - \tau)\,d\tau$$

$$= \int_{0}^{\infty} e^{-k\tau} u(t - \tau)\,d\tau.$$

3.1.2   Transfer Functions and Frequency Response

An immediate consequence of convolution is that an input of the form e^{st} results in an output $H(s)e^{st}$. Note that both input and output are exponential time functions, and that the output differs from the input only in the amplitude $H(s)$. $H(s)$ is the **transfer function** of the system. The constant s may be complex, expressed as $s = \sigma + j\omega$. Thus both the input and the output may be complex. If we let $u(t) = e^{st}$ in Eq. (3.3), then

Transfer function

$$y(t) = \int_{-\infty}^{\infty} h(\tau)u(t-\tau)\,d\tau$$

$$= \int_{-\infty}^{\infty} h(\tau)e^{s(t-\tau)}\,d\tau$$

$$= \int_{-\infty}^{\infty} h(\tau)e^{st}e^{-s\tau}\,d\tau$$

$$= \int_{-\infty}^{\infty} h(\tau)e^{-s\tau}\,d\tau\, e^{st}$$

$$= H(s)e^{st}, \tag{3.4}$$

where[1]

$$H(s) = \int_{-\infty}^{\infty} h(\tau)e^{-s\tau}d\tau. \tag{3.5}$$

The integral in Eq. (3.5) does not need to be computed to find the transfer function of a system. Instead, one can assume a solution of the form of Eq. (3.4), substitute that into the differential equation of the system, then solve for the transfer function $H(s)$.

EXAMPLE 3.3

Transfer Function

Compute the transfer function for the system of Example 3.2, and find the output y for the input $u = e^{st}$.

Solution.   The system equation from Example 3.2 is

$$\dot{y}(t) + ky(t) = u(t) = e^{st}. \tag{3.6}$$

We assume that we can express $y(t)$ as $H(s)e^{st}$. With this form, we have $\dot{y} = sH(s)e^{st}$, and Eq. (3.6) reduces to

$$sH(s)e^{st} + kH(s)e^{st} = e^{st}. \tag{3.7}$$

---

[1] Notice that this input is exponential for all time and that Eq. (3.5) represents the response for all time. If the system is causal, then $h(t) = 0$ for $t < 0$, and the integral reduces to $H(s) = \int_{0}^{\infty} h(\tau)e^{-s\tau}\,d\tau$.

Solving for the transfer function $H(s)$, we get

$$H(s) = \frac{1}{s + k}.$$

Substituting this back into Eq. (3.4) yields the output

$$y = \frac{e^{st}}{s + k}.$$

Frequency response

A very common way to use the exponential response of linear time-invariant systems is in finding the **frequency response,** or response to a sinusoid. First we express the sinusoid as a sum of two exponential expressions (Euler's relation):

$$A\cos(\omega t) = \frac{A}{2}(e^{j\omega t} + e^{-j\omega t}).$$

If we let $s = j\omega$ in the basic response formula Eq. (3.4), then the response to $u(t) = e^{j\omega t}$ is $y(t) = H(j\omega)e^{j\omega t}$; similarly, the response to $u(t) = e^{-j\omega t}$ is $H(-j\omega)e^{-j\omega t}$. By superposition, the response to the sum of these two exponentials, which make up the cosine signal, is the sum of the responses:

$$y(t) = \frac{A}{2}[H(j\omega)e^{j\omega t} + H(-j\omega)e^{-j\omega t}]. \tag{3.8}$$

The transfer function $H(j\omega)$ is a complex number that can be represented in polar form or in magnitude-and-phase form as $H(j\omega) = M(\omega)e^{j\varphi(\omega)}$ or simply $H = Me^{j\varphi}$. With this substitution, Eq. (3.8) becomes

$$y(t) = \frac{A}{2}M(e^{j(\omega t + \varphi)} + e^{-j(\omega t + \varphi)})$$

$$= AM\cos(\omega t + \varphi), \tag{3.9}$$

where

$$M = |H(j\omega)|, \quad \varphi = \angle H(j\omega).$$

This means that if a system represented by the transfer function $H(s)$ has a sinusoidal input with magnitude A, the output will be sinusoidal at the same frequency with magnitude AM and will be shifted in phase by the angle φ.

EXAMPLE 3.4

Frequency Response

For the system in Example 3.2, find the response to the sinusoidal input $u = A\cos(\omega t)$. That is,

(a) find the frequency response and plot the response for $k = 1$, and

(b) determine the complete response due to the sinusoidal input $u(t) = \sin(10t)$ again with $k = 1$.

Solution.

(a) In Example 3.3 we found the transfer function. To find the frequency response, we let $s = j\omega$ so that

$$H(s) = \frac{1}{s + k} \Longrightarrow H(j\omega) = \frac{1}{j\omega + k}.$$

From this we get

$$M = \frac{1}{\sqrt{\omega^2 + k^2}} \quad \text{and} \quad \varphi = -\tan^{-1}\left(\frac{\omega}{k}\right).$$

Therefore, the response of this system to a sinusoid will be

$$y(t) = AM\cos(\omega t + \varphi). \tag{3.10}$$

M is usually referred to as the **amplitude ratio** and φ is referred to as the **phase**, and they are both functions of the input frequency, ω.

The MATLAB program below is used to compute the amplitude ratio and phase for $k = 1$ as shown in Fig. 3.1. The logspace command is used to set the frequency range (on a logarithmic scale), and the bode command is used to compute the frequency response in MATLAB. Presenting frequency response in this manner (i.e., on log–log scale) was originated by H. W. Bode; thus, these plots are referred to as "Bode plots."[2] (see Chapter 6, Section 6.1).

Figure 3.1
Frequency response for $k = 1$

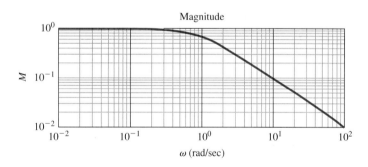

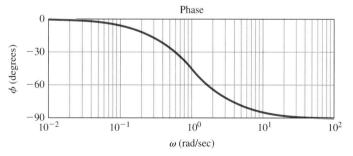

---

[2] Note that % is used in MATLAB to denote comments.

```
k = 1;
num = 1;                        % form numerator
den = [1 k];                    % form denominator
sysH = tf(numH,denH);          % define system by its numerator and denominator
w = logspace(-2,2);            % set frequency w to 50 values from 10^-2 to 10^+2
[mag,phase] = bode(sysH,w);    % compute frequency response
loglog(w,mag);                 % log–log plot of magnitude
semilogx(w,phase);             % semi log plot of phase
```

(b) To determine the response to an input that begins at $t = 0$ as $u(t) = \sin(10t)1(t)$, notice that from Laplace transform tables (Appendix A, Table A.2),

$$\mathcal{L}\{u(t)\} = \mathcal{L}\{\sin(10t)\} = \frac{10}{s^2 + 100},$$

where $\mathcal{L}$ denotes the Laplace transform, and the output of the system using partial fraction expansion (see Section 3.1.5) is given by

$$
\begin{aligned}
Y(s) &= H(s)U(s) \\
&= \frac{1}{s+1}\frac{10}{s^2 + 100} \\
&= \frac{\alpha_1}{s+1} + \frac{\alpha_0}{s + j10} + \frac{\alpha_0^*}{s - j10} \\
&= \frac{\frac{10}{101}}{s+1} + \frac{\frac{j}{2(1 - j10)}}{s + j10} + \frac{\frac{-j}{2(1 + j10)}}{s - j10}.
\end{aligned}
$$

The inverse Laplace transform of the output is given by (see Appendix A)

$$
\begin{aligned}
y(t) &= \frac{10}{101}e^{-t} + \frac{1}{\sqrt{101}}\sin(10t + \varphi) \\
&= y_1(t) + y_2(t),
\end{aligned}
$$

where

$$\varphi = \tan^{-1}(-10) = -84.2°.$$

The component $y_1(t)$ is called the *transient* response because it decays to zero as time goes on and the component $y_2(t)$ is called the *steady-state* and equals the response given by Eq. (3.10). Figure 3.2(a) is a plot of the time history of the output showing the different components (y_1, y_2) and the composite (y) output response. The output frequency is 10 rad/sec and the steady-state phase difference measured from Fig. 3.2(b) is approximately $10 * \delta t = 1.47$ rad $= 84.2°$.[3] Figure 3.2(b) shows the output lags the input by $84.2°$. Figure 3.2(b) shows that the steady-state amplitude of the output is the amplitude ratio $1/\sqrt{101} = 0.0995$—that is, the amplitude of the input signal times the magnitude of the transfer function evaluated at $\omega = 10$ rad/sec.

---

[3] The phase difference may also be determined by a Lissajous pattern.

Figure 3.2
(a) Complete transient
response, (b) Phase lag
between output and input

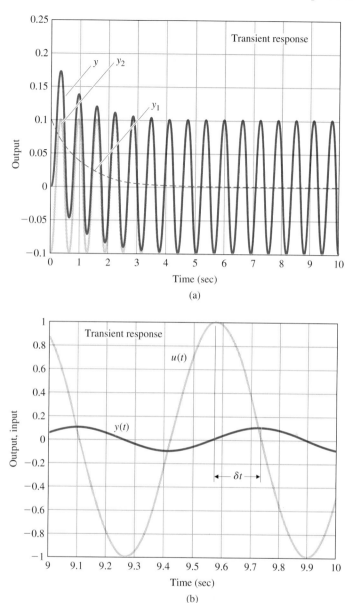

(a)

(b)

This example illustrates that the response of a linear time-invariant system to
a sinusoid of frequency ω is a sinusoid with the *same* frequency and with an amplitude ratio equal to the magnitude of the transfer function evaluated at the input
frequency. Furthermore, the phase difference between input and output signals is
given by the phase of the transfer function evaluated at the input frequency. The
magnitude ratio and phase difference can be computed from the transfer function
as just discussed; it can also be measured experimentally quite easily in the lab-

oratory by driving the system with a known sinusoidal input and measuring the steady-state amplitude and phase of the system's output. The input frequency is set to sufficiently many values so that curves such as the one in Fig. 3.1 are obtained.

We can generalize the frequency response by defining the **Laplace transform** of a signal $f(t)$ as

$$F(s) = \int_{-\infty}^{\infty} f(t)e^{-st}\, dt. \tag{3.11}$$

The key property of Laplace transforms

If we apply this definition to both $u(t)$ and $y(t)$ and use the convolution integral Eq. (3.3), we find that

$$Y(s) = H(s)U(s), \tag{3.12}$$

where $Y(s)$ and $U(s)$ are the Laplace transforms of $y(t)$ and $u(t)$, respectively. We prove this result in Appendix A.

Transient response

Laplace transforms such as Eq. (3.11) can be used to study the complete response characteristics of feedback systems, including the **transient response**; that is, the time response to an initial condition or suddenly applied signal. This is in contrast to the use of Fourier transforms, where the steady-state response is the main concern. A standard problem in control is to find the response, $y(t)$, of a system given the input, $u(t)$, and a model of the system. With Eq. (3.11) we have a means for computing the response of linear time-invariant systems to quite general inputs. Given any input into a system, we compute the transform of the input and the transfer function for the system. The transform of the output is then given by Eq. (3.12) as the product of these two. If we wanted the time function of the output, we would need to "invert" $Y(s)$ to get what is called the **inverse transform**; this step is typically not carried out explicitly. Nevertheless, understanding the process necessary for deriving $y(t)$ from $Y(s)$ is important because it leads to insight into the behavior of linear systems. Hence given a general linear system with transfer function $H(s)$ and an input signal $u(t)$, the procedure for determining y(t) using the Laplace transform is given by the following steps:

STEP 1. Determine the transfer function: $H(s) = \mathcal{L}\{\text{impulse response of the system}\}$. Compute H(s) by the following steps:

(a) Take the Laplace transform of the equations of motion. A table of transform properties is frequently useful in this process.

(b) Solve the resulting algebraic equations. Often this step is greatly helped by drawing the corresponding block diagram and solving the equations by graphical manipulation of the blocks or using MATLAB.

STEP 2. Determine the Laplace transform of the input signal: $U(s) = \mathcal{L}\{u(t)\}$.

STEP 3. Determine the Laplace transform of the output: $Y(s) = H(s)U(s)$.

STEP 4. Break up $Y(s)$ by partial-fraction expansion.

STEP 5. Find the output of the system by computing the inverse Laplace transform of $Y(s)$ in Step 4: $y(t) = \mathcal{L}^{-1}\{Y(s)\}$; that is, invert $Y(s)$ to get $y(t)$.

(a) Look up the components of $y(t)$ in a table of transform-time function pairs.

(b) Combine the components to give the total solution in the desired form.

As already mentioned, Steps 4 and 5 are almost never carried out in practice and a modified solution for a *qualitative* rather than a quantitative solution is often adequate and *almost always used for control design purposes*. The process begins with the first three steps as before. However, rather than inverting $Y(s)$, one can use prior knowledge and intuition about the effects of pole and zero locations in $Y(s)$ on the response $y(t)$ to estimate key parameters of $y(t)$. That is we get information about $y(t)$ from the pole-zero constellation of $Y(s)$ without actually inverting it as discussed in the rest of this chapter.

While it is possible to determine the transient response properties of the system using Eq. (3.11), it is generally more useful to use a simpler version of the Laplace transform based on the input beginning at time zero.

3.1.3   The $\mathcal{L}_-$ Laplace Transform

In many applications it is useful to define a **one-sided** (or **unilateral**) **Laplace transform**, which uses 0^- (that is, a value just before $t = 0$) as the lower limit of integration in Eq. (3.11). The $\mathcal{L}_-$ Laplace transform of $f(t)$, denoted by $\mathcal{L}_-\{f(t)\} = F(s)$, is a function of the complex variable $s = \sigma + j\omega$, where

Definition of Laplace transform

$$F(s) \stackrel{\triangle}{=} \int_{0^-}^{\infty} f(t)e^{-st}\, dt. \tag{3.13}$$

The decaying exponential term in the integrand in effect provides a built-in convergence factor if $\sigma > 0$. This means that even if $f(t)$ does not vanish as $t \to \infty$, the integrand will vanish for sufficiently large values of σ if f does not grow at a faster than exponential rate. The fact that the lower limit of integration is at 0^- allows the use of an impulse function at $t = 0$ as illustrated in Example 3.2; however, this distinction between $t = 0^-$ and $t = 0$ does not usually come up in practice. We will therefore for the most part drop the minus superscript on $t = 0$; however, we will return to using the notation $t = 0^-$ when an impulse at $t = 0$ is involved and the distinction is of practical value.

If Eq. (3.13) is a one-sided transform, then by extension, Eq. (3.11) is a **two-sided Laplace transform**.[4] We will use the $\mathcal{L}$ symbol from here on to mean $\mathcal{L}_-$.

Based on the formal definition in Eq. (3.13), we can ascertain the properties of Laplace transforms and compute the transforms of common time functions. The analysis of linear systems by means of Laplace transforms usually involves using tables of common properties and time functions, so we have provided this information in Appendix A. The tables of time functions and their Laplace transforms, together with the table of properties, permit us to find transforms of complex signals from simpler ones. For a thorough study of Laplace transforms and extensive tables, see Churchill (1972) and Campbell and Foster (1948). For more study of the two-sided transform, see Van der Pol and Bremmer (1955). These authors show that the time function can be obtained from the Laplace transform by the inverse relation

$$f(t) = \frac{1}{2\pi j} \int_{\sigma_c - j\infty}^{\sigma_c + j\infty} F(s)e^{st}\, ds, \tag{3.14}$$

---

[4]   The other possible one-sided transform is, of course, $\mathcal{L}_+$, in which the lower limit of the integral is 0^+. It is sometimes used in other applications.

where σ_c is a selected value to the right of all the singularities of $F(s)$ in the s-plane. In practice this relation is seldom used. Instead, complex Laplace transforms are broken down into simpler ones that are listed in the tables along with their corresponding time responses.

Let us compute a few Laplace transforms of some typical time functions.

EXAMPLE 3.5   *Step and Ramp Transforms*

Find the Laplace transform of the step, $a1(t)$, and ramp, $bt1(t)$, functions.

Solution.   For a step of size a, $f(t) = a1(t)$ and from Eq. (3.13) we have,

$$F(s) = \int_0^\infty ae^{-st}\, dt = \left.\frac{-ae^{-st}}{s}\right|_0^\infty = 0 - \frac{-a}{s} = \boxed{\frac{a}{s}}$$

For the ramp signal $f(t) = bt1(t)$, again from Eq. (3.13), we have

$$F(s) = \int_0^\infty bte^{-st}\, dt = \left[-\frac{bte^{-st}}{s} - \frac{be^{-st}}{s^2}\right]_0^\infty = \boxed{\frac{b}{s^2},}$$

where we employed the technique of integration by parts,

$$\int u\, dv = uv - \int v\, du,$$

with $u = bt$ and $dv = e^{-st}\, dt$.

A more subtle example is that of the impulse function.

EXAMPLE 3.6   *Impulse Function Transform*

Find the Laplace transform of the unit impulse function.

Solution.   From Eq. (3.13) we get

$$F(s) = \int_{0^-}^\infty \delta(t)e^{-st}\, dt = \int_{0^-}^{0^+} \delta(t)\, dt = 1. \tag{3.15}$$

It is the transform of the unit impulse function that led us to choose the $\mathcal{L}_-$ transform rather than the $\mathcal{L}_+$ transform.

EXAMPLE 3.7

Sinusoid Transform

Find the Laplace transform of the sinusoid function.

Solution.   Again we use Eq. (3.13) to get

$$\mathcal{L}\{\sin \omega t\} = \int_0^\infty (\sin \omega t) e^{-st}\, dt. \qquad (3.16)$$

If we substitute the relation

$$\sin \omega t = \frac{e^{j\omega t} - e^{-j\omega t}}{2j}$$

into Eq. (3.16), we find that

$$\mathcal{L}\{\sin \omega t\} = \int_0^\infty \left(\frac{e^{j\omega t} - e^{-j\omega t}}{2j} \right) e^{-st}\, dt$$

$$= \frac{1}{2j} \int_0^\infty (e^{(j\omega - s)t} - e^{-(j\omega + s)t})\, dt$$

$$= \frac{\omega}{s^2 + \omega^2}.$$

Table A.2 in Appendix A lists Laplace transforms for elementary time functions. Each entry in the table follows from direct application of the transform definition of Eq. (3.13) as demonstrated by Examples 3.5 to 3.7 above.

3.1.4   Properties of Laplace Transforms

In this section we will address each of the significant properties of the Laplace transform listed in Table A.1. For the proofs of these properties and related examples as well as the Initial Value Theorem, the reader is referred to Appendix A.

1. Superposition

One of the more important properties of the Laplace transform is that it is linear:

$$\mathcal{L}\{\alpha f_1(t) + \beta f_2(t)\} = \alpha F_1(s) + \beta F_2(s). \qquad (3.17)$$

The scaling property is a special case of this; that is,

$$\mathcal{L}\{\alpha f(t)\} = \alpha F(s). \qquad (3.18)$$

2. Time Delay

Suppose a function $f(t)$ is delayed by $\lambda > 0$ units of time. Its Laplace transform is

$$F_1(s) = \int_0^\infty f(t - \lambda)e^{-st}\, dt = e^{-s\lambda} F(s). \tag{3.19}$$

From this result we see that a time delay of λ corresponds to multiplication of the transform by $e^{-s\lambda}$.

3. Time Scaling

It is sometimes useful to time-scale equations of motion. For example, in the control system of a tape drive it is meaningful to measure time in milliseconds (see also Chapters 2 and 9). If the time t is scaled by a factor a, then the Laplace transform of the time-scaled signal is

$$F_1(s) = \int_0^\infty f(at)e^{-st}\, dt = \frac{1}{|a|} F\left(\frac{s}{a}\right). \tag{3.20}$$

4. Shift in Frequency

Multiplication (modulation) of $f(t)$ by an exponential expression in the time domain corresponds to a shift in frequency:

$$F_1(s) = \int_0^\infty e^{-at} f(t)e^{-st}\, dt = F(s + a). \tag{3.21}$$

5. Differentiation

The transform of the derivative of a signal is related to its Laplace transform and its initial condition as follows:

$$\mathcal{L}\left\{\frac{df}{dt}\right\} = \int_{0^-}^\infty \left(\frac{df}{dt}\right)e^{-st}\, dt = -f(0^-) + sF(s). \tag{3.22}$$

Another application of Eq. (3.22) leads to

$$\mathcal{L}\{\ddot{f}\} = s^2 F(s) - sf(0^-) - \dot{f}(0^-). \tag{3.23}$$

Repeated application of Eq. (3.22) leads to

$$\mathcal{L}\{f^m(t)\} = s^m F(s) - s^{m-1} f(0^-) - s^{m-2} \dot{f}(0^-) - \cdots - f^{(m-1)}(0^-), \tag{3.24}$$

where $f^m(t)$ denotes the mth derivative of $f(t)$ with respect to time.

6. Integration

Let us assume we wish to determine the Laplace transform of the integral of a time function $f(t)$:

$$F_1(s) = \mathcal{L}\left\{\int_0^t f(\xi)d\xi\right\} = \frac{1}{s} F(s). \tag{3.25}$$

which means we simply multiply its Laplace transform by $1/s$.

7. Convolution

We have seen previously that the response of a system is determined by convolving the input with the impulse response of the system, or by forming the product of the transfer function and the Laplace transform of the input. The following discussion extends this concept to various time functions.

Convolution in the time domain corresponds to multiplication in the frequency domain. Assume that $\mathcal{L}\{f_1(t)\} = F_1(s)$ and $\mathcal{L}\{f_2(t)\} = F_2(s)$. Then

$$\mathcal{L}\{f_1(t) * f_2(t)\} = \int_0^\infty f_1(t) * f_2(t)e^{-st}\,dt = F_1(s)F_2(s). \tag{3.26}$$

This implies that

$$\mathcal{L}^{-1}\{F_1(s)F_2(s)\} = f_1(t) * f_2(t). \tag{3.27}$$

A similar, or dual, of this result is discussed next.

8. Time Product

Multiplication in the time domain corresponds to convolution in the frequency domain:

$$\mathcal{L}\{f_1(t)f_2(t)\} = \frac{1}{2\pi j}F_1(s) * F_2(s). \tag{3.28}$$

9. Multiplication by Time

Multiplication by time corresponds to differentiation in the frequency domain:

$$\mathcal{L}\{tf(t)\} = -\frac{d}{ds}F(s). \tag{3.29}$$

3.1.5   Inverse Laplace Transform by Partial-Fraction Expansion

The easiest way to find $f(t)$ from its Laplace transform $F(s)$, if $F(s)$ is rational, is to expand $F(s)$ as a sum of simpler terms that can be found in the tables. The basic tool for performing this operation is called **partial-fraction expansion**. Consider the general form for the rational function $F(s)$ consisting of the ratio of two polynomials:

$$F(s) = \frac{b_1 s^m + b_2 s^{m-1} + \cdots + b_{m+1}}{s^n + a_1 s^{n-1} + \cdots + a_n}. \tag{3.30}$$

By factoring the polynomials this same function could also be expressed in terms of the product of factors as

$$F(s) = K\frac{\Pi_{i=1}^m (s - z_i)}{\Pi_{i=1}^n (s - p_i)}. \tag{3.31}$$

Zeros and poles

We will discuss the simple case of distinct poles here. For a transform $F(s)$ representing the response of any physical system, $m \leq n$. When $s = z_i$, s is referred to as a **zero** of the function, and when $s = p_i$, s is referred to as a **pole** of the function. Assuming for now that the poles $\{p_i\}$ are real or complex but distinct, we rewrite $F(s)$ as the partial fraction

$$F(s) = \frac{C_1}{s - p_1} + \frac{C_2}{s - p_2} + \cdots + \frac{C_n}{s - p_n}. \tag{3.32}$$

Next we determine the set of constants $\{C_i\}$. We multiply both sides of Eq. (3.32) by the factor $s - p_1$ to get

$$(s - p_1)F(s) = C_1 + \frac{s - p_1}{s - p_2}C_2 + \cdots + \frac{(s - p_1)C_n}{s - p_n}. \tag{3.33}$$

If we let $s = p_1$ on both sides of Eq. (3.33), then all the C_i terms will equal zero except for the first one. For this term,

$$C_1 = (s - p_1)F(s)|_{s=p_1}. \tag{3.34}$$

The other coefficients can be expressed in a similar form:

$$C_i = (s - p_i)F(s)|_{s=p_i}.$$

The cover-up method of determining coefficients

This process is called the **cover-up method** because in the factored form of $F(s)$ [Eq. (3.31)], we can cover up the individual denominator terms, evaluate the rest of the expression with $s = p_i$, and determine the coefficients C_i. Once this has been completed, the time function becomes

$$f(t) = \sum_{i=1}^{n} C_i e^{p_i t} 1(t)$$

because, as entry 7 in Table A.2 shows, if

$$F(s) = \frac{1}{s - p_i},$$

then

$$f(t) = e^{p_i t} 1(t).$$

For the cases of quadratic factors or repeated roots in the denominator, see Appendix A.

EXAMPLE 3.8

Partial-Fraction Expansion: Distinct Real Roots

Suppose you have computed $Y(s)$ and found that

$$Y(s) = \frac{(s + 2)(s + 4)}{s(s + 1)(s + 3)}.$$

Find $y(t)$.

Solution.   We may write $Y(s)$ in terms of its partial-fraction expansion:

$$Y(s) = \frac{C_1}{s} + \frac{C_2}{s+1} + \frac{C_3}{s+3}.$$

Using the cover-up method, we get

$$C_1 = \left.\frac{(s+2)(s+4)}{(s+1)(s+3)}\right|_{s=0} = \frac{8}{3}.$$

In a similar fashion,

$$C_2 = \left.\frac{(s+2)(s+4)}{s(s+3)}\right|_{s=-1} = -\frac{3}{2}$$

and

$$C_3 = \left.\frac{(s+2)(s+4)}{s(s+1)}\right|_{s=-3} = -\frac{1}{6}.$$

We can check the correctness of the result by adding the components again to verify that the original function has been recovered. With the partial fraction the solution can be looked up in the tables at once to be

$$y(t) = \frac{8}{3}1(t) - \frac{3}{2}e^{-t}1(t) - \frac{1}{6}e^{-3t}1(t).$$

The partial fraction expansion may be computed using the residue function in MATLAB:

```
num = conv([1 2],[1 4]);     % form numerator polynomial
den = conv([1 1 0],[1 3]);   % form denominator polynomial
[r,p,k] = residue(num,den);  % compute the residues
```

which yields the following result

```
r = [-0.1667 −1.5000 2.6667]';     p = [-3 −1 0]';     k = [];
```

and agrees with the hand calculations. Note that the conv function in MATLAB is used to multiply two polynomials (the arguments of the function are the polynomial coefficients).

3.1.6   The Final Value Theorem

An especially useful property of the Laplace transform in control known as the **Final Value Theorem** allows us to compute the constant steady-state value of a time function given its Laplace transform. The theorem follows from the development of partial-fraction expansion. Suppose we have a transform $Y(s)$ of a signal $y(t)$ and wish to know the final value $y(t)$ from $Y(s)$. There are three possibilities for the limit. It can be constant, undefined, or unbounded. If $Y(s)$ has any poles (i.e., denominator roots, see Section 3.1.5) in the right half of the s-plane—that is, if the real part of any $p_i > 0$—then $y(t)$ will grow and the limit will be unbounded. If $Y(s)$ has a pair of poles on the imaginary axis of the s-plane—that is, $p_i = \pm j\omega$—then $y(t)$ will contain a sinusoid that persists forever and the final value will not be defined. Only one case can provide a nonzero constant final value: If all poles of $Y(s)$ are in the left half of the s-plane except for one at $s = 0$, then all terms of $y(t)$ will decay to zero except the term corresponding to the pole at $s = 0$, and that term corresponds to a constant in time. Thus the final value is given by the coefficient associated with the pole at $s = 0$. Therefore, the Final Value Theorem is as follows:

The Final Value Theorem

If all poles of $sY(s)$ are in the left half of the s-plane, then

$$\lim_{t \to \infty} y(t) = \lim_{s \to 0} sY(s).$$ \hfill (3.35)

This relationship is proved in Appendix A.

EXAMPLE 3.9

Final Value Theorem

Find the final value of the system corresponding to

$$Y(s) = \frac{3(s+2)}{s(s^2 + 2s + 10)}.$$

Solution.   Applying the Final Value Theorem, we obtain

$$y(\infty) = sY(s)|_{s=0} = \frac{3 \cdot 2}{10} = 0.6.$$

Thus, after the transients have decayed to zero, $y(t)$ will settle to a constant value of 0.6.

Use the Final Value Theorem on stable systems only

Care must be taken to apply the Final Value Theorem only to stable systems (see Section 3.6). While one could use Eq. (3.35) on any $Y(s)$, doing so could result in erroneous results, as shown in the next example.

EXAMPLE 3.10

Incorrect Use of the Final Value Theorem

Find the final value of the signal corresponding to

$$Y(s) = \frac{3}{s(s-2)}.$$

Solution.   If we blindly apply Eq. (3.35), we obtain

$$y(\infty) = sY(s)|_{s=0} = -\frac{3}{2}.$$

However,

$$y(t) = \left(-\frac{3}{2} + \frac{3}{2}e^{2t}\right)1(t),$$

and Eq. (3.35) yields the constant term only. Of course, the true final value is unbounded.

  The theorem can also be used to find the DC gain of a system. The **DC gain** is the ratio of the output of a system to its input (presumed constant) after all transients have decayed. To find the DC gain, we assume there is a unit-step input [$U(s) = 1/s$] and use the Final Value Theorem to compute the steady-state value of the output. Therefore, for a system transfer function $G(s)$,

$$\text{DC gain} = \lim_{s \to 0} sG(s)\frac{1}{s} = \lim_{s \to 0} G(s). \tag{3.36}$$

EXAMPLE 3.11   *DC Gain*

Find the DC gain of the system whose transfer function is

$$G(s) = \frac{3(s+2)}{(s^2 + 2s + 10)}.$$

Solution.   Applying Eq. (3.36), we get

$$\text{DC gain} = G(s)|_{s=0} = \frac{3 \cdot 2}{10} = 0.6.$$

3.1.7   Using Laplace Transforms to Solve Problems

Laplace transforms can be used to solve differential equations using the properties described in Appendix A. First, we find the Laplace transform of the differential equation using the differentiation properties in Eqs. (A.12) and (A.13) in Appendix A. Then we find the Laplace transform of the output; using partial-fraction expansion and Table A.2, this can be converted to a time response function. We will illustrate this with three examples.

EXAMPLE 3.12   *Homogeneous Differential Equation Solution*

Find the solution to the differential equation

$$\ddot{y}(t) + y(t) = 0 \qquad \text{where} \quad y(0) = \alpha, \ \dot{y}(0) = \beta.$$

Solution.   Using Eq. (3.22), the Laplace transform of the differential equation is

$$s^2 Y(s) - \alpha s - \beta + Y(s) = 0,$$

$$(s^2 + 1)Y(s) = \alpha s + \beta,$$

$$Y(s) = \frac{\alpha s}{s^2 + 1} + \frac{\beta}{s^2 + 1}.$$

After looking up in the transform tables (Table A.2, Appendix A) the two terms on the right side of the above equation, we get

$$y(t) = [\alpha \cos t + \beta \sin t]1(t).$$

where $1(t)$ denotes a unit step function. We can verify that this solution is correct by substituting it back into the differential equation.

Another example will illustrate the solution when the equations are not homogeneous—that is, when the system is forced.

EXAMPLE 3.13   *Forced Differential Equation Solution*

Find the solution to the differential equation $\ddot{y}(t) + 5\dot{y}(t) + 4y(t) = 3$, where $y(0) = \alpha$, $\dot{y}(0) = \beta$.

Solution.   Taking the Laplace transform of both sides using Eqs. (3.21) and (3.22), we get

$$s^2 Y(s) - s\alpha - \beta + 5[sY(s) - \alpha] + 4Y(s) = \frac{3}{s}.$$

Solving for $Y(s)$ yields

$$Y(s) = \frac{s(s\alpha + \beta + 5\alpha) + 3}{s(s+1)(s+4)}.$$

The partial-fraction expansion using the cover-up method is

$$Y(s) = \frac{\frac{3}{4}}{s} - \frac{\frac{3-\beta-4\alpha}{3}}{s+1} + \frac{\frac{3-4\alpha-4\beta}{12}}{s+4}.$$

Therefore, the time function is given by

$$y(t) = \left(\frac{3}{4} + \frac{-3+\beta+4\alpha}{3}e^{-t} + \frac{3-4\alpha-4\beta}{12}e^{-4t} \right)1(t)$$

By differentiating this solution twice and substituting the result in the original differential equation, we can verify that this solution satisfies the differential equation.

The solution is especially simple if the initial conditions are all zero.

EXAMPLE 3.14   *Forced Equation Solution with Zero Initial Conditions*

Find the solution to $\ddot{y}(t) + 5\dot{y}(t) + 4y(t) = u(t)$, $y(0) = 0$, $\dot{y}(0) = 0$, $u(t) = 2e^{-2t}1(t)$, using

(a) partial-fraction expansion and

(b) MATLAB.

Solution.

(a) Taking the Laplace transform of both sides, we get

$$s^2 Y(s) + 5s Y(s) + 4Y(s) = \frac{2}{s+2}.$$

Solving for $Y(s)$ yields

$$Y(s) = \frac{2}{(s+2)(s+1)(s+4)}.$$

The partial-fraction expansion using the cover-up method is

$$Y(s) = -\frac{1}{s+2} + \frac{\frac{2}{3}}{s+1} + \frac{\frac{1}{3}}{s+4}.$$

Therefore, the time function is given by

$$y(t) = \left(-1e^{-2t} + \frac{2}{3}e^{-t} + \frac{1}{3}e^{-4t} \right) 1(t).$$

(b) The partial fraction expansion may also be computed using the MATLAB residue function:

```
num = 2;                    % form numerator
den = poly([-2;-1;-4]);     % form denominator polynomial from its roots
[r,p,k] = residue(num,den); % compute the residues
```

which results in the desired answer

r = [0.3333 −1 0.6667]';   p = [-4 −2 −1]';   k = [];

and agrees with the hand calculations.

Poles indicate response character

The primary value of using the Laplace transform method of solving differential equations is that it provides information concerning the qualitative characteristic behavior of the response. Once we know the values of the poles of $Y(s)$, we know what kind of characteristic terms will appear in the response. In the last example the pole at $s = -1$ produced a decaying $y = Ce^{-t}$ term in the response. The pole at $s = -4$ produced a $y = Ce^{-4t}$ term in the response, which decays faster. If there had been a pole at $s = +1$, there would have been a growing $y = Ce^{+t}$ term in the response. Using the pole locations to understand in essence how the system will respond is a powerful tool and will be developed further in Section 3.3. Control systems designers often manipulate design parameters so that the poles have values that would give acceptable responses, and they skip the steps associated with converting those poles to actual time responses until the final stages of the design. They use trial-and-error design methods (as described in Chapter 5) that graphically present how changes in design parameters affect the pole locations. Once a design has been obtained with pole locations predicted to give acceptable responses, the control designer determines a time response to verify that the design is satisfactory. This is typically done by computer which solves the differential equations directly by using numerical computer methods.

3.1.8   Poles and Zeros

A rational transfer function can be described either as a ratio of two polynomials in s,

$$H(s) = \frac{b_1 s^m + b_2 s^{m-1} + \cdots + b_{m+1}}{s^n + a_1 s^{n-1} + \cdots + a_n} = \frac{N(s)}{D(s)}, \qquad (3.37)$$

or in factored zero-pole form

$$H(s) = K \frac{\prod_{i=1}^{m}(s - z_i)}{\prod_{i=1}^{n}(s - p_i)}. \qquad (3.38)$$

Zeros

K is called the transfer function gain. The roots of the numerator $z_1, z_2, \ldots, z_m$ are called the finite **zeros** of the system. The zeros are locations in the s-plane where the transfer function is zero. If $s = z_i$ then,

$$H(s)|_{s=z_i} = 0.$$

The zeros also correspond to the signal transmission-blocking properties of the system and are also called the transmission zeros of the system. The system has the inherent capability to block frequencies coinciding with its zero locations. If we excite the system with the nonzero input, $u = u_0 e^{s_0 t}$, where s_0 is not a pole of the system, then the output is identically zero,[5] $y \equiv 0$, for frequencies where $s_0 = z_i$. The zeros also have a significant effect on the transient properties of the system (see Section 3.5).

Poles

The roots of the denominator, $p_1, p_2, \ldots, p_n$ are called the **poles**[6] of the system. The poles are locations in the s-plane where the magnitude of the transfer function becomes infinite. If $s = p_i$, then

$$|H(s)|_{s=p_i} = \infty.$$

The poles of the system determine its stability properties as we shall see in Section 3.6. The poles of the system also determine the natural or unforced behavior of the system referred to as the **modes** of the system. The zeros and poles may be complex quantities, and we may display their locations in a complex plane which we refer to as the s-plane. The locations of the poles and zeros lie at the heart of feedback control design and have significant practical implications for control system design. The system is said to have $n - m$ zeros at infinity if $m < n$ because the transfer function approaches zero as s approaches infinity. If the zeros at infinity are also counted, the system will have the same

---

[5] Identically zero means that the output and all of its derivatives are zero for $t > 0$.

[6] The meaning of the pole can also be appreciated by visualizing a three-dimensional plot of the transfer function, where the real and imaginary parts of s are plotted on the x and y axes, and the magnitude of the transfer function is plotted on the vertical z axis. For a single pole, the resulting three-dimensional plot will look like a tent with the "tent-pole" being located at the pole of the transfer function!

number of poles and zeros. No physical system can have $n < m$; otherwise, it would have an infinite response at $\omega = \infty$. If $z_i = p_j$ then there are *cancellations* in the transfer function which may lead to undesirable system properties as discussed in Chapter 7.

3.1.9   Linear System Analysis Using MATLAB

The first step in analyzing a system is to write down (or generate) the set of time-domain differential equations representing the dynamical behavior of the physical system. These equations are generated from the physical laws governing the system behavior—for example, rigid body dynamics, thermo-fluid mechanics, and electro-mechanics—as described in Chapter 2. The next step in system analysis is to determine and designate inputs and outputs of the system and then to compute the transfer function characterizing the input–output behavior of the dynamic system. In Chapter 2 we saw that a linear dynamic system may be represented by its differential equations, possibly in state-variable form,

$$\dot{\mathbf{x}} = \mathbf{Fx} + \mathbf{G}u,$$

$$y = \mathbf{Hx} + Ju.$$

Three ways of representing linear systems

Earlier in this chapter we saw that a linear dynamic system may also be represented by the Laplace transform of its differential equation—that is, its transfer function. The transfer function may be expressed as a ratio of two polynomials Eq. (3.37) or in factored zero-pole form Eq. (3.38). By analyzing the transfer function, we can determine the dynamic properties of the system both in a qualitative and quantitative manner. One way of extracting useful system information is to simply determine the pole-zero locations and deduce the essential characteristics of the dynamic properties of the system. Another way is to determine the time-domain properties of the system by determining the response of the system to typical excitation signals such as impulses, steps, ramps, and sinusoids. Yet another way is to determine the time response analytically by computing the inverse Laplace transform using partial fraction expansions and Tables A.1 and A.2 in Appendix A. Of course, it is also possible to determine the system response to an arbitrary input.

We will now illustrate this type of analysis by carrying out the above calculations for some of the physical systems addressed in the examples in Chapter 2 in order of increasing degree of difficulty. We will go back and forth between the different representations of the system—transfer function, pole-zero, state-space, and so on—using MATLAB as our computational engine. MATLAB typically accepts the specification of a system in any of the three forms—state-space, transfer function, and zero-pole and refers to the descriptions as `ss`, `tf`, and `zp`, respectively. Furthermore, it can transform the system description from any one form to another. Therefore, by specifying a system by its state-space differential equation description to MATLAB and then using MATLAB to transform the system description to the polynomial transfer-function form, we get the Laplace transform of the system. The analytical relationship between the state-space description ($\mathbf{F}$, $\mathbf{G}$, $\mathbf{H}$, and J) and the transfer-function description $H(s)$ is derived in Section 7.2 and shown by Eq. (7.36).

EXAMPLE 3.15

Cruise-Control Transfer Function Using MATLAB

Find the transfer function between the input u and the position of the car x in the cruise-control system in Example 2.8.

Solution.   In Example 2.8 we found the matrices $\mathbf{F}$, $\mathbf{G}$, and $\mathbf{H}$ and the value of J to be

$$\mathbf{F} = \begin{bmatrix} 0 & 1 \\ 0 & -0.05 \end{bmatrix}, \qquad \mathbf{G} = \begin{bmatrix} 0 \\ 0.001 \end{bmatrix}, \qquad \mathbf{H} = [1 \quad 0], \qquad J = 0.$$

The MATLAB statement

$$[\text{num, den}] = \text{ss2tf(F, G, H, J)}$$

will compute and display the coefficients of the numerator polynomial as the row vector num and the denominator as den. The results for this example are

$$\text{num} = [0 \quad 0 \quad 0.001] \quad \text{and} \quad \text{den} = [1 \quad 0.05 \quad 0].$$

This means that the transfer function of the system is

$$H(s) = \frac{0s^2 + 0s + 0.001}{s^2 + 0.05s + 0} = \frac{0.001}{s(s + 0.05)},$$

MATLAB printsys

and can be returned by MATLAB in this form using the printsys(num,den) command.

EXAMPLE 3.16

DC Motor Transfer Function Using MATLAB

In Example 2.15, assume that $J_m = 0.01$ kg·m^2, $b = 0.001$ N·m·sec, $K_t = K_e = 1$, $R_a = 10\ \Omega$, and $L_a = 1$ H. Find the transfer function between the input v_a and

(a) the output θ_m,

(b) the output $\omega = \dot{\theta}_m$.

Solution.

(a) Substituting the above parameters in Example 2.15, we find the matrices $\mathbf{F}$, $\mathbf{G}$, and $\mathbf{H}$ and the value of J to be

$$\mathbf{F} = \begin{bmatrix} 0 & 1 & 0 \\ 0 & -0.1 & 100 \\ 0 & -1 & -10 \end{bmatrix}, \qquad \mathbf{G} = \begin{bmatrix} 0 \\ 0 \\ 1 \end{bmatrix}, \qquad \mathbf{H} = [1 \quad 0 \quad 0], \qquad J = 0.$$

The MATLAB statement

$$[\text{numa, dena}] = \text{ss2tf}(F, G, H, J)$$

will compute and display the coefficients of the numerator polynomial as the row vector numa and the denominator as dena. The results for this example are

$$\text{numa} = [0 \quad 0 \quad 0 \quad 100] \quad \text{and} \quad \text{dena} = [1 \quad 10.1 \quad 101 \quad 0].$$

This means that the transfer function of the system is

$$H(s) = \frac{100}{s^3 + 10.1s^2 + 101s}.$$

The pole-zero description computed using the MATLAB command [z, p, k] = tf2zp (numa, dena) would result in

$$z = [] \quad p = [0 \quad -5.0500 \quad +8.6889j \quad -5.0500 \quad -8.6889j\,]' , \qquad k = 100,$$

and yields the transfer function in factored form,

$$H(s) = \frac{100}{s(s + 5.05 + j8.6889)(s + 5.05 - j8.6889)}.$$

(b) If we consider the velocity $\dot{\theta}_m$ as the output, $\mathbf{H1} = [0 \quad 1 \quad 0]$, then we find numb = [0 0 100], denb = [1 10.1 101], which tells us that the transfer function is

$$G(s) = \frac{100s}{s^3 + 10.1s^2 + 101s} = \frac{100}{s^2 + 10.1s + 101}.$$

This is as expected because $\dot{\theta}_m$ is simply the derivative of θ_m; thus $\mathcal{L}\{\dot{\theta}_m\} = s\mathcal{L}\{\theta_m\}$. For a unit step command in v_a, we can compute the step response in MATLAB (recall Example 2.8),

```
numb = [0 0 100];          % form numerator
denb = [1 10.1 101];       % form denominator
sysb = tf(numb,denb);      % define system by its numerator and denominator
t = 0:0.01:5;              % form time vector
y = step(sysb,t)           % compute step response;
plot(t,y)                  % plot step response
```

and the system yields a steady-state constant angular velocity as shown in Fig. 3.3. Note that there is a slight offset as the system does not have unity DC gain.

Figure 3.3
Transient response for DC motor

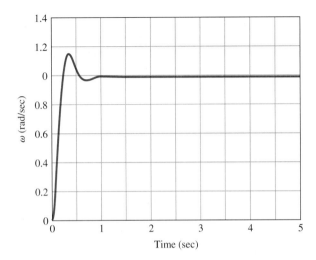

When a dynamic system is represented by a single differential equation of any order, finding the polynomial form of the transfer function from that differential equation is usually easier than finding the state-variable description from that differential equation. Therefore, you will find it best in these cases to specify a system directly in terms of its transfer function.

EXAMPLE 3.17   *Transformations Using* MATLAB

Find the transfer function and state-variable description of the system whose differential equation is

$$\ddot{y} + 6\dot{y} + 25y = 9u + 3\dot{u}.$$

Solution.   Using the differentiation rules given by Eqs. (3.22) and (3.23), we see by inspection that

$$\frac{Y(s)}{U(s)} = \frac{3s + 9}{s^2 + 6s + 25}.$$

The MATLAB statements

```
numG = [3 9];                  % form numerator
denG = [1 6 25];               % form denominator
[F,G,H,J] = tf2ss(numG,denG)   % convert to a state variable realization
```

will provide values of a state-variable description of the system:

$$\mathbf{F} = \begin{bmatrix} -6 & -25 \\ 1 & 0 \end{bmatrix}, \qquad \mathbf{G} = \begin{bmatrix} 1 \\ 0 \end{bmatrix}, \qquad \mathbf{H} = [3 \quad 9], \qquad J = 0.$$

Note that because of the derivatives of u in the differential equations, it is not as straightforward to write the state-variable description of this system as it was for Example 3.15.

Note also that the system output y is not one of the state elements, as evidenced by the fact that **H** contains two nonzero elements.

If the transfer function was desired in factored form, it could be obtained by transforming either the ss or tf description. Therefore, either of the following MATLAB statements

```
[z,p,k] = ss2zp(F,G,H,J)      % convert from state variable to pole-zero form
% convert from numerator-denominator polynomials to pole-zero form
[z,p,k] = tf2zp(numG,denG)
```

would result in $z = -3 \quad p = [\,-3+4j \quad -3-4j\,]'$, $k = 3$. This means the transfer function could also be written as

$$\frac{Y(s)}{U(s)} = \frac{3(s+3)}{(s+3-4j)(s+3+4j)}.$$

We may also convert from zero-pole representation to the transfer function representation using the MATLAB zp2tf command,

```
% convert from pole-zero form to numerator-denominator polynomials
[numG,denG] = zp2tf(z,p,k)
```

For this example, z = [-3], p = [-3+i4;-3-i*4], k = [3], will yield the numerator and denominator polynomials.

EXAMPLE 3.18

Satellite Transfer Function Using MATLAB

(a) Find the transfer function between the input F_c and the satellite attitude θ in Example 2.7 and

(b) determine the response of the system to a 25-N pulse of 0.1-sec duration starting at $t = 5$ sec. Let $d = 1$ m and $I = 5000$ kg-m^2.

Solution.

(a) From Example 2.7,

$$\frac{d}{I} = \frac{1}{5000} = 0.0002[\frac{m}{kg-m^2}]$$

and the matrices **F**, **G**, and **H** and the value of J is

$$\mathbf{F} = \begin{bmatrix} 0 & 1 \\ 0 & 0 \end{bmatrix}, \qquad \mathbf{G} = \begin{bmatrix} 0 \\ 0.0002 \end{bmatrix}, \qquad \mathbf{H} = [1 \quad 0], \qquad J = 0.$$

The MATLAB statement

$$[\text{numG, denG}] = \text{ss2tf}(F, G, H, J)$$

will compute and display the coefficients of the numerator polynomial as the row vector num and the denominator as the row vector den. The results for this example are

$$\text{numG} = [0 \quad 0 \quad 0.0002] \quad \text{and} \quad \text{denG} = [1 \quad 0 \quad 0].$$

This means that the transfer function of the system is

$$H(s) = \frac{0.0002}{s^2},$$

which can also be determined by inspection for this particular case.

(b) The following MATLAB statements compute the response of the system to a 25-N, 0.1-sec duration thrust pulse-input,

```
dl = 1/5000;
F = [0 1;0 0];                               % form state variable matrices
G = [0;dl];
H = [1 0];
J = [0];
sysG = ss(F,G,H,J);                          % define system by its
                                             state variable realization

t = 0:0.01:10;                               % set up time vector with
                                             dt = 0.01 sec

% pulse of 25N, at 5 sec, for 0.1 sec duration
u1 = [zeros(1,500) 25*ones(1,10) zeros(1,491)];   % pulse input
[y1] = lsim(sysG,u1,t);                      % linear simulation
ff = 180/pi;                                 % conversion factor from
                                             radians to degrees

y1 = ff*y1;                                  % output in degrees
plot(t,u1);                                  % plot input signal
plot(t,y1);                                  % plot output response
```

The system is excited with a short pulse (an impulsive input) that has the effect of imparting a nonzero angle θ_0 at time $t = 5$ sec on the system. Because the system is undamped, in the absence of any control it drifts with constant angular velocity with a value imparted by the impulse at $t = 5$ sec. The time response of the input in is shown in Fig. 3.4(a) along with the drift in angle θ in Fig. 3.4(b).

If we now excite the system with the same positive-magnitude thrust pulse at time $t = 5$ sec but follow that with a negative pulse with the same magnitude and duration at time at $t = 6.1$ sec [see Figure 3.5(a) for the input thrust], then the attitude response of the system is as shown in Figure 3.5(b). This is actually how the satellite attitude angle is controlled in practice. The additional relevant MATLAB statements are

```
% double pulse input
u2 = [zeros(1,500) 25*ones(1,10) zeros(1,100) -25*ones(1,10) zeros(1,381)];
[y2] = lsim(sysG,u2,t);          % linear simulation
plot(t,u2);                      % plot input signal
ff = 180/pi;                     % conversion factor from radians to degrees
y2 = ff*y2;                      % output in degrees
plot(t,y2);                      % plot output response
```

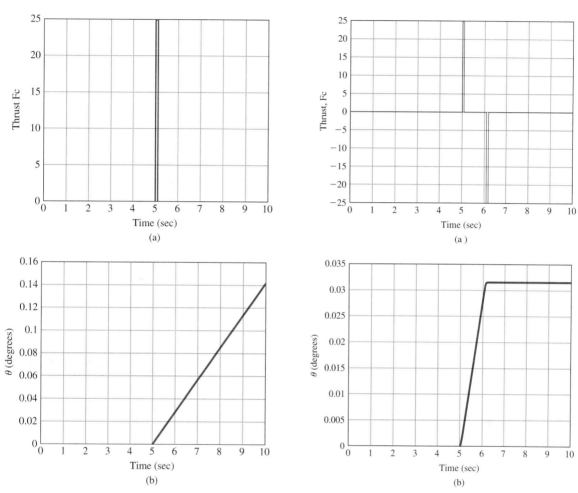

Figure 3.4 Transient response for satellite, (a) thrust input, (b) satellite attitude

Figure 3.5 Transient response for satellite (double-pulse), (a) thrust input (b) satellite attitude

3.2   System Modeling Diagrams

3.2.1   The Block Diagram

To obtain the transfer function, we need to find the Laplace transform of the equations of motion and solve the resulting algebraic equations for the relationship between the input and the output. In many control systems the system equations can be written so that their components do not interact except by having the input of one part be the output of another part. In these cases it is easy to draw a block diagram that represents the mathematical relationships in a similar manner to that used for the component block diagram in Fig. 1.2, Chapter 1. The transfer function of each component is placed in a box, and the input–output relationships between components are indicated by lines and arrows. We can then solve the equations by graphical simplification, which is often easier and more informative than algebraic manipulation, even though the methods are in every way equivalent. Drawings of three elementary block diagrams are seen in Fig. 3.6. It is convenient to think of each block as representing an electronic amplifier with the transfer function printed inside. The interconnections of blocks include summing points, where any number of signals may be added together. These are represented by a circle with the symbol Σ inside. In Fig. 3.6(a) the block with transfer function $G_1(s)$ is in series with the block with transfer function $G_2(s)$, and the overall transfer function is given by the product G_2G_1. In Fig. 3.6(b), two systems are in parallel with their outputs added, and the overall transfer function is given by the sum $G_1 + G_2$. These diagrams derive simply from the equations that describe them.

Figure 3.6(c) shows a more complicated case. Here the two blocks are connected in a feedback arrangement so that each feeds into the other. When the feedback $Y_2(s)$ is *subtracted*, as shown in the figure, we call it **negative feedback**. As you will see, negative feedback is usually required for system stability. For now we will simply solve the equations and then relate them back to the diagram. The equations are

$$U_1(s) = R(s) - Y_2(s),$$

$$Y_2(s) = G_2(s)G_1(s)U_1(s),$$

$$Y_1(s) = G_1(s)U_1(s),$$

and their solution is

$$Y_1(s) = \frac{G_1(s)}{1 + G_1(s)G_2(s)} R(s). \tag{3.39}$$

We can express the solution by the following rule:

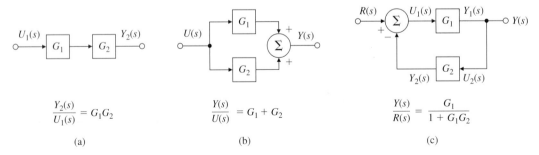

Figure 3.6 Three examples of elementary block diagrams

The gain of a single-loop negative feedback system is given by the forward gain divided by the sum of 1 plus the loop gain.

Positive feedback

When the feedback is added instead of subtracted, we call it **positive feedback**. In this case the gain is given by the forward gain divided by the sum of 1 minus the loop gain.

The three elementary cases given in Fig. 3.6 can be used in combination to solve by repeated reduction any transfer function defined by a block diagram. However, the manipulations can be tedious and subject to error when the topology of the diagram is complicated. Figure 3.7 shows examples of block-diagram algebra that complement those shown in Fig. 3.6. Figures 3.7(a) and (b) show how the interconnections of a block diagram can be manipulated without affecting the mathematical relationships. Figure 3.7(c) shows how the manipulations can be used to convert a general system (on the left) to a system without a component in the feedback path, usually referred to as a **unity feedback system**.

Unity feedback system

In all cases the basic principle is to simplify the topology while maintaining exactly the same relationships among the remaining variables of the block diagram. In relation to the algebra of the underlying linear equations, block-diagram reduction is a pictorial way to solve equations by eliminating variables.

Figure 3.7

Examples of block-diagram algebra

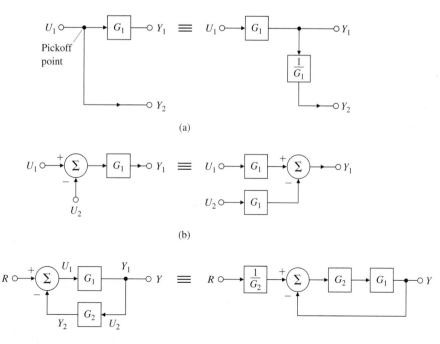

(a)

(b)

(c)

EXAMPLE 3.19   ***Transfer Function from a Simple Block Diagram***

Find the transfer function of the system shown in Fig. 3.8(a).

Solution.   First we simplify the block diagram by reducing the parallel combination of the controller path. This results in the diagram of Fig. 3.8(b) and we use the feedback rule to obtain the closed-loop transfer function.

$$T(s) = \frac{Y(s)}{R(s)} = \frac{\dfrac{2s+4}{s^2}}{1 + \dfrac{2s+4}{s^2}} = \frac{2s+4}{s^2 + 2s + 4}.$$

Figure 3.8
Block diagram of a
second-order system

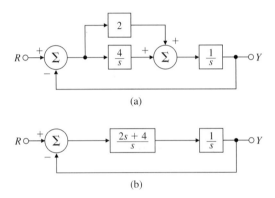

(a)

(b)

EXAMPLE 3.20   ***Transfer Function from the Block Diagram***

Find the transfer function of the system shown in Fig. 3.9(a).

Solution.   First we simplify the block diagram. Using the principles of Eq. (3.39) we replace the feedback loop involving G_1 and G_3 by its equivalent transfer function, noting that it is a positive feedback loop. The result is Fig. 3.9(b). The next step is to move the pick-off point preceding G_2 to its output [see Fig. 3.7(a)], as shown in Fig. 3.9(c). The negative feedback loop on the left is in series with the subsystem on the right, which is comprised of the two parallel blocks G_5 and G_6/G_2. The overall transfer function can be written using all three rules for reduction given by Fig. 3.6:

$$T(s) = \frac{Y(s)}{R(s)} = \frac{\dfrac{G_1 G_2}{1 - G_1 G_3}}{1 + \dfrac{G_1 G_2 G_4}{1 - G_1 G_3}} \left(G_5 + \frac{G_6}{G_2} \right)$$

$$= \frac{G_1 G_2 G_5 + G_1 G_6}{1 - G_1 G_3 + G_1 G_2 G_4}.$$

Figure 3.9

Example for block-diagram simplification

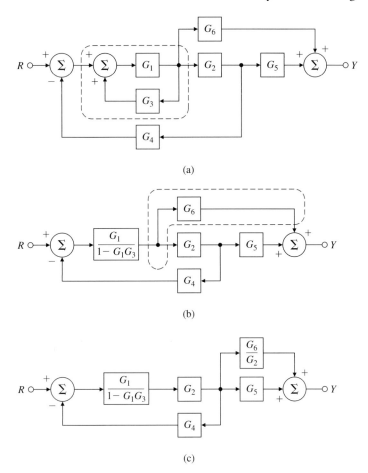

(a)

(b)

(c)

As we have seen, a system of algebraic equations may be represented by a block diagram that represents individual transfer functions by blocks and has interconnections that correspond to the system equations. A block diagram is a convenient tool to visualize the system as a collection of interrelated subsystems that emphasize the relationships among the system variables.

3.2.2   Block-Diagram Reduction Using MATLAB

If the individual transfer functions are available for components in a control system, it is possible to use MATLAB commands to compute the transfer functions of interconnected systems. The three commands series, parallel, and feedback can be used for this purpose. They compute the transfer functions of two component block transfer functions in series, parallel, and feedback configurations respectively. The following simple example illustrates their use.

EXAMPLE 3.21   *Transfer Function of a Simple System Using* MATLAB

Repeat computation of the transfer function for the block diagram in Fig. 3.8(a) using MATLAB.

Solution.   We label the transfer function of the separate blocks shown in Fig. 3.8(a) as shown in Fig. 3.10. Then we combine the two parallel blocks G1 and G2 by

```
num1 = [2];                                    % form G1
den1 = [1];
sysG1 = tf(num1,den1);                         % define subsystem G1
num2 = [4];                                    % form G2
den2 = [1 0];
sysG2 = tf(num2,den2);                         % define subsystem G2
% parallel combination of G1 and G2 to form subsystem G3
sysG3 = parallel(sysG1,sysG2);
```

then we combine the result G3, with the G4 in series by

```
num4 = [1];              % form G4
den4 = [1 0];
sysG4 = tf(num4,den4);   % define subsystem G4
sysG5 = series(sysG3,sysG4);   % series combination of G3 and G4
```

and complete the reduction of the feedback system by

```
num6 = [1];                    % form G6
den6 = [1];
sysG6 = tf(num6,den6)          % define subsystem G6
[sysCL] = feedback(sysG5,sysG6,-1)   % feedback combination of G5 and G6
```

The MATLAB results are sysCL of the form

$$\frac{Y(s)}{R(s)} = \frac{2s + 4}{s^2 + 2s + 4}$$

and is the same as the one obtained by block diagram reduction.

Figure 3.10

Example for block-diagram simplification

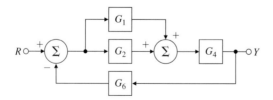

3.2.3   Mason's Rule and the Signal-Flow Graph

A compact alternative notation to the block diagram is given by the **signal-flow graph** introduced by S. J. Mason (1953, 1956). As with the block diagram, the signal-flow graph offers a visual tool for representing the causal relationships between the components of the system. The method consists of characterizing the system by a network of directed branches and associated gains (transfer functions) connected at nodes. Several block diagrams and their corresponding signal-flow graphs are shown in Fig. 3.11. The two ways of depicting a system are equivalent, and you can use either diagram to apply Mason's rule (to be defined shortly).

In a signal-flow graph the internal signals in the diagram, such as the common input to several blocks or the output of a summing junction, are called **nodes**. The system input point and the system output point are also nodes; the input node has outgoing branches only, and the output node has incoming

Figure 3.11

Block diagrams and corresponding signal flow graphs

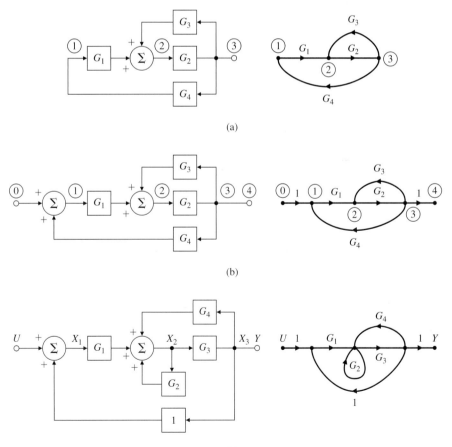

(a)

(b)

(c)

branches only. Mason defined a **path** through a block diagram as a sequence of connected blocks, the route passing from one node to another *in the direction of signal flow of the blocks* without including any block more than once. A **forward path** is a path from the input to output such that no node is included more than once. If the nodes are numbered in a convenient order, then a forward path can be identified by the numbers that are included. Any closed path that returns to its starting node without passing through any node more than once is a **loop**, and a path that leads from a given variable back to the same variable is a **loop path**. The **path gain** is the product of component gains (transfer functions) making up the path. Similarly, the **loop gain** is the path gain associated with a loop, that is, the product of gains in a loop. If two paths have a common component, they are said to touch. Notice particularly in this connection that the input and the output of a summing junction are not the same and that the summing junction is a one-way device from its inputs to its output.

Mason's rule relates the graph to the algebra of the simultaneous equations they represent.[7] Consider Fig. 3.11(c), where the signal at each node has been given a name and the gains are marked. Then the block diagram (or the signal-flow graph) represents the following system of equations:

$$X_1(s) = X_3(s) + U(s),$$

$$X_2(s) = G_1(s)X_1(s) + G_2(s)X_2(s) + G_4(s)X_3(s),$$

$$Y(s) = 1X_3(s).$$

Mason's rule states that the input–output transfer function associated with a signal-flow graph is given by

$$G(s) = \frac{Y(s)}{U(s)} = \frac{1}{\Delta} \sum_i G_i \Delta_i,$$

where

$G_i =$ path gain of the ith forward path.

$\Delta =$ the system determinant $= 1 - \sum$ (all individual loop gains) $+ \sum$ (gain products of all possible two loops that do not touch) $- \sum$ (gain products of all possible three loops that do not touch) $+ \ldots$,

$\Delta_i = i$th forward path determinant

$=$ value of Δ for that part of the block diagram that does *not* touch the ith forward path.

We will now illustrate the use of Mason's rule by several examples.

---

[7] The derivation is based on Cramer's rule for solving linear equations by determinants and is described in Mason's papers.

EXAMPLE 3.22

Mason's Rule in a Simple System

Find the transfer function for the block diagram in Fig. 3.12.

Solution.   From the block diagram shown in Fig. 3.12 we have

Forward Path	*Path Gain*
1236	$G_1 = 1 \left(\frac{1}{s} \right) (b_1)(1)$
12346	$G_2 = 1 \left(\frac{1}{s} \right) \left(\frac{1}{s} \right) (b_2)(1)$
123456	$G_3 = 1 \left(\frac{1}{s} \right) \left(\frac{1}{s} \right) \left(\frac{1}{s} \right) (b_3)(1)$

	Loop Path Gain
232	$l_1 = -a_1/s$
2342	$l_2 = -a_2/s^2$
23452	$l_3 = -a_3/s^3$

and the determinants are

$$\Delta = 1 - \left(-\frac{a_1}{s} - \frac{a_2}{s^2} - \frac{a_3}{s^3} \right) + 0$$

$$\Delta_1 = 1 - 0$$

$$\Delta_2 = 1 - 0$$

$$\Delta_3 = 1 - 0.$$

Applying Mason's rule, we find the transfer function to be

$$G(s) = \frac{Y(s)}{U(s)} = \frac{(b_1/s) + (b_2/s^2) + (b_3/s^3)}{1 + (a_1/s) + (a_2/s^2) + (a_3/s^3)}$$

$$= \frac{b_1 s^2 + b_2 s + b_3}{s^3 + a_1 s^2 + a_2 s + a_3}$$

Figure 3.12
Block diagram for
Example 3.22

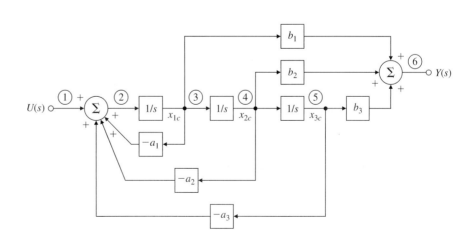

Mason's rule is particularly useful for more complex systems where there are several loops, some of which do not sum into the same point.

EXAMPLE 3.23   *Mason's Rule in a Complex System*

Find the transfer function for the system shown in Fig. 3.13.

Solution.   From the block diagram, we find that

Forward Path	*Path Gain*
12456	$G_1 = H_1 H_2 H_3$
1236	$G_2 = H_4$

	Loop Path Gain
242	$l_1 = H_1 H_5$ (does not touch l_3)
454	$l_2 = H_2 H_6$
565	$l_3 = H_3 H_7$ (does not touch l_1)
236542	$l_4 = H_4 H_7 H_6 H_5$

and the determinants are

$$\Delta = 1 - (H_1 H_5 + H_2 H_6 + H_3 H_7 + H_4 H_7 H_6 H_5) + (H_1 H_5 H_3 H_7),$$

$$\Delta_1 = 1 - 0,$$

$$\Delta_2 = 1 - H_2 H_6.$$

Therefore,

$$G(s) = \frac{Y(s)}{U(s)} = \frac{H_1 H_2 H_3 + H_4 - H_4 H_2 H_6}{1 - H_1 H_5 - H_2 H_6 - H_3 H_7 - H_4 H_7 H_6 H_5 + H_1 H_5 H_3 H_7}.$$

Figure 3.13
Block diagram for
Example 3.23

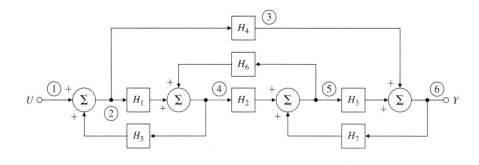

Mason's rule is useful for solving relatively complicated block diagrams by hand. It yields the solution in the sense that it provides an explicit input–output relationship for the system represented by the diagram. The advantage compared with path-by-path block-diagram reduction is that it is systematic and algorithmic rather than problem-dependent. MATLAB and other control systems computer-aided software allow you to specify a system in terms of individual blocks in an overall system, and the software algorithms perform the required block-diagram reduction; therefore, Mason's rule is less important today than in the past. However, there are some derivations that rely on the concepts embodied by the rule, so it still has a role in the control designer's toolbox.

3.2.4   Simulation

Perhaps the most effective way of understanding the state-variable equations is via an analog computer, block-diagram representation. The structure of the representation uses integrators as the central element, which are quite suitable for first-order, state-variable representation of equations of motion for a system. Even though the analog computers are almost extinct, analog computer implementation is still a useful concept for state-variable design, and in the circuit design of analog compensation.[8]

The analog computer was a device composed of electric components designed to simulate ordinary differential equations. The basic dynamic component of the analog computer is an **integrator**, constructed from an operational amplifier with a capacitor feedback and a resistor feed-forward as shown in Fig. 2.21. Because an integrator is a device whose input is the derivative of its output, as shown in Fig. 3.14, if, in an analog-computer simulation, we identify the outputs of the integrators as the state, we will then automatically have the equations in state-variable form. Conversely, if a system is described by state-variables, we can construct an analog-computer simulation of that system by taking one integrator for each state variable and connecting its input according to the given equation for that state variable as expressed in the state variable equations. The analog-computer diagram is a picture of the state equations.

The components of a typical analog computer used to accomplish these functions are shown in Fig. 3.15. Notice that the operational amplifier has a sign change that gives it a negative gain.

Figure 3.14
An integrator

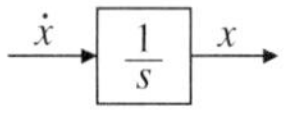

---

[8] As well as due to its historical significance.

Figure 3.15
Components of an analog
computer

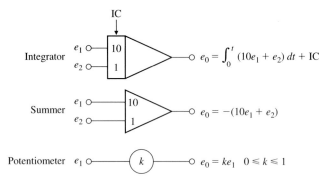

Analog-Computer Implementation

Find a state-variable description and the transfer function of the third-order system shown in Fig. 3.16 whose differential equation is

$$\dddot{y} + 6\ddot{y} + 11\dot{y} + 6y = 6u.$$

Solution. We solve for the highest derivative term in the ordinary differential equation to obtain

$$\dddot{y} = -6\ddot{y} - 11\dot{y} - 6y + 6u. \tag{3.40}$$

Now we assume that we have this highest derivative and note that the lower order terms can be obtained by integration as shown in Fig. 3.17(a). Finally, we apply Eq. (3.40) to complete the realization shown in Fig. 3.17(b). To obtain the state description, we simply define the state variables as the output of the integrators $x_1 = \ddot{y}$, $x_2 = \dot{y}$, $x_3 = y$, to obtain

$$\dot{x}_1 = -6x_1 - 11x_2 - 6x_3 + 6u,$$

$$\dot{x}_2 = x_1,$$

$$\dot{x}_3 = x_2,$$

which provides the state-variable description

$$\mathbf{F} = \begin{bmatrix} -6 & -11 & -6 \\ 1 & 0 & 0 \\ 0 & 1 & 0 \end{bmatrix}, \qquad \mathbf{G} = \begin{bmatrix} 6 \\ 0 \\ 0 \end{bmatrix}, \qquad \mathbf{H} = \begin{bmatrix} 0 & 0 & 1 \end{bmatrix}, \qquad J = 0.$$

Figure 3.16
Third-order system

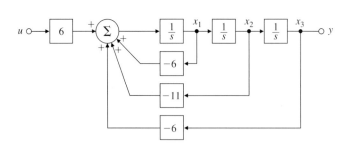

Figure 3.17
Block diagram of a
system to solve,
$\dddot{y} + 6\ddot{y} + 11\dot{y} + 6y = 6u$,
using only integrators
as dynamic elements
(a) intermediate diagram,
(b) final diagram

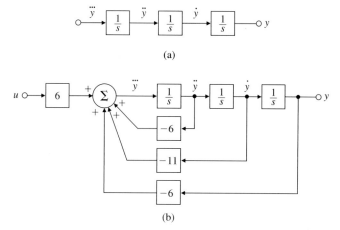

(a)

(b)

The MATLAB statement

$$[\text{num, den}] = \text{ss2tf}(F, G, H, J);$$

will yield the transfer function

$$\frac{Y(s)}{U(s)} = \frac{6}{s^3 + 6s^2 + 11s + 6}.$$

If the transfer function were desired in factored form, it could be obtained by transforming either the ss or tf description. Therefore, either of the following MATLAB statements

```
% convert state variable realization to pole-zero form
[z,p,k] = ss2zp(F,G,H,J)
% convert numerator-denominator to pole-zero form
[z,p,k] = tf2zp(num,den)
```

would result in $z = [\,]$, $p = [-3 \quad -2 \quad -1\,]'$, $k = 6$.
This means that the transfer function could also be written in factored form as

$$\frac{Y(s)}{U(s)} = \frac{6}{(s+1)(s+2)(s+3)}.$$

3.3   Effect of Pole Locations

Once the transfer function has been determined by any of the available methods, we can start to analyze the response of the system it represents. When the system equations are simultaneous ordinary differential equations (ODEs), the transfer function that results will be a ratio of polynomials; that is,

$$H(s) = b(s)/a(s).$$

If we assume that b and a have no common factors (as is usually the case), then values of s such that $a(s) = 0$ will represent points where $H(s)$ is infinity. As we saw in Section 3.1.5, these s-values are called poles of $H(s)$. Values of s such that $b(s) = 0$ are points where $H(s) = 0$, and the corresponding s-locations are called zeros. The effect of zeros on the transient response will be discussed in Section 3.5. These poles and zeros completely describe $H(s)$ except for a constant multiplier. Because the impulse response is given by the time function corresponding to the transfer function, we call the impulse response the **natural response** of the system. We can use the poles and zeros to compute the corresponding time response and thus identify time histories with pole locations in the s-plane. For example, the poles identify the classes of signals contained in the impulse response, as may be seen by a partial-fraction expansion of $H(s)$. For a first-order pole,

$$H(s) = \frac{1}{s + \sigma}.$$

Table A.2, entry 7, indicates that the impulse response will be an exponential function; that is,

$$h(t) = e^{-\sigma t} 1(t).$$

When $\sigma > 0$, the pole is located at $s < 0$, the exponential expression decays, and we say the impulse response is **stable**. If $\sigma < 0$, the pole is to the right of the origin. Because the exponential expression here grows with time, the impulse response is referred to as **unstable** (see Section 3.6). Figure 3.18(a) shows a typical stable response and defines the **time constant**

$$\tau = 1/\sigma \qquad (3.41)$$

as the time when the response is $1/e$ times the initial value. Hence it is a measure of the rate of decay. The straight line is tangent to the exponential curve at $t = 0$ and terminates at $t = \tau$. This characteristic of an exponential expression is useful in sketching a time plot or checking computer results.

Figure 3.18(b) shows the impulse and step response for a first-order system computed using MATLAB.

EXAMPLE 3.25   *Response Versus Pole Locations, Real Roots*

Compare the time response with the pole locations for the system with a transfer function between input and output given by

$$H(s) = \frac{2s + 1}{s^2 + 3s + 2}. \qquad (3.42)$$

Figure 3.18
First-order system response: (a) impulse response, (b) impulse and step response using MATLAB

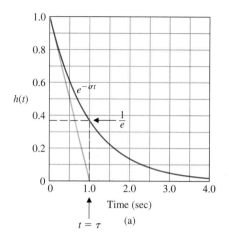

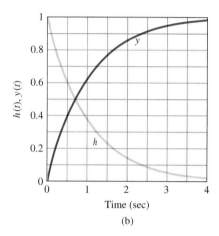

Solution.   The numerator is

$$b(s) = 2\left(s + \frac{1}{2}\right),$$

and the denominator is

$$a(s) = s^2 + 3s + 2 = (s + 1)(s + 2).$$

The poles of $H(s)$ are therefore at $s = -1$ and $s = -2$, and the one (finite) zero is at $s = -\frac{1}{2}$. A complete description of this transfer function is shown by the plot of the locations of the poles and the zeros in the s-plane using the MATLAB pzmap(num,den) function with

num = [2 1];
den = [1 3 2];

Figure 3.19
Sketch of s-plane showing
poles as crosses and zeros
as circles

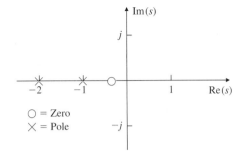

(see Fig. 3.19). A partial-fraction expansion of $H(s)$ results in

$$H(s) = -\frac{1}{s+1} + \frac{3}{s+2}.$$

From Table A.2 we can look up the inverse of each term in $H(s)$, which will give us the time function $h(t)$ that would result if the system input were an impulse. In this case,

$$h(t) = \begin{cases} -e^{-t} + 3e^{-2t}, & t \geq 0, \\ 0, & t < 0. \end{cases} \tag{3.43}$$

We see that the shape of the component parts of $h(t)$, which are e^{-t} and e^{-2t}, are determined by the poles at $s = -1$ and -2. This is true of more complicated cases as well: In general, the shapes of the components of the natural response are determined by the locations of the poles of the transfer function.

"Fast poles" and "slow poles" refer to relative rate of signal decay.

A sketch of these pole locations and corresponding natural responses is given in Fig. 3.20, along with other pole locations including complex ones, which will be discussed shortly.

The role of the numerator in the process of partial-fraction expansion is to influence the size of the coefficient that multiplies each component. Because e^{-2t} decays faster than e^{-t}, the signal corresponding to the pole at -2 decays faster than the signal corresponding to the pole at -1. For brevity we simply say that the pole at -2 is faster than the pole at -1. In general, poles farther to the left in the s-plane are associated with natural signals that decay faster than those associated with poles closer to the imaginary axis. If the poles had been located with positive values of s (in the right half of the s-plane), the response would have been a growing exponential function and thus unstable. Figure 3.21 shows that the fast $3e^{-2t}$ term dominates the early part of the time history and that the $-e^{-t}$ term is the primary contributor later on.

Impulse response using MATLAB

The purpose of this example is to illustrate the relationship between the poles and the character of the response, which can only be done exactly by finding the inverse Laplace transform and examining each term as above. However, if we simply wanted to plot the impulse response for this example, the expedient way would be to use the MATLAB sequence

```
numH = [2 1];          % form numerator
denH = [1 3 2];        % form denominator
sysH = tf(numH,denH);  % define system from its numerator and denominator
impulse(sysH);         % compute impulse response
```

and is shown in Fig. 3.21.

Figure 3.20

Time functions associated with points in the s-plane (LHP, left half-plane; RHP, right half-plane)

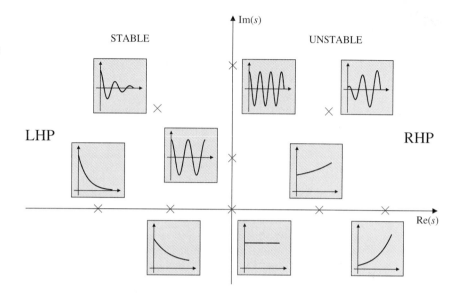

Figure 3.21

Impulse response of Example 3.25 (Eq. 3.42)

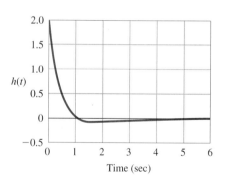

Complex poles can be defined in terms of their real and imaginary parts, traditionally referred to as

$$s = -\sigma \pm j\omega_d.$$

This means that a pole has a negative real part if σ is positive. Since complex poles always come in complex conjugate pairs, the denominator corresponding to a complex pair will be

$$a(s) = (s + \sigma - j\omega_d)(s + \sigma + j\omega_d) = (s + \sigma)^2 + \omega_d^2. \qquad (3.44)$$

When finding the transfer function from differential equations, we typically write the result in the polynomial form

$$H(s) = \frac{\omega_n^2}{s^2 + 2\zeta\omega_n s + \omega_n^2}. \tag{3.45}$$

By multiplying out the form given by Eq. (3.44) and comparing it with the coefficients of the denominator of $H(s)$ in Eq. (3.45), we find the correspondence between the parameters to be

$$\sigma = \zeta\omega_n \quad \text{and} \quad \omega_d = \omega_n\sqrt{1 - \zeta^2}, \tag{3.46}$$

Damping ratio; damped and undamped natural frequency

where the parameter ζ is the **damping ratio**[9] and ω_n is the **undamped natural frequency**. The poles of this transfer function are located at a radius ω_n in the s-plane and at an angle $\theta = \sin^{-1}\zeta$, as shown in Fig. 3.22. Therefore, the damping ratio reflects the level of damping as a fraction of the critical damping value where the poles become real. In rectangular coordinates the poles are at $s = -\sigma \pm j\omega_d$. When $\zeta = 0$, we have no damping, $\theta = 0$, and the damped natural frequency $\omega_d = \omega_n$, the undamped natural frequency.

For purposes of finding the time response from Table A.2 corresponding to a complex transfer function, it is easiest to manipulate the $H(s)$ so that the complex poles fit the form of Eq. (3.44), because then the time response can be found directly from the table. Equation (3.45) can be rewritten as

$$H(s) = \frac{\omega_n^2}{(s + \zeta\omega_n)^2 + \omega_n^2(1 - \zeta^2)}.$$

Figure 3.22
s-plane plot for a pair of complex poles

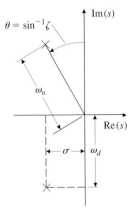

---

[9] In communications and filter engineering, the standard second-order transfer function is written $H = 1/[1 + Q(s/\omega_n + \omega_n/s)]$. Here ω_n is called the **band center** and Q is the **quality factor**. Comparison with Eq. (3.45) shows that $Q = 1/2\zeta$.

Therefore, from entry number 20 in Table A.2 and the definitions in Eq. (3.46), we see that the impulse response is

$$h(t) = \frac{\omega_n}{\sqrt{1 - \zeta^2}} e^{-\sigma t} (\sin \omega_d t) 1(t).$$

Figure 3.23(a) plots $h(t)$ for several values of ζ such that time has been normalized to the undamped natural frequency ω_n. Note that the actual frequency

Figure 3.23
Responses of second-order systems versus ζ;
(a) impulse responses;
(b) step responses

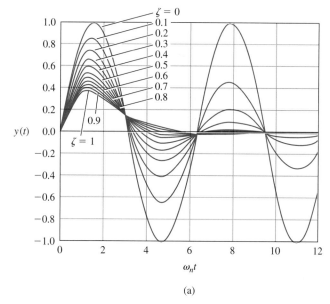

(a)

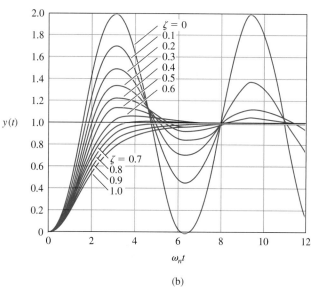

(b)

Figure 3.24
Pole locations
corresponding to three
values of ζ

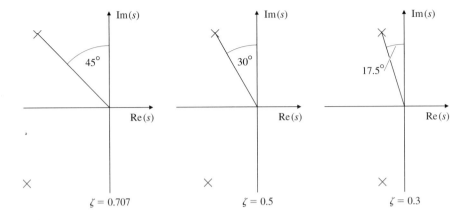

$\zeta = 0.707$ $\zeta = 0.5$ $\zeta = 0.3$

ω_d decreases slightly as the damping ratio increases. Also note that for very low damping the response is oscillatory, while for large damping (ζ near 1) the response shows no oscillation. A few of these responses are sketched in Fig. 3.20 to show qualitatively how changing pole locations in the s-plane affect impulse responses. You will find it useful as a control designer to commit the image of Fig. 3.20 to memory so that you can understand instantly how changes in pole locations influence the time response.

Stability depends on whether
natural response grows or
decays.

     Three pole locations are shown in Fig. 3.24 for comparison with the corresponding impulse responses in Fig. 3.23(a). The negative real part of the pole, σ, determines the decay rate of an exponential envelope that multiplies the sinusoid, as shown in Fig. 3.25. Note that if $\sigma < 0$ (and the pole is in the right half-plane, RHP), then the natural response will grow with time, so, as defined earlier, the system is said to be unstable. If $\sigma = 0$, the natural response neither grows nor decays, so stability is open to debate. If $\sigma > 0$, the natural response decays, so the system is stable.

Figure 3.25
Second-order system
response with an
exponential envelope

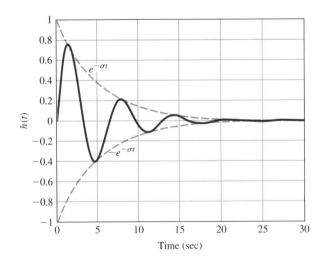

Step response

It is also interesting to examine the step response of $H(s)$—that is, the response of the system $H(s)$ to a unit step input $u = 1(t)$, where $U(s) = 1/s$. The step-response transform is given by $Y(s) = H(s)U(s)$, which is found in Table A.2, entry 21. Figure 3.23(b), which plots $y(t)$ for several values of ζ, shows that the basic transient response characteristics from the impulse response carries over quite well to the step response; the difference between the two responses is that the step response's final value is the commanded unit step.

EXAMPLE 3.26

Oscillatory Time Response

Discuss the correlation between the poles of

$$H(s) = \frac{2s + 1}{s^2 + 2s + 5} \qquad (3.47)$$

and the impulse response of the system, and find the exact impulse response.

Solution.   From the form of $H(s)$ given by Eq. (3.45), we see that

$$\omega_n^2 = 5 \Rightarrow \omega_n = \sqrt{5} = 2.24 \text{ rad/sec}$$

and

$$2\zeta\omega_n = 2 \Rightarrow \zeta = \frac{1}{\sqrt{5}} = 0.447.$$

This indicates we should expect a frequency around 2 rad/sec with very little oscillatory motion. In order to obtain the exact response, we manipulate $H(s)$ until the denominator is in the form of Eq. (3.44):

$$H(s) = \frac{2s + 1}{s^2 + 2s + 5} = \frac{2s + 1}{(s + 1)^2 + 2^2}.$$

From this equation we see that the poles of the transfer function are complex, with real part -1 and imaginary parts $\pm 2j$. Table A.2 has two entries, numbers 19 and 20, that match the denominator. The right side of the above equation needs to be broken into two parts so that they match the numerators of the entries in the table:

$$H(s) = \frac{2s + 1}{(s + 1)^2 + 2^2} = 2\frac{s + 1}{(s + 1)^2 + 2^2} - \frac{1}{2}\frac{2}{(s + 1)^2 + 2^2}$$

thus, the impulse response is

$$h(t) = \left(2e^{-t}\cos 2t - \frac{1}{2}e^{-t}\sin 2t\right)1(t).$$

Figure 3.26 is a plot of the response and shows how the envelope attenuates the sinusoid, the domination of the $2\cos 2t$ term, and the small phase shift caused by the $-\frac{1}{2}\sin 2t$ term.

Figure 3.26
System response for
Example 3.26

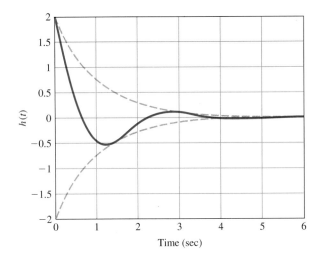

Time (sec)

Impulse response by MATLAB

As in the previous example, the expedient way of determining the impulse response would be to use the MATLAB sequence

```
numH = [2 1];            % form numerator
denH = [1 2 5];          % form denominator
sysH = tf(numH,denH);    % define system by its numerator and denominator
t = 0:0.1:6;             % form time vector
y = impulse(sysH,t);     % compute impulse response
plot(t,y);               % plot impulse response
```

as shown in Fig. 3.26.

3.4 Time-Domain Specifications

Definitions of rise time, settling time, overshoot, and peak time

Specifications for a control system design often involve certain requirements associated with the time response of the system. The requirements for a step response are expressed in terms of the standard quantities illustrated in Fig. 3.27:

1. The **rise time** t_r is the time it takes the system to reach the vicinity of its new set point.

2. The **settling time** t_s is the time it takes the system transients to decay.

3. The **overshoot** M_p is the maximum amount the system overshoots its final value divided by its final value (and often expressed as a percentage).

4. The **peak time** t_p is the time it takes the system to reach the maximum overshoot point.

3.4.1   Rise Time

For a second-order system the time responses shown in Fig. 3.23(b) yield information about the specifications that is too complex to be remembered unless converted to a simpler form. By examining these curves in light of the definitions given in Fig. 3.27, we can relate the curves to the pole-location parameters ζ and ω_n. For example, all the curves rise in roughly the same time. If we consider the curve for $\zeta = 0.5$ to be an average, the rise time from $y = 0.1$ to 0.9 is approximately $\omega_n t_r = 1.8$. Thus we can say that

$$t_r \cong \frac{1.8}{\omega_n}. \tag{3.48}$$

Rise time t_r

Although this relationship could be embellished by including the effect of the damping ratio, it is important to keep in mind how Eq. (3.48) is typically used. It is only accurate for a second-order system with no zeros; for all other systems it is a rough approximation to the relationship between t_r and ω_n. Most systems being analyzed for control systems design are more complicated than the pure second-order system, so designers use Eq. (3.48) with the knowledge that it is a rough approximation only.

3.4.2   Overshoot and Peak Time

For the overshoot M_p we can be more analytical. This value occurs when the derivative is zero, which can be found from calculus. The time history of the curves in Fig. 3.23(b), found from the inverse Laplace transform of $H(s)/s$, is

$$y(t) = 1 - e^{-\sigma t}\left(\cos \omega_d t + \frac{\sigma}{\omega_d}\sin \omega_d t\right),$$

Figure 3.27
Definition of rise time
t_r, settling time t_s, and
overshoot M_p

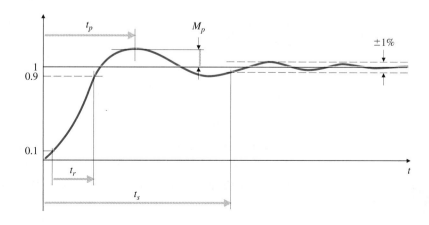

where $\omega_d = \omega_n\sqrt{1-\zeta^2}$ and $\sigma = \zeta\omega_n$. We may rewrite the above equation using the trigonometric identity

$$A\sin(\alpha) + B\cos(\alpha) = C\cos(\alpha - \beta),$$
$$C = \sqrt{A^2 + B^2}, \qquad \beta = \tan^{-1}\left(\frac{A}{B}\right),$$

with $A = \sigma/\omega_d$, $B = 1$, and $\alpha = \omega_d t$, in a more compact form as

$$y(t) = 1 - e^{-\sigma t}\sqrt{1 + \frac{\sigma^2}{\omega_d^2}}\cos(\omega_d t - \beta).$$

When $y(t)$ reaches its maximum value, its derivative will be zero:

$$\dot{y}(t) = \sigma e^{-\sigma t}\left(\cos\omega_d t + \frac{\sigma}{\omega_d}\sin\omega_d t\right) - e^{-\sigma t}(-\omega_d\sin\omega_d t + \sigma\cos\omega_d t) = 0$$

$$= e^{-\sigma t}\left(\frac{\sigma^2}{\omega_d}\sin\omega_d t + \omega_d\sin\omega_d t\right) = 0.$$

This occurs when $\sin\omega_d t = 0$, so

$$\omega_d t_p = \pi$$

and thus

$$t_p = \frac{\pi}{\omega_d}. \tag{3.49}$$

Peak time t_p

Substituting Eq. (3.49) into the expression for $y(t)$, we compute

$$y(t_p) \overset{\triangle}{=} 1 + M_p = 1 - e^{-\sigma\pi/\omega_d}\left(\cos\pi + \frac{\sigma}{\omega_d}\sin\pi\right)$$

$$= 1 + e^{-\sigma\pi/\omega_d}.$$

Overshoot M_p

Thus we have the formula

$$M_p = e^{-\pi\zeta/\sqrt{1-\zeta^2}}, \qquad 0 \le \zeta < 1, \tag{3.50}$$

which is plotted in Fig. 3.28. Two frequently used values from this curve are $M_p = 0.16$ for $\zeta = 0.5$ and $M_p = 0.05$ for $\zeta = 0.7$.

Figure 3.28
Overshoot M_p versus
damping ratio ζ for the
second-order system

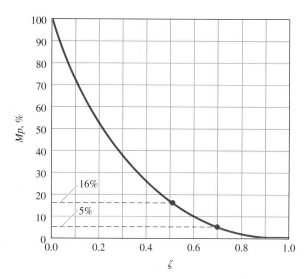

3.4.3   Settling Time

The final parameter of interest from the transient response is the settling time t_s. This is the time required for the transient to decay to a small value so that $y(t)$ is almost in the steady-state. Various measures of smallness are possible. For illustration we will use 1% as a reasonable measure; in other cases 2% or 5% are used. As an analytic computation we notice that the deviation of y from 1 is the product of the decaying exponential $e^{-\sigma t}$ and the circular functions sine and cosine. The duration of this error is essentially decided by the transient exponential, so we can define the settling time as that value of t_s when the decaying exponential reaches 1%:

$$e^{-\zeta \omega_n t_s} = 0.01.$$

Therefore,

$$\zeta \omega_n t_s = 4.6$$

or

$$t_s = \frac{4.6}{\zeta \omega_n} = \frac{4.6}{\sigma}, \tag{3.51}$$

Settling time t_s

where σ is the negative real part of the pole, as may be seen from Fig. 3.22.
Equations (3.48), (3.50), and (3.51) characterize the transient response of a system having no finite zeros and two complex poles with undamped natural frequency ω_n, damping ratio ζ, and negative real part σ. In analysis and design, they are used to estimate rise time, overshoot, and settling time, respectively, for just about any system. In design synthesis we wish to specify t_r, M_p, and

Design synthesis

t_s and to ask where the poles need to be so that the actual responses are less than or equal to these specifications. For specified values of t_r, M_p, and t_s, the synthesis form of the equation is then

$$\omega_n \geq \frac{1.8}{t_r}, \tag{3.52}$$

$$\zeta \geq \zeta(M_p) \qquad \text{(from Fig. 3.28),} \tag{3.53}$$

$$\sigma \geq \frac{4.6}{t_s}. \tag{3.54}$$

These equations, which can be graphed in the s-plane as shown in Fig. 3.29(a–c), will be used in later chapters to guide the selection of pole and zero locations to meet control system specifications for dynamic response.

It is important to keep in mind that Eqs. (3.52)–(3.54) are qualitative guides and not precise design formulas. They are meant to provide only a starting point for the design iteration. After the control design is complete, the time response should always be checked by an exact calculation, usually by numerical simulation, to verify whether the time specifications have actually been met. If not, another iteration of the design is required.

For a first-order system,

$$H(s) = \frac{\sigma}{s + \sigma},$$

the transform of the step response is

$$Y(s) = \frac{\sigma}{s(s + \sigma)}.$$

We see from entry 11 in Table A.2 that $Y(s)$ corresponds to

$$y(t) = (1 - e^{-\sigma t})1(t). \tag{3.55}$$

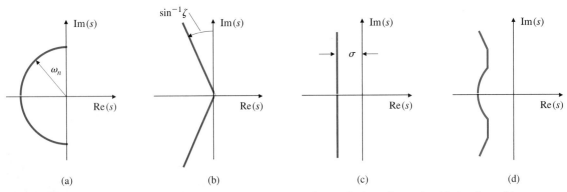

(a)   (b)   (c)   (d)

Figure 3.29 Graphs of regions in the s-plane delineated by certain transient requirements: (a) rise time; (b) overshoot; (c) settling time; and (d) composite of all three requirements

Comparison with the development for Eq. (3.51) shows that the value of t_s for a first-order system is the same:

$$t_s = \frac{4.6}{\sigma}.$$

No overshoot is possible, so $M_p = 0$. The rise time from $y = 0.1$ to $y = 0.9$ can be seen from Fig. 3.18 to be

$$t_r = \frac{\ln 0.9 - \ln 0.1}{\sigma} = \frac{2.2}{\sigma}.$$

Time constant τ

However, it is more typical to describe a first-order system in terms of its time constant, which was defined in Fig. 3.18 to be $\tau = 1/\sigma$.

EXAMPLE 3.27   *Transformation of the Specifications to the s-Plane*

Find the allowable regions in the s-plane for the poles of a transfer function of a system if the system response requirements are $t_r \leq 0.6$ sec, $M_p \leq 10\%$, and $t_s \leq 3$ sec.

Solution.   Without knowing whether or not the system is second-order with no zeros, it is impossible to find the allowable region accurately. Regardless of the system, we can obtain a first approximation using the relationships for a second-order system. Equation (3.52) indicates that

$$\omega_n \geq \frac{1.8}{t_r} = 3.0 \text{ rad/sec,}$$

Eq. (3.53) and Fig. 3.28 indicate that

$$\zeta \geq 0.6,$$

and Eq. (3.54) indicates that

$$\sigma \geq \frac{4.6}{3} = 1.5 \text{ sec.}$$

The allowable region is anywhere to the left of the solid line in Fig. 3.30. Note that any pole meeting the ζ and ω_n restrictions will automatically meet the σ restriction.

Figure 3.30
Allowable region in s-plane
for Example 3.27

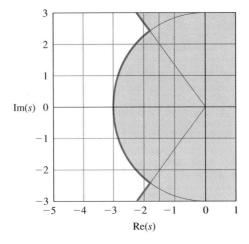

3.5   Effects of Zeros and Additional Poles

Relationships such as those shown in Fig. 3.29 are correct for the simple second-order system; for more complicated systems they can be used only as guidelines. If a certain design has an inadequate rise time (is too slow), we must raise the natural frequency; if the transient has too much overshoot, then the damping needs to be increased; if the transient persists too long, the poles need to be moved to the left in the s-plane.

Effect of zeros

Thus far only the poles of $H(s)$ have entered into the discussion. There may also be zeros of $H(s)$.[10] At the level of transient analysis, the zeros exert their influence by modifying the coefficients of the exponential terms whose shape is decided by the poles as seen in Example 3.26. To illustrate this further, consider the following two transfer functions, which have the same poles but different zeros:

$$H_1(s) = \frac{2}{(s+1)(s+2)} = \frac{2}{s+1} - \frac{2}{s+2}, \tag{3.56}$$

$$H_2(s) = \frac{2(s+1.1)}{1.1(s+1)(s+2)} = \frac{2}{1.1}\left(\frac{0.1}{s+1} + \frac{0.9}{s+2}\right) = \frac{0.18}{s+1} + \frac{1.64}{s+2}. \tag{3.57}$$

---

[10] We assume that $b(s)$ and $a(s)$ have no common factors. If this is not so, it is possible for $b(s)$ and $a(s)$ to be zero at the same location and for $H(s)$ to not equal zero there. The implications of this case will be discussed in Chapter 7, when we have a state-space description.

They are normalized to have the same DC gain—that is, gain at $s = 0$. Notice that the coefficient of the $(s + 1)$ term has been modified from 2 in $H_1(s)$ to 0.18 in $H_2(s)$. This dramatic reduction is brought about by the zero at $s = -1.1$ in $H_2(s)$, which almost cancels the pole at $s = -1$. If we put the zero exactly at $s = -1$, this term will vanish completely. In general, a zero near a pole reduces the amount of that term in the total response. From the equation for the coefficients in a partial-fraction expansion, Eq. (3.32),

$$C_1 = (s - p_1)F(s)|_{s=p_1}$$

we can see that, if $F(s)$ has a zero near the pole at $s = p_1$, the value of $F(s)$ will be small because the value of s is near the zero. Therefore, the coefficient C_1, which reflects how much of that term appears in the response, will be small.

In order to take into account how zeros affect the transient response when designing a control system, we consider transfer functions with two complex poles and one zero. To expedite the plotting for a wide range of cases, we write the transform in a form with normalized time and zero locations:

$$H(s) = \frac{(s/\alpha\zeta\omega_n) + 1}{(s/\omega_n)^2 + 2\zeta(s/\omega_n) + 1}. \tag{3.58}$$

The zero is located at $s = -\alpha\zeta\omega_n = -\alpha\sigma$. If α is large, the zero will be far removed from the poles and the zero will have little effect on the response. If $\alpha \cong 1$, the value of the zero will be close to that of the real part of the poles and can be expected to have a substantial influence on the response. The step-response curves for $\zeta = 0.5$ and for several values of α are plotted in Fig. 3.31. We see that the major effect of the zero is to increase the overshoot M_p, whereas it has very little influence on the settling time. A plot of M_p versus α is given

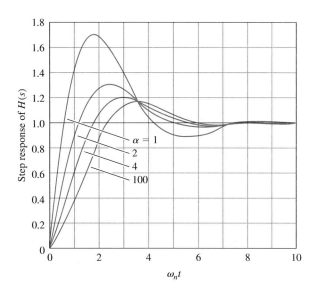

Figure 3.31

Plots of the step response of a second-order system with a zero ($\zeta = 0.5$)

Figure 3.32
Plot of overshoot M_p as a function of normalized zero location α. At $\alpha = 1$, the real part of the zero equals the real part of the poles

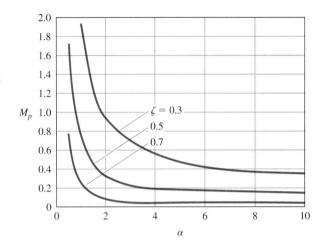

in Fig. 3.32. The plot shows that the zero has very little effect on M_p if $\alpha > 3$, but as α decreases below 3, it has an increasing effect, especially when $\alpha = 1$ or less.

Figure 3.31 can be explained in terms of Laplace-transform analysis. First we replace s/ω_n with s:

$$H(s) = \frac{s/\alpha\zeta + 1}{s^2 + 2\zeta s + 1}.$$

This has the effect of normalizing frequency in the transfer function and normalizing time in the corresponding step responses; thus $\tau = \omega_n t$. We then rewrite the transfer function as the sum of two terms:

$$H(s) = \frac{1}{s^2 + 2\zeta s + 1} + \frac{1}{\alpha\zeta}\frac{s}{s^2 + 2\zeta s + 1}. \tag{3.59}$$

The first term, which we will call $H_0(s)$, is the original term (having no finite zero); and the second term $H_d(s)$, which is introduced by the zero, is a product of a constant $(1/\alpha\zeta)$ times s times the original term. The Laplace transform of df/dt is $sF(s)$, so $H_d(s)$ corresponds to a product of a constant times the *derivative* of the original term. The step responses of $H_0(s)$ and $H_d(s)$ are plotted in Fig. 3.33. Looking at these curves, we can see why the zero increased the overshoot: The derivative has a large hump in the early part of the curve, and adding this to the $H_0(s)$ response lifts up the total response of $H(s)$ to produce the overshoot. This analysis is also very informative for the case when $\alpha < 0$ and the zero is in the RHP where $s > 0$. (This is typically called a **RHP zero** and sometimes referred to as a **nonminimum-phase zero**, a topic to be discussed in more detail in Section 6.1.1.) In this case the derivative term is subtracted rather than added. A typical case is sketched in Fig. 3.34.

RHP or non-minimum-phase zero

Figure 3.33
Second-order step
responses $y(t)$ of the
transfer functions $H(s)$,
$H_0(s)$, and $H_d(s)$

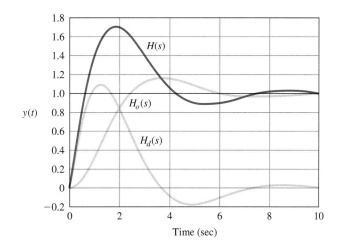

Figure 3.34
Step responses $y(t)$ of
a second-order system
with a zero in the RHP: a
nonminimum-phase system

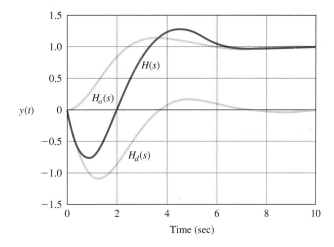

EXAMPLE 3.28

Aircraft Response Using MATLAB

The transfer function between the elevator and altitude of the Boeing 747 aircraft de-scribed in Section 9.3.2 can be approximated as

$$\frac{h(s)}{\delta_e(s)} = \frac{30(s-6)}{s(s^2 + 4s + 13)}.$$

(a) Use MATLAB to plot the altitude time history for a $1°$ impulsive elevator input. Ex-plain the response, noting the physical reasons for the non-minimum-phase nature of the response.

(b) Examine the accuracy of the approximations for t_r, t_s, and M_p [Eqs. (3.48) and (3.51) and Fig. 3.28].

Solution.

(a) The MATLAB statements to create the impulse response for this case are

```
u = -1;                          % u = delta e
numG = u*30*[1 -6];              % form numerator
denG = [1 4 13 0];               % form denominator
sysG = tf(numG,denG)             % define system by its numerator and denominator
y = impulse(sysG);               % compute impulse response; y = h
plot(y);                         % plot impulse response
```

The result is the plot shown in Fig. 3.35. Notice how the altitude drops initially and then rises to a new final value. The final value is predicted by the Final Value Theorem:

$$h(\infty) = s \left. \frac{30(s-6)(-1)}{s(s^2+4s+13)} \right|_{s=0} = \frac{30(-6)(-1)}{13} = +13.8.$$

Response of a non-minimum-phase system

The fact that the response has a finite final value for an impulsive input is due to the s-term in the denominator. This represents a pure integration, and the integral of an impulse function is a finite value. If the input had been a step, the altitude would have continued to increase with time; in other words the integral of a step function is a ramp function.

The initial drop is predicted by the RHP zero in the transfer function. The negative elevator deflection is defined to be upward by convention (see Fig. 9.30). The upward deflection of the elevators drives the tail down, which rotates the craft nose up and produces the climb. The deflection at the initial instant causes a downward force before the craft has rotated; therefore, the initial altitude response is down. After rotation the increased lift resulting from the increased angle of attack of the wings causes the airplane to climb.

(b) The rise time from Eq. (3.48) is

$$t_r = \frac{1.8}{\omega_n} = \frac{1.8}{\sqrt{13}} = 0.5 \text{ sec.}$$

Figure 3.35
Response of an airplane's altitude to an impulsive elevator input

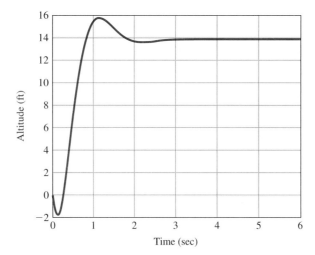

We find the damping ratio ζ from the relation

$$2\zeta\omega_n = 4 \Rightarrow \zeta = \frac{2}{\sqrt{13}} = 0.55.$$

From Fig. 3.28 we find the overshoot M_p to be 0.14. Because $2\zeta\omega_n = 2\sigma = 4$, Eq. (3.51) shows that

$$t_s = \frac{4.6}{\sigma} = \frac{4.6}{2} = 2.3 \text{ sec.}$$

Detailed examination of the time history $h(t)$ from MATLAB output shows that $t_r \cong 0.43$ sec, $M_p \cong 0.14$, and $t_s \cong 2.6$ sec, which are reasonably close to the estimates. The only significant effect of the nonminimum phase zero was to cause the initial response to go the "wrong direction" and make the response somewhat sluggish.

Effect of extra pole

In addition to studying the effects of zeros, it is useful to consider the effects of an extra pole on the standard second-order step response. In this case we take the transfer function to be

$$H(s) = \frac{1}{(s/\alpha\zeta\omega_n + 1)[(s/\omega_n)^2 + 2\zeta(s/\omega_n) + 1]}, \tag{3.60}$$

Plots of the step response for this case are shown in Fig. 3.36 for $\zeta = 0.5$ and several values of α. In this case the major effect is to increase the rise time. A plot of the rise time versus α is shown in Fig. 3.37 for several values of ζ.

Figure 3.36
Step responses for several third-order systems with $\zeta = 0.5$

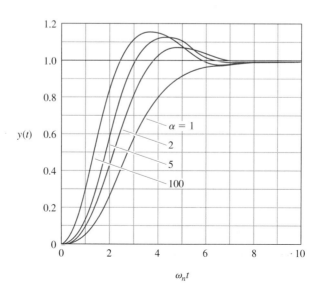

Figure 3.37
Normalized rise time for
several locations of an
additional pole

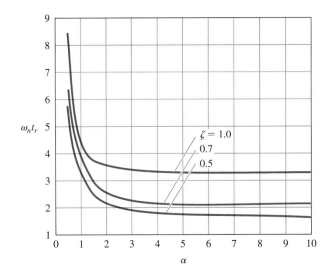

From this discussion we can draw several conclusions about the dynamic response of a simple system as revealed by its pole-zero patterns:

Effects of Pole-Zero Patterns on Dynamic Response

1. For a second-order system with no finite zeros, the transient response parameters are approximated by

$$\text{Rise time}: \quad t_r \cong \frac{1.8}{\omega_n},$$

$$\text{Overshoot}: \quad M_p \cong \begin{cases} 5\%, & \zeta = 0.7, \\ 16\%, & \zeta = 0.5 \quad \text{(see Fig. 3.28)}, \\ 35\%, & \zeta = 0.3, \end{cases}$$

$$\text{Settling time}: \quad t_s \cong \frac{4.6}{\sigma}.$$

2. A zero in the left half-plane (LHP) will increase the overshoot if the zero is within a factor of 4 of the real part of the complex poles. A plot is given in Fig. 3.32.

3. A zero in the right half-plane will depress the overshoot (and may cause the step response to start out in the wrong direction).

4. An additional pole in the left half-plane will increase the rise time significantly if the extra pole is within a factor of 4 of the real part of the complex poles. A plot is given in Fig. 3.37.

3.6 Stability

For nonlinear and time varying systems the study of stability is complex and often difficult subject. In this section we will consider only linear time-invariant systems for which we have the following condition for stability.

> A linear time-invariant system is said to be stable if all the roots of the transfer function denominator polynomial have negative real parts (i.e., they are *all* in the left-hand s-plane) and unstable otherwise.

Stable system

A system is stable if initial conditions decay to zero and unstable if they diverge. As stated above, a linear time-invariant (constant parameter) system is **stable** if *all* the poles of the system are strictly inside the left half s-plane; that is, all the poles must have negative real parts ($\sigma < 0$). If *any* pole of the system is in the right half s-plane [i.e., has a positive real part ($\sigma > 0$)], then the system is **unstable** as shown in Fig. 3.20. With any simple pole on the $j\omega$ axis ($\sigma = 0$), small initial conditions will persist. For any other pole with $\sigma = 0$, oscillatory motion will persist. Therefore, a system is stable if its transient response decays and unstable if it does not. Figure 3.20 shows the time response of a system due to its pole locations.

In later chapters we will address more advanced notions of stability, such as Nyquist's frequency-response stability test (Chapter 6) and Lyapunov stability (Chapter 7).

3.6.1 Stability of Linear Time-Invariant Systems

Consider the linear time-invariant system whose transfer function denominator polynomial leads to the characteristic equation

$$s^n + a_1 s^{n-1} + a_2 s^{n-2} + \cdots + a_n = 0. \tag{3.61}$$

Assume that the roots $\{p_i\}$ of the characteristic equation are real or complex but are distinct. Note that Eq. (3.61) shows up as the denominator in the transfer function for the system as follows *before any cancellations of poles by zeros are made:*

$$T(s) = \frac{Y(s)}{R(s)} = \frac{b_0 s^m + b_1 s^{m-1} + \cdots + b_m}{s^n + a_1 s^{n-1} + \cdots + a_n}$$

$$= \frac{K \prod_{i=1}^{m} (s - z_i)}{\prod_{i=1}^{n} (s - p_i)}, \qquad m \le n. \tag{3.62}$$

The solution to the differential equation whose characteristic equation is given by Eq. (3.61) may be written using partial fraction expansion as

$$y(t) = \sum_{i=1}^{n} K_i e^{p_i t}, \tag{3.63}$$

where $\{p_i\}$ are the roots of Eq. (3.61) and $\{K_i\}$ depend on the initial conditions and zero locations. If a zero were to cancel a pole in the RHP for the transfer function, the corresponding K_i would equal zero in the output but the unstable transient would appear in some internal variable.

The system is stable if and only if (necessary and sufficient condition) every term in Eq. (3.63) goes to zero as $t \to \infty$:

$$e^{p_i t} \to 0 \qquad \text{for all } p_i.$$

This will happen if all the poles of the system are strictly in the LHP, where

$$\text{Re}\{p_i\} < 0. \tag{3.64}$$

If any poles are repeated, the response must be changed from that of Eq. (3.63) by including a polynomial in t in place of K_i, but the conclusion is the same. This is called **internal stability**. Therefore, the stability of a system can be

Internal stability occurs when all poles are strictly in the LHP.

determined by computing the location of the roots of the characteristic equation and determining whether they are all in the LHP. If the system has any poles in the RHP, it is **unstable**. Hence the $j\omega$ axis is the stability boundary between asymptotically stable and unstable response. If the system has nonrepeated $j\omega$-axis poles, then it is said to be **neutrally stable**. For example, a pole at the origin (an integrator) results in a nondecaying transient. A pair of complex $j\omega$-axis poles results in an oscillating response (with constant amplitude). If the system has repeated poles on the $j\omega$ axis, then it is **unstable** [as it results in $te^{\pm j\omega_i t}$ terms in Eq. (3.63)]. For example, a pair of poles at the origin (double integrator) results in an unbounded response. MATLAB software makes the computation of the poles and therefore determination of the stability of the system relatively easy.

An alternative to locating the roots of the characteristic equation is given by Routh's stability criterion, which we will discuss next.

3.6.2   Routh's Stability Criterion

There are several methods of obtaining information about the locations of the roots of a polynomial without actually solving for the roots. These methods were developed in the 19th century and were especially useful before the availability of MATLAB software. They are still useful for determining the ranges of coefficients of polynomials for stability, especially when the coefficients are in symbolic (nonnumerical) form. Consider the characteristic equation of an nth-order system:[11]

$$a(s) = s^n + a_1 s^{n-1} + a_2 s^{n-2} + \cdots + a_{n-1} s + a_n. \tag{3.65}$$

---

[11] Without loss of generality we can assume the polynomial to be monic (that is, the coefficient of the highest power of s is 1).

It is possible to make certain statements about the stability of the system without actually solving for the roots of the polynomial. This is a classical problem, and several methods exist for the solution.

A *necessary condition for stability* of the system is that all of the roots of Eq. (3.65) have negative real parts, which in turn requires that all the $\{a_i\}$ be positive.[12]

A necessary (but not sufficient) condition for stability is that *all* the coefficients of the characteristic polynomial be positive.

If any of the coefficients are missing (are zero) or are negative, then the system will have poles located outside the LHP. This condition can be checked by inspection. Once the elementary necessary conditions have been satisfied, we need a more powerful test. Equivalent tests were independently proposed by Routh in 1874 and Hurwitz in 1895; we will discuss the former version. Routh's formulation requires the computation of a triangular array that is a function of the $\{a_i\}$. He showed that a *necessary and sufficient condition for stability* is that all of the elements in the first column of this array be positive.

A system is stable if and only if *all* the elements in the first column of the Routh array are positive.

To determine the Routh array, we first arrange the coefficients of the characteristic polynomial in two rows, beginning with the first and second coefficients and followed by the even-numbered and odd-numbered coefficients:

$$
\begin{array}{llllll}
s^n & : & 1 & a_2 & a_4 & \cdots \\
s^{n-1} & : & a_1 & a_3 & a_5 & \cdots
\end{array}
$$

We then add subsequent rows to complete the **Routh array**:

Row	n	s^n:	1	a_2	a_4	$\cdots$
Row	$n-1$	s^{n-1}:	a_1	a_3	a_5	$\cdots$
Row	$n-2$	s^{n-2}:	b_1	b_2	b_3	$\cdots$
Row	$n-3$	s^{n-3}:	c_1	c_2	c_3	$\cdots$
	$\vdots$	$\vdots$	$\vdots$	$\vdots$	$\vdots$	
Row	2	s^2:	*	*		
Row	1	s:	*			
Row	0	s^0:	*			

_____

[12] This is easy to see if we construct the polynomial as a product of first- and second-order factors.

Margin notes:
A necessary condition for Routh stability

A necessary and sufficient condition for stability

Routh array

We compute the elements from the $(n-2)$th and $(n-3)$th rows as follows:

$$b_1 = -\frac{\det \begin{bmatrix} 1 & a_2 \\ a_1 & a_3 \end{bmatrix}}{a_1} = \frac{a_1 a_2 - a_3}{a_1},$$

$$b_2 = -\frac{\det \begin{bmatrix} 1 & a_4 \\ a_1 & a_5 \end{bmatrix}}{a_1} = \frac{a_1 a_4 - a_5}{a_1},$$

$$b_3 = -\frac{\det \begin{bmatrix} 1 & a_6 \\ a_1 & a_7 \end{bmatrix}}{a_1} = \frac{a_1 a_6 - a_7}{a_1},$$

$$c_1 = -\frac{\det \begin{bmatrix} a_1 & a_3 \\ b_1 & b_2 \end{bmatrix}}{b_1} = \frac{b_1 a_3 - a_1 b_2}{b_1},$$

$$c_2 = -\frac{\det \begin{bmatrix} a_1 & a_5 \\ b_1 & b_3 \end{bmatrix}}{b_1} = \frac{b_1 a_5 - a_1 b_3}{b_1},$$

$$c_3 = -\frac{\det \begin{bmatrix} a_1 & a_7 \\ b_1 & b_4 \end{bmatrix}}{b_1} = \frac{b_1 a_7 - a_1 b_4}{b_1}.$$

Note that the elements of the $(n-2)$th row and the rows beneath it are formed from the two previous rows using determinants, with the two elements in the first column and other elements from successive columns. Normally these are $n+1$ elements in the first column when the array terminates. If these are all positive, then all the roots of the characteristic polynomial are in the LHP. However, if the elements of the first column are not all positive, then the number of roots in the RHP equals the number of sign changes in the column. A pattern of $+$, $-$, $+$ is counted as *two* sign changes: one change from $+$ to $-$ and another from $-$ to $+$. For a simple proof of the Routh test, the reader is referred to Ho et al. (1998).

EXAMPLE 3.29   *Routh's Test*

The polynomial

$$a(s) = s^6 + 4s^5 + 3s^4 + 2s^3 + s^2 + 4s + 4$$

satisfies the necessary condition for stability because all the $\{a_i\}$ are positive and nonzero. Determine whether any of the roots of the polynomial are in the RHP.

Solution.   The Routh array for this polynomial is

$$
\begin{array}{llll}
s^6: & 1 & 3 & 1 \qquad\qquad 4
\end{array}
$$

s^6:   1   3   1   4

s^5:   4   2   4   0

s^4: $\dfrac{5}{2} = \dfrac{4\cdot 3 - 1\cdot 2}{4}$ $0 = \dfrac{4\cdot 1 - 4\cdot 1}{4}$ $4 = \dfrac{4\cdot 4 - 1\cdot 0}{4}$

s^3: $2 = \dfrac{\dfrac{5}{2}\cdot 2 - 4\cdot 0}{\dfrac{5}{2}}$ $-\dfrac{12}{5} = \dfrac{\dfrac{5}{2}\cdot 4 - 4\cdot 4}{\dfrac{5}{2}}$ 0

s^2: $3 = \dfrac{2\cdot 0 - \dfrac{5}{2}(-\dfrac{12}{5})}{2}$ $4 = \dfrac{2\cdot 4 - (\dfrac{5}{2}\cdot 0)}{2}$

s: $-\dfrac{76}{15} = \dfrac{3(-\dfrac{12}{5}) - 8}{3}$ 0

s^0: $4 = \dfrac{-\dfrac{76}{15}\cdot 4 - 0}{-\dfrac{76}{15}}.$

We conclude that the polynomial has RHP roots, since the elements of the first column are not all positive. In fact, there are two poles in the RHP because there are two sign changes.[13]

Note that in computing the Routh array we can simplify the rest of the calculations by multiplying or dividing a row by a positive constant. Also note that the last two rows each have one nonzero element.

Routh's method is also useful in determining the range of parameters for which a feedback system remains stable.

EXAMPLE 3.30

Stability Versus Parameter Range

Consider the system shown in Fig. 3.38. The stability properties of the system are a function of the proportional feedback gain K. Determine the range of K over which the system is stable.

Figure 3.38
A feedback system for testing stability

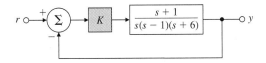

---

[13] The actual roots of the polynomial computed with the MATLAB roots command are -3.2644, $0.7797 \pm 0.7488j$, $-0.6046 \pm 0.9935j$, and -0.8858, which, of course, agree with our conclusion.

Solution.   The characteristic equation for the system is given by

$$1 + K\frac{s+1}{s(s-1)(s+6)} = 0,$$

or

$$s^3 + 5s^2 + (K-6)s + K = 0.$$

The corresponding Routh array is

$$
\begin{array}{lcc}
s^3: & 1 & K-6 \\
s^2: & 5 & K \\
s: & (4K-30)/5 & \\
s^0: & K &
\end{array}
$$

For the system to be stable, it is necessary that

$$\frac{4K-30}{5} > 0 \quad \text{and} \quad K > 0,$$

or

$$K > 7.5 \quad \text{and} \quad K > 0.$$

Thus Routh's method provides an analytical answer to the stability question. Although any gain satisfying this inequality stabilizes the system, the dynamic response could be quite different depending on the specific value of K. Given a specific value of the gain, we may compute the closed-loop poles by finding the roots of the characteristic polynomial. The characteristic polynomial has the coefficients represented by the row vector (in descending powers of s)

denT = [1 5 K–6 K],

Computing roots by MATLAB   and we may compute the roots using the MATLAB function

roots(denT).

For $K = 7.5$ the roots are at -5 and $\pm 1.22j$, and the system is neutrally stable. Note that Routh's method predicts the presence of poles on the $j\omega$ axis for $K = 7.5$. If we set $K = 13$, the closed-loop poles are at -4.06 and $-0.47 \pm 1.7j$, and for $K = 25$ they are at -1.90 and $-1.54 \pm 3.27j$. In both these cases the system is stable as predicted by Routh's method. Figure 3.39 shows the transient responses for the three gain values. To obtain these transient responses, we compute the closed-loop transfer function

$$T(s) = \frac{Y(s)}{R(s)} = \frac{K(s+1)}{s^3 + 5s^2 + (K-6)s + K},$$

so that the numerator polynomial is expressed as

numT = [K K];   % form numerator

and denT is as before. The MATLAB commands

sysT = tf(numT,denT);   % define system by its numerator and denominator
step(sysT);   % compute step response

produces a plot of the (unit) step response.

Figure 3.39
Transient responses for the
system in Figure 3.38

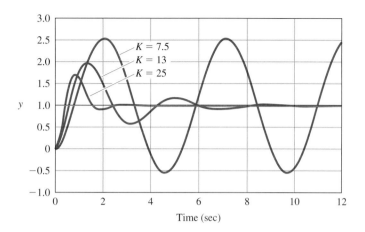

EXAMPLE 3.31

Stability Versus Two Parameter Ranges

Find the range of the controller gains (K, K_I) so that the PI (proportional-integral, see Chapter 4) feedback system in Fig. 3.40 is stable.

Solution.   The characteristic equation of the closed-loop system is

$$1 + \left(K + \frac{K_I}{s} \right) \frac{1}{(s+1)(s+2)} = 0,$$

which we may rewrite as

$$s^3 + 3s^2 + (2 + K)s + K_I = 0.$$

The corresponding Routh array is

$$
\begin{array}{ccc}
s^3 : & 1 & 2 + K \\
s^2 : & 3 & K_I \\
s : & (6 + 3K - K_I)/3 & \\
s^0 : & K_I &
\end{array}
$$

For asymptotic stability we must have

$$K_I > 0 \quad \text{and} \quad K > \frac{1}{3}K_I - 2.$$

Figure 3.40
System with
proportional-integral (PI)
control

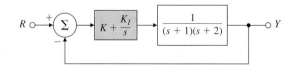

Figure 3.41
Allowable region for stability

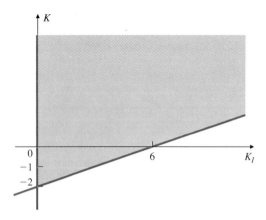

The allowable region is the shaded area in the (K_I, K) plane shown in Fig. 3.41, which represents an analytical solution to the stability question. This example illustrates the real value of Routh's approach and why it is superior to the numerical approaches. It would have been more difficult to arrive at these bounds on the gains using numerical search techniques. The closed-loop transfer function is

$$T(s) = \frac{Y(s)}{R(s)} = \frac{Ks + K_I}{s^3 + 3s^2 + (2 + K)s + K_I}.$$

MATLAB roots

As in Example 3.30, we may compute the closed-loop poles for different values of the dynamic compensator gains by using the MATLAB function `roots` on the denominator polynomial

```
denT = [1 3 2+K KI];    % form denominator
```

Similarly, we may find the zero by finding the root of the numerator polynomial

```
numT = [K KI];    % form numerator
```

The closed-loop zero of the system is at $-K_I/K$. Figure 3.42 shows the transient response for three sets of feedback gains. For $K = 1$ and $K_I = 0$ the closed-loop poles are at 0 and $-1.5 \pm 0.86j$, and there is a zero at the origin. For $K = K_I = 1$ the poles and zeros are all at -1. For $K = 10$ and $K_I = 5$ the closed-loop poles are at -0.46 and $-1.26 \pm 3.03j$, and the zero is at -0.5. The step responses were again obtained using the MATLAB function

```
sysT = tf(numT,denT);    % define system by its numerator and denominator
step(sysT);              % compute step response
```

There is a large steady-state error in this case when $K_I = 0$ (see Chapter 4).

Figure 3.42
Transient response for the
system in Fig. 3.41

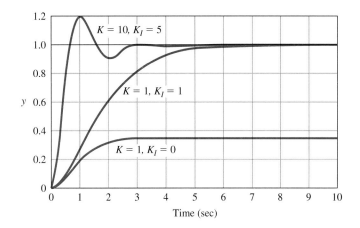

If the first term in one of the rows is zero or if an entire row is zero, then the standard Routh array cannot be formed, so we have to use one of the special techniques described next.

Special Cases

Special case I

If only the first element in one of the rows is zero, then we can replace the zero with a small positive constant $\epsilon > 0$ and proceed as before. We then apply the stability criterion by taking the limit as $\epsilon \to 0$.

EXAMPLE 3.32

Routh's Test for Special Case I

Consider the polynomial

$$a(s) = s^5 + 3s^4 + 2s^3 + 6s^2 + 6s + 9.$$

Determine whether any of the roots are in the RHP.

Solution.   The Routh array is

$s^5:$	1	2	6
$s^4:$	3	6	9
$s^3:$	0	3	0
New $s^3:$	ϵ	3	0	$\leftarrow$ Replace zero by ϵ
$s^2:$	$\dfrac{2\epsilon - 3}{\epsilon}$	3	0
$s:$	$3 - \dfrac{3\epsilon^2}{2\epsilon - 3}$	0	0
$s^0:$	3	0	

There are two sign changes in the first column of the array, which means there are two poles not in the LHP.[14]

---

[14] The actual roots computed with MATLAB are at $-2.9043, 0.6567 \pm 1.2881j, -0.7046 \pm 0.9929j$.

Another special case occurs when an entire row of the Routh array is zero. This indicates that there are complex conjugate pairs of roots that are mirror images of each other with respect to the imaginary axis. If the ith row is zero, we form the following auxiliary equation from the previous (nonzero) row:

$$a_1(s) = \beta_1 s^{i+1} + \beta_2 s^{i-1} + \beta_3 s^{i-3} + \cdots, \qquad (3.66)$$

where $\{\beta_i\}$ are the coefficients of the $(i+1)$th row in the array. We then replace the ith row by the coefficients of the *derivative* of the auxiliary polynomial, and complete the array. However, the roots of the auxiliary polynomial in Eq. (3.66) are also roots of the characteristic equation, and these must be tested separately.

EXAMPLE 3.33   *Routh Test for Special Case II*

For the polynomial

$$a(s) = s^5 + 5s^4 + 11s^3 + 23s^2 + 28s + 12,$$

determine whether there are any roots on the $j\omega$ axis or in the RHP.

Solution.   The Routh array is

s^5:	1	11	28
s^4:	5	23	12
s^3:	6.4	25.6	0
s^2:	3	12	
s:	0	0	$\leftarrow a_1(s) = 3s^2 + 12$
New s:	6	0	$\leftarrow \dfrac{da_1(s)}{ds} = 6s$
s^0:	12		

There are no sign changes in the first column. Hence all the roots have negative real parts except for a pair on the imaginary axis. We may deduce this as follows. When we replace the zero in the first column by $\epsilon > 0$, there are no sign changes. If we let $\epsilon < 0$, then there are two sign changes. Thus, if $\epsilon = 0$, there are two poles on the imaginary axis, which are the roots of

$$a_1(s) = s^2 + 4 = 0,$$

or

$$s = \pm j2.$$

This agrees with the fact that the actual roots are at -3, $\pm 2j$, -1, and -1 as computed using the roots command in MATLAB.

▲ 3.7 Numerical Simulation

Numerical solution of differential equations has become common with the widespread access to computers. Examination of the Laplace transform in Section 3.1 led us to insight into the character of solutions that we exploited in Sections 3.3 to 3.5. The Laplace transform also leads to a method of solving linear ODEs; however, this method is not used by MATLAB software to compute the solution of differential equations and is almost never used to compute an actual time history by hand. Numerical solution methods are capable of solving linear as well as nonlinear differential equations. However, some efficiencies are possible if the equations are linear; and typically, MATLAB software exploits these. Often a control design is carried out using linear analysis to arrive at a candidate feedback control scheme based solely on examination of pole locations in the s-plane. Then the design engineer examines a nonlinear simulation of the time response to acquire information that either confirms the acceptability of the design or indicates the need for more iteration.

3.7.1 Solution of Nonlinear Differential Equations

In Chapter 2 we gave the state-variable form of the differential equations for the nonlinear case (Section 2.6.1):

$$\dot{\mathbf{x}} = \mathbf{f}(\mathbf{x}, u). \tag{3.67}$$

The task at hand is to solve for $\mathbf{x}(t)$ given the initial condition $\mathbf{x}(0)$ and the forcing function $u(t)$.

    The fundamental ideas behind numerical integration are illustrated by deriving a particularly simple scheme called **Euler's method**. It follows from the definition of a derivative that

Euler's method

$$\frac{dx}{dt} = \lim_{\Delta t \to 0} \frac{\Delta x}{\Delta t}. \tag{3.68}$$

Even if Δt is not quite equal to zero, this relationship will be approximately true. Let us examine the differential equation for two values of time, t_i and t_{i+1}, where Δt is sufficiently close to zero that Eq. (3.68) is approximately correct. In this case, Eq. (3.67) becomes

$$\frac{\mathbf{x}_{i+1} - \mathbf{x}_i}{\Delta t} = \mathbf{f}(\mathbf{x}_i, u_i), \tag{3.69}$$

which can be solved to obtain

$$\mathbf{x}_{i+1} = \mathbf{x}_i + \Delta t\, \mathbf{f}(\mathbf{x}_i, u_i). \tag{3.70}$$

Many repeated evaluations of Eq. (3.70) lead to the desired solution. At the initial point $\mathbf{x}_i = \mathbf{x}(0)$ and for subsequent values of the index i, $\mathbf{x}_i$ takes on the value from the previous calculation of $\mathbf{x}_{i+1}$. Any arbitrary time history of the input u_i can be used: steps, ramps, sinusoids, random sequences, or stock market indices. As the step size Δt decreases, the accuracy of the method improves and the required computation time increases.

It is instructive to examine Euler's method from a graphical viewpoint. Figure 3.43 shows a time history of $\dot{x}$. Because

$$x(t) = x(0) + \int_0^t \dot{x}\, dt, \tag{3.71}$$

the value of $x(t)$ at time t_i is equal to $x(0)$ plus the shaded area under the $\dot{x}$ curve up until t_i. Euler's method approximates the area under the curve by assuming that $\dot{x}$ remains constant through the interval:

$$\int_{t_i}^{t_{i+1}} \dot{x}\, dt \cong \Delta t \dot{x}(t_i). \tag{3.72}$$

Although the graph can only be drawn for a scalar variable x, the ideas are the same for each element of a state vector $\mathbf{x}$, which is the case for the integration formula in Eq. (3.70).

Figure 3.43 suggests that a more accurate formula is to use the average of the values of the derivative at t_i and t_{i+1}. Because this is essentially a straight-line approximation to the $\dot{x}$ curve between t_i and t_{i+1}, it is called the **trapezoidal rule**. Unfortunately, it is impossible to implement for numerical integration of nonlinear systems because the derivative at t_{i+1} depends on $x(t_{i+1})$, which is not known yet. Many different numerical integration schemes have been devised to approximate the area under the curve; however, they are all iterative because the derivative at the endpoint is not initially known.

Runge–Kutta methods comprise one popular set of integration schemes. The second-order Runge–Kutta method obtains an approximate value of the endpoint $\mathbf{x}_{i+1}$ using the Euler method, estimates the derivative at the endpoint using the approximate $\mathbf{x}_{i+1}$, and then arrives at the final value for $\mathbf{x}_{i+1}$ using

Trapezoidal rule

Runge–Kutta methods

Figure 3.43
Graphical interpretation of numerical integration: Euler's method (solid blue), trapezoidal rule (dashed line)

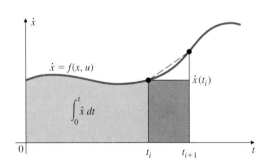

an average of the two derivatives:

$$\mathbf{k}_1 = \mathbf{f}(\mathbf{x}_i, u_i),$$

$$\mathbf{k}_2 = \mathbf{f}(\mathbf{x}_i + \mathbf{k}_1 \Delta t, u_{i+1}), \tag{3.73}$$

$$\mathbf{x}_{i+1} = \mathbf{x}_i + \frac{\Delta t}{2}(\mathbf{k}_1 + \mathbf{k}_2).$$

Third- and higher-order Runge–Kutta methods use this same basic idea; they differ from the second-order formula by using estimates of derivatives at midpoints as well as endpoints and including them in a weighted average to arrive at the final estimate of $\mathbf{x}_{i+1}$.

Integration errors will decrease as step size decreases.

Most computer-aided control system design software includes some form of numerical integration capability such as the Runge–Kutta method and most will include some sort of automatic Δt step size determination. Any method will become more accurate as the step size decreases; however, initially, neither the computer algorithms nor the user knows what step size is the best compromise between accuracy and speed. A commonly used scheme is to integrate using two different methods (perhaps a second- and third-order Runge–Kutta formula), compare the difference, and then cut the step size in half if the error exceeds a certain tolerance. The step size will continue to be cut in half until the error tolerance is met. Numerous books devoted entirely to the subject of numerical integration can provide more detail; see, for example, Shampine and Gordon (1975).

3.7.2   Solution of Linear Differential Equations

When the differential equations are linear, they can be written in the form

$$\dot{\mathbf{x}} = \mathbf{F}\mathbf{x} + \mathbf{G}u. \tag{3.74}$$

The solution to the homogeneous portion,

$$\dot{\mathbf{x}} = \mathbf{F}\mathbf{x}, \tag{3.75}$$

is found by assuming a solution of the form

$$\mathbf{x}(t) = e^{\mathbf{F}t}\mathbf{x}(0), \qquad t \geq 0, \tag{3.76}$$

where the matrix exponential $e^{\mathbf{F}t}$ is a power series exactly like a scalar exponential, that is,

$$e^{\mathbf{F}t} = \mathbf{I} + \mathbf{F}t + \frac{\mathbf{F}^2 t^2}{2!} + \frac{\mathbf{F}^3 t^3}{3!} + \cdots. \tag{3.77}$$

By differentiating this expression with respect to time, we find that, just like the scalar exponential,

$$\frac{d}{dt}(e^{\mathbf{F}t}) = \mathbf{F}e^{\mathbf{F}t}.$$

Substituting this back into Eq. (3.75) along with Eq. (3.76) shows that the assumed solution was correct. This solution is also the impulse response of the system in Eq. (3.74); therefore, the convolution integral in Eq. (3.3) can be used to write the total solution as

$$\mathbf{x}(t) = e^{\mathbf{F}(t-t_0)}\mathbf{x}(t_0) + \int_{t_0}^{t} e^{\mathbf{F}(t-\tau)}\mathbf{G}u(\tau)\,d\tau. \tag{3.78}$$

This equation can be specialized for solving between two time points t_i and t_{i+1}:

$$\mathbf{x}(t_{i+1}) = e^{\mathbf{F}(t_{i+1}-t_i)}\mathbf{x}(t_i) + \int_{t_i}^{t_{i+1}} e^{\mathbf{F}(t_{i+1}-\tau)}\mathbf{G}u(\tau)\,d\tau. \tag{3.79}$$

By assuming that $\Delta t = t_{i+1} - t_i$ is small and that $u(t)$ does not change over that time interval, the solution simplifies to

$$\mathbf{x}(t_{i+1}) = e^{\mathbf{F}\Delta t}\mathbf{x}(t_i) + \left[\int_0^{\Delta t} e^{\mathbf{F}\eta}\,d\eta\right]\mathbf{G}u(t_i). \tag{3.80}$$

Difference equation

Equation (3.80) is called a **difference equation** and usually written as

$$\mathbf{x}(t_{i+1}) = \mathbf{\Phi}\mathbf{x}(t_i) + \mathbf{\Gamma}u(t_i), \tag{3.81}$$

Discrete standard form

where

$$\mathbf{\Phi} = e^{\mathbf{F}\Delta t} \quad \text{and} \quad \mathbf{\Gamma} = \left[\int_0^{\Delta t} e^{\mathbf{F}\eta}\,d\eta\right]\mathbf{G}. \tag{3.82}$$

This is the standard form for discrete systems and corresponds to the standard form for continuous systems given by Eqs. (2.39)–(2.40).

Efficiency of linear versus nonlinear numerical simulation

The $\mathbf{\Phi}$ and $\mathbf{\Gamma}$ matrices are constant for *all* time steps. Therefore, to compute the response of a linear system at times that are spaced Δt seconds apart, the only quantities that require extensive computations, as in the Runge–Kutta method described in Section 3.7.1, are the $\mathbf{\Phi}$ and $\mathbf{\Gamma}$ matrices.[15] The numerical computation of the solution of $\mathbf{x}(t_i)$ is obtained by repeated evaluation of Eq. (3.81). The entire process is much more efficient than the nonlinear case because $\mathbf{\Phi}$ and $\mathbf{\Gamma}$ are only calculated once and the time step Δt can be larger than for nonlinear equations.

EXAMPLE 3.34

Numerical Simulation of a Pendulum Using MATLAB

Find and plot the first 10 sec of the pendulum time history using the equations from Example 2.5 using MATLAB. Assume the control torque is a pulse for the first second; that is, $T_c = 1$ N for $0 < t \leq 1$ sec and zero thereafter. The length $l = 1$ m, and its mass $m = 0.2$ kg.

---

[15] Note that the computation of $\mathbf{\Phi}$ by Eq. (3.77) is treacherous. MATLAB uses a sophisticated scheme to compute $e^{\mathbf{F}t}$.

(a) Use the linearized equation of motion [Eq. (2.23)].

(b) Use the nonlinear equation of motion [Eq. (2.22)].

(c) Repeat parts (a) and (b) for a smaller pulse of $T_c = 0.3$ N.

Solution.

(a) The most expedient method for entering this system into MATLAB for linear analysis is to write the equation of motion as a transfer function:

$$\frac{\theta(s)}{T_c(s)} = \frac{1/ml^2}{s^2 + g/l},$$

where $l/ml^2 = 5$ and $g/l = 9.82$. The lsim routine is required rather than the step routine used in Example 2.8 because the input time history is constant for 1 sec and then becomes zero thereafter. lsim will accept any input time history. Therefore, the statements in MATLAB

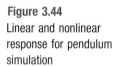
Linear simulation using
MATLAB

```
numG = 5;                              % form numerator
denG = [1 0 9.82];                     % form denominator
t = 0:.05:10;                          % form time vector
u = [ones(21, 1); zeros(180, 1)];      % form input vector
sysG = tf(numG,denG);                  % define system by its numerator and
                                       denominator

lsim(sysG,u,t)                         % compute system response
```

will produce the desired time history shown in Fig. 3.44. The index of the input time history vector u corresponds to the time history vector t and produces the square pulse of control for the first second.

Figure 3.44
Linear and nonlinear
response for pendulum
simulation

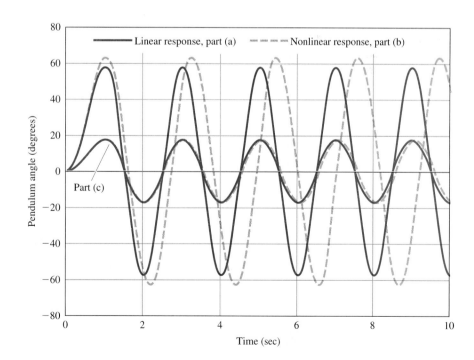

Nonlinear simulation using
MATLAB

(b) For the nonlinear solution, the differential equations need to be put into the standard first-order form of Eq. (3.67). We accomplish this by defining the state $\mathbf{x} = [\theta \quad \dot{\theta}]^T$ and rewriting Eq. (2.21) as

$$\dot{x}_1 = x_2,$$

$$\dot{x}_2 = -\frac{g}{l} \sin x_1 + \frac{T_c}{ml^2}.$$

A Runge–Kutta integration routine in MATLAB that performs both second- and third-order calculations and compares the two to determine the step size is called ode23. This routine requires that a separate function be provided that computes the derivatives. The following MATLAB statements

```
global Wn2 Uo                        % declare global variables
Wn2 = 9.82                           % g/l
Uo = 5                               % input
to = 0                               % initial time
tf = 10                              % final time
tspan = [to tf]                      % time span
Xo = [0;0]                           % initial condition
% call integration routine, ode23, that calls the function pendot
[t,x] = ode23('pendot',tspan,Xo)
```

call the integration routine, which in turn calls the derivative function, labeled pendot for this case. The derivative function is created by the following MATLAB statements:

```
function xdot = pendot(t, x)    % function pendot
global Wn2 Uo                   % declare global variables
if t < = 1,
U = Uo
  else U = 0
  end
xdot(1) = x(2)                  % compute derivative of first state variable
xdot(2) = -Wn2*sin (x(1)) + U   % compute derivative of second state variable
```

This produces the large-amplitude (dashed) time histories shown in Fig. 3.44. For this level of the control input, the nonlinear response exhibits a slightly longer period and larger response amplitude than the linear case. This is because $\sin \theta$ is less than θ at the high angles excited by the 1-N control pulse, thus lowering the restoring gravity torque.

(c) To repeat part (a) for $T_c = 0.3$ N, we need to change the num $= 5$ statement to num $= 1.5$ in order to reduce the magnitude of the input impulse. To repeat part (b) for $T_c = 0.3$ N, we need to change Uo $= 5$ to Uo $= 1.5$. In this case, because the angles never exceed $20°$, the linear approximation in Fig. 3.44 is almost indistinguishable from the nonlinear one.

▲ 3.8 Obtaining Models from Experimental Data

There are several reasons for using experimental data to obtain a model of the dynamic system to be controlled. In the first place, the best theoretical model built from equations of motion is still only an approximation of reality. Sometimes, as in the case of a very rigid spacecraft, the theoretical model is extremely good. Other times, as with many chemical processes such as paper-making or metalworking, the theoretical model is very approximate. In every case, before the final control design is done, it is important and prudent to verify the theoretical model with experimental data. Secondly, in situations where the theoretical model is especially complicated or the physics of the process is poorly understood, the only reliable information on which to base the control design is the experimental data. Finally, the system is sometimes subject to on-line changes, which occur when the environment of the system changes. Examples include when an aircraft changes altitude or speed, a paper machine is given a different composition of fiber, or a nonlinear system moves to a new operating point. On these occasions we need to "retune" the controller by changing the control parameters. This requires a model for the new conditions, and experimental data are often the most effective, if not the only, information available for the new model.

There are four kinds of experimental data for generating a model.

Four sources of experimental data

1. **Transient response**, such as comes from an impulse or a step.

2. **Frequency response data**, which result from exciting the system with sinusoidal inputs at many frequencies.

3. **Stochastic steady-state information**, as might come from flying an aircraft through turbulent weather or from some other natural source of randomness.

4. **Pseudorandom-noise data**, as may be generated in a digital computer.

Each class of experimental data has its properties, advantages, and disadvantages.

Transient response

Transient response data are quick and relatively easy to obtain. They are also often representative of the natural signals to which the system is subjected. Thus a model derived from such data can be reliable for designing the control system. On the other hand, in order for the signal-to-noise ratio to be sufficiently high, the transient response must be highly noticeable. Thus the method is rarely suitable for normal operations, so the data must be collected as part of special tests. A second disadvantage is that the data do not come in a form suitable for standard control systems designs, and some parts of the model, such as poles and zeros, must be computed from the data.[16] This computation can be simple in special cases or complex in the general case.

---

[16] Ziegler and Nichols (1943), building on the earlier work of Callender et al. (1936), use the step response directly in designing the controls for certain classes of processes. See Chapter 4 for details.

Frequency-response data (see Chapter 6) is simple to obtain but substantially more time-consuming than transient-response information. This is especially so if the time constants of the process are large, as often occurs in chemical processing industries. As with the transient-response data, it is important to have a good signal-to-noise ratio, so obtaining frequency-response data can be very expensive. On the other hand, as we will see in Chapter 6, frequency-response data are exactly in the right form for frequency-response design methods, so once the data have been obtained, the control design can proceed immediately.

Normal operating records from a natural stochastic environment at first appear to be an attractive basis for modeling systems since such records are by definition nondisruptive and inexpensive to obtain. Unfortunately, the quality of such data is inconsistent, tending to be worst just when the control is best, because then the upsets are minimal and the signals are smooth. At such times some or even most of the system dynamics are hardly excited. Because they contribute little to the system output, they will not be found in the model constructed to explain the signals. The result is a model that represents only part of the system and is sometimes unsuitable for control. In some instances, as occurs when trying to model the dynamics of the electroencephalogram (brain waves) of a sleeping or anesthetized person to locate the frequency and intensity of alpha waves, normal records are the only possibility. Usually they are the last choice for control purposes.

Finally, the pseudorandom signals that can be constructed using digital logic have much appeal. Especially interesting for model making is the pseudorandom binary signal (PRBS). The PRBS takes on the value $+A$ or $-A$ according to the output (1 or 0) of a feedback shift register. The feedback to the register is a binary sum of various states of the register that have been selected to make the output period (which must repeat itself in finite time) as long as possible. For example, with a register of 20 bits, $2^{20} - 1$ (over a million) steps are produced before the pattern repeats. Analysis beyond the scope of this text has revealed that the resulting signal is almost like a broad-band random signal. Yet this signal is entirely under the control of the engineer who can set the level (A) and the length (bits in the register) of the signal. The data obtained from tests with a PRBS must be analyzed by computer, and both special-purpose hardware and programs for general-purpose computers have been developed to perform this analysis.

3.8.1   Models from Transient-Response Data

To obtain a model from transient data we assume that a step response is available. If the transient is a simple combination of elementary transients, then a reasonable low-order model can be estimated using hand calculations. For example, consider the step response shown in Fig. 3.45. The response is monotonic and smooth. If we assume that it is given by a sum of exponentials, we can write

$$y(t) = y(\infty) + Ae^{-\alpha t} + Be^{-\beta t} + Ce^{-\gamma t} + \dots . \tag{3.83}$$

Figure 3.45
A step response
characteristic of many
chemical processes

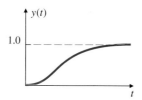

Subtracting off the final value and assuming that $-\alpha$ is the slowest pole, we write

$$y - y(\infty) \cong Ae^{-\alpha t}$$

$$\log_{10}[y - y(\infty)] \cong \log_{10} A - \alpha t \log_{10} e$$

$$\cong \log_{10} A - 0.4343\alpha t. \tag{3.84}$$

This is the equation of a line whose slope determines α and intercept determines A. If we fit a line to the plot of $\log_{10}[y - y(\infty)]$ (or $\log_{10}[y(\infty) - y]$ if A is negative), then we can estimate A and α. Once these are estimated, we plot $y - [y(\infty) + Ae^{-\alpha t}]$, which as a curve approximates $Be^{-\beta t}$ and on the log plot is equivalent to $\log_{10} B - 0.4345\beta t$. We repeat the process, each time removing the slowest remaining term, until the data stop being accurate. Then we plot the final model step response and compare it with data so we can assess the quality of the computed model. It is possible to get a good fit to the step response and yet be far off from the true time constants (poles) of the system. However, the method gives a good approximation for control of processes whose step responses look like Fig. 3.45.

EXAMPLE 3.35   *Determining the Model from Time-Response Data*

Find the transfer function that generates the data given in Table 3.1 and which are plotted in Fig. 3.46.

TABLE 3.1   **Step-Response Data**

t	$y(t)$	t	$y(t)$
0.1	0.000	1.0	0.510
0.1	0.005	1.5	0.700
0.2	0.034	2.0	0.817
0.3	0.085	2.5	0.890
0.4	0.140	3.0	0.932
0.5	0.215	4.0	0.975
		∞	1.000

Source: Sinha and Kuszta (1983).

Figure 3.46

Step response data in Table 3.1

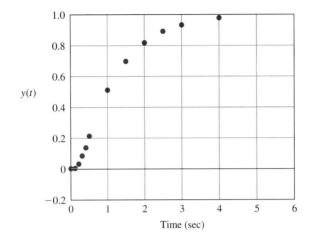

Solution.   Table 3.1 shows and Fig. 3.46 implies that the final value of the data is $y(\infty) = 1$. We know that A is negative because $y(\infty)$ is greater than $y(t)$. Therefore, the first step in the process is to plot $\log_{10}[y(\infty) - y]$, which is shown in Fig. 3.47. From the line (fitted by eye) the values are

$$\log_{10}|A| = 0.125,$$

$$0.4343\alpha = \frac{1.602 - 1.167}{\Delta t} = \frac{0.435}{1} \Rightarrow \alpha \cong 1.$$

Figure 3.47

$\log_{10}[y(\infty) - y]$ versus t

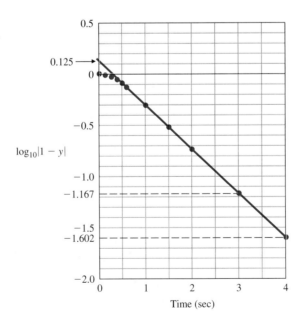

Figure 3.48
$\log_{10}[y - (1 + Ae^{-\alpha t})]$
versus t

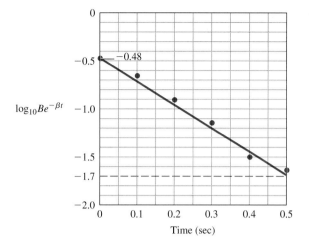

Thus

$$A = -1.33,$$

$$\alpha = 1.0.$$

If we now subtract $1 + Ae^{\alpha t}$ from the data and plot the log of the result, we find the plot of Fig. 3.48. Here we estimate

$$\log_{10} B = -0.48,$$

$$0.4343\beta = \frac{-0.48 - (-1.7)}{0.5} = 2.5,$$

$$\beta \cong 5.8,$$

$$B = 0.33.$$

Combining these results, we arrive at the y estimate

$$\hat{y}(t) \cong 1 - 1.33e^{-t} + 0.33e^{-5.8t}. \tag{3.85}$$

Equation (3.85) is plotted as the colored line in Fig. 3.49 and shows a reasonable fit to the data, although some error is noticeable near $t = 0$.

From $\hat{y}(t)$ we compute

$$\hat{Y}(s) = \frac{1}{s} - \frac{1.33}{s+1} + \frac{0.33}{s+5.8}$$

$$= \frac{(s+1)(s+5.8) - 1.33s(s+5.8) + 0.33s(s+1)}{s(s+1)(s+5.8)}$$

$$= \frac{-0.58s + 5.8}{s(s+1)(s+5.8)}.$$

Figure 3.49
Model fits to the
experimental data

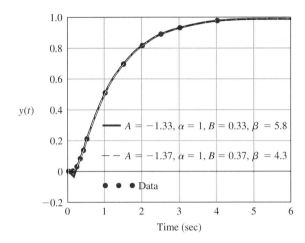

The resulting transfer function is

$$G(s) = \frac{-.58(s-10)}{(s+1)(s+5.8)}.$$

Notice that this method has given us a system with a zero in the RHP, even though the data showed no values of y that were negative. Very small differences in the estimated value for A, all of which approximately fit the data, can cause values of β to range from 4 to 6. This illustrates the sensitivity of pole locations to the quality of the data and emphasizes the need for a good signal-to-noise ratio.

By using a computer to perform the plotting, we are better able to iterate the four parameters to achieve the best overall fit. The data presentation in Figs. 3.47 and 3.48 can be obtained directly by using a semilog plot. This eliminates having to calculate $\log_{10}$ and the exponential expression to find the values of the parameters. The equations of the lines to be fit to the data are $y(t) = Ae^{\alpha t}$ and $y(t) = Be^{\beta t}$, which are straight lines on a semilog plot. The parameters A and α, or B and β, are iteratively selected so that the straight line comes as close as possible to passing through the data. This process produces the improved fit shown by the dashed black line in Fig. 3.49. The revised parameters, $A = -1.37$, $B = 0.37$, and $\beta = 4.3$ result in the transfer function

$$G(s) = \frac{-0.22s + 4.3}{(s+1)(s+4.3)}.$$

The RHP zero is still present, but it is now located at $s \cong +20$ and has no noticeable effect on the time response.

This set of data was fitted quite well by a second-order model. In many cases a higher-order model is required to explain the data and the modes may not be as well-separated.

If the transient response has oscillatory modes, then these can sometimes be estimated by comparing them with the standard plots of Fig. 3.23. The period will give the frequency ω_d, and the decay from one period to the next will afford an estimate of the damping ratio. If the response has a mixture of modes not well separated in frequency, then more sophisticated methods need to be used. One such is **least-squares identification**, in which a numerical optimization routine selects the best combination of parameters so as to minimize the fit error. The fit error is defined to be a scalar **cost function**

Least-squares identification

$$J = \sum_i (y_{\mathrm{data}} - y_{\mathrm{model}})^2, \qquad i = 1, 2, 3, \ldots \text{ for each data point,}$$

so that fit errors at all data points are taken into account in determining the best value for the parameters.

3.8.2   Models from Other Data

As mentioned early in Section 3.8, we can also generate a model using frequency-response data, which are obtained by exciting the system with a set of sinusoids and plotting $H(j\omega)$. In Chapter 6 we will show how such plots can be used directly for design. Alternatively, we can use the frequency response to estimate the poles and zeros of a transfer function using straight-line asymptotes on a logarithmic plot.

The construction of dynamic models from normal stochastic operating records or from the response to a PRBS can be based either on the concept of cross-correlation or on the least-squares fit of a discrete equivalent model, both topics in the field of **system identification**. They require substantial presentation and background that are beyond the scope of this text. An introduction to system identification can be found in Chapter 8 of Franklin et al. (1998), and a comprehensive treatment is given in Ljüng (1999). Based largely on the work of Professor Ljüng, the MATLAB Toolbox on Identification provides substantial software to perform system identification and to verify the quality of the proposed models.

SUMMARY

- The Laplace transform is the primary tool used to determine the behavior of linear systems. The Laplace transform of a time function $f(t)$ is given by

$$\mathcal{L}[f(t)] = F(s) = \int_{0^-}^{\infty} f(t)e^{-st}\,dt. \tag{3.86}$$

- This relationship leads to the key property of Laplace transforms, namely,

$$\mathcal{L}[\dot{f}(t)] = sF(s) - f(0^-) \tag{3.87}$$

- This property allows us to find the transfer function of a linear ODE. Given the transfer function $G(s)$ of a system and the input $u(t)$, with transform $U(s)$, the system output transform is

$$Y(s) = G(s)U(s).$$

- Normally, inverse transforms are found by referring to tables such as Table A.2 in Appendix A or by computer. Properties of Laplace transforms and their inverses are summarized in Table A.1 in Appendix A.

- The Final Value Theorem is useful in finding steady-state errors for stable systems: If all the poles of $s\,Y(s)$ are in the LHP, then

$$\lim_{t \to \infty} y(t) = \lim_{s \to 0} s\,Y(s). \tag{3.88}$$

- Block diagrams are a convenient way to show the relationships between the components of a system. They can usually be simplified using the relations in Fig. 3.7 and Eq. (3.39), that is the transfer function of the block diagram below

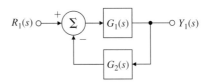

is equivalent to

$$Y_1(s) = \frac{G_1(s)}{1 + G_1(s)G_2(s)} R_1(s). \tag{3.89}$$

- The locations of poles in the s-plane determine the character of the response, as shown in Figure 3.20.

- The location of a pole in the s-plane is defined by the parameters shown in Fig. 3.27. These parameters are related to the time-domain quantities of rise time t_r, settling time t_s, and overshoot M_p, which are defined in Fig. 3.27. The correspondences between them, for a second-order system with no zeros, are given by

$$t_r \cong \frac{1.8}{\omega_n}, \tag{3.90}$$

$$M_p = e^{-\pi\zeta/\sqrt{1-\zeta^2}}, \tag{3.91}$$

$$t_s = \frac{4.6}{\zeta\omega_n}. \tag{3.92}$$

- When a zero in the left half plane is present, the overshoot increases. This effect is summarized in Figs. 3.31 and 3.32.

- When an additional stable pole is present, the system response is more sluggish. This effect is summarized in Figs. 3.36 and 3.37.

- For a stable system, all the closed-loop poles must be in the LHP.

- A system is stable if and only if all the elements in the first column of the Routh array are positive. To determine the Routh array, refer to the formulas in Section 3.6.2.

- Numerical simulation is usually used to find the time history of a solution to a linear or nonlinear ODE. This can be carried out using computer-aided control system design software. In MATLAB the step, lsim, or ODE23 functions accomplish the task.

- Determining a model from experimental data, or verifying an analytically based model by experiment, is an important step in system design.

Review Questions

1. What is the definition of "transfer function"?

2. What are the properties of systems whose responses can be described by transfer functions?

3. What is the Laplace transform of $f(t - \lambda)1(t - \lambda)$ if the transform of $f(t)$ is $F(s)$?

4. State the Final Value Theorem (FVT).

5. What is the most common use of the FVT in control?

6. Given a second-order transfer function with damping ratio ζ and natural frequency ω_n, what is the estimate of the step response rise time? What is the estimate of the percent overshoot in the step response? What is the estimate of the settling time?

7. What is the major effect of a zero in the left half plane on the second-order step response?

8. What is the most noticeable effect of a zero in the RHP on the step response of the second-order system?

9. What is the main effect of an extra real pole on the second-order step response?

10. Why is stability an important consideration in control system design?

11. What is the main use of Routh's criterion?

12. Under what conditions might it be important to know how to estimate a transfer function from experimental data?

Problems

Problems for Section 3.1

3.1. Show that, in a partial-fraction expansion, complex conjugate poles have coefficients that are also complex conjugates. (The result of this relationship is that whenever complex conjugate pairs of poles are present, only one of the coefficients needs to be computed.)

3.2. Find the Laplace transform of the following time functions:

    (a) $f(t) = 1 + 2t$

    (b) $f(t) = 3 + 7t + t^2 + \delta(t)$

    (c) $f(t) = e^{-t} + 2e^{-2t} + te^{-3t}$

    (d) $f(t) = (t + 1)^2$

    (e) $f(t) = \sinh t$

3.3. Find the Laplace transform of the following time functions:

    (a) $f(t) = 3 \cos 6t$

    (b) $f(t) = \sin 2t + 2 \cos 2t + e^{-t} \sin 2t$

    (c) $f(t) = t^2 + e^{-2t} \sin 3t$

3.4. Find the Laplace transform of the following time functions:

    (a) $f(t) = t \sin t$

    (b) $f(t) = t \cos 3t$

    (c) $f(t) = te^{-t} + 2t \cos t$

    (d) $f(t) = t \sin 3t - 2t \cos t$

    (e) $f(t) = 1(t) + 2t \cos 2t$

3.5. Find the Laplace transform of the following time functions (* denotes convolution):

    (a) $f(t) = \sin t \sin 3t$

    (b) $f(t) = \sin^2 t + 3 \cos^2 t$

    (c) $f(t) = (\sin t)/t$

    (d) $f(t) = \sin t * \sin t$

    (e) $f(t) = \int_0^t \cos(t - \tau) \sin \tau d\tau$

3.6. Given that the Laplace transform of $f(t)$ is $F(s)$, find the Laplace transform of the following:

    (a) $g(t) = f(t) \cos t$

    (b) $g(t) = \int_0^t \int_0^{t_1} f(\tau) d\tau dt_1$

3.7. Find the time function corresponding to each of the following Laplace transforms using partial fraction expansions:

    (a) $F(s) = \dfrac{2}{s(s + 2)}$

    (b) $F(s) = \dfrac{10}{s(s + 1)(s + 10)}$

    (c) $F(s) = \dfrac{3s + 2}{s^2 + 4s + 20}$

    (d) $F(s) = \dfrac{3s^2 + 9s + 12}{(s + 2)(s^2 + 5s + 11)}$

    (e) $F(s) = \dfrac{1}{s^2 + 4}$

    (f) $F(s) = \dfrac{2(s + 2)}{(s + 1)(s^2 + 4)}$

    (g) $F(s) = \dfrac{s + 1}{s^2}$

(h) $F(s) = \dfrac{1}{s^6}$

(i) $F(s) = \dfrac{4}{s^4 + 4}$

(j) $F(s) = \dfrac{e^{-s}}{s^2}$

3.8. Find the time function corresponding to each of the following Laplace transforms:

(a) $F(s) = \dfrac{1}{s(s+2)^2}$

(b) $F(s) = \dfrac{2s^2 + s + 1}{s^3 - 1}$

(c) $F(s) = \dfrac{2(s^2 + s + 1)}{s(s+1)^2}$

(d) $F(s) = \dfrac{s^3 + 2s + 4}{s^4 - 16}$

(e) $F(s) = \dfrac{2(s+2)(s+5)^2}{(s+1)(s^2+4)^2}$

(f) $F(s) = \dfrac{(s^2 - 1)}{(s^2 + 1)^2}$

(g) $F(s) = \tan^{-1}\left(\dfrac{1}{s}\right)$

3.9. Solve the following ordinary differential equations using Laplace transforms:

(a) $\ddot{y}(t) + \dot{y}(t) + 3y(t) = 0;\ y(0) = 1,\ \dot{y}(0) = 2$

(b) $\ddot{y}(t) - 2\dot{y}(t) + 4y(t) = 0;\ y(0) = 1,\ \dot{y}(0) = 2$

(c) $\ddot{y}(t) + \dot{y}(t) = \sin t;\ y(0) = 1,\ \dot{y}(0) = 2$

(d) $\ddot{y}(t) + 3y(t) = \sin t;\ y(0) = 1,\ \dot{y}(0) = 2$

(e) $\ddot{y}(t) + 2\dot{y}(t) = e^t;\ y(0) = 1,\ \dot{y}(0) = 2$

(f) $\ddot{y}(t) + y(t) = t;\ y(0) = 1,\ \dot{y}(0) = -1$

3.10. Write the dynamic equations describing the circuit in Fig. 3.50. Write the equations in both state-variable form and as a second-order differential equation in $y(t)$. Assuming a zero input, solve the differential equation for $y(t)$ using Laplace-transform methods for the parameter values and initial conditions shown in the figure. Verify your answer using the initial command in MATLAB.

Figure 3.50
Circuit for Problem 3.10

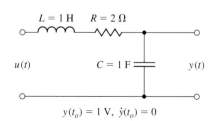

$L = 1\,\text{H} \qquad R = 2\,\Omega$

$u(t)$

$C = 1\,\text{F}$

$y(t)$

$y(t_o) = 1\,\text{V},\ \dot{y}(t_o) = 0$

3.11. Consider the standard second-order system

$$G(s) = \frac{\omega_n^2}{s^2 + 2\zeta\omega_n s + \omega_n^2}.$$

a) Write the Laplace transform of the signal in Fig. 3.51. b). What is the transform of the output if this signal is applied to G(s). c) Find the output of the system for the input shown in Fig. 3.51.

Figure 3.51
Plot of input for
Problem 3.11

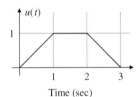

Time (sec)

3.12. A rotating load is connected to a field-controlled DC motor with negligible field inductance. A test results in the output load reaching a speed of 1 rad/sec within 1/2 sec when a constant input of 100 V is applied to the motor terminals. The output steady-state speed from the same test is found to be 2 rad/sec. Determine the transfer function $\theta(s)/V_f(s)$ of the motor.

3.13. For the tape drive shown in Fig. 2.48, compute the following, using the numbers given in Problem 2.20(a):

(a) The transfer function from the motor current to the tape position.

(b) The poles and zeros for the transfer function in part (a).

3.14. For the system in Fig. 2.50, compute the transfer function from the motor voltage to position θ_2.

3.15. Compute the transfer function for the two-tank system in Fig. 2.54 with holes at A and C.

3.16. For a second-order system with transfer function

$$G(s) = \frac{3}{s^2 + 2s - 3},$$

determine the following:

(a) DC gain.

(b) The final value to a step input.

3.17. Consider the continuous rolling mill depicted in Fig. 3.52. Suppose that the motion of the adjustable roller has a damping coefficient b, and that the force exerted the rolled material on the adjustable roller is proportional to the material's change in thickness: $F_s = c(T-x)$. Suppose further that the DC motor has a torque constant K_t and a back-emf constant K_e, and that the rack-and-pinion has effective radius of R.

(a) What are the inputs to this system? The output?

(b) Without neglecting the effects of gravity on the adjustable roller, draw a block diagram of the system that explicitly shows the following quantities: $V_s(s)$, $I_0(s)$, $F(s)$ (the force the motor exerts on the adjustable roller), and $X(s)$.

(c) Simplify your block diagram as much as possible while still identifying output and each input separately.

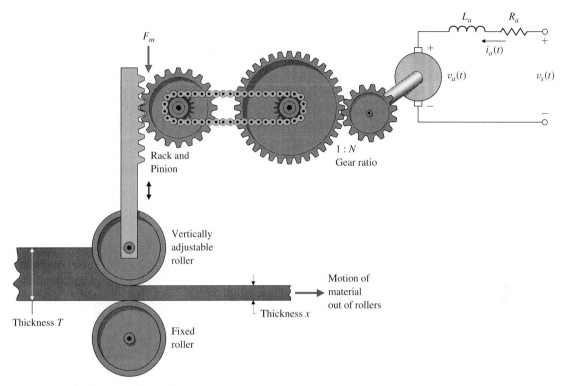

Figure 3.52 Continuous rolling mill

Problems for Section 3.2

3.18. Compute the transfer function for the block diagram shown in Fig. 3.53. Note that a_i and b_i are constants.

  (a) Write the third-order differential equation that relates y and u. (*Hint*: Consider the transfer function.)

Figure 3.53
Block diagram for
Problem 3.18

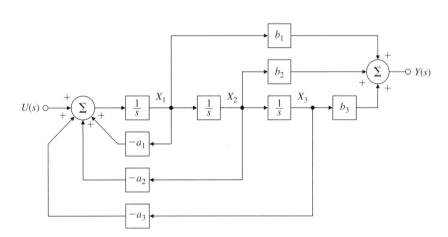

(b) Write three simultaneous first-order (state-variable) differential equations using variables x_1, x_2, and x_3, as defined on the block diagram in Fig. 3.53. Notice how the same constant parameters enter the transfer function, the differential equations, and the matrices of the state-variable form. (This special structure is called the control canonical form and will be discussed further in Chapter 7.) Repeat for the block diagram of Fig. 3.55(b). This is the "observer canonical form" for a third-order system.

3.19. Find the transfer functions for the block diagrams in Fig. 3.54.

Figure 3.54
Block diagrams for
Problem 3.19

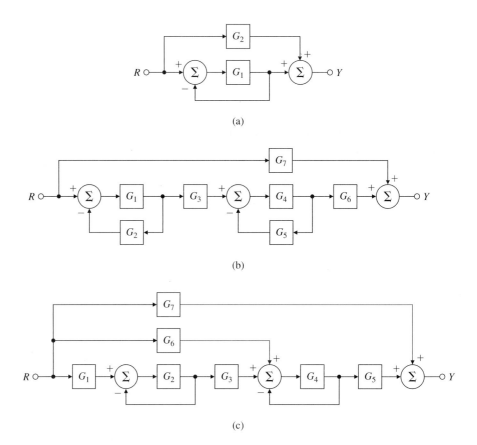

(a)

(b)

(c)

3.20. Find the transfer functions for the block diagrams in Fig. 3.55, using the following:

(a) The ideas of Figs. 3.6 and 3.7.

(b) Mason's rule.

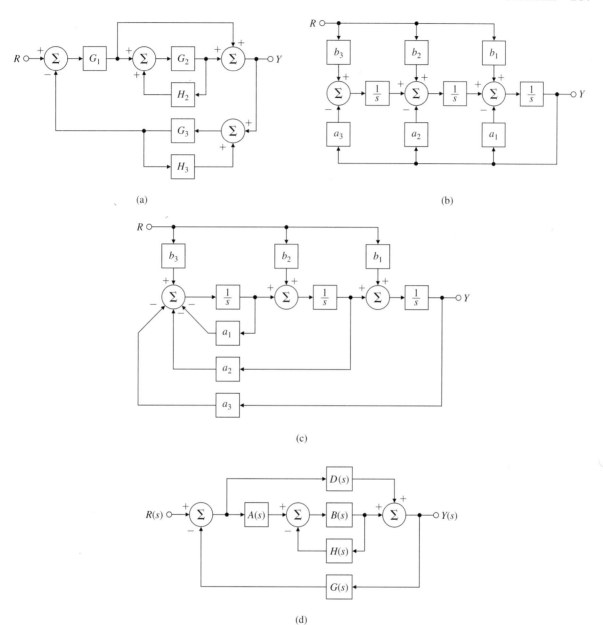

Figure 3.55 Block diagrams for Problem 3.20

3.21. Use block-diagram algebra or Mason's rule to determine the transfer function between $R(s)$ and $Y(s)$ in Fig. 3.56.

Figure 3.56
Block diagram for
Problem 3.21

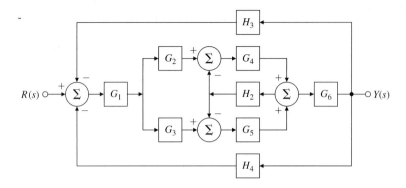

3.22. Use block-diagram algebra to determine the transfer function between $R(s)$ and $Y(s)$ in Fig. 3.57.

Figure 3.57
Block diagram for
Problem 3.22

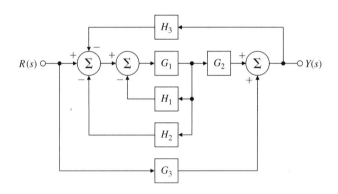

3.23. For the electric circuit shown in Fig. 3.58, find the following:

(a) The time-domain equation relating $i(t)$ and $v_1(t)$.

(b) The time-domain equation relating $i(t)$ and $v_2(t)$.

(c) Assuming all initial conditions are zero, the transfer function $V_2(s)/V_1(s)$ and the damping ratio ζ and undamped natural frequency ω_n of the system.

(d) The values of R that will result in $v_2(t)$ having an overshoot of no more than 25%, assuming $v_1(t)$ is a unit step, $L = 10$ mH, and $C = 4\ \mu\text{F}$.

Figure 3.58
Circuit for Problem 3.23

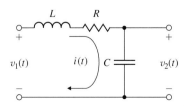

Problems for Section 3.3

3.24. For the unity feedback system shown in Fig. 3.59, specify the gain K of the proportional controller so that the output $y(t)$ has an overshoot of no more than 10% in response to a unit step.

Figure 3.59
Unity feedback system for
Problem 3.24

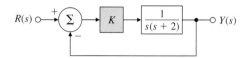

3.25. For the unity feedback system shown in Fig. 3.60, specify the gain and pole location of the compensator so that the overall closed-loop response to a unit-step input has an overshoot of no more than 25%, and a 1% settling time of no more than 0.1 sec. Verify your design using MATLAB.

Figure 3.60
Unity feedback system for
Problem 3.25

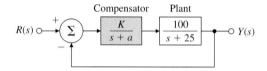

Problems for Section 3.4

3.26. Suppose you desire the peak time of a given second-order system to be less than t_p'. Draw the region in the s-plane that corresponds to values of the poles that meet the specification $t_p < t_p'$.

3.27. Suppose you are to design a unity feedback controller for a first-order plant depicted in Fig. 3.61. (As you will learn in Chapter 4, the configuration shown is referred to as a proportional-integral controller.) You are to design the controller so that the closed-loop poles lie within the shaded regions shown in Fig. 3.62.

Figure 3.61
Unity feedback system for
Problem 3.27

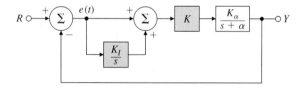

Figure 3.62
Desired closed-loop pole
locations for Problem 3.27

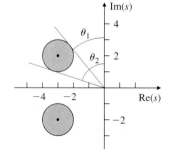

(a) What values of ω_n and ζ correspond to the shaded regions in Fig. 3.62? (A simple estimate from the figure is sufficient.)

(b) Let $K_\alpha = \alpha = 2$. Find values for K and K_1 so that the poles of the closed-loop system lie within the shaded regions.

(c) Prove that no matter what the values of K_α and α are, the controller provides enough flexibility to place the poles anywhere in the complex (left-half) plane.

3.28. The open-loop transfer function of a unity feedback system is

$$G(s) = \frac{K}{s(s+2)}.$$

The desired system response to a step input is specified as peak time $t_p = 1$ sec and overshoot $M_p = 5\%$.

(a) Determine whether both specifications can be met simultaneously by selecting the right value of K.

(b) Sketch the associated region in the s-plane where both specifications are met, and indicate what root locations are possible for some likely values of K.

(c) Pick a suitable value for K, and use MATLAB to verify that the specifications are satisfied.

3.29. The equations of motion for the DC motor shown in Fig. 2.26 were given in Eqs. (2.63)–(2.64) as

$$J_m\ddot{\theta}_m + \left(b + \frac{K_t K_e}{R_a}\right)\dot{\theta}_m = \frac{K_t}{R_a}v_a.$$

Assume that

$$J_m = 0.01 \text{ kg} \cdot \text{m}^2,$$

$$b = 0.001 \text{ N} \cdot \text{m} \cdot \text{sec},$$

$$K_e = 0.02 \text{ V} \cdot \text{sec},$$

$$K_t = 0.02 \text{ N} \cdot \text{m/A},$$

$$R_a = 10 \ \Omega.$$

(a) Find the transfer function between the applied voltage v_a and the motor speed $\dot{\theta}_m$.

(b) What is the steady-state speed of the motor after a voltage $v_a = 10$ V has been applied?

(c) Find the transfer function between the applied voltage v_a and the shaft angle θ_m.

(d) Suppose feedback is added to the system in part (c) so that it becomes a position servo device such that the applied voltage is given by

$$v_a = K(\theta_r - \theta_m),$$

where K is the feedback gain. Find the transfer function between θ_r and θ_m.

(e) What is the maximum value of K that can be used if an overshoot $M_p < 20\%$ is desired?

(f) What values of K will provide a rise time of less than 4 sec? (Ignore the M_p constraint.)

(g) Use MATLAB to plot the step response of the position servo system for values of the gain $K = 0.5, 1$, and 2. Find the overshoot and rise time of the three step responses by examining your plots. Are the plots consistent with your calculations in parts (e) and (f)?

3.30. You wish to control the elevation of the satellite-tracking antenna shown in Figs. 3.63 and 3.64. The antenna and drive parts have a moment of inertia J and a damping B; these arise to some extent from bearing and aerodynamic friction but

Figure 3.63
Satellite-tracking antenna *(Courtesy Space Systems/Loral)*

Figure 3.64
Schematic of antenna for Problem 3.30

mostly from the back emf of the DC drive motor. The equations of motion are

$$J\ddot{\theta} + B\dot{\theta} = T_c,$$

where T_c is the torque from the drive motor. Assume that

$$J = 600,000 \text{ kg·m}^2 \quad \text{and} \quad B = 20,000 \text{ N·m·sec.}$$

(a) Find the transfer function between the applied torque T_c and the antenna angle θ.

(b) Suppose the applied torque is computed so that θ tracks a reference command θ_r according to the feedback law

$$T_c = K(\theta_r - \theta),$$

where K is the feedback gain. Find the transfer function between θ_r and θ.

(c) What is the maximum value of K that can be used if you wish to have an overshoot $M_p < 10\%$?

(d) What values of K will provide a rise time of less than 80 sec? (Ignore the M_p constraint.)

(e) Use MATLAB to plot the step response of the antenna system for $K = 200$, 400, 1000, and 2000. Find the overshoot and rise time of the four step responses by examining your plots. Do the plots confirm your calculations in parts (c) and (d)?

3.31. (a) Show that the second-order system

$$\ddot{y} + 2\zeta\omega_n\dot{y} + \omega_n^2 y = 0, \quad y(0) = y_o, \quad \dot{y}(0) = 0,$$

has the response

$$y(t) = y_o \frac{e^{-\sigma t}}{\sqrt{1-\zeta^2}} \sin(\omega_d t + \cos^{-1}\zeta).$$

(b) Prove that, for the underdamped case ($\zeta < 1$), the response oscillations decay at a predictable rate (see Fig. 3.65) called the **logarithmic decrement** δ, where

$$\delta = \ln \frac{y_o}{y_1} = \sigma\tau_d$$

$$= \ln \frac{\Delta y_1}{y_1} \cong \ln \frac{\Delta y_i}{y_i},$$

and τ_d is the damped natural period of vibration

$$\tau_d = \frac{2\pi}{\omega_d}.$$

Figure 3.65
Definition of logarithmic
decrement

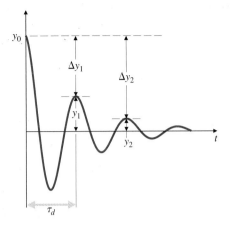

Problems for Section 3.5

3.32. In aircraft control systems, an ideal pitch response (q_o) versus a pitch command (q_c) is described by the transfer function

$$\frac{Q_o(s)}{Q_c(s)} = \frac{\tau\omega_n^2(s+1/\tau)}{s^2+2\zeta\omega_n s+\omega_n^2}.$$

The actual aircraft response is more complicated than this ideal transfer function; nevertheless, the ideal model is used as a guide for autopilot design. Assume that t_r is the desired rise time, and that

$$\omega_n = \frac{1.789}{t_r},$$

$$\frac{1}{\tau} = \frac{1.6}{t_r},$$

$$\zeta = 0.89.$$

Show that this ideal response possesses a fast settling time and minimal overshoot by plotting the step response for $t_r = 0.8, 1.0, 1.2,$ and 1.5 sec.

3.33. Consider the system shown in Fig. 3.66, where

$$G(s) = \frac{1}{s(s+3)} \quad \text{and} \quad D(s) = \frac{K(s+z)}{s+p}. \tag{3.93}$$

Find K, z, and p so that the closed-loop system has a 10% overshoot to a step input and a settling time of 1.5 sec (1% criterion).

Figure 3.66
Unity feedback system for
Problem 3.33

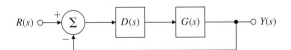

3.34. Sketch the step response of a system with the transfer function

$$G(s) = \frac{s/2 + 1}{(s/40 + 1)[(s/4)^2 + s/4 + 1]}.$$

Justify your answer based on the locations of the poles and zeros (do not find inverse Laplace transform). Then compare your answer with the step response computed using MATLAB.

3.35. Consider the two nonminimum phase systems,

$$G_1(s) = -\frac{2(s - 1)}{(s + 1)(s + 2)}; \tag{3.94}$$

$$G_2(s) = \frac{3(s - 1)(s - 2)}{(s + 1)(s + 2)(s + 3)}. \tag{3.95}$$

(a) Sketch the unit step responses for $G_1(s)$ and $G_2(s)$, paying close attention to the transient part of the response.

(b) Explain the difference in the behavior of the two responses as it relates to the zero locations.

(c) Consider a stable, strictly proper system (that is, m zeros and n poles, where $m < n$). Let $y(t)$ denote the step response of the system. The step response is said to have an undershoot if it initially starts off in the "wrong" direction. Prove that a stable, strictly proper system has an undershoot if and only if its transfer function has an *odd* number of *real* RHP zeros.

3.36. Consider the following second-order system with an extra pole:

$$H(s) = \frac{\omega_n^2 p}{(s + p)(s^2 + 2\zeta\omega_n s + \omega_n^2)}.$$

Show that the unit step response is

$$y(t) = 1 + Ae^{-pt} + Be^{-\sigma t} \sin(\omega_d t - \theta),$$

where

$$A = \frac{-\omega_n^2}{\omega_n^2 - 2\zeta\omega_n p + p^2},$$

$$B = \frac{p}{\sqrt{(p^2 - 2\zeta\omega_n p + \omega_n^2)(1 - \zeta^2)}},$$

$$\theta = \tan^{-1}\frac{\sqrt{1 - \zeta^2}}{-\zeta} + \tan^{-1}\frac{\sqrt{1 - \zeta^2}}{p - \zeta\omega_n}.$$

(a) Which term dominates $y(t)$ as p gets large?

(b) Give approximate values for A and B for small values of p.

(c) Which term dominates as p gets small? (Small with respect to what?)

(d) Using the explicit expression for $y(t)$ above or the step command in MATLAB, and assuming $\omega_n = 1$ and $\zeta = 0.7$, plot the step response of the system above for several values of p ranging from very small to very large. At what point does the extra pole cease to have much effect on the system response?

3.37. The block diagram of an autopilot designed to maintain the pitch attitude θ of an aircraft is shown in Fig. 3.67. The transfer function relating the elevator angle δ_e and the pitch attitude θ is

$$\frac{\theta(s)}{\delta_e(s)} = G(s) = \frac{50(s+1)(s+2)}{(s^2 + 5s + 40)(s^2 + 0.03s + 0.06)},$$

where θ is the pitch attitude in degrees and δ_e is the elevator angle in degrees. The autopilot controller uses the pitch attitude error ε to adjust the elevator according to the following transfer function:

$$\frac{\delta_e(s)}{\varepsilon(s)} = D(s) = \frac{K(s+3)}{s+10}.$$

Using MATLAB, find a value of K that will provide an overshoot of less than 10% and a rise time faster than 0.5 sec for a unit step change in θ_r. After examining the step response of the system for various values of K, comment on the difficulty associated with making rise-time and overshoot measurements for complicated systems.

Figure 3.67
Block diagram of autopilot

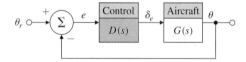

Problems for Section 3.6

3.38. A measure of the degree of instability in an unstable aircraft response is the amount of time it takes for the *amplitude* of the time response to double (see Fig. 3.68) given some nonzero initial condition.

   (a) For a first-order system, show that the **time to double** τ_2 is

$$\tau_2 = \frac{\ln 2}{p},$$

   where p is the pole location in the RHP.

   (b) For a second-order system (with two complex poles in the RHP), show that

$$\tau_2 = \frac{\ln 2}{-\zeta \omega_n}.$$

Figure 3.68
Time to double

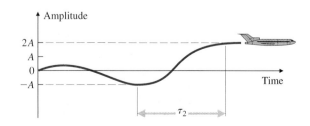

3.39. Suppose that unity feedback is to be applied around the following open-loop systems. Use Routh's stability criterion to determine whether the resulting closed-loop systems will be stable.

(a) $KG(s) = \dfrac{4(s+2)}{s(s^3 + 2s^2 + 3s + 4)}$

(b) $KG(s) = \dfrac{2(s+4)}{s^2(s+1)}$

(c) $KG(s) = \dfrac{4(s^3 + 2s^2 + s + 1)}{s^2(s^3 + 2s^2 - s - 1)}$

3.40. Use Routh's stability criterion to determine how many roots with positive real parts the following equations have.

(a) $s^4 + 8s^3 + 32s^2 + 80s + 100 = 0$.

(b) $s^5 + 10s^4 + 30s^3 + 80s^2 + 344s + 480 = 0$.

(c) $s^4 + 2s^3 + 7s^2 - 2s + 8 = 0$.

(d) $s^3 + s^2 + 20s + 78 = 0$.

(e) $s^4 + 6s^2 + 25 = 0$.

3.41. Find the range of K for which all the roots of the following polynomial are in the LHP.

$$s^5 + 5s^4 + 10s^3 + 10s^2 + 5s + K = 0.$$

Use MATLAB to verify your answer by plotting the roots of the polynomial in the s-plane for various values of K.

3.42. The transfer function of a typical tape-drive system is given by

$$G(s) = \frac{K(s+4)}{s[(s+0.5)(s+1)(s^2 + 0.4s + 4)]},$$

where time is measured in milliseconds. Using Routh's stability criterion, determine the range of K for which this system is stable when the characteristic equation is $1 + G(s) = 0$.

3.43. Consider the system shown in Fig. 3.69.

(a) Compute the closed-loop characteristic equation.

(b) For what values of (T, A) is the system stable? *Hint:* An approximate answer may be found using

$$e^{-Ts} \cong 1 - Ts$$

or

$$e^{-Ts} \cong \frac{1 - \dfrac{T}{2}s}{1 + \dfrac{T}{2}s}$$

for the pure delay. As an alternative, you could use the computer MATLAB (Simulink) to simulate the system or to find the roots of the system's characteristic equation for various values of T and A.

Figure 3.69
Control system for
Problem 3.43

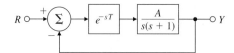

3.44. Modify the Routh criterion so that it applies to the case where all the poles are to be to the left of $-\alpha$ when $\alpha > 0$. Apply the modified test to the polynomial

$$s^3 + (6+K)s^2 + (5+6K)s + 5K = 0,$$

finding those values of K for which all poles have a real part less than -1.

3.45. Suppose the characteristic polynomial of a given closed-loop system is computed to be

$$s^4 + (11+K_2)s^3 + (121+K_1)s^2 + (K_1 + K_1 K_2 + 110K_2 + 210)s + 11K_1 + 100 = 0.$$

Find constraints on the two gains K_1 and K_2 that guarantee a stable closed-loop system, and plot the allowable region(s) in the (K_1, K_2) plane. You may wish to use the computer to help solve this problem.

3.46. Overhead electric power lines sometimes experience a low-frequency, high-amplitude vertical oscillation, or **gallop**, during winter storms when the line conductors become covered with ice. In the presence of wind, this ice can assume aerodynamic lift and drag forces that result in a gallop up to several meters in amplitude. Large-amplitude gallop can cause clashing conductors and structural damage to the line support structures caused by the large dynamic loads. These effects in turn can lead to power outages. Assume that the line conductor is a rigid rod, constrained to vertical motion only and suspended by springs and dampers as shown in Fig. 3.70. A simple model of this conductor galloping is

$$m\ddot{y} + \frac{D(\alpha)\dot{y} - L(\alpha)v}{(\dot{y}^2 + v^2)^{1/2}} + T\left(\frac{n\pi}{\ell}\right)y = 0,$$

where

$$m = \text{mass of conductor,}$$

$$y = \text{conductor's vertical displacement,}$$

$$D = \text{aerodynamic drag force,}$$

$$L = \text{aerodynamic lift force,}$$

$$v = \text{wind velocity,}$$

$$\alpha = \text{aerodynamic angle of attack} = -\tan^{-1}(\dot{y}/v),$$

$$T = \text{conductor tension,}$$

$$n = \text{number of harmonic frequencies,}$$

$$\ell = \text{length of conductor.}$$

Assume that $L(0) = 0$ and $D(0) = D_0$ (a constant), and linearize the equation around the value $y = \dot{y} = 0$. Use Routh's stability criterion to show that galloping can occur whenever

$$\frac{\partial L}{\partial \alpha} + D_0 < 0.$$

Figure 3.70
Electric power-line
conductor

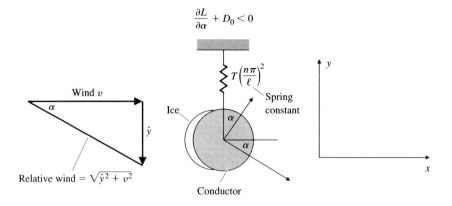

Problem for Section 3.7

▲ **3.47.** Repeat Example 3.34 using Simulink.

Problem for Section 3.8

▲ **3.48.** Samples from a step response are given in Table 3.2. Plot this data on a linear scale $[y(t)$ versus $t]$ and semilog scale $[\log(y - y_\infty)$ versus $t]$, and obtain an estimate of the transfer function.

TABLE 3.2   **Step-Response Data for Problem 3.48**

t	$y(t)$	t	$y(t)$	t	$y(t)$
0	0	0.20	0.0138	0.90	0.4409
0.02	0.0001	0.22	0.0395	1.00	0.4924
0.04	0.0005	0.24	0.0480	1.50	0.6904
0.06	0.0014	0.26	0.0571	2.00	0.8121
0.08	0.0031	0.28	0.0668	2.50	0.8860
0.10	0.0057	0.30	0.0771	3.00	0.9309
0.12	0.0091	0.50	0.1979	3.50	0.9581
0.14	0.0135	0.60	0.2624	4.00	0.9746
0.16	0.0187	0.70	0.3253	5.00	0.9907
0.18	0.0248	0.80	0.3851		

4 Basic Properties of Feedback

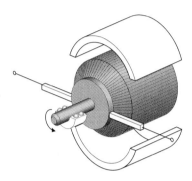

Chapter Overview

The chapter begins with a case study using speed control of an electric motor to illustrate a number of advantages of feedback control compared with open-loop control. First the transfer functions of the two systems are developed and they are compared with respect to steady-state errors in **disturbance rejection**, **reference tracking**, and **sensitivity** to simple process model changes. In an optional subsection the sensitivity of the dynamic response to parameter changes is computed. In Section 4.2 the elementary types of dynamic feedback are considered in a more general setting, beginning with simple proportional control, followed by integral control or "automatic reset" to improve the steady-state error properties. Next we introduce derivative control in order to improve the system's transient properties. Together these three control actions—called **proportional-integral-derivative** (PID) control—constitute a heuristic approach to controller design that has found wide acceptance, especially in the chemical process industries. As part of the consideration of PID control, we discuss a few methods that have been proposed for setting the controller parameters, known as tuning the controller and also **anti-windup** methods to prevent the integral action from continuing or "winding up" when the actuator saturates. In Section 4.3 we present a general analysis of steady-state tracking of polynomial inputs and introduce the concepts of **system type** and error constants to summarize these results. Finally, in Section 4.4 we introduce the important issue of controller implementation by digital logic in recognition of the increasing use of flexible, high-performance, cost-effective digital processors in feedback control.

A Perspective on Properties of Feedback

Control of a dynamic process begins with a model, often based on the principles outlined in Chapter 2, and a description of what the control is required to do. In this book it will usually be assumed that the models of the processes being controlled can be approximated as linear and time-invariant, and they will be described by transfer functions. The control specifications usually include both static and dynamic requirements, such as:

- The permissible steady-state error to a constant or "bias" disturbance signal.
- The permissible steady-state tracking error to a polynomial reference signal such as a step or a ramp.
- The sensitivity of the system to changes in model parameters.
- The dynamic properties such as rise time and overshoot in response to a step in either the reference or the disturbance input.
- Stability of the closed-loop system.

Open-loop and closed-loop control

The two fundamental structures for realizing controls are open-loop control (Fig. 4.1) and closed-loop control, also known as feedback control (Fig. 4.2). Open-loop control is generally simpler, does not require a sensor to measure the output, and does not, of itself, introduce stability problems. Feedback control is more complex and may cause stability problems but also has the potential to give much better performance than is possible with open-loop control. If the process is naturally unstable, feedback control is the only possibility to obtain a stable system.

Figure 4.1
Open-loop control system

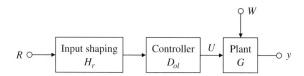

Figure 4.2
Feedback control system

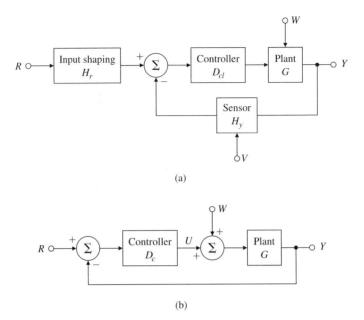

(a)

(b)

4.1 A Case Study of Speed Control

It is very important to maintain constant speed of the spindle in high-perform-ance hard-disk drives used for computer data storage and, in fact, speed control with an electric motor is a generic problem calling for control. As derived in Chapter 2, Eqs. (4.1) and (4.2) below describe the dynamics of a DC motor with negligible armature inductance and including a term, $T_\ell(t)$, for load-torque disturbance.

$$J_m \ddot{\theta}_m + b\dot{\theta}_m = K_t i_a + \tau_\ell, \tag{4.1}$$

$$K_e \dot{\theta}_m + R_a i_a = v_a, \tag{4.2}$$

If we take the Laplace transform of Eqs. (4.1) and (4.2) and let the velocity $\dot{\theta}_m = \omega_m$ with transform $\Omega_m(s) \triangleq s\Theta(s)$ and the transform of the load torque be $T_\ell(s)$, we obtain the transformed equations of speed control as

$$s J_m \Omega_m(s) + b\Omega_m(s) = K_t I_a(s) + \tau_\ell(s), \tag{4.3}$$

$$K_e \Omega_m(s) + R_a I_a(s) = V_a(s). \tag{4.4}$$

After solving for I_a from Eq. (4.4) and substituting it into Eq. (4.3), the equation for motor speed becomes

$$(J_m R_a s + b R_a + K_t K_e)\Omega_m(s) = K_t V_a(s) + R_a T_\ell(s), \tag{4.5}$$

or

$$\left(\frac{J_m R_a}{b R_a + K_t K_e} s + 1\right) \Omega_m(s) = \frac{K_t}{b R_a + K_t K_e} V_a(s) + \frac{R_a}{(b R_a + K_t K_e)} T_\ell(s). \quad (4.6)$$

We can simplify the coefficients in this equation by defining new parameters as

$$(\tau s + 1)\Omega_m(s) = A V_a(s) + B T_\ell(s), \quad (4.7)$$

where the time constant and gains are given by

$$\tau = \frac{J_m R_a}{b R_a + K_t K_e} \qquad \text{(time constant, sec)},$$

$$A = \frac{K_t}{b R_a + K_t K_e} \qquad \text{(voltage gain, rad/volt·sec)}, \quad (4.8)$$

$$B = \frac{R_a}{b R_a + K_t K_e} \qquad \text{(torque gain, rad/N·m·sec)}.$$

We may also rewrite Eq. (4.7) in transfer-function form:

$$\Omega_m(s) = \frac{A}{\tau s + 1} V_a(s) + \frac{B}{\tau s + 1} T_\ell(s). \quad (4.9)$$

If the inputs are constants, say $v_a = a$ and $\tau_\ell = b$, then the steady-state solution of these equations is[1]

$$\omega_{ss} = Aa + Bb. \quad (4.10)$$

The transfer functions from Eq. (4.9) are shown in Fig. 4.3 for open-loop control with a controller having transfer function $D_{ol}(s)$. For this configuration we

Figure 4.3
Open-loop speed-control
system

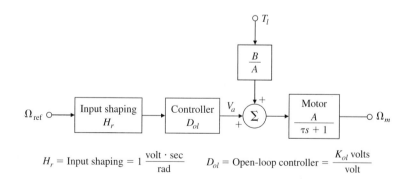

$H_r = $ Input shaping $= 1 \dfrac{\text{volt} \cdot \text{sec}}{\text{rad}}$ $D_{ol} = $ Open-loop controller $= \dfrac{K_{ol} \text{ volts}}{\text{volt}}$

---

[1] From the final value theorem presented in Section 3.1.6, if $Y(s) = G(s)U(s)$ and $U(s) = 1/s$, the final value of $y(t)$ is $y_{ss} = G(0)$. Thus we just have to set $s = 0$ in a *stable* transfer function to find the final or steady-state value to a unit step input.

will refer to the transfer function from Ω_{ref} to Ω_m as the **open-loop transfer function**,

$$T_{ol}(s) = H_r D_{ol} \frac{A}{\tau s + 1}.$$

Modeling speed control by a block diagram

In contrast, the block diagram in Fig. 4.4 shows a closed-loop system which requires an output sensor. In this case the sensor is a tachometer, which is usually a small permanent-magnet DC machine which produces a voltage proportional to the shaft speed $(= \omega_m)$. In this case, the **closed-loop transfer function** is

$$T_{cl}(s) = \frac{\Omega_m}{\Omega_{ref}} = H_r \frac{D_{cl} \dfrac{A}{\tau s + 1}}{1 + D_{cl} H_y \dfrac{A}{\tau s + 1}}.$$

We will often refer to the transfer function of the elements around a closed loop as the **loop gain**, which in this case is

$$D_{cl} H_y \frac{A}{\tau s + 1}.$$

In these figures and in the analysis below, the gain or scale factor of the sensor is taken to be $H_y = 1$ volt·sec/rad. Techniques for treating more practical values will be described shortly. The input-shaping filter, H_r, is a units conversion device also taken here to have gain 1 volt·sec/rad. The controllers in the figures are taken to be electronic circuits with transfer functions having dimensions *volts/volt*.

Figure 4.4
Feedback speed control system

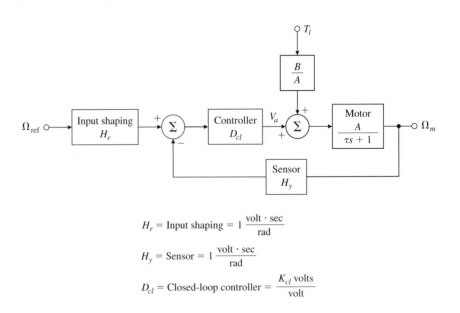

$$H_r = \text{Input shaping} = 1 \ \frac{\text{volt} \cdot \text{sec}}{\text{rad}}$$

$$H_y = \text{Sensor} = 1 \ \frac{\text{volt} \cdot \text{sec}}{\text{rad}}$$

$$D_{cl} = \text{Closed-loop controller} = \frac{K_{cl} \text{ volts}}{\text{volt}}$$

4.1.1 Disturbance Rejection

Now let us compare open-loop control with feedback control with respect to how well each system maintains a constant steady-state reference speed in the face of load or disturbance torques. For the open-loop controller (Fig. 4.3), the amplifier voltage is taken to be

$$v_a = D_{ol}\omega_{ref} = K_{ol}\omega_{ref}. \tag{4.11}$$

Open-loop speed control

The constant gain $D_{ol}(s) = K_{ol}$ is determined so that in steady-state the output speed $\omega_m = \omega_{ss}$ equals the constant reference speed ω_{ref} when the load torque τ_ℓ is zero. From Eq. (4.10) this value is found to be

$$K_{ol} = \frac{1}{A}.$$

In the steady state ($s = 0$) and with no load torque ($\tau_\ell = 0$), Eq. (4.10) gives the output speed as

$$\omega_m = \omega_{ss} = Av_a = A\frac{1}{A}\omega_{ref} = \omega_{ref}.$$

With the feedback controller (Fig. 4.4) the controller transfer function is taken to be $D_{cl} = K_{cl}$ so that

$$v_a = K_{cl}(\omega_{ref} - \omega_m). \tag{4.12}$$

Closed-loop speed control

Combining the motor model given by Eq. (4.7) with the transform of the feedback controller equation described by Eq. (4.12), the closed-loop system equations are

$$[(\tau s + 1) + AK_{cl}]\Omega_m(s) = AK_{cl}\Omega_{ref}(s) + BT_\ell(s). \tag{4.13}$$

For no load torque ($\tau_\ell = 0$) and in the steady state ($s = 0$) the speed is given by

$$\omega_{ss} = \frac{AK_{cl}}{1 + AK_{cl}}\omega_{ref}. \tag{4.14}$$

If the gain K_{cl} is selected large so that $AK_{cl} \gg 1$, then, although some error remains, $\omega_{ss} \cong \omega_{ref}$.

With these controls, both approaches provide the desired result of good tracking, $\omega_{ss} \cong \omega_{ref}$, although on the face of it the open-loop seems to be more accurate. However, the gains have been set with zero load torque; now let us compute what effect a nonzero torque will have on the steady-state speed in the two cases. Equation (4.10) shows that in the open-loop system the steady-state speed is

$$\omega_{ss} = AK_{ol}\omega_{ref} + B\tau_\ell. \tag{4.15}$$

Here, with $K_{ol} = 1/A$, the speed with load torque for the open-loop case is

$$\omega_{ss} = \omega_{ref} + B\tau_\ell,$$

and the variation in speed caused by the load torque for the open-loop case is $\omega_{ss} - \omega_{ref} = \delta\omega$, where

$$\delta\omega = B\tau_\ell.$$

So we see that the speed error is proportional to the disturbing load torque, and the control designer has no influence over the parameter B which determines the size of the error.

For the feedback case [Eq. (4.13)], the steady-state speed for constant values of ω_m and τ_ℓ becomes

$$\omega_{ss} = \frac{AK_{cl}}{1 + AK_{cl}}\omega_{ref} + \frac{B}{1 + AK_{cl}}\tau_\ell. \qquad (4.16)$$

If it is possible for the designer to pick values of K_{cl} so that $AK_{cl} \gg 1$ and $AK_{cl} \gg B$, a very small error will result with or without a disturbing load torque. In fact, as long as $AK_{cl} > 0$, comparison of Eqs. (4.15) and (4.16) reveals that the feedback-controller speed errors due to a load torque will be less than the open-loop errors by exactly $1 + AK_{ol}$. This is our first result on the advantages of feedback:

Advantage of feedback

> System errors to constant disturbances can be made smaller with feedback than they are in open-loop systems by a factor of $1 + AK_{cl}$ where AK_{cl} is the loop gain at $s = 0$.

EXAMPLE 4.1   *Steady-State Error to Disturbance, Open-Loop Case*

For a certain small servomotor and load, it is determined that $\tau = \frac{1}{60}$ sec, $A = 10$ rad/V·sec, and $B = 50$ rad/N·m·sec. The reference speed is $\omega_{ref} = 100$ rad/sec. Find the steady-state open-loop armature voltage needed to get this speed with zero load torque. What is the steady-state speed error with this load voltage if the load torque is $\tau_\ell = -0.1$ N·m?

Solution.   For the open-loop system with an armature voltage input v_a and a load torque τ_ℓ, the steady-state output is given by Eq. (4.10) with $K_{ol} = \frac{1}{10}$ and thus $v_a = \omega_{ref}/10$.

$$\omega_{ss} = 10v_a + 50\tau_\ell.$$

Assuming $\tau_\ell = 0$, the output is

$$\omega_{ss} = 100 \text{ rad/sec} = \omega_{\text{ref}}.$$

Now suppose a constant load torque of $\tau_\ell = -0.1$ N·m is applied. The steady-state value of the output with the disturbance is

$$\omega_{ss} = 100 + 50(-0.1) = 95 \text{ rad/sec}.$$

Thus the error is 5 rad/sec or 5%.

Now consider the same example using proportional feedback of the output.

EXAMPLE 4.2   *Steady-State Error to Disturbance, Closed-Loop Case*

Consider the feedback control structure shown in Fig. 4.4, where tachometer feedback and proportional control are employed. You are to improve the ability of the system to reject steady-state disturbances by a factor of at least 100 compared with the open-loop system. The parameter values are the same as those given in Example 4.1.

Solution.   From Eq. (4.16) we see that to get an improvement factor of 100 we require $1 + AK_{cl} = 100$, from which $K_{cl} = 9.9$. With this value of K_{cl} the steady-state value of the tracking output assuming that the load torque $T_\ell = 0$, is[2]

$$\omega_{ss} = \frac{AK_{cl}}{1 + AK_{cl}}\omega_{ref} = \frac{99}{100}100$$

$$= 99 \text{ rad/sec.}$$

The steady-state output due to the input of a constant disturbance torque is [Eq. 4.16)]

$$\omega_{ss} = \frac{B\,\tau_\ell}{1 + K_{cl}A} = \frac{50\,\tau_\ell}{1 + 10K_{cl}}.$$

The output with both reference and disturbance is, $\omega_{ss} = 99 - 0.05 = 98.95$ rad/sec which is a change of only 0.051% from the speed without disturbance. Thus the feedback has improved the disturbance rejection by a factor of about 100 over the open-loop system.

4.1.2   Sensitivity of System Gain to Parameter Changes

As another comparison of open- and closed-loop control, we consider the sensitivity of the steady-state gain or transfer function at $s = 0$ to parameter changes. The change might come about because of external effects such as temperature changes or might simply be due to an error in the value of the parameter from the start. Suppose that the motor gain in operation differs from its original design value of A to be $A + \delta A$. In the open-loop case the nominal gain is[3] $T_{ol} = K_{ol}A$, and the new overall system gain would be

$$T_{ol} + \delta T_{ol} = K_{ol}(A + \delta A) = K_{ol}A + K_{ol}\delta A = T_{ol} + K_{ol}\delta A,$$

where T_{ol} is the open-loop transfer function at $s = 0$. Using the fact that $K_{ol} = 1/A$, the normalized error in the gain is

$$\frac{\delta A}{A}.$$

---

[2] In process control it is common practice to note that if the set point were *reset* to $\omega'_{ref} = 100/99\omega_{ref}$, then the output will be equal to the reference value.

[3] We use T_{ol} and T_{cl} for the open-loop and closed-loop transfer functions, respectively. These are not to be confused with the transform of the disturbance torque, T_ℓ used earlier.

In terms of change in the normalized gain, defined as $\delta T / T$, we find that

$$\frac{\delta T_{ol}}{T_{ol}} = \frac{K_{ol}\delta A}{K_{ol} A} = \frac{\delta A}{A}. \tag{4.17}$$

This means that a 10% error in A would yield a 10% error in T_{ol}. H. W. Bode called the ratio of $\delta T / T$ to $\delta A / A$ the **sensitivity,** S, of the gain from Ω_{ref} to Ω_m with respect to the parameter A. In the open-loop case, $S = 1$.

The same change in A in the feedback case (Eq. 4.14) yields the new steady-state feedback gain

$$T_{cl} + \delta T_{cl} = \frac{(A + \delta A)K_{cl}}{1 + (A + \delta A)K_{cl}},$$

where T_{cl} is the closed-loop gain. We can compute the sensitivity of this closed-loop gain directly using differential calculus. The closed-loop steady-state gain is

$$T_{cl} = \frac{AK_{cl}}{1 + AK_{cl}}.$$

The first-order variation is proportional to the derivative and is given by

$$\delta T_{cl} = \frac{dT_{cl}}{dA}\delta A.$$

The general expression for sensitivity of a transfer function T to a parameter A is thus given by

$$\frac{\delta T}{T} = \left(\frac{A}{T}\frac{dT}{dA}\right)\frac{\delta A}{A}$$

$$= (\text{sensitivity})\,\frac{\delta A}{A}.$$

From this formula the sensitivity is

$$S_A^{T_{cl}} \triangleq \text{sensitivity of } T_{cl} \text{ with respect to } A$$

$$\triangleq \frac{A}{T_{cl}}\frac{dT_{cl}}{dA},$$

so

$$S_A^{T_{cl}} = \frac{A}{AK_{cl}/(1 + AK_{cl})}\frac{(1 + AK_{cl})K_{cl} - K_{cl}(AK_{cl})}{(1 + AK_{cl})^2}$$

$$= \frac{1}{1 + AK_{cl}}. \tag{4.18}$$

This result exhibits another major advantage of feedback:

In feedback control, the error in the overall transfer function gain at $s = 0$ is less sensitive to variations in the plant gain by a factor of $1 + AK_{cl}$ compared to errors in open-loop control.

We can conclude that even if the system gain is subject to change, it is still possible to achieve precise control using feedback if the loop gain can be made sufficiently large. This is often very important when the system being controlled—for example, an electric motor or steam engine—is a large, high-powered device whose gain is not only difficult to compute but naturally subject to substantial variation.

The superiority of feedback control is equally apparent when we consider errors in the control gain K_{cl}. Because K_{cl} always appears as a product with A, the analysis for an error in A applies directly to an error in K_{cl}.

EXAMPLE 4.3

Open- and Closed-Loop Sensitivity to Parameter Changes

Compare the sensitivity of the open-loop and closed-loop transfer functions to changes in the parameter A for the DC motor of Example 4.1.

Solution.   For the open-loop case, Eq. (4.17) shows immediately that $S_A^{T_{ol}} = 1$. For the closed-loop situation we use Eq. (4.18) to find that

$$S_A^{T_{cl}} = \frac{1}{1 + 10(9.9)} = 0.01.$$

Thus, for the closed-loop case a 10% change in motor gain A will cause only a 0.1% change in the steady-state speed. The open-loop controller is 100 times more sensitive to gain changes than this closed-loop system whose loop gain is 100.

The results in this section so far have been computed for the steady-state error in the presence of constant inputs, either reference or disturbance. Very similar results can be obtained for the steady-state behavior in the presence of sinusoidal reference and disturbance signals. This is important because there are times when such signals naturally occur as with a disturbance of 60 Hz due to power-line interference in an electronic system, for example. The concept is also important because more complex signals can be described as containing sinusoidal components over a band of frequencies and analyzed using superposition one frequency at a time. For example, it is well known that human hearing is restricted to signals in the frequency range of about 60 to 15,000 Hz. A feedback amplifier and loudspeaker system designed for high-fidelity sound must accurately track any sinusoidal (pure tone) signal in this range. If we take the controller in the feedback system shown in Fig. 4.2 to have the transfer function $D(s)$ and we take the process to have the transfer function $G(s)$, then the steady-state open-loop gain at the sinusoidal signal of frequency ω_o will be $|D(j\omega_o)G(j\omega_o)|$ and the error of the feedback system will be

$$|E(j\omega_o)| = |R(j\omega_o)| \left| \frac{1}{1 + D(j\omega_o)G(j\omega_o)} \right|.$$

Thus, to reduce errors to 1% of the input at the frequency ω_o, we must make $|1 + DG| \geq 100$ or $|D(j\omega_o)G(j\omega_o)| \gtrsim 100$, and a good audio amplifier must have this loop gain over the range $2\pi 60 \leq \omega \leq 2\pi 15{,}000$. We will revisit this concept in Chapter 6 as part of the design based on frequency response techniques.

4.1.3   Effect of Feedback on Time Response

Thus far we have looked at steady-state properties in the presence of a constant reference or a constant disturbance. However, the systems we are interested in are dynamic, and tracking time-varying inputs is also an important role for control. A constant-gain open-loop controller has no effect on the dynamics of the system for either reference or disturbance inputs. If an open-loop controller includes a dynamic input-shaping filter, $H_r(s)$, this can change the dynamic response to the reference signal but the plant dynamics will still determine the system's response to disturbances. On the other hand, feedback of any kind changes the dynamics of the system for both reference and disturbance inputs. In the case of open-loop speed control, Eq. (4.9) shows that the plant dynamics are described by the (open loop) time constant τ. The dynamics with proportional feedback control are described by Eq. (4.13), and the root of the characteristic equation of this system is

$$s = \frac{1 + AK_{cl}}{\tau}. \tag{4.19}$$

Therefore, the closed-loop time constant is a function of the feedback gain K_{cl} and is given by

$$\tau_{cl} = \frac{\tau}{1 + AK_{cl}}$$

and is decreased as compared to the open-loop value. It is typically the case that closed-loop systems have a faster response as the feedback gain is increased and, if there were no other effects, this is generally desirable. As we will see, however, the responses of higher-order systems typically become less well damped and eventually will become unstable as the gain is steadily increased. Thus a definite limit exists on how large we can make the gain in our efforts to reduce the effects of disturbances and the sensitivity to changes in plant parameters. Attempts to resolve the conflict between small steady-state errors and good dynamic response will characterize a large fraction of control design problems. The conclusion is as follows:

Important property of
feedback

Feedback changes the dynamic response and often makes the system faster and less stable.

▲ 4.1.4   Sensitivity of Time Response to Parameter Change

We have considered the effects of errors on the steady-state gain of a dynamic system and showed how feedback control can reduce these errors. Because many control specifications are in terms of the step response, the sensitivity of the time response to parameter changes is sometimes very useful to explore. For example, by looking at the sensitivity plot we can tell if increasing a particular parameter will increase or decrease the overshoot of the response.[4] The following analysis is also a good exercise in small-signal linearization.

To consider the sensitivity of the output $y(t, \theta)$ of a system having a parameter of interest, θ, we compute the effect of a perturbation in the parameter, $\delta\theta$, on the nominal response by using Taylor's series expansion

$$y(t, \theta + \delta\theta) = y(t, \theta) + \frac{\partial y}{\partial \theta}\delta\theta + \cdots. \tag{4.20}$$

The first-order approximation of the parameter perturbation effect is the term

$$\delta y(t) = \frac{\partial y}{\partial \theta}\delta\theta. \tag{4.21}$$

This function can be generated from the system itself as shown by Perkins et al. (1991). We assume that the response depends linearly on the parameter and therefore that the overall transfer function $T(s, \theta)$ is composed of component transfer functions which can be defined to bring out the dependence on the parameter explicitly. A block diagram of the transfer function in terms of the components $T_{ij}(s)$ can be expressed as shown in Fig. 4.5 where we have labeled the parameter as θ and its input signal as Z. In terms of this block diagram, the equations relating Y and Z to the reference input can be written immediately.

$$Y = T_{11}R + T_{21}\theta Z \tag{4.22}$$

$$Z = T_{12}R + T_{22}\theta Z \tag{4.23}$$

Figure 4.5
Block diagram showing the dependence of output y on parameter θ

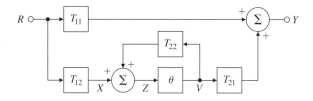

<hr />

[4] As we will see in the next chapter, the development of the MATLAB root locus interface rltool gives the designer a computer aid to this result.

The perturbed equations are

$$Y + \delta Y = T_{11}R + T_{21}(\theta + \delta\theta)(Z + \delta Z) \tag{4.24}$$

$$Z + \delta Z = T_{12}R + T_{22}(\theta + \delta\theta)(Z + \delta Z) \tag{4.25}$$

Multiplying these out and ignoring the small term $\delta\theta\delta Z$, the expressions for the perturbations in Y and Z are given by

$$\delta Y = T_{21}(Z\delta\theta + \theta\delta Z) \tag{4.26}$$

$$\delta Z = T_{22}(Z\delta\theta + \theta\delta Z) \tag{4.27}$$

The solutions to these equations can be best presented as a block diagram, shown in Fig. 4.6(a). The output of this figure is

$$\delta Y = \frac{\partial y}{\partial\theta}\delta\theta$$

and we notice that the input Z is multiplied by a gain of $\delta\theta$. Therefore, if we drop the block $\delta\theta$ the output will be simply $\frac{\partial y}{\partial\theta}$ as shown in Fig. 4.6(b). Finally,

Figure 4.6

Block diagrams showing the generation of (a) δY and δZ; (b) $\frac{\partial y}{\partial\theta}$; and (c) $\theta\frac{\partial y}{\partial\theta}$

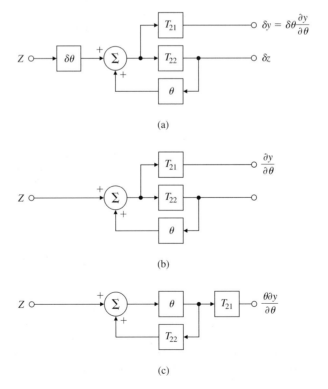

Figure 4.7

Block diagram showing the computation of $\theta \frac{\partial y}{\partial \theta}$ from the original transfer function

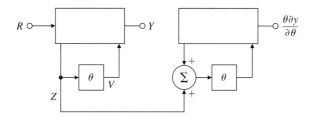

to compute the sensitivity as the variation to a percent change in the parameter which is

$$\frac{\partial y}{\partial \ln \theta} = \frac{\dfrac{\partial y(t, \theta)}{\partial \theta}}{\dfrac{\partial \ln \theta}{\partial \theta}} = \theta \frac{\partial y}{\partial \theta},$$

we need only shift the input Z from the output side of the θ block to its input as shown in Fig. 4.6(c). We are now in a position to give the final block diagram of the system as it is to be implemented, shown in Fig. 4.7.

In this figure it is clear that to compute the sensitivity of the output to a parameter, one needs to simulate two copies of the system. The input to the first system is the reference input of interest and the input to the second system is at the input to the parameter of interest of the variable Z taken from the input to the parameter in the original system. The transfer function from the reference input to the output sensitivity is readily computed to be

$$\frac{T_{12} \theta T_{21}}{(1 - \theta T_{22})^2}. \tag{4.28}$$

From this function it is clear that:

Response sensitivity

> To keep the sensitivity of the output signal to a parameter change low, it is important to have feedback with high gain around the parameter in question.

EXAMPLE 4.4   *Time-Domain Sensitivity*

Compute the sensitivity of the output of the speed control example shown in Fig. 4.4 with respect to the control gain, K_{cl}. Take the nominal values to be $K_{cl} = 9$, $\tau = 0.01$ sec, and $A = 1$ rad/volt-sec.

Solution.   The required block diagram for the computation is given in Fig. 4.8 based on Fig. 4.4 and Fig. 4.7. In MATLAB, we will construct the several transfer functions with $T_{ij} = n_{ij}/d_{ij}$ and implement Eq. (4.28). For comparison, we compute the nominal response from Fig. 4.5 and add 10% of the sensitivity to the nominal response. The instructions to do the computation in MATLAB are:

Figure 4.8
Block diagram showing
the computation of the
sensitivity of the output of
the speed control example

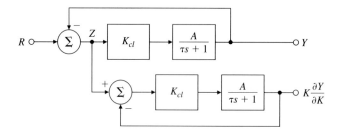

% script to compute sensitivity for Fig. 4.8

% First input the data for the component transfer functions Tij

% and the nominal parameter, Kcl for this problem

Kcl = 9; tau = .01;

n11 = 0; d11 = 1;

n12 = 1; d12 = 1;

n22 = [0 −1]; d22 = [tau 1];

n21 = 1; d21 = [tau 1];

% Now compute the numerator and denominator polynomials of the transfer functions

% using the convolution function conv to multiply the polynomials

% and put them into system transfer function forms with the MATLAB function tf.

% The over-all transfer function is

% Y/R = n11/d11 + (n12*n21* d22)/(d12*d21* [d22-Kcl*n22]) = sysy

% The transfer function from the reference input to the sensitivity is

% Kcl*(dy/dKcl)/R = sysdy

% Now define the numerators and denominators of several intermediate transfer functions

n1 = Kcl*conv(n21,n12);

d1 = conv(d21,d12);

n2 = d22;

d2 = [d22-Kcl*n22];

ny = conv(n1,n2);

dy = conv(d1,d2);

% Now put these together to form two intermediate transfer functions

sysy1 = tf(ny,dy);

sysy2 = tf(n11,d11);

% Now construct the final transfer functions

% The overall transfer function Y/R

sysy = sysy1+sysy2;

% The sensitivity transfer function

```
ndy = conv(ny,n2);
ddy = conv(dy,d2);
sysdy = tf(ndy,ddy);
% Now use these to compute the step responses and
% plot the output, the sensitivity and a perturbed response
[y,t] = step(sysy);
[yd,t] = step(sysdy);
plot(t,[y yd y+.1*yd]);
```

These instructions are constructed to compute the sensitivity for any system, given the several transfer functions. The script input is for the specific example. Plots of the output, its sensitivity, and the result of a 10% change in the parameter value are given in Fig. 4.9.

Figure 4.9
Plots of the output, the sensitivity, and the result of a 10% change in the parameter value for the speed control example

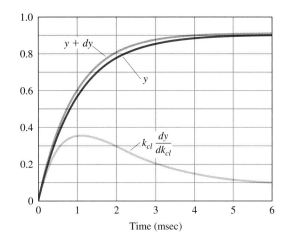

4.2   The Classical Three-Term Controller

We have seen in Section 4.1 that proportional feedback control (**P**) can reduce error responses to disturbances but that it still allows a nonzero steady-state error to constant inputs. As we will see in this section, when the controller includes a term proportional to the integral of the error (**I**), then the steady-state error can be eliminated, although typically at the cost of deterioration in the dynamic response. Addition of a term proportional to the derivative of the error (**D**) can often improve the dynamic response. Combined, these three terms form the classical **PID** controller which is widely used in the process and robotics industries.

The PID (proportional-integral-derivative) controller

4.2.1   Proportional Feedback Control (P)

As we saw earlier, when the feedback control signal is made to be linearly proportional to the system error, we call the result **proportional feedback**. This was the case for the feedback [Eq. (4.12)] used in the controller of speed in Section 4.1. In the speed control test case, the controller input was $\omega_{ref} - \omega_m$ and its output was the motor voltage, v_m. In the general form of proportional control we take the input to the controller to be the tracking error $e(t)$ and the output control variable to be $u(t)$. The control law is

$$u = k_p e, \tag{4.29}$$

and the controller transfer function $D_c(s)$ as shown in Fig. 4.2(b) would be[5]

$$\frac{U(s)}{E(s)} = D_c(s) = k_p. \tag{4.30}$$

The typical proportional controller as simply an amplifier with a knoblike volume control to adjust its gain up or down. The system with proportional control will usually have a steady-state offset (or droop) in response to a constant reference input or to a constant disturbance input which gets smaller as the gain is increased, as we saw in the speed control example. For higher-order systems, large values of the proportional feedback gain will almost always lead to instability, and there is an upper limit on the size of the proportional feedback gain if we are to achieve a well-damped stable response. This limit may still have an unacceptable steady-state error. The standard way to improve the steady-state accuracy of control without adding extremely high proportional gains is to introduce integral control.

Limits of proportional feedback

4.2.2   Proportional Plus Integral Control (PI)

Adding an integral term to the controller results in the **proportional plus integral (PI)** control equation

Proportional plus integral control

$$u(t) = k_p e + k_I \int_{t_0}^{t} e(\tau)\, d\tau; \tag{4.31}$$

for which the $D_c(s)$ in Fig. 4.2 becomes

$$\frac{U(s)}{E(s)} = D_c(s) = k_p + \frac{k_I}{s}, \tag{4.32}$$

This feedback has the primary virtue that in the steady state its control output can be a *nonzero* constant value even when the error signal at its input is *zero*.

---

[5]   For this section, we take $H_r = H_y = 1.0$ (unity feedback) and $D_{cl}(s) = D_c(s)$.

This comes about because the integral term in the control signal is a summation of all past values of $e(t)$. In fact the integral term will not stop changing until its input is zero and therefore if the system reaches a stable steady state the input signal to the integrator will of necessity be zero. This feature means that a constant disturbance w (see Fig. 4.2) can be canceled by the controller integrator's output even while the system error is zero.

EXAMPLE 4.5

Integral Control of Motor Speed

Consider again the speed controller discussed in Section 4.1. Compute the effects of adding integral control on the steady-state error to a step disturbance input.

Solution.   For PI control, the voltage v_a in Eq. (4.7) will include a term proportional to the integral of the error between ω_m and ω_{ref} for which the transform equation is

$$V_a = k_p(\Omega_{ref} - \Omega_m) + k_I \frac{\Omega_{ref} - \Omega_m}{s}. \qquad (4.33)$$

The system transform equation with this controller is

$$(\tau s + 1)\Omega_m = A\left(k_p + \frac{k_I}{s}\right)(\Omega_{ref} - \Omega_m) + BT_\ell. \qquad (4.34)$$

and, if we multiply by s and collect terms,

$$(\tau s^2 + (Ak_p + 1)s + Ak_I)\Omega_m = A(k_p s + k_I)\,\Omega_{ref} + Bs T_\ell. \qquad (4.35)$$

We can analyze this equation in several ways to answer the problem. The most direct way is to assume that $\Omega_{ref} = 0$, $T_\ell = T_o/s$ and apply the final value theorem as follows:

$$\lim_{t \to \infty} \omega_m = \lim_{s \to 0} s\Omega_m(s) \qquad (4.36)$$

$$= \lim_{s \to 0} s \frac{Bs}{\tau s^2 + (Ak_p + 1)s + Ak_I} \frac{T_o}{s} \qquad (4.37)$$

$$= 0. \qquad (4.38)$$

A second way to analyze the equation is to recognize that for constant inputs, the steady-state transfer gains are obtained simply by setting $s = 0$. If we do this in Eq. (4.35), then

$$Ak_I\Omega_{mo} = Ak_I\Omega_{refo} + 0T_{\ell o} \qquad (4.39)$$

$$= Ak_I\Omega_{refo} \qquad (4.40)$$

$$\Omega_{mo} = \Omega_{refo} \qquad (4.41)$$

with the result that we have the steady-state motor speed equaling the reference speed *regardless of the presence of a constant disturbance or of the actual value of either the motor gain A or the controller gains, k_p and k_I as long as k_I is different from zero.*[6]

---

[6]   This result should be contrasted to the open-loop case where although the nominal steady-state error is also zero, it would become nonzero if either the process gain or the controller gain were to change.

Because the PI controller includes dynamics, use of this controller will change the dynamic response in more complicated ways than the simple speed up we saw with proportional control. This we can understand by considering the characteristic equation of the speed control with PI control as seen in Eq. (4.35):

$$\tau s^2 + (Ak_p + 1)s + Ak_I = 0 \tag{4.42}$$

The two roots of this equation may be complex and, if so, the natural frequency is $\omega_n = \sqrt{Ak_I/\tau}$ and the damping ratio is $\zeta = (Ak_p + 1)/2\tau\omega_n$. The point is that these parameters are determined by the controller gains and unless care is taken, the design may result in an unsatisfactory lightly damped response. To explore this phenomenon further, we consider a higher-order case and the introduction of derivative control.

4.2.3   Proportional-Integral-Derivative Feedback Control (PID)

The final term in the classical controller is derivative control, **D**, and the complete three-term controller is described by the transform equation we will use, namely,

$$D_c(s) = \frac{U(s)}{E(s)} = k_p + \frac{k_I}{s} + k_D s \tag{4.43}$$

or, equivalently, by the equation often used in the process industries:

$$D_c(s) = k_p \left[1 + \frac{1}{T_I s} + T_D s \right], \tag{4.44}$$

where the "reset rate" in seconds, T_I, and the "derivative rate," T_D also in seconds can be given physical meaning to the operator who must select values for them to "tune" the controller. For our purposes, Eq. (4.43) is simpler to use. The effect of the derivative control term depends on the rate of change of the error. As a result, a controller with derivative control exhibits an anticipatory response as illustrated by the fact that the output of a PID controller having a ramp error $e(t) = t1(t)$ input would *lead* the output of a proportional controller having the same input by $k_D/k_p \overset{\triangle}{=} T_D$ seconds, as shown in Fig. 4.10.

Figure 4.10
Anticipatory nature of derivative control

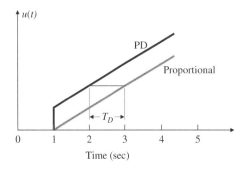

Because of the sharp effect of derivative control on suddenly changing signals, the D term is sometimes introduced into the feedback path as shown in Fig. 4.11(a), which would describe for example a tachometer on the shaft of a motor. The closed-loop characteristic equation is the same as if the term were in the forward path as given by Eq. (4.43) and drawn in Fig. 4.11(b) if the derivative gain is $k_D = k_p k_t$ but the zeros from the reference to the output are different are different in the two cases. With the derivative in the feedback path, the reference is not differentiated, which may be a desirable result if the reference is subject to sudden changes. With the derivative in the forward path a step change in the reference input will cause an intense initial pulse[7] in the control signal which may be very undesirable.

To illustrate the effect of a derivative term on PID control, the speed control equations will be reconsidered but with a non-negligible armature inductance. Simplification of the resulting motor equations leads to a process with a second-order characteristic equation having the form

$$[(L_a s + R_a)(J_m s + b) + K_t K_e]\Omega_m = K_t V_a - (L_a s + R_a)T_\ell \qquad (4.45)$$

$$(a_o s^2 + a_1 s + a_2)\Omega_m(s) = K_t V_a - (L_a s + R_a)T_\ell \qquad (4.46)$$

Applying PID according to Eq. (4.43) to this example amounts to letting

$$V_a = \left[k_p + \frac{k_I}{s} + k_D s\right](\Omega_{ref} - \Omega_m) \qquad (4.47)$$

If we substitute this expression for V_a into Eq. (4.46), we obtain

$$\left[a_o s^2 + a_1 s + a_2 + K_t\left(k_p + \frac{k_I}{s} + k_D s\right)\right]\Omega_m$$

$$= K_t\left[k_p + \frac{k_I}{s} + k_D s\right]\Omega_{ref} - (L_a s + R_a)T_\ell. \qquad (4.48)$$

Figure 4.11
Alternative ways of
configuring rate feedback

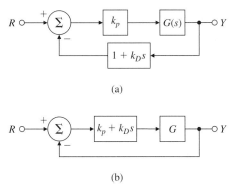

(a)

(b)

---

[7] An impulse in theory.

For this system, the characteristic equation is given by multiplying through by s and collecting terms as

$$a_o s^3 + (a_1 + K_t k_D)s^2 + (a_2 + K_t k_p)s + K_t k_I = 0 \qquad (4.49)$$

The point here is that this equation, whose three roots determine the nature of the dynamic response of the system, has three free parameters in k_p, k_I, and k_D and by selection of these parameters, the roots can be uniquely and, in theory, arbitrarily determined. Without the derivative term, there would be only two free parameters and three roots so the choice of roots of the characteristic equation would be restricted. To illustrate the effect more concretely, a numerical example is useful.

EXAMPLE 4.6   *PID Control of Motor Speed*

Consider the DC motor speed control described in Eq. (4.46) with parameters[8]

$J_m = 1.13 \times 10^{-2}$ N-m- sec 2/rad, $b = 0.028$ N-m-sec/rad, $L_a = 10^{-1}$ henry

$R_a = 0.45$ ohms, $K_t = 0.067$ N-m/amp, $K_e = 0.067$ V-sec/rad. (4.50)

Use the controller parameters

$$k_p = 3, \qquad k_I = 15 \text{ sec}^{-1}, \qquad k_D = 0.3 \text{ sec}. \qquad (4.51)$$

Discuss the effects of P, PI, and PID control on the responses of this system to steps in the disturbance and steps in the reference input. Let the unused controller parameters be zero.

Solution.   Figure 4.12(a) illustrates the effects of P, PI, and PID feedback on the step disturbance response of the system. Note that adding the integral term increases the oscillatory behavior but eliminates the steady-state error and that adding the derivative term reduces the oscillation while maintaining zero steady-state error. Figure 4.12(b) illustrates the effects of P, PI, and PID feedback on the step reference response with similar results. The step responses can be computed by forming the numerator and denominator coefficient vectors (in descending powers of s) and using the step function in MATLAB. For example, after the values for the parameters are entered, the following commands produce a plot of the response of PID control to a disturbance step.

```
numG = [La Ra 0];
denG = [Jm*La Ra*b + Ke*Ke + Ke*kD Ra*Ke*Ke + Ke*kp Ke*ki];
sysG = tf(numG,denG);
y = step(sysG).
```

---

[8] These values have been scaled to measure time in milliseconds by multiplying the true L_a and J_m by 1000 each.

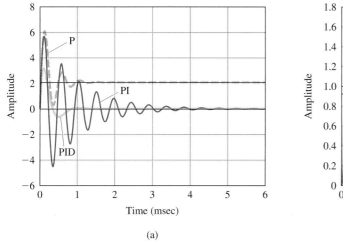
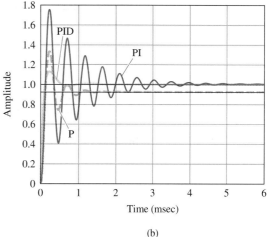

(a)
(b)

Figure 4.12 Responses of P, PI, and PID control to (a) step disturbance input (b) step reference input

▲ 4.2.4   Ziegler–Nichols Tuning of PID Regulators

As we will see in later chapters, sophisticated methods are available to develop a controller that will meet steady-state and transient specifications for both tracking input references and rejecting disturbances. These methods require that the designer has either a dynamic model of the process in the form of equations of motion or a detailed frequency response over a substantial range of frequencies. Either of these data can be quite difficult to obtain and the difficulty has led to the development of sophisticated techniques of system model identification. Engineers early on explored ways to avoid these requirements.

Callender et al. (1936) proposed a design for the widely used PID controller by specifying satisfactory values for the controller settings based on estimates of the plant parameters that an operating engineer could make from experiments on the process itself. The approach was extended by J. G. Ziegler and N. B. Nichols (1942, 1943), who recognized that the step responses of a large number of process control systems exhibit a **process reaction curve** like that shown in Fig. 4.13 which can be generated from experimental step response data. The S-shape of the curve is characteristic of many systems and can be approximated by the step response of

Transfer function for a high-order system with a characteristic process reaction curve

$$\frac{Y(s)}{U(s)} = \frac{Ae^{-st_{cl}}}{\tau s + 1}, \tag{4.52}$$

which is a first-order system with a time delay of t_d seconds. The constants in Eq. (4.52) can be determined from the unit step response of the process. If a tangent is drawn at the inflection point of the reaction curve, then the slope of the line is $R = A/\tau$ and the intersection of the tangent line with the time axis identifies the time delay $L = t_d$.

Figure 4.13
Process reaction curve

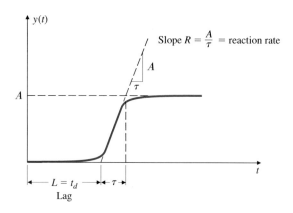

Ziegler and Nichols gave two methods for tuning the PID controller for

Tuning by decay ratio of 0.25

such a model. In the first method the choice of controller parameters is designed to result in a closed-loop step response transient with a decay ratio of approximately 0.25. This means that the transient decays to a quarter of its value after one period of oscillation, as shown in Fig. 4.14. A quarter decay corresponds to $\zeta = 0.21$ and is a reasonable compromise between quick response and adequate stability margins. The authors simulated the equations for the system on an analog computer and adjusted the controller parameters until the transients showed the decay of 25% in one period. The regulator parameters suggested by Ziegler and Nichols for the controller terms defined by

$$D_c(s) = k_p \left(1 + \frac{1}{T_I s} + T_D s \right) \tag{4.53}$$

are given in Table 4.1.

Figure 4.14
Quarter decay ratio

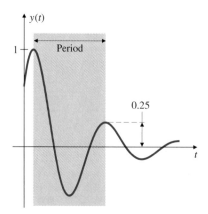

TABLE 4.1

Ziegler–Nichols Tuning for the Regulator
$D(s) = K(1 + 1/T_I s + T_D s)$, **for a Decay Ratio of 0.25**

Type of Controller	Optimum Gain
Proportional	$k_p = 1/RL$
PI	$\begin{cases} k_p = 0.9/RL, \\ T_I = L/0.3 \end{cases}$
PID	$\begin{cases} k_p = 1.2/RL, \\ T_I = 2L, \\ T_D = 0.5L \end{cases}$

Tuning by evaluation at limit of stability (ultimate sensitivity method)

In the **ultimate sensitivity method** the criteria for adjusting the parameters are based on evaluating the amplitude and frequency of the oscillations of the system at the limit of stability rather than on taking a step response. To use the method, the proportional gain is increased until the system becomes marginally stable and continuous oscillations just begin with amplitude limited by the saturation of the actuator. The corresponding gain is defined as K_u (called the **ultimate gain**) and the period of oscillation is P_u (called the **ultimate period**). These are determined as shown in Figs. 4.15 and 4.16. P_u should be measured when the amplitude of oscillation is as small as possible. Then the tuning parameters are selected as shown in Table 4.2.

TABLE 4.2

Ziegler–Nichols Tuning for the Regulator
$D_c(s) = k_p(1 + 1/T_I s + T_D s)$, **Based on the Ultimate Sensitivity Method**

Type of Controller	Optimum Gain
Proportional	$k_p = 0.5K_u$
PI	$\begin{cases} k_p = 0.45K_u, \\ T_I = \dfrac{P_u}{1.2} \end{cases}$
PID	$\begin{cases} k_p = 0.6K_u, \\ T_I = \dfrac{1}{2}P_u, \\ T_D = \dfrac{1}{8}P_u \end{cases}$

Figure 4.15
Determination of the ultimate gain and period

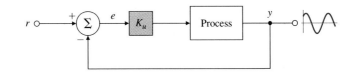

Figure 4.16
Neutrally stable system

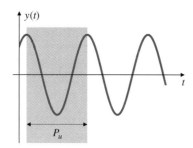

Experience has shown that the controller settings according to Ziegler–Nichols rules provide acceptable closed-loop response for many systems. The process operator will often do final tuning of the controller iteratively on the actual process to yield satisfactory control.[9]

EXAMPLE 4.7

Tuning of a Heat Exchanger: Quarter Decay Ratio

Consider the heat exchanger of Example 2.17. The process reaction curve of this system is shown in Fig. 4.17. Determine proportional and PI regulator gains for the system using the Zeigler–Nichols rules to achieve a quarter decay ratio. Plot the corresponding step responses.

Figure 4.17
A measured process
reaction curve

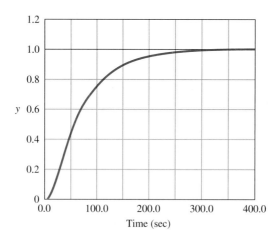

---

[9] Tuning of PID controllers has been the subject of continuing study since 1936. A recent publication on the topic is H. Panagopoulous, K. J. Astrom, and T. Hagglund, *Proceedings of the American Control Conference*, San Diego, CA, June 1999.

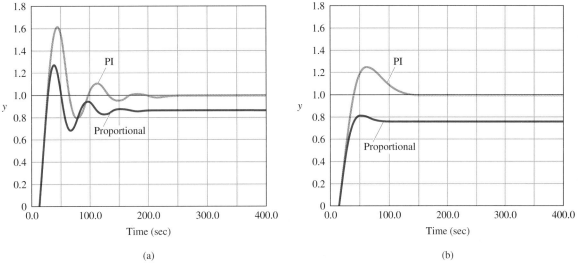

Figure 4.18 Closed-loop step responses

Solution.   From the process reaction curve we measure the maximum slope to be $R \cong \frac{1}{90}$ and the time delay to be $L \cong 13$ sec. According to the Zeigler–Nichols rules of Table 4.1 the gains are

$$\text{Proportional}: \quad k_p = \frac{1}{RL} = \frac{90}{13} = 6.92,$$

$$\text{PI}: \quad k_p = \frac{0.9}{RL} = 6.22 \quad \text{and} \quad T_I = \frac{L}{0.3} = \frac{13}{0.3} = 43.3.$$

Figure 4.18(a) shows the step responses of the closed-loop system to these two regulators. Note that the proportional regulator results in a steady state offset, while the PI regulator tracks the step exactly in the steady state. Both regulators are rather oscillatory and have considerable overshoot. If we arbitrarily reduce the gain k_p by a factor of 2 in each case, the overshoot and oscillatory behaviors are substantially reduced, as shown in Fig. 4.18(b).

EXAMPLE 4.8

Tuning of a Heat Exchanger: Oscillatory Behavior

Proportional feedback was applied to the heat exchanger in the previous example until the system showed nondecaying oscillations in response to a short pulse (impulse) input, as shown in Fig. 4.19. The ultimate gain was $K_u = 15.3$, and the period was measured at $P_u = 42$ sec. Determine the proportional and PI regulators according to the Zeigler–Nichols rules based on the ultimate sensitivity method. Plot the corresponding step responses.

Figure 4.19
Ultimate period of Heat
Exchanger

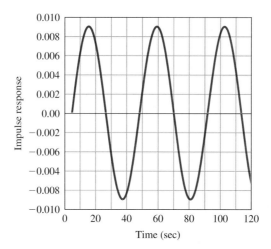

Solution.   The regulators from Table 4.2 are

$$\text{Proportional}: \quad k_p = 0.5, \quad K_u = 7.65,$$

$$\text{PI}: \quad k_p = 0.45 K_u = 6.885 \quad \text{and} \quad T_I = \frac{1}{1.2} P_u = 35.$$

The step responses of the closed-loop system are shown in Fig. 4.20(a). Note that
the responses are similar to those in Example 4.7. If we reduce k_p by 50%, then the
overshoot is substantially reduced, as shown in Fig. 4.20(b).

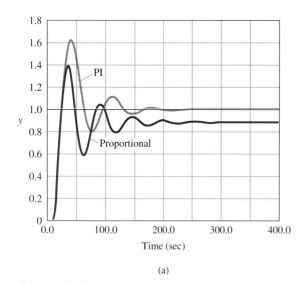

(a)

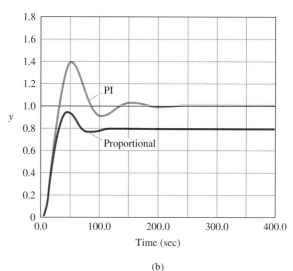

(b)

Figure 4.20 Closed-loop step response

▲ 4.2.5   Integrator Anti-Windup

In any control system the output of the actuator can saturate because the dynamic range of all real actuators is limited. For example, a valve saturates when it is fully open or closed, the control surfaces on an aircraft cannot be deflected beyond certain angles from their nominal positions, electronic amplifiers can only produce finite voltage outputs, and so on. Whenever actuator saturation happens, the control signal to the process stops changing and the feedback path is effectively opened. If the error signal continues to be applied to the integrator input under these conditions, the integrator stored value will grow (wind up) until the sign of the error changes and the integration turns around. The result can be a very large overshoot because the output must grow to produce the necessary unwinding error and poor transient response is the result. In effect, the integrator is an unstable element in open loop and must be stabilized when saturation occurs.[10]

Consider the feedback system shown in Fig. 4.21. Suppose a given reference step is more than large enough to cause the actuator to saturate at u_{max}. The integrator continues integrating the error e, and the signal u_c keeps growing. However, the input to the plant is stuck at its maximum value, namely, $u = u_{max}$, so the error remains large until the plant output exceeds the reference and the error changes sign. The increase in u_c is not helpful because the input to the plant is not changing but u_c may become quite large if saturation lasts a long time. It will then take a considerable negative error e and the resulting poor transient response to bring the integrator output back to within the linear band where the control is not saturated.

The solution to this problem is an **integrator anti-windup** circuit, which "turns off" the integral action when the actuator saturates. (This can be done quite easily with logic if the controller is implemented digitally by including a statement such as "if $|u| = u_{max}$, $k_I = 0$"; see Chapter 8.) Two equivalent anti-windup schemes are shown in Fig. 4.22(a, b) for a PI controller. The method in Fig. 4.22(a) is somewhat easier to understand, whereas the one in Fig. 4.22(b) is easier to implement because it does not require a separate nonlinearity but

Figure 4.21

Feedback system with actuator saturation

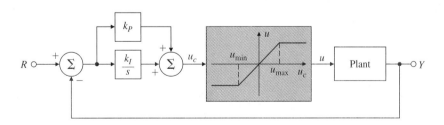

---

[10] In process control, integral control is usually called **reset control**, and so integrator windup is often called **reset windup**. Without integral control, a given setpoint of, say, 10 results in a response of less value, say 9.9. The operator must then *reset* to 10.1 to bring the output to the desired value of 10. With integral control the controller automatically brings the output to 10 with a setpoint of 10; hence the integrator does *automatic reset*.

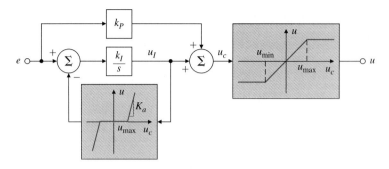

(a) PI controller with anti-windup.

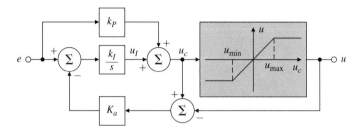

(b) Implementation of anti-windup with a single non-linearity

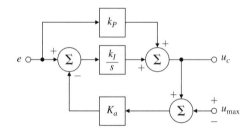

(c) Equivalent block diagram of integrator loop during saturation.

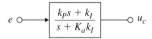

(d) First order lag equivalent of an anti-windup integrator during saturation.

Figure 4.22 Integrator anti-windup techniques: (a) PI controller with anti-windup; (b) implementation of antiwindup with a single nonlinearity; (c) equivalent block diagram during saturation; (d) first-order lag equivalent of an antiwindup integrator during saturation

uses the saturation itself.[11] In these schemes, as soon as the actuator saturates, the feedback loop around the integrator becomes active and acts to keep the input to the integrator at e_1 small. During this time the integrator essentially becomes a fast first-order lag. To see this, note that we can redraw the portion of the block diagram in Fig. 4.22(a) from e to u_c as shown in Fig. 4.22(c). The integrator part then becomes the first-order lag shown in Fig. 4.22(d). The anti-windup gain, K_a, should be chosen to be large enough that the anti-windup circuit keeps the input to the integrator small under all error conditions.

The effect of the anti-windup is to reduce both the overshoot and the control effort in the feedback system. Implementation of such anti-windup schemes is a necessity in any practical application of integral control, and omission of this technique may lead to serious deterioration of the response. From the point of view of stability, the effect of the saturation is to open the feedback loop and leave the open-loop plant with a constant input and leave the controller as an open-loop system with the system error as input.

Purpose of anti-windup

> The purpose of the anti-windup is provide local feedback to make the controller stable alone when the main loop is opened by signal saturation, and any circuit which does this will perform as anti-windup.[12]

EXAMPLE 4.9   *Anti-Windup Compensation for a PI Controller*

Consider a plant with the following transfer function for small signals,

$$G(s) = \frac{1}{s},$$

and a PI controller,

$$D_c(s) = k_p + \frac{k_I}{s} = 2 + \frac{4}{s},$$

in the unity feedback configuration. The input to the plant is limited to ± 1.0. Study the effect of anti-windup on the response of the system.

---

[11] In some cases, especially with mechanical actuators such as an aircraft control surface or a flow control valve it is not desirable and may cause damage to have the physical device bang against its stops. In such cases it is common practice to include an *electronic* saturation with lower limits than the physical device so the system hits the electrical stops just before the physical device would saturate.

[12] A more sophisticated scheme might use an anti-windup feedback at a lower level of saturation than that imposed by the actuator so PD control continues for a time after integration has been stopped. Any such scheme needs to be analyzed carefully to evaluate its performance and to assure stability.

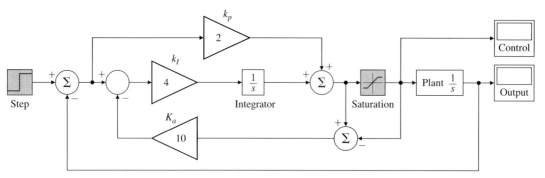

Figure 4.23 Antiwindup example

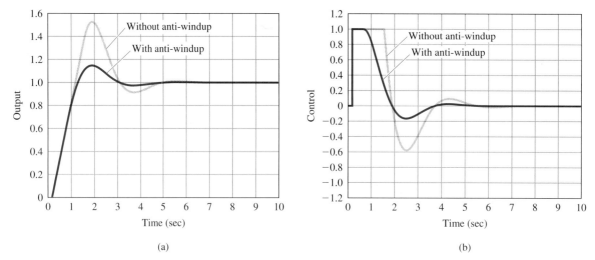

(a)   (b)

Figure 4.24 Integrator antiwindup: (a) step response; (b) control effort

Solution.   Suppose we use an anti-windup circuit with a feedback gain of $K_a = 10$, as shown in the Simulink block diagram of Fig. 4.23. Figure 4.24(a) shows the step response of the system with and without the anti-windup element. Figure 4.24(b) shows the corresponding control effort. Note that the system with anti-windup has substantially less overshoot and less control effort.

4.3   Steady-State Tracking and System Type

In the speed-control case study in Section 4.1 we considered constant reference inputs and constant disturbances. We found in Section 4.2 that for a system with such signals, integral control could keep the steady-state error at zero

Figure 4.25
Signal for satellite tracking

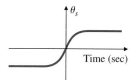

even when the motor gain differed from the one used in the design. In a number of important cases the reference input will not be constant but can be approximated as a polynomial in time long enough for the system to effectively reach steady state. For example, when an antenna is tracking the elevation angle to a satellite, the time history as the satellite approaches overhead is an S-shaped curve as sketched in Fig. 4.25. This signal may be approximated by a linear function of time (called a ramp function or velocity input) for a significant time relative to the speed of response of the servomechanism. In the position control of an elevator, a ramp function reference input will direct the elevator to move with constant speed until it comes near the next floor. In rare cases, the input can be approximated over a substantial period as having a constant acceleration. In this section we consider steady-state errors in stable systems with such polynomial inputs.

   The general method is to represent the input as a polynomial in time and to consider the resulting steady-state tracking errors for polynomials of different degrees. As we will see, the error will be zero for input polynomials below a certain degree and unbounded for inputs of higher degrees. A stable system

Definition of system type

can be classified as a **system type** *defined to be the degree of the polynomial for which the steady-state system error is a nonzero finite constant.* In the speed control example proportional control was used and the system had a constant finite error to a step input, which is an input polynomial of zero degree and therefore this system is called a type zero (type 0). With integral control, the error to a step input was zero and we will see that the error to a ramp or first-degree polynomial is a finite nonzero constant. Such a system is called type one (type 1), and so on. System types can be defined with regard to either reference inputs or disturbance inputs, and in this section we will consider both classifications. Determining the system type involves calculating the transform of the system error and then applying the Final Value Theorem. As we will see, determination of system type is easiest for the case of unity feedback and we will begin with that case, but first it is useful to develop the equations for the basic feedback block diagrams for future reference.

4.3.1   The Equations of Feedback

Considering the advantages of feedback seen in the case study of motor speed control in Section 4.1, the remainder of our attention will focus on the feedback structure for control. We begin by collecting the basic equations and transfer functions that will be used throughout the rest of the text. Figure 4.2 gives the basic structure of interest but with the disturbance and the sensor noise entering

Figure 4.26

Basic feedback control block diagram

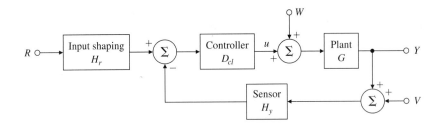

in unspecific ways. For analysis it is necessary to be more specific, and for the purpose we will take these signals to be at the inputs of the process and the sensor, respectively, as shown in Fig. 4.26. The sensor transfer function is H_y and may show important dynamics. However, the sensor can often be selected to be fast and accurate and with an electrical output so the signals may be easily manipulated in the controller. If this is the case, its transfer function can be taken to be a constant H_y with units of *volts/unit-of-output*. The reference input, r, has the same units as the output, of course, and the input shaping element's transfer function is H_r also with units of *volts/unit-of-output*. An equivalent block diagram is drawn in Fig. 4.27 with controller transfer function $D(s) = H_r D_{cl}$ and with the feedback transfer function as the ratio $H = H_y/H_r$. It is standard practice, especially if H_y is constant, to select equal scale factors so that $H_r = H_y$, and the block diagram can be drawn as a unity feedback structure as shown in Fig. 4.28. We will develop the equations and transfer functions for this standard structure. When we use these equations it will be important to be sure that the assumptions given above actually apply. If the sensor has dynamics which cannot be ignored, for example, then the equations will need to be modified accordingly.

For the block diagram of Fig. 4.28, the equations for the output and the control are

$$Y = \frac{DG}{1 + DG} R + \frac{G}{1 + DG} W - \frac{DG}{1 + DG} V \qquad (4.54)$$

$$U = \frac{D}{1 + DG} R - \frac{DG}{1 + DG} W - \frac{D}{1 + DG} V \qquad (4.55)$$

Figure 4.27

Equivalent feedback block diagram with H_r included inside the loop

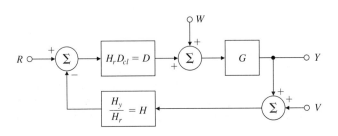

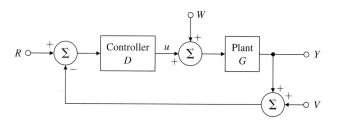

Figure 4.28
Unity feedback system when $H_r = H_y$ and letting $D = H_r D_{ol}$

Perhaps more important than these is the equation of the error, $E = R - Y$.

$$E = R - \left[\frac{DG}{1+DG} R + \frac{G}{1+DG} W - \frac{DG}{1+DG} V \right] \tag{4.56}$$

$$= \frac{1}{1+DG} R - \frac{G}{1+DG} W + \frac{DG}{1+DG} V. \tag{4.57}$$

This equation is simplified by the definition of the sensitivity $\mathcal{S}$ and the complementary sensitivity $\mathcal{T}$ as

$$\mathcal{S} = \frac{1}{1+DG}, \tag{4.58}$$

$$\mathcal{T} = 1 - \mathcal{S} = \frac{DG}{1+DG}. \tag{4.59}$$

In terms of these, the error is given by

$$E = \mathcal{S}R - \mathcal{S}GW + \mathcal{T}V. \tag{4.60}$$

A major goal of control is to keep the error small for any input and in the face of parameter changes. Although we will focus on the selection of the controller transfer function, $D_c(s)$, the control engineer must be aware that changes to the plant may be possible which will greatly help the process. It is also possible that the selection and location of the sensor can be very helpful. Given a fixed plant described by G, Eq. (4.60) would seem to require that both $\mathcal{S}$ and $\mathcal{T}$ should be kept small. However, *Eq. (4.59) shows that this is not possible.* The most common approach here is to recognize that the reference and the disturbance are not large for all frequencies but are typically restricted in content to a band of frequencies below some limit. On the other hand, the sensor can be selected so the sensor noise, V, is held small in the (low-frequency) band where R and W are substantial. If these assumptions hold, then it is sufficient if the controller keeps $\mathcal{S}$ small where R and W are large and arranges to have $\mathcal{T}$ be small (and $\mathcal{S}$ necessarily larger) for higher frequencies where sensor noise is unavoidable. It is these assumptions and compromises that will occupy most of our attention in the design of controllers. However, we first consider the simpler case of the steady-state error to polynomial inputs.

4.3.2   System Type for Reference Tracking: The Unity Feedback Case

In the unity feedback case drawn in Fig. 4.28 the system error is given by Eq. (4.57). If we consider only the reference input alone and set $W = V = 0$, then the equation is simply

$$E = \frac{1}{1 + DG} R = \mathcal{S} R \tag{4.61}$$

To consider polynomial inputs, we let $r(t) = t^k 1(t)$ for which the transform is $R = 1/s^{k+1}$. As a generic reference nomenclature, step inputs for which $k = 0$ are called "position" inputs, ramp inputs for which $k = 1$ are called "velocity" inputs and if $k = 2$ the inputs are called "acceleration" inputs, regardless of the units of the actual signals. Application of the final value theorem to the error gives the formula

$$\lim_{t \to \infty} e(t) = e_{ss} = \lim_{s \to 0} E(s) \tag{4.62}$$

$$= \lim_{s \to 0} s \frac{1}{1 + DG} R(s) \tag{4.63}$$

$$= \lim_{s \to 0} s \frac{1}{1 + DG} \frac{1}{s^{k+1}}. \tag{4.64}$$

We consider first a system for which DG has no pole at the origin and a step input for which $R(s) = 1/s$. In this case, Eq. (4.64) reduces to

$$e_{ss} = \lim_{s \to 0} s \frac{1}{1 + DG} \frac{1}{s} \tag{4.65}$$

$$= \frac{1}{1 + DG(0)}. \tag{4.66}$$

We define the system to be *type 0* and we define the constant, $DG(0) \stackrel{\triangle}{=} K_p$, as the "position error constant." If DG has one pole at the origin, we could consider both steps and ramp inputs, but it is quite straightforward to evaluate Eq. (4.64) in a general setting. For this case, it is useful to be able to describe the behavior of the controller and plant as s approaches 0. For this purpose, we collect all the terms except the pole(s) at the origin into a function $G_o(s)$ which is thus finite at $s = 0$ so that we can define the constant $G_o(0) = K_n$ and write the loop transfer function as

$$DG = \frac{G_o(s)}{s^n}. \tag{4.67}$$

For example, if DG has no integrator, then $n = 0$. If the system has one integral, then $n = 1$, and so forth. Substituting this expression into Eq. (4.64), we obtain

$$e_{ss} = \lim_{s \to 0} s \frac{1}{1 + \dfrac{G_o(s)}{s^n}} \frac{1}{s^{k+1}} \tag{4.68}$$

$$= \lim_{s \to 0} \frac{s^n}{s^n + K_n} \frac{1}{s^k}. \tag{4.69}$$

From this equation we can see at once that if $n > k$, then $e = 0$ and if $n < k$, then $e \to \infty$. If $n = k = 0$, then $e_{ss} = 1/(1 + K_0)$ and if $n = k \neq 0$, then $e_{ss} = 1/K_n$. If $n = k = 0$, the input is a zero-degree polynomial otherwise known as a step or position, the constant K_o is called the "position constant" written as K_p, and the system is classified as "type 0" as we saw above. If $n = k = 1$, the input is a first-degree polynomial otherwise known as a ramp or velocity, the constant K_1 is called the "velocity constant" written as K_v, and the system is classified "type 1." In a similar way, systems of type 2 and higher types may be defined. The type information can be usefully gathered in a table of errors as shown in Table 4.3.

The most common case is that of simple integral control leading to a type 1 system. In this case, the relationship between K_v and the steady-state error to a ramp input is shown in Fig. 4.29. Looking back at the expression given for $D_c G$ in Eq. (4.67), we can readily see that the several error constants can be calculated by counting the degree, n, of the poles of DG at the origin (the number of integrators in the loop with unity gain feedback) and applying the appropriate one of the following simple formulas:

Type II systems

$$K_p = \lim_{s \to 0} DG(s), \qquad n = 0, \tag{4.70}$$

$$K_v = \lim_{s \to 0} s\,DG(s), \qquad n = 1, \tag{4.71}$$

$$K_a = \lim_{s \to 0} s^2 DG(s), \qquad n = 2.$$

TABLE 4.3   **Errors as a Function of System Type**

Type	Input		
	Step (Position)	Ramp (Velocity)	Parabola (Acceleration)
Type 0	$\dfrac{1}{1 + K_p}$	∞	∞
Type 1	0	$\dfrac{1}{K_v}$	∞
Type 2	0	0	$\dfrac{1}{K_a}$

Figure 4.29

Relationship between ramp response and K_v

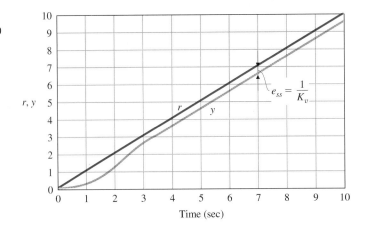

EXAMPLE 4.10

System Type with Proportional Control

Determine the system type and the relevant error constant for the speed control example shown in Fig. 4.4 with proportional feedback given by $D_{cl} = k_p$. The plant transfer function is $G = A/\tau s + 1$.

Solution.   In this case, $D_c G = k_p A/\tau s + 1$ and applying Eq. (4.70) we see that $n = 0$ in this case because there is no pole at $s = 0$. Thus the system is type 0 and the error constant is a position constant given by $K_p = k_p A$.

EXAMPLE 4.11

System Type with PI Control

Determine the system type and the relevant error constant for the speed control example shown in Fig. 4.4 with PI feedback. The plant transfer function is $G = A/(\tau s + 1)$, and in this case the controller transfer function is $D_c = k_p + k_I/s$.

Solution.   In this case, the transfer function $D_c G = A (k_p s + k_I)/s(\tau s + 1)$, and as a unity feedback system with a single pole at $s = 0$ the system is immediately seen as type 1. The velocity constant is given by Eq. (4.71) to be $K_v = \lim_{s \to 0} s D_c G = A k_I$.

The definition of system type helps us to identify quickly the ability of a system to track polynomials. In the unity feedback structure, if the process parameters change without removing the pole at the origin in a type 1 system, the velocity constant will change but the system will still have zero steady-state error in response to a constant input and will still be type 1. Similar statements can be made for systems of type 2 or higher. Thus we can say that system type is a **robust property** with respect to parameter changes in the unity feedback structure. Robustness is the major reason for preferring unity feedback over other kinds of control structure.

Robustness of system type

4.3.3   System Type for Reference Tracking: The General Case

If the feedback $H = H_y/H_r$ in Fig. 4.27 is different from unity, the formulas given in the unity feedback case do not apply and a more general approach is needed. There are two immediate possibilities. In the first instance, if one adds and subtracts 1.0 from H as shown by block diagram manipulation in Fig. 4.30, the general case is reduced to the unity feedback case and the formulas can be applied to the redefined transfer function $L = DG/1 + (H-1)DG$ for which the error equation is $E = [1/(1+L)]R$.

Another possibility is to develop formulas directly in terms of the closed-loop transfer function which is the complementary sensitivity function, $\mathcal{T}(s)$. From Fig. 4.27 the transfer function is

$$\frac{Y(s)}{R(s)} = \mathcal{T}(s) = \frac{DG}{1 + HDG},\tag{4.72}$$

and therefore the error is

$$E(s) = R(s) - Y(s) = R(s) - \mathcal{T}(s)R(s).$$

The reference-to-error transfer function is thus

$$\frac{E(s)}{R(s)} = 1 - \mathcal{T}(s),$$

and the system error transform is

$$E(s) = [1 - \mathcal{T}(s)]R(s) = \mathcal{S}R.$$

We assume that the conditions of the Final Value Theorem are satisfied—that is, that all poles of $sE(s)$ are in the left half-plane. In that case the steady-state error is given by applying the Final Value Theorem to get

$$e_{ss} = \lim_{t\to\infty} e(t) = \lim_{s\to 0} sE(s) = \lim_{s\to 0} s[1 - \mathcal{T}(s)]R(s).\tag{4.73}$$

With a polynomial test input the error transform becomes

$$E(s) = \frac{1}{s^{k+1}}[1 - \mathcal{T}(s)],$$

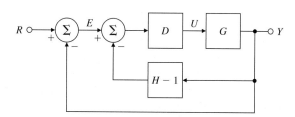

Figure 4.30
Block diagram reduction to an equivalent unity feedback system

and the steady-state error is given again by the Final Value Theorem:

$$e_{ss} = \lim_{s \to 0} s \frac{1 - \mathcal{T}(s)}{s^{k+1}} = \lim_{s \to 0} \frac{1 - \mathcal{T}(s)}{s^k}. \tag{4.74}$$

The result of evaluating the limit in Eq. (4.74) can be zero, a nonzero constant, or infinite. If the solution to Eq. (4.74) is a nonzero constant, the system is referred to as type k. For example, if $k = 0$ and the solution to Eq. (4.74) is a nonzero constant equal, *by definition*, to $1/(1 + K_p)$, then the system is type 0. Similarly, if $k = 1$ and the solution to Eq. (4.74) is a nonzero constant, then the system is type 1 and has a zero steady-state error to a position input and a constant steady-state error equal *by definition* to $1/K_v$ to a unit velocity reference input. Type 1 systems are by far the most common in practice. A system of type 1 or higher has a closed-loop DC gain of 1.0, which means that $\mathcal{T}(0) = 1$.

EXAMPLE 4.12   *System Type for a Servo with Tachometer Feedback*

Consider the electric motor position control problem including a non-unity feedback system caused by having a tachometer fixed to the motor shaft and its voltage (which is proportional to shaft speed) is fed back as part of the control. The parameters corresponding to Fig. 4.27 are

$$G(s) = \frac{1}{s(\tau s + 1)},$$

$$D(s) = k_p,$$

$$H(s) = 1 + k_t s.$$

Determine the system type and relevant error constant with respect to reference inputs.

Solution.   The system error is

$$E(s) = R(s) - Y(s)$$

$$= R(s) - \mathcal{T}(s)R(s)$$

$$= R(s) - \frac{DG(s)}{1 + HDG(s)} R(s)$$

$$= \frac{1 + (H(s) - 1)DG(s)}{1 + HDG(s)} R(s).$$

The steady-state system error from Eq. (4.74) is

$$e_{ss} = \lim_{s \to 0} s R(s)[1 - \mathcal{T}(s)].$$

For a polynomial reference input, $R(s) = 1/s^{k+1}$ and hence

$$e_{ss} = \lim_{s \to 0} \frac{[1 - \mathcal{T}(s)]}{s^k} = \lim_{s \to 0} \frac{1}{s^k} \frac{s(\tau s + 1) + (1 + k_t s - 1)k_p}{s(\tau s + 1) + (1 + k_t s)k_p}$$

$$= 0, \qquad k = 0$$

$$= \frac{1 + k_t k_p}{k_p}, \qquad k = 1;$$

therefore the system is type 1 and the velocity constant is $K_v = k_p/(1 + k_t k_p)$. Notice that if $k_t > 0$, this velocity constant is smaller than the unity feedback value of k_p.

4.3.4   System Type with Respect to Disturbance Inputs

In most control systems, disturbances of one type or another exist. In practice, these disturbances can sometimes be usefully approximated by polynomial time functions such as steps or ramps. This would suggest that system type can also be defined with respect to the system's ability to reject disturbance inputs in a way analogous to the classification scheme based on reference inputs. System type with regard to disturbance inputs specifies the degree of the polynomial expressing those input disturbances that the system can reject in the steady state. Knowing the system type, we know the qualitative steady-state response of the system to polynomial disturbance inputs such as step or ramp signals. The system type depends on exactly where the disturbance enters into the control system.

To determine the system type with respect to disturbances, we form the transfer function from the disturbance input $W(s)$ to the error $E(s)$ as

$$\frac{E(s)}{W(s)} = \frac{-Y(s)}{W(s)} = T_w(s). \tag{4.75}$$

Equation (4.75) describes the system error resulting from a disturbance input because with the reference equal to zero, the output is the error. In a similar way as for reference inputs, the system is type 0 if a step disturbance input results in a nonzero constant steady-state error and is type 1 if a ramp disturbance input results in a steady-state value of the error that is a nonzero constant. In general, following the same approach used in developing Eq. (4.69), we assume that a constant n and a function $T_{o,w}(s)$ can be defined with the properties that $T_{o,w}(0) = 1/K_{n,w}$ and that the disturbance-to-error transfer function can be written as

$$T_w(s) = s^n T_{o,w}(s). \tag{4.76}$$

Then the steady-state error to a disturbance input which is a polynomial of degree k is

$$y_{ss} = \lim_{s \to 0} \left[s T_w(s) \frac{1}{s^{k+1}} \right]$$

$$= \lim_{s \to 0} \left[T_{o,w}(s) \frac{s^n}{s^k} \right]. \tag{4.77}$$

From Eq. (4.77), if $n > k$, then the error is zero and if $n < k$, the error is unbounded. If $n = k$, the system is type k and the error is given by $1/K_{n,w}$.

EXAMPLE 4.13   *System Type for a DC Motor Position Control*

Consider the simplified model of a DC motor in unity feedback as shown in Fig. 4.31 where the disturbance torque is labeled $W(s)$.

(a) Use the proportional controller

$$D(s) = k_p, \qquad (4.78)$$

and determine the system type and steady-state error properties with respect to disturbance inputs.

(b) Let the control be PI as given by

$$D(s) = k_p + \frac{k_I}{s} \qquad (4.79)$$

and determine the system type and the steady-state error properties for disturbance inputs.

Solution.

(a) The closed-loop transfer function from W to E (where $R = 0$) is

$$T_w(s) = \frac{-B}{s(\tau s + 1) + A k_p}$$
$$= s^0 T_{o,w},$$
$$n = 0,$$
$$K_{o,w} = \frac{-A k_p}{B}.$$

Applying Eq. (4.77), we see that the system is **type 0** and the steady-state error to a unit step torque input is $e_{ss} = -B/A k_p$. From the earlier section, this system is seen to be type 1 for reference inputs and illustrates that system type can be different for different inputs to the same system.

Figure 4.31
DC motor with unity feedback

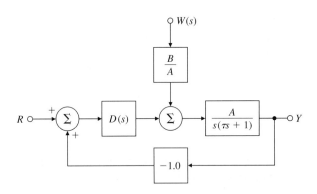

(b) If the controller is PI, the disturbance error transfer function is

$$T_w(s) = \frac{-Bs}{s^2(\tau s + 1) + (k_p s + k_I)A},$$ (4.80)

$$n = 1,$$ (4.81)

$$K_{n,w} = \frac{Ak_I}{-B},$$ (4.82)

and therefore the system is **type 1** and the error to a unit ramp disturbance input will be

$$e_{ss} = \frac{-B}{Ak_I}.$$ (4.83)

EXAMPLE 4.14   *Satellite Attitude Control*

Consider the model of a satellite attitude control system shown in Fig. 4.32(a) where

$$J = \text{moment of inertia,}$$

$$W = \text{disturbance torque,}$$

$$H_y = \text{sensor gain,}$$

$$D_c(s) = \text{the compensator.}$$

With equal input filter and sensor scale factors, the system with PD control can be redrawn with unity feedback as in Fig. (4.32)(b) and with PID control drawn as in Fig. 4.32(c). Assume that the control results in a stable system and determine the system types and error responses to disturbances of the control system for

(a) System Fig. 4.32(b) PD control

(b) System Fig. 4.32(c) PID control

Solution.

(a) We see from inspection of Fig. 4.32(b) that the system is type 2 with respect to reference inputs. The transfer function from disturbance to error is

$$T_w(s) = \frac{1}{Js^2 + k_D s + k_p}$$ (4.84)

$$= T_{o,w}(s),$$ (4.85)

for which $n = 0$ and $K_{o,w} = k_p$. The system is **type 0** and the error constant is k_p so the error to a unit step is $1/k_p$.

Figure 4.32

Model of a satellite attitude control (a) Basic system (b) PD control; (c) PID control

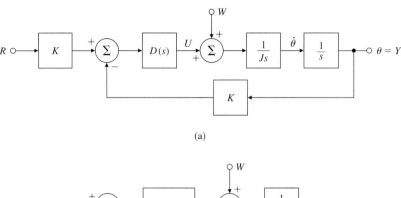

(a)

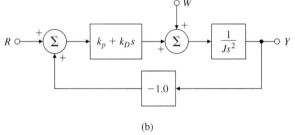

(b)

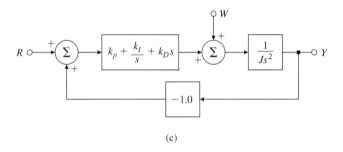

(c)

(b) With PID control, the forward loop has three poles at the origin, so this system is type 3 for reference inputs but the disturbance transfer function is

$$T_w(s) = \frac{s}{Js^3 + k_D s^2 + k_p s + k_I}, \tag{4.86}$$

$$n = 1, \tag{4.87}$$

$$T_{o,w}(s) = \frac{1}{Js^3 + k_D s^2 + k_p s + k_I}, \tag{4.88}$$

from which the system is **type 1** and the error constant is k_I so the error to a disturbance ramp of unit slope will be $1/k_I$.

▲ 4.3.5   Truxal's Formula

In this chapter we have derived formulas for the error constants in terms of the system transfer function. The most common case is the type 1 system whose error constant is K_v, the velocity error constant. Truxal (1955) derived a formula for the velocity constant in terms of the closed-loop poles and zeros, a formula that connects the steady-state error to the dynamic response. Because control design often requires a trade-off between these two characteristics, Truxal's formula can be useful to know. Its derivation is quite direct. Suppose the closed-loop transfer function $\mathcal{T}(s)$ of a type 1 system is

$$\mathcal{T}(s) = K \frac{(s - z_1)(s - z_2) \cdots (s - z_m)}{(s - p_1)(s - p_2) \cdots (s - p_n)}. \tag{4.89}$$

Because the steady-state error in response to a step input in a type 1 system is zero, the DC gain is unity; thus

$$\mathcal{T}(0) = 1. \tag{4.90}$$

The system error is given by

$$E(s) \triangleq R(s) - Y(s) = R(s)\left[1 - \frac{Y(s)}{R(s)}\right] = R(s)[1 - \mathcal{T}(s)]. \tag{4.91}$$

The system error due to a unit ramp input is given by

$$E(s) = \frac{1 - \mathcal{T}(s)}{s^2}. \tag{4.92}$$

Using the Final Value Theorem, we get

$$e_{ss} = \lim_{s \to 0} \frac{1 - \mathcal{T}(s)}{s}. \tag{4.93}$$

Using L'Hôpital's rule we rewrite Eq. (4.93) as

$$e_{ss} = -\lim_{s \to 0} \frac{d\mathcal{T}}{ds} \tag{4.94}$$

or

$$e_{ss} = -\lim_{s \to 0} \frac{d\mathcal{T}}{ds} = \frac{1}{K_v}. \tag{4.95}$$

Equation (4.95) implies that $1/K_v$ is related to the slope of the transfer function at the origin, a result that will also be shown in Section 6.1.2. Using Eq. (4.90), we can rewrite Eq. (4.95) as

$$e_{ss} = -\lim_{s \to 0} \frac{d\mathcal{T}}{ds} \frac{1}{\mathcal{T}} \tag{4.96}$$

or

$$e_{ss} = -\lim_{s \to 0} \frac{d}{ds} [\ln \mathcal{T}(s)]. \tag{4.97}$$

Substituting Eq. (4.89) into Eq. (4.97), we get

$$e_{ss} = -\lim_{s \to 0} \frac{d}{ds} \left\{ \ln \left[K \frac{\prod_{i=1}^{m}(s - z_i)}{\prod_{i=1}^{n}(s - p_i)} \right] \right\} \tag{4.98}$$

$$= -\lim_{s \to 0} \frac{d}{ds} \left[K + \sum_{i=1}^{m} \ln(s - z_i) - \sum_{i=1}^{m} \ln(s - p_i) \right], \tag{4.99}$$

or

$$\frac{1}{K_v} = -\frac{d \ln \mathcal{T}}{ds} \bigg|_{s=0} = \sum_{i=1}^{n} -\frac{1}{p_i} + \sum_{i=1}^{m} \frac{1}{z_i}. \tag{4.100}$$

Truxal's formula

We observe from Eq. (4.100) that K_v increases as the closed-loop poles move away from the origin. Similar relationships exist for other error coefficients, and these are explored in the problems.

EXAMPLE 4.15   *Truxal's Formula*

A third-order type 1 system has closed-loop poles at $-2 \pm 2j$ and -0.1. The system has only one closed-loop zero. Where should the zero be if a $K_v = 10$ is desired?

Solution.   From Truxal's formula we have

$$\frac{1}{K_v} = -\frac{1}{-2 + 2j} - \frac{1}{-2 - 2j} - \frac{1}{-0.1} + \frac{1}{z}$$

or

$$0.1 = 0.5 + 10 + \frac{1}{z}.$$

Therefore, the closed-loop zero should be at $z = -0.1$.

4.4   Digital Implementation of Controllers

As a result of the revolution in the cost-effectiveness of digital computers there has been an increasing use of digital logic in embedded applications such as controllers in feedback systems. With the formula for calculating the control signal in software rather than hardware, a digital controller gives the designer much more flexibility in making modifications to the control law after the hardware design is fixed. In many instances, this means that the hardware and software designs can proceed almost independently, saving a great deal of time. Also, it is easy to include binary logic and nonlinear operations as part of the function of a digital controller. Special processors designed for real-time signal processing and known as digital signal processors (DSPs) are particularly well-suited for use as real-time controllers. While in general the design of systems to use a digital processor requires sophisticated use of new concepts to be introduced in Chapter 8 such as the $z-transform$ it is quite straight forward to translate a linear continuous analog design into a discrete equivalent. A digital controller differs from an analog controller in that the signals must be **sampled** and **quantized**.[13] A signal to be used in digital logic needs to be sampled first and then the samples need to be converted by an analog-to-digital converter or A/D[14] into a quantized digital number. Once the digital computer has calculated the proper next control signal value, this value needs to be converted back into a voltage and held constant or otherwise extrapolated by a digital-to-analog converter or D/A in order to be applied to the actuator of the process. The control signal is not changed until the next sampling period. As a result of the sampling, there are more strict limits on the speed or bandwidth of a digital controller than on analog devices. Discrete design methods are described in Chapter 8 which tend to minimize these limitations. A reasonable rule of thumb for selecting the sampling period is that during the rise time of the response to a step, the input to the discrete controller should be sampled approximately 6 times. By adjusting the controller for the effects of sampling, the sample period can be as large as 2 to 3 times per rise time. This corresponds to a sampling frequency that is 10 to 20 times the system's closed-loop bandwidth. The quantization of the controller signals introduces an equivalent extra noise into the system and to keep this interference at an acceptable level, the A/D converter usually has an accuracy of 10 to 12 bits. For a first analysis, the effects of the quantization are usually ignored. A simplified block diagram of a system with a digital controller is shown in Fig. 4.33.

For this introduction to digital control, we will describe a simplified technique for finding a discrete (sampled but not quantized) equivalent to a given continuous controller. The method depends on the sampling period, T_s, being

---

[13] A controller that operates on signals that are sampled but *not* quantized is called **discrete**, while one that operates on signals that are both sampled and quantized is called **digital**.

[14] Pronounced "A to D."

Figure 4.33
Block diagram of a digital controller

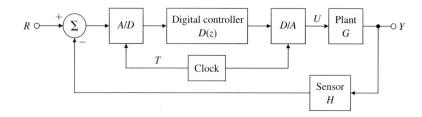

short enough that the reconstructed control signal is close to the signal that the original analog controller would have produced. We also assume that the numbers used in the digital logic have enough accurate bits so that the quantization implied in the A/D and D/A processes can be ignored. While there are good analysis tools to determine how well these requirements are met, here we will test our results by simulation, following the well-known advice that "The proof of the pudding is in the eating."

Finding a discrete equivalent to a given analog controller is equivalent to finding a recurrence equation for the samples of the control which will approximate the differential equation of the analog device as described in Section 3.7.2. The assumption is that we have the transfer function of an analog controller and wish to replace it with a discrete controller that will accept samples of the controller input, $e(kT_s)$, from a sampler and, using past values of the control signal, $u(kT_s)$, and present and past samples of the input, will compute the next control signal to be sent to the actuator. As an example, consider a PID controller with the transfer function

$$U(s) = \left(k_p + \frac{k_I}{s} + k_D s \right) E(s), \tag{4.101}$$

which is equivalent to the three terms of the time-domain expression

$$u(t) = k_p e(t) + k_I \int_0^t e(\tau)\, d\tau + k_D\, \dot{e}(t) \tag{4.102}$$

$$= u_P + u_I + u_D. \tag{4.103}$$

Based on these terms and the fact that the system is linear, the next control sample can be computed term-by-term. The proportional term is immediate:

$$u_P(kT_s + T_s) = k_p e(kT_s + T_s) \tag{4.104}$$

The integral term can be computed by breaking the integral into two parts and approximating the second part, which is the integral over one sample period, as follows:

$$u_I(kT_s + T_s) = k_I \int_0^{kT_s+T_s} e(\tau)\,d\tau \qquad (4.105)$$

$$= k_I \int_0^{kT_s} e(\tau)\,d\tau + k_I \int_{kT_s}^{kT_s+T_s} e(\tau)\,d\tau \qquad (4.106)$$

$$= u_I(kT_s) + \{\text{area under } e(\tau) \text{ over one period}\} \qquad (4.107)$$

$$\cong u_I(kT_s) + k_I \frac{T_s}{2}\{e(kT_s + T_s) + e(kT_s)\}. \qquad (4.108)$$

In Eq. (4.108) the area in question has been approximated by that of the trapezoid formed by the base T_s and vertices $e(kT_s + T_s)$ and $e(kT_s)$ as discussed in Section 3.7.1 and shown in Fig. 3.43. The area can also be approximated by the rectangle of amplitude $e(kT_s)$ and width T_s to give $u_I(kT_s + T_s) = u_I(kT_s) + k_I T_s e(kT_s)$. These and other possibilities are considered in Chapter 8.

In the derivative term, the roles of u and e are reversed from integration and the consistent approximation can be written down at once from Eq. (4.108) and Eq. (4.102) as

$$\frac{T_s}{2}\{u_D(kT_s + T_s) + u_D(kT_s)\} = k_D\{e(kT_s + T_s) - e(kT_s)\} \qquad (4.109)$$

As with linear analog transfer functions, these relations are greatly simplified and generalized by the use of transform ideas. At this time, the discrete transform will be introduced simply as a prediction operator z much as if we described the Laplace Transform variable, s, as a differential operator. Here we define the operator z as the forward shift operator in the sense that if $U(z)$ is the transform of $u(kT_s)$ then $zU(z)$ will be the transform of $u(kT_s + T_s)$. With this definition, the integral term can be written as

$$zU_I(z) = U_I(z) + k_I \frac{T_s}{2}[zE(z) + E(z)], \qquad (4.110)$$

$$U_I(z) = k_I \frac{T_s}{2} \frac{z+1}{z-1} E(z), \qquad (4.111)$$

and from Eq. (4.109) the derivative term becomes the inverse as

$$U_D(z) = k_D \frac{2}{T_s} \frac{z-1}{z+1} E(z) \qquad (4.112)$$

The complete discrete PID controller is thus described by

$$U(z) = \left(k_P + k_I \frac{T_s}{2} \frac{z+1}{z-1} + k_D \frac{2}{T_s} \frac{z-1}{z+1}\right) E(z) \qquad (4.113)$$

Comparing the two discrete equivalents of integration and differentiation with the corresponding analog terms, it is seen that the effect of the discrete approximation in the z domain is as if everywhere in the analog transfer function the operator s has been replaced by the composite operator

$$\frac{2}{T_s}\frac{z-1}{z+1}.$$

Trapezoid Rule

This is the trapezoid rule[15] of discrete equivalents:

The discrete equivalent to $D_a(s)$ is $D_d(z) = D_a\left(\frac{2}{T_s}\frac{z-1}{z+1}\right)$.   (4.114)

EXAMPLE 4.16   *Finding a Discrete Equivalent*

Find the discrete equivalent to the analog controller having transfer function

$$D(s) = \frac{U(s)}{E(s)} = \frac{11s+1}{3s+1}$$ (4.115)

using the sample period $T_s = 1$.

Solution.   The discrete operator is $\frac{2(z-1)}{z+1}$ and thus the discrete transfer function is

$$D_d(z) = \frac{U(z)}{E(z)} = D(s)\bigg|_{s=\frac{2}{T_s}\frac{z-1}{z+1}}$$ (4.116)

$$= \frac{11\left[\dfrac{2(z-1)}{z+1}\right]+1}{3\left[\dfrac{2(z-1)}{z+1}\right]+1}.$$ (4.117)

Clearing fractions, the discrete transfer function is

$$D_d(z) = \frac{U(z)}{E(z)} = \frac{23z-21}{7z-5}.$$ (4.118)

Converting the discrete transfer function to a discrete difference equation using the definition of z as the forward shift operator is done as follows. First we cross-multiply in Eq. (4.118) to obtain

$$(7z-5)U(z) = (23z-21)E(z)$$ (4.119)

and, interpreting z as a shift operator, this is equivalent to the difference equation[16]

$$7u(k+1) - 5u(k) = 23e(k+1) - 21e(k),$$ (4.120)

---

[15] The formula is also called Tustin's Method after the English engineer who used the technique to study the responses of nonlinear circuits.

[16] The process is entirely similar to that used in Chapter 3 to find the ordinary differential equation to which a rational Laplace Transform corresponds.

where we have replaced $kT_s + T_s$, with $k + 1$ to simplify the notation. To compute the next control at time $kT_s + T_s$, therefore, we solve the difference equation

$$u(k + 1) = \frac{5}{7}u(k) + \frac{23}{7}e(k + 1) - \frac{21}{7}e(k). \qquad (4.121)$$

Now let's apply these results to a control problem. Fortunately, MATLAB provides us with the Simulink capability to simulate both continuous and discrete systems allowing us to compare the responses of the systems with continuous and discrete controllers.

EXAMPLE 4.17   *Discrete Control of Motor Speed*

A motor speed control is found to have the plant transfer function

$$\frac{Y}{U} = \frac{45}{(s + 9)(s + 5)}. \qquad (4.122)$$

A PI controller designed for this system has the transfer function

$$D(s) = \frac{U}{E} = 1.4\frac{s + 6}{s}. \qquad (4.123)$$

The closed-loop system has a rise time of about 0.2 sec and an overshoot of about 20%. Design a discrete equivalent to this controller and compare the step responses and control signals of the two systems.

(a) Compare the responses if the sample period is 0.07, which is about 3 samples per rise time.

(b) Compare the responses with a sample period of $T_s = 0.035$, which corresponds to about 6 samples per rise time.

Solution.

(a) Using the substitution given by Eq. (4.114), the discrete equivalent for $T_s = 0.07$ is given by replacing s by

$$s \leftarrow \frac{2}{0.07}\frac{z - 1}{z + 1}$$

in $D(s)$ as follows:

$$D_d(z) = 1.4\frac{\dfrac{2}{.07}\dfrac{z - 1}{z + 1} + 6}{\dfrac{2}{.07}\dfrac{z - 1}{z + 1}} \qquad (4.124)$$

$$= 1.4\frac{2(z - 1) + 6 * 0.07(z + 1)}{2(z - 1)} \qquad (4.125)$$

$$= 1.4\frac{1.21z - 0.79}{(z - 1)}. \qquad (4.126)$$

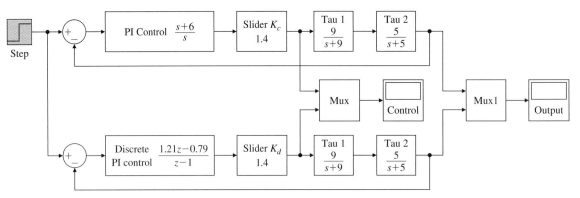

Figure 4.34 Simulink block diagram to compare continuous and discrete controllers

Based on this expression, the equation for the control is (the sample period is suppressed)

$$u(k + 1) = u(k) + 1.4 * [1.21e(k + 1) - 0.79e(k)]. \tag{4.127}$$

(b) For $T_s = 0.035$, the discrete transfer function is

$$D_d = 1.4\frac{1.105z - 0.895}{z - 1}, \tag{4.128}$$

for which the difference equation is

$$u(k + 1) = u(k) + 1.4[1.105\,e(k + 1) - 0.895\,e(k)].$$

A Simulink block diagram for simulating the two systems is given in Fig. 4.34, and plots of the step responses are given in Fig. 4.35(a). The respective control signals are plotted in Fig. 4.35(b). Notice that the discrete controller for $T_s = 0.07$ results in a substantial increase in the overshoot in the step response, while with $T_s = 0.035$ the digital controller matches the performance of the analog controller fairly well.

For controllers with many poles and zeros, making the continuous-to-discrete substitution called for in Eq. (4.114) can be very tedious. Fortunately, MATLAB provides a command that does all the work. If one has a continuous transfer function given by $D_c(s) = num D/den D$ represented in MATLAB as sysDa = tf(numD,denD), then the discrete equivalent with sampling period T_s is given by

$$sysDd = c2d(sysDa, T_s ,\text{'t'}) \tag{4.129}$$

In this expression, of course, the polynomials are represented in MATLAB form. The last parameter in the c2d function given by "t" calls for the conversion to be done using the trapezoid method. The alternatives can be found by asking

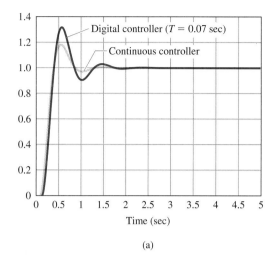
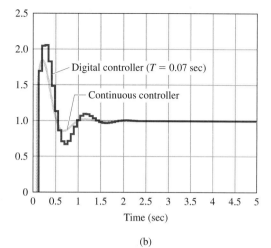

(a)   (b)

Figure 4.35 Comparison plots of a speed control system with continuous and discrete controllers. (a) Output responses. (b) Control signals

MATLAB for help c2d. For example, to compute the polynomials for $T_s = 0.07$ for the example above, the commands would be

```
numDa = [1 6];
denDa = [1 −1];
sysDa = tf(numD,denD)
sysDd = c2d( sysDa,0.07,'t')
```

SUMMARY

- Compared to open-loop control, feedback can be used to reduce steady-state error to disturbances, reduce the system's transfer function sensitivity to parameter variations, speed up the transient response, and reduce the sensitivity of the output signal to parameter changes.
- Increasing the proportional feedback gain reduces steady-state errors, but high gains almost always destabilize the system. Integral control provides robust reduction in steady-state errors, but often makes the system less stable. Derivative control usually increases damping and improves stability. These three kinds of control combined form the classical PID controller.
- The standard PID controller is described by the equation

$$U(s) = \left(k_p + \frac{k_I}{s} + k_D s \right) E(s)$$

or

$$U(s) = k_p \left(1 + \frac{1}{T_I s} + T_D s \right) E(s) = D(s) E(s).$$

This form is ubiquitous in the process-control industry and is the basic controller in many control systems.

- Useful guidelines for tuning PID controllers were presented in Tables 4.1 and 4.2.
- Classifying a system as type k indicates the ability of the system to achieve zero steady-state error to polynomials of degree less than but not equal to k. A stable *unity* feedback system is type k with respect to reference inputs if the loop transfer function has k poles at the origin in which case we can write

$$D(s)G(s) = \frac{(s + z_1)(s + z_2) \cdots}{s^k(s + p_1)(s + p_2) \cdots}.$$

and the error constant is given by

$$K_k = \lim_{s \to 0} s^k D(s)G(s). \qquad (4.130)$$

- A table of steady-state errors for systems of types 0, 1 and 2 to reference inputs is given in Table 4.3.
- Systems can be classified as to type for rejecting disturbances by computing the system error to polynomial disturbance inputs. The system is type k if the error is zero to all polynomials of degree less than k but nonzero for a polynomial of degree k.
- A difference equation describing a digital controller to be used to replace a given analog controller can be found by replacing s with

$$\frac{2}{T_s} \frac{z - 1}{z + 1}$$

in the analog's transfer function and using z as a forward shift operator in the sense that if $U(z)$ corresponds to $u(kT_s)$ then $zU(z)$ corresponds to $u(kT_s + T_s)$.
- MATLAB can compute a discrete equivalent with the command c2d.

Review Questions

1. Give three advantages of feedback in control.

2. Give two disadvantages of feedback in control.

3. What is the main objective of introducing integral control?

4. What is the major objective of adding derivative control?

5. Why might a designer wish to put the derivative term in the feedback rather than in the error path?

▲   6. What is integrator windup?

▲   7. Why is an anti-windup circuit important?

▲   8. Using the nonlinear saturation function having gain 1 and limits ±1, sketch the block diagram of saturation for an actuator which has gain 7 and limits of ±20.

9. A temperature control system is found to have zero error to a constant tracking input and an error of $0.5°$C to a tracking input that is linear in time, rising at the rate of $40°$C/sec. What is the system type of this control system and what is the relevant error constant $[K_p$ or K_v or etc.]?

10. What are the units of K_p, K_v, and K_a?

11. What is the definition of system type with respect to reference inputs?

12. What is the definition of system type with respect to disturbance inputs?

13. Why does system type depend on where the external signal enters the system?

14. Give two reason to use a digital controller rather than an analog controller.

15. Give two disadvantages to using a digital controller.

16. Give the substitution in the discrete operator z for the Laplace operator s if the approximation to the integral in Eq. (4.108) is taken to be the rectangle of height $e(kT_s)$ and base T_s.

Problems

Problems for Section 4.1

4.1. Consider a system with the configuration of Fig. 4.28, where D is the constant gain of the controller and G is that of the process. The nominal values of these gains are $D = 5$ and $G = 7$. Suppose a constant disturbance w is added to the control input u before the signal goes to the process.

   (a) Compute the gain from w to y in terms of D and G.

   (b) Suppose the system designer knows that an increase by a factor of 6 in the loop gain DG can be tolerated before the system goes out of specification. Where should the designer place the extra gain if the objective is to minimize the system error $r - y$ due to the disturbance? For example, either D or G could be increased by a factor of 6, or D could be doubled and G tripled, and so on. Which choice is the best?

4.2. Bode defined the sensitivity function relating a transfer function G to one of its parameters k as the ratio of percent change in k to percent change in G. We define the reciprocal of Bode's function as

$$S_k^G = \frac{dG/G}{dk/k} = \frac{d\ln G}{d\ln k} = \frac{k}{G}\frac{dG}{dk}.$$

Thus, when the parameter k changes by a certain percentage, S tells us what percent change to expect in G. In control systems design we are almost always interested in the sensitivity at zero frequency, or when $s = 0$. The purpose of this exercise is to examine the effect of feedback on sensitivity. In particular, we would like to compare the topologies shown in Fig. 4.36 for connecting three amplifier stages with a gain of $-K$ into a single amplifier with a gain of -10.

   (a) For each topology in Fig. 4.36, compute β_i so that if $K = 10$, $Y = -10R$.

   (b) For each topology, compute S_k^G when $G = Y/R$. [Use the respective β_i values found in part (a).] Which case is the *least* sensitive?

   (c) Compute the sensitivities of the systems in Fig. 4.36(b, c) to β_2 and β_3. Using your results, comment on the relative need for precision in sensors and actuators.

Figure 4.36
Three-amplifier topologies
for Problem 4.2

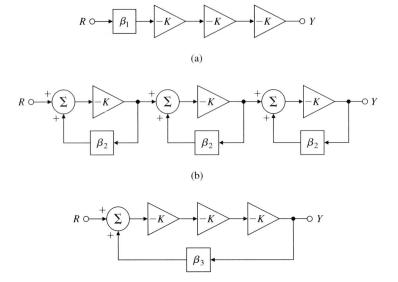

4.3.   Compare the two structures shown in Fig. 4.37 with respect to sensitivity to changes in the overall gain due to changes in the amplifier gain. Use the relation

$$S = \frac{d \ln F}{d \ln K} = \frac{K}{F} \frac{dF}{dK}$$

as the measure. Select H_1 and H_2 so that the nominal system outputs satisfy $F_1 = F_2$, and assume $K H_1 > 0$.

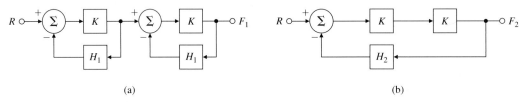

Figure 4.37 Block diagrams for Problem 4.3

4.4.   The DC-motor speed control shown in Fig. 4.38 is described by the differential equation

$$\dot{y} + 60y = 600v_a - 1500w,$$

where y is the motor speed, v_a is the armature voltage, and w is the load torque. Assume the armature voltage is computed using the PI control law

$$v_a = \left(k_p e + k_I \int_0^t e \, dt \right),$$

where $e = r - y$.

Figure 4.38
Unity feedback system with
prefilter for Problem 4.4

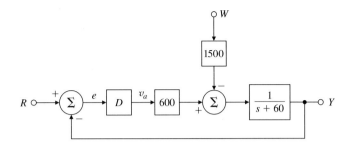

(a) Compute the transfer function from W to Y as a function of k_p and k_I.

(b) Compute values for k_p and k_I so that the characteristic equation of the closed-loop system will have roots at $-60 \pm 60j$.

4.5. Consider the system shown in Fig. 4.39, which consists of a prefilter and a unity feedback system.

(a) Determine the transfer function from R to Y.

(b) Determine the steady-state error due to a step input.

(c) Discuss the effect of different values of (K_r, a) on the system's response.

(d) For each of the following three cases,

$$(1)\ A = 1,\ \tau = 1, \quad (2)\ A = 10,\ \tau = 1, \quad (3)\ A = 1,\ \tau = 2,$$

use MATLAB to find values for K_r and a so that (if possible)

  i. the rise time is less than 1.5 sec,

  ii. the overshoot is less than 20%,

  iii. the settling time is less than 10 sec, and

  iv. the steady-state error is less than 5%.

In cases where the specifications are easily met, try to make the rise time as small as possible. If the specifications cannot be met, find the design to meet as many of the specifications as possible, in the order given.

Figure 4.39
Block diagrams for
Problem 4.5

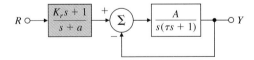

4.6. A unity feedback control system has the open-loop transfer function

$$G(s) = \frac{A}{s(s + a)}.$$

(a) Compute the sensitivity of the closed-loop transfer function to changes in the parameter A.

(b) Compute the sensitivity of the closed-loop transfer function to changes in the parameter a.

(c) If the unity gain in the feedback changes to a value of $\beta \neq 1$, compute the sensitivity of the closed-loop transfer function with respect to β.

(d) Assuming $A = 1$ and $a = 1$, plot the magnitude of each of the above sensitivity functions for $s = j\omega$ using semilogy command in MATLAB. Comment on the relative effect of parameter variations in A, a, and β at different frequencies ω, paying particular attention to DC (when $\omega = 0$).

▲ **4.7.** For the unity feedback system with proportional control $D = k_p$ and process transfer function

$$G(s) = \frac{A}{s(\tau s + 1)},$$

(a) Draw the block diagram from which to compute the sensitivity to changes in the parameter τ of the output response to a reference step input. Let the parameter be $\theta = 1/\tau$.

(b) Use MATLAB to compute and plot the sensitivity computed from the block diagram of part (a) if $A = \tau = k_p = 1$.

Problems for Section 4.2

4.8. Consider the satellite-attitude control problem shown in Fig. 4.40 where the normalized parameters are

$$J = 10 \quad \text{spacecraft inertia, N-m-sec}^2/\text{rad},$$

$$\theta_r = \text{reference satellite attitude, rad},$$

$$\theta = \text{actual satellite attitude, rad},$$

$$H_y = 1 \quad \text{sensor scale, factor volts/rad},$$

$$H_r = 1 \quad \text{reference sensor scale factor, volts/rad},$$

$$w = \text{disturbance torque, N-m}.$$

(a) Use proportional control, **P**, with $D(s) = k_p$, and give the range of values for k_p for which the system will be stable.

(b) Use **PD** control, let $D(s) = (k_p + k_D s)$, and determine the system type and error constant with respect to reference inputs.

Figure 4.40
Satellite attitude control

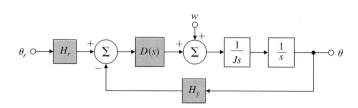

(c) Use **PD** control, let $D(s) = (k_p + k_D s)$, and determine the system type and error constant with respect to disturbance inputs.

(d) Use **PI** control, let $D(s) = (k_p + k_I/s)$, and determine the system type and error constant with respect to reference inputs.

(e) Use **PI** control, let $D(s) = (k_p + k_I/s)$, and determine the system type and error constant with respect to disturbance inputs.

(f) Use **PID** control, let $D(s) = (k_p + k_I/s + k_D s)$, and determine the system type and error constant with respect to reference inputs.

(g) Use **PID** control, let $D(s) = (k_p + k_I/s + k_D s)$, and determine the system type and error constant with respect to disturbance inputs.

▲   **4.9.**   The unit-step response of a paper machine is shown in Fig. 4.41(a), where the input into the system is stock flow onto the wire and the output is basis weight (thickness). The time delay and slope of the transient response may be determined from the figure.

(a) Find the proportional, PI, and PID-controller parameters using the Zeigler–Nichols transient-response method.

(b) Using proportional feedback control, control designers have obtained a closed-loop system with the unit impulse response shown in Fig. 4.41(b). When the gain $K_u = 8.556$, the system is on the verge of instability. Determine the proportional-, PI-, and PID-controller parameters according to the Zeigler–Nichols ultimate sensitivity method.

Figure 4.41
Paper-machine response data for problem 4.9

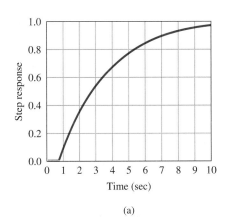

(a)

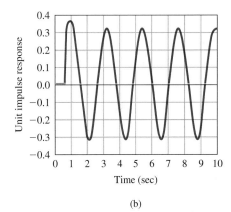

(b)

▲   **4.10.**   A paper machine has the transfer function

$$G(s) = \frac{e^{-2s}}{3s + 1},$$

where the input is stock flow onto the wire and the output is basis weight or thickness.

(a) Find the PID-controller parameters using the Zeigler–Nichols tuning rules.

(b) The system becomes marginally stable for a proportional gain of $K_u = 3.044$ as shown by the unit impulse response in Fig. 4.42. Find the optimal PID-controller parameters according to the Zeigler–Nichols tuning rules.

Figure 4.42
Unit impulse response
for paper-machine in
Problem 4.10

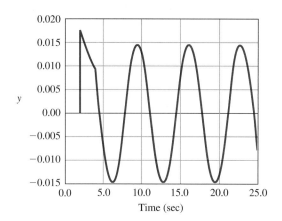

4.11. Consider the system with the plant transfer function

$$G(s) = \frac{1}{s^2 + 1}.$$

We would like to use PID control on this system. It is known that the system's actuator is a saturation nonlinearity with a slope of 1 and $|u| \le 10$.

(a) Design the values of k_p, k_D, and k_I so that the closed-loop characteristic equation has roots at $s = -1, -1 \pm j1$. Connect the derivative term to the output, not to the error with the other terms, and use the modified form $\frac{-k_D s}{0.1s + 1}$ because MATLAB cannot realize the unrealistic pure derivative term.

(b) Using Simulink as in Fig. 4.23, add an anti-windup system using the techniques discussed in this chapter. Experiment with different values for the anti-windup feedback gain K_a, and select a value that gives good response to large steps.

(c) Plot both the step tracking response and the control effort for steps that cause the actuator to saturate. Qualitatively describe the effect of the anti-windup on both the output response and the control effort.

Problems for Section 4.3

4.12. Consider the second-order plant

$$G(s) = \frac{1}{(s + 1)(5s + 1)}.$$

(a) Determine the system type and error constant with respect to tracking polynomial reference inputs of the system for P, PD, and PID controllers (as configured in Fig. 4.28). Let $k_p = 19$, $k_I = 0.5$, and $k_D = \frac{4}{19}$.

(b) Determine the system type and error constant of the system with respect to disturbance inputs for each of the three regulators in part (a) with respect to rejecting polynomial disturbances $w(t)$ at the *input* to the plant.

(c) Is this system better at tracking references or rejecting disturbances? Explain your response briefly.

(d) Verify your results for parts (a) and (b) using MATLAB by plotting unit step and ramp responses for both tracking and disturbance rejection.

4.13. Consider a system with the plant transfer function $G(s) = 1/s(s+1)$. You wish to add a dynamic controller so that $\omega_n = 2$ rad/sec and $\zeta \geq 0.5$. Several dynamic controllers have been proposed:

1. $D(s) = (s+2)/2$,

2. $D(s) = 2\dfrac{s+2}{s+4}$,

3. $D(s) = 5\dfrac{(s+2)}{s+10}$,

4. $D(s) = 5\dfrac{(s+2)(s+0.1)}{(s+10)(s+0.01)}$.

(a) Using MATLAB, compare the resulting transient and steady-state responses to reference step inputs for each controller choice. Which controller is best for the smallest rise time and smallest overshoot?

(b) Which system would have the smallest steady-state error to a ramp reference input?

(c) Compare each system for peak control effort, that is, measure the peak magnitude of the plant input $u(t)$ for a unit reference step input.

(d) Based on your results from parts (a) to (c), recommend a dynamic controller for the system from the four candidate designs.

4.14. A certain control system has the following specifications: rise time $t_r \leq 0.010$ sec, overshoot $M_p \leq 16\%$, and steady-state error to unit ramp $e_{ss} \leq 0.005$.

(a) Sketch the allowable region in the s-plane for the dominant second-order poles of an acceptable system.

(b) If $Y/R = G/(1+G)$, what condition must $G(s)$ satisfy near $s = 0$ for the closed-loop system to meet specifications; that is, what is the required asymptotic low-frequency behavior of $G(s)$?

4.15. For the system in Problem 4.4, compute the following steady-state errors:

(a) To a unit-step reference input.

(b) To a unit-ramp reference input.

(c) To a unit-step disturbance input.

(d) For a unit-ramp disturbance input.

(e) Verify your answers to parts (a) to (d) using MATLAB. Note that a ramp response can be generated as the step response of a system modified by an added integrator at the reference input.

4.16. Consider the system shown in Fig. 4.43. Show that the system is type 1 and compute the K_v,

Figure 4.43
Control system for
Problem 4.16

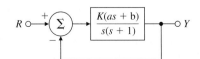

4.17. Consider the DC-motor control system with rate (tachometer) feedback shown in Fig. 4.44(a).

(a) Find values for K' and k'_t so that the system of Fig. 4.44(b) has the same transfer function as the system of Fig. 4.44(a).

(b) Determine the system type with respect to tracking θ_r and compute the system K_v in terms of parameters K' and k'_t.

(c) Does the addition of tachometer feedback with positive k_t increase or decrease K_v?

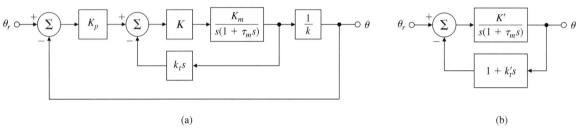

(a)   (b)

Figure 4.44 Control system for Problem 4.17

4.18. Consider the system shown in Fig. 4.45, where

$$D(s) = K\frac{(s+\alpha)^2}{s^2 + \omega_o^2}.$$

(a) Prove that if the system is stable, it is capable of tracking a sinusoidal reference input $r = \sin \omega_o t$ with zero steady-state error. (Look at the transfer function from R to E and consider the gain at ω_o.)

(b) Use Routh's criteria to find the range of K such that the closed-loop system remains stable if $\omega_o = 1$ and $\alpha = 0.25$.

Figure 4.45
Control system for
Problem 4.18

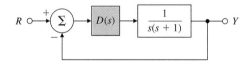

4.19. Consider the system shown in Fig. 4.46 which represents control of the angle of a pendulum which has no damping.

(a) What condition must $D(s)$ satisfy so that the system can track a ramp reference input with constant steady-state error?

(b) For a transfer function $D(s)$ that stabilizes the system and satisfies the condition in part (a), find the class of disturbances $w(t)$ that the system can reject with zero steady-state error.

(c) Show that although a PI controller satisfies the condition derived in part (a), it will not yield a stable closed-loop system. Will a PID controller work; that is, satisfy part (a) *and* stabilize the system? If so, what constraints must k_p, k_I, and k_D satisfy?

(d) Discuss qualitatively and briefly the effects of small variations on the controller parameters k_p, k_I, and k_D on the system's step response rise time and overshoot.

Figure 4.46
Control system for
Problem 4.19

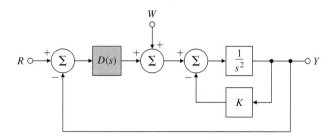

4.20. A unity feedback system has the overall transfer function

$$\frac{Y(s)}{R(s)} = \mathcal{T}(s) = \frac{\omega_n^2}{s^2 + 2\zeta\omega_n s + \omega_n^2}.$$

Give the system type and corresponding error constant for tracking polynomial reference inputs in terms of ζ and ω_n.

4.21. Consider the second-order system

$$G(s) = \frac{1}{s^2 + 2\zeta s + 1}.$$

We would like to add a transfer function of the form $D(s) = K(s + a)/(s + b)$ in series with $G(s)$ in a unity-feedback structure.

(a) Ignoring stability for the moment, what are the constraints on K, a, and b so that system type 1?

(b) What are the constraints placed on K, a, and b so that the system is stable and type 1?

(c) What are the constraints on a and b so that the system is type 1 and remains stable for every positive value for K?

4.22. The transfer function for the plant in a motor position control is given by

$$G(s) = \frac{A}{s(s + a)}.$$

If we were able to select values for both A and a, what would they be to result in a system with $K_v = 20$ and $\varsigma = 0.707$?

4.23. Consider the system shown in Fig. 4.47(a).

(a) What is the system type? Compute the steady-state tracking error due to a ramp input $r(t) = r_o t 1(t)$.

(b) For the modified system shown in Fig. 4.47(b), give the values of H_f and H_r so the system is type 2 for reference inputs and compute the K_a in this case.

(c) Is the resulting type 2 property of this system robust with respect to changes in H_f? Will the system remain type 2 if H_f changes slightly?

Figure 4.47
Control system for
Problem 4.23

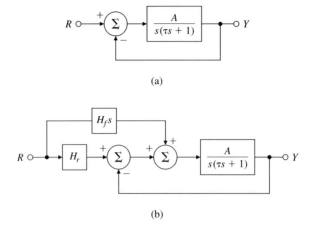

(a)

(b)

4.24. A controller for a satellite attitude control with transfer function $G = 1/s^2$ has been designed with a unity feedback structure and has the transfer function

$$D(s) = \frac{10(s+2)}{s+5}.$$

    (a) Find the system type for reference tracking and the corresponding error constant for this system.

    (b) If a disturbance torque adds to the control so that the input to the process is $u + w$, what is the system type and corresponding error constant with respect to disturbance rejection?

4.25. A compensated motor position control system is shown in Fig. 4.48. Assume that the sensor dynamics are $H(s) = 1$.

    (a) Can the system track a step reference input r with zero steady-state error? If yes, give the value of the velocity constant. 0

    (b) Can the system reject a step disturbance w with zero steady-state error? If yes, give the value of the velocity constant.

Figure 4.48
Control system for
Problem 4.25

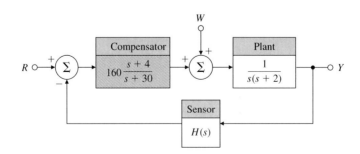

(c) Compute the sensitivity of the closed-loop transfer function to changes in the plant pole at -2.

(d) In some instances there are dynamics in the sensor. Repeat parts (a) to (c) for $H(s) = 20/(s + 20)$ and compare the corresponding velocity constants.

4.26. Consider the system shown in Fig. 4.49 with PI control.

(a) Determine the transfer function from R to Y.

(b) Determine the transfer function from W to Y.

(c) Use Routh's criteria to find the range of (k_p, k_I) for which the system is stable.

(d) What is the system type and error constant with respect to reference tracking?

(e) What is the system type and error constant with respect to disturbance rejection?

Figure 4.49
Control system for
Problem 4.26

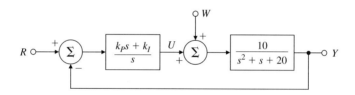

4.27. The general unity feedback system shown in Fig. 4.50 has disturbance inputs w_1, w_2, and w_3 and is asymptotically stable. Also,

$$G_1(s) = \frac{K_1 \prod_{i=1}^{m_1}(s + z_{1i})}{s^{l_1} \prod_{i=1}^{m_1}(s + p_{1i})}, \qquad G_2(s) = \frac{K_2 \prod_{i=1}^{m_1}(s + z_{2i})}{s^{l_2} \prod_{i=1}^{m_1}(s + p_{2i})}.$$

(a) Show that the system is of type 0, type l_1, and type $(l_1 + l_2)$ with respect to disturbance inputs w_1, w_2, and w_3.

(b) Consider the multivariable system shown in Fig. 4.51. Assume that the system is stable. Find the transfer functions from each disturbance input to each output and determine the steady-state values of y_1 and y_2 for constant disturbances. We define a multivariable system to be type k with respect to polynomial inputs at w_i if the steady-state value of *every* output is zero for any combination of inputs of degree less than k and at least one input is a nonzero constant for an input of degree k. What is the system type with respect to disturbance rejection at w_1? At w_2?

Figure 4.50
Single input-single output
unity feedback system with
disturbance inputs

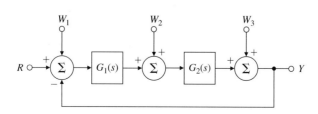

Figure 4.51
Multivariable system

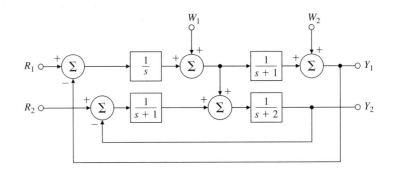

4.28. One possible representation of an automobile speed-control system with integral control is shown in Fig. 4.52.

   (a) With a zero reference velocity input ($v_c = 0$), find the transfer function relating the output speed v to the wind disturbance w.

   (b) What is the steady-state response of v if w is a unit ramp function?

   (c) What type is this system in relation to reference inputs? What is the value of the corresponding error constant?

   (d) What is the type and corresponding error constant of this system in relation to tracking the disturbance w?

Figure 4.52
System using integral control

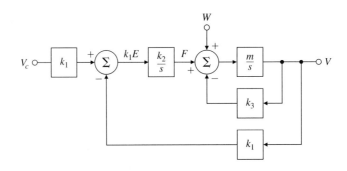

4.29. For the feedback system shown in Fig. 4.53, find the value of α that will make the system type 1 for $K = 5$. Give the corresponding velocity constant. Show that the system is not robust by using this value of α and computing the tracking error $e = r - y$ to a step reference for $K = 4$ and $K = 6$.

Figure 4.53
Control system for
Problem 4.29

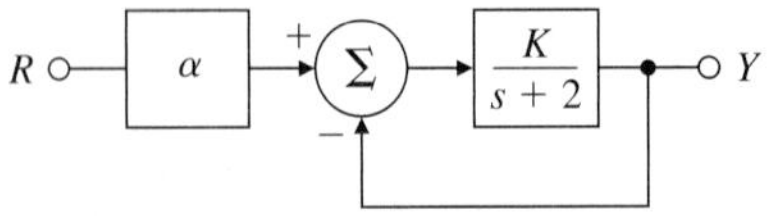

$$R \; \circ \!\!-\!\! \boxed{\alpha} \longrightarrow \overset{+}{\underset{-}{\Sigma}} \longrightarrow \boxed{\dfrac{K}{s+2}} \!\!-\!\! \circ \; Y$$

4.30. A position control system has the closed-loop transfer function (meter/meter) given by

$$\frac{Y(s)}{R(s)} = \frac{b_0 s + b_1}{s^2 + a_1 s + a_2}.$$

(a) Choose the parameters (a_1, a_2, b_0, b_1) so that the following specifications are satisfied simultaneously:

    i. The rise time $t_r < 0.1$ sec.

    ii. The overshoot $M_p < 20\%$.

    iii. The settling time $t_s < 0.5$ sec.

    iv. The steady-state error to a step reference is zero.

    v. The steady-state error to a ramp reference input of 0.1 m/sec is not more than 1 mm.

(b) Verify your answer via MATLAB simulation.

4.31. Suppose you are given the system depicted in Fig. 4.54(a), where the plant parameter a is subject to variations.

(a) Find $G(s)$ so that the system shown in Fig. 4.54(b) has the same transfer function from r to y as the system in Fig. 4.54(a).

(b) Assume that $a = 1$ is the nominal value of the plant parameter. What is the system type and the error constant in this case?

(c) Now assume that $a = 1 + \delta a$, where δa is some perturbation to the plant parameter. What is the system type and the error constant for the perturbed system?

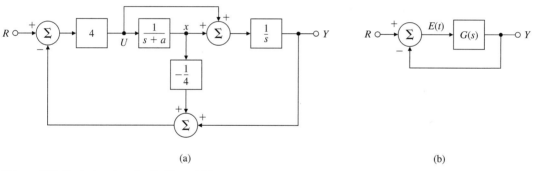

(a)　　　　　　　　　　　　　　　　(b)

Figure 4.54 Control system for Problem 4.31

4.32. Two feedback systems are shown in Fig. 4.55.

(a) Determine values for K_1, K_2, and K_3 so that both systems:

    i. Exhibit zero steady-state error to step inputs (that is, both are type 1).

    ii. Have static velocity error constant $K_v = 1$ when $K_0 = 1$.

(b) Suppose K_0 undergoes a small perturbation: $K_0 \rightarrow K_0 + \delta K_0$. What effect does this have on the system type in each case? Which system has a type which is robust? Which system do you think would be preferred?

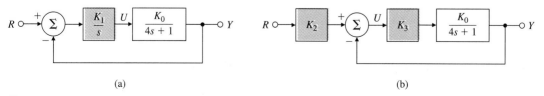

(a)　　　　　　　　　　　　　　　　(b)

Figure 4.55 Two feedback systems for Problem 4.32

(c) Estimate the transient response of both systems to a step reference input, and give estimates for t_s, t_r, and M_p. In your opinion, which system has a better transient response at the nominal parameter values?

4.33. You are given the system shown in Fig. 4.56, where the feedback gain β is subject to variations. You are to design a controller for this system so that the output $y(t)$ accurately tracks the reference input $r(t)$.

(a) Let $\beta = 1$. You are given the following three options for the controller $D_i(s)$:

$$D_1(s) = k_p, \qquad D_2(s) = \frac{k_p s + k_I}{s}, \qquad D_3(s) = \frac{k_p s^2 + k_I s + k_2}{s^2}.$$

Choose the controller (including particular values for the controller constants) that will result in a type 1 system with a steady-state error to a unit reference ramp of less than $\frac{1}{10}$.

(b) Next, suppose that there is some attenuation in the feedback path that is modeled by $\beta = 0.9$. Find the steady-state error due to a ramp input for your choice of $D_i(s)$ in part (a).

(c) If $\beta = 0.9$, what is the system type for part (b)? What are the values of the appropriate error constant?

Figure 4.56

Control system for
Problem 4.33

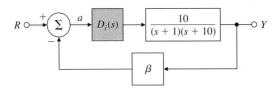

4.34. Consider the system shown in Fig. 4.57.

(a) Find the transfer function from the reference input to the tracking error.

(b) For this system to respond to inputs of the form $r(t) = t^n 1(t)$ (where $n < q$) with zero steady-state error, what constraint is placed on the open-loop poles $p_1, p_2, \ldots, p_q$?

Figure 4.57

Control system for
Problem 4.34

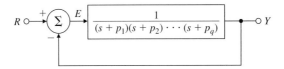

4.35. The feedback control system shown in Fig. 4.58 is to be designed to satisfy the following specifications: (1) steady-state error of less than 10% to a ramp reference input, (2) maximum overshoot for a unit step input of less than 5%, and (3) 1% settling time of less than 3 sec.

Figure 4.58

Control system for
Problem 4.35

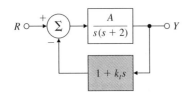

(a) Compute the closed-loop transfer function.

(b) Sketch the region in the complex plane where the closed-loop poles may lie.

(c) What does specification (1) imply about the possible values of A?

(d) What does specification (3) imply about the closed-loop poles?

(e) Find the error due to a unit ramp input in terms of A and k_t.

(f) Suppose $A = 32$. Find the value of k_t that yields closed-loop poles on the right-hand boundary of the feasible region. Use MATLAB to check whether this choice for k_t satisfies the desired specifications. If not, adjust k_t until it does.

(g) Using $A = 32$ and the value for k_t computed in part (f), estimate the settling time of the system. Use MATLAB to check your answer.

4.36. The transfer functions of speed control for a magnetic tape-drive system are shown in Fig. 4.59. The speed sensor is fast enough that its dynamics can be neglected and the diagram shows the equivalent unity feedback system.

(a) Assuming $\omega_r = 0$, what is the steady-state error due to a step disturbance torque of $1 \text{ N} \cdot \text{m}$? What must the amplifier gain K be in order to make the steady-state error $e_{ss} \leq 0.001$ rad/sec?

(b) Plot the roots of the closed-loop system in the complex plane, and accurately sketch the time response $\omega(t)$ for a step input ω_r using the gain K computed in part (a). Are these roots satisfactory? Why or why not?

(c) Plot the region in the complex plane of acceptable closed-loop poles corresponding to the specifications of a 1% settling time of $t_s \leq 0.1$ sec. and an overshoot $M_p \leq 5\%$.

(d) Give values for k_p and k_D for a PD controller which will meet the specifications.

(e) How would the disturbance-induced steady-state error change with the new control scheme in part (d)? How could the steady-state error to a disturbance torque be eliminated entirely?

Figure 4.59
Speed-control system for a magnetic tape drive

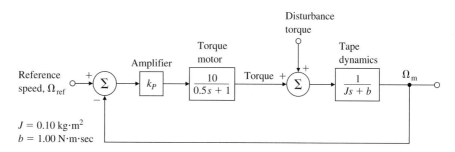

$J = 0.10$ kg·m^2
$b = 1.00$ N·m·sec

4.37. A linear ODE model of the DC motor with negligible armature inductance ($L_a = 0$) and disturbance torque w was given earlier in the chapter; it is restated here, in slightly different form, as

$$\frac{J R_a}{K_t}\ddot{\theta}_m + K_e\dot{\theta}_m = v_a + \frac{R_a}{K_t}w,$$

where θ_m is measured in radians. Dividing through by the coefficient of $\ddot{\theta}_m$, we obtain

$$\ddot{\theta}_m + a_1\dot{\theta}_m = b_0v_a + c_0w,$$

where

$$a_1 = \frac{K_t K_e}{J R_a}, \qquad b_0 = \frac{K_t}{J R_a}, \qquad c_0 = \frac{1}{J}.$$

With rotating potentiometers, it is possible to measure the positioning error between θ and the reference angle θ_r or $e = \theta_{ref} - \theta_m$. With a tachometer we can measure the motor speed $\dot{\theta}_m$. Consider using feedback of the error e and the motor speed $\dot{\theta}_m$ in the form

$$v_a = K(e - T_D\dot{\theta}_m),$$

where K and T_D are controller gains to be determined.

(a) Draw a block diagram of the resulting feedback system showing both θ_m and $\dot{\theta}_m$ as variables in the diagram representing the motor.

(b) Suppose the numbers work out so that $a_1 = 65$, $b_0 = 200$, and $c_0 = 10$. If there is no load torque ($w = 0$), what speed (in rpm) results from $v_a = 100$ V?

(c) Using the parameter values given in part (b), find k_p and k_D so that a step change in θ_{ref} with zero load torque results in a transient that has an approximately 17% overshoot and that settles to within 5% of steady-state in less than 0.05 sec.

(d) Derive an expression for the steady-state error to a reference angle input, and compute its value for your design in part (c) assuming $\theta_{ref} = 1$ rad.

(e) Derive an expression for the steady-state error to a constant disturbance torque when $\theta_{ref} = 0$, and compute its value for your design in part (c) assuming $w = 1.0$.

4.38. We wish to design an automatic speed control for an automobile. Assume that (1) the car has a mass m of 1000 kg, (2) the accelerator is the control U and supplies a force on the automobile of 10 N per degree of accelerator motion, and (3) air drag provides a friction force proportional to velocity of 10 N · sec/m.

(a) Obtain the transfer function from control input U to the velocity of the automobile.

(b) Assume the velocity changes are given by

$$V(s) = \frac{1}{s + 0.002}U(s) + \frac{0.05}{s + 0.02}W(s),$$

where V is given in meters per second, U is in degrees, and W is the percent grade of the road. Design a proportional control law $U = -k_p V$ that will maintain a velocity error of less than 1 m/sec in the presence of a constant 2% grade.

(c) Discuss what advantage (if any) integral control would have for this problem.

(d) Assuming that pure integral control (that is, no proportional term) is advantageous, select the feedback gain so that the roots have critical damping ($\zeta = 1$).

4.39. Consider the automobile speed control system depicted in Fig. 4.60.

    (a) Find the transfer functions from $W(s)$ and from $R(s)$ to $Y(s)$.

    (b) Assume that the desired speed is a constant reference r, so that $R(s) = r_o/s$. Assume that the road is level, so $w(t) = 0$. Compute values of the gains K, H_r, and H_f to guarantee that

$$\lim_{t \to \infty} y(t) = r_o.$$

    Include both the open-loop (assuming $H_y = 0$) and feedback cases ($H_y \neq 0$) in your discussion.

    (c) Repeat part (b) assuming that a constant grade disturbance $W(s) = w_o/s$ is present *in addition to* the reference input. In particular, find the variation in speed due to the grade change for both the feed forward and feedback cases. Use your results to explain (1) why feedback control is necessary and (2) how the gain k_p should be chosen to reduce steady-state error.

    (d) Assume that $w(t) = 0$ and that the gain A undergoes the perturbation $A + \delta A$. Determine the error in speed due to the gain change for both the feed forward and feedback cases. How should the gains be chosen in this case to reduce the effects of δA?

Figure 4.60
Automobile speed-control system

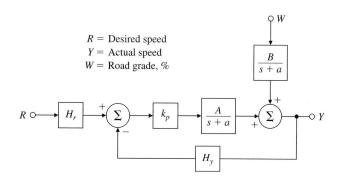

R = Desired speed
Y = Actual speed
W = Road grade, %

4.40. Prove that for a type 2 system, the acceleration error constant is given by

$$\frac{1}{K_a} = \frac{1}{2} \left(\sum_{i=1}^{m} \frac{1}{z_i^2} - \sum_{i=1}^{n} \frac{1}{p_i^2} \right),$$

where z_i and p_i are the closed-loop zeros and poles of the system.

4.41. For a system with impulse response $h(t)$, prove that the velocity constant is given by

$$\frac{1}{K_v} = \int_0^\infty t h(t)\, dt,$$

and the acceleration constant is given by $\dfrac{1}{K_a} = -\dfrac{1}{2} \int_0^\infty t^2 h(t)\, dt$.

5 The Root-Locus Design Method

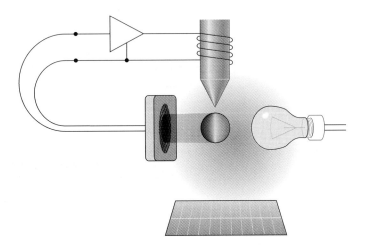

Chapter Overview

We open in Section 5.1 by illustrating the root-locus for some simple feedback systems for which the equations can be solved directly. In Section 5.2 we show how to put an equation into the proper form to plot a root locus by hand and introduce the main rules to aid such plotting, including the case where the parameter can be negative. In Section 5.3 this approach is applied to sketch the locus for a number of typical control problems which illustrate the factors which influence the final shape. MATLAB is used for detailed plotting of specific loci. In Section 5.4 we discuss how to select the particular value of the parameter that will result in root locations corresponding to the system dynamic response requirements. Again, MATLAB can be of substantial assistance in the process. When adjustment of the selected parameter alone cannot produce a satisfactory design, changes in other parameters can be studied or new zeros and poles can be used to introduce new dynamics such as lead, lag, or notch compensation, as described in Section 5.5. In Section 5.6 the uses of the root locus for design is summarized by a comprehensive design for the attitude control of a small airplane. Finally, in Section 5.7 the root-locus method is extended to guide the design of systems with more than one variable parameter, systems with simple time delay, and elementary nonlinear systems.

A Perspective on the Root-locus Design Method

In Chapter 3 we related the features of a step response, such as rise time, overshoot, and settling time, to pole locations in the s-plane of the transform of a second order system characterized by the natural frequency ω_n, the damping ratio ζ, and the real part σ. We also examined the changes in these transient-response features when a pole or a zero is added to the transfer function. In Chapter 4 we saw how feedback can improve steady-state errors and can also influence dynamic response by changing the system's pole locations. In this chapter we present a specific technique which shows how changes in one of a system's parameters will modify the *roots of the characteristic equation* which are the closed-loop poles and thus change the system's dynamic response. The method was developed by W. R. Evans, who gave rules for plotting the paths of the roots, a plot he called the **Root locus**.

The root locus is most commonly used to study the effect of loop gain variations; however, the method is general and can be used to plot the roots of any polynomial with respect to any parameter which enters the equation linearly. For example, the root-locus method can be used to plot the roots of a characteristic equation as the gain of velocity sensor feedback changes or the parameter can be a physical parameter such as motor inertia or armature inductance. Finally, a root locus can be plotted for the characteristic equation that results from the analysis of digital control systems using the z-transform, a topic we introduced in Chapter 4 and will discuss further in Chapter 8.

5.1 Root Locus of a Basic Feedback System

We begin with the basic feedback system shown in Fig. 5.1. For this system the closed-loop transfer function is

$$\frac{Y(s)}{R(s)} = T(s) = \frac{D(s)G(s)}{1 + D(s)G(s)H(s)}, \tag{5.1}$$

and the characteristic equation, whose roots are the poles of this transfer function, is

$$1 + D(s)G(s)H(s) = 0. \tag{5.2}$$

To put the equation in a form suitable for study of the roots as a parameter changes, we first put the equation in polynomial form and select the parameter of interest which we will call K. We assume that we can define component polynomials $a(s)$ and $b(s)$ so that the characteristic polynomial is in the form $a(s) + Kb(s)$. We then define the transfer function $L(s) = b(s)/a(s)$ so that the characteristic equation can be written as[1]

$$1 + KL(s) = 0 \tag{5.3}$$

If, as is often the case, the parameter is the gain of the controller, then $L(s)$ is simply proportional to $D(s)G(s)H(s)$. Evans suggested that we plot the locus of *all possible* roots of Eq. (5.3) as K varies from zero to infinity and then use the resulting plot to aid us in selecting the best value of K. Furthermore, by studying the effects of additional poles and zeros on this graph, we can determine the consequences of additional dynamics added to $D(s)$ as compensation in the loop. We thus have a tool not only for selecting the specific parameter value but for designing the dynamic compensation as well. The graph of all possible roots of Eq. (5.3) relative to parameter K is called the **root locus,**

Figure 5.1
Basic closed loop block diagram

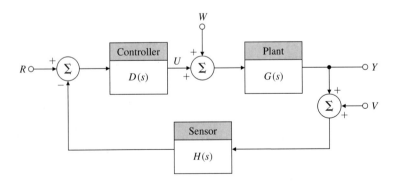

---

[1] In the most common case, $L(s)$ is the loop transfer function of the feedback system and K is the gain of the controller–plant combination. However, the root locus is a general method suitable for the study of any polynomial and any parameter that can be put in the form of Eq. (5.3).

Evans's method

and the rules to construct this graph is called the **root-locus method of Evans**. We begin our discussion of the method with the mechanics of constructing a root locus, using the equation in the form of Eq. (5.3) and K as the variable parameter.

To set the notation for our study, we assume here that the transfer function $L(s)$ is a rational function whose numerator is a monic[2] polynomial $b(s)$ of degree m and whose denominator is a monic polynomial $a(s)$ of degree n such that[3] $n \geq m$. We can factor these polynomials as

$$b(s) = s^m + b_1 s^{m-1} + \cdots + b_m$$

$$= (s - z_1)(s - z_2) \cdots (s - z_m)$$

$$= \prod_{i=1}^{m} (s - z_i), \tag{5.4}$$

$$a(s) = s^n + a_1 s^{n-1} + \cdots + a_n$$

$$= \prod_{i=1}^{n} (s - p_i).$$

The roots of $b(s) = 0$ are the zeros of $L(s)$ and are labeled z_i and the roots of $a(s) = 0$ are the poles of $L(s)$ and are labeled p_i. The roots of the characteristic equation itself are r_i from the factored form

$$a(s) + Kb(s) = (s - r_1)(s - r_2) \cdots (s - r_n). \tag{5.5}$$

Root-locus forms

We may now state the root-locus problem expressed in Eq. (5.3) in several equivalent but useful ways. Each of the following equations has the same roots:

$$1 + KL(s) = 0, \tag{5.6a}$$

$$1 + K\frac{b(s)}{a(s)} = 0, \tag{5.6b}$$

$$a(s) + Kb(s) = 0, \tag{5.6c}$$

$$L(s) = -\frac{1}{K}. \tag{5.6d}$$

Equations (5.6) are sometimes referred to as the **root-locus form** or Evans form of a characteristic equation. The root locus is the set of values of s for which

---

[2] Monic means the coefficient of the highest power of s is 1.

[3] If $L(s)$ is the transfer function of a physical system, it is necessary that $n \geq m$ or else the system would have an infinite response to a finite input. If the parameter should be chosen so that $n < m$, then we can consider the equivalent equation $1 + K^{-1}L(s)^{-1} = 0$.

Eqs. (5.6) hold for some positive real value[4] of K. Because the solutions to Eqs. (5.6) are the roots of the closed-loop system characteristic equation and are thus closed-loop poles of the system, the root-locus method can be thought of as a method for inferring dynamic properties of the closed-loop system as the parameter K changes.

EXAMPLE 5.1

Root Locus of a Motor Position Control

In Chapter 2 we saw that a normalized transfer function of a DC motor voltage-to-position can be

$$\frac{\Theta_m(s)}{V_a(s)} = \frac{Y(s)}{U(s)} = G(s) = \frac{A}{s(s+c)}.$$

Solve for the root locus of closed-loop poles of the system created by feeding back the output Θ_m as shown in Fig. 5.1 with respect to the parameter A if $D(s) = H(s) = 1$ and also $c = 1$.

Solution.   In terms of our notation, the values are

$$L(s) = \frac{1}{s(s+1)}, \qquad b(s) = 1, \qquad m = 0, \qquad z_i = \{empty\}, \tag{5.7}$$

$$K = A, \qquad a(s) = s^2 + s, \qquad n = 2, \qquad p_i = 0, -1.$$

From Eq. (5.6) the root locus is a graph of the roots of the quadratic equation

$$a(s) + Kb(s) = s^2 + s + K = 0. \tag{5.8}$$

Using the quadratic formula, we can immediately express the roots of Eq. (5.8) as

$$r_1, r_2 = -\frac{1}{2} \pm \frac{\sqrt{1-4K}}{2}. \tag{5.9}$$

A plot of the corresponding root locus is shown in Fig. 5.2. For $0 \le K \le \frac{1}{4}$, the roots are real between -1 and 0. At $K = \frac{1}{4}$ there are two roots at $-\frac{1}{2}$, and for $K > \frac{1}{4}$ the roots become complex with real parts constant at $-\frac{1}{2}$ and imaginary parts that increase essentially in proportion to the square root of K. The dashed lines in Fig. 5.2 correspond to roots with a damping ratio $\zeta = 0.5$. The poles of $L(s)$ at $s = 0$ and $s = -1$ are marked by the symbol $\times$, and the points where the locus crosses the lines where the damping ratio equals 0.5 are marked with dots ($\bullet$). We can compute K at the point where the locus crosses $\zeta = 0.5$ because we know that, if $\zeta = 0.5$, then $\theta = 30°$ and the magnitude of the imaginary part of the root is $\sqrt{3}$ times the magnitude of the real part. Because the size of the real part is $\frac{1}{2}$, then from Eq. (5.9) we have

$$\frac{\sqrt{4K-1}}{2} = \frac{\sqrt{3}}{2},$$

and, therefore, $K = 1$.

---

[4] If K is positive, the locus is called the "positive" locus. We will consider later the simple changes if $K < 0$, resulting in a "negative" locus.

Figure 5.2
Root locus for
$L(s) = 1/[s(s+1)]$

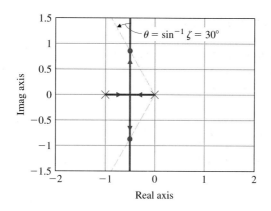

We can observe several features of this simple locus by looking at Eqs. (5.8) and (5.9) and Fig. 5.2. First, there are two roots and thus two branches of the root locus. At $K = 0$ these branches begin at the poles of $L(s)$ (which are at 0 and -1) as they should, since for $K = 0$ the system is open-loop and the characteristic equation is $a(s) = 0$. As K is increased, the roots move toward each other, coming together at $s = -\frac{1}{2}$ and at that point they break away from the real axis. After the **breakaway point** the roots move off to infinity with equal real parts, so the sum of the two roots is always -1. From the viewpoint of design, we see that by altering the value of the parameter K, we can cause the closed-loop poles to be at any point along the locus in Fig. 5.2. If some points along this locus correspond to a satisfactory transient response, then we can complete the design by choosing the corresponding value of K; otherwise, we are forced to consider a more complex controller. As we pointed out earlier, the root locus technique is not limited to focusing on the system gain ($K = A$ in Example 5.1); the same ideas are applicable for finding the locus with respect to *any* parameter that enters linearly in the characteristic equation.

Breakaway points are where roots move away from the real axis

EXAMPLE 5.2   *Root Locus with Respect to a Plant Open-Loop Pole*

Consider the characteristic equation as in Example 5.1, except that now let $D(s) = H(s) = 1$ and also $A = 1$. Select c as the parameter of interest in the equation:

$$1 + G(s) = 1 + \frac{1}{s(s+c)}. \tag{5.10}$$

Find the root locus of the characteristic equation with respect to c.

Solution.   The corresponding closed-loop characteristic equation in polynomial form is

$$s^2 + cs + 1 = 0. \tag{5.11}$$

Figure 5.3
Root locus vs. damping factor c for $1 + G(s) = 1 + 1/[s(s + c)] = 0$

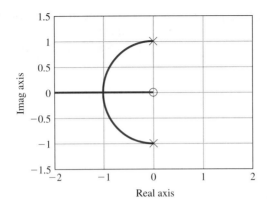

The alternatives of Eq. (5.6) with the associated definitions of poles and zeros will apply if we let

$$L = \frac{s}{s^2 + 1}, \qquad b(s) = s, \qquad m = 1, \qquad z_i = 0,$$

$$K = c, \qquad a(s) = s^2 + 1, \qquad n = 2, \qquad p_i = +j, -j. \tag{5.12}$$

Thus, the root-locus form of the characteristic equation is

$$1 + c\frac{s}{s^2 + 1} = 0.$$

The solutions to Eq. (5.11) are easily computed as

$$r_1, r_2 = -\frac{c}{2} \pm \frac{\sqrt{c^2 - 4}}{2}. \tag{5.13}$$

The locus of solutions is shown in Fig. 5.3, with the poles [roots of $a(s)$] again indicated by ×s and the zero [root of $b(s)$] by the circle ○ symbol. Note that when $c = 0$, the roots are at the ×s on the imaginary axis and the corresponding response would be oscillatory. The damping ratio ζ grows as c increases from 0. At $c = 2$, there are two roots at $s = -1$ and the two locus segments abruptly change direction and move in opposite directions along the real axis; this point of multiple roots where two or more roots come into the real axis is called a **break-in point**.

Break-in point

Of course, computing the root locus for a quadratic equation is easy to do because we can solve the characteristic equation for the roots, as was done in Eq. (5.9) and (5.13), and directly plot these as a function of the parameter K. To be useful the method must be suitable for higher-order systems where explicit solutions are difficult to obtain and rules for the construction of a general root locus were developed by Evans. With the availability of MATLAB, these rules are no longer necessary to plot a specific locus. The command rlocus(sys) will do that. However, in control design we are interested not only in a specific locus but also in how to modify the dynamics in such a way as to propose a system that will meet the dynamic response specifications for good control performance.

For this purpose, it is very useful to be able to roughly sketch a locus so as to be able to evaluate the consequences of possible compensation alternatives. It is also important to be able to quickly evaluate the correctness of a computer generated locus to verify that what is plotted by MATLAB is in fact what was meant to be plotted. It is easy to get a constant wrong or to leave out a term and GIGO[5] is the well known first rule of computation.

5.2   Guidelines for Sketching a Root Locus

We begin with a formal definition of a root locus. From the form of Eq. (5.6a) we define the root locus this way:

Definition I

> The root locus is the set of values of s for which $1 + KL(s) = 0$ is satisfied as the real parameter K varies from 0 to $+\infty$. Typically, $1 + KL(s) = 0$ is the characteristic equation of the system and in this case the roots on the locus are the closed-loop poles of that system.

Now suppose we look at Eq. (5.6d). If K is to be real and positive, $L(s)$ must be real and negative. In other words, if we arrange $L(s)$ in polar form as magnitude and phase, then the phase of $L(s)$ must be $180°$ in order to satisfy Eq. (5.6d). We can thus define the root locus in terms of this **phase condition** as follows.

Definition II

The basic root-locus rule; the phase of $L(s) = 180°$

> The root locus of $L(s)$ is the set of points in the s-plane where the phase of $L(s)$ is $180°$. If we define the angle to the test point from a zero as ψ_i and the angle to the test point from a pole as ϕ_i then Definition II is expressed as those points in the s-plane where, for integer l,
>
> $$\sum \psi_i - \sum \phi_i = 180° + 360°(l - 1) \qquad (5.14)$$

The immense merit of Definition II is that, while it is very difficult to solve a high-order polynomial by hand, computing the phase of a transfer function is relatively easy. The usual case is when K is real and positive, and we call this case the **positive** or **180° locus**. When K is real and negative, $L(s)$ must be real and positive with a phase of $0°$, and this case is called the **negative** or **0° locus**.

From Definition II we can in principle sketch a positive root locus for a complex transfer function by measuring the phase and marking those places where we find $180°$. This direct approach can be illustrated by considering the example

$$L(s) = \frac{s + 1}{s(s + 5)[(s + 2)^2 + 4]}. \qquad (5.15)$$

---

[5] Garbage in, garbage out.

Figure 5.4
Measuring the phase of
Eq. (5.15)

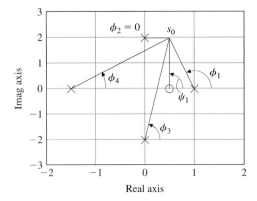

In Fig. 5.4 the poles of this $L(s)$ are marked $\times$ and the zero is marked $\circ$. Suppose we select the test point $s_0 = -1 + 2j$. We would like to test whether or not s_0 lies on the root locus for some value of K. For this point to be on the locus, we must have $\angle L(s_0) = 180° + 360°(l - 1)$ for some integer l, or, equivalently, from Eq. (5.15)

$$\angle(s_0 + 1) - \angle s_0 - \angle(s_0 + 5) - \angle[(s_0 + 2)^2 + 4] = 180° + 360°(l - 1) \quad (5.16)$$

The angle from the zero term $s_0 + 1$ can be computed[6] by drawing a line from the location of the zero at -1 to the test point s_0. In this case the line is vertical and has a phase angle, marked $\psi_1 = 90°$ on Fig. 5.4. In similar fashion, the vector from the pole at $s = 0$ to the test point s_0 is shown with angle ϕ_1, and the angles of the two vectors from the complex poles at $-2 \pm 2j$ to s_0 are shown with angles ϕ_2 and ϕ_3. The phase of the vector $s_0 + 5$ is shown with angle ϕ_4. From Eq. (5.16) we find the total phase of $L(s)$ at $s = s_0$ to be the sum of the phase of the numerator term corresponding to the zero minus the phases of the denominator terms corresponding to the poles:

$$\angle L = \psi_1 - \phi_1 - \phi_2 - \phi_3 - \phi_4$$
$$= 90° - 116.6° - 0° - 76° - 26.6°$$
$$= -129.2°.$$

Because the phase of $L(s)$ is not $180°$, we conclude that s_0 is *not* on the root locus, so we must select another point and try again. Although measuring phase is not particularly hard, measuring phase at every point in the s-plane is hardly practical. Therefore, to make the method practical we need some general guidelines for determining where the root locus is. Evans developed

---

[6] The graphical evaluation of the magnitude and phase of a complex number is reviewed in Appendix B, Section B.3.

a set of rules for the purpose which we will illustrate by applying them to the root locus for

$$L(s) = \frac{1}{s[(s+4)^2 + 16]}.$$ (5.17)

We begin by consideration of the positive locus, which is by far the most common case.[7] The first three rules are relatively simple to remember and are essential for any reasonable sketch. The last three are useful but are marked with the ▲ to show that they are used occasionally but are optional. This judgment is based on our assumption that MATLAB or its equivalent is always available to make an accurate plot of a promising locus.

5.2.1   Rules for Plotting a Positive $(180°)$ Root Locus

RULE 1. The n branches of the locus start at the poles of $L(s)$ and m of these branches end on the zeros of $L(s)$.

  From the equation $a(s) + Kb(s) = 0$, if $K = 0$ the equation reduces to $a(s) = 0$ whose roots are the poles. When K approaches infinity, s must be such that either $b(s) = 0$ or $s \to \infty$. Because there are m zeros where $b(s) = 0$, m branches can end in these places. The case for $s \to \infty$ is considered in Rule 3.

RULE 2. The loci are on the real axis to the left of an odd number of poles and zeros.

  If we take a test point on the real axis such as s_0 in Fig. 5.5, we find that the angles ϕ_1 and ϕ_2 of the two complex poles cancel each other as would the angles from complex conjugate zeros. Angles from real poles or zeros are $0°$ if the test point is to the right and $180°$ if the test point is to the *left* of a given pole or zero. Therefore, for the total angle to add to $180° + 360°l$ the test point

Figure 5.5
Rule 2. The real-axis parts of the locus are to the left of an odd number of poles and zeros.

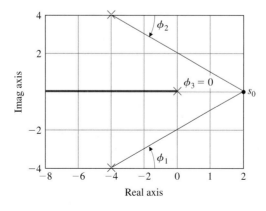

---

[7] The negative locus will be considered shortly.

must be to the left of an *odd* number of real-axis poles plus zeros as shown in Fig. 5.5.

RULE 3. For large s and K, $n - m$ of the loci are asymptotic to lines at angles ϕ_l radiating out from the point $s = \alpha$ on the real axis where

$$\phi_l = \frac{180° + 360°(l - 1)}{n - m}, \qquad l = 1, 2, \ldots, n - m,$$

$$\alpha = \frac{\sum p_i - \sum z_i}{n - m}.$$

(5.18)

As $K \to \infty$, the equation

$$L(s) = -\frac{1}{K}$$

can be satisfied only if $L(s) = 0$. This can occur in two apparently different ways. In the first instance, as discussed in Rule 1, m roots will be found to approach the zeros of $L(s)$. The second manner in which $L(s)$ may go to zero is if $s \to \infty$ because, by assumption, n is larger than m. The asymptotes describe how these $n - m$ roots approach $s \to \infty$. For large s, the equation

$$1 + K \frac{s^m + b_1 s^{m-1} + \cdots + b_m}{s^n + a_1 s^{n-1} + \cdots + a_n} = 0.$$

(5.19)

can be approximated[8] by

$$1 + K \frac{1}{(s - \alpha)^{n-m}} = 0.$$

(5.20)

This is the equation for a system where there are $n - m$ poles, all clustered at $s = \alpha$. Another way to visualize this same result is to consider the picture we would see if we could observe the locations of poles and zeros from a vantage point of very large s: They would appear to cluster near the s-plane origin. Thus m zeros would cancel the effects of m of the poles, and the other $n - m$ poles would appear to be in the same place. We say that the locus of Eq. (5.19) is asymptotic to the locus of Eq. (5.20) for large values of K and s. We need to compute α and to find the locus for the resulting asymptotic system. To find the locus, we choose our search point s_0 such that $s_0 = R e^{j\phi}$ for some large fixed value of R and variable ϕ. Because all poles of this simple system are in the same place, the angle of its transfer function is $180°$ if all $n - m$ angles, each equal to ϕ_l, sum to $180°$. Therefore, ϕ_l is given by

$$(n - m)\phi_l = 180° + 360°(l - 1)$$

---

[8] This approximation can be obtained by dividing $a(s)$ by $b(s)$ and matching the dominant two terms (highest powers in s) to the expansion of $(s - \alpha)^{n-m}$.

for some integer l. Thus, the asymptotic root locus consists of radial lines at the $n - m$ distinct angles given by

$$\phi_l = \frac{180° + 360°(l - 1)}{n - m}, \qquad l = 1, 2, \ldots, n - m. \qquad (5.21)$$

For the system described by Eq. (5.17), $n - m = 3$ and $\phi_{1,2,3} = 60°$, $180°$, and $300°$ or $\pm 60°$, $180°$.

The lines of the asymptotic locus come from $s_0 = \alpha$ on the real axis. To determine α we make use of a simple property of polynomials. Suppose we consider the monic polynomial $a(s)$ with coefficients a_i and roots p_i, as in Eq. (5.4), and we equate the polynomial form with the factored form

$$s^n + a_1 s^{n-1} + a_2 s^{n-2} + \cdots + a_n = (s - p_1)(s - p_2) \cdots (s - p_n).$$

If we multiply out the factors on the right side of this equation, we see that the coefficient of s^{n-1} is $-p_1 - p_2 - \cdots - p_n$. On the left side of the equation we see that this term is a_1. Thus $a_1 = -\sum p_i$; in other words, the coefficient of the *second* highest term in a monic polynomial is the negative sum of its roots—in this case, the poles of $L(s)$. Applying this result to the polynomial $b(s)$, we find the negative sum of the zeros to be b_1. These results can be written as

$$\begin{aligned} -b_1 &= \sum z_i, \\ -a_1 &= \sum p_i. \end{aligned} \qquad (5.22)$$

Finally, we apply this result to the closed-loop characteristic polynomial obtained from Eq. (5.19):

$$s^n + a_1 s^{n-1} + \cdots + a_n + K(s^m + b_1 s^{m-1} + \cdots + b_m)$$

$$= (s - r_1)(s - r_2) \cdots (s - r_n) = 0. \qquad (5.23)$$

Note that the sum of the roots is the negative of the coefficient of s^{n-1}, and **is independent of K** if $m < n - 1$. Therefore if $L(s)$ has at least two more poles than zeros, we have $a_1 = -\sum r_i$. We have thus shown that the center point of the roots *does not change with K* if $m < n - 1$ and that the open-loop and closed-loop sum is the same and is $-a_1$, which can be expressed as

$$-\sum r_i = -\sum p_i. \qquad (5.24)$$

For large values of K, we have seen that m of the roots r_i approach the zeros z_i and $n - m$ of the roots approach the branches of the asymptotic system $1/(s - \alpha)^{n-m}$ whose poles add up to $(n - m)\alpha$. Combining these results we conclude that the sum of all the roots equals the sum of those roots that go to infinity plus the sum of those roots that go to the zeros of $L(s)$:

$$-\sum r_i = -(n - m)\alpha - \sum z_i = -\sum p_i.$$

The center of the asymptotes   Solving for α, we get

$$\alpha = \frac{\sum p_i - \sum z_i}{n - m}. \tag{5.25}$$

Notice that in the sums $\sum p_i$ and $\sum z_i$ the imaginary parts *always* add to zero because complex poles and zeros always occur in complex conjugate pairs. Thus Eq. (5.25) requires information about the real parts only. For Eq. (5.25) we obtain

$$\alpha = \frac{-4 - 4 + 0}{3 - 0}$$

$$= -\frac{8}{3} = -2.67.$$

The asymptotes at $\pm 60°$ are shown dashed in Fig. 5.6. Notice that they cross the imaginary axis at $\pm (2.67)j\sqrt{3} = \pm 4.62j$. The asymptote at $180°$ was already found on the real axis by Rule 2.

▲   RULE 4. The angle(s) of departure of a branch of the locus from a pole of multiplicity q is given by

$$q\phi_{l,\text{dep}} = \sum \psi_i - \sum_{i \neq l} \phi_i - 180° - 360°(l - 1), \tag{5.26}$$

and the angle(s) of arrival of a branch at a zero of multiplicity q is given by

$$q\psi_{l,\text{arr}} = \sum \phi_i - \sum_{i \neq l} \psi_i + 180° + 360°(l - 1). \tag{5.27}$$

If a system has poles near the imaginary axis it can be important to know if the locus which starts at such a pole starts off toward the stable left half-plane or heads toward the unstable right half-plane. To compute the angle by which a branch of the locus departs from one of the poles we take a test point s_o very

Figure 5.6
The asymptotes are $n - m$ radial lines from α at equal angles.

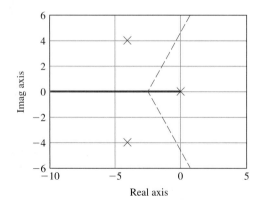

near the pole in question, define the angle from that pole to the test point as $\phi_{l,\text{dep}}$ and transpose all other terms of Eq. (5.14) to the right-hand side. We can illustrate the process by taking the test point s_o to be near the pole at $-4 + 4j$ of our example and computing the angle of $L(s_0)$. The situation is sketched in Fig. 5.7 and the angle from $-4 + 4j$ to the test point we define as ϕ_1. We select the test point close enough to the pole that the angles ϕ_2 and ϕ_3 to the test point can be considered the same as those angles to the pole. Thus, $\phi_2 = 90°$, $\phi_3 = 135°$, and ϕ_1 can be calculated from the angle condition as *whatever it takes* to make the total be $180°$. The calculation is ($l = 1$)

$$\phi_1 = -90° - 135° - 180° \tag{5.28}$$

$$= -405 \tag{5.29}$$

$$= -45 \tag{5.30}$$

By the complex conjugate symmetry of the plots, the angle of departure of the locus near the pole at $-4 - 4j$ will be $+45°$.

If there had been zeros in $L(s)$, the angles from the pole to the zeros would have been added to the right side of Eq. (5.28). For the general case, we can see from Eq. (5.28) that the angle of departure from a single pole is

$$\phi_{1,\text{dep}} = \sum \psi_i - \sum_{i \neq 1} \phi_i - 180°, \tag{5.31}$$

where $\sum \phi_i$ is the sum of the angles to the remaining poles and $\sum \psi_i$ is the sum of the angles to all the zeros. For a multiple pole of order q, we must count the angle from the pole q times. This alters Eq. (5.31) to

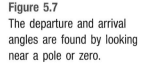

Rule for departure angles

$$q\phi_{l,\text{dep}} = \sum \psi_i - \sum_{i \neq l} \phi_i - 180° - 360°(l - 1), \tag{5.32}$$

where l takes on q values because there are q branches of the locus that depart from such a multiple pole.

Figure 5.7
The departure and arrival angles are found by looking near a pole or zero.

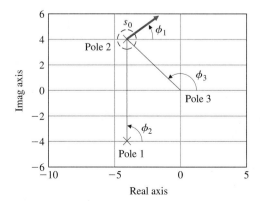

The process of calculating a departure angle for small values of K, as shown in Fig. 5.7, is also valid for computing the angle by which a root locus arrives at a zero of $L(s)$ for large values of K. The general formula that results is

$$q\psi_{l,\text{arr}} = \sum \phi_i - \sum_{i \neq l} \psi_i + 180° + 360°(l - 1) \qquad (5.33)$$

where $\sum \phi_i$ is the sum of the angles to all the poles, $\sum \psi_i$ is the sum of the angles to the remaining zeros, and l is an integer as before.

▲   **RULE 5.** The locus crosses the $j\omega$ axis at points where the Routh criterion shows a transition from roots in the left half-plane to roots in the right half-plane.

In all cases where $n - m > 2$ at least one branch of the locus will cross the imaginary axis and knowing these values can help sketch the locus. A root of the characteristic equation in the right half-plane (RHP) implies that the closed-loop system is unstable, a fact that can be tested by Routh's stability criterion. If we compute the Routh array using K as a parameter, we can locate those values of K for which an incremental change will cause the number of roots in the RHP to change. Such values must correspond to a root locus crossing the imaginary axis. Given the facts that K is known and that a root exists for $s_0 = j\omega_0$, for low order systems it is not hard to solve for the frequency of crossing, ω_0. For the third-order example the characteristic equation is

$$1 + \frac{K}{s[(s + 4)^2 + 16]} = 0,$$

which is equivalent to

$$s^3 + 8s^2 + 32s + K = 0. \qquad (5.34)$$

The Routh array for this polynomial is

s^3 :	1	32
s^2 :	8	K
s^1 :	$\dfrac{8 \cdot 32 - K}{8}$	0
s^0 :	K	

In this case we see that the equation has no roots in the RHP for $0 < K < 8 \cdot 32 = 256$. For $K = 0$, there is obviously a root at $s = 0$ on the axis. For $K < 256$ there are no roots in the RHP, and if $K > 256$, the Routh test indicates that there are two roots in the RHP. Thus $K = 256$ must correspond to a solution at $s = j\omega_0$ for some ω_0. Substituting this data into Eq. (5.34), we find

$$(j\omega_0)^3 + 8(j\omega_0)^2 + 32(j\omega_0) + 256 = 0. \qquad (5.35)$$

For Eq. (5.35) to be true, both real and imaginary parts must equal zero.[9] This gives us

$$-8\omega_0^2 + 256 = 0,$$

---

[9]   Notice that we could have substituted an unknown value of K into Eq. (5.35) and computed $K = 256$ without using the Routh array in this case.

Figure 5.8
A locus crosses the imaginary axis as determined by Routh's criterion for stability.

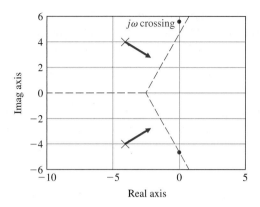

and

$$-\omega_0^3 + 32\omega_0 = 0. \tag{5.36}$$

The solution to Eq. (5.36) is $\omega_0 = \pm\sqrt{32} = \pm 5.66$, which is plotted in Fig. 5.8. We note that the asymptotic locus crosses the imaginary axis at $s = 4.62j$, which is below the actual locus crossing.

▲   RULE 6. The locus will have multiple roots at points *on the locus* where

$$\left(b\frac{da}{ds} - a\frac{db}{ds}\right) = 0. \tag{5.37}$$

and the branches will approach a point of q roots at angles separated by

$$\frac{180° + 360°(l-1)}{q} \tag{5.38}$$

and will depart at angles with the same separation.

As with any polynomial, it is possible for a characteristic polynomial of a degree greater than 1 to have multiple roots. For example, in the second-order locus of Fig. 5.2, there are two roots at $s = -\frac{1}{2}$ when $K = \frac{1}{4}$. Here the horizontal branches of the locus come together and the vertical branches break away from the real axis, becoming complex for $K > \frac{1}{4}$. The locus arrives at $0°$ and $180°$ and departs at $+90°$ and $-90°$. The general situation is that when $K = K_1$ such that there are multiple roots at the point $s = r_1$, the characteristic equation can be written as

$$a(s) + K_1 b(s) = (s - r_1)^q f(s) \tag{5.39}$$

where $f(s)$ represents the product of all the other root factors. If we differentiate this equation with respect to s we find

$$\frac{da}{ds} + K_1\frac{db}{ds} = q(s - r_1)^{q-1}(s - r_2)(\cdots) + (s - r_1)^q\frac{df(s)}{ds}. \tag{5.40}$$

Now if we set $s = r_1$ in this equation, the result is zero. In other words, at a point of multiple roots, the derivative of the polynomial is zero as well as the polynomial being zero at this point. In terms of the original parameters, at this point

$$K_1 = -\left.\frac{a(s)}{b(s)}\right|_{s=r_1} \tag{5.41}$$

Making this substitution, the requirement for a zero derivative is that

$$\left[b\frac{da}{ds} - a\frac{db}{ds}\right]_{s=r_1} = 0. \tag{5.42}$$

For the second-order case in Example 5.1, we have

$$b = 1, \qquad \frac{db}{ds} = 0;$$

$$a = s^2 + s, \qquad \frac{da}{ds} = 2s + 1.$$

Substituting these expressions into Eq. (5.42), we find

$$2s_0 + 1 = 0 \tag{5.43}$$

$$s_0 = -\frac{1}{2}, \tag{5.44}$$

which confirms our previous observation of the breakaway point in Fig. 5.2. For the third-order example given by Eq. (5.17) the polynomials are

$$b = 1, \qquad \frac{db}{ds} = 0;$$

$$a = s[(s + 4)^2 + 16] = s^3 + 8s^2 + 32s,$$

$$\frac{da}{ds} = 3s^2 + 16s + 32.$$

The points of possible multiple roots are therefore given by

$$3s^2 + 16s + 32 = 0,$$

or

$$s_0 = -2.67 \pm 1.89j. \tag{5.45}$$

However, Fig. 5.9 shows that these points are *not* on the root locus and are therefore extraneous. This emphasizes that Eq. (5.42) does not indicate whether s_0 is a multiple root *on the locus*. The derivative condition is necessary but not sufficient to indicate a multiple-root situation. It is also necessary for the point to be on the locus for which $a(s) + Kb(s) = 0$ at $s = r_1$. Referring again to Eq. (5.39), it is easy to see that a similar argument would show that the first $q - 1$ derivatives of the polynomial will be zero at a point of a root of multiplicity q.

Figure 5.9
Root locus for $L(s) = 1/[s(s^2 + 8s + 32)]$

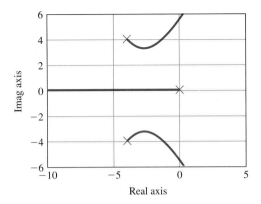

Continuation locus

In order to compute the angles of arrival and departure from a point of multiple roots, it is useful to use a trick we call the **continuation locus**. We can imagine plotting a root locus for an initial range of K, perhaps for $0 \leq K \leq K_1$. If we let $K = K_1 + K_2$, we can then plot a new locus with parameter K_2, a locus which is the *continuation of the original locus* and whose starting poles are the roots of the original system at $K = K_1$. To see how this works, we return to the second-order root locus of Eq. (5.8) and let K_1 be the value corresponding to the breakaway point $K_1 = \frac{1}{4}$. If we let $K = \frac{1}{4} + K_2$, we have the locus equation $s^2 + s + \frac{1}{4} + K_2 = 0$, or

$$\left(s + \frac{1}{2}\right) 2 + K_2 = 0. \tag{5.46}$$

The steps for plotting this locus are, of course, the same as for any other, except that now the initial departure of the locus of Eq. (5.46) corresponds to the breakaway point of the original locus of Eq. (5.8). Applying the rule for departure angles [Eq. (5.32)] from the double pole at $s = -\frac{1}{2}$, we find that

$$2\phi_{\mathrm{dep}} = -180° - 360°(l - 1), \tag{5.47}$$

$$\phi_{\mathrm{dep}} = -90° - 180°(l - 1), \tag{5.48}$$

$$\phi_{\mathrm{dep}} = \pm 90° \text{(departure angles at breakaway)}. \tag{5.49}$$

In this case the arrival angles at $s = -\frac{1}{2}$ are, from the original root locus, along the real axis and are clearly 0° and 180°.

The complete locus for our third-order example is drawn in Fig. 5.9. It combines all the results found so far—that is, the real-axis segment, the center of the asymptotes and their angles, the angles of departure from the poles, and the imaginary-axis crossing points. It is usually sufficient to sketch the locus by using only Rules 1 to 3, which should be memorized. Rule 4 and Rule 5 can be used to refine a hand-drawn locus, especially if it appears that a selected root will be near the $j\omega$ axis. Rule 6 is rarely used, but the information about multiple roots is useful to help interpret plots that come from the computer and, as we will see in the next section, to explain qualitative changes in some loci as a pole or zero is moved. The actual locus in Fig. 5.9 was drawn using the MATLAB commands

```
numL = [1];
denL = [1 8 32 0];
sysL = tf(numL,denL);
rlocus(sysL)
```

5.2.2   Rules for Plotting a Negative ($0°$) Root Locus

We now consider modifying the root-locus procedure to permit analysis of negative values of the parameter. In a number of important cases the transfer function of the plant has a zero in the right half plane and is said to be non-minimum phase. The result is often a locus of the form $1 + A(z_i - s)G'(s) = 1+(-A)(s-z_i)G'(s) = 0$ and in the standard form, the parameter $K = -A$ must be negative. Another important issue calling for understanding the negative locus arises in building a control system. In any physical implementation of a control system there are inevitably a number of amplifiers and components whose gain sign must be selected. By Murphy's Law[10] when the loop is first closed, the sign will be wrong and the behavior will be unexpected unless the engineer understands how the response will go if the gain which should be positive is instead negative. So what are the rules for a negative locus (a root locus relative to a negative parameter)? First of all, Eq. (5.6) must be satisfied for negative values of K, which implies that $L(s)$ is real and *positive*. In other words, for the negative locus the phase condition is:

The angle of $L(s)$ is $0° + 360°(l - 1)$ for s on the negative locus.

The steps for plotting a negative locus are essentially the same as for the positive locus except that we search for places where the angle of $L(s)$ is $0°+360°(l-1)$ instead of $180°+360°(l-1)$. For this reason a negative locus is also referred to as a $0°$ root locus. This time we find that the locus is to the left of an *even* number of real poles plus zeros (zero being even). Computation of the asymptotes for large values of s is, as before, given by

$$\alpha = \frac{\sum p_i - \sum z_i}{n - m},$$ (5.50)

but we modify the angles to be

$$\phi_l = \frac{360°(l - 1)}{n - m}, \qquad l = 1, 2, 3, \ldots, n - m$$

---

[10] Any thing that can go wrong, will go wrong.

[shifted by $180°/(n - m)$ from the $180°$ locus]. Following are the guidelines for plotting a $0°$ locus.

RULE 1. (As before) The n branches of the locus leave the poles and m approach the zeros and $n - m$ approach asymptotes to infinity.

RULE 2. The locus is on the real axis to the left of an *even* number of real poles plus zeros.

RULE 3. The asymptotes are described by

$$\alpha = \frac{\sum p_i - \sum z_i}{n - m} = \frac{-a_1 + b_1}{n - m},$$

$$\phi_l = \frac{360°(l - 1)}{n - m}, \qquad l = 1, 2, 3, \ldots, n - m.$$

Notice that the angle condition here is measured from $0°$ rather than from $180°$ as in the positive locus.

▲   RULE 4. Departure angles from poles and arrival angles to zeros are found by searching for points in the near neighborhood of the pole or zero for points where the phase of $L(s)$ is $0°$, so that

$$q\phi_{\text{dep}} = \sum \psi_i - \sum \phi_i - 360°(l - 1),$$

$$q\psi_{\text{arr}} = \sum \phi_i - \sum \psi_i + 360°(l - 1),$$

where q is the order of the pole or zero, and l takes on q integer values such that the angles are between $\pm 180°$.

▲   RULE 5. The locus crosses the imaginary axis where either by letting $s = j\omega_0$ or applying Routh's criterion shows a change between stability and instability.

▲   RULE 6. The equation has multiple roots at points *on the locus* [for which the angle of $L(s)$ is $0° + 360°(l - 1)$], where

$$b\frac{da}{ds} - a\frac{db}{ds} = 0.$$

The result of extending the guidelines for constructing root loci to include negative parameters is that we can visualize the root locus as a set of continuous curves showing the location of possible solutions to the equation $1 + KL(s) = 0$ for *all real values of* K, both positive and negative. One branch of the locus departs from every pole in one direction for positive values of K, and another branch departs from the same pole in another direction for negative K. Likewise, all zeros will have two branches arriving, one with positive and the other with negative values of K. For the $n - m$ excess poles, there will be $2(n - m)$ branches of the locus asymptotically approaching infinity as K approaches respectively positive and negative infinity. For a single pole or zero the angles of departure or arrival for the two locus branches will be $180°$ apart.

For a double pole or zero the two positive branches will be 180° apart, and the two negative branches will be at 90° to the positive branches.

The negative locus is often required when studying a nonminimum phase transfer function. A well-known example is that of the control of liquid level in the boiler of a steam power plant. If the level is too low, the actuator valve adds (relatively) cold water to the boiling water in the vessel. The initial effect of the addition is to slow down the rate of boiling which reduces the number and size of the bubbles and causes the level to fall momentarily before the added volume and heat cause it to rise again to the new increased level. This initial underflow is typical of nonminimum phase systems. Another typical nonminimum phase transfer function is of the altitude control of an airplane. To make the plane climb the upward deflection of the elevators initially causes the plane to drop before it rotates and climbs. A Boeing 747 in this mode can be described by the scaled and normalized transfer function

$$G(s) = \frac{6 - s}{s(s^2 + 4s + 13)}. \tag{5.51}$$

To put $1 + KG(s)$ in root locus form we need to multiply by -1 to get

$$G(s) = -\frac{s - 6}{s(s^2 + 4s + 13)}. \tag{5.52}$$

EXAMPLE 5.3   *Negative Root Locus for an Airplane*

Sketch the negative root locus for the equation

$$1 + K \frac{s - 6}{s(s^2 + 4s + 13)} = 0. \tag{5.53}$$

Solution.

RULE 1. There are three branches and two asymptotes.

RULE 2. A real axis segment is to the right of $s = 6$ and a segment is to the left of $s = 0$.

RULE 3. The angles of the asymptotes are

$$\phi_l = \frac{(l - 1)360°}{2} = 0° \quad \text{and} \quad 180°,$$

and the center of the asymptotes is at

$$\alpha = \frac{-2 - 2 - (6)}{3 - 1} = -5.$$

▲   RULE 4. The branch departs the pole at $s = -2 + j3$ at the angle

$$\phi = \tan^{-1}\left(\frac{3}{-6}\right) - \tan^{-1}\left(\frac{3}{-2}\right) - 90° + 360°(l - 1),$$

$$\phi = 159.4 - 123.7 - 90 + 360°(l - 1).$$

$$\phi = -54.3°.$$

Figure 5.10
Negative root locus
corresponding to $L(s) = (s - 6)/s(s^2 + 4s + 13)$

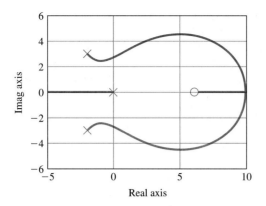

The locus is plotted in Fig. 5.10 by MATLAB, which is seen to be consistent with these values.

5.2.3 Summary of the Rules for Plotting a Root Locus

Primary Rules

RULE 1. The n branches of the locus start at the poles of $L(s)$ and m branches end on the zeros of $L(s)$.

RULE 2. The loci are on the real axis to the left of an odd number of poles and zeros.

RULE 2a. A negative locus is on the real axis to the left of an even number of poles and zeros.

RULE 3. For large s and K, $n - m$ of the loci are asymptotic to lines at angles ϕ_l radiating out from the center point $s = \alpha$ on the real axis where

$$\phi_l = \frac{180° + 360°(l - 1)}{n - m}, \qquad l = 1, 2, \ldots, n - m, \tag{5.54}$$

$$\alpha = \frac{\sum p_i - \sum z_i}{n - m}. \tag{5.55}$$

RULE 3a. The negative locus asymptotes are described by

$$\phi_l = \frac{360°(l - 1)}{n - m}, \qquad l = 1, 2, \ldots, n - m, \tag{5.56}$$

$$\alpha = \frac{\sum p_i - \sum z_i}{n - m}. \tag{5.57}$$

RULE 4. The angle(s) of departure of a branch of the locus from a pole of multiplicity q is given by

$$q\phi_{l,\text{dep}} = \sum \psi_i - \sum \phi_i - 180° - 360°(l-1), \tag{5.58}$$

and the angle(s) of arrival of a branch at a zero of multiplicity q is given by

$$q\psi_{l,\text{arr}} = \sum \phi_i - \sum \psi_i + 180° + 360°(l-1). \tag{5.59}$$

RULE 4a. The angles of arrival and departure for a negative locus satisfy

$$q\phi_{l,\text{dep}} = \sum \psi_i - \sum \phi_i - 360°(l-1), \tag{5.60}$$

$$q\psi_{l,\text{arr}} = \sum \phi_i - \sum \psi_i + 360°(l-1). \tag{5.61}$$

Secondary Rules

RULE 5. The locus crosses the $j\omega$ axis at points where the Routh criterion shows a transition from roots in the left half-plane to roots in the right half-plane.

RULE 6. The locus will have multiple roots at points *on the locus* where the derivative is zero as

$$\left(b\frac{da}{ds} - a\frac{db}{ds}\right) = 0. \tag{5.62}$$

The point will be of multiplicity q if it is on the locus and the first $q-1$ derivatives are zero there. The branches will approach a point of q roots at angles separated by

$$\frac{180° + 360°(l-1)}{q} \tag{5.63}$$

and will depart at angles with the same separation forming an array of $2q$ rays equally spaced. If the point is on the real axis then the orientation of this array is given by the real axis rule. If the point is in the complex plane, then the angle of departure rule must be applied.

RULE 6a. The angles of arrival of q roots on a negative locus to a point of multiple roots are separated by

$$\frac{360°(l-1)}{q}. \tag{5.64}$$

▲ 5.3 Selected Illustrative Root Loci

A number of important control problems are characterized by a process with the simple "double integrator" transfer function

$$G(s) = \frac{1}{s^2}. \tag{5.65}$$

The control of attitude of a satellite is described by this equation. Also the read/write head assembly of a computer hard disk drive is typically floating on an air bearing so that friction is negligible for all but the smallest motion. The motor is typically driven by a current source so the back emf does not affect the torque. The result is a plant described by Eq. (5.65). If we form a unity feedback system with this plant, and a proportional controller, the root locus with respect to controller gain is

$$1 + k_p \frac{1}{s^2} = 0. \tag{5.66}$$

If we apply the rules to this (trivial) case, the results are as follows:

RULE 1. The locus has two branches that start at $s = 0$.

RULE 2. There are no parts of the locus on the real axis.

RULE 3. The two asymptotes have origin at $s = 0$ and are at the angles of $\pm 90°$.

RULE 4. The loci depart from $s = 0$ at the angles of $\pm 90°$.

RULE 5. The loci remain on the imaginary axis for all values of k_p.

RULE 6. The breakaway point is at $s = 0$.

Conclusion: The locus consists of the imaginary axis and the transient would be oscillatory for any value of k_p. A more useful design results with the use of proportional plus derivative control.

EXAMPLE 5.4

Root Locus for Satellite Attitude Control with PD Control

The characteristic equation with PD control is

$$1 + [k_p + k_D s] \frac{1}{s^2} = 0. \tag{5.67}$$

To put the equation in root locus form, we define $K = k_D$, and for the moment arbitrarily select the gain ratio[11] as $k_p/k_D = 1$ which results in the root locus form

$$1 + K \frac{s + 1}{s^2} = 0. \tag{5.68}$$

---

[11] Given a specific physical system, this number would be selected with consideration of the specified rise time of the design or the maximum control signal (control authority) of the actuator.

Solution.   Again we compute the results of the rules.

RULE 1. There are two branches which start at $s = 0$, one of which terminates on the zero at $s = -1$ and the other approaches infinity.

RULE 2. The real axis to the left of $s = -1$ is on the locus.

RULE 3. Because $n - m = 1$, there is one asymptote along the negative real axis.

RULE 4. The angles of departure from the double pole at $s = 0$ are $\pm 90°$.

RULE 5. Applying Routh's criterion, we find the array

$$
\begin{array}{cc}
1 & K \\
K & \\
K & \\
\end{array}
\tag{5.69}
$$

and conclude that the locus does not cross the imaginary axis.

RULE 6. The points of multiple roots are found from

$$
\begin{array}{ll}
b = s + 1 & \dfrac{db}{ds} = 1 \\
a = s^2 & \dfrac{da}{ds} = 2s
\end{array}
\tag{5.70}
$$

and are the solutions to

$$
b\frac{da}{ds} - a\frac{db}{ds} = (s + 1)2s - s^2 = 0
\tag{5.71}
$$

$$
s^2 + 2s = 0
\tag{5.72}
$$

$$
s_i = 0, \quad -2
\tag{5.73}
$$

Because the point $s = -2$ is on the locus from Rule 2, it is a point of multiple roots, in this case a point of break-in. We conclude that two branches of the locus leave the origin going North and South, curve around[12] without passing into the right half-plane, and break into the real axis at $s = -2$, from which point one branch goes West toward infinity while the other goes East to rendezvous with the zero at $s = -1$. The locus is plotted in Fig. 5.11 with the commands

```
numS = [1 1];
denS = [1 0 0];
sysS = tf(numS,denS);
rlocus( sysS)
```

---

[12] You can prove that the path is a circle by assuming that $s + 1 = e^{j\theta}$ and showing that the equation has a solution for a range of positive K and real θ under this assumption.

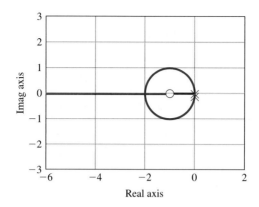

Figure 5.11
Root locus for $L(s) = G(s) = (s+1)/s^2$

Comparing this case with that for the simple $1/s^2$, we see the following:

Effect of a zero in the left half-plane

> The addition of the zero has pulled the locus into the left half-plane, a point of general importance in constructing a compensation.

In the previous case, we considered pure PD control. However, as we have mentioned earlier, the physical operation of differentiation is not practical and in practice PD control is approximated by

$$D(s) = k_p + \frac{k_D s}{s/p + 1}, \qquad (5.74)$$

which can be put in root locus form by defining $K = k_p + p k_D$ and $z = p k_p / K$ so that[13]

$$D(s) = K \frac{s+z}{s+p}. \qquad (5.75)$$

For reasons we will see when we consider design by frequency response, this controller transfer function is called a "lead compensator" or, referring to the frequent implementation by electrical components, a "lead network." The characteristic equation for the $1/s^2$ plant with this controller is

$$1 + D(s)G(s) = 1 + KL(s) = 0,$$
$$1 + K \frac{s+z}{s^2(s+p)} = 0.$$

---

[13] The use of z here for zero is not to be confused with the use of the operator z used in defining the discrete transfer function needed to describe digital controllers.

EXAMPLE 5.5

Root Locus of the Satellite Control with Modified PD or Lead Compensation

To evaluate the effect of the added pole, we will again set $z = 1$ and consider three different values for p. We begin with a somewhat large value, $p = 12$ and consider the root locus for

$$1 + K\frac{s+1}{s^2(s+12)}. \tag{5.76}$$

Solution.   Again we apply the rules for plotting a root locus.

RULE 1. There are now three branches to the locus, two starting at $s = 0$ and one starting at $s = -12$.

RULE 2. The real axis segment between $-12 \le s \le -1$ is part of the locus.

RULE 3. There are $n - m = 3 - 1 = 2$ asymptotes centered at

$$\alpha = \frac{-12 - (-1)}{2} = -11/2$$

and at the angles $\pm 90°$.

RULE 4. The angles of departure of the branches at $s = 0$ are again $\pm 90°$.

RULE 5. The Routh array will again confirm that no branch of the locus crosses the imaginary axis for positive K.

RULE 6. The points of multiple roots are found from

$$b = s + 1, \qquad \frac{db}{ds} = 1$$

$$a = s^3 + 12s^2, \qquad \frac{da}{ds} = 3s^2 + 24s$$

and are the solutions of

$$b\frac{da}{ds} - a\frac{db}{ds} = (s+1)(3s^2 + 24s) - (s^3 + 12s^2) = 0,$$

$$2s^3 + 15s^2 + 24s = 0,$$

$$s_i = 0, -2.31, -5.18.$$

In this case the equation has an obvious solution at $s = 0$ and the resulting quadratic can be readily solved by the quadratic formula. However, MATLAB is quicker. Try roots([2 15 24 0]). From the point of view of the root locus, the solution at $s = 0$ is an obvious point of multiple roots and the other two solutions are also on the locus based on the results of Rule 2. We conclude that two branches of this locus break away vertically from the poles at $s = 0$, curve around to the left without passing into the right half plane, and break in at $s = -2.31$ where one branch goes right to meet the zero at $s = -1$ and the other goes left where it is met by the root which left the pole at $s = -12$. These two form a multiple root at $s = -5.18$ and break away there and approach the vertical asymptotes located at $s = -5.5$. The locus is plotted in Fig. 5.12 by the MATLAB commands

```
numL = [1 1];
denL = [1 12 0 0];
sysL = tf(numL,denL);
rlocus(sysL)
```

Figure 5.12
Root locus for
$L(s) = (s+1)/s^2(s+12)$

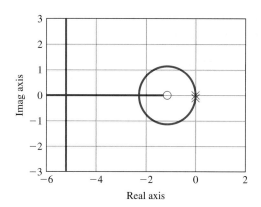

Considering this locus, we see that the effect of the added pole has been to distort the simple circle of the PD control but, for points near the origin, the locus is quite similar to the earlier case. The situation changes when the pole is brought closer in.

EXAMPLE 5.6

Root Locus of the Satellite Control with Lead Having a Relatively Small Value for the Pole

Now consider $p = 4$ and sketch the root locus for

$$1 + K\frac{s+1}{s^2(s+4)} = 0. \tag{5.77}$$

Solution.   Again, by the rules.

RULE 1. There are again three branches to the locus two starting from $s = 0$ and one from $s = -4$.

RULE 2. The segment of the real axis from $-4 \le s \le -1$ is part of the locus.

RULE 3. There are two asymptotes centered at $\alpha = -3/2$ and at the angles $\pm 90°$.

RULE 4. The branches again depart from the poles at $s = 0$ at $\pm 90°$.

RULE 5. The Routh array again shows no crossing of the imaginary axis.

RULE 6. The equation for multiple roots is, in this case,

$$b\frac{da}{ds} - a\frac{db}{ds} = (s+1)(3s^2 + 8s) - (s^3 + 4s^2) = 0,$$
$$2s^3 + 7s^2 + 8s = 0,$$
$$s = 0, \qquad -1.75 \pm j0.97,$$

The solution at $s = 0$ is an obvious point of a double root. The other two are not on the locus for, with only three roots and one of these on the real axis it is not possible

Figure 5.13
Root locus for
$L(s) = (s+1)/s^2(s+4)$

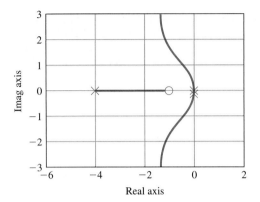

with a real polynomial to have two roots in the upper half-plane and none in the lower half-plane. In this case, the locus differs from the case when $p = -12$ in that there are no break-in or breakaway points on the real axis as part of the locus. The MATLAB plot is given in Fig. 5.13.

In these two cases we have similar systems, but in one case, $p = -12$, there were both a break-in and a breakaway on the real axis, whereas for $p = -4$ these features have disappeared. A logical question might be to ask at what point they went away. As a matter of fact, it happens at $p = 9$ and we'll look at that locus next.

The Root Locus for the Satellite with a Transition Value for the Pole

EXAMPLE 5.7

Plot the root locus for

$$1 + K \frac{s+1}{s^2(s+9)} = 0. \tag{5.78}$$

RULE 1. The locus has three branches, starting from $s = 0$ and $s = -9$.

RULE 2. The real axis segment from $-9 \le s \le -1$ is part of the locus.

RULE 3. The two asymptotes are centered at $\alpha = -8/2 = -4$.

RULE 4. The departures are as before, at $\pm 90°$ from $s = 0$.

RULE 5. The roots do not cross the imaginary axis.

RULE 6. The equation for multiple roots is

$$b\frac{da}{ds} - a\frac{db}{ds} = (s+1)(3s^2 + 18s) - (s^3 + 9s^2) = 0,$$

$$2s^3 + 12s^2 + 18s = 0,$$

$$s(s+3)^2 = 0,$$

$$s = 0, \qquad -3, \qquad -3.$$

$$\tag{5.79}$$

Now we have a solution that is on the locus, but what are we to make of the fact that the solution is itself a repeated root? The fact is that in this special case the break-in and the breakaway are at the same point where not two but three roots break in. From the rule considering the continuation locus, the angles of break-in are at $-180°$, $60°$, and $-60°$ and the breakaway angles are at $0°$ and $\pm120°$. The locus is plotted in Fig. 5.14.

From Figs. 5.12 to 5.14, it is evident that when the third pole is near the zero, (p near 1) there is only a modest distortion of the locus that would result for $D(s)G(s) = K1/s^2$, which consists of two straight-line locus branches departing at $\pm90°$ from the two poles at $s = 0$. Then, as we increase p, the locus changes until at $p = 9$ the locus breaks in at -3 in a triple multiple root. As the pole p is moved to the left beyond -9, the locus exhibits distinct break-in and break-away points, approaching, as p gets very large, the circular locus of one zero and two poles. Figure 5.14, when $p = 9$, is thus a transition locus between the two second-order extremes, which occur at $p = 1$ (when the zero is canceled) and $p \to \infty$ (where the extra pole has no effect). A useful conclusion drawn from this example is the following:

> An additional pole moving in from the far left tends to push the locus branches to the right as it approaches a given locus.

The double integrator is the simplest model of the examples assuming a rigid body with no friction. A more realistic case would include the affects of flexibility in either the satellite attitude control where at least the solar panels would be flexible. In the case of the disk drive read/write mechanism, the head and supporting arm assembly always has flexibility and usually a very complex behavior with a number of lightly damped modes which can often be usefully approximated by a single dominant mode. In Section 2.1 it was showed that flexibility in the disk drive added a set of complex poles to the $1/s^2$ model. Generally there are two possibilities depending on whether the sensor is on

Figure 5.14
Root locus for
$L(s) = (s + 1)/s^2(s + 9)$

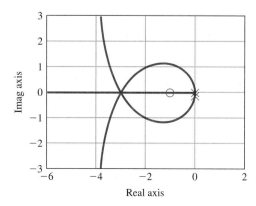

Imag axis

Real axis

the same rigid body as the actuator, which is called the collocated case,[14] or if the sensor is on another body, in which case we have the noncollocated case.[15] We begin with consideration of the collocated case similar to that given by Eq. (2.19). As we saw in Chapter 2, the transfer function in the collocated case has not only a pair of complex poles but also a pair of nearby complex zeros located at a lower natural frequency than the poles. The numbers in the following examples are chosen more to illustrate the root locus properties than to represent particular physical models.

EXAMPLE 5.8

Root Locus of the Satellite Control with a Collocated Flexibility

Plot the root locus of the characteristic equation $1 + G(s)D(s) = 0$ where

$$G(s) = \frac{(s+0.1)^2 + 6^2}{s^2[(s+0.1)^2 + 6.6^2]} \tag{5.80}$$

is in a unity feedback structure with the controller transfer function

$$D(s) = K\frac{s+1}{s+12}. \tag{5.81}$$

Solution.   In this case

$$L(s) = \frac{s+1}{s+12}\frac{(s+0.1)^2 + 6^2}{s^2[(s+0.1)^2 + 6.6^2]}$$

has both poles and zeros near the imaginary axis and we should expect to find the departure angles of particular importance.

RULE 1. There are five branches to the locus, three of which approach finite zeros and two of which approach asymptotes.

RULE 2. The real axis segment between $-12 \le s \le -1$ is part of the locus.

---

[14] Typical of the satellite attitude control where the flexibility arises from solar panels, and both actuator and sensor act on the main body of the satellite.

[15] Typical of the satellite where the flexibility arises from a scientific package whose attitude is to be controlled from a command body coupled to the package by a flexible strut. This case is also typical of computer hard disk read/write head control where the motor is on one end of the arm and the head is on the other.

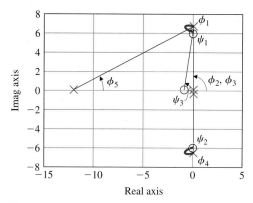

Figure 5.15 Figure for computing a departure angle

for $L(s) = \dfrac{s+1}{s+12}\dfrac{(s+0.1)^2 + 6^2}{s^2[(s+0.1)^2 + 6.6^2]}$

RULE 3. The center of the two asymptotes is at

$$\alpha = \frac{-12 - .1 - .1 - (-.1 - .1 - 1)}{5 - 3} = \frac{11}{2}.$$

The angle of the asymptotes is $\pm 90°$.

RULE 4. We compute the departure angle from the pole at $s = -0.1 + j6.6$. The angle at this pole we will define to be ϕ_1. The other angles are marked on Fig. 5.15. The root locus condition is

$$\phi_1 = \psi_1 + \psi_2 + \psi_3 - (\phi_2 + \phi_3 + \phi_4 + \phi_5) - 180°,$$

$$\phi_1 = 90° + 90° + \tan^{-1}(6.6) - \left[90° + 90° + 90° + \tan^{-1}\left(\frac{6.6}{12}\right)\right] - 180°,$$

(5.82)

$$\phi_1 = 81.4° - 90° - 28.8° - 180°$$

$$= -217.4° = 142.6°,$$

so the root leaves this pole up and to the left, into the stable region of the plane. An interesting exercise would be to compute the arrival angle at the zero located at $s = -0.1 + j6$.

RULE 5. Application of Routh's criterion reveals that there is no crossing of the imaginary axis by this locus.

RULE 6. Although straightforward but tedious, applying this rule is not worth the calculation in this case.

The locus is plotted in Fig. 5.16.

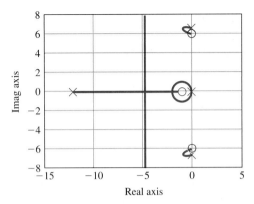

Figure 5.16 Root locus for
$$L(s) = \frac{s+1}{s+12}\frac{(s+0.1)^2 + 6^2}{s^2[(s+0.1)^2 + 6.6^2]}$$

The previous example showed the following:

> In the collocated case, the presence of a single flexible mode introduces a lightly damped root to the characteristic equation but does not cause the system to be unstable.

The departure angle calculation showed that the root departs from the pole introduced by the flexible mode toward the left half-plane. Next, let's consider the noncollocated case for which we take the plant transfer function to be[16]

$$G(s) = \frac{1}{s^2[(s+0.1)^2 + 6.6^2]} \tag{5.83}$$

compensated again by the lead

$$D(s) = K\frac{s+1}{s+12}. \tag{5.84}$$

As these equations show, the noncollocated transfer function has the complex poles but does not have associated complex zeros as shown in Eqs. (2.12) and (2.18). This will have a substantial effect as illustrated by the following example.

---

[16] In a physical structure of two masses with noncollocated actuator and sensor and some damping, there would be a real zero far to the left of the main singularities as shown in Eq. (2.12). Its omission here results in a MATLAB plot with very little distortion and more accuracy in the areas of interest.

EXAMPLE 5.9   *Root Locus for the Noncollocated Case*

Apply the rules and sketch the root locus for

$$KL(s) = DG = K\frac{s+1}{s+12}\frac{1}{s^2[(s+0.1)^2 + 6.6^2]},\qquad (5.85)$$

paying special attention to the departure angles from the complex poles.

RULE 1. There are five branches to the root locus, of which one approaches the zero and four approach the asymptotes.

RULE 2. The real axis segment defined by $-12 \le s \le -1$ is part of the locus.

RULE 3. The center of the asymptotes is located at

$$\alpha = \frac{-12 - 0.2 - (-1)}{5 - 2} = \frac{-11.2}{3}$$

and the angles for the four asymptotic branches are at $\pm 45°, \pm 135°$.

RULE 4. We again compute the departure angle from the pole at $s = -0.1 + j6.6$. The angle at this pole we will define to be ϕ_1. The other angles are marked in Fig. 5.17. The root locus condition is

$$\phi_1 = \psi_1 - (\phi_2 + \phi_3 + \phi_4 + \phi_5) - 180°,$$

$$\phi_1 = \tan^{-1}(6.6) - (90° + 90° + 90° + \tan^{-1}\left(\frac{6.6}{12}\right) - 180°,$$

$$\phi_1 = 81.4° - 90° - 90° - 90° - 28.8° - 180°,\qquad (5.86)$$

$$\phi_1 = 81.4° - 90° - 28.8° - 360°,$$

$$\phi_1 = -37.4°.$$

In this case, the root leaves the pole down and to the *right*, toward the unstable region. We would expect the system to soon become unstable as gain is increased.

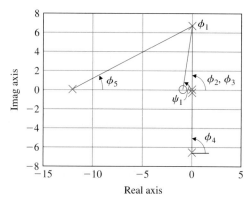

Figure 5.17 Figure to compute a departure angle
for $L(s) = \dfrac{s+1}{s+12}\dfrac{1}{s^2[(s+0.1)^2 + 6.6^2]}$

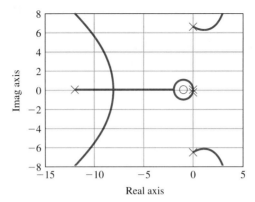

Figure 5.18 Root locus for
$$L(s) = \frac{s+1}{s+12}\frac{1}{s^2[(s+0.1)^2 + 6.6^2]}$$

RULE 5. The first two rows of the Routh array are

$$\begin{matrix} 1 & 45.97 & K \\ 12.2 & 522.84 & K \end{matrix} \qquad (5.87)$$

At this point it is time to admit that applying the Routh criterion in a case of this complexity is of doubtful value. The information given already permits a rough sketch to be made, showing that the locus branches cross the imaginary axis very near the complex poles at $\omega = \pm j6.6$. MATLAB provides a command to verify the value. If the root locus is drawn by rlocus(sysL), then the command [K,P] = rlocfind(sysL) places a crosshair on the plot and a bit of iteration finds that the locus crosses the axis at $\omega = \pm j6.525$ where the parameter is $K = 144.4$.

RULE 6. Again, computing the exact points of breakin and breakaway is of limited value.

The locus is plotted in Fig. 5.18 with the commands

```
numG = 1;
denG = [1.0  0.20  43.57  0];
sysG = tf(numS,denS);
numD = [1   1];
denD = [1  12];
sysD = tf(numD,denD);
sysL = sysD*sysS;
rlocus(sysL)
axis([-18  2   -7.5  7.5])
```

A Locus with Complex Multiple Roots

We have seen loci with break-in and breakaway points on the real axis. Of course an equation of fourth or higher order can have multiple roots that are complex. Although such a feature of a root locus is a rare event, it is an interesting curiosity that is illustrated by the next example.

EXAMPLE 5.10   ***Root Locus Having Complete Multiple Roots***

Sketch the root locus of $1 + KL(s) = 0$, where

$$L(s) = \frac{1}{s(s+2)[(s+1)^2+4]}.$$

Solution.

RULE 1. There are four branches of the locus, all of which approach asymptotes.

RULE 2. The real axis segment $-2 \leq s \leq 0$ is on the locus.

RULE 3. The center of the asymptotes is at

$$\alpha = \frac{-2-1-1-0+0}{4-0} = -1$$

and the angles are $\phi_l = 45°, 135°, -45°, -135°$.

RULE 4. The departure angle ϕ_{dep} from the pole at $= -1+2j$ based on Fig. 5.19 is

$$\phi_{\text{dep}} = \phi_3 = -\phi_1 - \phi_2 - \phi_4 + 180°$$
$$= -\tan^{-1}\left(\frac{2}{-1}\right) - \tan^{-1}\left(\frac{2}{1}\right) - 90° + 180°$$
$$= -116.6° - 63.4° - 90° + 180°$$
$$= -90°$$

We can observe at once that, along the line $s = -1 + j\omega$, ϕ_2, and ϕ_1 are angles of an isosceles triangle and always add to $180°$. Hence, the entire line from one complex pole to the other is on the locus *in this special case*.

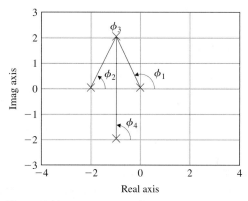

Figure 5.19 Figure to compute departure angle for $L(s) = \dfrac{1}{s(s+2)[(s+1)^2+4]}$

RULE 5. Crossings of the imaginary axis are solutions for $s = j\omega_0$ of the equation

$$s^4 + 4s^3 + 9s^2 + 10s + K = 0.$$

Making the substitution, we find that ω_0 and K must satisfy the equation

$$\omega_0^4 - 4j\omega_0^3 - 9\omega_0^2 + 10j\omega_0 + K = 0.$$

Therefore

$$\omega_0^4 - 9\omega_0^2 + K = 0, \tag{5.88}$$

$$-4\omega_0^3 + 10\omega_0 = 0. \tag{5.89}$$

From Eq.(5.89) we get $\omega_0^2 = \frac{5}{2}$ and thus $\omega_0 = 1.58$. From both Eqs. (5.88) and Eq. (5.89), it follows that

$$K = 9\left(\frac{5}{2}\right) - \frac{25}{4}$$

$$= \frac{90 - 25}{4} = 16.25.$$

This value of K can also be verified by the Routh array.

RULE 6. We locate possible multiple roots from the data:

$$b = 1, \qquad a = s^4 + 4s^3 + 9s^2 + 10s,$$
$$\frac{db}{ds} = 0, \qquad \frac{da}{ds} = 4s^3 + 12s^2 + 18s + 10.$$

The condition reduces to $da/ds = 0$. We could find solutions to this cubic using MATLAB, but there is also another way: From Rule 4 we notice that the line at $s = -1 + j\omega$ is on the locus, so there must be a multiple root breakaway point at $s = -1$, which can be divided out. That is, we can easily show that

$$4s^3 + 12s^2 + 18s + 10 = (s + 1)(4s^2 + 8s + 10).$$

The quadratic has roots $-1 \pm j\sqrt{\frac{3}{2}} = -1 \pm 1.22j$. Because these points are on the line between the complex poles, they are true points of multiple roots on the locus.

The locus plotted with MATLAB is shown in Fig. 5.20. Notice that we have complex multiple roots at $s = -1 \pm 1.22j$. and branches of the locus come together at $-1 \pm 1.22j$ and break away at $0°$ and $180°$.

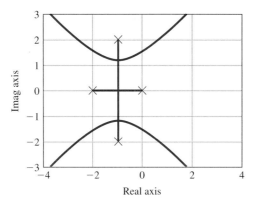

Figure 5.20 Root locus for
$$L(s) = \frac{1}{s(s+2)[(s+1)^2+4]}$$

The locus in this example is a transition between two types of loci: one where the complex poles are to the left of the example case and approach the asymptotes at $\pm 135°$ and another where the complex poles are to the right of their positions in the example and approach the asymptotes at $\pm 45°$.

▲ 5.4 Selecting the Parameter Value

The positive root locus is a plot of *all possible locations* for roots to the equation $1 + KL(s) = 0$ for some real positive value of K. The purpose of design is to select a particular value of K that will meet the specifications for static and dynamic response. We now turn to the issue of selecting K from a particular locus so that the roots are at specific places. Although we will show how the gain selection can be made by hand calculations from a plot of the locus, this is rarely done by hand because the calculation requires considerable care to be done accurately; and because we are dealing with a specific locus, MATLAB has a function which solves the problem easily and accurately. We present the hand calculation first to show what MATLAB is doing and then illustrate the calculation both by hand and by computer.

Using Definition II of the locus, we developed rules to sketch a root locus from the phase of $L(s)$ alone. If the equation is actually to have a root at a particular place when the phase of $L(s)$ is $180°$, then a **magnitude condition** must also be satisfied. This condition is given by Eq. (5.6d), rearranged as

$$K = -\frac{1}{L(s)}.$$

For values of s on the root locus the phase of $L(s)$ is $180°$, so we can write the magnitude condition as

$$K = \frac{1}{|L|}. \tag{5.90}$$

Equation (5.90) has both an algebraic and a graphical interpretation. To see the latter, consider the locus of $1 + KL(s)$ where

$$L(s) = \frac{1}{s[(s+4)^2 + 16]}. \tag{5.91}$$

For this transfer function, the locus is plotted in Fig. 5.21. In Fig. 5.21 the lines corresponding to a damping ratio of $\zeta = 0.5$ are sketched, and the points where the locus crosses these lines are marked with dots ($\bullet$). Suppose we wish to set the gain so that the roots are located at the dots. This corresponds to selecting the gain so that two of the closed-loop system poles have a damping ratio of $\zeta = 0.5$. (We will find the third pole shortly.) What is the value of K when a root is at the dot? From Eq. (5.90), the value of K is given by 1 over the magnitude of $L(s_0)$, where s_0 is the coordinate of the dot. In the figure we have plotted three vectors marked $s_0 - s_1$, $s_0 - s_2$, and $s_0 - s_3$, which are the vectors from the poles of $L(s)$ to the point s_0. (Because $s_1 = 0$, the first vector equals s_0.) Algebraically, we have

$$L(s_0) = \frac{1}{s_0(s_0 - s_2)(s_0 - s_3)}. \tag{5.92}$$

Using Eq. (5.90), this becomes

$$K = \frac{1}{|L(s_0)|} = |s_0||s_0 - s_2||s_0 - s_3|. \tag{5.93}$$

Figure 5.21
Root locus for $L(s) = 1/\{s[(s+4)^2 + 16]\}$ showing calculations of gain K

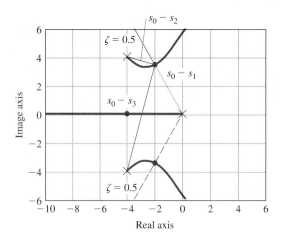

Graphical calculation of the
desired gain

The graphical interpretation of Eq.(5.93) shows that its three magnitudes are the lengths of the corresponding vectors drawn on Fig. 5.21 (see Appendix B). Hence we can compute the gain to place the roots at the dot ($s = s_0$) by measuring the lengths of these vectors and multiplying the lengths together, *provided that the scales of the imaginary and real axes are identical.* Using the scale of the figure we estimate that

$$|s_0| \cong 4.0,$$

$$|s_0 - s_2| \cong 2.1,$$

$$|s_0 - s_3| \cong 7.7.$$

Thus the gain is estimated to be

$$K = 4.0(2.1)(7.7) \cong 65.$$

We conclude that if K is set to the value 65, then a root of $1 + KL$ will be at s_0, which has the desired damping ratio of 0.5. Another root is at the conjugate of s_0. Where is the third root? The third branch of the locus lies along the negative real axis. Ordinarily, we would need to take a test point, compute a trial gain, and repeat this process until we found the point where $K = 65$. In this case it is more convenient to use the property of the root locus expressed in Eq. (5.24)—that the sum of the roots is constant (does not change as K changes) if $m < n - 1$. Thus, the unknown root must be moved far enough to the left to keep the sum fixed. From Fig. 5.21 we estimate that $s_0 = -2 + 3.4j$. Because the starting point was at $s = -4 + 4j$, this root has moved two units to the right. The conjugate has moved an equal distance to the right so the third root must have moved $2 + 2$ units to the left of where it began at $s = 0$. We have marked the new location at -4 with the third dot.

The same calculation can be done with MATLAB. If the locus has been plotted using the command rlocus(sysL), for example, then the command [K,p] = rlocfind(sysL) will produce a crosshair on the plot; and, when spotted at the desired location of the root and selected with a mouse click, the value of the gain K is returned as well as the roots corresponding to that K in the variable p.

Finally, with the parameter selected, it is possible to compute the error constant of the control system. A process with the transfer function given by Eq. (5.91) has one integrator and, in a unity feedback configuration, will be a type 1 control system. In this case the steady-state error in tracking a ramp input is given by the velocity constant:

$$K_v = \lim_{s \to 0} s K L(s) \tag{5.94}$$

$$= \lim_{s \to 0} s \frac{K}{s[(s + 4)^2 + 16]} \tag{5.95}$$

$$= \frac{K}{32}. \tag{5.96}$$

With the gain set for complex roots at a damping $\zeta = 0.5$, the root-locus gain is $K = 65$, so from Eq. (5.96) we get $K_v = 65/32 \cong 2$. If the closed-loop dynamic response as determined by the root locations is satisfactory and the steady-state accuracy as measured by K_v is good enough, then the design can be completed by gain selection alone. However, if no value of K satisfies all the constraints, as is typically the case, then additional modifications are necessary to meet the system specifications.

5.5   Dynamic Compensation

Consideration of control design begins with the design of the process itself. The importance of early consideration of potential control problems in the design of the process and selection of the actuator and sensor cannot be over em-phasized. It is not uncommon for a first study of the control to suggest that the process itself can be changed by, for example, adding damping or stiffness to a structure to make a flexibility easier to control. Once these factors have been taken into account, the design of the controller begins. If the process dynamics are of such a nature that a satisfactory design cannot be obtained by adjustment of the proportional gain alone, then some modification or com-pensation of the process dynamics is indicated. While the variety of possible compensation schemes is great, three categories have been found to be par-ticularly simple and effective. These are lead, lag, and notch compensation.[17]

Lead and lag compensation   **Lead compensation** approximates the function of PD control and acts mainly to speed up a response by lowering rise time and to decrease the transient over-shoot. **Lag compensation** approximates the function of PI control and is usually used to improve the steady-state accuracy of the system. **Notch compensation** will be used to achieve stability for systems with lightly damped flexible modes as we saw with the satellite attitude control having noncollocated actuator and sensor. In this section we will examine techniques to select the parameters of these three schemes. Lead, lag, and notch compensations have historically been implemented using analog electronics and hence were often referred to as networks. Today, however, most new control system designs use digital com-puter technology, in which the compensation is implemented in the software. In this case, one needs to compute discrete equivalents to the analog transfer functions as described in Chapter 4 and discussed further in Chapter 8 and in Franklin et al. (1998).

Compensation with a transfer function of the form

$$D(s) = K\frac{s+z}{s+p} \tag{5.97}$$

---

[17] The names of these compensation schemes derives from their frequency (sinusoidal) responses, where the output leads the input in one case (a positive phase shift) and lags the input in another (a negative phase shift). The frequency response of the third looks as if a notch had been cut in an otherwise flat frequency response. See Chapter 6.

Figure 5.22
Feedback system with compensation

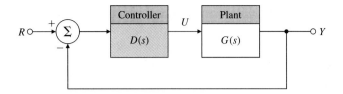

is called lead compensation if $z < p$ and lag compensation if $z > p$. Compensation is typically placed in series with the plant in the feed forward path, as shown in Fig. 5.22. It can also be placed in the feedback path and in that location has the same effect on the overall system poles but results in different transient responses from reference inputs.. The characteristic equation of the system in Fig. 5.22 is

$$1 + D(s)G(s) = 0,$$
$$1 + KL(s) = 0,$$

where K and $L(s)$ are selected to put the equation in root locus form as before.

5.5.1   Lead Compensation

To explain the basic stabilizing effect of lead compensation on a system, we first consider proportional control for which $D(s) = K$. If we apply this compensation to a second-order position control system with normalized transfer function

$$G(s) = \frac{1}{s(s+1)},$$

the root locus with respect to K is shown as the solid-line portion of the locus in Fig. 5.23. Also shown in Fig. 5.23 is the locus produced by proportional plus derivative control where $D(s) = K(s + 2)$. The modified locus is the circle

Figure 5.23
Root loci for
$1 + D(s)G(s) = 0$,
$G(s) = 1/[s(s+1)]$:
with compensation
$D(s) = K$ (solid
lines) and with
$D(s) = K(s+2)$
(dashed lines)

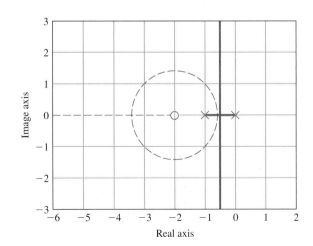

sketched with dashed lines. As we saw in the examples, the effect of the zero is to move the locus to the left, toward the more stable part of the s-plane. If, now, our speed-of-response specification calls for $\omega_n \cong 2$, then proportional control alone ($D = K$) can produce only a very low value of damping ratio ζ when the roots are put at the required value of ω_n hence at the required gain the transient overshoot will be substantial. However, by adding the zero of PD control we can move the locus to a position having closed-loop roots at $\omega_n = 2$ and damping ratio $\zeta \geq 0.5$. We have "compensated" the given dynamics by using $D(s) = K(s + 2)$.

As we observed earlier, pure derivative control is not normally practical because of the amplification of sensor noise implied by the differentiation and must be approximated. If the pole of the lead network is placed well outside the range of the design ω_n then we would not expect it to upset the dynamic response of the design in a serious way. For example, consider the lead compensation

$$D(s) = K \frac{s + 2}{s + p}.$$

The root loci for two cases with $p = 10$ and $p = 20$ are shown in Fig. 5.24 along with the locus for PD control. The important fact about these loci is that for small gains, before the real root departing from $-p$ approaches -2, the loci with lead control are almost identical with the locus where $D(s) = K(s + 2)$. Note that, as expected from the example loci, the effect of the pole is to press the locus toward the right, but for the early part of the locus, the effect of the pole is not great if $p > 10$.

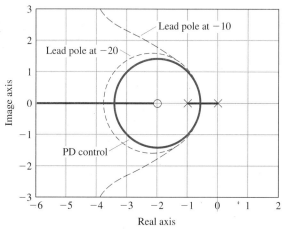

Figure 5.24 Root loci for three cases with
$G(s) = 1/[s(s + 1)]$: (a) $D(s) = (s + 2)/(s + 20)$;
(b) $D(s) = (s + 2)/(s + 10)$; (c) $D(s) = s + 2$
(solid lines)

Selecting exact values of z and p in Eq. (5.97) for particular cases is usually done by trial and error, which can be minimized with experience. In general, the zero is placed in the neighborhood of the closed-loop ω_n, as determined by rise-time or settling-time requirements, and the pole is located at a distance 5 to 20 times the value of the zero location. The choice of the exact pole location is a compromise between the conflicting effects of noise suppression, for which one wants a small value for p, and compensation effectiveness for which one wants a large p. In general, if the pole is too close to the zero, then, as seen in Fig. 5.24, the root locus moves back too far toward its uncompensated shape and the zero is not successful in doing its job. On the other hand, for reasons that are perhaps easier to understand from the frequency response, when the pole is too far to the left the magnification of sensor noise appearing at the output of $D(s)$ is too great and the motor or other actuator of the process can be overheated by noise energy in the control signal, $u(t)$. With a large value of p the lead compensation approaches pure PD control. A simple example will illustrate the approach.

EXAMPLE 5.11 ⟶ *Design of Lead Compensation*

Find a compensation for $G(s) = 1/[s(s + 1)]$ that will provide overshoot no more than 20% and rise time no more than 0.25 sec.

Solution.   From Chapter 3, we estimate that a damping ratio of $\zeta \geq 0.5$ and a natural frequency of $\omega_n \cong 1.8/0.25 \cong 7.2$ will satisfy the requirements. Considering the root loci plotted in Fig. 5.24 we will first try

$$D(s) = K\frac{s+2}{s+10}$$

Figure 5.25 shows that $K = 70$ will comfortably meet both specifications, resulting in $\zeta = 0.56$ and $\omega_n = 7.7$ rad/sec. The real parts of the complex poles are at $\zeta\omega_n = -4.3$

Figure 5.25
Root locus for lead design

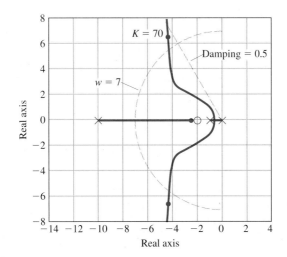

Figure 5.26
Step response for
Example 5.11

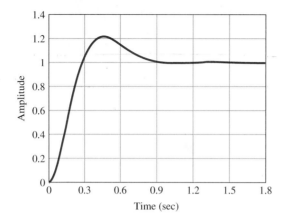

and the third pole will be at $-11 + 8.6 = -2.4$, to keep the sum of the roots constant at -11. Because this third pole is so near the lead zero at -2 the overshoot is not greatly increased from the second order case. Figure 5.26 shows the step response of the system. Typically, lead compensation in the feed-forward path will increase the step-response overshoot because the zero of the compensation has a differentiating effect, as discussed in Chapter 3.

We can make the method for selecting the pole and zero of lead compensation more analytical if we select the desired closed-loop pole location first. With r_o selected, we can use the root-locus angle condition to select z. This method will be illustrated with the same plant.

EXAMPLE 5.12   *A Second Lead Compensation Design*

Suppose we require the closed-loop system to have a pole at $r_o = -3.5 + j3.5\sqrt{3}$ corresponding exactly to $\omega_n = 7$ and $\zeta = 0.5$ so that, as discussed in Chapter 3, we should expect a rise time $t_r \cong 0.26$ sec and an overshoot $M_p \cong 20\%$. The noise suppression requirements require that the lead pole be no larger than 20. Design a lead compensation to meet these specifications.

Solution.   In order to keep the lead zero as far to the left as possible to minimize the effect it will have on the step response, we will take the lead pole to be at the limit, $p = 20$. The corresponding s-plane is sketched in Fig. 5.27. Because $\angle L(s) = 107.4°$ at the desired point for the root, $r_o = -3.5 + j3.5\sqrt{3}$, the root-locus angle condition will be satisfied if the angle ψ from the lead zero is $72.6°$. The location of the zero is found to be $z = -5.4$ at a gain of 127 and thus with the compensation

$$D(s) = 127\frac{s + 5.4}{s + 20} \tag{5.98}$$

the point $r_o = -3.5 + j3.5\sqrt{3}$ will be on the locus as shown in Fig. 5.28. Other combinations can also be selected that yield the required angle from $D(s)$, perhaps with a

Figure 5.27
Construction for placing a specific point on the root locus

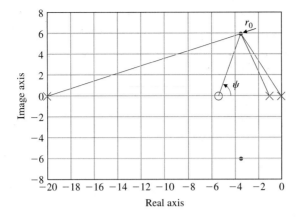

larger magnitude for the zero, however, the limit on how far the zero can be placed in the left half-plane is found as p goes to infinity and the point of diminishing returns is quickly found as p gets much larger than 20. The step response of the design is plotted in Fig. 5.29.

As stated earlier, the name *lead compensation* is a reflection of the fact that to sinusoidal signals these transfer functions impart phase lead. For example, the phase of Eq. (5.97) at $s = j\omega$ is given by

$$\phi = \tan^{-1}\left(\frac{\omega}{z}\right) - \tan^{-1}\left(\frac{\omega}{p}\right). \tag{5.99}$$

If $z < p$, then ϕ is positive, which by definition indicates phase lead. The details of design using the phase angle of the lead compensation will be treated in Chapter 6.

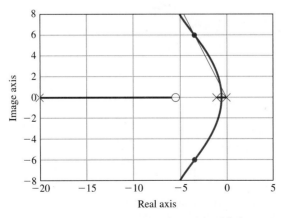

Figure 5.28 Root locus for $L(s) = \dfrac{s + 5.4}{s(s + 20)(s + 1)}$

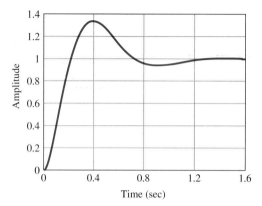

Figure 5.29 Step response for $K = 127$ and
$L(s) = \dfrac{s + 5.4}{(s + 20)} \dfrac{1}{s(s + 1)}$

5.5.2 Lag Compensation

Once satisfactory dynamic response has been obtained, perhaps by using one or more lead compensations, we may discover that the low-frequency gain—the value of the relevant steady-state error constant such as K_v—is still too low. As we saw in Chapter 4, the system type, which determines the degree of the polynomial the system is capable of following, is determined by the order of the pole of the transfer function $D(s)G(s)$ at $s = 0$. If the system is type 1, the velocity error constant, which determines the magnitude of the error to a ramp input, is given by $\lim_{s \to 0} D(s)G(s)$. In order to increase this constant, it is necessary to increase this constant in such a way that does not upset the already satisfactory dynamic response. Thus, we want an expression for $D(s)$ that will yield a significant gain at $s = 0$ to raise K_v (or other steady-state error constant) but is nearly unity (no effect) at the higher frequency ω_n where dynamic response is determined. The result is

$$D(s) = \frac{s + z}{s + p}, \qquad z > p, \tag{5.100}$$

where the values of z and p are small compared to ω_n, yet $D(0) = z/p = 3$ to 10 (the value depending on the extent to which the steady-state gain requires boosting). Because $z > p$, the phase ϕ given by Eq. (5.99) is negative, corresponding to phase lag. Hence a device with this transfer function is called lag compensation.

An example of lag compensation

The effects of lag compensation on dynamic response can be studied by looking at the corresponding root locus. Again we take $G(s) = 1/[s(s + 1)]$, include the lead compensation $K D_1(s) = K(s+5.4)/(s+20)$ that produced the locus in Fig. 5.25(a), and raised the gain until the closed-loop roots correspond to a damping ratio of $\zeta = 0.5$. At this point, the root-locus gain of about $K = 127$ can be found using MATLAB's rlocfind or by the methods described in Section 5.5. Thus the velocity constant is

$$K_v = \lim_{s \to 0} s K D G$$

$$= \lim_{s \to 0} s (127) \frac{s + 5.4}{s + 20} \frac{1}{s(s + 1)}$$

$$= \frac{127 * 5.4}{20} = 34.3.$$

Suppose we require a $K_v = 100$. To obtain this, we add a lag compensation with $z/p = 3$ in order to increase the velocity constant by 3 and add a pole at $p = -0.01$ to keep the values of both z and p very small so that $D_2(s)$ would have little effect on the portions of the locus representing the dominant dynamics at $\omega_n = 7$. The result is a lag compensation with the transfer function of $D_2(s) = (s + 0.03)/(s + 0.01)$ The root locus with both lead and lag compensation is plotted in Fig. 5.30(a) showing the dominant roots at $-3.5 \pm j6.1$ and the region around the origin is expanded in Fig. 5.30(b).

In Fig. 5.30(b), we can see that the circle is a result of the small pole and zero of the lag compensation. Notice that a closed-loop root remains very near the lag-compensation zero at $-0.03 + 0j$. The transient response corresponding to this root will be a very slowly decaying term which, however, will have a small magnitude because the zero will almost cancel the pole in the transfer function. Still, the decay is so slow that this term may seriously influence the settling time. Furthermore, the zero will *not* be present in the step response to a disturbance torque and the slow transient will be much more evident there. Because of this effect it is important to place the lag pole-zero combination at as high a frequency as possible without causing major shifts in the dominant root locations.

5.5.3   Notch Compensation

Suppose the design has been completed with lead and lag compensation given by

$$K D(s) = 127 \frac{s + 5.4}{s + 20} \frac{s + .03}{s + .01} \tag{5.101}$$

but is found to have a substantial oscillation at about 50 rad/sec when tested because there was an unsuspected flexibility of the non-collocated type at a natural frequency of $\omega_n = 50$. On reexamination, the plant transfer function including the effect of the flexibility is estimated to be

$$G(s) = \frac{2500}{s(s + 1)(s^2 + s + 2500)}. \tag{5.102}$$

Figure 5.30
Root locus with both lead and lag compensations: (a) whole locus; (b) portion of part (a) expanded to show the root locus near the lag compensation

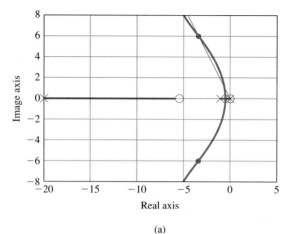

(a)

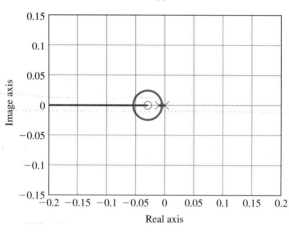

(b)

A mechanical engineer claims that some of the "control energy" has spilled over into the flexible mode and caused it to be excited. How can this instability be corrected? There are two possibilities. An additional lag compensation might lower the loop gain at $\omega = 50$ far enough that there is greatly reduced spillover and the oscillation eliminated. Reducing the gain at the high frequency is called **gain stabilization**. If the response time resulting from gain stabilization is too long, a second alternative is to add a zero near the resonance so as to shift the phase of the loop at ω_n and cause the closed loop root to move into the left half-plane and thus cause the associated transient to die out. This approach is called **phase stabilization** and its action is similar to that of flexibility in the collocated motion control discussed earlier. For phase stabilization, the result

Gain and phase stabilization are discussed in Chapter 6.

Figure 5.31
Root locus with lead, lag, and notch compensations

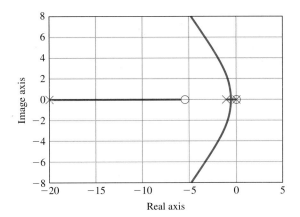

is called a notch compensation and an example has a transfer function

$$D_{notch}(s) = \frac{s^2 + 2\zeta\omega_o s + \omega_o^2}{(s + \omega_o)^2}.$$ (5.103)

A design decision is whether to place the notch frequency above or below that of the natural resonance of the flexibility in order to get the necessary phase. A check of the angle of departure shows that with the plant as compensated by Eq. (5.101) and the notch as given it is necessary to place the frequency of the notch *above* that of the resonance to get the departure angle to point toward the left half-plane. Thus the compensation is added with the transfer function

$$D_{notch}(s) = \frac{s^2 + .8s + 3600}{(s + 60)^2}.$$ (5.104)

The gain of the notch at $s = 0$ has been kept at 1 so as not to change the K_v. The new root locus to the same scale is shown in Fig. 5.31, and the step response is shown in Fig. 5.32.

Figure 5.32
Step response with lead, lag, and notch compensations

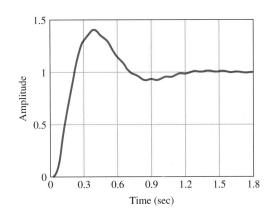

When considering notch or phase stabilization it is important to understand that its success depends on maintaining the correct phase at the frequency of the resonance. If that frequency is subject to significant change, which is common in many cases, then the notch needs to be removed far enough from the nominal frequency in order to work for all cases. The result may be interference of the notch with the rest of the dynamics and poor performance. As a general rule, gain stabilization is substantially more robust to plant changes than is phase stabilization.

5.5.4   Analog and Digital Implementations

Lead compensation can be implemented using analog electronics, but digital computers are preferred.

Lead compensation can be physically realized in many ways. In analog electronics a common method is to use an operational amplifier, an example of which is shown in Fig. 5.33. The transfer function of the circuit in Fig. 5.33 is readily found by the methods of Chapter 3 to be

$$D_{lead}(s) = -a\frac{s+z}{s+p}, \tag{5.105}$$

where

$$a = \frac{p}{z}, \text{ if } R_f = R_1 + R_2$$

$$z = \frac{1}{R_1 C},$$

$$p = \frac{R_1 + R_2}{R_2}\frac{1}{R_1 C}.$$

If a design for $D(s)$ is complete and a digital implementation is desired, then the technique of Chapter 4 can be used by first selecting a sampling period T_s and then making substitution of $\frac{2}{T_s}\frac{z-1}{z+1}$ for s. For example, consider the lead compensation $D(s) = \frac{s+5.4}{s+20}$ and because the rise time is about 0.25, a sampling period of 5 samples per rise time results in the selection of $T_s = 0.05$. With the substitution of $\frac{2}{0.05}\frac{z-1}{z+1}$ for s into this transfer function, the discrete transfer function is

$$\frac{U(z)}{E(z)} = \frac{40\dfrac{z-1}{z+1} + 5.4}{40\dfrac{z-1}{z+1} + 20}$$

$$= \frac{2.27z - 1.73}{3z - 1}. \tag{5.106}$$

Figure 5.33
Possible circuit of a lead
compensation

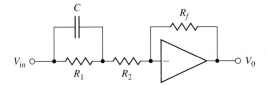

Clearing of fractions and using the fact that operating on the time functions $zu(kT_s) = u(kT_s + T_s)$, Eq. (5.106) is equivalent to the formula for the controller given by

$$u(kT_s + T_s) = \frac{1}{3}u(kT_s) + \frac{2.27}{3}e(kT_s + T_s) - \frac{1.73}{3}e(kT_s). \qquad (5.107)$$

The control and step responses with this controller are plotted on Fig. 5.34 along with the results of using the analog controller. As with lead compensation, lag

Figure 5.34
Comparison of analog and
digital control (a) Output
responses (b) Control
responses

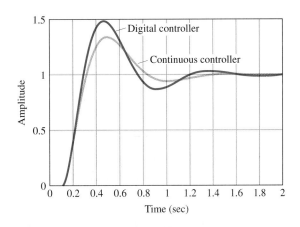

(a)

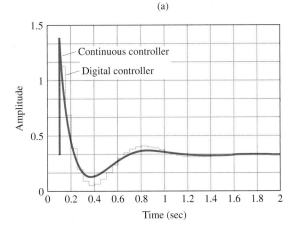

(b)

Figure 5.35
Possible circuit of lag compensation

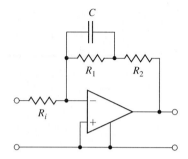

or notch compensation can implemented using a digital computer following the same procedure. However, they, too, can be implemented using analog electronics and a circuit diagram of a lag network is given in Fig. 5.35. The transfer function of this circuit can be shown to be

$$D(s) = -a\frac{s + z}{s + p},$$

where

$$a = \frac{R_2}{R_i},$$

$$z = \frac{R_1 + R_2}{R_1 R_2 C},$$

$$p = \frac{1}{R_1 C}.$$

Usually $R_i = R_2$, so the high-frequency gain is unity or $a = 1$ and the low-frequency increase in gain to enhance K_v or other error constant is set by $k = a\frac{z}{p} = \frac{R_1 + R_2}{R_2}$.

5.6   A Design Example Using the Root Locus

EXAMPLE 5.13   *Control of a Small Airplane*

For the Piper Dakota shown in Fig. 5.36, the transfer function between the elevator input and the pitch attitude is

$$G(s) = \frac{\theta(s)}{\delta_e(s)} = \frac{160(s + 2.5)(s + 0.7)}{(s^2 + 5s + 40)(s^2 + 0.03s + 0.06)}, \tag{5.108}$$

where

$$\theta = \text{pitch attitude, degrees (see Fig. 9.29)},$$

$$\delta_e = \text{elevator angle, degrees.}$$

(For a more detailed discussion of longitudinal aircraft motion, refer to Chapter 9.)

Figure 5.36
Auto-pilot design in the
Piper Dakota, showing
elevator and trim tab
*(Photo courtesy of Denise
Freeman)*

(a)

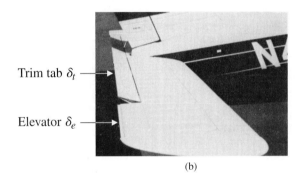

Trim tab δ_t —→

Elevator δ_e —→

(b)

(a) Design an auto pilot so that the response to a step elevator input has a rise time of 1 sec or less and an overshoot less than 10%.

(b) When there is a constant disturbing moment acting on the aircraft so that the pilot must supply a constant force on the controls for steady flight, it is said to be out of trim. The transfer function between the disturbing moment and the attitude is the same as that due to the elevator; that is,

$$\frac{\theta(s)}{M_d(s)} = \frac{160(s + 2.5)(s + 0.7)}{(s^2 + 5s + 40)(s^2 + 0.03s + 0.06)}, \tag{5.109}$$

where M_d is the moment acting on the aircraft. There is a separate aerodynamic surface for trimming, δ_t, that can be actuated and will change the moment on the aircraft. It is shown in the close-up of the tail in Fig. 5.36. Its influence is depicted in the block diagram shown in Fig. 5.37(a). For both manual and auto-pilot flight it is desirable to adjust the trim so that there is no steady-state control effort required from the elevator (that is, so $\delta_e = 0$). In manual flight this means no force is required by the pilot to keep the aircraft at a constant attitude, whereas in auto-pilot control it means reducing the amount of electrical power required and saving wear and tear on the servomotor that drives the elevator. Design an auto pilot that will command the trim δ_t so as to drive the steady-state value of δ_e to zero for an arbitrary moment M_d as well as meet the specifications in part (a).

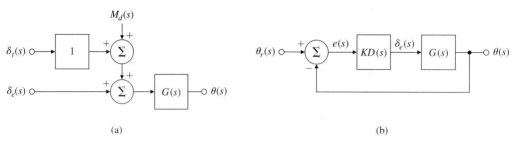

Figure 5.37 Block diagrams for auto-pilot design: (a) open loop; (b) feedback scheme excluding trim control

Solution.

(a) To satisfy the requirement that the rise time $t_r \le 1$ sec, Eq. 3.48 indicates that, for the ideal second-order case, ω_n must be greater than 1.8 rad/sec. And to provide an overshoot of less than 10%, Fig. 3.28 indicates that ζ should be greater than 0.6, again, for the ideal second-order case. In the design process we can examine a root locus for a candidate for feedback compensation and then look at the resulting time response when the roots appear to satisfy the design guidelines. However, because this is a fourth-order system, the design guidelines might not be sufficient, or they might be overly restrictive.

To initiate the design process, it is often instructive to look at the system characteristics with proportional feedback, that is, where $D(s) = 1$ in Fig. 5.37(b). The statements in MATLAB to create a root locus with respect to K and a time response for the proportional feedback case with $K = 0.3$ are:

```
numG = 160*conv ([1    2.5],[1    0.7]);
denG = conv([1    5    40],[1  0.03    0.06]);
sysG = tf(numG,denG);
rlocus(sysG)
K = 0.3
sysL = K*sysG
sysH = tf(1,1);
[sysT] = feedback (sysL,sysH)
step(sysT)
```

The resulting root locus and time response are shown with dashed lines in Figs. 5.38 and 5.39. Notice from Fig. 5.38 that the two faster roots will always have a damping ratio ζ that is less than 0.4 and the maximum damping of both sets can be increased only to about 0.3; therefore, proportional feedback will not be acceptable. Also, the slower roots have some effect on the time response shown in Fig. 5.39 (dashed curve) with $K = 0.3$ in that they cause a long-term settling. However, the dominating characteristic of the response that determines whether or not the compensation meets the specifications is the behavior in the first few seconds, which is dictated by the fast roots. The low damping of the fast roots causes the time response to be oscillatory, which leads to excess overshoot and a longer settling time than desired.

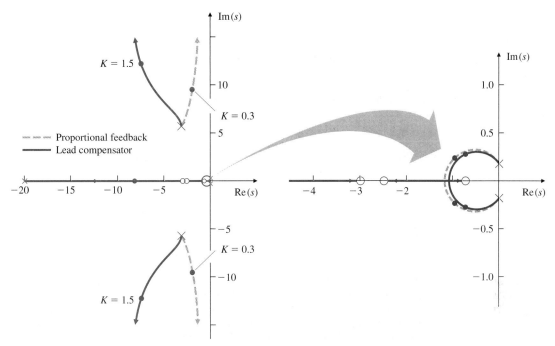

Figure 5.38 Root loci for auto-pilot design

We saw in Section 5.5.1 that lead compensation causes the locus to shift to the left, a change needed here to increase the damping. Some trial and error will be required to arrive at a suitable pole and zero location. Values of $z = 3$ and $p = 20$ in Eq. (5.97) have a substantial effect in moving the fast branches of the locus to the left; thus

$$D(s) = \frac{s+3}{s+20}.$$

Figure 5.39
Time-response plots for auto-pilot design

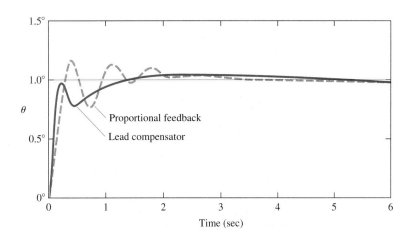

Trial and error is also required to arrive at a value of K that meets the specifications. The statements in MATLAB to add this compensation are

```
numD = [1  3];
denD = [1  20];
sysD = tf(numD,denD);
sysDG = sysD*sysG
rlocus(sysDG)
K = 1.5;
sysKDG = K*sysDG;
sysH = tf(1,1)
sysT = feedback(sysKDG,sysH)
step(sysT)
```

The root locus for this case and the corresponding time response are also shown in Figs. 5.38 and 5.39 by the solid lines. Note that the damping of the fast roots that corresponds to $K = 1.5$ is $\zeta = 0.52$, which is slightly lower than we would like; also, the natural frequency is $\omega_n = 15$ rad/sec, much faster than we need. However, these values are close enough to meeting the guidelines to suggest a look at the time response. In fact, the time response shows that $t_r \cong 0.9$ sec and $M_p \cong 8\%$, both within the specifications, although by a very slim margin.

In summary, the primary design path consisted of adjusting the compensation to influence the fast roots, examining their effect on the time response, and continuing the design iteration until the time specifications were satisfied.

(b) The purpose of the trim is to provide a moment that will eliminate a steady-state nonzero value of the elevator. Therefore, if we integrate the elevator command δ_e and feed this integral to the trim device, the trim should eventually provide the moment required to hold an arbitrary attitude, thus eliminating the need for a steady-state δ_e. This idea is shown in Fig. 5.40(a). If the gain on the integral term K_I is small enough, the destabilizing effect of adding the integral should be small and the system should behave approximately as before since that feedback loop has been left intact. The block diagram in Fig. 5.40(a) can be reduced to that in Fig. 5.40(b) for analysis purposes by defining the compensation to include the PI form

$$D_I(s) = KD(s)\left(1 + \frac{K_I}{s}\right).$$

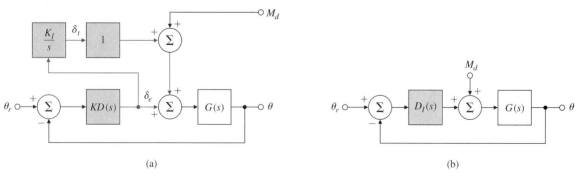

(a)   (b)

Figure 5.40 Block diagram showing the trim-command loop

However, it is important to keep in mind that, physically, there will be two outputs from the compensation: δ_e (used by the elevator servomotor) and δ_t (used by the trim servomotor).

The characteristic equation of the system with the integral term is

$$1 + KDG + \frac{K_I}{s}KDG = 0.$$

To aid in the design process, it is desirable to find the locus of roots with respect to K_I, but the characteristic equation is not in any of the root-locus forms given by Eqs. (5.6d). Therefore, dividing by $1 + KDG$ yields

$$1 + \frac{(K_I/s)KDG}{1 + KDG} = 0.$$

To put this system in root locus form, we define

$$L(s) = \frac{1}{s}\frac{KDG}{1 + KDG}. \qquad (5.110)$$

In MATLAB, with $\frac{KDG}{1+KGD}$ already computed as sysT, we construct the integrator as sysI = tf(1,[1 0]), the loop gain of the system with respect to K_I as sysL = sysI*sysT and the root locus with respect to K_I is found with rlocus(sysL).

It can be seen from the locus in Fig. 5.41 that the damping of the fast roots decreases as K_I increases, as is typically the case when integral control is added.

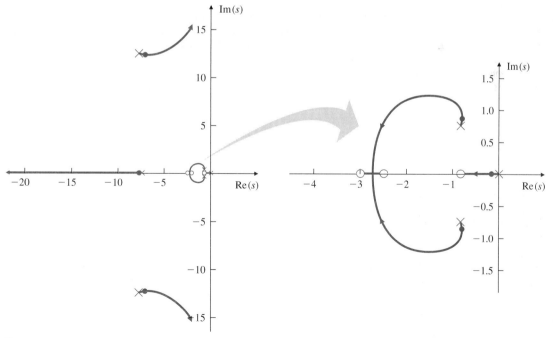

Figure 5.41 Root locus versus K_I: assumes an added integral term and lead compensation with a gain $K = 1.5$; roots for $K_I = 0.15$ marked with •

This shows the necessity for keeping the value of K_I as low as possible. After some trial and error we select $K_I = 0.15$. This value has little effect on the roots—note the roots are virtually on top of the previous roots obtained without the integral term—and little effect on the short-term behavior of the step response, as shown in Fig. 5.42(a), so the specifications are still met. $K_I = 0.15$ does cause the longer-term attitude behavior to approach the commanded value with no error, as we would expect with integral control. It also causes δ_e to approach zero [Fig. 5.42(b) shows it settling in approximately 30 sec], which is good because this is the reason for choosing integral control in the first place. The time for the integral to reach the correct value is predicted by the new, slow real root that is added by the integral term at $s = -0.14$. The time constant associated with this root is $\tau = 1/0.14 \cong 7$ sec. The settling time to 1% for a root with $\sigma = 0.14$ is shown by Eq. (3.51) to be $t_s = 33$ sec, which agrees with the behavior in Fig. 5.42(b).

Figure 5.42

Step response for the case with an integral term and 5° command

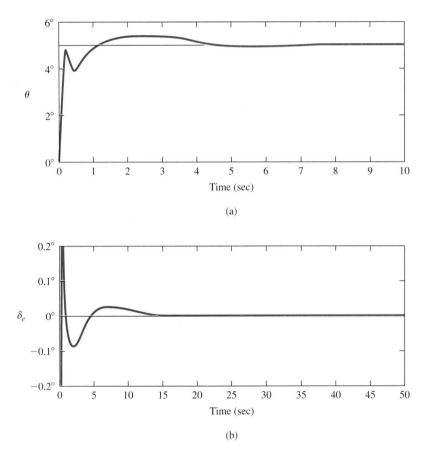

5.7 Extensions of the Root-Locus Method

As we have seen in this chapter, the root-locus technique is a graphical scheme to show locations of possible roots of an algebraic equation as a single real parameter varies. The method can be extended to a sequential consideration of more than one parameter and can also be useful in the study of simple transcendental equations as well as certain nonlinear systems. In this section we examine these possibilities.

5.7.1 Consideration of Two Parameters

Successive loop closure

An important technique for practical control is to consider a structure with two loops, an inner loop around an actuator or part of the process dynamics and an outer loop around the entire plant-plus-inner-controller. The process is called **successive loop closure**. A controller is selected for the inner loop to be robust and give good response alone and then the outer loop can be designed to be simpler and more effective than if the entire control was done without the aid of the inner loop. The use of the root locus to study such a system with two parameters can be illustrated by a simple example.

EXAMPLE 5.14

A block diagram of a relatively common servomechanism structure is shown in Fig. 5.43. Here a speed-measuring device (a tachometer) is available and the problem is to use the root locus to guide the selection of the tachometer gain K_T as well as the amplifier gain K_A. The characteristic equation of the system in Fig. 5.43 is

$$1 + \frac{K_A}{s(s+1)} + \frac{K_T}{s+1} = 0,$$

which is not in the standard $1 + KL(s)$ form. After clearing fractions, the characteristic equation becomes

$$s^2 + s + K_A + K_T s = 0 \tag{5.111}$$

which is a function of two parameters whereas the root locus technique can consider only one parameter at a time. In this case, we set the gain K_A to a nominal value of 4 and consider first the locus with respect to K_T. With $K_A = 4$ Eq. (5.111) can be put into root locus form for a root-locus study with respect to K_T with $L(s) = \frac{s}{s^2+s+4}$ or

$$1 + K_T \frac{s}{s^2 + s + 4} = 0. \tag{5.112}$$

Figure 5.43

Block diagram of a
servomechanism structure
including tachometer
feedback

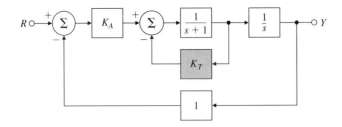

For this root-locus the zero is at $s = 0$, and the poles are at the roots of $s^2 + s + 4 = 0$, or $s = -\frac{1}{2} \pm 1.94j$. A sketch of the locus using the rules as before is shown in Fig. 5.44.

From this locus, we can select K_T so the complex roots have a specific damping ratio or take any other value of K_T that would result in satisfactory roots for the characteristic equation. Consider $K_T = 1$. Having selected a trial value of K_T, we can now reform the equation to consider the effects of changing from $K_A = 4$ by taking the new parameter to be K_1 so that $K_A = 4 + K_1$. The locus with respect to K_1 is governed by Eq. (5.93) with now $L(s) = \frac{1}{s^2+2s+4}$ so that the locus is for the equation

$$1 + K_1 \frac{1}{s^2 + 2s + 4} = 0. \tag{5.113}$$

Note that the *poles* of the new locus corresponding to Eq. (5.113) are the *roots* of the previous locus, which was drawn versus K_T and the roots were taken at $K_T = 1$. The locus is sketched in Fig. 5.45, with the previous locus versus K_T left dashed. We could draw a locus with respect to K_1 for a while, stop, resolve the equation, and continue the locus with respect to K_T; in a sort of seesaw between the parameters K_A and K_T and thus use the root locus to study the effects of two parameters on the roots of a characteristic equation. Notice, of course, that we can also plot the root locus for negative values of K_1 and thus consider values of K_A less than 4.

Figure 5.44

Root locus of closed-loop
poles of the system in
Fig. 5.42 vs. K_T

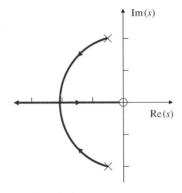

Figure 5.45
Root locus vs.
$K_1 = K_A + 4$ after
choosing $K_T = 1$

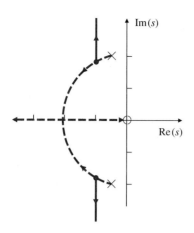

▲ 5.7.2 Time Delay

Time delays often arise in control systems, both from delays in the process itself and from delays in the processing of sensed signals. Chemical plants often have processes with a time delay representing the time material takes to be transported via pipes or other conveyer. In measuring the attitude of a spacecraft en route to Mars, there is a significant time delay for the sensed quantity to arrive back on Earth due to the speed of light. There is also a small time delay in any digital control system due to the cycle time of the computer and the fact that data is processed at discrete intervals. Time delay *always* reduces the stability of a system; therefore, it is important to be able to analyze its effect. In this section we discuss how to use the root locus for such analysis. Although an exact method of analyzing time delay is available in the frequency-response methods to be described in Chapter 6, knowing several different ways to analyze a design allows the control designer to do a better job.

Consider the problem of designing a control system for the temperature of the heat exchanger described in Chapter 2. The transfer function between the control A_s and the measured output temperature T_m is described by two first-order terms plus a time delay T_d of 5 sec. The time delay results because the temperature sensor is physically located downstream from the exchanger so that there is a delay in its reading. The transfer function is

$$G(s) = \frac{e^{-5s}}{(10s + 1)(60s + 1)}, \qquad (5.114)$$

where the e^{-5s} term arises from the time delay.[18]

Time delays always reduce the stability of a system.

An example of a root locus with time delay

---

[18] Time delay is often referred to as "transportation lag" in the process industries

The corresponding root-locus equations with respect to proportional gain K are

$$1 + KG(s) = 0,$$

$$1 + K\frac{e^{-5s}}{(10s + 1)(60s + 1)} = 0,$$

$$600s^2 + 70s + 1 + Ke^{-5s} = 0. \tag{5.115}$$

How would we plot the root locus corresponding to Eq. (5.115)? Because it is not a polynomial, we cannot proceed with the methods used in previous examples. Instead, there are two basic approaches we will describe: approximation and direct application of the phase condition.

In the first approach we reduce the given problem to one we have previously solved by approximating the nonrational function e^{-5s} with a rational function. Since we are concerned with control systems and hence typically with low frequencies, we want an approximation that will be good for small s.[19] The most common means for finding such an approximation is attributed to H. Padé. It consists of matching the series expansion of the transcendental function e^{-5s} with the series expansion of a rational function whose numerator is a polynomial of degree p and whose denominator is a polynomial of degree q. The result is called a **(p, q) Padé approximant**[20] to e^{-5s}. We will initially compute the approximants to e^{-s}, and in the final result we will substitute $T_d s$ for s to allow for any desired delay.

Padé approximant

To illustrate the process, we begin with the (1, 1) approximant. In this case we wish to select b_0, b_1, and a_0 so that the error

$$e^{-s} - \frac{b_0 s + b_1}{a_0 s + 1} = \varepsilon \tag{5.116}$$

is small. For the Padé approximant we expand both e^{-s} and the rational function given by Eq.(5.116) into a McLauren series and match as many of the initial terms as possible. The two series are

$$e^{-s} = 1 - s + \frac{s^2}{2} - \frac{s^3}{3!} + \frac{s^4}{4!} - \cdots,$$

$$\frac{b_0 s + b_1}{a_0 s + 1} = b_1 + (b_0 - a_0 b_1)s - a_0(b_0 - a_0 b_1)s^2 + a_0^2(b_0 - a_0 b_1)s^3 + \cdots.$$

---

[19] The nonrational function e^{-5s} is analytic for all finite values of s and so may be approximated by a rational function. If nonanalytic functions such as $\sqrt{s}$ were involved, great caution would be needed in selecting an approximation valid near $s = 0$.

[20] The (p,p) Padé approximant for a delay of T seconds is most commonly used and is computed by the MATLAB command [num,den] = pade(T, p).

Matching coefficients of the first four terms, we get the following equations to solve:

$$b_1 = 1,$$

$$b_0 - a_0 b_1 = -1,$$

$$-a_0(b_0 - a_0 b_1) = \frac{1}{2},$$

$$a_0^2(b_0 - a_0 b_1) = -\frac{1}{6}.$$

There are an infinite number of equations but only three parameters. We match the first three coefficients and then substitute $T_d s$ for s. The resulting $(1,1)$ Padé approximant is

$$e^{-T_d s} \cong \frac{1 - (T_d s/2)}{1 + (T_d s/2)}. \tag{5.117}$$

If we assume $p = q = 2$, we have five parameters, and a better match is possible. In this case we have the $(2, 2)$ approximant which has the transfer function

$$e^{-T_d s} \cong \frac{1 - T_d s/2 + (T_d s)^2/12}{1 + T_d s/2 + (T_d s)^2/12}. \tag{5.118}$$

The comparison of these approximants is best seen from their pole-zero configurations as plotted in Fig. 5.46. The locations of the poles are in the LHP and the zeros are in the RHP at the reflections of the poles.

In some cases a very crude approximation is acceptable. For small delays the $(0, 1)$ approximant can be used which is simply a first-order lag given by

$$e^{-T_d s} \cong \frac{1}{1 + T_d s}. \tag{5.119}$$

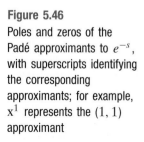

Figure 5.46
Poles and zeros of the Padé approximants to e^{-s}, with superscripts identifying the corresponding approximants; for example, x^1 represents the $(1, 1)$ approximant

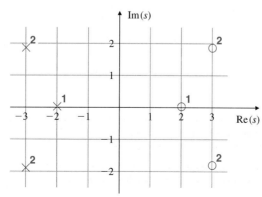

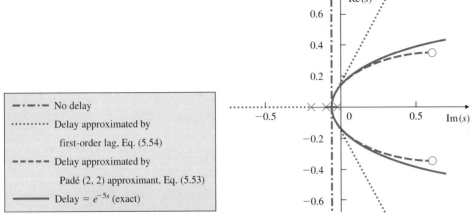

Legend:

- – · – · – No delay
- ·········· Delay approximated by first-order lag, Eq. (5.54)
- – – – – Delay approximated by Padé (2, 2) approximant, Eq. (5.53)
- ——— Delay $= e^{-5s}$ (exact)

Figure 5.47 Root loci for the heat exchanger with and without time delay

Contrasting methods of approximating delay

To illustrate the effect of a delay and the accuracy of the different approximations, root loci for the heat exchanger are drawn in Fig. 5.47 for four cases. Notice that for low gains and up to the point where the loci cross the imaginary axis, the approximate curves are very close to the exact. However, the (2,2) Padé curve follows the exact curve much further than does the first-order lag, and its increased accuracy would be useful if the delay were larger. All analyses of the delay show its destabilizing effect and how it limits the achievable response time of the system.

While the Padé approximation leads to a rational transfer function, in theory it is not necessary for plotting a root locus. A direct application of the phase condition can be used to plot portions of an exact locus of a system with time delay. The phase-angle condition does not change if the transfer function of the process is nonrational, so we still must search for values of s for which the phase is $180° + 360°l$. If we write the transfer function as

$$G(s) = e^{-T_d s} \bar{G}(s),$$

the phase of $G(s)$ is the phase of $\bar{G}(s)$ minus $\lambda\omega$ for $s = \sigma + j\omega$. Thus we can formulate a root-locus problem as searching for locations where the phase of $\bar{G}(s)$ is $180° + T_d\omega + 360°l$. To plot such a locus, we would fix ω and search along a horizontal line in the s-plane until we found a point on the locus, then raise the value of ω, change the target angle, and repeat. Similarly, the departure angles are modified by $T_d\omega$, where ω is the imaginary part of the pole from which the departure is being computed. Unfortunately, MATLAB does not provide a program to plot the root locus of systems with delay and we must be satisfied here with Padé approximants. Because it is possible to plot the frequency response (or Bode plot) of delay exactly and easily, if the designer feels that the Padé approximant is not satisfactory, the expedient approach is to use the frequency-response design methods described in Chapter 6.

▲ 5.7.3 Nonlinear Systems

As we have tried to make clear, every real control system is nonlinear, and the linear analysis and design methods we have described use linear approximations to the real models. There is one important category of nonlinear systems for which some significant analysis (and design) can be done. This comprises the systems in which the nonlinearity has no dynamics and is well-approximated as a gain that varies as the size of its input signal varies. Sketches of a few such nonlinear system elements and their common names are shown in Fig. 5.48.

The behavior of systems containing any one of these nonlinearities can be qualitatively described by considering the nonlinear element as a varying signal-dependent gain. For example, with the saturation element (Fig. 5.48a), it is clear that for input signals with magnitudes of less than a, the nonlinearity is linear with the gain N/a. However, for signals larger than a, the output size is bounded by N, while the input size can get much larger than a, so once the input exceeds a, the ratio of output to input goes down. Thus, saturation has the gain characteristics shown in Fig. 5.49. *All* actuators saturate at some level. If they did not, their output would increase to infinity, which is physically impossible. An important aspect of control system design is **sizing the actuator**, which means picking the size, weight, power required, cost, and saturation level of the device. Generally, higher saturation levels require bigger, heavier, and more costly actuators. From the control point of view, the key factor that enters into the sizing is the effect of the saturation on the control system's performance.

Figure 5.48

Nonlinear elements with no dynamics: (a) saturation, (b) relay, (c) relay with dead zone, (d) gain with dead zone, (e) preloaded spring, or coulomb plus viscous friction, and (f) quantization

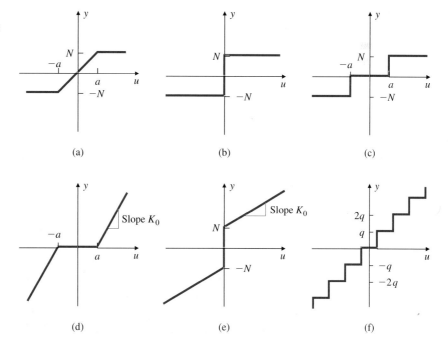

Figure 5.49

General shape of the effective gain of saturation

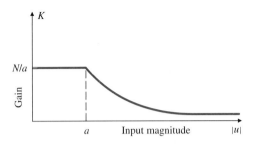

To illustrate saturation with an example, consider the system shown in Fig. 5.50. The root locus of this system versus K with the saturation removed is given by Fig. 5.51. At $K = 1$ the damping ratio is $\zeta = 0.5$. As the gain is reduced, the locus shows that the roots move toward the origin of the s-plane with less and less damping. Plots of the step responses of this system were obtained using the Simulink program. A series of different size step inputs r were introduced to the system with magnitudes $r_0 = 2$, 4, 6, 8, 10, and 12, and the results are shown in Fig. 5.52. As long as the signal entering the saturation remains less than 0.4, the system will be linear and should behave according to the roots at $\zeta = 0.5$. However, notice that as the input gets larger, the response has more and more overshoot and slower and slower recovery. This can be explained by noting that larger and larger input signals correspond to smaller and smaller effective gain K, as seen in Fig. 5.49. From the root-locus plot of Fig. 5.51, we see that as K decreases, the closed-loop poles move closer to the origin and have a smaller damping ζ. This results in the longer rise and settling times, increased overshoot, and greater oscillatory response.

Figure 5.50

Dynamic system with saturation

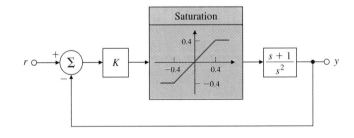

Figure 5.51

Root locus of $(s + 1)/s^2$, the system in Fig. 5.50 with the saturation removed

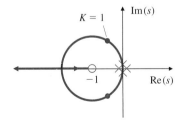

Figure 5.52
Step responses of system
in Fig. 5.50 for various
input step sizes

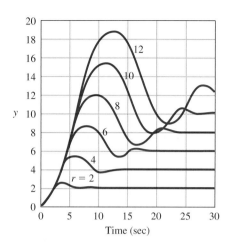

A nonlinear example: stability
depends on input magnitude.

As a second example of a nonlinear response described by signal-dependent gain, consider the system whose block diagram is drawn in Fig. 5.53 and whose root locus, excluding the saturation, is plotted in Fig. 5.54. From this locus we can readily calculate that the imaginary axis crossing occurs at $\omega_0 = 1$ and $K = \frac{1}{2}$. Systems such as this, which are stable for (relatively) large gains but

Conditional stability

unstable for smaller gains, are called **conditionally stable systems**. If $K = 2$,

Figure 5.53
Block diagram of a
conditionally stable system

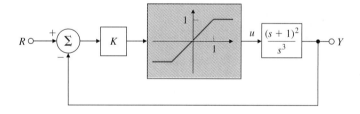

Figure 5.54
Root locus for
$G(s) = (s + 1)^2/s^3$ from
system in Fig. 5.53

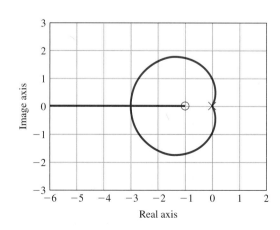

Figure 5.55

Step responses of system
in Fig. 5.52

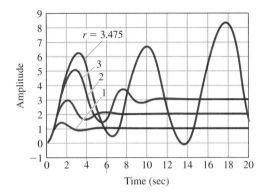

Figure 5.55

Step responses of system
in Fig. 5.52

which corresponds to $\zeta = 0.5$ on the locus, the system would be expected to show responses consistent with $\zeta = 0.5$ for small reference input signals. However, as the reference input size gets larger, the equivalent gain would get smaller due to the saturation and the system would be expected to become less well damped. Finally, the system would be expected to become unstable at some point for large inputs. Step responses from nonlinear simulation of the system with $K = 2$ for input steps of size $r_0 = 1.0, 2.0, 3.0$, and 3.4 are shown in Fig. 5.55. These responses confirm our predictions. Furthermore, the marginally stable case shows oscillations near 1 rad/sec, which is predicted by the frequency at the point at which the root locus crosses into the RHP.

A nonlinear example: an oscillatory system with saturation

The final illustration of the use of the root locus to give a qualitative description of the response of a nonlinear system is based on the block diagram in Fig. 5.56. This system is typical of electromechanical control problems where the designer perhaps at first is not aware of the resonant mode corresponding to the denominator term $s^2 + 0.2s + 1, (\omega = 1, \zeta = 0.1)$. The root locus for this system versus K, excluding the saturation, is sketched in Fig. 5.57. The imaginary-axis crossing can be verified to be at $\omega_0 = 1, K = 0.2$; thus a gain of $K = 0.5$ is enough to force the roots of the resonant mode into the RHP, as shown by the dots. If the system gain is set at $K = 0.5$, our analysis predicts a system that is initially unstable but becomes stable as the gain decreases. Thus we would expect the response of the system with the saturation to build up due to the instability until the magnitude is sufficiently large that the effective gain is lowered to $K = 0.2$ *and then stop growing!*

Figure 5.56

Block diagram of a system
with an oscillatory mode

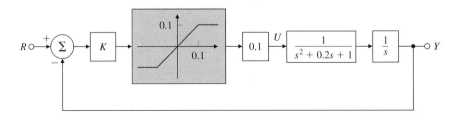

Figure 5.57
Root locus for the system in Fig. 5.56

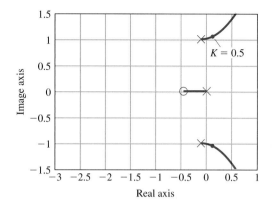

Plots of the step responses with $K = 0.5$ for three steps of size $r_0 = 1$, 4, and 8 are shown in Fig. 5.58, and again our heuristic analysis is exactly correct: The error builds up to a fixed amplitude and then starts to oscillate at a fixed amplitude. The oscillations have a frequency of ≈ 1 rad/sec and hold constant amplitude regardless of the step sizes of the input. In this case the response always approaches a periodic solution of fixed amplitude known as a **limit cycle**, so-called because the response is cyclic and is approached in the limit as time grows large. A nonlinear analysis method known as **describing functions** based on the assumption that the input to the nonlinearity during the limit cycle is sinusoidal can be used to predict the amplitude and frequency of a limit cycle for systems like these (Khalil, 1996).

Describing functions

In order to prevent the limit cycle, the locus has to be modified by compensation so that no branches cross into the right half-plane. One common method to do this for a lightly damped oscillatory mode is to place compensation zeros near the poles at a frequency such that the angle of departure of the root locus branch from these poles is toward the left half-plane, a procedure

Figure 5.58
Step responses of system in Fig. 5.56

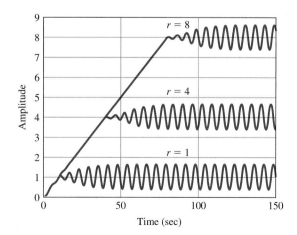

Figure 5.59

Root locus including compensation

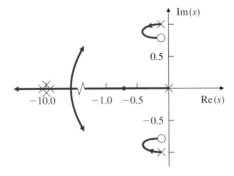

called phase stabilization earlier. Example 5.10 for collocated mechanical motion demonstrated that a pole-zero pair located in this manner will often cause a locus branch to go from the pole to the zero, looping to the left, and thus staying away from the RHP. Figure 5.59 shows the root locus for the system, $1/[s(s^2 + 0.2s + 1.0)]$, including a **notch compensation** with zeros located as just discussed. In addition, the compensation also includes two poles to make the compensation physically realizable. In this case, both poles were placed at $s = -10$, fast enough to not cause stability problems with the system, yet slow enough that high-frequency noise would not be amplified too much. Thus the compensation used for the root locus is

$$D(s) = 123\frac{s^2 + 0.18s + 0.81}{(s + 10)^2},$$

where the gain of 123 has been selected to make the compensation's DC gain equal to unity. This notch filter compensation attenuates inputs in the vicinity of $\omega_n^2 = 0.81$ or $\omega_n = 0.9$ rad/sec, so that any input from the plant resonance is attenuated and is thus prevented from detracting from the stability of the system. Figure 5.60 shows the system including the notch filter, and Fig. 5.61 shows the time response for two step inputs. Both inputs $r_0 = 2$ and 4 are sufficiently large so that the nonlinearity is saturated initially; however, because the system is unconditionally stable, the saturation results only in lowering the gain, so the response is slower than predicted by linear analysis but still stable, as also predicted by our piecewise linear analysis. In both cases the

Figure 5.60

Block diagram of the system with a notch filter

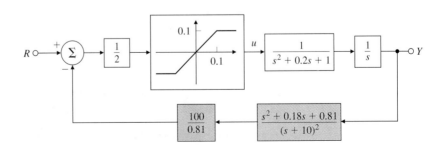

Figure 5.61

Step responses of the system in Fig. 5.60

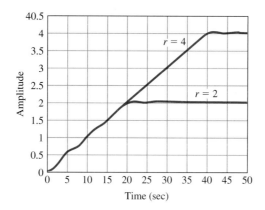

nonlinearity eventually becomes unsaturated and the system stabilizes to its new commanded value of r.

SUMMARY

- A root locus is a graph of the values of s that are solutions to the equation

$$1 + KL(s) = 0$$

  with respect to a real parameter K.

  1. When $K > 0$, s is on the locus if $\angle L(s) = 180°$, producing a $180°$ of positive K locus.

  2. When $K < 0$, s is on the locus if $\angle L(s) = 0°$, producing a $0°$ or negative K locus.

- If $KL(s)$ is the loop transfer function of a system with negative feedback, then the characteristic equation of the closed-loop system is

$$1 + KL(s) = 0,$$

  and the root-locus method displays the effect of changing the gain K on the closed-loop system roots.

- A *specific* locus for a system sysL in MATLAB notation can be plotted by rlocus(sysL).

- A working knowledge of how to sketch a root locus by hand quickly is useful for verifying computer results and for suggesting design alternatives.

- The key features for the aid in sketching a $180°$ locus are as follows:

1. The locus is on the real axis to the left of an odd number of poles plus zeros.

2. Of the n branches, m approach the zeros of $L(s)$ and $n - m$ branches approach asymptotes centered at α and leaving at angles ϕ_l:

$$n = \text{number of poles,}$$

$$m = \text{number of zeros,}$$

$$n - m = \text{number of asymptotes,}$$

$$\alpha = \frac{\sum p_i - \sum z_i}{n - m},$$

$$\phi_l = \frac{180° + 360°(l - 1)}{n - m}, \qquad l = 1, 2, \ldots, n - m.$$

3. Branches of the locus depart from the poles of order q and arrive at the zeros of order q with angles

$$\phi_{l,\text{dep}} = \frac{1}{q}\left(\sum \psi_i - \sum_{i \neq l} \phi_i - 180° - 360°(l - 1)\right),$$

$$\psi_{l,\text{arr}} = \frac{1}{q}\left(\sum \phi_i - \sum_{i \neq l} \psi_i + 180° + 360°(l - 1)\right),$$

where

$$q = \text{order of the pole or zero,}$$

$$\psi_i = \text{angles from the zeros,}$$

$$\phi_i = \text{angles from the poles.}$$

- The parameter K corresponding to a root at a particular point s_0 on the locus can be found from

$$K = \frac{1}{|L(s_0)|},$$

where $|L(s_0)|$ can be found graphically by measuring the distances from s_0 to each of the poles and zeros.

- For a locus drawn with rlocus(sysL), the parameter and corresponding roots can be found with
  [K,p] = rlocfind(sysL)

- Lead compensation, given by

$$D(s) = \frac{s+z}{s+p}, \qquad z < p,$$

approximates proportional-derivative (PD) control. For a fixed error co-efficient, it generally moves the locus to the left and improves the system damping.

- Lag compensation, given by

$$D(s) = \frac{s+z}{s+p}, \qquad z > p,$$

approximates proportional-integral (PI) control. It generally improves the steady-state error for fixed speed of response by increasing the low-frequency gain and typically degrades stability.

- The root locus can be used to analyze successive loop closures by studying two (or more) parameters in succession.

- The root locus can be used to analyze the effect of time delay.

- Nonlinearities with no dynamics, such as saturation, can be analyzed using the root locus by considering the nonlinearity to be a variable gain.

Review Questions

1. Give two definitions for the root locus.

2. Define the negative root locus.

3. Where are the sections of the (positive) root locus on the real axis?

4. What are the angles of departure from two coincident poles at $s = -a$ on the real axis. There are no poles or zeros to the right of $-a$.

5. What are the angles of departure from *three* coincident poles at $s = -a$ on the real axis. There are no poles or zeros to the right of $-a$.

6. What is the requirement on the location of a lead compensation zero that will cause the locus to pass through a desired location, r_o?

7. What is the value of the compensator gain that will cause a closed-loop pole to be at r_o?

8. What is the principal effect of a lead compensation on a root locus?

9. What is the principal effect of a lag compensation on a root locus in the vicinity of the dominant closed-loop roots?

10. What is the principal effect of a lag compensation on the steady-state error to a polynomial reference input?

11. Why is the angle of departure from a pole near the imaginary axis especially important?

12. Define a conditionally stable system.

13. Show, with a root-locus argument, that a system having three poles at the origin *must* be conditionally stable.

Problems

Problems for Section 5.1

5.1. Set up the following characteristic equations in the form suited to Evans's root-locus method. Give $L(s)$, $a(s)$, and $b(s)$ and the parameter, K, in terms of the original parameters in each case. Be sure to select K so that $a(s)$ and $b(s)$ are monic in each case and the degree of $b(s)$ is not greater than that of $a(s)$.

(a) $s + (1/\tau) = 0$ versus parameter τ.

(b) $s^2 + cs + c + 1 = 0$ versus parameter c.

(c) $(s + c)^3 + A(Ts + 1) = 0$.

    i. versus parameter A,

    ii. versus parameter T,

    iii. versus the parameter c, if possible. Say why you can or cannot. Can a plot of the roots be drawn versus c for given constant values of A and T by any means at all?

(d) $1 + \left[k_p + \frac{k_I}{s} + \frac{k_D s}{\tau s + 1} \right] G(s) = 0$.

Assume that $G(s) = A \frac{c(s)}{d(s)}$, where $c(s)$ and $d(s)$ are monic polynomials with the degree of $d(s)$ greater than that of $c(s)$.

    i. versus k_p,

    ii. versus k_I,

    iii. versus k_D,

    iv. versus τ.

Problems for Section 5.2

5.2. Roughly sketch the root loci for the pole-zero maps as shown in Fig. 5.62. Show your estimates of the center and angles of the asymptotes, a rough evaluation of arrival and departure angles for complex poles and zeros, and the loci for positive values of the parameter K. Each pole-zero map is from a characteristic equation of the form

$$1 + K \frac{b(s)}{a(s)} = 0,$$

where the roots of the numerator $b(s)$ are shown as small circles ∘ and the roots of the denominator $a(s)$ are shown as ×'s on the s-plane. Note that in Fig. 5.62(c), there are two poles at the origin.

Figure 5.62
Pole-zero maps

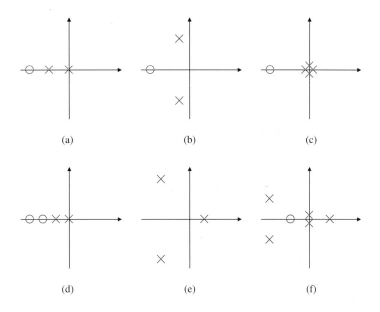

(a)   (b)   (c)

(d)   (e)   (f)

5.3. For the characteristic equation

$$1 + \frac{K}{s(s+1)(s+5)} = 0$$

do the following:

(a) Draw the real-axis segments of the corresponding root locus.

(b) Sketch the asymptotes of the locus for $K \to \infty$.

(c) For what value of K are the roots on the imaginary axis?

(d) Verify your sketch with a MATLAB plot.

5.4. *Real poles and zeros.* Sketch the root locus with respect to K for the equation $1 + KL(s) = 0$ and the following choices for $L(s)$. Be sure to give the asymptotes, arrival, and departure angles at any complex zero or pole, and give the frequency of any imaginary-axis crossing. After completing each hand sketch, verify your results using MATLAB. Turn in your hand sketches and the MATLAB results on the same scales.

(a) $L(s) = \dfrac{1}{s(s+1)(s+5)(s+10)}.$

(b) $L(s) = \dfrac{s+2}{s(s+1)(s+5)(s+10)}.$

(c) $L(s) = \dfrac{(s+2)(s+6)}{s(s+1)(s+5)(s+10)}.$

(d) $L(s) = \dfrac{(s+2)(s+4)}{s(s+1)(s+5)(s+10)}.$

5.5. *Complex poles and zeros.* Sketch the root locus with respect to K for the equation $1 + KL(s) = 0$ and the following choices for $L(s)$. Be sure to give the asymptotes, arrival and departure angles at any complex zero or pole, and the frequency of any imaginary-axis crossing. After completing each hand sketch verify your results using MATLAB. Turn in your hand sketches and the MATLAB results on the same scales.

(a) $L(s) = \dfrac{1}{s^2 + 3s + 10}$.

(b) $L(s) = \dfrac{1}{s(s^2 + 3s + 10)}$.

(c) $L(s) = \dfrac{s^2 + 2s + 8}{s(s^2 + 2s + 10)}$.

(d) $L(s) = \dfrac{s^2 + 2s + 12}{s(s^2 + 2s + 10)}$.

(e) $L(s) = \dfrac{s^2 + 1}{s(s^2 + 4)}$.

(f) $L(s) = \dfrac{s^2 + 4}{s(s^2 + 1)}$.

5.6.   *Multiple poles at the origin.*   Sketch the root locus with respect to K for the equation $1 + KL(s) = 0$ and the following choices for $L(s)$. Be sure to give the asymptotes, arrival and departure angles at any complex zero or pole, and the frequency of any imaginary-axis crossing. After completing each hand sketch verify your results using MATLAB. Turn in your hand sketches and the MATLAB results on the same scales.

(a) $L(s) = \dfrac{1}{s^2(s + 8)}$.

(b) $L(s) = \dfrac{1}{s^3(s + 8)}$.

(c) $L(s) = \dfrac{1}{s^4(s + 8)}$.

(d) $L(s) = \dfrac{s + 3}{s^2(s + 8)}$.

(e) $L(s) = \dfrac{s + 3}{s^3(s + 4)}$.

(f) $L(s) = \dfrac{(s + 1)^2}{s^3(s + 4)}$.

(g) $L(s) = \dfrac{(s + 1)^2}{s^3(s + 10)^2}$.

5.7.   *Mixed real and complex poles.*   Sketch the root locus with respect to K for the equation $1 + KL(s) = 0$ and the following choices for $L(s)$. Be sure to give the asymptotes, arrival and departure angles at any complex zero or pole, and the frequency of any imaginary-axis crossing. After completing each hand sketch verify your results using MATLAB. Turn in your hand sketches and the MATLAB results on the same scales.

(a) $L(s) = \dfrac{s + 2}{s(s + 10)(s^2 + 2s + 2)}$.

(b) $L(s) = \dfrac{s + 2}{s^2(s + 10)(s^2 + 6s + 25)}$.

(c) $L(s) = \dfrac{(s + 2)^2}{s^2(s + 10)(s^2 + 6s + 25)}$.

(d) $L(s) = \dfrac{(s + 2)(s^2 + 4s + 68)}{s^2(s + 10)(s^2 + 4s + 85)}$.

(e) $L(s) = \dfrac{[(s + 1)^2 + 1]}{s^2(s + 2)(s + 3)}$.

5.8. *Right half-plane poles and zeros.*   Sketch the root locus with respect to K for the equation $1 + KL(s) = 0$ and the following choices for $L(s)$. Be sure to give the asymptotes, arrival and departure angles at any complex zero or pole, and the frequency of any imaginary-axis crossing. After completing each hand sketch verify your results using MATLAB. Turn in your hand sketches and the MATLAB results on the same scales.

(a) $L(s) = \dfrac{s+2}{s+10} \dfrac{1}{s^2-1}$;

the model for a case of magnetic levitation with lead compensation.

(b) $L(s) = \dfrac{s+2}{s(s+10)} \dfrac{1}{(s^2-1)}$;

the magnetic levitation system with integral control and lead compensation.

(c) $L(s) = \dfrac{s-1}{s^2}$.

(d) $L(s) = \dfrac{s^2+2s+1}{s(s+20)^2(s^2-2s+2)}$.

What is the largest value that can be obtained for the damping ratio of the stable complex roots on this locus?

(e) $L(s) = \dfrac{s+2}{s(s-1)(s+6)^2}$.

(f) $L(s) = \dfrac{1}{(s-1)[(s+2)^2+3]}$.

5.9. Plot the loci for the $0°$ locus or negative K for

(a) The examples given in Problem 5.3.

(b) The examples given in Problem 5.4.

(c) The examples given in Problem 5.5.

(d) The examples given in Problem 5.6.

(e) The examples given in Problem 5.7.

(f) The examples given in Problem 5.8.

Problems for Section 5.3

5.10. A simplified model of the longitudinal motion of a certain helicopter near hover has the transfer function

$$G(s) = \frac{9.8(s^2-0.5s+6.3)}{(s+0.66)(s^2-0.24s+0.15)}$$

and the characteristic equation $1 + D(s)G(s) = 0$. Let $D(s) = k_p$ at first.

(a) Compute the departure and arrival angles at the complex poles and zeros.

(b) Sketch the root locus for this system for parameter $K = 9.8k_p$. Use axes $-4 \le x \le 4$. $-3 \le y \le 3$;

(c) Verify your answer using MATLAB. Use the command axes ([-4 4 −3 3]) to get the right scales.

(d) Suggest a practical (at least as many poles as zeros) alternative compensation $D(s)$ which will at least result in a stable system.

Figure 5.63

Control system for
Problem 11

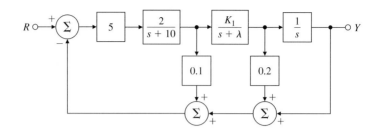

5.11. **(a)** For the system given in Fig. 5.63, plot the root locus of the characteristic equation as the parameter K_1 is varied from 0 to ∞ with $\lambda = 2$. Give the corresponding $L(s)$, $a(s)$, and $b(s)$.

(b) Repeat part (a) with $\lambda = 5$. Is there anything special about this value?

(c) Repeat part (a) for fixed $K_1 = 2$ with the parameter $K = \lambda$ varying from 0 to ∞.

5.12. For the system shown in Fig. 5.64, determine the characteristic equation and sketch the root locus of it with respect to positive values of the parameter c. Give $L(s)$, $a(s)$, and $b(s)$ and be sure to show with arrows the direction in which c increases on the locus.

Figure 5.64

Control system for
Problem 12

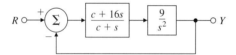

5.13. Suppose you are given a system with the transfer function

$$L(s) = \frac{s + z}{(s + p)^2},$$

where z and p are real and $z > p$. Show that the root locus for $1 + KL(s) = 0$ with respect to K is a circle centered at z with radius given by

$$r = (z - p).$$

Hint: Assume $s + z = re^{j\phi}$ and show that $L(s)$ is real and negative for real ϕ under this assumption.

5.14. The loop transmission of a system has two poles at $s = -1$ and a zero at $s = -2$. There is a third real-axis pole p located somewhere to the *left* of the zero. Several different root loci are possible, depending on the exact location of the third pole. The extreme cases occur when the pole is located at infinity or when it is located at $s = -2$. Give values for p and sketch the three distinct types of loci.

5.15. For the feedback configuration of Fig. 5.65, use asymptotes, center of asymptotes, angles of departure and arrival, and the Routh array to sketch root loci for the characteristic equations of the following feedback control systems versus the parameter K. Use MATLAB to verify your results.

Figure 5.65
Feedback system for
Problem 5.15

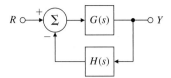

$R \circ \!\!\! \xrightarrow{+} \!\!\! \Sigma \xrightarrow{} G(s) \xrightarrow{} \circ Y$

(a) $G(s) = \dfrac{1}{s(s+1+3j)(s+1-3j)}$, $H(s) = \dfrac{s+2}{s+8}$

(b) $G(s) = \dfrac{1}{s^2}$, $H(s) = \dfrac{s+1}{s+3}$

(c) $G(s) = \dfrac{(s+5)}{(s+1)}$, $H(s) = \dfrac{s+7}{s+3}$

(d) $G(s) = \dfrac{(s+3+4j)(s+3-4j)}{s(s+1+2j)(s+1-2j)}$, $H(s) = 1 + 3s$

5.16. Consider the system in Fig. 5.66.

   (a) Using Routh's stability criterion, determine all values of K for which the system is stable.

   (b) Sketch the root locus of the characteristic equation versus K. Include angles of departure and arrival, and find the values for K and s at all breakaway points, break-in points, and imaginary-axis crossings.

Figure 5.66
Feedback system for
Problem 5.16

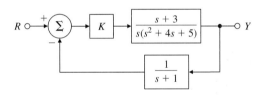

Problems for Section 5.4

5.17. Put the characteristic equation of the system shown in Fig. 5.67 in root locus form with respect to the parameter α and identify the corresponding $L(s)$, $a(s)$, and $b(s)$. Sketch the root locus with respect to the parameter α, estimate the closed-loop pole locations and sketch the corresponding step responses when $\alpha = 0, 0.5$, and 2. Use MATLAB to check the accuracy of your approximate step responses.

Figure 5.67
Control system for
Problem 5.17

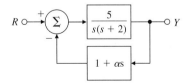

5.18. Suppose you are given the plant

$$L(s) = \dfrac{1}{s^2 + (1+\alpha)s + (1+\alpha)},$$

where α is a system parameter that is subject to variations. Use both positive and negative root-locus methods to determine what variations in α can be tolerated before instability occurs.

5.19. Use the MATLAB function rltool to study the behavior of the root locus of $1+KL(s)$ for

$$L(s) = \frac{(s+a)}{s(s+1)(s^2+8s+52)}$$

as the parameter a is varied from 0 to 10, paying particular attention to the region between 2.5 and 3.5. Verify that a multiple root occurs at a complex value of s for some value of a in this range.

5.20. Using root-locus methods, find the range of the gain K for which the systems in Fig. 5.68 are stable and use the root locus to confirm your calculations.

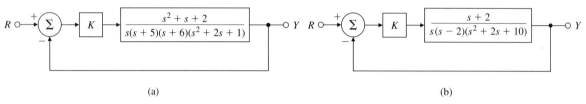

(a)   (b)

Figure 5.68 Feedback systems for Problem 5.20

5.21. Sketch the root locus for the characteristic equation of the system for which

$$L(s) = \frac{(s+1)}{s(s+1)(s+2)},$$

and determine the value of the root-locus gain for which the complex conjugate poles have a damping ratio of 0.5.

5.22. For the system in Fig. 5.69:

(a) Find the locus of closed-loop roots with respect to K.

(b) Is there a value of K that will cause all roots to have a damping ratio greater than 0.5?

(c) Find the values of K that yield closed-loop poles with the damping ratio $\zeta = 0.707$.

(d) Use MATLAB to plot the response of the resulting design to a reference step.

Figure 5.69
Feedback system for
Problem 5.22

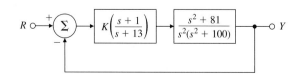

5.23. For the feedback system shown in Fig. 5.70, find the value of the gain K that results in dominant closed-loop poles with a damping ratio $\zeta = 0.5$.

Figure 5.70
Feedback system for
Problem 5.23

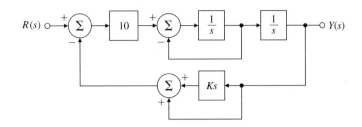

Problems for Section 5.5

5.24. Let

$$G(s) = \frac{1}{(s+2)(s+3)} \quad \text{and} \quad D(s) = K\frac{s+a}{s+b}.$$

Using root-locus techniques, find values for the parameters a, b, and K of the compensation $D(s)$ that will produce closed-loop poles at $s = -1 \pm j$ for the system shown in Fig. 5.71.

Figure 5.71
Unity feedback system for
Problems 5.24 to 5.30 and
5.35

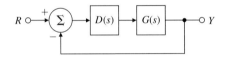

5.25. Suppose that in Fig. 5.71,

$$G(s) = \frac{1}{s(s^2 + 2s + 2)} \quad \text{and} \quad D(s) = \frac{K}{s+2}.$$

Sketch the root-locus with respect to K of the characteristic equation for the closed-loop system, paying particular attention to points that generate multiple roots if $KL(s) = D(s)G(s)$.

5.26. Suppose the unity feedback system of Fig. 5.71 has an open-loop plant given by $G(s) = 1/s^2$. Design a lead compensation $D(s) = K\frac{s+z}{s+p}$ to be added in series with the plant so that the dominant poles of the closed-loop system are located at $s = -2 \pm 2j$.

5.27. Assume that the unity feedback system of Fig. 5.71 has the open-loop plant

$$G(s) = \frac{1}{s(s+3)(s+6)}.$$

Design a lag compensation to meet the following specifications:

- The step response settling time is to be less than 5 sec.
- The step response overshoot is to be less than 17%.
- The steady-state error to a unit ramp input must not exceed 10%.

5.28. A numerically controlled machine tool positioning servomechanism has a normalized and scaled transfer function given by

$$G(s) = \frac{1}{s(s+1)}.$$

Performance specifications of the system in the unity feedback configuration of Fig. 5.71 are satisfied if the closed-loop poles are located at $s = -1 \pm j\sqrt{3}$.

(a) Show that this specification cannot be achieved by choosing proportional control alone, $D(s) = k_p$.

(b) Design a lead compensator $D(s) = K\frac{s+z}{s+p}$ that will meet the specification.

5.29. A servomechanism position control has the plant transfer function

$$G(s) = \frac{10}{s(s+1)(s+10)}.$$

You are to design a series compensation transfer function $D(s)$ in the unity feedback configuration to meet the following closed-loop specifications:

- The response to a reference step input is to have no more than 16% overshoot.

- The response to a reference step input is to have a rise time of no more than 0.4 sec.

- The steady-state error to a unit ramp at the reference input must be less than 0.02

(a) Design a lead compensation that will cause the system to meet the dynamic response specifications.

(b) If $D(s)$ is proportional control, $D(s) = k_p$, what is the velocity constant K_v?

(c) Design a lag compensation to be used in series with the lead you have designed to cause the system to meet the steady-state error specification.

(d) Give the MATLAB plot of the root locus of your final design.

(e) Give the MATLAB response of your final design to a reference step.

5.30. Assume the closed-loop system of Fig. 5.71 has a feed-forward transfer function $G(s)$ given by

$$G(s) = \frac{1}{s(s+2)}.$$

Design a lag compensation so that the dominant poles of the closed-loop system are located at $s = -1 \pm j$ and the steady-state error to a unit ramp input is less than 0.2.

5.31. An elementary magnetic suspension scheme is depicted in Fig. 5.72. For small motions near the reference position, the voltage e on the photo detector is related to the ball displacement x (in meters) by $e = 100x$. The upward force (in newtons) on the ball caused by the current i (in amperes) may be approximated by $f = 0.5i + 20x$. The mass of the ball is 20 g, and the gravitational force is 9.8 N/kg. The power amplifier is a voltage-to-current device with an output (in amperes) of $i = u + V_0$.

Figure 5.72
Elementary magnetic
suspension

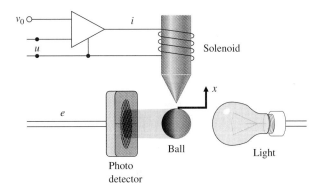

(a) Write the equations of motion for this setup.

(b) Give the value of the bias V_0 that results in the ball being in equilibrium at $x = 0$.

(c) What is the transfer function from u to e?

(d) Suppose the control input u is given by $u = -Ke$. Sketch the root locus of the closed-loop system as a function of K.

(e) Assume that a lead compensation is available in the form $\frac{U}{E} = D(s) = K\frac{s+z}{s+p}$. Give values of K, z, and p that yields improved performance over the one proposed in part (d).

5.32. A certain plant with the nonminimum phase transfer function

$$G(s) = \frac{4 - 2s}{s^2 + s + 9},$$

is in a unity positive feedback system with the controller transfer function $D(s)$.

(a) Use root-locus techniques to determine a (negative) value for $D(s) = K$ so that the closed-loop system with negative feedback has a damping ratio $\zeta = 0.707$.

(b) Use MATLAB to plot the system's response to a reference step.

(c) Give the value of a constant input filter H_r such that the system has zero steady-state error to a reference step.

5.33. Consider the rocket-positioning system shown in Fig. 5.73.

Figure 5.73
Block diagram for
rocket-positioning control
system

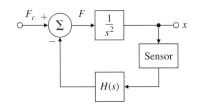

(a) Show that if the sensor that measures x has a unity transfer function, the lead compensator

$$H(s) = K\frac{s+2}{s+4}$$

stabilizes the system.

(b) Assume that the sensor transfer function is modeled by a single pole with a 0.1-sec time constant and unity DC gain. Using the root-locus procedure, find a value for the gain K that will provide the maximum damping ratio.

5.34. For the system in Fig. 5.74:

(a) Sketch the locus of closed-loop roots with respect to K.

(b) Find the maximum value of K for which the system is stable. Assume $K = 2$ for the remaining parts of this problem.

(c) What is the steady-state error ($e = r - y$) for a step change in r?

(d) What is the steady-state error in y for a constant disturbance w_1?

(e) What is the steady-state error in y for a constant disturbance w_2?

(f) If you wished to have more damping, what changes would you make to the system?

Figure 5.74
Control system for
Problem 5.34

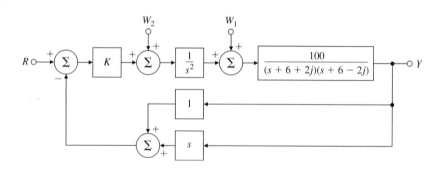

5.35. Consider the plant transfer function

$$G(s) = \frac{bs + k}{s^2[mMs^2 + (M+m)bs + (M+m)k]}$$

to be put in the unity feedback loop of Fig. 5.71. This is the transfer function relating the input force $u(t)$ on mass m and the position $y(t)$ of mass M in the non-collocated sensor and actuator similar to Example 2.2. In this problem we will use root-locus techniques to design a controller $D(s)$ so that the closed-loop step response has a rise time of less than 0.1 sec and an overshoot of less than 10%. You may use MATLAB for any of the following questions.

(a) Approximate $G(s)$ by assuming that $m \cong 0$, and let $M = 1$, $k = 1$, $b = 0.1$, and $D(s) = K$. Can K be chosen to satisfy the performance specifications? Why or why not?

(b) Repeat part (a) assuming $D(s) = K(s+z)$, and show that K and z can be chosen to meet the specifications.

(c) Repeat part (b) but with a practical controller given by the transfer function

$$D(s) = K \frac{p(s+z)}{s+p},$$

and pick p so that the values for K and z computed in part (b) remain more or less valid.

(d) Now suppose that the small mass m is not negligible, but is given by $m = M/10$. Check to see if the controller you designed in part (c) still meets the given specifications. If not, adjust the controller parameters so that the specifications are met.

5.36. Consider the type 1 system drawn in Fig. 5.75. We would like to design the compensation $D(s)$ to meet the following requirements: (1) The steady-state value of y due to a constant unit disturbance w should be less than $\frac{4}{5}$, and (2) the damping ratio $\zeta = 0.7$. Using root-locus techniques:

(a) Show that proportional control alone is not adequate.

(b) Show that proportional-derivative control will work.

(c) Find values of the gains k_p and k_D for $D(s) = k_p + k_D s$ that meet the design specifications.

Figure 5.75
Control system for
Problem 5.36

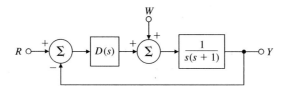

Problems for Section 5.6

5.37. Consider the positioning servomechanism system shown Fig. 5.76, where

$$e_i = K_{\text{pot}}\theta_i, \qquad e_o = K_{\text{pot}}\theta_o, \qquad K_{\text{pot}} = 10 \text{ V/rad},$$

$$T = \text{motor torque} = k_m i_a,$$

$$k_m = \text{torque constant} = 0.1 \text{ N} \cdot \text{m/A},$$

$$R_a = \text{armature resistance} = 10 \text{ }\Omega,$$

$$\text{Gear ratio} = 1:1,$$

$$J_L + J_m = \text{total inertia} = 10^{-3} \text{ kg} \cdot \text{m}^2,$$

$$C = 200 \text{ }\mu\text{F},$$

$$v_a = K_A(e_i - e_f).$$

Figure 5.76
Positioning servomechanism
(*Reprinted from Clark,
1962, with permission*)

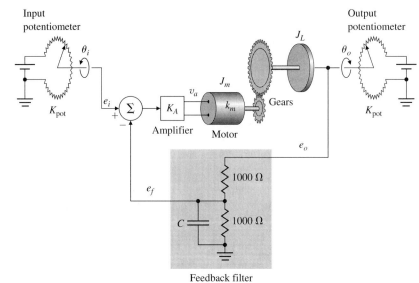

(a) What is the range of the amplifier gain K_A for which the system is stable? Estimate the upper limit graphically using a root-locus plot.

(b) Choose a gain K_A that gives roots at $\zeta = 0.7$. Where are all three closed-loop root locations for this value of K_A?

5.38. We wish to design a velocity control for a tape-drive servomechanism. The transfer function from current $I(s)$ to tape velocity $\Omega(s)$ (in millimeters per millisecond per ampere) is

$$\frac{\Omega(s)}{I(s)} = \frac{15(s^2 + 0.9s + 0.8)}{(s+1)(s^2 + 1.1s + 1)}.$$

We wish to design a type 1 feedback system so that the response to a reference step satisfies

$$t_r \le 4 \text{ msec}, \quad t_s \le 15 \text{ msec}, \quad M_p \le 0.05.$$

(a) Use the integral compensator k_I/s to achieve type 1 behavior, and sketch the root locus with respect to k_I. Show on the same plot the region of acceptable pole locations corresponding to the specifications.

(b) Assume a proportional-integral compensator of the form $k_p(s + \alpha)/s$, and select the best possible values of k_p and α you can find. Sketch the root-locus plot of your design, giving values for k_p and α, and the velocity constant K_v your design achieves. On your plot, indicate the closed-loop poles with a dot •, and include the boundary of the region of acceptable root locations.

5.39. The normalized, scaled equations of a cart as drawn in Fig. 5.77 of mass m_c holding an inverted uniform pendulum of mass m_p and length ℓ with no friction are

$$\ddot{\theta} - \theta = -v,$$

$$\ddot{y} + \beta\theta = v,$$

(5.120)

Figure 5.77
Figure of cart-pendulum for
problem 5.39

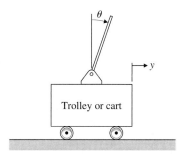

where $\beta = \frac{3m_p}{4(m_c+m_p)}$ is a mass ratio bounded by $0 < \beta < 0.75$. Time is measured in terms of $\tau = \omega_o t$ where $\omega_o^2 = \frac{3g(m_c+m_p)}{\ell(4m_c+m_p)}$. The cart motion, y, is measured in units of pendulum length as $y = \frac{3x}{4\ell}$ and the input is force normalized by the system weight, $v = \frac{u}{g(m_c+m_p)}$. These equations can be used to compute the transfer functions

$$\frac{\Theta}{V} = -\frac{1}{s^2 - 1}, \tag{5.121}$$

$$\frac{Y}{V} = \frac{s^2 - 1 + \beta}{s^2(s^2 - 1)}. \tag{5.122}$$

In this problem you are to design a control for this system by first closing a loop around the pendulum, Eq. (5.121) and then, with this loop closed, closing a second loop around the cart plus pendulum Eq. (5.122). For this problem, let the mass ratio be $m_c = 5m_p$.

(a) Draw a block diagram for the system with V input and both Y and Θ as outputs.

(b) Design a lead compensation $D_p(s) = K_p\frac{s+z}{s+p}$ for the Θ loop to cancel the pole at $s = -1$ and place the two remaining poles at $-4 \pm j4$. The new control is $U(s)$, where the force is $V(s) = U(s) + D(s)\Theta(s)$. Draw the root locus of the angle loop.

(c) Compute the transfer function of the new plant from U to Y with $D(s)$ in place.

(d) Design a controller $D_c(s)$ for the cart position with the pendulum loop closed. Draw the root locus with respect to the gain of $D_c(s)$

(e) Use MATLAB to plot the control, cart position, and pendulum position for a unit step change in cart position.

5.40. Consider the 270-ft U.S. Coast Guard cutter *Tampa* (902) shown in Fig. 5.78. Parameter identification based on sea-trials data (Trankle, 1987) was used to estimate the hydrodynamic coefficients in the equations of motion. The result is that the response of the heading angle of the ship ψ to rudder angle δ and wind changes w can be described by the second-order transfer functions

$$G_\delta(s) = \frac{\psi(s)}{\delta(s)} = \frac{-0.0184(s + 0.0068)}{s(s + 0.2647)(s + 0.0063)},$$

$$G_w(s) = \frac{\psi(s)}{w(s)} = \frac{0.0000064}{s(s + 0.2647)(s + 0.0063)},$$

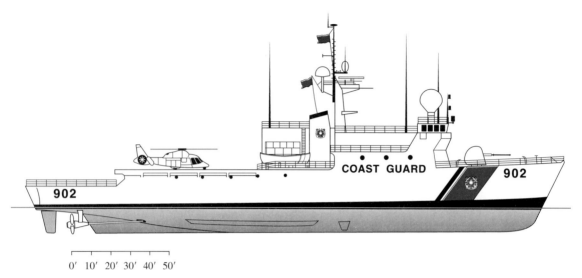

Figure 5.78 USCG cutter *Tampa* (902)

where

$$\psi = \text{heading angle, rad,}$$

$$\psi_r = \text{reference heading angle, rad,}$$

$$r = \dot{\psi} \text{ yaw rate, rad/sec,}$$

$$\delta = \text{rudder angle, rad,}$$

$$w = \text{wind speed, m/sec.}$$

(a) Determine the open-loop settling time of r for a step change in δ.

(b) In order to regulate the heading angle ψ, design a compensator that uses ψ and the measurement provided by a yaw-rate gyroscope (that is, by $\dot{\psi} = r$). The settling time of ψ to a step change in ψ_r is specified to be less than 50 sec, and, for a 5° change in heading the maximum allowable rudder angle deflection is specified to be less than 10°.

(c) Check the response of the closed-loop system you designed in part (b) to a wind gust disturbance of 10 m/sec (model the disturbance as a step input). If the *steady-state* value of the heading due to this wind gust is more than 0.5°, modify your design so that it meets this specification as well.

5.41. Golden Nugget Airlines has opened a free bar in the tail of their airplanes in an attempt to lure customers. In order to automatically adjust for the sudden weight shift due to passengers rushing to the bar when it first opens, the airline is mechanizing a pitch-attitude autopilot. Figure 5.79 shows the block diagram of the proposed arrangement. We will model the passenger moment as a step disturbance $M_p(s) = M_0/s$, with a maximum expected value for M_0 of 0.6.

Figure 5.79
Golden Nugget Airlines
autopilot

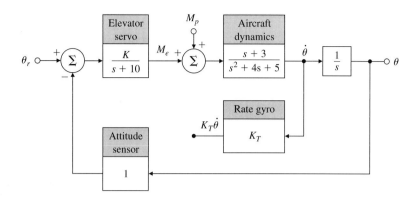

(a) What value of K is required to keep the steady-state error in θ to less than 0.02 rad($\cong 1°$)? (Assume the system is stable.)

(b) Draw a root locus with respect to K.

(c) Based on your root locus, what is the value of K when the system becomes unstable?

(d) Suppose the value of K required for acceptable steady-state behavior is 600. Show that this value yields an unstable system with roots at

$$s = -3, -14, +1 \pm 6.5j.$$

(e) You are given a black box with *rate gyro* written on the side and told that when installed, it provides a perfect measure of $\dot{\theta}$, with output $K_T\dot{\theta}$. Assume $K = 600$ as in part (d) and draw a block diagram indicating how you would incorporate the rate gyro into the auto pilot. (Include transfer functions in boxes.)

(f) For the rate gyro in part (e), sketch a root locus with respect to K_T.

(g) What is the maximum damping factor of the complex roots obtainable with the configuration in part (e)?

(h) What is the value of K_T for part (g)?

(i) Suppose you are not satisfied with the steady-state errors and damping ratio of the system with a rate gyro in parts (e) through (h). Discuss the advantages and disadvantages of adding an integral term and extra lead networks in the control law. Support your comments using MATLAB or with rough root-locus sketches.

5.42. Consider the instrument servomechanism with the parameters given in Fig. 5.80. For each of the following cases, draw a root locus with respect to the parameter K, and indicate the location of the roots corresponding to your final design.

(a) *Lead network*: Let

$$H(s) = 1, \qquad D(s) = K\frac{s+z}{s+p}, \qquad \frac{p}{z} = 6.$$

Select z and K so that the roots nearest the origin (the dominant roots) yield

$$\zeta \geq 0.4, \qquad -\sigma \leq -7, \qquad K_v \geq 16\frac{2}{3}\ \sec^{-1}.$$

Figure 5.80

Control system for
Problem 5.42

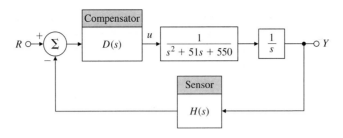

(b) *Output-velocity (tachometer) feedback*: Let

$$H(s) = 1 + K_T s \quad \text{and} \quad D(s) = K.$$

Select K_T and K so that the dominant roots are in the same location as those of part (a). Compute K_v. If you can, give a physical reason explaining the reduction in K_v when output derivative feedback is used.

(c) *Lag network*: Let

$$H(s) = 1 \quad \text{and} \quad D(s) = K \frac{s+1}{s+p}.$$

Using proportional control, it is possible to obtain a $K_v = 12$ at $\zeta = 0.4$. Select K and p so that the dominant roots correspond to the proportional-control case but with $K_v = 100$ rather than $K_v = 12$.

Problems for Section 5.7

5.43. Consider the system in Fig. 5.81.

(a) Use Routh's criterion to determine the regions in the (K_1, K_2) plane for which the system is stable.

(b) Use root-locus methods to verify your answer to part (a).

Figure 5.81

Feedback system for
Problem 5.43

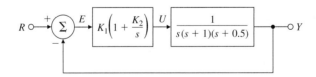

▲   **5.44.** Consider the third-order system shown in Fig. 5.82.

Figure 5.82

Control system for
Problem 5.44

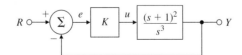

(a) Sketch the root locus for this system with respect to K, showing your calculations for the asymptote angles, departure angles, and so on.

(b) Using graphical techniques, locate carefully the point at which the locus crosses the imaginary axis. What is the value of K at that point?

(c) Assume that, due to some unknown mechanism, the amplifier output is given by the following saturation non linearity (instead of by a proportional gain K):

$$u = \begin{cases} e, & |e| \leq 1; \\ 1, & e > 1; \\ -1, & e < -1. \end{cases}$$

Qualitatively describe how you would expect the system to respond to a unit step input.

5.45. The block diagram of a positioning servomechanism is shown in Fig. 5.83.

(a) Sketch the root locus with respect to K when no tachometer feedback is present ($K_T = 0$).

(b) Indicate the root locations corresponding to $K = 4$ on the locus of part (a). For these locations, estimate the transient-response parameters t_r, M_p, and t_s. Compare your estimates to measurements obtained using the step command in MATLAB.

(c) For $K = 4$, draw the root locus with respect to K_T.

(d) For $K = 4$ and with K_T set so that $M_p = 0.05(\zeta = 0.707)$, estimate t_r and t_s. Compare your estimates to the actual values of t_r and t_s obtained using MATLAB.

(e) For the values of K and K_T in part (d), what is the velocity constant K_v of this system?

Figure 5.83
Control system for
Problem 5.45

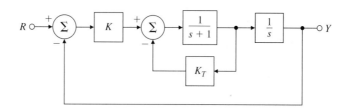

5.46. Consider the mechanical system shown in Fig. 5.84, where g and a_0 are gains. The feedback path containing gs controls the amount of rate feedback. For a fixed value of a_0, adjusting g corresponds to varying the location of a zero in the s-plane.

Figure 5.84
Control system for
Problem 5.46

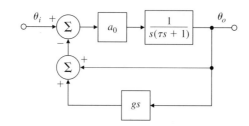

(a) With $g = 0$ and $\tau = 1$, find a value for a_0 such that the poles are complex.

(b) Fix a_0 at this value, and construct a root locus that demonstrates the effect of varying g.

▲   **5.47.** Sketch the root locus with respect to K for the system in Fig. 5.85. What is the range of values of K for which the system is unstable?

Figure 5.85

Control system for Problem 5.47

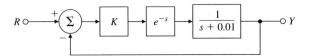

5.48. Prove that the plant $G(s) = 1/s^3$ *cannot* be made unconditionally stable if pole cancellation is forbidden.

5.49. For the equation $1 + KG(s)$ where

$$G(s) = \frac{1}{s(s - p)[(s + 1)^2 + 4]},$$

use MATLAB to examine the root locus as a function of K for p in the range from $p = 1$ to $p = 10$, making sure to include the point $p = 2$.

6 The Frequency-Response Design Method

Chapter Overview

The chapter opens with a discussion of how to obtain the frequency response of a system by analyzing its poles and zeros. An important extension of this discussion is how to use Bode plots to graphically display the frequency response. In Sections 6.2 and 6.3 we discuss stability briefly, and then in more depth the use of the Nyquist stability criterion. In Sections 6.4 through 6.6 we introduce the notion of stability margins, discuss Bode's gain-phase relationship, and study the closed-loop frequency response of dynamic systems. The gain-phase relationship suggests a very simple rule for compensation design: Shape the frequency response magnitude so that it crosses magnitude 1 with a slope of -1. As with our treatment of the root-locus method, we describe how adding dynamic compensation can adjust the frequency response (Section 6.7) and improve system stability and/or error characteristics. We also show how to implement compensation digitally in an example.

Several alternate methods of displaying frequency-response data have been developed over the years; we present two of them—the Nichols chart and the inverse Nyquist plot—in optional Section 6.8. In optional Section 6.9 we discuss issues of sensitivity that relate to the frequency response, including optional material on sensitivity functions and stability robustness. The final two sections—on analyzing time delays in the system and obtaining a pole-zero model from frequency response data—represent additional, somewhat advanced material that may also be considered optional.

A Perspective on the Frequency-response Design Method

The design of feedback control systems in industry is probably accomplished using frequency-response methods more often than any other. Frequency-response design is popular primarily because it provides good designs in the face of uncertainty in the plant model. For example, for systems with poorly known or changing high-frequency resonances, we can temper their feedback compensation to alleviate the effects of those uncertainties. Currently, this tempering is carried out more easily using frequency-response design than any other method.

Another advantage of using frequency response is the ease with which experimental information can be used for design purposes. Raw measurements of the output amplitude and phase of a plant undergoing a sinusoidal input excitation are sufficient to design a suitable feedback control. No intermediate processing of the data (such as finding poles and zeros or determining system matrices) is required to arrive at the system model. The wide availability of computers has rendered this advantage less important now than it was years ago; however, for relatively simple systems, frequency response is often still the most cost-effective design method. The method is most effective for systems that are stable in open-loop.

Yet another advantage is that it is the easiest method to use for designing compensation. A simple rule can be used to provide reasonable designs with a minimum of trial and error.

Although the underlying theory is somewhat challenging and requires a rather broad knowledge of complex variables, the methodology of frequency-response design is easy, and the insights gained by learning the theory are well worth the struggle.

6.1 Frequency Response

The basic concepts of frequency response were discussed in Section 3.1.2. In this section we will review those ideas and extend the concepts for use in control system design.

Frequency response

A linear system's response to sinusoidal inputs—called the system's **frequency response**—can be obtained from knowing its pole and zero locations.

To review the ideas, we consider a system described by

$$\frac{Y(s)}{U(s)} = G(s),$$

where the input $u(t)$ is a sine wave with an amplitude A:

$$u(t) = A \sin(\omega_o t)1(t),$$

which has a Laplace transform

$$U(s) = \frac{A\omega_o}{s^2 + \omega_o^2}.$$

With zero initial conditions, the Laplace transform of the output is

$$Y(s) = G(s)\frac{A\omega_o}{s^2 + \omega_o^2}. \tag{6.1}$$

Partial fraction expansion

A partial-fraction expansion of Eq. (6.1) [assuming the poles of $G(s)$ are distinct] will result in an equation of the form

$$Y(s) = \frac{\alpha_1}{s - p_1} + \frac{\alpha_2}{s - p_2} + \cdots + \frac{\alpha_n}{s - p_n} + \frac{\alpha_o}{s + j\omega_o} + \frac{\alpha_o^*}{s - j\omega_o}, \tag{6.2}$$

where $p_1, p_2, \ldots, p_n$ are the poles of $G(s)$, α_o would be found by performing the partial-fraction expansion and α_o^* is the complex conjugate of α_o. The time response that corresponds to $Y(s)$ is

$$y(t) = \alpha_1 e^{p_1 t} + \alpha_2 e^{p_2 t} \cdots + \alpha_n e^{p_n t} + 2|\alpha_o| \sin(\omega_o t + \phi), \qquad t \geq 0, \tag{6.3}$$

where

$$\phi = \tan^{-1}\left[\frac{\mathrm{Im}(\alpha_o)}{\mathrm{Re}(\alpha_o)}\right].$$

If all the poles of the system represent stable behavior (the real parts of $p_1, p_2, \ldots, p_n < 0$), the natural unforced response will die out eventually and therefore the steady-state response of the system will be due solely to the sinusoidal term in Eq. (6.3) which is caused by the sinusoidal excitation. Example 3.4 determined the response of the system $G(s) = 1/(s+1)$ to the input $u = \sin 10t$; and it showed that response in Fig. 3.2, which is repeated here as Fig. 6.1. It shows that e^{-t}, the natural part of the response associated with $G(s)$, disappears after several time constants and the pure sinusoidal response is essentially all that remains. Example 3.4 showed that the remaining sinusoidal term in Eq. (6.3) can be expressed as

$$y(t) = AM \sin(\omega_o t + \phi), \qquad plus \quad \angle G(j\omega) \qquad (6.4)$$

where

$$M = |G(j\omega_o)| = |G(s)|_{s=j\omega_o} = \sqrt{\{\text{Re}[G(j\omega_o)]\}^2 + \{\text{Im}[G(j\omega_o)]\}^2}, \qquad (6.5)$$

$$\phi = \tan^{-1}\left[\frac{\text{Im}[G(j\omega_o)]}{\text{Re}[G(j\omega_o)]}\right] = \angle G(j\omega_o). \qquad (6.6)$$

In polar form,

$$G(j\omega_o) = Me^{j\phi}. \qquad (6.7)$$

Equation (6.4) shows that a stable system with transfer function $G(s)$ excited by a sinusoid with unit amplitude and frequency ω_o will, after the response has reached steady state, exhibit a sinusoidal output with a magnitude $M(\omega_o)$ and a phase $\phi(\omega_o)$ at the frequency ω_o. The facts that the output y is a sinusoid with the *same* frequency as the input u and that the magnitude ratio M and phase ϕ of the output are independent of the amplitude A of the input are a consequence of $G(s)$ being a linear constant system. If the system being excited were a nonlinear or time-varying system, the output might contain frequencies other than the input frequency and the output-input ratio might be dependent on the input magnitude.

More generally, the **magnitude** M is given by $|G(j\omega)|$, and the **phase** ϕ is given by $\angle[G(j\omega)]$; that is, the magnitude and angle of the complex quantity $G(s)$ are evaluated with s taking on values along the imaginary axis ($s = j\omega$). The frequency response of a system consists of these functions of frequency that

Frequency response plot

Magnitude and phase

Figure 6.1
Response of
$G(s) = 1/(s + 1)$ to
$\sin 10t$

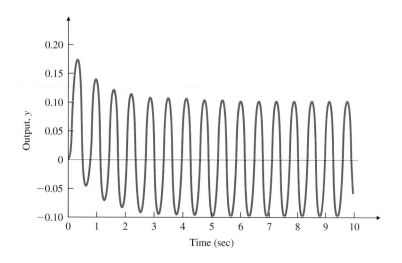

Time (sec)

tell us how a system will respond to a sinusoidal input of any frequency. We are interested in analyzing the frequency response not only because it will help us understand how a system responds to a sinusoidal input, but also because evaluating $G(s)$ with s taking on values along the $j\omega$ axis will prove to be very useful in determining the stability of a closed-loop system. As we saw in Chapter 3, the $j\omega$ axis is the boundary between stability and instability; we will see in Section 6.4 that evaluating $G(j\omega)$ provides information that allows us to determine closed-loop stability from the open-loop $G(s)$.

EXAMPLE 6.1

Frequency-Response Characteristics of a Capacitor

Consider the capacitor described by the equation

$$i = C\frac{dv}{dt},$$

where v is the input and i is the output. Determine the sinusoidal steady-state response of the capacitor.

Solution.   The transfer function of this circuit is

$$\frac{I(s)}{V(s)} = G(s) = Cs,$$

so

$$G(j\omega) = Cj\omega.$$

Computing the magnitude and phase, we find that

$$M = |Cj\omega| = C\omega \quad \text{and} \quad \phi = \angle(Cj\omega) = 90°.$$

For a unit-amplitude sinusoidal input v the output i will be a sinusoid with magnitude $C\omega$ and the phase of the output will lead the input by $90°$. Note for this example that the magnitude is proportional to the input frequency while the phase is independent of frequency.

EXAMPLE 6.2

Frequency-Response Characteristics of a Lead Compensator

Recall from Eq. (5.97) in Chapter 5 the transfer function of the lead compensation, which is equivalent to

$$D(s) = K\frac{Ts + 1}{\alpha Ts + 1}, \qquad \alpha < 1. \tag{6.8}$$

(a) Analytically determine its frequency-response characteristics and discuss what you would expect from the result.

(b) Use MATLAB to plot $D(j\omega)$ with $K = 1$, $T = 1$, and $\alpha = 0.1$ for $0.1 \le \omega \le 10$ and verify the features predicted from the analysis in (a).

Solution.

(a) **Analytical evaluation**: Substituting $s = j\omega$ into Eq. (6.8), we get

$$D(j\omega) = K\,\frac{Tj\omega + 1}{\alpha Tj\omega + 1}.$$

From Eqs. (6.5) and (6.6) the amplitude is

$$M = |D| = |K|\,\frac{\sqrt{1 + (\omega T)^2}}{\sqrt{1 + (\alpha\omega T)^2}}$$

and the phase is given by

$$\phi = \angle(1 + j\omega T) - \angle(1 + j\alpha\omega T)$$

$$= \tan^{-1}(\omega T) - \tan^{-1}(\alpha\omega T).$$

At very low frequencies, the amplitude is just $|K|$ and at very high frequencies it is $|K/\alpha|$. Therefore, the amplitude increases as a function of frequency. The phase is zero at very low frequencies and goes back to zero at very high frequencies. At intermediate frequencies, evaluation of the $\tan^{-1}(\cdot)$ functions would reveal that ϕ becomes positive. These are the general characteristics of lead compensation.

(b) **Computer evaluation**: A MATLAB script for frequency response evaluation was shown for Example 3.4. A similiar script for the lead compensation is

```
num = [1 1];
den = [0.1 1];
sysD = tf(num,den);
[mag,phase,w] = bode(sysD);      % computes magnitude, phase, and frequen-
                                   cies over range of interest
loglog(w,mag)
semilogx(w,phase)
```

produces the frequency response magnitude and phase plots shown in Fig. 6.2.

Figure 6.2
(a) Magnitude and (b) phase for the lead compensation in Example 6.2

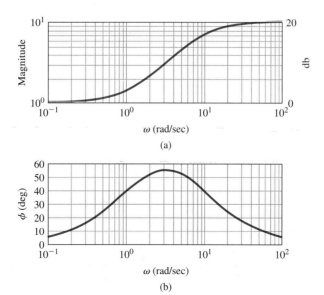

The analysis indicated that the low-frequency magnitude should be $K(= 1)$ and the high frequency magnitude should be K/α $(= 10)$, which are both verified by the magnitude plot. The phase plot also verifies that the value approaches zero at high and low frequencies and that the intermediate values are positive.

In the cases where we do not have a good model of the system and wish to determine the frequency-response magnitude and phase experimentally, we can excite the system with a sinusoid varying in frequency. The magnitude $M(\omega)$ is obtained by measuring the ratio of the output sinusoid to input sinusoid in the steady-state at each frequency. The phase $\phi(\omega)$ is the measured difference in phase between input and output signnals.[1]

A great deal can be learned about the dynamic response of a system from knowledge of the magnitude $M(\omega)$ and the phase $\phi(\omega)$ of its transfer function. In the obvious case, if the signal is a sinusoid, then M and ϕ completely describe the response. Furthermore, if the input is periodic, then a Fourier series can be constructed to decompose the input into a sum of sinusoids, and again $M(\omega)$ and $\phi(\omega)$ can be used with each component to construct the total response. For transient inputs, our best path to understanding the meaning of M and ϕ is to relate the frequency response $G(j\omega)$ to the transient responses calculated by the Laplace transform. For example, in Fig. 3.23(b) we plotted the step response of a system having the transfer function

$$G(s) = \frac{1}{(s/\omega_n)^2 + 2\zeta(s/\omega_n) + 1}, \tag{6.9}$$

for various values of ζ. These transient curves were normalized with respect to time as $\omega_n t$. In Fig. 6.3 we plot $M(\omega)$ and $\phi(\omega)$ for these same values of ζ to help us see what features of the frequency response correspond to the transient-response characteristics. Specifically, Figs. 3.23(b) and 6.3 indicate the effect of damping on system time response and the corresponding effect on the frequency response. They show that the damping of the system can be determined from the transient response overshoot or from the peak in the magnitude of the frequency response [Fig. 6.3(a)]. Furthermore, from the frequency response we see that ω_n is approximately equal to the bandwidth—the frequency where the magnitude starts to fall off from its low-frequency value. (We will define bandwidth more formally in the next paragraph.) Therefore, the rise time can be estimated from the bandwidth. We also see that the peak overshoot in frequency is approximately $1/2\zeta$ for $\zeta < 0.5$, so the peak over-shoot in the step response can be estimated from the peak overshoot in the frequency response. Thus we see that essentially the same information is contained in the frequency-response curve as is found in the transient-response curve.

---

[1] Agilent Technologies produces instruments called spectral analyzers that automate this experimental procedure and greatly speed up the process.

Figure 6.3
(a) Magnitude and
(b) phase of Eq. (6.9)

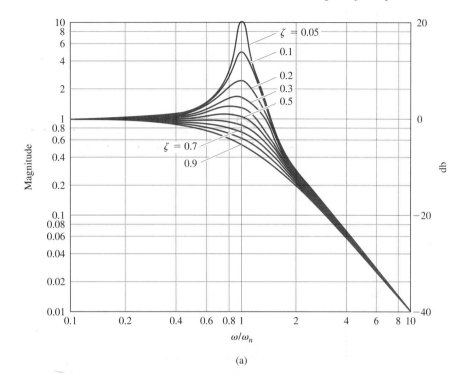

(a)

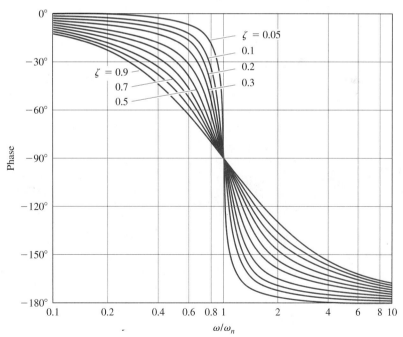

(b)

Figure 6.4
Simplified system definition

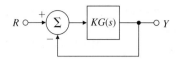

Bandwidth

A natural specification for system performance in terms of frequency response is the **bandwidth,** defined to be the maximum frequency at which the output of a system will track an input sinusoid in a satisfactory manner. By convention, for the system shown in Fig. 6.4 with a sinusoidal input r, the bandwidth is the frequency of r at which the output y is attenuated to a factor of 0.707 times the input.[2]   Figure 6.5 depicts the idea graphically for the frequency response of the *closed-loop* transfer function

$$\frac{Y(s)}{R(s)} \triangleq T(s) = \frac{KG(s)}{1 + KG(s)}.$$

The plot is typical of most closed-loop systems in that (1) the output follows the input $[|T| \cong 1]$ at the lower excitation frequencies and (2) the output ceases to follow the input $[|T| < 1]$ at the higher excitation frequencies. The maximum value of the frequency-response magnitude is referred to as the **resonant peak** M_r.

Bandwidth is a measure of speed of response and is therefore similar to time-domain measures like rise time and peak time or the s-plane measure of dominant-root(s) natural frequency. In fact, if the $KG(s)$ in Fig. 6.4 is such that the closed-loop response is given by Fig. 6.3, we can see that the bandwidth will equal the natural frequency of the closed-loop root (that is, $\omega_{BW} = \omega_n$ for a closed-loop damping ratio of $\zeta = 0.7$). For other damping ratios the bandwidth is approximately equal to the natural frequency of the closed-loop roots, with an error typically less than a factor of 2.

Figure 6.5
Definitions of bandwidth and resonant peak

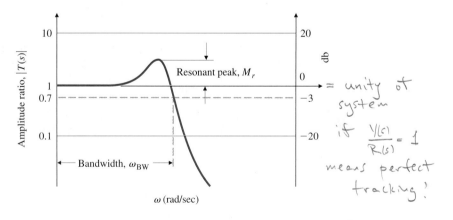

---

[2]   If the output is a voltage across a 1-Ω resistor, the power is v^2; and when $|v| = 0.707$, the power is a factor of 2 reduced. By convention, this is called the half-power point.

The definition of the bandwidth stated here is meaningful for systems that have a low-pass filter behavior, as is the case for any physical control system. In other applications the bandwidth may be defined differently. Also, if the ideal model of the system does not have a high-frequency roll-off (e.g., if it has an equal number of poles and zeros), the bandwidth is infinite; however, this does not occur in nature because nothing responds well at infinite frequencies.

In order to use the frequency response for control systems design, we need to consider an efficient and meaningful form in which to make frequency-response plots as well as find methods to relate the open-loop characteristics of KG to the closed-loop characteristics of $T(s)$. These are the concerns of Section 6.1.1.

6.1.1   Bode Plot Techniques

Display of frequency response is a problem that has been studied for a long time. Before computers, this was accomplished by hand; therefore, it was useful to be able to accomplish this quickly. The most useful technique for hand plotting was developed by H. W. Bode at Bell Laboratories between 1932 and 1942. This technique allows plotting that is quick and yet sufficiently accurate for control systems design. Most control systems designers now have access to computer programs that diminish the need for hand plotting; however, it is still important to develop good intuition so that you can quickly identify erroneous computer results, and for this you need the ability to check results by hand.

The idea in Bode's method is to plot magnitude curves using a logarithmic scale and phase curves using a linear scale. This strategy allows us to plot a high-order $G(j\omega)$ by simply adding the separate terms graphically, as discussed in Appendix B. This addition is possible because a complex expression with zero and pole factors can be written in polar (or phasor) form as

$$G(j\omega) = \frac{\vec{s}_1 \vec{s}_2}{\vec{s}_3 \vec{s}_4 \vec{s}_5} = \frac{r_1 e^{j\theta_1} r_2 e^{j\theta_2}}{r_3 e^{j\theta_3} r_4 e^{j\theta_4} r_5 e^{j\theta_5}} = \left(\frac{r_1 r_2}{r_3 r_4 r_5} \right) e^{j(\theta_1 + \theta_2 - \theta_3 - \theta_4 - \theta_5)}. \qquad (6.10)$$

Composite plot from individual terms

(The overhead arrow indicates a phasor.) Note from Eq. (6.10) that the phases of the individual terms are added directly to obtain the phase of the **composite** expression, $G(j\omega)$. Furthermore, because

$$|G(j\omega)| = \frac{r_1 r_2}{r_3 r_4 r_5},$$

then

$$\log_{10} |G(j\omega)| = \log_{10} r_1 + \log_{10} r_2 - \log_{10} r_3 - \log_{10} r_4 - \log_{10} r_5. \qquad (6.11)$$

We see that addition of the logarithms of the individual terms provides the logarithm of the magnitude of the composite expression. The frequency response is typically presented as two curves: the logarithm of magnitude versus log ω,

Bode plot

and the phase versus log ω. Together these two curves comprise a **Bode plot** of the system. Because

$$\log_{10} Me^{j\phi} = \log_{10} M + j\phi \log_{10} e, \tag{6.12}$$

Decibel

we see that the Bode plot shows the real and imaginary parts of the logarithm of $G(j\omega)$. In communications it is standard to measure the power gain in decibels (db):[3]

$$|G|_{db} = 10 \log_{10} \frac{P_2}{P_1}, \tag{6.13}$$

where P_1 and P_2 are the input and output powers. Because power is proportional to the square of the voltage, the power gain is also given by

$$|G|_{db} = 20 \log_{10} \frac{V_2}{V_1}. \tag{6.14}$$

Hence we can present a Bode plot as the magnitude in decibels versus log ω and the phase in degrees versus log ω.[4] In this book we give Bode plots in the form $\log |G|$ versus $\log \omega$; also we mark an axis in decibels on the right-hand side of the magnitude plot to give you the choice of working with the representation you prefer. However, for frequency response plots, we are not actually plotting power and use of Eq. (6.14) can be somewhat misleading. If the magnitude data are derived in terms of $\log |G|$, it is conventional to plot them on a log scale but identify the scale in terms of $|G|$ only (without "log"). If the magnitude data are given in decibels, the vertical scale is linear such that each decade of $|G|$ represents 20 db.

Advantages of Working with Frequency Response in terms of Bode Plots

Advantages of Bode plots

1. Dynamic compensator design can be based entirely on Bode plots.

2. Bode plots can be determined experimentally.

3. Bode plots of systems in series (or tandem) simply add, which is quite convenient.

4. The use of a log scale permits a much wider range of frequencies to be displayed on a single plot than is possible with linear scales.

---

[3] Researchers at Bell Laboratories first defined the unit of power gain as a **bel** (named for Alexander Graham Bell, the founder of the company). However, this unit proved to be too large, and hence a **decibel or db** (1/10 of a bel) was selected as a more useful unit. The abbreviation dB is also sometimes used; however, Bode used db and we choose to follow his lead.

[4] Henceforth we will drop the base of the logarithm; it is understood to be 10.

It is important for the control systems engineer to be able to hand-plot frequency responses for several reasons: This skill not only allows the engineer to deal with simple problems but also to perform a sanity check on computer results for more complicated cases. Often approximations can be used to quickly sketch the frequency response and deduce stability as well as determine the form of the needed dynamic compensations. Finally, hand plotting is useful in interpreting frequency-response data that have been generated experimentally.

In Chapter 5 we wrote the open-loop transfer function in the form

$$KG(s) = K\frac{(s - z_1)(s - z_2)\cdots}{(s - p_1)(s - p_2)\cdots} \tag{6.15}$$

because it was the most convenient form for determining the degree of stability from the root locus with respect to the gain K. In working with frequency response, it is more convenient to replace s with $j\omega$ and to write the transfer functions in the **Bode form**

Bode form of the transfer function

$$KG(j\omega) = K_o\frac{(j\omega\tau_1 + 1)(j\omega\tau_2 + 1)\cdots}{(j\omega\tau_a + 1)(j\omega\tau_b + 1)\cdots} \tag{6.16}$$

because the gain K_o in this form is directly related to the transfer-function magnitude at very low frequencies. In fact, for type 0 systems, K_o is the gain at $\omega = 0$ in Eq. (6.16) and is also equal to the DC gain of the system. Although a straightforward calculation will convert a transfer function in the form of Eq. (6.15) to an equivalent transfer function in the form of Eq. (6.16), note that K and K_o will not usually have the same value in the two expressions.

Transfer functions can also be rewritten according to Eqs. (6.10) and (6.11). As an example, suppose that

$$KG(j\omega) = K_o\frac{j\omega\tau_1 + 1}{(j\omega)^2(j\omega\tau_a + 1)}. \tag{6.17}$$

Then

$$\angle KG(j\omega) = \angle K_o + \angle(j\omega\tau_1 + 1) - \angle(j\omega)^2 - \angle(j\omega\tau_a + 1) \tag{6.18}$$

and

$$\log|KG(j\omega)| = \log|K_o| + \log|j\omega\tau_1 + 1| - \log|(j\omega)^2| - \log|j\omega\tau_a + 1|. \tag{6.19}$$

In decibels, Eq. (6.19) becomes

$$|KG(j\omega)|_{\mathrm{db}} = 20\,\log|K_o| + 20\,\log|j\omega\tau_1 + 1| - 20\,\log|(j\omega)^2|$$
$$- 20\,\log|j\omega\tau_a + 1|. \tag{6.20}$$

All transfer functions for the kinds of systems we have talked about so far are composed of three classes of terms:

Classes of terms of transfer functions

1. $K_o(j\omega)^n$,

2. $(j\omega\tau + 1)^{\pm 1}$,

3. $\left[\left(\dfrac{j\omega}{\omega_n}\right)^2 + 2\zeta\dfrac{j\omega}{\omega_n} + 1\right]^{\pm 1}$.

First we will discuss the plotting of each individual term and how the terms affect the composite plot including all the terms; then we will discuss how to draw the composite curve

Class 1: singularities at the origin

1. $K_o(j\omega)^n$. Because

$$\log K_o|(j\omega)^n| = \log K_o + n \log|j\omega|,$$

the magnitude plot of this term is a straight line with a slope $n \times$ (20 db/decade). Examples for different values of n are shown in Fig. 6.6. $K_o(j\omega)^n$ is the only class of term that affects the slope at the lowest frequencies because all other terms are constant in that region. The easiest way to draw the curve is to locate $\omega = 1$ and plot $\log K_o$ at that frequency. Then draw the line with slope n through that point.[5] The phase of $(j\omega)^n$ is $\phi = n \times 90°$; it is independent of frequency and is thus a horizontal line: $-90°$ for $n = -1$, $-180°$ for $n = -2$, $+90°$ for $n = +1$, and so forth.

Class 2: first-order term

2. $j\omega\tau + 1$. The magnitude of this term approaches one asymptote at very low frequencies and another asymptote at very high frequencies:

(a) For $\omega\tau \ll 1$, $j\omega\tau + 1 \cong 1$.

(b) For $\omega\tau \gg 1$, $j\omega\tau + 1 \cong j\omega\tau$.

Break point

If we call $\omega = 1/\tau$ the **break point**, then we see that below the break point the magnitude curve is approximately constant ($= 1$), while above the break point the magnitude curve behaves approximately like the class 1

Figure 6.6
Magnitude of $(j\omega)^n$

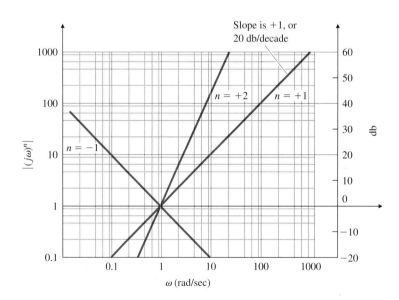

---

[5] In decibels the slopes are $n \times 20$ dB per decade or $n \times 6$ dB per octave (an octave is a change in frequency by a factor of 2).

Figure 6.7
Magnitude plot for
$j\omega\tau + 1j$; $\tau = 0.1$

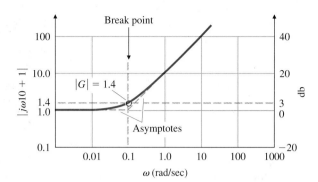

term $K_o(j\omega)^n$. The example plotted in Fig. 6.7, $G(s) = 10s + 1$, shows how the two asymptotes cross at the break point and how the actual magnitude curve lies above that point by a factor of 1.4 (or +3 db). (If the term were in the denominator, it would be below the break point by a factor of 0.707 or -3 db.) Note that this term will have only a small effect on the composite magnitude curve below the break point because its value is equal to 1 (= 0 db) in this region. The slope at high frequencies is +1 (or +20 db/decade). The phase curve can also be easily drawn by using the following low- and high-frequency asymptotes:

(a) For $\omega\tau \ll 1$, $\angle 1 = 0°$.

(b) For $\omega\tau \gg 1$, $\angle j\omega\tau = 90°$.

(c) For $\omega\tau \cong 1$, $\angle(j\omega\tau + 1) \cong 45°$.

For $\omega\tau \cong 1$, the $\angle(j\omega + 1)$ curve is tangent to an asymptote going from 0° at $\omega\tau = 0.2$ to 90° at $\omega\tau = 5$ as shown in Fig. 6.8. The figure also illustrates the three asymptotes (dashed lines) used for the phase plot and how the actual curve deviates from the asymptotes by 11° at their intersections. Both the composite phase and magnitude curves are unaffected by this class of term

Figure 6.8
Phase plot for $j\omega\tau + 1j$;
$\tau = 0.1$

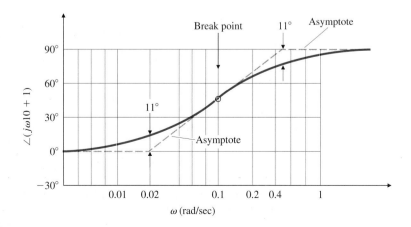

at frequencies below the break point by more than a factor of 10 because the term's magnitude is 1 (or 0 db) and its phase is $0°$.

Class 3: second-order term

3. $[(\mathbf{j}\omega/\omega_{\mathbf{n}})^2 + 2\zeta(\mathbf{j}\omega/\omega_{\mathbf{n}}) + 1]^{\pm 1}$. This term behaves in a manner similar to the class 2 term, with differences in detail: The break point is now $\omega = \omega_n$. The magnitude changes slope by a factor of $+2$ (or $+40$ db per decade) at the break point, (and -2, or -40 db per decade, when the term is in the denominator). The phase changes by $\pm 180°$, and the transition through the break point region varies with the damping ratio ζ. Figure 6.3 shows the magnitude and phase for several different damping ratios when the term is in the denominator. Note that the magnitude asymptote for frequencies above the break point has a slope of -2 (or -40 db per decade), and that the transition through the break-point region has a large dependence on the damping ratio. A rough sketch of this transition can be made by noting that

Peak amplitude

$$|G(j\omega)| = \frac{1}{2\zeta} \quad \text{at} \quad \omega = \omega_n \tag{6.21}$$

for this class of second-order term in the denominator. If the term was in the numerator, the magnitude would be the reciprocal of the curve plotted in Fig. 6.3(a).

No such handy rule as Eq. (6.21) exists for sketching in the transition for the phase curve; therefore, we would have to resort to Fig. 6.3(b) for an accurate plot of the phase. However, a very rough idea of the transition can be gained by noting that it is a step function for $\zeta = 0$, while it obeys the rule for two first-order (class 2) terms when $\zeta = 1$ with simultaneous break-point frequencies. All intermediate values of ζ fall between these two extremes. The phase of a second-order term is always $\pm 90°$ at ω_n.

Composite curve

When the system has several poles and several zeros, plotting the frequency response requires that the components be combined into a composite curve. To plot the composite magnitude curve, it is useful to note that the slope of the asymptotes is equal to the sum of the slopes of the individual curves. Therefore, the composite asymptote curve has integer slope changes at each break point frequency: $+1$ for a first-order term in the numerator, -1 for a first-order term in the denominator, and ± 2 for second-order terms. Furthermore, the lowest-frequency portion of the asymptote has a slope determined by the value of n in the $(j\omega)^n$ term and is located by plotting the point $K_o\omega^n$ at $\omega = 1$. Therefore, the complete procedure consists of plotting the lowest-frequency portion of the asymptote, then sequentially changing the asymptote's slope at each break point in order of ascending frequency, and finally drawing the actual curve by using the transition rules discussed earlier for classes 2 and 3.

The composite phase curve is the sum of the individual curves. Adding of the individual phase curves graphically is made possible by locating the curves so that the composite phase approaches the individual curve as closely as possible. A quick but crude sketch of the composite phase can be found by starting the phase curve below the lowest break point and setting it equal to $n \times 90°$. The phase is then stepped at each break point in order of ascending frequency. The amount of the phase step is $\pm 90°$ for a first-order term and $\pm 180°$ for a second-order term. Break points in the numerator indicate a positive step in phase, while break points in the denominator indicate a negative phase step.[6] The plotting rules so far have only considered poles and zeros in the LHP. Changes for singularities in the RHP will be discussed at the end of the section.

Summary of Bode Plot Rules

1. Manipulate the transfer function into the Bode form given by Eq. (6.16).

2. Determine the value of n for the $K_o(j\omega)^n$ term (class 1). Plot the low-frequency magnitude asymptote through the point K_o at $\omega = 1$ with a slope of n (or $n \times 20$ db per decade).

3. Complete the composite magnitude asymptotes: Extend the low-frequency asymptote until the first frequency break point. Then step the slope by ± 1 or ± 2, depending on whether the break point is from a first- or second-order term in the numerator or denominator. Continue through all break points in ascending order.

4. Sketch in the approximate magnitude curve: Increase the asymptote value by a factor of 1.4 ($+3$ db) at first-order numerator break points, and decrease it by a factor of 0.707 (-3 db) at first-order denominator break points. At second-order break points, sketch in the resonant peak (or valley) according to Fig. 6.3(a) using the relation $|G(j\omega)| = 1/2\zeta$ at denominator (or $|G(j\omega)| = 2\zeta$ at numerator) break points.

5. Plot the low-frequency asymptote of the phase curve, $\phi = n \times 90°$.

6. As a guide, sketch in the approximate phase curve by changing the phase by $\pm 90°$ or $\pm 180°$ at each break point in ascending order. For first-order terms in the numerator, the change of phase is $+90°$; for those in the denominator the change is $-90°$. For second-order terms, the change is $\pm 180°$.

7. Locate the asymptotes for each individual phase curve so that their phase change corresponds to the steps in the phase toward or away from the approximate curve indicated by Step 6. Sketch in each individual phase curve as indicated by Fig. 6.8 or Fig. 6.3(b).

8. Graphically add each phase curve. Use grids if an accuracy of about $\pm 5°$ is desired. If less accuracy is acceptable, the composite curve can be done by eye. Keep in mind that the curve will start at the lowest-frequency asymptote and end on the highest-frequency asymptote and will approach the intermediate asymptotes to an extent that is determined by how close the break points are to each other.

---

[6] This approximate method was pointed out to us by our Parisian colleagues.

EXAMPLE 6.3   *Bode Plot for Real Poles and Zeros*

Plot the Bode magnitude and phase for the system with the transfer function

$$KG(s) = \frac{2000(s + 0.5)}{s(s + 10)(s + 50)}.$$

Solution.

STEP 1. We convert the function to the Bode form of Eq. (6.16):

$$KG(j\omega) = \frac{2[(j\omega/0.5) + 1]}{j\omega[(j\omega/10) + 1][(j\omega/50) + 1]}.$$

STEP 2. We note that the term in $j\omega$ is first-order and in the denominator, so $n = -1$. Therefore, the low-frequency asymptote is defined by the first term:

$$KG(j\omega) = \frac{2}{j\omega}.$$

This asymptote is valid for $\omega < 0.1$ because the lowest break point is at $\omega = 0.5$. The magnitude plot of this term has the slope of -1 (or -20 db per decade). We locate the magnitude by passing through the value 2 at $\omega = 1$ even though the composite curve will not go through this point because of the break point at $\omega = 0.5$. This is shown in Fig. 6.9(a).

STEP 3. We obtain the remainder of the asymptotes, also shown in Fig. 6.9(a): The first breakpoint is at $\omega = 0.5$ and is a first order term in the numerator, thus calls for a change in slope of $+1$. We therefore draw a line with 0 slope that intersects the original -1 slope. Then we draw a -1 slope line that intersects the previous one at $\omega = 10$. Finally, we draw a -2 slope line that intersects the previous -1 slope at $\omega = 50$.

STEP 4. We sketch in the actual curve so that it is approximately tangent to the asymptotes when far away from the break points, a factor of 1.4 ($+3$ db) above the asymptote at the $\omega = 0.5$ break point, and a factor of 0.7 (-3 db) below the asymptote at the $\omega = 10$ and $\omega = 50$ break points.

STEP 5. Because the phase of $\frac{2}{j\omega}$ is $-90°$, the phase curve in Fig. 6.9(b) starts at $-90°$ at the lowest frequencies.

STEP 6. The result is shown in Fig. 6.9(c).

STEP 7. The individual phase curves, shown dashed in Fig. 6.9(b), have the correct phase change for each term and are aligned vertically so that their phase change corresponds to the steps in the phase from the approximate curve in Fig. 6.9(c). Note that the composite curve approaches each individual term.

STEP 8. The graphical addition of each dashed curve results in the solid composite curve in Fig. 6.9(b). As can be seen from the figure, the vertical placement of each individual phase curve makes the required graphical addition particularly easy because the composite curve approaches each individual phase curve in turn.

Figure 6.9
Composite plots:
(a) magnitude; (b) phase;
(c) approximate phase

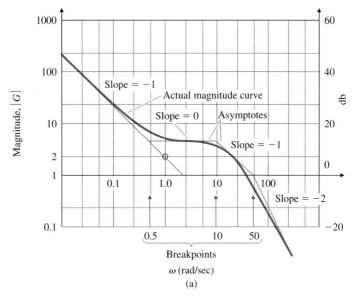

(a)

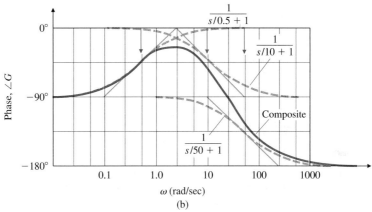

(b)

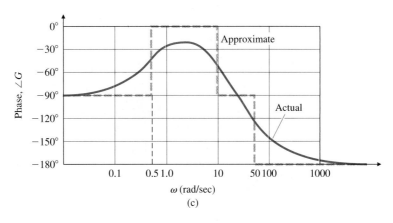

(c)

EXAMPLE 6.4   ***Bode Plot with Complex Poles***

As a second example, draw the frequency response for the system

$$KG(s) = \frac{10}{s(s^2 + 0.4s + 4)}.$$ (6.22)

Solution.   A system like this is more difficult to plot than the one in the previous example because the transition between asymptotes is dependent on the damping ratio; however, the same basic ideas illustrated in Example 6.3 apply.

This system contains a second-order term in the denominator. Proceeding through the steps, we convert Eq. (6.22) to the Bode form of Eq. (6.16):

$$KG(s) = \frac{10}{4} \frac{1}{s(s^2/4 + 2(0.1)s/2 + 1)}.$$

Starting with the low-frequency asymptote, we have $n = -1$ and $|G(j\omega)| \cong 2.5/\omega$. The magnitude plot of this term has a slope of -1 (-20 db per decade) and passes through the value of 2.5 at $\omega = 1$ as shown in Fig. 6.10(a). For the second-order pole, note that $\omega_n = 2$ and $\zeta = 0.1$. At the break-point frequency of the poles, $\omega = 2$, the slope

Figure 6.10
Bode plot for a transfer function with complex poles: (a) magnitude; (b) phase

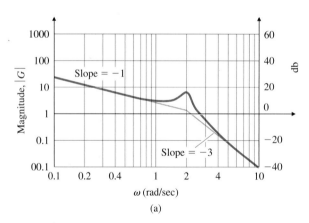

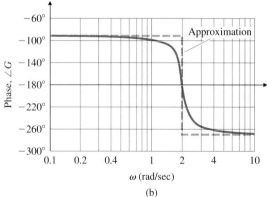

shifts to -3 (-60 db per decade). At the pole break point the magnitude ratio above the asymptote is $1/2\zeta = 1/0.2 = 5$. The phase curve for this case starts at $\phi = -90°$ corresponding to the $1/s$ term, falls to $\phi = -180°$ at $\omega = 2$ due to the pole as shown in Fig. 6.10(b), and then approaches $\phi = -270°$ for higher frequencies. Because the damping is small, the stepwise approximation is a very good one. The true composite phase curve is shown in Fig. 6.10(b).

EXAMPLE 6.5

Bode Plot for Complex Poles and Zeros: Satellite with Flexible Appendages

As a third example, draw the Bode plots for a system with second-order terms. The transfer function represents a mechanical system with two equal masses coupled with a lightly damped spring. The applied force and position measurement are collocated on the same mass. For the transfer function the time scale has been chosen so that the resonant frequency of the complex zeros is equal to 1. The transfer function is

$$KG(s) = \frac{0.01(s^2 + 0.01s + 1)}{s^2[(s^2/4) + 0.02(s/2) + 1]}.$$

Solution.   Proceeding through the steps, we start with the low-frequency asymptote, $0.01/\omega^2$. It has a slope of -2 (-40 db per decade) and passes through magnitude $= 0.01$ at $\omega = 1$ as shown in Fig. 6.11(a). At the break-point frequency of the zero, $\omega = 1$, the slope shifts to zero until the break point of the pole, which is located at $\omega = 2$, when the slope returns to a slope of -2. To interpolate the true curve, we plot the point at the zero break point, $\omega = 1$, with a magnitude ratio below the asymptote of $2\zeta = 0.01$. At the pole break point the magnitude ratio above the asymptote is $1/2\zeta = 1/0.02 = 50$. The magnitude curve is a "doublet" of a negative pulse followed by a positive pulse. Figure 6.11(b) shows that the phase curve for this system starts at $-180°$ (corresponding to the $1/s^2$ term), jumps $180°$ to $\phi = 0$ at $\omega = 1$ due to the zeros, and then falls $180°$ back to $\phi = -180°$ at $\omega = 2$ due to the pole. With such small damping ratios the stepwise approximation is quite good. [We haven't drawn this on Fig. 6.3(b) because it would not be easily distinguishable from the true phase curve.] Thus the true composite phase curve is a nearly square pulse between $\omega = 1$ and $\omega = 2$.

In actual designs, most Bode plots are made with the aid of a computer. However, acquiring the ability to quickly sketch Bode plots by hand is a useful skill because it gives the designer insight into how changes in the compensation parameters will affect the frequency response. This allows the designer to iterate to the best designs more quickly.

EXAMPLE 6.6

Computer-Aided Bode Plot for Complex Poles and Zeros

Repeat Example 6.5 using MATLAB.

Figure 6.11
Bode plot for a transfer
function with complex poles
and zeros: (a) magnitude;
(b) phase

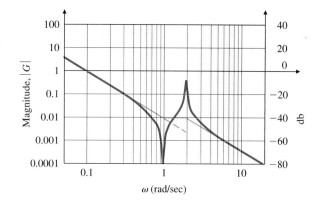

(a)

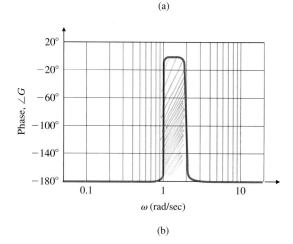

(b)

Solution. To obtain Bode plots using MATLAB, we call the function **bode** as follows:

```
numG = 0.01*[1  0.01  1];
denG = [0.25  0.01  1  0  0];
sysG = tf(numG,denG);
[mag, phase, w] = bode(sysG);
loglog(w,mag)
semilogx(w,phase)
```

These commands will result in a Bode plot that matches that in Fig. 6.11 very closely.
To obtain the magnitude plot in decibels, the last three lines can be replaced with

```
bode(sysG)
```

Nonminimum-Phase Systems

A system with a zero in the right half-plane (RHP) undergoes a net change in
phase when evaluated for frequency inputs between zero and infinity, which,

for an associated magnitude plot, is greater than if all poles and zeros were in the left half-plane (LHP). Such a system is called **nonminimum phase**. As can be seen from the construction in Fig. B3, if the zero is in the RHP, then the phase *decreases* at the zero break point instead of exhibiting the usual phase increase that occurs for an LHP zero. Consider the transfer functions

$$G_1(s) = 10\frac{s+1}{s+10},$$

$$G_2(s) = 10\frac{s-1}{s+10}.$$

Both transfer functions have the same magnitude for all frequencies; that is,

$$|G_1(j\omega)| = |G_2(j\omega)|,$$

as shown in Fig. 6.12(a). But the phases of the two transfer functions are drastically different [Fig. 6.12(b)]. A mimimum phase system (all zeros in the LHP) with a given magnitude curve will produce the smallest net change in the

Figure 6.12
Bode plot for minimum and nonminimum phase systems: (a) magnitude; (b) phase

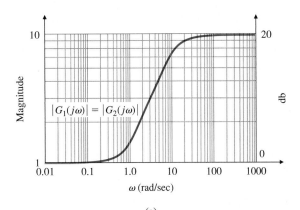

(a)

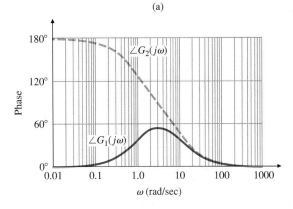

(b)

associated phase as shown in G_1 compared to what the nonminimum phase system will produce as shown by the phase of G_2. Hence, G_2 is nonminimum phase. The discrepancy between G_1 and G_2 in regard to the phase change would be greater if two or more zeros of the plant were in the RHP.

6.1.2   Steady-State Errors

We saw in Section 4.3 that the steady-state error of a feedback system decreases as the gain of the open-loop transfer function increases. In plotting a composite magnitude curve, we saw in Section 6.1.1 that the open-loop transfer function, at very low frequencies, is approximated by

$$KG(j\omega) \cong K_o(j\omega)^n. \tag{6.23}$$

Therefore, we can conclude that the larger the value of the magnitude on the low-frequency asymptote, the lower the steady-state errors will be for the closed-loop system. This relationship is very useful in the design of compensation. Often we want to evaluate several alternate ways to improve stability and to do so we want to be able to see quickly how changes in the compensation will affect the steady-state errors.

Position error constant

For a system of the form given by Eq. (6.16)—that is, where $n = 0$ in Eq. (6.23) (a type 0 system)—the low-frequency asymptote is a constant and the gain K_o of the open-loop system is equal to the position-error constant K_p. For a unity feedback system with a unit step input, the Final Value Theorem (Section 3.1.6) was used in Section 4.3.2 to show that the steady-state error is given by

$$e_{ss} = \frac{1}{1 + K_p}.$$

Velocity error coefficient

For a unity-feedback system in which $n = -1$ in Eq. (6.23), defined to be a type 1 system in Section 4.3.2, the low-frequency asymptote has a slope of -1. The magnitude of the low-frequency asymptote is related to the gain according to Eq. (6.23); therefore, we can again read the gain, K_o/ω, directly from the Bode magnitude plot. Equation (4.71) tells us that the velocity-error constant

$$K_v = K_o,$$

where, for a unity feedback system with a unit ramp input, the steady-state error is

$$e_{ss} = \frac{1}{K_v}.$$

The easiest way of determining the value of K_v in a type 1 system is to read the magnitude of the low-frequency asymptote at $\omega = 1$ rad/sec because this asymptote is $A(\omega) = K_v/\omega$. In some cases the lowest-frequency break point will be below $\omega = 1$ rad/sec; therefore, the asymptote needs to extend to $\omega = 1$ rad/sec in order to read K_v directly. Alternately, we could read the magnitude at any frequency on the low-frequency asymptote and compute it from $K_v = \omega A(\omega)$.

EXAMPLE 6.7   *Computation of K_v*

As an example of the determination of steady-state errors, a Bode magnitude plot of an open-loop system is shown in Fig. 6.13. Assuming there is unity feedback as in Fig. 6.4, find the velocity-error constant, K_v.

Solution.   Because the slope at the low frequencies is -1, we know that the system is type 1. The extension of the low-frequency asymptote crosses $\omega = 1$ rad/sec at a magnitude of 10. Therefore, $K_v = 10$ and the steady-state error to a unit ramp for a unity feedback system would be 0.1. Alternatively, at $\omega = 0.01$ we have $|A(\omega)| = 1000$; therefore, from Eq. (6.23) we have

$$K_o = K_v \cong \omega |A(\omega)| = 0.01(1000) = 10.$$

Figure 6.13

Determination of K_v from the Bode plot for the system $KG(s) = 10/[s(s+1)]$

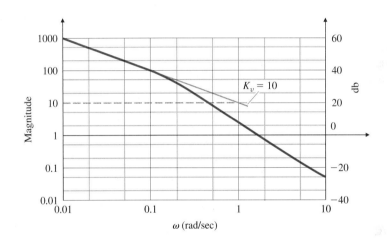

6.2   Neutral Stability

In the early days of electronic communications, most instruments were judged in terms of their frequency response. It is therefore natural that when the feedback amplifier was introduced, techniques to determine stability in the presence of feedback were based on this response.

Suppose the closed-loop transfer function of a system is known. We can determine the stability of a system by simply inspecting the denominator in factored form (because the factors give the system roots directly) to observe whether the real parts are positive or negative. However, the closed-loop transfer function is usually not known; in fact, the whole purpose behind understanding the root-locus technique is to be able to find the factors of the denominator in the closed-loop transfer function, given only the open-loop transfer function. Another way to determine closed-loop stability is to evaluate the frequency response of the *open-loop* transfer function $KG(j\omega)$ and then perform a test on that response. Note that this method does not require factoring the denominator of the closed-loop transfer function. In this section we will explain the principles of this method.

Suppose we have a system defined by Fig. 6.14(a) and whose root locus behaves as shown in Fig. 6.14(b); that is, instability results if K is larger than 2. The neutrally stable points lie on the imaginary axis, that is, where $K = 2$ and $s = j1.0$. Furthermore, we saw in Section 5.1 that all points on the locus have the property that

$$|KG(s)| = 1 \quad \text{and} \quad \angle G(s) = 180°.$$

At the point of neutral stability we see that these root-locus conditions hold for $s = j\omega$, so

$$|KG(j\omega)| = 1 \quad \text{and} \quad \angle G(j\omega) = 180°. \tag{6.24}$$

Thus a Bode plot of a system that is neutrally stable (that is, with K defined such that a closed-loop root falls on the imaginary axis) will satisfy the conditions of Eq. (6.24). Figure 6.15 shows the frequency response for the system whose root locus is plotted in Fig. 6.14 for various values of K. The magnitude response corresponding to $K = 2$ passes through 1 at the same frequency ($\omega = 1$ rad/sec) at which the phase passes through 180°, as predicted by Eq. (6.24).

Having determined the point of neutral stability, we turn to a key question: Does increasing the gain increase or decrease the system's stability? We can see from the root locus in Fig. 6.14(b) that any value of K less than the value at the neutrally stable point will result in a stable system. At the frequency ω where the phase $\angle G(j\omega) = -180°$ ($\omega = 1$ rad/sec), the magnitude $|KG(j\omega)| < 1.0$ for stable values of K and >1 for unstable values of K. Therefore, we have

Figure 6.14

Stability example:
(a) system definition;
(b) root locus

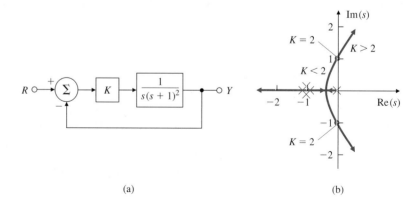

(a)   (b)

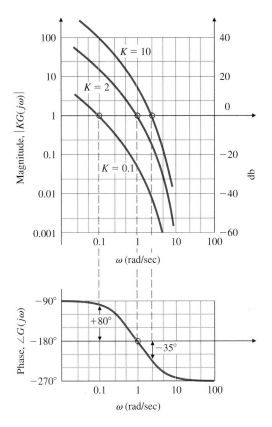

Figure 6.15
Frequency response
magnitude and phase for
the system in Fig. 6.14

Stability condition

the following trial stability condition based on the character of the open-loop frequency response:

$$|KG(j\omega)| < 1 \quad \text{at} \quad \angle G(j\omega) = -180°. \tag{6.25}$$

This stability criterion holds for all systems where increasing gain leads to instability and $|KG(j\omega)|$ crosses the magnitude $= 1$ once, the most common situation. However, there are systems where an increasing gain can lead from instability to stability; in this case, the stability condition is

$$|KG(j\omega)| > 1 \quad \text{at} \quad \angle G(j\omega) = -180°. \tag{6.26}$$

There are also cases when $|KG(j\omega)|$ crosses magnitude $= 1$ more than once. One way to resolve the ambiguity that is usually sufficient is to perform a rough sketch of the root locus. Another more rigorous way to resolve the ambiguity is to use the Nyquist stability criterion, the subject of the next section. However, because the Nyquist criterion is fairly complex, it is important while studying it to bear in mind the theme of this section, namely, that for most systems a simple relationship exists between closed-loop stability and the open-loop frequency response.

6.3   The Nyquist Stability Criterion

For most systems, as we saw in the previous section, an increasing gain eventually causes instability. In the very early days of feedback control design, this relationship between gain and stability margins was assumed to be universal. However, designers found occasionally that in the laboratory the relationship reversed itself; that is, the amplifier would become unstable when the gain was decreased. The confusion caused by these conflicting observations motivated Harry Nyquist of the Bell Telephone Laboratories to study the problem in 1932. His study explained the occasional reversals and resulted in a more sophisticated analysis with no loopholes. Not surprisingly, his test has come to be called the **Nyquist stability criterion**. It is based on a result from complex variable theory known as the **argument principle**,[7] as we briefly explain in this section and in more detail in Appendix B.

The Nyquist stability criterion relates the open-loop frequency response to the number of closed-loop poles of the system in the RHP. Study of the Nyquist criterion will allow you to determine stability from the frequency response of a complex system, perhaps with one or more resonances, where the magnitude curve crosses 1 several times and/or the phase crosses $180°$ several times. It is also very useful in dealing with open-loop, unstable systems, nonminimum-phase systems, and systems with pure delays (transportation lags).

6.3.1   The Argument Principle

Consider the transfer function $H_1(s)$ whose poles and zeros are indicated in the s-plane in Fig. 6.16(a). We wish to evaluate H_1 for values of s on the clockwise contour C_1. (Hence this is called a **contour evaluation**.) We choose the test point s_o for evaluation. The resulting complex quantity has the form $H_1(s_o) = \vec{v} = |\vec{v}|e^{j\alpha}$. The value of the argument of $H_1(s_o)$ is

$$\alpha = \theta_1 + \theta_2 - (\phi_1 + \phi_2).$$

As s traverses C_1 in the clockwise direction starting at s_o, the angle α of $H_1(s)$ in Fig. 6.16(b) will change (decrease or increase), but it will not undergo a net change of $360°$ as long as there are no poles or zeros within C_1. This is because none of the angles that make up α go through a net revolution. The angles θ_1, θ_2, ϕ_1, and ϕ_2 increase or decrease as s traverses around C_1, but they return to their original values as s returns to s_o without rotating through $360°$. This means that the plot of $H_1(s)$ [Fig. 6.16(b)] will not encircle the origin. This conclusion follows from the fact that α is the sum of the angles indicated in Fig. 6.16(a), so the only way that α can be changed by $360°$ after s executes one full traverse of C_1 is for C_1 to contain a pole or zero.

---

[7] Sometimes referred to as "Cauchy's Principle of the Argument."

Figure 6.16

Contour evaluations:
(a) s-plane plot of poles and zeros of $H_1(s)$ and the contour C_1; (b) $H_1(s)$ for s on C_1; (c) s-plane plot of poles and zeros of $H_2(s)$ and the contour C_1; (d) $H_2(s)$ for s on C_1

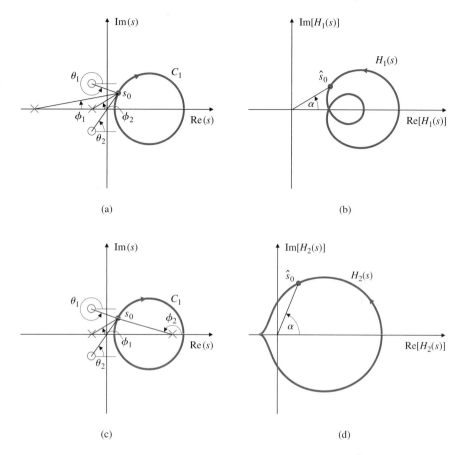

(a)

(b)

(c)

(d)

Now consider the function $H_2(s)$, whose pole-zero pattern is shown in Fig. 6.16(c). Note that it has a singularity (pole) within C_1. Again we start at the test point s_o. As s traverses in the clockwise direction around C_1, the contributions from the angles θ_1, θ_2, and ϕ_1 change, but they return to their original values as soon as s returns to s_o. In contrast, ϕ_2, the angle from the pole within C_1, undergoes a net change of $-360°$ after one full traverse of C_1. Therefore, the argument of $H_2(s)$ undergoes the same change, causing H_2 to encircle the origin in the counterclockwise direction, as shown in Fig. 6.16(d). The behavior would be similar if the contour C_1 had enclosed a zero instead of a pole. The mapping of C_1 would again enclose the origin once in the $H_2(s)$-plane except it would do so in the clockwise direction.

Thus we have the essence of the argument principle:

Argument principle

A contour map of a complex function will encircle the origin $Z - P$ times where Z is the number of zeros and P is the number of poles of the function inside the contour.

For example, if the number of poles and zeros within C_1 is the same, the net angles cancel and there will be no net encirclement of the origin.

6.3.2   Application to Control Design

To apply the principle to control design, we let the C_1 contour in the s-plane encircle the entire RHP, the region in the s-plane where a pole would cause an unstable system (Fig. 6.17). The resulting evaluation of $H(s)$ will only encircle the origin if $H(s)$ has a RHP pole or zero.

    As stated earlier, what makes all this contour behavior useful is that a contour evaluation of an *open-loop* $KG(s)$ can be used to determine stability of the *closed-loop* system. Specifically, for the system in Fig. 6.18, the closed-loop transfer function is

$$\frac{Y(s)}{R(s)} = T(s) = \frac{KG(s)}{1 + KG(s)}.$$

Therefore, the closed-loop roots are the solutions of

$$1 + KG(s) = 0$$

and we apply the principle of the argument to the function $1 + KG(s) = 0$. If the evaluation contour of this function of s enclosing the entire RHP contains a zero or pole of $1 + KG(s)$, then the evaluated contour of $1 + KG(s)$ will encircle the origin. Notice that $1 + KG(s)$ is simply $KG(s)$ shifted to the right 1 unit as shown in Fig. 6.19. Therefore, if the plot of $1 + KG(s)$ encircles the origin, the plot of $KG(s)$ will encircle -1 on the real axis. Therefore, we can plot the contour evaluation of the open-loop $KG(s)$, examine its encirclements of -1, and draw conclusions about the origin encirclements of the closed-loop function $1 + KG(s)$. Presentation of the evaluation of $KG(s)$ in this manner is often referred to as a **Nyquist plot** or **polar plot** because we plot the magnitude of $KG(s)$ versus the angle of $KG(s)$.

Nyquist plot; polar plot

Figure 6.17
An s-plane plot of a contour C_1 that encircles the entire RHP

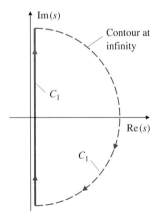

Figure 6.18
Block diagram for
$Y(s)/R(s) =$
$KG(s)/[1 + KG(s)]$

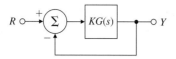

To determine whether an encirclement is due to a pole or zero, we write $1 + KG(s)$ in terms of poles and zeros of $KG(s)$:

$$1 + KG(s) = 1 + K\frac{b(s)}{a(s)} = \frac{a(s) + Kb(s)}{a(s)}. \tag{6.27}$$

Equation (6.27) shows that the poles of $1 + KG(s)$ are also the poles of $G(s)$. Because it is safe to assume that the poles of $G(s)$ [or factors of $a(s)$] are known, the (rare) existence of any of these poles in the RHP can be accounted for. Assuming for now that there are no poles of $G(s)$ in the RHP, an encirclement of -1 by $KG(s)$ indicates a zero of $1 + KG(s)$ in the RHP and thus an unstable root of the closed-loop system.

We can generalize this basic idea by noting that a clockwise contour C_1 enclosing a zero of $1 + KG(s)$—that is, a closed-loop system root—will result in $KG(s)$ encircling the -1 point in a clockwise direction. Likewise, if C_1 encloses a pole of $1 + KG(s)$; that is, if there is an unstable open-loop pole, there will be a counterclockwise $KG(s)$ encirclement of -1. Furthermore, if two poles or two zeros are in the RHP, $KG(s)$ will encircle -1 twice, and so on. The net number of clockwise encirclements, N, equals the number of zeros (closed-loop system roots) in the RHP, Z, minus the number of open-loop poles in the RHP, P:

$$N = Z - P.$$

This is the key concept of the Nyquist stability criterion.

A simplification in the plotting of $KG(s)$ results from the fact that any $KG(s)$ that represents a physical system will have zero response at infinite

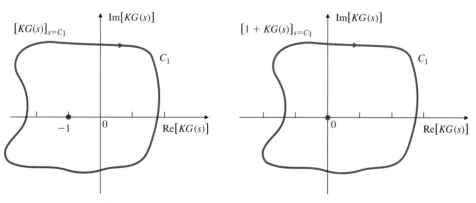

Figure 6.19 Evaluations of $KG(s)$ and $1 + KG(s)$: Nyquist plots

frequency (i.e., has more poles than zeros). This means that the big arc of C_1 corresponding to s at infinity (Fig. 6.19) results in $KG(s)$ being a point of infinitesimally small value near the origin for that portion of C_1. Therefore, we accomplish a complete evaluation of a physical system $KG(s)$ by letting s traverse the imaginary axis from $-j\infty$ to $+j\infty$ (actually from $-j\omega_h$ to $+j\omega_h$ where ω_h is large enough that $|KG(j\omega)|$ is much less than 1 for all $\omega > \omega_h$). The evaluation of $KG(s)$ from $s = 0$ to $s = +j\infty$ has already been discussed in Section 6.1 under the context of finding the frequency response of $KG(s)$. Because $G(-j\omega)$ is the complex conjugate of $G(j\omega)$, we can easily obtain the entire plot of $KG(s)$ by reflecting the $0 \le s \le +j\infty$ portion about the real axis, to get the $(-j\infty \le s < 0)$ portion. Hence we see that closed-loop stability can be determined in all cases by examination of the frequency response of the open-loop transfer function on a polar plot. In some applications, models of physical systems are simplified so as to eliminate some high-frequency dynamics. The resulting reduced-order transfer function might have an equal number of poles and zeros. In that case the big arc of C_1 at infinity needs to be considered.

In practice, many systems behave like those discussed in Section 6.2, so you need not carry out a complete evaluation of $KG(s)$ with subsequent inspection of the -1 encirclements; a simple look at the frequency response may suffice to determine stability based on the gain and phase curves. However, in the case of a complex system for which the simplistic rules given in Section 6.2 become ambiguous, you will want to perform the complete analysis, summarized as follows:

Procedure for Plotting the
Nyquist Plot

1. Plot $KG(s)$ for $-j\infty \le s \le +j\infty$. Do this by first evaluating $KG(j\omega)$ for $\omega = 0$ to ω_h, where ω_h is so large that the magnitude of $KG(j\omega)$ is negligibly small for $\omega > \omega_h$, then reflecting the image about the real axis and adding it to the preceding image. The magnitude of $KG(j\omega)$ will be small at high frequencies for any physical system. The Nyquist plot will always be symmetric with respect to the real axis.

2. Evaluate the number of clockwise encirclements of -1, and call that number N. Do this by drawing a straight line in *any* direction from -1 to ∞. Then count the net number of left-to-right crossings of the straight line by $KG(s)$. If encirclements are in the counterclockwise direction, N is negative.

3. Determine the number of unstable (RHP) poles of $G(s)$, and call that number P.

4. Calculate the number of unstable closed-loop roots, Z:

$$Z = N + P. \tag{6.28}$$

For stability we wish to have $Z = 0$; that is, no characteristic equation roots in the RHP.

Let us now examine a rigorous application of the procedure for drawing Nyquist plots for some examples.

EXAMPLE 6.8   *Nyquist Plot for a Second-Order System*

Determine the stability properties of the system defined in Fig. 6.20.

Figure 6.20
Control system for
Example 6.8

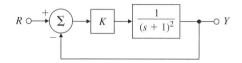

Solution.   The root locus of the system in Fig. 6.20 is shown in Fig. 6.21. It shows that the system is stable for all values of K. The magnitude of the frequency response of $KG(s)$ is plotted in Fig. 6.22(a) for $K = 1$, and the phase is plotted in Fig. 6.22(b); this is the typical Bode method of presenting frequency response and represents the evaluation of $G(s)$ over the interesting range of frequencies. The same information is replotted in Fig. 6.23 in the Nyquist (polar) plot form. Note how the points A, B, C, D, and E are mapped from the Bode plot to the Nyquist plot in Fig. 6.23. The arc from $G(s) = +1$ ($\omega = 0$) to $G(s) = 0$ ($\omega = \infty$) that lies below the real axis is derived from Fig. 6.22. The portion of the C_1 arc at infinity from Fig. 6.17 transforms into $G(s) = 0$ in Fig. 6.23; therefore, a continuous evaluation of $G(s)$ with s traversing C_1 is completed by simply reflecting the lower arc about the real axis. This creates the portion of the contour above the real axis and completes the Nyquist (polar) plot. Because the plot does not encircle -1, $N = 0$. Also, there are no poles of $G(s)$ in the RHP, so $P = 0$. From Eq. (6.28), we conclude that $Z = 0$, which indicates there are no unstable roots of the closed-loop system for $K = 1$. Furthermore, different values of K would simply change the magnitude of the polar plot, but no positive value of K would cause the plot to encircle -1 because the polar plot will always cross the negative real axis when $KG(s) = 0$. Thus the Nyquist stability criterion confirms what the root locus indicated: The closed-loop system is stable for all $K > 0$.

The MATLAB statements that will produce this Nyquist plot are

```
numG = 1;
denG = [1  2  1];
sysG = tf(numG,denG);
nyquist(sysG);
```

Figure 6.21
Root locus of
$G(s) = 1/(s + 1)^2$ with
respect to K

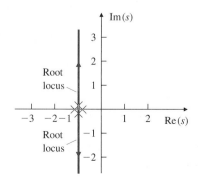

Figure 6.22
Open loop Bode plot for
$G(s) = 1/(s+1)^2$

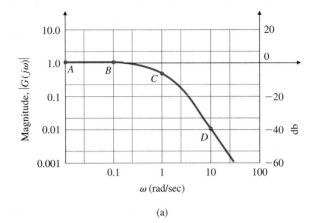

(a)

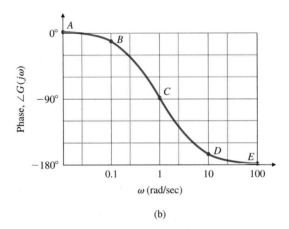

(b)

Figure 6.23
Nyquist plot of the
evaluation of $KG(s)$ for
$s = C_1$ and $K = 1$

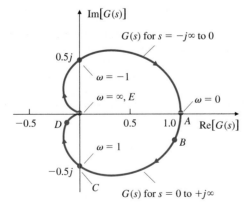

Often the control systems engineer is more interested in determining a range of gains K for which the system is stable than in testing for stability at a specific value of K. To accommodate this requirement, but to avoid drawing multiple Nyquist plots for various values of the gain, the test can be slightly modified. To do so we scale $KG(s)$ by K and examine $G(s)$ to determine stability for a range of gains K. This is possible because an encirclement of -1 by $KG(s)$ is equivalent to an encirclement of $-1/K$ by $G(s)$. Therefore, instead of having to deal with $KG(s)$, we need only consider $G(s)$, and count the number of the encirclements of the $-1/K$ point.

Applying this idea to Example 6.8, we see that the Nyquist plot cannot encircle the $-1/K$ point. For positive K, the $-1/K$ point will move along the negative real axis, so there will not be an encirclement of $G(s)$ for any value of $K > 0$.

(There are also values of $K < 0$ for which the Nyquist plot shows the system to be stable; specifically, $-1 < K < 0$. This result may be verified by drawing the $0°$ locus.)

EXAMPLE 6.9

Nyquist Plot for a Third-Order System

As a second example, consider the system $G(s) = 1/s(s + 1)^2$ for which the closed-loop system is defined in Fig. 6.24. Determine its stability properties using the Nyquist criterion.

Solution.   This is the same system discussed in Section 6.2. The root locus in Fig. 6.14(b) shows that this system is stable for small values of K but unstable for large values of K. The magnitude and phase of $G(s)$ in Fig. 6.25 are transformed into the Nyquist plot shown in Fig. 6.26. Note how the points A, B, C, D, and E on the Bode plot of Fig. 6.25 map into those on the Nyquist plot of Fig. 6.26. Also note the large arc at infinity that arises from the open-loop pole at $s = 0$. This pole creates an infinite magnitude of $G(s)$ at $\omega = 0$; in fact, a pole or zero anywhere on the imaginary axis will create an arc at infinity. To correctly determine the number of $-1/K$ point encirclements, we must draw this arc in the proper half-plane: Should it cross the *positive* real axis, as shown in Fig. 6.26, or the negative one? It is also necessary to assess whether the arc should sweep out $180°$ (as in Fig. 6.26), $360°$, or $540°$.

A simple artifice suffices to answer these questions. We modify the C_1 contour to take a small detour around the pole either to the right (Fig. 6.27), or to the left. It makes no difference to the final stability question which way, but it is more convenient to go to the right because then no poles are introduced within the C_1 contour, keeping the value of P equal to 0. Because the phase of $G(s)$ is the negative of the sum of the angles from all of the poles, we see that the evaluation results in a Nyquist plot moving from $+90°$ for s just below the pole at $s = 0$, across the positive real axis to $-90°$ for s just above the pole. Had there been two poles at $s = 0$, the Nyquist plot at infinity would have executed a full $360°$ arc, and so on for three or more poles. Furthermore, for a pole

Figure 6.24

Control system for Example 6.9

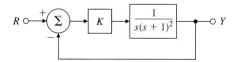

Figure 6.25
Bode plot for
$G(s) = 1/[s(s+1)^2]$

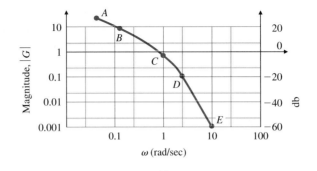

(a)

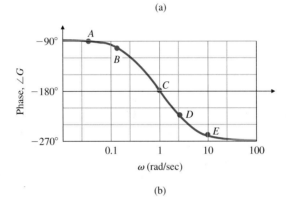

(b)

Figure 6.26
Nyquist plot for
$G(s) = 1/[s(s+1)^2]$

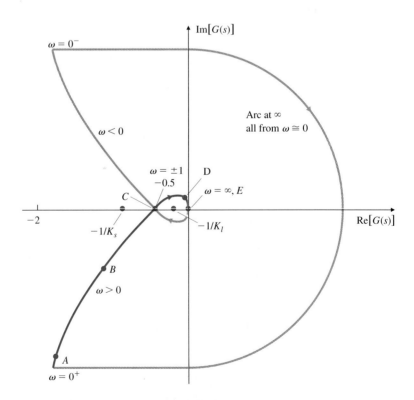

Figure 6.27
C_1 contour enclosing
the RHP for system in
Example 6.9

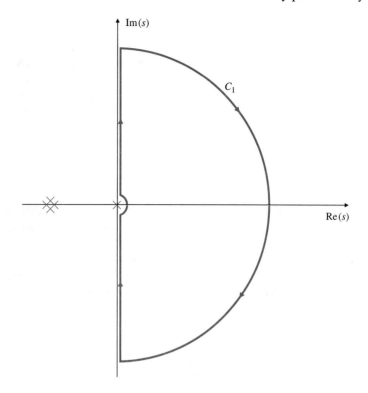

elsewhere on the imaginary axis, a 180° clockwise arc would also result but would be oriented differently than the example in Fig. 6.26.

The Nyquist plot crosses the real axis at $\omega = 1$ with $|G| = 0.5$, as indicated by the Bode plot. For $K > 0$, there are two possibilities for the location of $-1/K$: inside the two loops of the Nyquist plot, or outside the Nyquist contour completely. For large values of K (K_l in Fig. 6.26), $-0.5 < -1/K_l < 0$ will lie inside the two loops; hence $N = 2$ and therefore $Z = 2$, indicating there are two unstable roots. This happens for $K > 2$. For small values of K (K_s in Fig. 6.26), $-1/K$ lies outside the loops; thus $N = 0$, and all roots are stable. All this information is in agreement with the root locus in Figure 6.14(b). (When $K < 0$, $-1/K$ lies on the positive real axis, then $N = 1$, which means $Z = 1$ and the system has one unstable root. The 0° root locus will verify this result.)

For this and many similar systems, we can see that the encirclement criterion reduces to a very simple test for stability based on the open-loop frequency response: The system is stable if $|KG(j\omega)| < 1$ when the phase of $G(j\omega)$ is 180°. Note that this relation is identical to the stability criterion given in Eq. (6.25); however, by using the Nyquist criterion, we don't require the root locus to determine whether $|KG(j\omega)| < 1$ or $|KG(j\omega)| > 1$.

Nyquist plot via MATLAB   We draw the Nyquist plot using MATLAB with

```
numG = 1;
denG = [1  2  1  0];
sysG = tf(numG,denG);
axis([-5  5  −5  5])
nyquist(sysG);
```

The **axis** command scaled the plot so that only points between $+5$ and -5 on the real and imaginary axes were included. Without manual scaling, the plot would be scaled between $\pm\infty$ and the essential features in the vicinity of the the -1 region would be lost.

For systems that are open-loop unstable, care must be taken because now $P \neq 0$ in Eq. (6.28). We will see that the simple rules from Section 6.2 will need to be revised in this case.

EXAMPLE 6.10   *Nyquist Plot for an Open-Loop Unstable System*

The third example is defined in Fig. 6.28. Determine its stability properties using the Nyquist criterion.

Figure 6.28
Control system for
Example 6.10

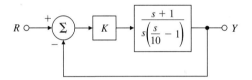

Solution.   The root locus for this system is sketched in Fig. 6.29. The open-loop system is unstable since it has a pole in the RHP. The open-loop Bode plot is shown in Fig. 6.30. Note in the Bode that $|KG(j\omega)|$ behaves exactly the same as if the pole had been in the LHP. However, $\angle G(j\omega)$ increases by 90° instead of the usual decrease at a pole. Any system with a pole in the RHP is unstable; hence it would be impossible to determine its frequency response experimentally because the system would never reach a steady-state sinusoidal response for a sinusoidal input. It is, however, possible to compute the magnitude and phase of the transfer function according to the rules in Section 6.1. The pole in the RHP affects the Nyquist encirclement criterion because the value of P in Eq. (6.28) is $+1$.

Figure 6.29
Root locus for $G(s) =$
$(s + 1)/[s(s/10 - 1)]$

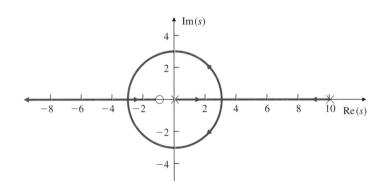

Figure 6.30
Bode plot for $G(s) = (s+1)/[s(s/10 - 1)]$

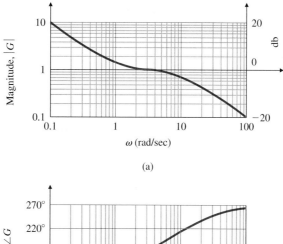

(a)

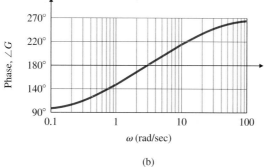

(b)

We convert the frequency-response information of Fig. 6.30 into the Nyquist plot (Fig. 6.31) as in the previous examples. As before, the C_1 detour around the pole at $s = 0$ in Fig. 6.32 creates a large arc at infinity in Fig. 6.31. This arc crosses the *negative* real axis because of the 180° phase contribution of the pole in the RHP.

The real-axis crossing occurs at $|G(s)| = 1$ because in the Bode plot $|G(s)| = 1$ when $\angle G(s) = 180°$, which happens to be at $\omega \cong 3$ rad/sec.

The contour shows two different behaviors depending on the values of $K \; (> 0)$. For large values of $K \; (K_1$ in Fig. 6.31), there is one counterclockwise encirclement; hence $N = -1$. However, since $P = 1$ from the RHP pole, $Z = N + P = 0$, so there are no unstable system roots and the system is stable for $K > 1$. For small values of K (K_s in Fig. 6.31), $N = +1$ because of the clockwise encirclement and $Z = 2$, indicating two unstable roots. This happens if $K < 1$. These results can be verified qualitatively by the root locus in Fig. 6.29. (If $K < 0$, $-1/K$ is on the positive real axis so that $N = 0$ and $Z = 1$, indicating the system will have one unstable closed-loop pole which can be verified by a 0° root locus.)

As with all systems, the stability boundary occurs at $|KG(j\omega)| = 1$ for the phase of $\angle G(j\omega) = 180°$. However, in this case, $|KG(j\omega)|$ must be greater than 1 to yield the correct number of -1 point encirclements to achieve stability.

To draw the Nyquist plot using MATLAB, use the following commands:

```
numG = [1  1];
denG = [0.1  -1  0];
sysG = tf(numG,denG);
axis([-5  5  -5  5]);
nyquist(sysG)
```

Figure 6.31
Nyquist plot for $G(s) = (s+1)/[s(s/10-1)]$

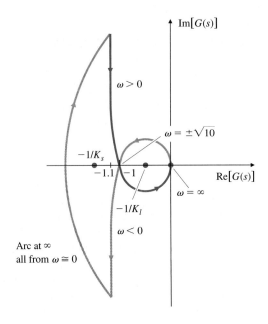

Figure 6.32
C_1 contour for Example 6.10

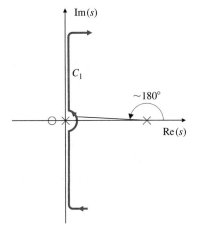

The existence of the RHP pole in Example 6.10 affected the Bode plotting rules of the phase curve and affected the relationship between encirclements and unstable closed-loop roots because $P = 1$ in Eq. (6.28). But we apply the Nyquist stability criterion without any modifications. The same is true for systems with a RHP zero; that is, a nonminimum-phase zero has no effect on the Nyquist stability criterion, but the Bode plotting rules are affected.

6.4   Stability Margins

A large fraction of control system designs behave in a pattern roughly similar to that of the system in Section 6.2 and Example 6.9 in Section 6.3; that is, the system is stable for all small gain values and becomes unstable if the gain increases past a certain critical point. Two commonly used quantities that measure the stability margin for such systems are directly related to the stability criterion of Eq. (6.25): gain margin and phase margin. In this section we will define and use these two concepts to study system design. Another measure of stability, originally defined by O. J. M. Smith (1958), combines these two margins into one and gives a better indication of stability for complicated cases.

Gain margin

The **gain margin (GM)** is the factor by which the gain can be raised before instability results. For the typical case it can be read directly from the Bode plot (for example, see Fig. 6.15) by measuring the vertical distance between the $|KG(j\omega)|$ curve and the $|KG(j\omega)| = 1$ line at the frequency where $\angle G(j\omega) = 180°$. We see from the figure that when $K = 0.1$, the system is stable and GM $= 20$ (or 26 db). When $K = 2$, the system is neutrally stable with GM $= 1$ (0 db), while $K = 10$ results in an unstable system with GM $= 0.125$ (-18 db). Note that GM is the *factor* by which the gain K can be raised before instability results; therefore, $|GM| < 1$ (or $|GM| < 0$ db) indicates an unstable system. The GM can also be determined from a root locus with respect to K by noting two values of K: 1) at the point where the locus crosses the $j\omega$-axis, and 2) at the nominal closed-loop poles. The GM is the ratio of these two values.

Phase margin

Another measure that is used to indicate the stability margin in a system is the **phase margin (PM)**. It is the amount by which the phase of $G(j\omega)$ exceeds $-180°$ when $|KG(j\omega)| = 1$, which is an alternate way of measuring the degree to which the stability conditions of Eq. (6.25) are met. For the case in Fig. 6.15 we see that PM $\cong 80°$ for $K = 0.1$, PM $= 0°$ for $K = 2$, and PM $= -35°$ for $K = 10$. A positive PM is required for stability.

Note that the two stability measures, PM and GM, together determine how far the complex quantity $G(j\omega)$ passes from the -1 point, which is another way of stating the neutral-stability point specified by Eq. (6.24).

The stability margins may also be defined in terms of the Nyquist plot. Figure 6.33 shows that GM and PM are measures of how close the Nyquist plot

Figure 6.33
Nyquist plot for defining GM and PM

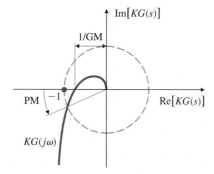

comes to encircling the -1 point. Again we can see that the GM indicates how much the gain can be raised before instability results in a system like the one in Example 6.9. The PM is the difference between the phase of $G(j\omega)$ and $180°$ when $KG(j\omega)$ crosses the circle $|KG(s)| = 1$; the positive value of PM is assigned to the stable case (that is, with no Nyquist encirclements).

Crossover frequency

It is easier to determine these margins directly from the Bode plot than from the Nyquist plot. The term **crossover frequency**, ω_c, is often used to refer to the frequency at which the gain is unity, or 0 db. Figure 6.34 shows the same data plotted in Fig. 6.25. The same values of PM $(= 22°)$ and GM $(= 2)$ may be obtained from the Nyquist plot shown in Fig. 6.26: The real-axis crossing at -0.5 corresponds to a GM of $1/0.5$ or 2 and the PM could be computed graphically by measuring the angle of $G(j\omega)$ as it crosses $|G(j\omega)| = 1$ circle.

One of the useful aspects of frequency-response design is the ease with which we can evaluate the effects of gain changes. In fact, we can determine the PM from Fig. 6.34 for any value of K without redrawing the magnitude or

Figure 6.34
GM and PM from the magnitude and phase plots

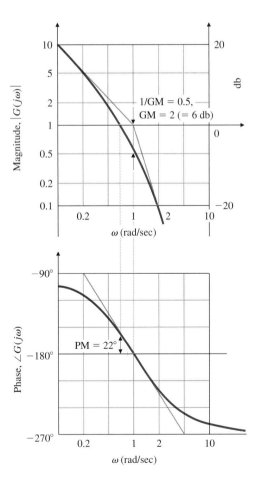

phase information. We need only indicate on the figure where $|KG(j\omega)| = 1$ for selected trial values of K, as has been done with dashed lines in Fig. 6.35. Now we can see that $K = 5$ yields an unstable PM of $-22°$, while a gain of $K = 0.5$ yields a PM of $+45°$. Furthermore, if we wish a certain PM (say 70°), we simply read the value of $|G(j\omega)|$ corresponding to the frequency that would create the desired PM (here $\omega = 0.2$ rad/sec yields 70°, where $|G(j\omega)| = 5$), and note that the magnitude at this frequency is $1/K$. Therefore, a PM of 70° will be achieved with $K = 0.2$.

The PM is more commonly used to specify control system performance because it is most closely related to the damping ratio of the system. This can be seen easily for the open-loop second-order system

$$G(s) = \frac{\omega_n^2}{s(s + 2\zeta\omega_n)},$$
(6.29)

Figure 6.35
PM vs. K from the frequency-response data

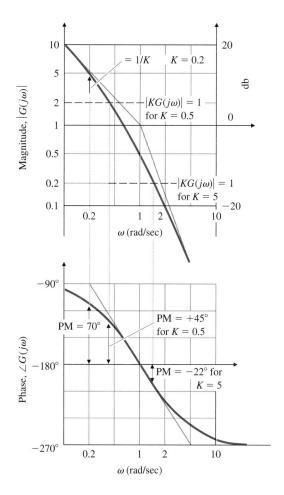

which, with unity feedback produces the closed-loop system

$$T(s) = \frac{\omega_n^2}{s^2 + 2\zeta\omega_n s + \omega_n^2}.$$ (6.30)

It can be shown that the relationship between the PM and ζ in this system is

$$PM = \tan^{-1}\left[\frac{2\zeta}{\sqrt{\sqrt{1+4\zeta^4} - 2\zeta^2}}\right],$$ (6.31)

and this function is plotted in Fig. 6.36. Note that the function is approximately a straight line up to about PM $= 60°$. The dashed line shows a straight-line approximation to the function where

$$\zeta \cong \frac{PM}{100}.$$ (6.32)

It is clear that the approximation only holds for phase margins below about 70°. Furthermore, Eq. (6.31) is only accurate for the second-order system of Eq. (6.30). In spite of these limitations, Eq. (6.32) is often used as a rule of thumb for relating the closed-loop damping ratio to PM. It is useful as a starting point; however, it is important always to check the actual damping of a design as well as other aspects of the performance before calling the design complete.

The gain margin for the second-order system [given by Eq. (6.29)] is infinite (GM $= \infty$) because the phase curve does not cross $-180°$ as the frequency increases. This would also be true for any first- or second-order system.

Additional data to aid in evaluating a control system based on its PM can be derived from the relationship between the resonant peak M_r and ζ seen in Fig. 6.3. Note that this figure was derived for the same system Eq. (6.9) as Eq. (6.30). We can convert the information in Fig. 6.36 into a form relating M_r to the PM. This is depicted in Fig. 6.37, along with the step-response overshoot M_p. Therefore, we see that, given the PM, one can infer information about what the overshoot of the closed-loop step response would be.

Figure 6.36
Damping ratio vs. phase margin (PM)

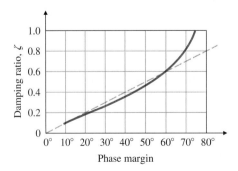

Figure 6.37

Transient-response overshoot (M_p) and frequency response resonant peak (M_r) versus phase margin (PM) for $T(s) = \omega_n^2/(s^2 + 2\zeta\omega_n s + \omega_n^2)$

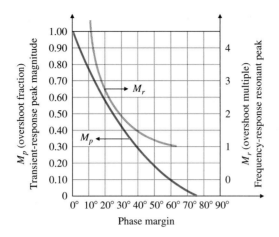

Many engineers think directly in terms of the PM when judging whether a control system is adequately stabilized. In these terms, a PM = $30°$ is often judged to be the lowest adequate value. In addition to testing the stability of a system design using the PM, a designer would typically also be concerned with meeting a speed-of-response specification like bandwidth, as discussed in Section 6.1. In terms of the frequency-response parameters discussed so far, the crossover frequency would best describe a system's speed of response. This idea will be discussed further in Sections 6.6 and 6.7.

In some cases the PM and GM are not helpful indicators of stability. For first- and second-order systems, the phase never crosses the $180°$ line; hence the GM is always ∞ and not a useful design parameter. For higher-order systems it is possible to have more than one frequency where $|KG(j\omega)| = 1$ or where $\angle KG(j\omega) = 180°$, and the margins as previously defined need clarification. An example of this can be seen in Fig. 9.12, where the magnitude crosses 1 three times. A decision was made to define PM by the first crossing, because the PM at this crossing was the smallest of the three values and thus the most conservative assessment of stability. A Nyquist plot based on the data in Fig. 9.12 would show that the portion of the Nyquist curve closest to the -1 point was the critical indicator of stability, and therefore use of the crossover frequency yielding the minimum value of PM was the logical choice. At best, a designer needs to be judicious when applying the margin definitions described in Fig. 6.33. In fact, the actual stability margin of a system can only be rigorously assessed by examining the Nyquist plot to determine its closest approach to the -1 point.

Vector margin

To aid in this analysis, O. J. M. Smith (1958) introduced the **vector margin** which he defined to be the distance to the -1 point from the closest approach of the Nyquist plot.[8] Figure 6.38 illustrates the idea graphically. Because the

---

[8]   This value is closely related to the use of the sensitivity function for design and the concept of stability robustness, to be discussed in optional Section 6.9.

Figure 6.38
Definition of the vector
margin on the Nyquist plot

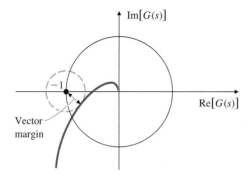

vector margin is a single margin parameter, it removes all the ambiguities in assessing stability that come with using GM and PM in combination. In the past it has not been used extensively due to difficulties in computing it. However, with the widespread availability of computer aids, the idea of using the vector margin to describe the degree of stability is much more feasible.

Conditionally stable systems

There are certain practical examples where an increase in the gain can make the system stable. As we saw in Chapter 5, these systems are called **conditionally stable**. A representative root-locus plot for such systems is shown in Fig. 6.39. For a point on the root locus such as A, an increase in the gain would make the system stable by bringing the unstable roots into the LHP. For point B, either a gain increase or decrease could make the system become unstable. Therefore, several gain margins exist that correspond to either gain reduction or gain increase and the definition of the GM in Fig. 6.33 is not valid.

Figure 6.39
Root locus for a
conditionally stable system

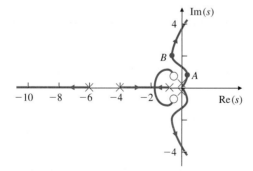

EXAMPLE 6.11

Stability Properties for a Conditionally Stable System

Determine the stability properties as a function of the gain K for the system with the open-loop transfer function

$$KG(s) = \frac{K(s+10)^2}{s^3}.$$

Figure 6.40
System in which increasing gain leads from instability to stability: (a) root locus; (b) Nyquist plot

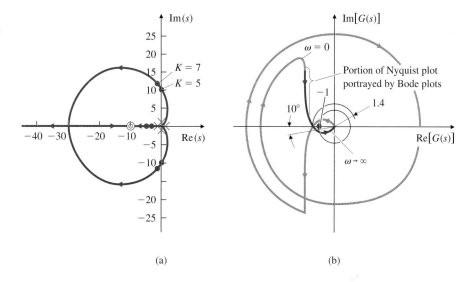

(a)   (b)

Solution.   This is a system for which increasing gain causes a transition from instability to stability. The root locus in Fig. 6.40(a) shows that the system is unstable for $K < 5$ and stable for $K > 5$. The Nyquist plot in Fig. 6.40(b) was drawn for the stable value $K = 7$. Determination of the margins according to Fig. 6.33 yields PM = $+10°$ (stable) and GM = 0.7 (unstable). According to the rules for stability discussed earlier, these two margins yield conflicting signals on the system's stability.

   We resolve the conflict by counting the Nyquist encirclements in Fig. 6.40(b). There is one clockwise encirclement and one counterclockwise encirclement of the -1 point. Hence there are no net encirclements, which confirms that the system is stable for $K = 7$. For systems like this it is best to resort to the root locus and/or Nyquist plot (rather than the Bode plot) to determine stability.

EXAMPLE 6.12   *Nyquist Plot for a System with Multiple Crossover Frequencies*

Draw the Nyquist plot for the system

$$G(s) = \frac{85(s + 1)(s^2 + 2s + 43.25)}{s^2(s^2 + 2s + 82)(s^2 + 2s + 101)}$$

$$= \frac{85(s + 1)(s + 1 \pm 6.5j)}{s^2(s + 1 \pm 9j)(s + 1 \pm 10j)},$$

and determine the stability margins.

Solution.   The Nyquist plot (Fig. 6.41) shows there are three crossover frequencies ($\omega = 0.7, 8.5,$ and 9.8 rad/sec) with three corresponding PM values of $37°$, $80°$, and $40°$, respectively. However, the key indicator of stability is the proximity of the Nyquist plot as it approaches the -1 point while crossing the real axis. In this case only the GM indicates the poor stability margins of this system. The Bode plot for this system (Fig. 6.42)

Figure 6.41
Nyquist plot of the complex system in Example 6.12

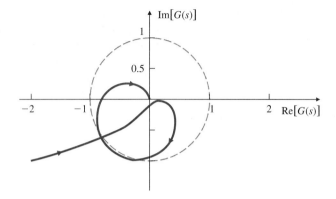

shows the same three crossings of magnitude $= 1$ at 0.7, 8.5, and 9.8 rad/sec. The GM value of 1.26 from the Bode plot corresponding to $\omega = 10.4$ rad/sec qualitatively agrees with the GM from the Nyquist plot and would be the most useful and unambiguous margin for this example.

Figure 6.42
Bode plot of the system in Example 6.12

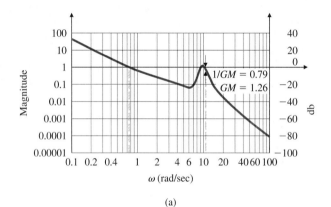

(a)

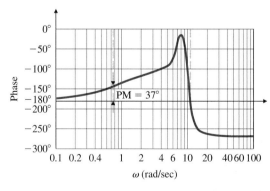

(b)

In summary, many systems behave roughly like Example 6.9 and, for them, the GM and PM are well-defined and useful. There are also frequent instances of more complicated systems with multiple magnitude 1 crossovers or unstable open-loop systems where the stability criteria defined by Fig. 6.33 are ambiguous or incorrect; therefore, we need to verify the GM and PM as previously defined, and/or modify them by reverting back to the Nyquist stability criterion.

6.5   Bode's Gain-Phase Relationship

One of Bode's important contributions is the following theorem:

> For any stable minimum-phase system (that is, one with no RHP zeros *or poles*), the phase of $G(j\omega)$ is uniquely related to the magnitude of $G(j\omega)$.

Bode's gain-phase relationship

When the slope of $|G(j\omega)|$ versus ω on a log-log scale persists at a constant value for approximately a decade of frequency, the relationship is particularly simple:

$$\angle G(j\omega) \cong n \times 90°, \tag{6.33}$$

where n is the slope of $|G(j\omega)|$ in units of decade of amplitude per decade of frequency. For example, in considering the magnitude curve alone in Fig. 6.43,

Figure 6.43

An approximate gain-phase relationship demonstration

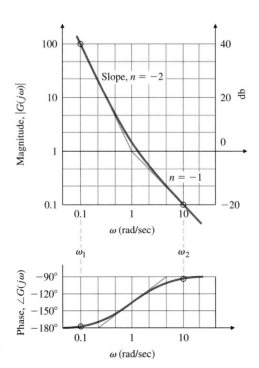

we see that Eq. (6.33) can be applied to the two frequencies $\omega_1 = 0.1$ (where $n = -2$) and $\omega_2 = 10$ (where $n = -1$) which are a decade removed from the change in slope, to yield the approximate values of phase, $-180°$ and $-90°$. The exact phase curve shown in the figure verifies that indeed the approximation is quite good. It also shows that the approximation will degrade if the evaluation is performed at frequencies closer to the change in slope.

An exact statement of the Bode gain-phase theorem is

$$\angle G(j\omega_o) = \frac{1}{\pi} \int_{-\infty}^{+\infty} \left(\frac{dM}{du}\right) W(u)\, du \qquad \text{(in radians)}, \qquad (6.34)$$

where

$$M = \log \text{ magnitude} = \ln |G(j\omega)|,$$

$$u = \text{normalized frequency} = \ln(\omega/\omega_o),$$

$$dM/du \cong \text{slope } n, \text{ as defined in Eq. (6.33)},$$

$$W(u) = \text{weighting function} = \ln(\coth|u|/2).$$

Figure 6.44 is a plot of the weighting function $W(u)$ and shows how the phase is most dependent on the slope at ω_o; it is also dependent, though to a lesser degree, on slopes at neighboring frequencies. The figure also suggests that the weighting could be approximated by an impulse function centered at ω_o. We may approximate the weighting function as

$$W(u) \cong \frac{\pi^2}{2}\delta(u),$$

which is precisely the approximation made to arrive at Eq. (6.33) using the "sifting" property of the impulse function (and conversion from radians to degrees).

Figure 6.44
Weighting function in Bode's gain-phase theorem

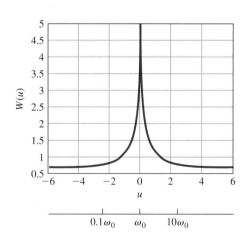

In practice, Eq. (6.34) is never used, but Eq. (6.33) *is* used as a guide to infer stability from $|G(\omega)|$ alone. When $|KG(j\omega)| = 1$,

$$\angle G(j\omega) \cong -90° \quad \text{if} \quad n = -1,$$
$$\angle G(j\omega) \cong -180° \quad \text{if} \quad n = -2.$$

For stability we want $\angle G(j\omega) > -180°$ for PM > 0. Therefore, we adjust the $|KG(j\omega)|$ curve so that it has a slope of -1 at the "crossover" frequency, ω_c, (that is, where $|KG(j\omega)| = 1$). If the slope is -1 for a decade above and below the crossover frequency, then PM $\cong 90°$; however, to ensure a reasonable PM, it is usually only necessary to insist that a -1 slope (-20 db per decade) persist for a decade in frequency that is centered at the crossover frequency. We therefore see that there is a very simple design criterion:

> Adjust the slope of the magnitude curve $|KG(j\omega)|$ so that it crosses over magnitude 1 with a slope of -1 for a decade around ω_c.

This criterion will usually be sufficient to provide an acceptable PM and hence provide adequate system damping. To achieve the desired speed of response, the system gain is adjusted so that the crossover point is at a frequency that will yield the desired bandwidth or speed of response as determined by Eq. (3.48). Recall that the natural frequency, ω_n, bandwidth, and crossover frequency are all approximately equal, as will be discussed further in Section 6.6.

Crossover frequency (margin note)

EXAMPLE 6.13   *Use of Simple Design Criterion for Spacecraft Attitude Control*

For the spacecraft-attitude-control problem defined in Fig. 6.45, find a suitable expression for $KD(s)$ that will provide good damping and a bandwidth of approximately 0.2 rad/sec.

Solution.   The magnitude of the frequency response of the spacecraft (Fig. 6.46) clearly requires some reshaping because it has a slope of -2 (or -40 db per decade) everywhere. The simplest compensation to do the job consists of using proportional and derivative terms (a PD compensator), which produces the relation

$$KD(s) = K(T_D s + 1). \tag{6.35}$$

We will adjust the gain K to produce the desired bandwidth, and we will adjust break point $\omega_1 = 1/T_D$ to provide the -1 slope at the crossover frequency. The actual design process to achieve the desired specifications is now very simple: We pick a value of K to provide a crossover at 0.2 rad/sec and choose a value of ω_1 that is about 4 times lower than the crossover frequency so the slope will be -1 in the vicinity of the crossover. Figure 6.47 shows the steps we take to arrive at the final compensation:

Figure 6.45
Spacecraft-attitude control system

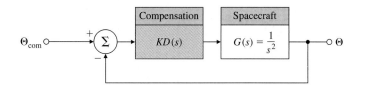

Figure 6.46
Magnitude of the
spacecraft's frequency
response

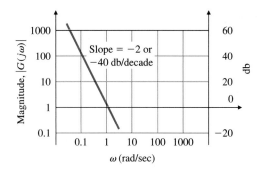

Figure 6.47
Compensated open-loop
transfer function

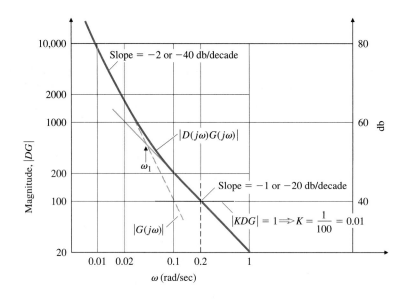

Figure 6.48
Closed-loop frequency
response

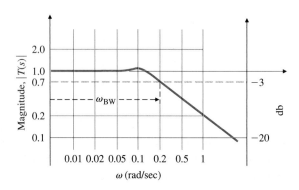

Figure 6.49

Step response for PD compensation

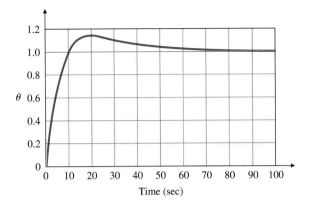

STEP 1. Plot $|G(j\omega)|$.

STEP 2. Modify the plot to include $|D(j\omega)|$, with $\omega_1 = 0.05$ rad/sec ($T_D = 20$), so that the slope will be $\cong -1$ at $\omega = 0.2$ rad/sec.

STEP 3. Determine that $|DG| = 100$ where the $|DG|$ curve crosses the line $\omega = 0.2$ rad/sec which is where we want magnitude 1 crossover to be.

STEP 4. In order for crossover to be at $\omega = 0.2$ rad/sec, compute

$$K = \frac{1}{[|DG|]_{\omega=0.2}} = \frac{1}{100} = 0.01.$$

Therefore,

$$KD(s) = 0.01(20s + 1)$$

will meet the specifications, thus completing the design.

If we were to draw the phase curve of KDG above, we would find that PM = 75°, which is certainly quite adequate. A plot of the closed-loop frequency-response magnitude (Fig. 6.48) shows that, indeed, the crossover frequency and the bandwidth are almost identical in this case. The step response of the closed-loop system is shown in Fig. 6.49, and its 14% overshoot confirms the adequate damping.

6.6   Closed-Loop Frequency Response

The closed-loop bandwidth was defined in Section 6.1 and in Fig. 6.5. Figure 6.3 showed that the natural frequency is always within a factor of two of the bandwidth for a second-order system. In Example 6.13, we designed the compensation so that the crossover frequency was at the desired bandwidth and verified by computation that the bandwidth was identical to the crossover frequency. Generally, the match between the crossover frequency and the bandwidth is not as good as in Example 6.13. We can help establish a more exact correspondence by making a few observations. Consider a system in which $|KG(j\omega)|$ shows the typical behavior

$$|KG(j\omega)| \gg 1 \qquad \text{for} \quad \omega \ll \omega_c,$$

$$|KG(j\omega)| \ll 1 \qquad \text{for} \quad \omega \gg \omega_c,$$

where ω_c is the crossover frequency. The closed-loop frequency-response magnitude is approximated by

$$|T(j\omega)| = \left| \frac{KG(j\omega)}{1 + KG(j\omega)} \right| \cong \begin{cases} 1, & \omega \ll \omega_c, \\ |KG|, & \omega \gg \omega_c \end{cases} \tag{6.36}$$

In the vicinity of crossover where $|KG(j\omega)| = 1$, $|T(j\omega)|$ depends heavily on the PM. A PM of 90° means that $\angle G(j\omega_c) = -90°$, and therefore $|T(j\omega_c)| = 0.707$. On the other hand, PM = 45° yields $|T(j\omega_c)| = 1.31$.

The approximations in Eq. (6.36) were used to generate the curves of $|T(j\omega)|$ in Fig. 6.50. It shows that the bandwidth for smaller values of PM is typically somewhat greater than ω_c, though usually it is less than $2\omega_c$; thus

$$\omega_c \leq \omega_{BW} \leq 2\omega_c.$$

Another specification related to the closed-loop frequency response is the resonant-peak magnitude M_r, defined in Fig. 6.5. Figures 6.3 and 6.37 show that, for linear systems, M_r is generally related to the damping of the system. In practice, M_r is rarely used; most designers prefer to use the PM to specify the damping of a system because the imperfections that make systems nonlinear or cause delays usually erode the phase more significantly than the magnitude.

Figure 6.50
Closed-loop bandwidth with respect to PM

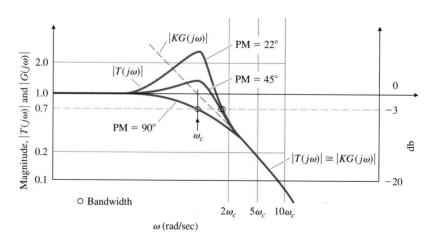

6.7 Compensation

As we discussed in Chapters 4 and 5, dynamic elements (or compensation) are typically added to feedback controllers to improve the system's stability and error characteristics because the process itself cannot be made to have acceptable characteristics with proportional feedback alone.

Section 4.2 discussed the basic types of feedback: proportional, derivative, and integral. Section 5.5 discussed three kinds of dynamic compensation: lead compensation, which approximates proportional-derivative (PD) feedback, lag compensation, which approximates proportional-integral (PI) control, and notch compensation, which has special characteristics for dealing with resonances. In this section we discuss these and other kinds of compensation in terms of their frequency-response characteristics. In most cases the compensation will be implemented in a microprocessor. Techniques for converting the continuous compensation $D(s)$ into a form that can be coded in the computer was briefly discussed in Section 4.4. It will be illustrated further in this section and will be discussed in more detail in Chapter 8.

The frequency response stability analysis to this point has considered the closed-loop system to have the characteristic equation $1 + KG(s) = 0$. With the introduction of compensation, the closed-loop characteristic equation becomes $1 + KD(s)G(s) = 0$ and all the previous discussion in this chapter pertaining to the frequency response of $KG(s)$ applies directly to the compensated case if we apply it to the frequency response of $KD(s)G(s)$. We call this quantity $L(s)$ the "loop gain," or open-loop transfer function of the system, where $L(s) = KD(s)G(s)$.

6.7.1 PD Compensation

PD compensation

We will start the discussion of compensation design by using the frequency response with PD control. The compensator transfer function, given by

$$D(s) = (T_D s + 1), \tag{6.37}$$

was shown in Fig. 5.23 to have a stabilizing effect on the root locus of a second-order system. The frequency-response characteristics of Eq. (6.37) are shown in Fig. 6.51. A stabilizing influence is apparent by the increase in phase and the corresponding +1 slope at frequencies above the break point $1/T_D$. We use this compensation by locating $1/T_D$ so that the increased phase occurs in the vicinity of crossover (that is, where $|KD(s)G(s)| = 1$), thus increasing the phase margin.

Note that the magnitude of the compensation continues to grow with increasing frequency. This feature is undesirable because it amplifies the high-frequency noise that is typically present in any real system and, as a continuous transfer function, cannot be realized with physical elements. It is also the reason we stated in Section 5.5 that pure derivative compensation gives trouble.

Figure 6.51
Frequency response of PD
control

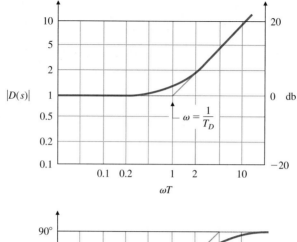

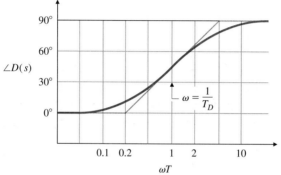

6.7.2   Lead Compensation

In order to alleviate the high-frequency amplification of the PD compensation,
a first-order pole is added in the denominator at frequencies well higher than
the breakpoint of the PD compensator. Thus the phase increase (or lead) still
occurs, but the amplification at high frequencies is limited. The resulting **lead
compensation** has a transfer function of

Lead compensation

$$D(s) = \frac{Ts + 1}{\alpha Ts + 1}, \qquad \text{where} \quad \alpha < 1 \qquad (6.38)$$

and $1/\alpha$ is the ratio between the pole-zero breakpoint frequencies. Figure 6.52
shows the frequency response of this lead compensation. Note that a significant
amount of phase lead is still provided, but with much less amplification at high
frequencies. A lead compensator is generally used whenever a substantial
improvement in damping of the system is required.

The phase contributed by the lead compensation in Eq. (6.38) is given by

$$\phi = \tan^{-1}(T\omega) - \tan^{-1}(\alpha T\omega).$$

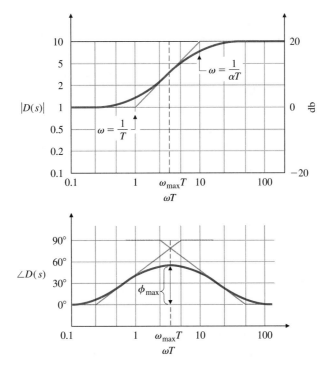

Figure 6.52
Lead-compensation
frequency response with
$1/\alpha = 10$

It can be shown (see Problem 6.43) that the frequency where the phase is maximum is given by

$$\omega_{\max} = \frac{1}{T\sqrt{\alpha}}. \tag{6.39}$$

The maximum phase contribution—that is, the peak of the $\angle D(s)$ curve in Fig. 6.52—corresponds to

$$\sin \phi_{\max} = \frac{1-\alpha}{1+\alpha} \tag{6.40}$$

or

$$\alpha = \frac{1 - \sin \phi_{\max}}{1 + \sin \phi_{\max}}.$$

Another way to look at this is the following: The maximum frequency occurs midway between the two break-point frequencies (sometimes called corner frequencies) on a logarithmic scale,

$$\log \omega_{\max} = \log \frac{1/\sqrt{T}}{\sqrt{\alpha T}} = \log \frac{1}{\sqrt{T}} + \log \frac{1}{\sqrt{\alpha T}}$$

$$= \frac{1}{2}\left[\log\left(\frac{1}{T}\right) + \log\left(\frac{1}{\alpha T}\right)\right], \tag{6.41}$$

as shown in Fig. 6.52. Alternatively, we may state these results in terms of the pole-zero locations. Rewriting $D(s)$ in the form used for root-locus analysis, we obtain

$$D(s) = \frac{s+z}{s+p}.$$ (6.42)

Problem 6.43 shows that

$$\omega_{max} = \sqrt{|z|\,|p|}$$ (6.43)

and

$$\log \omega_{max} = \frac{1}{2}(\log|z| + \log|p|).$$ (6.44)

These results agree with the previous ones if we let $z = -1/T$ and $p = -1/\alpha T$ in Eqs. (6.39) and (6.41).

For example, a lead compensator with a zero at $s = -2$ ($T = 0.5$) and a pole at $s = -10$ ($\alpha T = 0.1$ so that $\alpha = \frac{1}{5}$) would yield the maximum phase lead at

$$\omega_{max} = \sqrt{2 \cdot 10} = 4.47 \text{ rad/sec.}$$

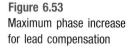
Lead ratio $= \frac{1}{\alpha}$

The amount of phase lead at the midpoint depends only on α in Eq. (6.40) and is plotted in Fig. 6.53. For $\alpha = \frac{1}{5}$, Fig. 6.53 shows that $\phi_{max} = 40°$. Note from the figure that we could increase the phase lead up to 90° using higher values of the **lead ratio**, $1/\alpha$; however, Fig. 6.52 shows that increasing values of $1/\alpha$ also produces higher amplifications at higher frequencies. Thus our task is to select a value of $1/\alpha$ that is a good compromise between an acceptable phase margin and an acceptable noise sensitivity at high frequencies. Usually the compromise suggests a lead compensation should contribute a maximum of 60° to the phase. If a greater phase lead is needed, then a double lead compensation would be suggested, where

$$D(s) = \left(\frac{Ts+1}{\alpha Ts+1}\right)^2.$$

Figure 6.53
Maximum phase increase
for lead compensation

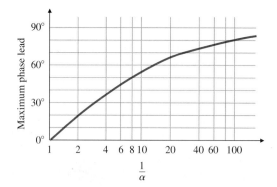

Even if a system had negligible amounts of noise present and the pure derivative compensation of Eq. (6.37) were acceptable, a continuous compensation would look more like Eq. (6.38) than Eq. (6.37) because of the impossibility of building a pure differentiator. No physical system—mechanical or electrical—responds with infinite amplitude at infinite frequencies, so there will be a limit in the frequency range (or bandwidth) for which derivative information (or phase lead) can be provided. This is also true with a digital implementation. Here, the sample rate limits the high-frequency amplification and essentially places a pole in the compensation transfer function.

EXAMPLE 6.14   *Lead Compensation for a DC Motor*

As an example of designing a lead compensator, let us repeat the design of compensation for the DC motor with the transfer function

$$G(s) = \frac{1}{s(s+1)}$$

that was carried out in Section 5.5.1. This also represents the model of a satellite tracking antenna (see Fig. 3.63). This time we wish to obtain a steady-state error of less than 0.1 for a unit ramp input. Furthermore, we desire an overshoot $M_p < 25\%$.

(a) Determine the lead compensation satisfying the specifications,

(b) determine the digital version of the compensation with $T_s = 0.05$ sec, and

(c) compare the step and ramp responses of both implementations.

Solution.

(a) The steady-state error is given by

$$e_{ss} = \lim_{s \to 0} s \left[\frac{1}{1 + KD(s)G(s)} \right] R(s), \tag{6.45}$$

where $R(s) = 1/s^2$ for a unit ramp, so Eq. (6.45) reduces to

$$e_{ss} = \lim_{s \to 0} \left\{ \frac{1}{s + KD(s)[1/(s+1)]} \right\} = \frac{1}{KD(0)}.$$

Therefore, we find that $KD(0)$, the steady-state gain of the compensation, cannot be less than 10 ($K_v \geq 10$) if it is to meet the error criterion, so we pick $K = 10$. To relate the overshoot requirement to phase margin, Fig. 6.37 shows that a PM of $45°$ should suffice. The frequency response of $KG(s)$ in Fig. 6.54 shows that the PM = $20°$ if no phase lead is added by compensation. If it were possible to simply add phase without affecting the magnitude, we would need an additional phase of only $25°$ at the $KG(s)$ crossover frequency of $\omega = 3$ rad/sec. However, maintaining the same low-frequency gain and adding a compensator zero would increase the crossover frequency; hence more than a $25°$ phase contribution will be required from the lead compensation. To be safe we will design the lead compensator so that it supplies a maximum phase lead of $40°$. Fig. 6.53 shows that $1/\alpha = 5$ will accomplish that goal. We will derive the greatest benefit from the compensation if

Figure 6.54
Frequency response for
lead-compensation design

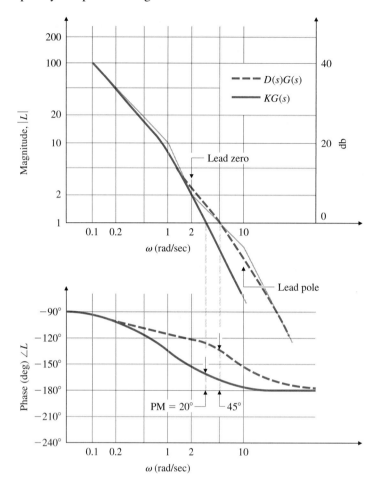

the maximum phase lead from the compensator occurs at the crossover frequency. With some trial and error, we determine that placing the zero at $\omega = 2$ rad/sec and the pole at $\omega = 10$ rad/sec, causes the maximum phase lead to be at the crossover frequency. The compensation, therefore, is

$$K D(s) = 10\frac{s/2 + 1}{s/10 + 1}.$$

The frequency-response characteristics of $L(s) = K D(s)G(s)$ in Fig. 6.54 can be seen to yield a phase margin of 45°, which satisfies the design goals.

The root locus for this design, originally given as Fig. 5.24, is repeated here as Fig. 6.55 with the root locations marked for $K = 10$. The locus is not needed for the frequency response design procedure; it is presented here only for comparison with the root locus design method presented in Chapter 5.

Figure 6.55
Root locus for lead
compensation design

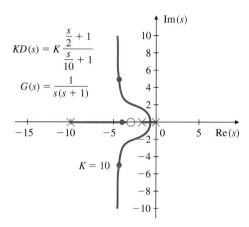

$$KD(s) = K \frac{\frac{s}{2} + 1}{\frac{s}{10} + 1}$$

$$G(s) = \frac{1}{s(s+1)}$$

(b) To find the discrete equivalent of $D(s)$, we use the trapezoidal rule given by Eq. (4.114). That is,

$$D_d(z) = \frac{\frac{2}{T_s} \frac{z-1}{z+1}/2 + 1}{\frac{2}{T_s} \frac{z-1}{z+1}/10 + 1},$$

(6.46)

which, with $T_s = 0.05$ sec, reduces to

$$D_d(z) = \frac{4.2z - 3.8}{z - 0.6}.$$

(6.47)

This same result can be obtained by the MATLAB statement,

```
sysD = tf([0.5  1],[0.1  1]);
sysDd = c2d(sysD, 0.05, 'tustin').
```

Because

$$\frac{U(z)}{E(z)} = K D_d(z),$$

(6.48)

the discrete control equations that result are

$$u(k+1) = 0.6u(k) + 10(4.2e(k+1) - 3.8e(k)).$$

(6.49)

(c) The Simulink block diagram of the continuous and discrete versions of $D(s)$ controlling the DC motor are is shown in Fig. 6.56. The plot of the step response of the two controllers are plotted together in Fig. 6.57(a) and are reasonably close to one another, however the discrete controller does exhibit slightly increased overshoot as is often the case. Both overshoots are less than 25% and thus meet the specifications. The ramp responses of the two controllers shown in Fig. 6.57(b) are essentially identical and both meet the 0.1 specified error.

Figure 6.56
Simulink block diagram
for transient response of
lead-compensation design

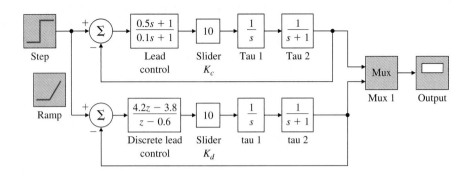

Figure 6.57
Lead-compensation design:
(a) step response; (b) ramp
response

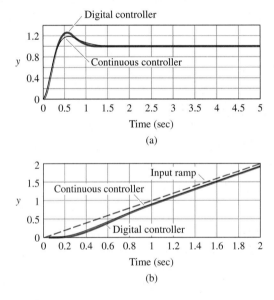

The design procedure used in Example 6.14 can be summarized as follows:

1. Determine the low-frequency gain so that the steady-state errors are within specification.

2. Select the combination of lead ratio $1/\alpha$ and zero values $(1/T)$ that achieves an acceptable phase margin at crossover.

3. The pole location is then at $(1/\alpha T)$.

This design procedure will apply to many cases; however, keep in mind that the specific procedure followed in any particular design may need to be tailored to its particular set of specifications.

In Example 6.14 there were two specifications: peak overshoot and steady-state error. We transformed the overshoot specification into a PM, but the steady-state error specification we used directly. No speed-of-response type of specification was given; however, it would have impacted the design in the same way that the steady-state error specification did. The speed of response or bandwidth of a system is directly related to the crossover frequency, as we pointed out earlier in Section 6.6. Figure 6.54 shows that the crossover frequency was ∼ 5 rad/sec. We could have increased it by raising the gain K and increasing the frequency of the lead compensator pole and zero in order to keep the slope of −1 at the crossover frequency. Raising the gain would also have decreased the steady-state error to be better than the specified limit. The gain margin was never introduced into the problem because the stability was adequately specified by the phase margin alone. Furthermore, the gain margin would not have been useful for this system because the phase never crossed the 180° line and the GM was always infinite.

In lead-compensation designs there are three primary design parameters:

Designs parameters for lead-networks

1. The crossover frequency ω_c, which determines bandwidth ω_{BW}, rise time t_r, and settling time t_s;

2. The phase margin (PM), which determines the damping coefficient ζ and the overshoot M_p;

3. The low-frequency gain, which determines the steady-state error characteristics.

The design problem is to find the best values for the parameters given the requirements. In essence, lead compensation increases the value of $\frac{\omega_c}{L(0)}$ (= $\frac{\omega_c}{K_v}$ for a type 1 system). That means that, if the low-frequency gain is kept the same, the crossover frequency will increase. Or if the crossover frequency is kept the same, the low-frequency gain will decrease. Keeping this interaction in mind, the designer can assume a fixed value of one of these three design parameters and then adjust the other two iteratively until the specifications are met. One approach is to set the low-frequency gain to meet the error specifications and add a lead compensator to increase the PM at the crossover frequency. An alternative is to pick the crossover frequency to meet a time response specification, then adjust the gain and lead characteristics so that the PM specification is met. A step-by-step procedure is outlined below for these two cases. They apply to a sizable class of problems for which a single lead is sufficient. As with all such design procedures, it provides only a starting point; the designer will typically find it necessary to go through several design iterations in order to meet all the specifications.

We now summarize a step-by-step procedure for lead-compensator design.

1. Determine open-loop gain K to satisfy error or bandwidth requirements:
   (a) To meet error requirement, pick K to satisfy error constants (K_p, K_v, or K_a) so that e_{ss} error specification is met.
   (b) Alternatively, to meet bandwidth requirement, pick K so that the open-loop crossover frequency is a factor of two below the desired closed-loop bandwidth.

2. Evaluate the phase margin (PM) of the uncompensated system using the value of K obtained from Step 1.

3. Allow for extra margin (about $10°$), and determine the needed phase lead ϕ_{max}.

4. Determine α from Eq. (6.40) or Fig. 6.53.

5. Pick ω_{max} to be at the crossover frequency; thus the zero is at $1/T = \omega_{max}\sqrt{\alpha}$ and the pole is at $1/\alpha T = \omega_{max}/\sqrt{\alpha}$.

6. Draw the compensated frequency response and check the PM.

7. Iterate on the design. Adjust compensator parameters (poles, zeros, and gain) until all specifications are met. Add an additional lead compensator (that is, a double lead compensation) if necessary.

While these guidelines will not apply to all the systems you will encounter in practice, they do suggest a systematic trial-and-error process to search for a satisfactory compensator that will often be successful.

EXAMPLE 6.15   *Lead Compensator for a Temperature Control System*

The third-order system

$$KG(s) = \frac{K}{(s/0.5 + 1)(s + 1)(s/2 + 1)}$$

is representative of a typical temperature control system. Design a lead compensator such that $K_p = 9$ and the phase margin is at least $25°$.

Solution.   Let us follow the design procedure:

STEP 1. Given the specification for K_p, we solve for K:

$$K_p = \lim_{s \to 0} KG(s) = K = 9.$$

STEP 2. The Bode plot of the uncompensated system with $K = 9$ can be created by the MATLAB statements below and is shown in Fig. 6.58.

```
numG = 9;
den2 = conv([1 0.5],[1 1]);
denG = conv(den2,[1 2]);
sysG = tf(numG,denG);
[mag,phas,w] = bode(sysG);
```

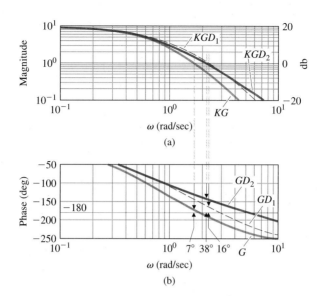

Figure 6.58
Bode plot for the
lead-compensation design
in Example 6.15

It is difficult to read the PM and crossover frequencies accurately from the Bode plots; therefore, the MATLAB command

[GM,PM,Wcg,Wcp] = margin(mag,phas,w)

can be invoked. The quantity PM is the phase margin and Wcp is the frequency at which the gain crosses magnitude 1. (GM and Wcg are the open loop gain margin and the frequency at which the phase crosses 180°.) For this example, the output is

GM =1.2500, PM = 7.1249, Wcg = 1.8708, Wcp = 1.6844,

which says that the PM of the uncompensated system is 7° and that this occurs at a crossover frequency of 1.7 rad/sec.

STEP 3. Allowing for 10° of extra margin, we want the lead compensator to contribute $25° + 10° - 7° = 28°$ at the crossover frequency. The extra margin is typically required because the lead will increase the crossover frequency from the open-loop case, at which point more phase increase will be required.

STEP 4. From Fig. 6.53 we see that $\alpha = 1/3$ will produce approximately 30° phase increase midway between the zero and pole.

STEP 5. As a first cut, let's place the zero at 1 rad/sec ($T = 1$) and the pole at 3 rad/sec ($\alpha T = 1/3$), thus bracketing the open-loop crossover frequency and preserving the factor of 3 between pole and zero as indicated by $\alpha = 1/3$. The lead compensator is

$$D_1(s) = \frac{s+1}{s/3+1} = \frac{1}{0.333}\left(\frac{s+1}{s+3}\right).$$

STEP 6. The Bode plot of the system with $D_1(s)$ (Fig. 6.58, middle curve) has a PM of 16°. We did not achieve the desired PM of 25° because the lead shifted the crossover frequency from 1.7 rad/sec to 2.3 rad/sec, thus increasing the required phase increase from the lead. The step response of the system with $D_1(s)$ (Fig. 6.59) shows a very oscillatory response, as we might expect from the low PM of 16°.

Figure 6.59
Step response for
lead-compensation design

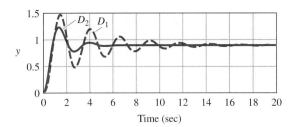

STEP 7. We repeat the design with extra phase increase and move the zero location slightly to the right so that the crossover frequency won't be shifted so much. We choose $\alpha = \frac{1}{10}$ with the zero at $s = -1.5$, so

$$D_2(s) = \frac{s/1.5 + 1}{s/15 + 1} = \frac{1}{0.1}\left(\frac{s + 1.5}{s + 15}\right).$$

This compensation produces a PM $= 38°$, and the crossover frequency lowered slightly to 2.2 rad/sec. Figure 6.58 (upper curve) shows the frequency response of the revised design. Figure 6.59 shows a substantial reduction in the oscillations, which you should expect from the higher PM value.

EXAMPLE 6.16

Lead-Compensator Design for a Type 1 Servomechanism System

Consider the third-order system

$$KG(s) = K\frac{10}{s(s/2.5 + 1)(s/6 + 1)}.$$

This type of system would result for a DC motor with a lag in the shaft position sensor. Design a lead compensator so that the PM $= 45°$ and $K_v = 10$.

Solution.   Again we follow the design procedure given earlier:

STEP 1. $KG(s)$ as given will yield $K_v = 10$ if $K = 1$. Therefore, the K_v requirement is met by $K = 1$ and the low-frequency gain of the compensation should be 1.

STEP 2. The Bode plot of the system is shown in Fig. 6.60. The phase margin of the uncompensated system (lower curve) is approximately $-4°$ and the crossover frequency is at $\omega_c \cong 4$ rad/sec.

STEP 3. Allowing for 5° of extra phase margin, we need PM $= 45° + 5° - (-4°) = 54°$ to be contributed by the lead compensator.

Figure 6.60
Bode plot for the
lead-compensation design
in Example 6.16

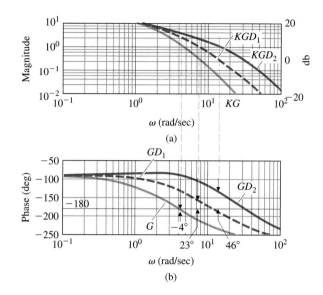

STEP 4. From Fig. 6.53 we find that α must be 0.1 to achieve a maximum phase lead of 54°.

STEP 5. The new gain crossover frequency will be higher than the open-loop value of $\omega_c = 4$ rad/sec, so let's select the pole and zero of the lead compensation to be at 20 and 2 rad/sec, respectively. So the candidate compensator is

$$D_1(s) = \frac{s/2+1}{s/20+1} = \frac{1}{0.1}\frac{s+2}{s+20}.$$

STEP 6. The Bode plot of the compensated system (Fig. 6.60, middle curve) shows a PM of 23°. Further iteration will show that a single lead compensator cannot meet the specification because of the high-frequency slope of -3.

STEP 7. We need a double lead compensator in this system. If we try a compensator of the form

$$D_2(s) = \frac{1}{(0.1)^2}\frac{(s+2)(s+4)}{(s+20)(s+40)} = \frac{(s/2+1)(s/4+1)}{(s/20+1)(s/40+1)},$$

we obtain PM $= 46°$. The Bode plot for this case is shown as the upper curve in Fig. 6.60.

Both Examples 6.15 and 6.16 are third order. Example 6.16 was more difficult to design compensation for because the error requirement, K_v, forced the crossover frequency, ω_c, to be so high that a single lead could not provide enough PM.

6.7.3   PI Compensation

PI compensation

In many problems it is important to keep the bandwidth low and also to reduce the steady-state error. For this purpose a proportional-integral (PI) or lag compensator is useful. By letting $k_D = 0$ in Eq. (4.43), we see that PI control has the transfer function

$$D(s) = \frac{K}{s}\left(s + \frac{1}{T_I}\right),\tag{6.50}$$

which results in the frequency-response characteristics shown in Fig. 6.61. The desirable aspect of this compensation is the infinite gain at zero frequency, which reduces the steady-state errors. This is accomplished, however, at the cost of a phase decrease at frequencies lower than the break point at $\omega = 1/T_I$. Therefore, $1/T_I$ is usually located at a frequency substantially less than the crossover frequency so that the system's phase margin is not affected significantly.

Figure 6.61
Frequency response of PI control

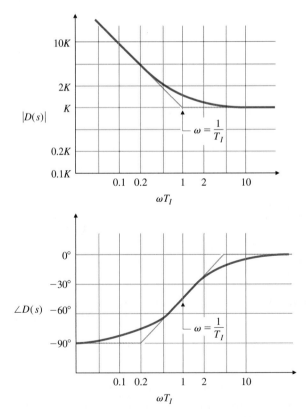

6.7.4   Lag Compensation

Lag compensation

As we discussed in Section 5.5, **lag compensation** approximates PI control. Its transfer function was given by Eq. (5.100) for root-locus design, but for frequency-response design, it is more convenient to write the transfer function of the lag compensation *alone* in the Bode form

$$D(s) = \alpha \frac{Ts + 1}{\alpha Ts + 1}, \qquad \text{where} \quad \alpha > 1; \qquad (6.51)$$

where α is the ratio between the zero-pole breakpoint frequencies. The complete controller will almost always include an overall gain K and perhaps other dynamics in addition to the lag compensation. Although Eq. (6.51) looks very similiar to the lead compensation in Eq. (6.38), the fact that $\alpha > 1$ causes the pole to have a lower break-point frequency than the zero. This relationship produces the low-frequency increase in amplitude and phase decrease (lag) apparent in the frequency-response plot in Fig. 6.62 and gives the compensation the essential feature of integral control: an increased low-frequency gain. The typical objective of lag-compensation design is to provide additional gain of α in the low-frequency range and to leave the system sufficient phase margin (PM). Of course, phase lag is not the useful effect, and the pole and zero

Figure 6.62

Frequency response of lag compensation with $\alpha = 10$

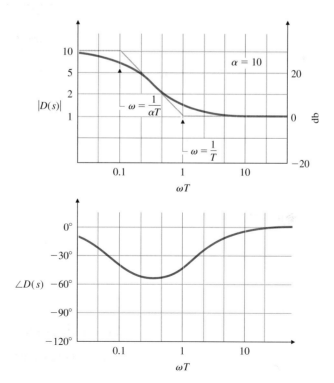

of the lag compensator are selected to be at much lower frequencies than the uncompensated system crossover frequency in order to keep the effect on the PM to a minimum. Thus, the lag compensator increases the open-loop DC gain thereby improving the steady-state response characteristics, without changing the transient response characteristics significantly. If the pole and zero are relatively close together and near the origin (that is, if the value of T is large), we can increase the low frequency gain (and thus K_p, K_v, or K_a) by a factor α without moving the closed-loop poles appreciably. Hence the transient response remains approximately the same while the steady-state response is improved.

We now summarize a step-by-step procedure for lag-compensator design.

Design Procedure for Lag Compensation

1. Determine the open-loop gain K that will meet the phase-margin requirement without compensation.

2. Draw the Bode plot of the uncompensated system with crossover frequency from step 1, and evaluate the low-frequency gain.

3. Determine α to meet the low-frequency gain error requirement.

4. Choose the corner frequency $\omega = 1/T$ (the zero of the lag compensator) to be one octave to one decade below the new crossover frequency ω_c.

5. The other corner frequency (the pole location of the lag compensator) is then $\omega = 1/\alpha T$.

6. Iterate on the design. Adjust compensator parameters (poles, zeros, and gain) to meet all the specifications.

EXAMPLE 6.17

Lag Compensator Design for Temperature Control System

Again consider the third-order system of Example 6.15:

$$KG(s) = \frac{K}{\left(\dfrac{1}{0.5}s + 1\right)(s + 1)\left(\dfrac{1}{2}s + 1\right)}.$$

Design a lag compensator so the phase margin is at least $40°$ and $K_p = 9$.

Solution.   We follow the design procedure enumerated above.

STEP 1. From the open-loop plot of $KG(s)$ shown for $K = 9$ in Fig. 6.58, it can be seen that a PM $> 40°$ will be achieved if the crossover frequency $\omega_c \lesssim 1$ rad/sec. This will be the case if $K = 3$. So we pick $K = 3$ in order to meet the PM specification.

STEP 2. The Bode plot of $KG(s)$ shown in Fig. 6.63 with $K = 3$ shows that the PM is $\approx 50°$ and the low-frequency gain is now 3. Exact calculation of the PM using MATLAB's margin shows that PM $= 53°$.

Figure 6.63

Frequency response of lag-compensation design in Example 6.17

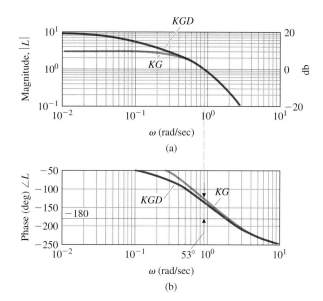

STEP 3. The low-frequency should be raised by a factor of 3, which means the lag compensation needs to have $\alpha = 3$.

STEP 4. We choose the corner frequency for the zero to be an octave slower than the crossover frequency—that is, at 0.2 rad/sec. So, $1/T = 0.2$, or $T = 5$.

STEP 5. We then have the value for the other corner frequency: $\omega = 1/\alpha T = \frac{1}{(3)(5)} = \frac{1}{15}$ rad/sec. The compensator is thus

$$D(s) = 3\frac{5s + 1}{15s + 1}.$$

The compensated frequency response is also shown in Fig. 6.63. The low-frequency gain of $L(0) = K D(0)G(0) = 3K = 9$, thus $K_p = 9$ and the PM lowers slightly to 44°, which satisfies the specifications. The step response of the system, shown in Fig. 6.64, illustrates the reasonable damping that we would expect from PM = 44°.

STEP 6. No iteration is required in this case.

Figure 6.64

Step response of lag-compensation design in Example 6.17

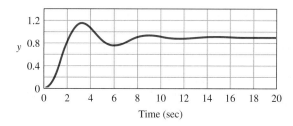

Note that Examples 6.15 and 6.17 are both for the same plant and both had the same steady-state error requirement. One was compensated with lead and one was compensated with lag. The result is that the bandwidth of the lead compensated design is approximately a factor of 3 higher than the lag compensated design. This result can be seen by comparing the crossover frequencies of the two designs.

A beneficial effect of lag compensation, an increase in the low-frequency gain for better error characteristics, was just demonstrated in Example 6.17. However, in essence, lag compensation reduces the value of $\frac{\omega_c}{L(0)}$ (= $\frac{\omega_c}{K_v}$ for a type 1 system). That means that, if the crossover frequency is kept the same, the low frequency gain will increase. Likewise, if the low frequency gain is kept the same, the crossover frequency will decrease. Therefore, lag compensation could also be interpreted to reduce the crossover frequency and thus obtain a better phase margin. The procedure for design in this case is partially modified. First, pick the low-frequency gain to meet error requirements, then locate the lag compensation pole and zero in order to provide a crossover frequency with adequate PM. The example below illustrates this design procedure. The end result of the design will be the same no matter what procedure is followed.

EXAMPLE 6.18   *Lag Compensation of the DC Motor*

Repeat the design of the DC motor control in Example 6.14, this time using lag compensation. Fix the low-frequency gain in order to meet the error requirement of $K_v = 10$; then use the lag compensation to meet the PM requirement of $45°$.

Solution.   The frequency response of the system $KG(s)$, with the required gain of $K = 10$, is shown in Fig. 6.65. The uncompensated system has a crossover frequency at approximately 3 rad/sec where the PM = $20°$. The designer's task is to select the lag compensation break points so that the crossover frequency is lowered and more favorable PM results. To prevent detrimental effects from the compensation phase lag, the pole and zero position values of the compensation need to be substantially lower than the new crossover frequency. One possible choice is shown in Fig. 6.65: The lag zero is at 0.1 rad/sec, and the lag pole is at 0.01 rad/sec. This selection of parameters produces a PM of $50°$, thus satisfying the specifications. Here the stabilization is achieved by keeping the crossover frequency to a region where $G(s)$ has favorable phase characteristics. The criterion for selecting the pole and zero locations $1/T$ is to make them low enough to minimize the effects of the phase lag from the compensation at the crossover frequency. Generally, however, the pole and zero are located no lower than necessary because the additional system root [compare with the root locus of a similar system design in Fig. 5.31(b)] introduced by the lag will be in the same frequency range as the compensation zero and will have some effect on the output response, especially the response to disturbance inputs.

The response of the system to a step reference input is shown in Fig. 6.66. It shows no steady-state error to a step input because this is a type 1 system. However, the introduction of the slow root from the lag compensation has caused the response to require about 25 sec to settle down to the zero steady-state value. The overshoot, M_p, is somewhat larger than you would expect from the guidelines based on a second-order system shown in Fig. 6.37 for a PM = $50°$; however, the performance is adequate.

Figure 6.65

Frequency response of
lag-compensation design in
Example 6.18

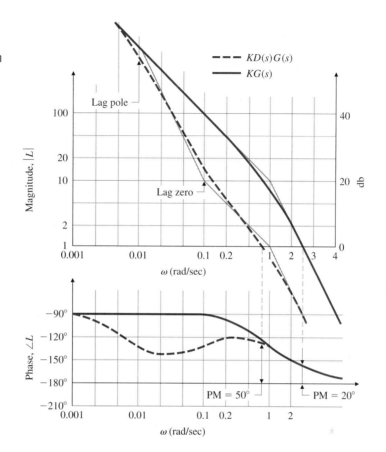

Figure 6.66

Step response of
lag-compensation design in
Example 6.18

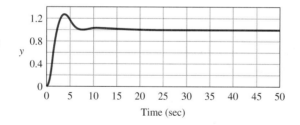

As we saw previously for a similar situation, Examples 6.14 and 6.18 meet
an identical set of specifications for the same plant in very different ways. In the
first case the specifications are met with a lead compensation, and a crossover
frequency $\omega_c = 5$ rad/sec ($\omega_{BW} \cong 6$ rad/sec) results. In the second case the
same specifications are met with a lag compensation, and $\omega_c \cong 0.8$ rad/sec

($\omega_{BW} \cong 1$ rad/sec) results. Clearly, had there been specifications for rise time or bandwidth, they would have influenced the choice of compensation (lead or lag). Likewise, if the slow settling to the steady-state value was a problem, it might have suggested the use of lead compensation instead of lag.

In more realistic systems, dynamic elements usually represent the actuator and sensor as well as the process itself, so it is typically impossible to raise the crossover frequency much beyond the value representing the speed of response of the components being used. Although linear analysis seems to suggest that almost any system can be compensated, in fact, if we attempt to drive a set of components much faster than their natural frequencies, the system will saturate, the linearity assumptions will be no longer valid, and the linear design will represent little more than wishful thinking. With this behavior in mind, we see that simply increasing the gain of a system and adding lead compensators to achieve an adequate PM may not always be possible. It may be preferable to satisfy error requirements by adding a lag network so that the closed-loop bandwidth is kept at a more reasonable frequency.

6.7.5   PID Compensation

PID compensation

For problems that need phase margin improvement at ω_c *and* low-frequency gain improvement, it is effective to use both derivative and integral control. By combining Eqs. (6.37) and (6.50), we obtain PID control. Its transfer function is

$$D(s) = \frac{K}{s}\left[(T_D s + 1)\left(s + \frac{1}{T_I} \right) \right], \tag{6.52}$$

and its frequency-response characteristics are shown in Fig. 6.67. This form is slightly different from that given by Eq. (4.44); however, the effect of the difference is inconsequential. This compensation is roughly equivalent to combining lead and lag compensators in the same design, and so it is sometimes referred to as a **lead-lag compensator**. Hence, it can provide simultaneous improvement in transient and steady-state responses.

EXAMPLE 6.19

PID Compensation Design for Spacecraft Attitude Control

A simplified design for spacecraft-attitude-control was presented in Section 6.5; however, here we have a more realistic situation that includes a sensor lag and a disturbing torque. Figure 6.68 defines the system. Design a PID controller to have zero steady-state error to a constant-disturbance torque, a phase margin of $65°$, and as high a bandwidth as is reasonably possible.

Solution.   First, let us take care of the steady-state error. For the spacecraft to be at a steady final value, the total input torque, $T_d + T_c$, must equal zero. Therefore, if $T_d \neq 0$, then $T_c = -T_d$. The only way this can be true with no error ($e = 0$) is for $D(s)$ to contain an integral term, hence including integral control in the compensation will meet the steady-state requirement. This could also be verified mathematically by use of the Final Value Theorem (see Problem 6.46).

Figure 6.67
Frequency response of
PID compensation with
$T_I/T_D = 20$

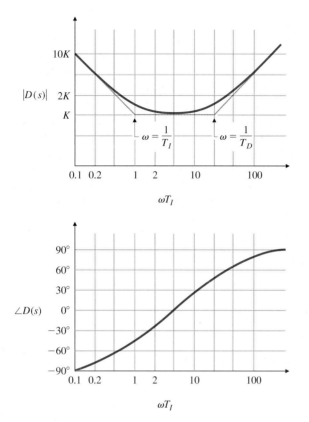

The frequency response of the spacecraft and sensor,

$$G(s) = \frac{0.9}{s^2}\left(\frac{2}{s+2}\right),\qquad(6.53)$$

is shown in Fig. 6.69. The slopes of -2 (that is, -40 db per decade) and -3 (-60 db per decade) show that the system would be unstable for any value of K if no derivative feedback were used. This is clear because of Bode's gain-phase relationship which shows

Figure 6.68
Block diagram of Spacecraft
Control using PID design,
Example 6.19

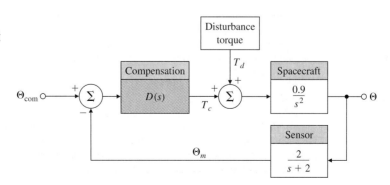

Figure 6.69
Compensation for PID
design in Example 6.19

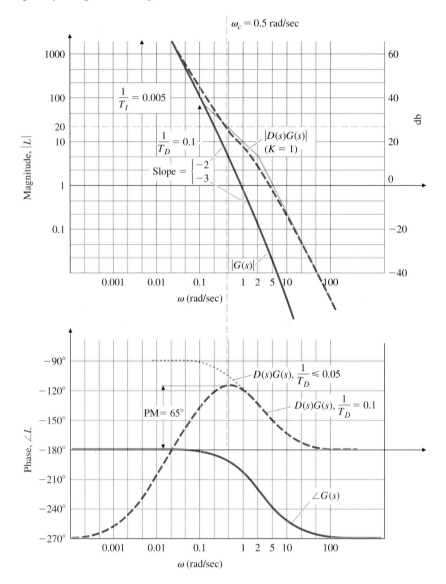

that the phase would be $-180°$ for the -2 slope and $-270°$ for the -3 slope and which would correspond to a PM of $0°$ or $-90°$. Therefore, derivative control is required to bring the slope to -1 at the crossover frequency that was shown in Section 6.5 to be a requirement for stability. The problem now is to pick values for the three parameters in Eq. (6.52)—K, T_D, and T_I—that will satisfy the specifications.

The easiest approach is to work first on the phase so that PM $= 65°$ is achieved at a reasonably high frequency. This can be accomplished primarily by adjusting T_D, noting that T_I has a minor effect if sufficiently larger than T_D. Once the phase is adjusted, we establish the crossover frequency and then can easily determine the gain K.

We examine the phase of the PID controller in Fig. 6.67, to determine what would happen to the compensated spacecraft system, $D(s)G(s)$, as T_D is varied. If $1/T_D \geq 2$ rad/sec, the phase lead from the PID control would simply cancel the sensor phase lag, and the composite phase would never exceed $-180°$, an unacceptable situation. If $1/T_D \leq 0.01$, the composite phase would approach $-90°$ for some range of frequencies and would exceed $-115°$ for an even wider range of frequencies; the latter threshold would provide a PM of 65°. In the compensated phase curve shown in Fig. 6.69, $1/T_D = 0.1$, which is the largest value of $1/T_D$ that could provide the required PM of 65°. The phase would never cross the $-115°$ (65°PM) line for any $1/T_D > 0.1$. For $1/T_D = 0.1$, the crossover frequency ω_c that produces the 65°PM is 0.5 rad/sec. For a value of $1/T_D \ll 0.05$, the phase essentially follows the dotted curve in Fig. 6.69, which indicates that the maximum possible ω_c is approximately 1 rad/sec and is provided by $1/T_D = 0.05$. Therefore, $0.05 < 1/T_D < 0.1$ is the only sensible range for $1/T_D$; anything less than 0.05 would provide no significant increase in bandwidth, while anything more than 0.1 could not meet the PM specification. Although the final choice is somewhat arbitrary, we have chosen $1/T_D = 0.1$ for our final design.

Our choice for $1/T_I$ is a factor of 20 lower than $1/T_D$; that is, $1/T_I = 0.005$. A factor less than 20 would negatively impact the phase at crossover, thus lowering the PM. Furthermore, it is generally desirable to keep the compensated magnitude as large as possible at frequencies below ω_c in order to have a faster transient response and smaller errors; maintaining $1/T_D$ and $1/T_I$ at the highest possible frequencies will bring this about.

The only remaining task is to determine the proportional part of the PID controller, or K. Unlike the system in Example 6.17, where we selected K in order to meet a steady-state error specification, here we select a value of K that will yield a crossover frequency at the point corresponding to the required PM of 65°. The basic procedure for finding K, discussed in Section 6.6, consists of plotting the compensated system amplitude with $K = 1$, finding the amplitude value at crossover, then setting $1/K$ equal to that value. Figure 6.69 shows that when $K = 1$, $|D(s)G(s)| = 20$ at the desired crossover frequency $\omega_c = 0.5$ rad/sec. Therefore,

$$\frac{1}{K} = 20, \quad \text{so} \quad K = \frac{1}{20} = 0.05.$$

The compensation equation that satisfies all of the specifications is now complete:

$$D(s) = \frac{0.05}{s}[(10s + 1)(s + 0.005)].$$

It is interesting to note that this system would become unstable if the gain were lowered so that $\omega_c \leq 0.02$ rad/sec, the region in Fig. 6.69 where the phase of the compensated system is less than $-180°$. As mentioned in Section 6.4, this situation is referred to as a conditionally stable system. A root locus with respect to K for this and any conditionally stable system would show the portion of the locus corresponding to very low gains in the RHP. The response of the system for a unit step θ_{com} is shown in Fig. 6.70(a) and exhibits well-damped behavior, as should be expected with a 65° PM.

The response of the system for a step disturbance torque $T_d = 0.1$ N is shown in Fig. 6.70(b). Note that the integral control term does eventually drive the error to zero; however, it is slow due to the presence of a closed-loop pole in the vicinity of the zero at $s = -0.005$. However, recall from the design process that this zero was located in order that the integral term not impact the PM unduly. So if the slow disturbance response is not acceptable, speeding up this pole will decrease the PM and damping of the system. Compromise is often a necessity in control system design!

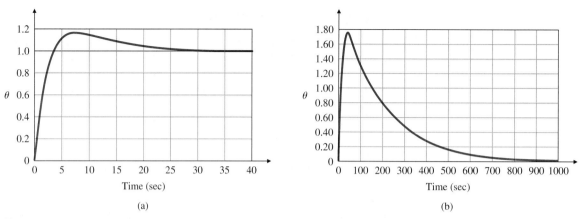

Figure 6.70 Transient response for PID example: (a) step response; (b) step-disturbance response

We now summarize conmpensation characteristics.

1. *PD Control* adds phase lead at all frequencies above the break point. If there is no change in gain on the low-frequency asymptote, PD compensation will increase the crossover frequency and the speed of response. The increase in magnitude of the frequency response at the higher frequencies will increase the system's sensitivity to noise.

2. *Lead Compensation* adds phase lead at a frequency band between the two break points, which are usually selected to bracket the crossover frequency. If there is no change in gain on the low-frequency asymptote, lead compensation will increase both the crossover frequency and the speed of response over the uncompensated system.

3. *PI Control* increases the frequency response magnitude at frequencies below the break point, thereby decreasing steady-state errors. It also contributes phase lag below the break point, which must be kept at a low enough frequency to avoid degrading the stability excessively.

4. *Lag Compensation* increases the frequency response magnitude at frequencies below the two break points, thereby decreasing steady-state errors. Alternatively, with suitable adjustments in K, lag compensation can be used to decrease the frequency-response magnitude at frequencies above the two break points so that ω_c yields an acceptable phase margin. Lag compensation also contributes phase lag between the two break points, which must be kept at frequencies low enough to keep the phase decrease from degrading the PM excessively. This compensation will typically provide a slower response than using lead compensation.

6.7.6   Design Considerations

We have seen in the preceding designs that characteristics of the open-loop Bode plot of the loop gain, $L(s)$ $(= KDG)$, determine performance with respect to steady-state errors and dynamic response. Other properties of feedback, developed in Chapter 4, include reducing the effects of sensor noise and parameter changes on the performance of the system.

The consideration of steady-state errors due to command inputs and disturbances has been an important design component in the different design methods presented. Design for acceptable steady-state errors can be thought of as placing a lower bound on the very-low-frequency gain of the system. Another aspect of the sensitivity issue concerns the high-frequency portion of the system. So far, Chapter 4 and Sections 5.5 and 6.7 have briefly discussed the idea that, to alleviate the effects of sensor noise, the gain of the system at high frequencies must be kept low. In fact, in the development of lead compensation, we added a pole to pure derivative control specifically to reduce the effects of sensor noise at the higher frequencies. It is not unusual for designers to place an extra pole in the compensation, that is, to use the relation

$$D(s) = \frac{Ts + 1}{(\alpha Ts + 1)^2},$$

in order to introduce even more attenuation for noise reduction.

A second consideration affecting high-frequency gains is that many systems have high-frequency dynamic phenomena such as mechanical resonances that could have an impact on the stability of a system. In very-high-performance designs, these high-frequency dynamics are included in the plant model, and a compensator is designed with a specific knowledge of those dynamics. A standard approach to designing for unknown high-frequency dynamics is to keep the high-frequency gain low, just as we did for sensor-noise reduction. The reason for this can be seen from the gain–frequency relationship of a typical system, shown in Fig. 6.71. The only way instability can result from high-frequency dynamics is if an unknown high-frequency resonance causes the magnitude to

Figure 6.71
Effect of high-frequency plant uncertainty

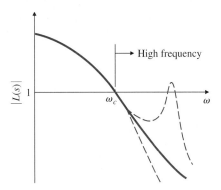

rise above 1. Conversely, if all unknown high-frequency phenomena are guaranteed to remain below a magnitude of 1, stability can be guaranteed. The likelihood of an unknown resonance in the plant G rising above 1 can be reduced if the nominal high-frequency loop gain (L) is lowered by the addition of extra poles in $D(s)$. When the stability of a system with resonances is assured by tailoring the high-frequency magnitude never to exceed 1, we refer to this process as **amplitude** or **gain stabilization**. Of course, if the resonance characteristics are known exactly, a specially tailored compensation, such as one with a notch at the resonant frequency, can be used to change the phase at a specific frequency to avoid encirclements of -1, thus stabilizing the system even though the amplitude does exceed magnitude 1. This method of stabilization is referred to as **phase stabilization**. A drawback to phase stabilization is that the resonance information is often not available with adequate precision or varies with time; therefore, the method is more susceptible to errors in the plant model used in the design. Thus we see that sensitivity to plant uncertainty and sensor noise are both reduced by sufficiently low loop gain at high frequency.

These two aspects of sensitivity—high- and low-frequency behavior—can be depicted graphically, as shown in Fig. 6.72. There is a minimum low-frequency gain allowable for acceptable steady-state error performance and a maximum high-frequency gain allowable for acceptable noise performance and for low probability of instabilities caused by plant-modeling errors. It is sometimes convenient to define the low-frequency lower bound on the frequency response as W_1 and the upper bound as W_2^{-1} as shown in the figure. Between these two bounds the control engineer must achieve a gain crossover near the required bandwidth; as we have seen, the crossover must occur at essentially a slope of -1 for good PM and hence damping.

For example, if a control system was required to follow a sine reference input with frequencies from 0 to ω_d with errors no greater than 1%, the function W_1 would be 100 from $\omega = 0$ to ω_d. Similiar ideas enter into defining possible values for the W_2^{-1} function. These ideas will be discussed further in Section 6.9.

Gain stabilization

Phase stabilization

Figure 6.72
Design criteria for low sensitivity

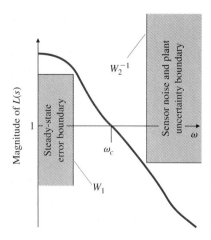

▲ 6.8 Alternate Presentations of Data

Before computers were widely available, other ways to present frequency-response data were developed to aid both in understanding design and in easing the designer's work load. The widespread availability of computers has reduced the need for these methods. Two techniques are the Nichols chart and the inverse Nyquist plot, both of which we examine in this section because of their place in history.

6.8.1 Nichols Chart

A rectangular plot of $\log |G(j\omega)|$ versus $\angle G(j\omega)$ can be drawn by simply transferring the information directly from the separate magnitude and phase portions in a Bode plot; one point on the new curve thus results from a given value of the frequency ω. This means that the new curve is parameterized as a function of frequency. As with the Bode plots, the magnitude information is plotted on a logarithmic scale, while the phase information is plotted on a linear scale. This template was suggested by N. Nichols and is usually referred to as a **Nichols chart**. The idea of plotting the magnitude of $G(j\omega)$ versus its phase is similar to the concept of plotting the real and imaginary parts of $G(j\omega)$, which formed the basis for the Nyquist plots shown in Sections 6.3 and 6.4. However, it is difficult to capture all the pertinent characteristics of $G(j\omega)$ on the linear scale of the Nyquist plot. The log scale for magnitude in the Nichols chart alleviates this difficulty, allowing this kind of presentation to be useful for design.

For any value of the complex transfer function $G(j\omega)$, Section 6.6 showed that there is a unique mapping to the unity-feedback closed-loop transfer function

$$T(j\omega) = \frac{G(j\omega)}{1 + G(j\omega)}, \tag{6.54}$$

or, in polar form,

$$T(j\omega) = M(\omega)e^{j\alpha(\omega)}, \tag{6.55}$$

where $M(\omega)$ is the magnitude of the closed-loop transfer function and $\alpha(\omega)$ is the phase of the closed-loop transfer function. Specifically,

$$M = \left| \frac{G}{1 + G} \right|, \tag{6.56}$$

$$\alpha = \tan^{-1}(N) = \angle \frac{G}{1 + G}. \tag{6.57}$$

It can be proven that the contours of constant closed-loop magnitude and phase are circles when $G(j\omega)$ is presented in the linear Nyquist plot. These circles are referred to as the **M and N circles**, respectively.

M and N circles

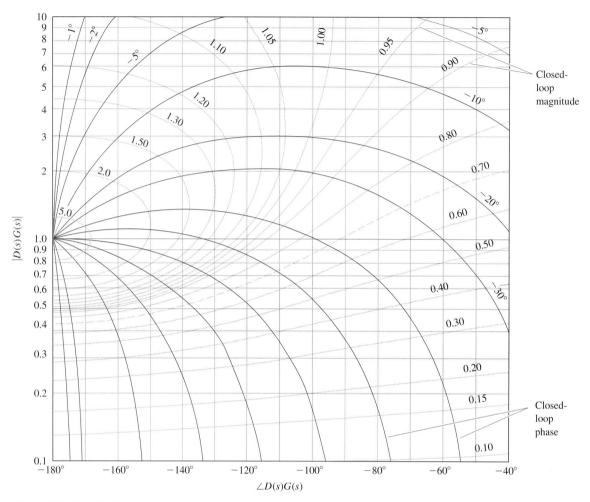

Figure 6.73 Nichols chart

The Nichols chart also contains contours of constant *closed-loop* magnitude and phase based on these relationships, as shown in Fig. 6.73; however, they are no longer circles because the Nichols charts are semilog plots of magnitude versus phase. A designer can therefore graphically determine the bandwidth of a closed-loop system from the plot of the open-loop data on a Nichols chart by noting where the open-loop curve crosses the 0.70 contour of the closed-loop magnitude and determining the frequency of the corresponding data point. Likewise, a designer can determine the resonant peak amplitude M_r by noting the value of the magnitude of the highest closed-loop contour tangent to the curve. The frequency associated with the magnitude and phase at the point of tangency is sometimes referred to as the **resonant frequency** ω_r. Similarly, a designer can determine the gain margin (GM) by observing the value of the gain

Resonant frequency

where the Nichols plot crosses the $-180°$ line, and he or she can determine the phase margin (PM) by observing the phase where the plot crosses the amplitude 1 line (James et al., 1947). MATLAB provides for easy drawing of a Nichols chart via the nichols function.

EXAMPLE 6.20   *Nichols Chart for PID Example*

Determine the bandwith and resonant peak magnitude of the compensated system whose frequency response is shown in Fig. 6.69.

Solution.   The magnitude and phase information of the compensated design example seen in Fig. 6.69 is shown on a Nichols chart in Fig. 6.74. When comparing the two figures, it is important to divide the magnitudes in Fig. 6.69 by a factor of 20 in order

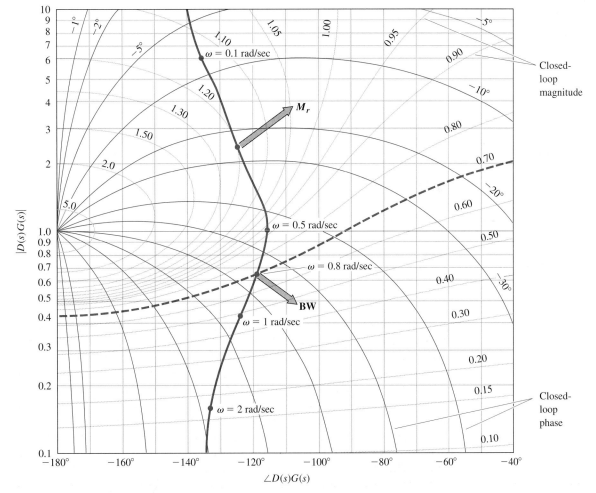

Figure 6.74 Example plot on the Nichols chart for determining bandwidth and M_r

to obtain $|D(s)G(s)|$ rather than the normalized values used in Fig. 6.69. Because the curve crosses the closed-loop magnitude 0.70 contour at $\omega = 0.8$ rad/sec, we see that the bandwidth of this system is 0.8 rad/sec. Because the largest-magnitude contour touched by the curve is 1.20, we also see that $M_r = 1.2$.

This presentation of data was particularly valuable when a designer had to generate plots and perform calculations by hand. A change in gain, for example, could be evaluated by sliding the curve vertically on transparent paper over a standard Nichols chart. The GM, PM, and bandwidth were then easy to read off the chart, thus allowing evaluations of several values of gain with a minimal amount of effort. With access to computer-aided methods, however, we can now calculate the bandwidth and perform many repetitive evaluations of the gain or any other parameter with a few key strokes. Today the Nichols chart is used primarily as an alternate way to present the information in a Nyquist plot. For complex systems where the -1 encirclements need to be evaluated, the magnitude log scale of the Nichols chart enables us to examine a wider range of frequencies than a Nyquist plot does, as well as allowing us to read the gain and phase margins directly.

EXAMPLE 6.21   *Stability Margins from Nichols Chart*

For the system of Example 6.12, whose Nyquist plot is shown in Fig. 6.41, determine the PM and GM using the Nyquist plot.

Solution.   Figure 6.75 shows a Nichols chart with the data from the same system shown in Fig. 6.41. Note that the PM for the magnitude 1 crossover frequency is 36° and the GM is 1.25 ($= 1/0.8$). It is clear from this presentation of the data that the most critical portion of the curve is where it crosses the $-180°$ line; hence the GM is the most relevant stability margin in this example.

6.8.2   Inverse Nyquist

The **inverse Nyquist plot** is simply the reciprocal of the Nyquist plot described in Section 6.3 and used in Section 6.4 for the definition and discussion of stability margins. It is obtained most easily by computing the inverse of the magnitude from the Bode plot and plotting that quantity at an angle in the complex plane as indicated by the phase from the Bode plot. It can be used to find the PM and GM in the same way that the Nyquist plot was used. When $|G(j\omega)| = 1$, $|G^{-1}(j\omega)| = 1$ also, so the definition of PM is identical on the two plots. However, when the phase is $-180°$ or $+180°$, the value of $|G^{-1}(j\omega)|$ is the GM directly; no calculation of an inverse is required, as was the case for the Nyquist plot.

The inverse Nyquist plot for the system in Fig. 6.24 (Example 6.9) is shown in Fig. 6.76 for the case where $K = 1$ and the system is stable. Note that

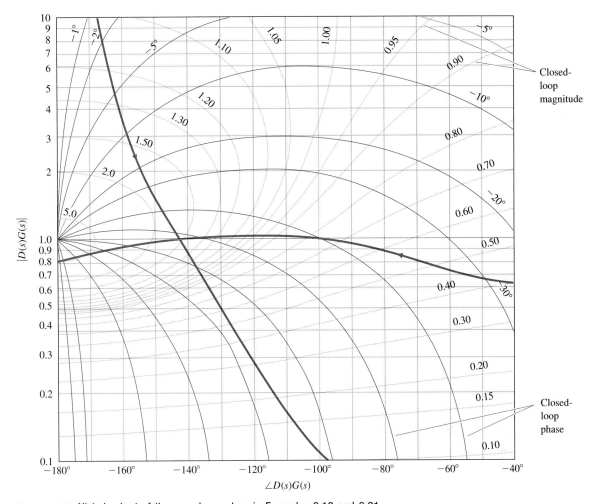

Figure 6.75 Nichols chart of the complex system in Examples 6.12 and 6.21

Figure 6.76
Inverse Nyquist plot for
Example 6.9

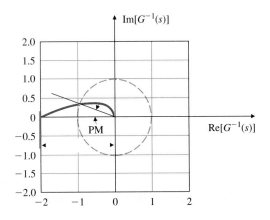

Figure 6.77
Inverse Nyquist plot of the
system whose Nyquist plot
is in Fig. 6.41

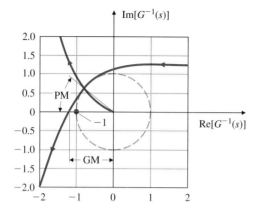

GM $= 2$ and PM $\cong 20°$. As an example of a more complex case, Fig. 6.77 shows an inverse Nyquist plot for the sixth-order case whose Nyquist plot was shown in Fig. 6.41 and whose Nichols chart was shown in Fig. 6.75. Note here that GM $= 1.2$ and PM $\cong 35°$. Had the two crossings of the unit circle not occurred at the same point, the crossing with the smallest PM would have been the appropriate one to use.

▲ 6.9   Specifications in Terms of the Sensitivity Function

We have seen how the gain and phase margins give useful information about the relative stability of nominal systems and can be used to guide the design of lead and lag compensations. However, the GM and PM are only two numbers and have limitations as guides to the design of realistic control problems. We can express more complete design specifications in the frequency domain if we first give frequency descriptions for the external signals such as the reference and disturbance and consider the sensitivity function defined in Section 4.3. For example, we have so far described dynamic performance by the transient response to simple steps and ramps. A more realistic description of the actual complex input signals is to represent them as random processes with corresponding frequency power density spectra. A less sophisticated description which is adequate for our purposes is to assume that the signals can be represented as a sum of sinusoids with frequencies in a specified range. For example, we can usually describe the frequency content of the reference input as a sum of sinusoids with relative amplitudes given by a magnitude function $|R|$ such as that plotted in Fig. 6.78 which represents a signal with sinusoidal components having about the same amplitudes up to some value ω_1 and very small amplitudes for frequencies above that. With this assumption the response tracking specification can be expressed by a statement such as "the magnitude of the system error is to be less than the bound e_b (a value such as 0.01) for any sinusoid of frequency ω_o in the range $0 \le \omega_o \le \omega_1$ and of amplitude given by

Figure 6.78
Plot of typical reference
spectrum

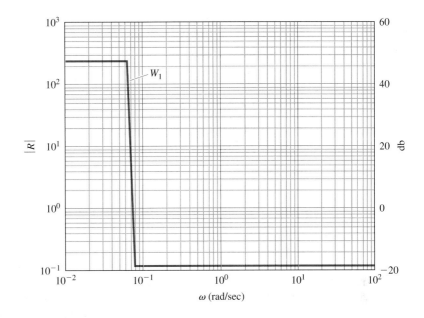

$|R(j\omega_o)|$. To express such a performance requirement in terms that can be used in design, we consider again the unity feedback system drawn in Fig. 6.79. For this system the error is given by

$$E(j\omega) = \frac{1}{1 + DG} R \triangleq S(j\omega)R \qquad (6.58)$$

Sensitivity function

where we have used the **sensitivity function** S, where

$$S(j\omega) \triangleq \frac{1}{1 + DG}. \qquad (6.59)$$

In addition to being the factor multiplying the system error, the sensitivity function is also the reciprocal of the distance of the Nyquist curve, DG, from the critical point -1. A large value for S indicates a Nyquist plot that comes close to the point of instability. The frequency-based error specification based on Eq. (6.58) can be expressed as $|E| = |S||R| \leq e_b$. In order to normalize

Figure 6.79
Closed-loop block diagram

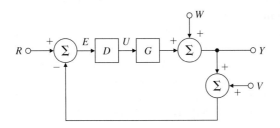

the problem and not need to define both the spectrum R and the error bound each time, we define the real function of frequency $W_1(\omega) = |R|/e_b$ and the requirement can be written as

$$\boxed{|S|\,W_1 \leq 1.}$$

(6.60)

EXAMPLE 6.22   **Performance Bound Function**

A unity feedback system is to have an error less than 0.005 for all unity amplitude sinusoids below frequency 100 Hz. Draw the performance frequency function $W_1(\omega)$ for this design.

Solution.   The spectrum, from the problem description, is unity for $0 \leq \omega \leq 200\pi$. Because $e_b = 0.005$, the required function is given by a rectangle of amplitude $1/0.005 = 200$ over the given range. The function is plotted in Fig. 6.80.

Figure 6.80
Plot of example
performance function, W_1

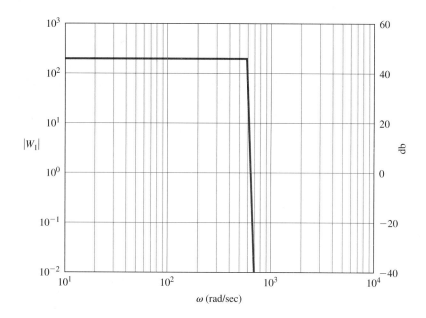

The expression in Eq. (6.60) can be translated to the more familiar Bode plot coordinates and given as a requirement on loop gain by observing that over the frequency range when errors are small, the loop gain is large. In that case $|S| \approx 1/|DG|$ and the requirement is approximately

$$\frac{W_1}{|DG|} \leq 1,$$

(6.61)

$$\boxed{|DG| \geq W_1.}$$

This requirement can be seen as an extension of the steady-state error requirement from just $\omega = 0$ to the range $0 \le \omega_o \le \omega_1$.

In addition to the requirement on dynamic performance, the designer is usually required to design for **stability robustness**. By this we mean that, while the design is done for a nominal plant transfer function, the actual system is expected to be stable for an entire class of transfer functions that represents the range of changes that are expected to be faced as temperature, age, and other operational and environmental factors vary the plant dynamics from the nominal case. A realistic way to express this uncertainty is to describe the plant transfer function as having a multiplicative uncertainty as

$$G(j\omega) = G_o(j\omega)[1 + W_2(\omega)\triangle(j\omega)]. \tag{6.62}$$

In Eq. (6.62), the real function W_2 is a magnitude function that expresses the size of changes as a function of frequency that the transfer function is expected to experience. In terms of G and G_o the expression is

$$W_2 = \left| \frac{G - G_o}{G_o} \right|. \tag{6.63}$$

The shape of W_2 is almost always very small for low frequencies (we know the model very well there) and increases substantially as we go to high frequencies where parasitic parameters come into play and unmodeled structural flexibility is common. A typical shape is sketched in Fig. 6.81. The complex function, $\triangle(j\omega)$, represents the uncertainty in phase and is restricted only by the constraint

$$0 \le |\triangle| \le 1. \tag{6.64}$$

Figure 6.81
Plot of typical plant uncertainty, W_2

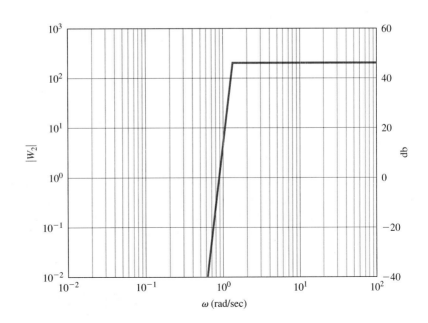

We assume that the nominal design has been done and is stable so that the Nyquist plot of DG_o satisfies the Nyquist stability criterion. In this case, the nominal characteristic equation $1 + DG_o = 0$ is never satisfied for any real frequency. If the system is to have stability robustness, the characteristic equation using the uncertain plant as described by (6.62) must not go to zero for any real frequency for any value of $\triangle$. The requirement can be written as

$$1 + DG \neq 0,$$

$$1 + DG_o[1 + W_2\triangle] \neq 0, \tag{6.65}$$

$$(1 + DG_o)(1 + TW_2\triangle) \neq 0,$$

where we have defined the **complementary sensitivity function** as

$$T(s) \overset{\triangle}{=} DG_o/(1 + DG_o) = 1 - S. \tag{6.66}$$

Because the nominal system is stable, the first term in Eq. (6.65), $(1 + DG_o)$, is never zero. Thus, if Eq. (6.65) is not to be zero for any frequency and any $\triangle$, then it is necessary and sufficient that

$$|TW_2\triangle| < 1,$$

which reduces to

$$|T| W_2 < 1, \tag{6.67}$$

making use of Eq. (6.64). As with the performance specification, for single-input–single-output unity feedback systems this requirement can be approximated by a more convenient form. Over the range of high frequencies where W_2 is non-negligible because there is significant model uncertainty, DG_o is small. Therefore we can approximate $T \approx DG_o$ and the constraint reduces to

$$|DG_o| W_2 < 1,$$

$$\boxed{|DG_o| < \frac{1}{W_2}.} \tag{6.68}$$

EXAMPLE 6.23   *Typical Plant Uncertainty*

The uncertainty in a plant model is described by a function W_2 which is zero until $\omega = 3000$ and increases linearly from there to a value of 100 at $\omega = 10{,}000$ and remains at 100 for higher frequencies. Plot the constraint on DG_o to meet this requirement.

Solution.   Where $W_2 = 0$, there is no constraint on the magnitude of loop gain; above $\omega = 3000$, $1/W_2 = DG_o$ is a hyperbola from ∞ to 0.01 at $\omega = 10{,}000$ and remains at 0.01 for $\omega > 10{,}000$. The bound is sketched in Fig. 6.82.

Figure 6.82

Plot of constraint on $|DG_o|\ (=|W_2^{-1}|)$

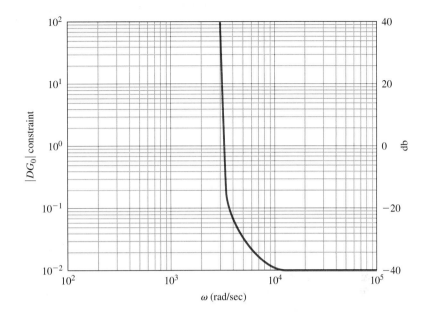

In practice, the magnitude of the loop gain is plotted on log–log (Bode) coordinates and the constraints of Eq. (6.61) and Eq. (6.68) are included on the same plot. A typical sketch is drawn in Fig. 6.83. The designer is expected to

Figure 6.83

Tracking and stability robustness constraints on the Bode plot

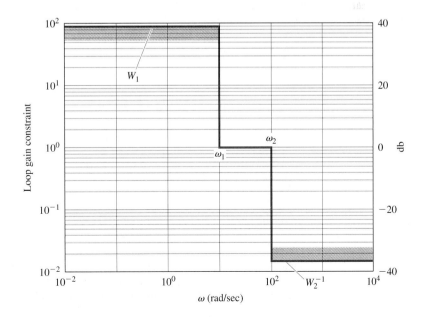

construct a loop gain that will stay above W_1 for frequencies below ω_1, cross over the magnitude 1 line ($|DG| = 0$) in the range $\omega_1 \leq \omega \leq \omega_2$, and stay below $1/W_2$ for frequencies above ω_2.

6.9.1   Limitations on Design in Terms of the Sensitivity Function

One of the major contributions of Bode was to derive important limitations on transfer functions that set limits on achievable design specifications. For example, one would like to have the system error kept small for the widest possible range of frequencies and yet to have a system that is robustly stable for a very uncertain plant. In terms of the plot in Fig. 6.83, we want W_1 and W_2 to be very large in their respective frequency ranges and for ω_1 to be pushed up close to ω_2. Thus the loop gain is expected to plunge with a large negative slope from being greater than W_1 to being less than $1/W_2$ in a very short span, while maintaining a good phase margin to assure stability and good dynamic performance. The Bode gain-phase formula given earlier shows that this is *impossible* with a linear controller by showing that the minimum possible phase is determined by an integral depending on the slope of the magnitude curve. If the slope is constant for a substantial range around ω_o, then the formula can be approximated by

$$\phi(\omega_o) \approx \frac{\pi}{2} \left. \frac{dM}{du} \right|_{u=0}, \tag{6.69}$$

where M is the log magnitude and $u = \log \frac{\omega}{\omega_o}$. If, for example, the phase is to be kept above $-150°$ to maintain a $30°$ phase margin, then the magnitude slope near ω_o is estimated to be

$$\frac{dM}{du} \approx \frac{2}{\pi} \left(-150 \frac{\pi}{180} \right)$$
$$\approx -1.667.$$

If we try to make the average slope faster (more negative) than this, we will lose the phase margin. From this condition there developed the design rule that the asymptotes of the Bode plot magnitude, which are restricted to be integral values for rational functions, should be made to cross over the 0-db line at a slope of -1 over a frequency range of about one decade around the crossover frequency as already discussed in Section 6.5. Modifications to this rule need of course be made in particular cases, but the limitation implied by Eq. (6.69) is a hard limit that cannot be avoided.

EXAMPLE 6.24   *Robustness Constraints*

If $W_1 = W_2 = 100$, and we want PM $= 30°$, what is the minimum ratio of ω_2/ω_1?

Solution.   The slope is

$$\frac{\log W_1 - \log \dfrac{1}{W_2}}{\log \omega_1 - \log \omega_2} = \frac{2+2}{\log \dfrac{\omega_1}{\omega_2}} = -1.667.$$

Thus the log of the ratio is $\log \frac{\omega_1}{\omega_2} = -2.40$ and $\omega_2 = 251\omega_1$.

An alternative to the standard Bode plot as a design guide can be based on a plot of the sensitivity function as a function of frequency. In this format, Eq. (6.60) requires that $|S| < 1/W_1$ over the range $0 \le \omega \le \omega_1$ for performance and Eq. (6.68) requires that $|S| \approx 1$ over the range $\omega_2 \le \omega$ for stability robustness. It should come as no surprise that Bode found a limitation on the possibilities in this case, too. The constraint, extended by Freudenberg and Looze (1985), shows that an integral of the sensitivity function is determined by the presence of poles in the right half-plane. Suppose the loop gain DG_o has n_p poles, p_i, in the right half-plane and "rolls off" at high frequencies at a slope faster than -1. For rational functions, this means that there is an excess of at least two more finite poles than zeros. Then it can be shown that

$$\int_0^\infty \ln(|S|)\, d\omega = \pi \sum_{i=1}^{n_p} \mathrm{Re}\{p_i\}. \tag{6.70}$$

If there are no right half-plane poles, then the integral is zero. This means that if we make the log of the sensitivity function very negative over some frequency band to reduce errors in that band, then *of necessity* $\ln|S|$ will be positive over another part of the band and errors will be amplified there. If there are unstable poles, the situation is worse because the positive area where sensitivity magnifies the error must *exceed* the negative area where the error is reduced by the feedback. If the system is minimum phase, then it is in principle possible to keep the magnitude of the sensitivity small by spreading the sensitivity increase over all positive frequencies to infinity but such a design requires an excessive bandwidth and is rarely practical. If a specific bandwidth is imposed, then the sensitivity function is constrained to take on a finite, possibly large, positive value at some point below the bandwidth. As implied by the definition of the vector margin (VM) in Section 6.4, (Fig. 6.38), a large $S_{\max}$ corresponds to a Nyquist plot that comes close to the -1 critical point and a system having a small vector margin since

Vector margin

$$VM = \frac{S_{\max}}{S_{\max} - 1}. \tag{6.71}$$

If the system is not minimum phase, the situation is worse. An alternative to Eq. (6.70) is true if there is a nonminimum-phase zero of DG_o, a zero in the right half-plane. Suppose the zero is located at $z_o = \sigma_o + j\omega_o$, where $\sigma_o > 0$. Again we assume there are n_p right half-plane poles at locations p_i

with conjugate values $\overline{p}_i$. Now the condition can be expressed as a two-sided weighted integral

$$\int_{-\infty}^{\infty} \ln(|S|) \frac{\sigma_o}{\sigma_o^2 + (\omega - \omega_o)^2} \, d\omega = \pi \sum_{i=1}^{n_p} \ln \left| \frac{\overline{p}_i + z_o}{p_i - z_o} \right|. \tag{6.72}$$

In this case, we do not have the "roll-off" restriction and there is no possibility of spreading the positive area over high frequencies because the weighting function goes to zero with frequency. The important point about this integral is that if the nonminimum-phase zero is close to a right half-plane pole, the right side of the integral can be very large and the excess of positive area is required to be correspondingly large. Based on this result, one expects especially great difficulty meeting both tracking and robustness specifications on sensitivity with a system having right half-plane poles and zeros close together.

EXAMPLE 6.25   *Sensitivity Function for Antenna*

Compute and plot the sensitivity function for the design of the antenna for which $G(s) = 1/s(s + 1)$ and $D(s) = 10(0.5s + 1)/(0.1s + 1)$.

Solution.   The sensitivity function for this case is

$$S = \frac{s(s + 1)(s + 10)}{s^3 + 11s^2 + 60s + 100} \tag{6.73}$$

and the plot shown in Fig. 6.84 is given by the MATLAB commands

```
numS = [1 11 10 0];
denS = [1 11 60 100];
sysS = tf(numS,denS);
[mag,ph,w] = bode(sysS);
semilogy(w,mag)
```

Figure 6.84
Sensitivity function for
Example 6.25

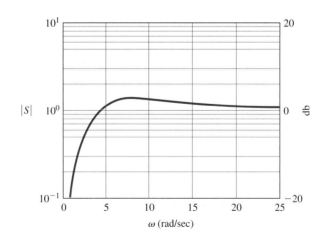

The largest value of S is given by $M = \max(\text{mag})$ and is given by 1.366 from which the vector margin is $VM = 3.73$.

▲ 6.10 Time Delay

The Laplace transform of a pure time delay is $G_D(s) = e^{-sT_d}$ and was approximated by a rational function (Padé approximate) in our earlier discussion of root-locus analysis. Although this same approximation could be used with frequency-response methods, an exact analysis of the delay is possible with the Nyquist criterion and Bode plots.

The frequency response of the delay is given by the magnitude and phase of $e^{-sT_d}|_{s=j\omega}$. The magnitude is

Time-delay magnitude

$$|G_D(j\omega)| = |e^{-j\omega T_d}| = 1 \qquad \text{for all } \omega. \qquad (6.74)$$

Time-delay phase

This result is expected, because a time delay merely shifts the signal in time and has no effect on its magnitude. The phase is

$$\angle G_D(j\omega) = -\omega T_d \qquad (6.75)$$

in radians, and it grows increasingly negative in proportion to the frequency. This, too, is expected since a fixed time delay T_d becomes a larger fraction or multiple of a sine wave as the period drops due to increasing frequency. A plot of $\angle G_D(j\omega)$ is drawn in Fig. 6.85. Note that the phase lag is greater than 270° for values of ωT_d greater than about 5 rad. This trend implies that it would be virtually impossible to stabilize a system (or to achieve a positive PM) with a crossover frequency greater than $\omega = 5/T_d$, and it would be difficult for frequencies greater than $\omega \cong 3/T_d$. These characteristics essentially place a constraint on the achievable bandwidth of any system with a time delay. (See Problem 6.68 for an illustration of this constraint.)

The frequency-domain concepts such as the Nyquist criterion apply directly to systems with pure time delay. This means that no approximations (Padé type or otherwise) are needed, as shown in the following example.

Figure 6.85
Phase lag due to pure time delay

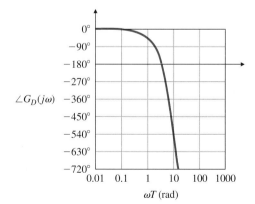

ωT (rad)

EXAMPLE 6.26   *Nyquist Plot for System with Time Delay*

Consider the system with

$$KG(s) = \frac{Ke^{-T_d s}}{s},$$

where $T_d = 1$ sec. Determine the range of K for which the system is stable.

Solution.   Because the Bode plotting rules do not apply for the phase of a time delay term, we will use an analytical approach to determine the key features of the frequency response plot. As just discussed, the magnitude of the frequency response of the delay term is unity and its phase is $-\omega$ radians. The magnitude of the frequency response of the pure integrator is $1/\omega$ with a constant phase of $-\pi/2$. Therefore,

$$G(j\omega) = \frac{1}{\omega} e^{-j(\omega + \pi/2)}$$

$$= \frac{1}{\omega}(-\sin\omega - j\cos\omega). \tag{6.76}$$

Using Eq. (6.76) and substituting in different values of ω, we can make the Nyquist plot, which is the spiral shown in Fig. 6.86.

Figure 6.86
Nyquist plot for
Example 6.26

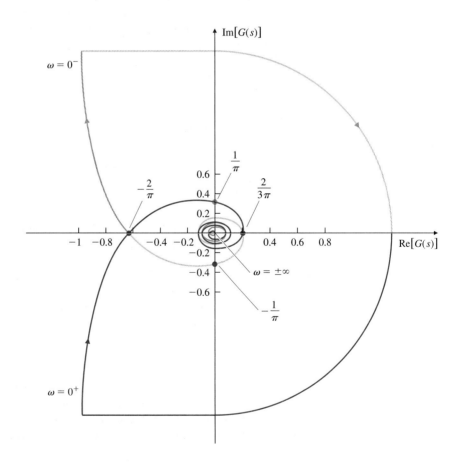

Let us examine the shape of the spiral in more detail. We pick a Nyquist path with a small detour to the right of the origin. The effect of the pole at the origin is the large arc at infinity with a $180°$ sweep, as shown in Fig. 6.86. From Eq. (6.76), for small values of $\omega > 0$, the real part of the frequency response is close to -1 because $\sin\omega \cong \omega$ and $\text{Re}[G(j\omega)] \cong -1$. Similarly, for small values of $\omega > 0$, $\cos\omega \cong 1$ and $\text{Im}[G(j\omega)] \cong -1/\omega$, that is, very large, negative values as shown in Fig. 6.86. To obtain the crossover points on the real axis, we set the imaginary part equal to zero:

$$\frac{\cos\omega}{\omega} = 0. \tag{6.77}$$

The solution is then

$$\omega_0 = \frac{(2n+1)\pi}{2}, \qquad n = 0, 1, 2, \ldots. \tag{6.78}$$

After substituting Eq. (6.78) back into Eq. (6.76), we find

$$G(j\omega_0) = \frac{(-1)^n}{(2n+1)}\left(\frac{2}{\pi}\right), \qquad n = 0, 1, 2, \ldots.$$

So the first crossover of the negative real axis is at $-2/\pi$ corresponding to $n = 0$. The first crossover of the positive real axis occurs for $n = 1$ and is located at $2/3\pi$. As we can infer from Fig. 6.86, there are an infinite number of other crossings of the real axis. Finally, for $\omega = \infty$, the Nyquist plot converges to the origin. Note that the Nyquist plot for $\omega < 0$ is the mirror image of the one for $\omega > 0$.

The number of poles in the RHP is zero ($P = 0$), so for closed-loop stability we need $Z = N = 0$. Therefore, the Nyquist plot cannot be allowed to encircle the $-1/K$ point. It will not do so as long as

$$-\frac{1}{K} < -\frac{2}{\pi}, \tag{6.79}$$

which means for stability we must have $0 < K < \pi/2$.

▲ 6.11   Obtaining a Pole-Zero Model from Frequency-Response Data

As we pointed out earlier, it is relatively easy to obtain the frequency response of a system experimentally. Sometimes it is desirable to obtain an approximate model, in terms of a transfer function, directly from the frequency response. The derivation of such a model can be done to various degrees of accuracy. The method described in this section is usually adequate and is widely used in practice.

There are two ways to obtain a model from frequency-response data. In the first case we can introduce a sinusoidal input, measure the gain (logarithm of the amplitude ratio of output to input) and the phase difference between output and input, and accept the curves plotted from this data as the model. Using the methods given in previous sections, we can derive the design directly from this information. In the second case, we wish to use the frequency data to verify a mathematical model obtained by other means. To do so we need to extract an approximate transfer function from the plots, again by fitting straight lines to the data, estimating break points (that is, finding the poles and zeros), and using Fig. 6.3 to estimate the damping ratios of complex factors from the frequency overshoot. The next example illustrates the second case.

EXAMPLE 6.27   *Transfer Function from Measured Frequency Response*

Determine a transfer function from the frequency response plotted in Fig. 6.87, where frequency f is plotted in hertz.

Figure 6.87
An experimental frequency response

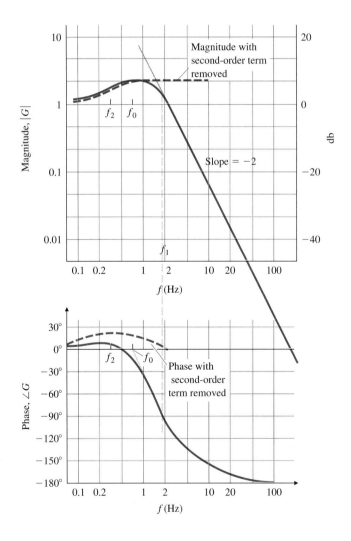

Solution.   Drawing an asymptote to the final slope of -2 (or -40 db per decade), we assume a break point at the frequency where the phase is $-90°$. This occurs at $f_1 \cong 1.66$ Hz ($\omega_1 = 2\pi f_1 = 10.4$ rad/sec). We need to know the damping ratio in order to subtract out this second-order pole. For this the phase curve may be of more help. Because the phase around the break-point frequency is symmetric, we draw a line at the slope of the phase curve at f_1 to find that the phase asymptote intersects the $0°$ line at $f_0 \cong 0.71$ Hz (or 4.46 rad/sec). This corresponds to $f_1/f_0 \cong 2.34$, which in time corresponds to $\zeta \cong 0.5$, as seen on the normalized response curves in Fig. 6.3(b). The magnitude curve with the second-order factor taken out shows an asymptotic amplitude gain of about 6.0 db, or a factor of $10^{6.0/20} = 2.0$. Because this is a gain rise, it occurs because of a lead compensation of the form

$$\frac{s/a + 1}{s/b + 1},$$

where $b/a = 2.0$. If we remove the second-order terms in the phase curve, we obtain a phase curve with a maximum phase of about $20°$, which also corresponds to a frequency separation of about 2. To locate the center of the lead compensation, we must estimate the point of maximum phase based on the lead term alone, which occurs at the geometric mean of the two break-point frequencies. The lead center seems to occur at $f_2 \cong 0.3$ Hz (or $\omega_2 = 1.88$ rad/sec).

Thus we have the relations

$$ab(1.88)^2 = 3.55,$$

$$\frac{b}{a} = 2,$$

from which we can solve

$$2a^2 = 3.55,$$

$$a = 1.33,$$

$$b = 2.66.$$

Model from measured response

Our final model is given by

$$\hat{G}(s) = \frac{(s/1.33) + 1}{[(s/2.66) + 1][(s/10.4)^2 + (s/10.4) + 1]}. \tag{6.80}$$

The actual data were plotted from

$$G(s) = \frac{(s/2) + 1}{[(s/4) + 1][(s/10)^2 + (s/10) + 1]}.$$

As can be seen, we found the second-order term quite easily, but the location of the lead compensation is off in center frequency by a factor of $4/2.66 \cong 1.5$. However, the subtraction of the second-order term from the composite curve was not done with great accuracy, rather, by reading the curves. Again, as with the transient response, we conclude that by a bit of approximate plotting we can obtain a crude model (usually within a factor of 1.4 (± 3 db) in amplitude and $\pm 10°$ in phase) that can be used for control design.

Refinements on these techniques with computer aids are rather obvious, and an interactive program for removing standard first- and second-order terms and accurately plotting the residual function would greatly improve the speed and accuracy of the process. It is also common to have computer tools that can find the parameters of an assumed model structure by minimizing the sum of squares of the difference between the model's frequency response and the experimental frequency response.

SUMMARY

- The frequency-response **Bode plot** is a graph of the transfer function magnitude in logarithmic scale and the phase in linear scale versus frequency in logarithmic scale. For a transfer function $G(s)$,

$$A = |G(j\omega)| = |G(s)|_{s=j\omega}$$

$$= \sqrt{\{\mathrm{Re}[G(j\omega)]\}^2 + \{\mathrm{Im}[G(j\omega)]\}^2},$$

$$\phi = \tan^{-1}\left[\frac{\mathrm{Im}[G(j\omega)]}{\mathrm{Re}[G(j\omega)]}\right] = \angle G(j\omega).$$

- For a transfer function in Bode form,

$$KG(\omega) = K_0 \frac{(j\omega\tau_1 + 1)(j\omega\tau_2 + 1)\cdots}{(j\omega\tau_a + 1)(j\omega\tau_b + 1)\cdots},$$

the Bode frequency response can be easily plotted by hand using the rules described in Section 6.1.1.

- Bode plots can be obtained using computer algorithms (bode in MATLAB), but hand-plotting skills are still extremely helpful.

- For a second-order system the peak magnitude of the Bode plot is related to the damping by

$$|G(j\omega)| = \frac{1}{2\zeta} \qquad \text{at } \omega = \omega_n.$$

- A method of determining the stability of a closed-loop system based on the frequency response of the system's open-loop transfer function is the **Nyquist stability criterion**. Rules for plotting the **Nyquist plot** are described in Section 6.3. The number of RHP closed-loop roots, Z, is given by

$$Z = N + P,$$

where

$$N = \text{number of clockwise encirclements of the } -1 \text{ point,}$$

$$P = \text{number of open} - \text{loop poles in the RHP.}$$

- The Nyquist plot may be obtained using computer algorithms (`nyquist` in MATLAB).

- The **gain margin** (GM) and **phase margin** (PM), can be determined directly by inspecting the open-loop Bode plot or the Nyquist plot. Also, use of MATLAB's `margin` function determines the values directly.

- For a standard second-order system, the PM is related to the closed-loop damping by Eq. (6.32):

$$\zeta \cong \frac{PM}{100}.$$

- The **bandwidth** of the system is a measure of speed of response. For control systems it is defined as the frequency corresponding to 0.707 (-3 db) in the closed-loop magnitude Bode plot and is approximately given by the crossover frequency, ω_c, which is the frequency at which the open-loop gain curve crosses magnitude 1.

- The **vector margin** is a single-parameter stability margin based on the closest point of the Nyquist plot to the critical point $-1/K$.

- For a stable minimum-phase system, Bode's gain-phase relationship uniquely relates the phase to the gain of the system and is approximated by Eq. (6.33):

$$\angle G(j\omega) \cong n \times 90°,$$

where n is the slope of $|G(j\omega)|$ in units of decade of amplitude per decade of frequency. The relationship shows that, in most cases, stability is ensured if the gain plot crosses the magnitude 1 line with a slope of -1.

- Experimental frequency response data of the open-loop system can be used directly for analysis and design of a closed-loop control system with no analytical model.

- For the system shown in Fig. 6.88, the open-loop Bode plot is the frequency response of GD, and the closed-loop frequency response is obtained from $T(s) = GD/(1 + GD)$.

- The frequency response characteristics of several types of compensation have been described, and examples of design using these characteristics have been discussed. Design procedures were given for lead and lag compensators in Section 6.7. The examples in that section show the ease of selecting specific values of design variables, a result of using frequency-response methods. A summary was provided at the end of Section 6.7.5 on page 440.

Figure 6.88
Typical system

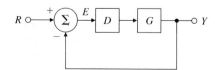

- **Lead compensation**, given by Eq. (6.38),

$$D(s) = \frac{Ts + 1}{\alpha Ts + 1}, \qquad \alpha < 1,$$

is a high-pass filter and approximates PD control. It is used whenever substantial improvement in damping of the system is required. It tends to increase the speed of response of a system for a fixed low frequency gain.

- **Lag compensation**, given by Eq. (6.51),

$$D(s) = \alpha \frac{Ts + 1}{\alpha Ts + 1}, \qquad \alpha > 1, \tag{6.81}$$

is a low-pass filter and approximates PI control. It is usually used to increase the low-frequency gain of the system so as to improve steady-state response for fixed bandwidth. For a fixed low-frequency gain, it will decrease the speed of response of a system.

- The **Nichols plot** is an alternate representation of the frequency response as a plot of gain versus phase and is parameterized as a function of frequency.

- Tracking-error reduction and disturbance rejection can be specified in terms of the low-frequency gain of the Bode plot. Sensor-noise rejection can be specified in terms of high-frequency attenuation of the Bode plot (see Fig. 6.72).

- Time delay can be analyzed directly by sketching the Nyquist plot.

- Transfer-function poles and zeros can be estimated from the measured frequency response.

Review Questions

1. Why did Bode suggest plotting the magnitude of a frequency response on log–log coordinates?

2. Define a decibel.

3. What is the transfer function magnitude if the gain is listed as 14 db?

4. Define gain crossover.

5. Define phase crossover.

6. Define phase margin, PM.

7. Define gain margin, GM.

8. What Bode plot characteristic is the best indicator of the closed-loop step response overshoot?

9. What Bode plot characteristic is the best indicator of the closed-loop step response rise time?

10. A certain control system is required to follow sinusoids which may be any frequency in the range $0 \leq \omega_\ell \leq 450$ rad/sec and have amplitudes up to 5 units with (sinusoidal) steady-state error to be never more than 0.01. Sketch (or describe) the corresponding performance function $W_1(\omega)$.

11. What is the principal effect of a lead compensation on Bode plot performance measures?

12. What is the principal effect of a lag compensation on Bode plot performance measures?

13. How do you find the K_v of a type 1 system from its Bode plot?

14. Why do we need to know before hand the number of open-loop unstable poles in order to tell stability from the Nyquist plot?

15. What is the main advantage in control design of counting the encirclements of $-1/K$ of $D(j\omega)G(j\omega)$ rather than encirclements of $-1/K$ of $D(j\omega)G(j\omega)$?

16. Define a conditionally stable feedback system. How can you identify one on a Bode plot?

Problems

Problems for Section 6.1

6.1. **(a)** Show that α_0 in Eq. (6.2) is given by

$$\alpha_0 = \left[G(s) \frac{U_0 \omega}{s - j\omega} \right]\Bigg|_{s = -j\omega} = -U_0 G(-j\omega) \frac{1}{2j}$$

and

$$\alpha_0^* = \left[G(s) \frac{U_0 \omega}{s + j\omega} \right]\Bigg|_{s = +j\omega} = U_0 G(j\omega) \frac{1}{2j}.$$

(b) By assuming the output can be written as

$$y(t) = \alpha_0 e^{-j\omega t} + \alpha_0^* e^{j\omega t},$$

derive Eqs. (6.4)–(6.6).

6.2. **(a)** Calculate the magnitude and phase of

$$G(s) = \frac{1}{s + 10}$$

by hand for $\omega = 1, 2, 5, 10, 20, 50,$ and 100 rad/sec.

(b) Sketch the asymptotes for $G(s)$ according to the Bode plot rules, and compare these with your computed results from part (a).

6.3. Sketch the asymptotes of the Bode plot magnitude and phase for each of the following open-loop transfer functions. After completing the hand sketches verify your result using MATLAB. Turn in your hand sketches and the MATLAB results on the same scales.

(a) $L(s) = \dfrac{4000}{s(s+400)}$.

(b) $L(s) = \dfrac{100}{s(0.1s+1)(0.5s+1)}$.

(c) $L(s) = \dfrac{1}{s(s+1)(0.02s+1)}$.

(d) $L(s) = \dfrac{1}{(s+1)^2(s^2+2s+4)}$.

(e) $L(s) = \dfrac{10(s+4)}{s(s+1)(s^2+2s+5)}$.

(f) $L(s) = \dfrac{1000(s+0.1)}{s(s+1)(s^2+8s+64)}$.

(g) $L(s) = \dfrac{(s+5)(s+3)}{s(s+1)(s^2+s+4)}$.

(h) $L(s) = \dfrac{4s(s+10)}{(s+100)(4s^2+5s+4)}$.

(i) $L(s) = \dfrac{s}{(s+1)(s+10)(s^2+2s+2500)}$.

6.4. *Real poles and zeros.* Sketch the asymptotes of the Bode plot magnitude and phase for each of the following open-loop transfer functions. After completing the hand sketches verify your result using MATLAB. Turn in your hand sketches and the MATLAB results on the same scales.

(a) $L(s) = \dfrac{1}{s(s+1)(s+5)(s+10)}$.

(b) $L(s) = \dfrac{(s+2)}{s(s+1)(s+5)(s+10)}$.

(c) $L(s) = \dfrac{(s+2)(s+6)}{s(s+1)(s+5)(s+10)}$.

(d) $L(s) = \dfrac{(s+2)(s+4)}{s(s+1)(s+5)(s+10)}$.

6.5. *Complex poles and zeros.* Sketch the asymptotes of the Bode plot magnitude and phase for each of the following open-loop transfer functions and approximate the transition at the second order break point based on the value of the damping ratio. After completing the hand sketches verify your result using MATLAB. Turn in your hand sketches and the MATLAB results on the same scales.

(a) $L(s) = \dfrac{1}{s^2+3s+10}$.

(b) $L(s) = \dfrac{1}{s(s^2+3s+10)}$.

(c) $L(s) = \dfrac{(s^2+2s+8)}{s(s^2+2s+10)}$.

(d) $L(s) = \dfrac{(s^2+2s+12)}{s(s^2+2s+10)}$.

(e) $L(s) = \dfrac{(s^2+1)}{s(s^2+4)}$.

(f) $L(s) = \dfrac{(s^2+4)}{s(s^2+1)}$.

6.6.   *Multiple poles at the origin.* Sketch the asymptotes of the Bode plot magnitude and phase for each of the following open-loop transfer functions. After completing the hand sketches verify your result using MATLAB. Turn in your hand sketches and the MATLAB results on the same scales.

(a) $L(s) = \dfrac{1}{s^2(s+8)}$.

(b) $L(s) = \dfrac{1}{s^3(s+8)}$.

(c) $L(s) = \dfrac{1}{s^4(s+8)}$.

(d) $L(s) = \dfrac{(s+3)}{s^2(s+8)}$.

(e) $L(s) = \dfrac{(s+3)}{s^3(s+4)}$.

(f) $L(s) = \dfrac{(s+1)^2}{s^3(s+4)}$.

(g) $L(s) = \dfrac{(s+1)^2}{s^3(s+10)^2}$.

6.7.   *Mixed real and complex poles.* Sketch the asymptotes of the Bode plot magnitude and phase for each of the following open-loop transfer functions. Embellish the asymptote plots with a rough estimate of the transitions for each break point. After completing the hand sketches verify your result using MATLAB. Turn in your hand sketches and the MATLAB results on the same scales.

(a) $L(s) = \dfrac{(s+2)}{s(s+10)(s^2+2s+2)}$.

(b) $L(s) = \dfrac{(s+2)}{s^2(s+10)(s^2+6s+25)}$.

(c) $L(s) = \dfrac{(s+2)^2}{s^2(s+10)(s^2+6s+25)}$.

(d) $L(s) = \dfrac{(s+2)(s^2+4s+68)}{s^2(s+10)(s^2+4s+85)}$.

(e) $L(s) = \dfrac{[(s+1)^2+1]}{s^2(s+2)(s+3)}$.

6.8.   *Right half-plane poles and zeros.* Sketch the asymptotes of the Bode plot magnitude and phase for each of the following open-loop transfer functions. Make sure the phase asymptotes properly take the RHP singularity into account by sketching the complex plane to see how the $\angle L(s)$ changes as s goes from 0 to $+j\infty$. After completing the hand sketches verify your result using MATLAB. Turn in your hand sketches and the MATLAB results on the same scales.

(a) $L(s) = \dfrac{s+2}{s+10}\dfrac{1}{s^2-1}$; the model for a case of magnetic levitation with lead compensation.

(b) $L(s) = \dfrac{s+2}{s(s+10)}\dfrac{1}{(s^2-1)}$; the magnetic levitation system with integral control and lead compensation.

(c) $L(s) = \dfrac{s-1}{s^2}$.

(d) $L(s) = \dfrac{s^2+2s+1}{s(s+20)^2(s^2-2s+2)}$.

(e) $L(s) = \dfrac{(s+2)}{s(s-1)(s+6)^2}$.

(f) $L(s) = \dfrac{1}{(s-1)[(s+2)^2+3]}$.

6.9. A certain system is represented by the asymptotic Bode diagram shown in Fig. 6.89. Find and sketch the response of this system to a unit step input (assuming zero initial conditions).

Figure 6.89
Magnitude portion of Bode plot for Problem 6.9

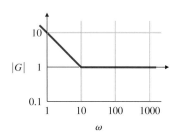

6.10. Prove that a magnitude slope of -1 in a Bode plot corresponds to -20 db per decade or -6 db per octave.

6.11. A normalized second-order system with a damping ratio $\zeta = 0.5$ and an additional zero is given by

$$G(s) = \frac{s/a+1}{s^2+s+1}.$$

Use MATLAB to compare the M_p from the step response of the system for $a = 0.01, 0.1, 1, 10$, and 100 with the M_r from the frequency response of each case. Is there a correlation between M_r and M_p?

6.12. A normalized second-order system with $\zeta = 0.5$ and an additional pole is given by

$$G(s) = \frac{1}{[(s/p)+1](s^2+s+1)}.$$

Draw Bode plots with $p = 0.01, 0.1, 1, 10$ and 100. What conclusions can you draw about the effect of an extra pole on the bandwidth compared to the bandwidth for the second-order system with no extra pole?

6.13. For the closed-loop transfer function

$$T(s) = \frac{\omega_n^2}{s^2+2\zeta\omega_n s+\omega_n^2},$$

derive the following expression for the bandwidth ω_{BW} of $T(s)$ in terms of ω_n and ζ:

$$\omega_{BW} = \omega_n\sqrt{1-2\zeta^2+\sqrt{2+4\zeta^4-4\zeta^2}}.$$

Assuming $\omega_n = 1$, plot ω_{BW} for $0 \leq \zeta \leq 1$.

6.14. Consider the system whose transfer function is

$$G(s) = \frac{A_0 \omega_0 s}{Q s^2 + \omega_0 s + \omega_0^2 Q}.$$

This is a model of a tuned circuit with *quality factor Q*.

(a) Compute the magnitude and phase of the transfer function analytically, and plot them for $Q = 0.5, 1, 2$, and 5 as a function of the normalized frequency ω/ω_0.

(b) Define the bandwidth as the distance between the frequencies on either side of ω_0 where the magnitude drops to 3 db below its value at ω_0 and show that the bandwidth is given by

$$\text{BW} = \frac{1}{2\pi} \left(\frac{\omega_0}{Q} \right).$$

(c) What is the relation between Q and ς?

6.15. A DC voltmeter schematic is shown in Fig. 6.90. The pointer is damped so that its maximum overshoot to a step input is 10%.

(a) What is the undamped natural frequency of the system?

(b) What is the damped natural frequency of the system?

(c) Plot the frequency response using MATLAB to determine what input frequency will produce the largest magnitude output?

(d) Suppose this meter is now used to measure a 1-V AC input with a frequency of 2 rad/sec. What amplitude will the meter indicate after initial transients have died out? What is the phase lag of the output with respect to the input? Use a Bode plot analysis to answer these questions. Use the lsim command in MATLAB to verify your answer in part (d).

Figure 6.90
Voltmeter schematic

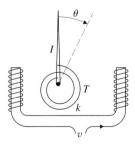

$I = 40 \times 10^{-6}\ \text{kg} \cdot \text{m}^2$
$k = 4 \times 10^{-6}\ \text{kg} \cdot \text{m}^2/\text{sec}^2$
$T = \text{input torque} = K_m v$
$v = \text{input voltage}$
$K_m = 1\ \text{N} \cdot \text{m/V}$

6.16. Determine the range of K for which each of the following systems is stable by making a Bode plot for $K = 1$ and imagining the magnitude plot sliding up or down until instability results. Verify your answers using a very rough sketch of a root-locus plot.

(a) $KG(s) = \dfrac{K(s+2)}{s+20}$.

(b) $KG(s) = \dfrac{K}{(s+10)(s+1)^2}$.

(c) $KG(s) = \dfrac{K(s+10)(s+1)}{(s+100)(s+5)^3}$.

6.17. Determine the range of K for which each of the following systems is stable by making a Bode plot for $K = 1$ and imagining the magnitude plot sliding up or down until instability results. Verify your answers using a very rough sketch of a root-locus plot.

(a) $KG(s) = \dfrac{K(s+1)}{s(s+5)}$.

(b) $KG(s) = \dfrac{K(s+1)}{s^2(s+10)}$.

(c) $KG(s) = \dfrac{K}{(s+2)(s^2+9)}$.

(d) $KG(s) = \dfrac{K(s+1)^2}{s^3(s+10)}$.

6.18. (a) Sketch the Nyquist plot for an open-loop system with transfer function $1/s^2$; that is, sketch

$$\left.\frac{1}{s^2}\right|_{s=C_1},$$

where C_1 is a contour enclosing the entire RHP, as shown in Fig. 6.17. (*Hint:* Assume C_1 takes a small detour around the poles at $s = 0$, as shown in Fig. 6.27.)

(b) Repeat part (a) for an open-loop system whose transfer function is $G(s) = 1/(s^2 + \omega_0^2)$.

6.19. Sketch the Nyquist plot based on the Bode plots for each of the following systems, then compare your result with that obtained using the MATLAB command nyquist:

(a) $KG(s) = \dfrac{K(s+2)}{s+10}$.

(b) $KG(s) = \dfrac{K}{(s+10)(s+2)^2}$.

(c) $KG(s) = \dfrac{K(s+10)(s+1)}{(s+100)(s+2)^3}$.

(d) Using your plots, estimate the range of K for which each system is stable, and qualitatively verify your result using a rough sketch of a root-locus plot.

6.20. Draw a Nyquist plot for

$$KG(s) = \frac{K(s+1)}{s(s+3)} \tag{6.82}$$

choosing the contour to be to the right of the singularity on the $j\omega$-axis and determine the range of K for which the system is stable using the Nyquist criterion. Then redo the Nyquist plot, this time choosing the contour to be to the left of the singularity on the imaginary axis and again check the range of K for which the system is stable using the Nyquist criterion. Are the answers the same? Should they be?

6.21. Draw the Nyquist plot for the system in Fig. 6.91. Using the Nyquist stability criterion, determine the range of K for which the system is stable. Consider both positive and negative values of K.

Figure 6.91
Control system for
Problem 6.21

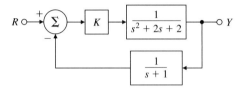

6.22. (a) For $\omega = 0.1$ to 100 rad/sec, sketch the phase of the minimum-phase system

$$G(s) = \frac{s+1}{s+10}\Big|_{s=j\omega}$$

and the nonminimum-phase system

$$G(s) = -\frac{s-1}{s+10}\Big|_{s=j\omega},$$

noting that $\angle(j\omega - 1)$ decreases with ω rather than increasing.

(b) Does a RHP zero affect the relationship between the -1 encirclements on a polar plot and the number of unstable closed-loop roots in Eq. (6.28)?

(c) Sketch the phase of the following unstable system for $\omega = 0.1$ to 100 rad/sec:

$$G(s) = \frac{s+1}{s-10}\Big|_{s=j\omega}.$$

(d) Check the stability of the systems in (a) and (c) using the Nyquist criterion on $KG(s)$. Determine the range of K for which the closed-loop system is stable, and check your results qualitatively using a rough root-locus sketch.

Problems for Section 6.4

6.23. The Nyquist plot for some actual control systems resembles the one shown in Fig. 6.92. What are the gain and phase margin(s) for the system of Fig. 6.92 given that $\alpha = 0.4$, $\beta = 1.3$, and $\phi = 40°$. Describe what happens to the stability of the system as the gain goes from zero to a very large value. Sketch what the corresponding root locus must look like for such a system. Also sketch what the corresponding Bode plots would look like for the system.

Figure 6.92
Nyquist plot for
Problem 6.23

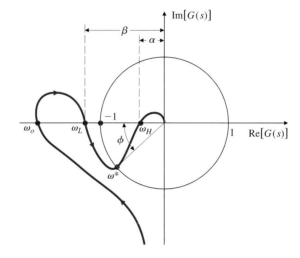

6.24. The Bode plot for

$$G(s) = \frac{100[(s/10) + 1]}{s[(s/1) - 1][(s/100) + 1]}$$

is shown in Fig. 6.93.

(a) Why does the phase start at 270° at the low frequencies?

(b) Sketch the Nyquist plot for $G(s)$.

(c) Is the closed-loop system derived from Fig. 6.93 stable?

(d) Will the system be stable if the gain is lowered by a factor of 100? Make a rough sketch of a root locus for the system and qualitatively confirm your answer.

6.25. Suppose that in Fig. 6.94,

$$G(s) = \frac{25(s + 1)}{s(s + 2)(s^2 + 2s + 16)}.$$

Use MATLAB's margin to calculate the PM and GM for $G(s)$ and, based on the Bode plots, conclude which margin would provide more useful information to the control designer for this system.

6.26. Consider the system given in Fig. 6.95.

(a) Use MATLAB to obtain Bode plots for $K = 1$ and use the plots to estimate the range of K for which the system will be stable.

(b) Verify the stable range of K by using margin to determine PM for selected values of K.

(c) Use rlocus and rlocfind to determine the values of K at the stability boundaries.

(d) Sketch the Nyquist plot of the system, and use it to verify the number of unstable roots for the unstable ranges of K.

(e) Using Routh's criterion, determine the ranges of K for closed-loop stability of this system.

Figure 6.93
Bode plot for Problem 6.24

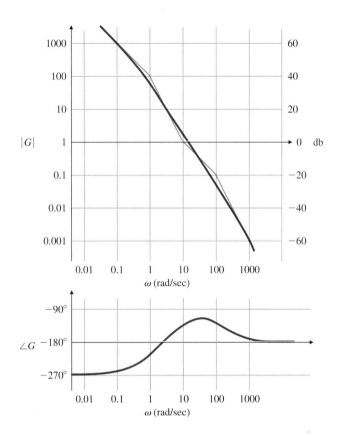

Figure 6.94
Control system for
Problem 6.25

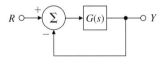

Figure 6.95
Control system for
Problem 6.26

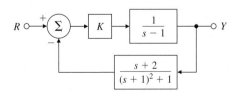

6.27. Suppose that in Fig. 6.94,

$$G(s) = \frac{3.2(s+1)}{s(s+2)(s^2 + 0.2s + 16)}.$$

Use MATLAB's margin to calculate the PM and GM for $G(s)$ and comment on whether you think this system will have well damped closed-loop roots.

6.28. For a given system, show that the ultimate period P_u and the corresponding ultimate gain K_u for the Zeigler-Nichols method can be found using the following:

(a) Nyquist diagram.

(b) Bode plot.

(c) Root locus.

6.29. If a system has the open-loop transfer function

$$G(s) = \frac{\omega_n^2}{s(s + 2\zeta\omega_n)}$$

with unity feedback, then the closed-loop transfer function is given by

$$T(s) = \frac{\omega_n^2}{s^2 + 2\zeta\omega_n s + \omega_n^2}.$$

Verify the values of the PM shown in Fig. 6.36 for $\zeta = 0.1, 0.4$, and 0.7.

6.30. Consider the unity feedback system with the open-loop transfer function

$$G(s) = \frac{K}{s(s + 1)[(s^2/25) + 0.4(s/5) + 1]}.$$

(a) Use MATLAB to draw the Bode plots for $G(j\omega)$ assuming $K = 1$.

(b) What gain K is required for a PM of $45°$? What is the GM for this value of K?

(c) What is K_v when the gain K is set for PM $= 45°$?

(d) Create a root locus with respect to K, and indicate the roots for a PM of $45°$.

6.31. For the system depicted in Fig. 6.96(a), the transfer-function blocks are defined by

$$G(s) = \frac{1}{(s + 2)^2(s + 4)} \quad \text{and} \quad H(s) = \frac{1}{s + 1}.$$

(a) Using rlocus and rlocfind, determine the value of K at the stability boundary.

(b) Using rlocus and rlocfind, determine the value of K that will produce roots with damping corresponding to $\zeta = 0.707$.

(c) What is the gain margin of the system if the gain is set to the value determined in part (b)? Answer this question *without* using any frequency response methods.

(d) Create the Bode plots for the system, and determine the gain margin that results for PM $= 65°$. What damping ratio would you expect for this PM?

(e) Sketch a root locus for the system shown in Fig. 6.96(b). How does it differ from the one in part (a)?

(f) For the systems in Figs. 6.96(a) and (b), how does the transfer function $Y_2(s)/R(s)$ differ from $Y_1(s)/R(s)$? Would you expect the step response to $r(t)$ be different for the two cases?

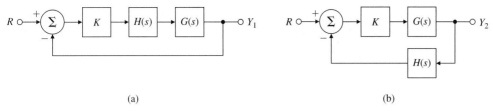

Figure 6.96 Block diagram for Problem 6.31. (a) unity feedback; (b) $H(s)$ in feedback

6.32. For the system shown in Fig. 6.97, use Bode and root-locus plots to determine the gain and frequency at which instability occurs. What gain (or gains) gives a PM of 20°? What is the gain margin when PM = 20°?

Figure 6.97
Control system for
Problem 6.32

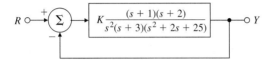

6.33. A magnetic tape-drive speed-control system is shown in Fig. 6.98. The speed sensor is slow enough that its dynamics must be included. The speed-measurement time constant is $\tau_m = 0.5$ sec; the reel time constant is $\tau_r = J/b = 4$ sec, where b = the output shaft damping constant = $1\,\text{N}\cdot\text{m}\cdot\text{sec}$; and the motor time constant is $\tau_1 = 1$ sec.

(a) Determine the gain K required to keep the steady-state speed error to less than 7% of the reference-speed setting.

(b) Determine the gain and phase margins of the system. Is this a good system design?

Figure 6.98
Magnetic tape-drive speed
control

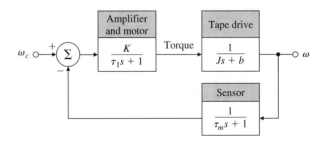

6.34. For the system in Fig. 6.99, determine the Nyquist plot and apply the Nyquist criterion

(a) to determine the range of values of K (positive and negative) for which the system will be stable and

(b) to determine the number of roots in the RHP for those values of K for which the system is unstable. Check your answer using a rough root-locus sketch.

Figure 6.99

Control system for
Problems 6.34, 6.61, and
6.62

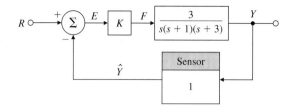

6.35. For the system shown in Fig. 6.100, determine the Nyquist plot and apply the
Nyquist criterion

(a) to determine the range of values of K (positive and negative) for which the
system will be stable and

(b) to determine the number of roots in the RHP for those values of K for which
the system is unstable. Check your answer using a rough root-locus sketch.

Figure 6.100

Control system for
Problem 6.35

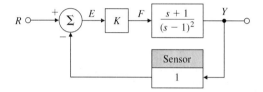

6.36. For the system shown in Fig. 6.101, determine the Nyquist plot and apply the
Nyquist criterion

(a) to determine the range of values of K (positive and negative) for which the
system will be stable and

(b) to determine the number of roots in the RHP for those values of K for which
the system is unstable. Check your answer using a rough root-locus sketch.

Figure 6.101

Control system for
Problem 6.36

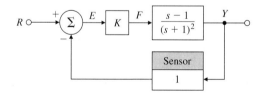

6.37. The Nyquist diagrams for two stable, open-loop systems are sketched in Fig. 6.102.
The proposed operating gain is indicated as K_0, and arrows indicate increasing
frequency. In each case give a rough estimate of the following quantities for the
closed-loop (unity feedback) system:

(a) Phase margin.

(b) Damping ratio.

(c) Range of gain for stability (if any).

(d) System type (0, 1, or 2).

Figure 6.102
Nyquist plots for
Problem 6.37

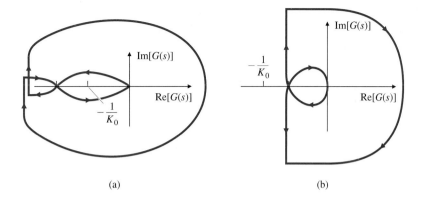

(a)   (b)

6.38. The steering dynamics of a ship are represented by the transfer function

$$\frac{V(s)}{\delta_r(s)} = G(s) = \frac{K[-(s/0.142) + 1]}{s(s/0.325 + 1)(s/0.0362) + 1)},$$

where v is the ship's lateral velocity in meters per second, and δ_r is the rudder angle in radians.

(a) Use the MATLAB command **bode** to plot the log magnitude and phase of $G(j\omega)$ for $K = 0.2$.

(b) On your plot, indicate the crossover frequency, PM, and GM.

(c) Is the ship steering system stable with $K = 0.2$?

(d) What value of K would yield a PM of $30°$ and what would the crossover frequency be?

6.39. For the open-loop system

$$KG(s) = \frac{K(s + 1)}{s^2(s + 10)^2},$$

determine the value for K at the stability boundary and the values of K at the points where PM $= 30°$.

Problems for Section 6.5

6.40. The frequency response of a plant in a unity feedback configuration is sketched in Fig. 6.103. Assume the plant is open-loop stable and minimum phase.

(a) What is the velocity constant K_v for the system as drawn?

(b) What is the damping ratio of the complex poles at $\omega = 100$?

(c) What is the PM of the system as drawn? (Estimate to within $\pm 10°$.)

Figure 6.103
Magnitude frequency
response for Problem 6.40

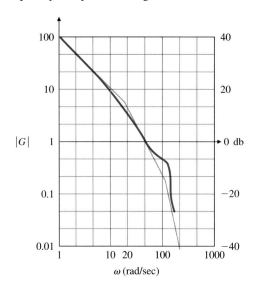

6.41. For the system

$$G(s) = \frac{100(s/a + 1)}{s(s + 1)(s/b + 1)},$$

where $b = 10a$, find the approximate value of a that will yield the best PM by sketching only candidate values of the frequency response magnitude.

Problem for Section 6.6

6.42. For the open-loop system

$$KG(s) = \frac{K(s + 1)}{s^2(s + 10)^2},$$

determine the value for K that will yield PM $\geq 30°$ and the maximum possible closed-loop bandwidth. Use MATLAB to find the bandwidth.

Problems for Section 6.7

6.43. For the lead compensator

$$D(s) = \frac{Ts + 1}{\alpha Ts + 1},$$

where $\alpha < 1$.

(a) Show that the phase of the lead compensator is given by

$$\phi = \tan^{-1}(T\omega) - \tan^{-1}(\alpha T\omega).$$

(b) Show that the frequency where the phase is maximum is given by

$$\omega_{max} = \frac{1}{T\sqrt{\alpha}},$$

and that the maximum phase corresponds to

$$\sin \phi_{max} = \frac{1 - \alpha}{1 + \alpha}.$$

(c) Rewrite your expression for ω_{max} to show that the maximum-phase frequency occurs at the geometric mean of the two corner frequencies on a logarithmic scale:

$$\log \omega_{max} = \frac{1}{2} \left(\log \frac{1}{T} + \log \frac{1}{\alpha T} \right).$$

(d) To derive the same results in terms of the pole-zero locations, rewrite $D(s)$ as

$$D(s) = \frac{s + z}{s + p},$$

and then show that the phase is given by

$$\phi = \tan^{-1} \left(\frac{\omega}{|z|} \right) - \tan^{-1} \left(\frac{\omega}{|p|} \right),$$

such that

$$\omega_{max} = \sqrt{|z||p|}.$$

Hence the frequency at which the phase is maximum is the square root of the product of the pole and zero locations.

6.44. For the third-order servo system

$$G(s) = \frac{50,000}{s(s + 10)(s + 50)}.$$

Design a lead compensator so that PM $\geq 50°$ and $\omega_{BW} \geq 20$ rad/sec using Bode plot sketches, then verify and refine your design using MATLAB.

6.45. For the system shown in Fig. 6.104, suppose that

$$G(s) = \frac{5}{s(s + 1)(s/5 + 1)}.$$

Design a lead compensation $D(s)$ with unity DC gain so that PM $\geq 40°$ using Bode plot sketches, then verify and refine your design using MATLAB. What is the approximate bandwidth of the system?

Figure 6.104
Control system for
Problem 6.45

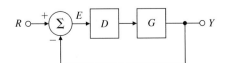

6.46. Derive the transfer function from T_d to θ for the system in Fig. 6.68. Then apply the Final Value Theorem (assuming T_d = constant) to determine whether $\theta(\infty)$ is nonzero for the following two cases:

(a) When $D(s)$ has no integral term, where $\lim_{s\to 0} D(s) =$ constant.

(b) When $D(s)$ has an integral term:

$$D(s) = \frac{D'(s)}{s},$$

where $\lim_{s\to 0} D'(s) =$ constant.

6.47. The inverted pendulum has a transfer function given by Eq. (2.35), which is similiar to

$$G(s) = \frac{1}{s^2 - 1}.$$

(a) Design a lead compensator to achieve a PM of $30°$ using Bode plot sketches, then verify and refine your design using MATLAB.

(b) Sketch a root locus and correlate it with the Bode plot of the system.

(c) Could you obtain the frequency response of this system experimentally?

6.48. The open-loop transfer function of a unity feedback system is

$$G(s) = \frac{K}{s(s/5 + 1)(s/50 + 1)}.$$

(a) Design a lag compensator for $G(s)$ using Bode plot sketches so that the closed-loop system satisfies the following specifications:

i. The steady-state error to a unit ramp reference input is less than 0.01.

ii. PM $\geq 40°$

(b) Verify and refine your design using MATLAB.

6.49. The open-loop transfer function of a unity feedback system is

$$G(s) = \frac{K}{s(s/5 + 1)(s/200 + 1)}.$$

(a) Design a lead compensator for $G(s)$ using Bode plot sketches so that the closed-loop system satisfies the following specifications:

i. The steady-state error to a unit ramp reference input is less than 0.01.

ii. For the dominant closed-loop poles the damping ratio $\zeta \geq 0.4$.

(b) Verify and refine your design using MATLAB including a direct computation of the damping of the dominant closed-loop poles.

6.50. A DC motor with negligible armature inductance is to be used in a position control system. Its open-loop transfer function is given by

$$G(s) = \frac{50}{s(s/5 + 1)}.$$

(a) Design a compensator for the motor using Bode plot sketches so that the closed-loop system satisfies the following specifications:

i. The steady-state error to a unit ramp input is less than 1/200.

ii. The unit step response has an overshoot of less than 20%.

iii. The bandwidth of the compensated system is no less than that of the uncompensated system.

(b) Verify and/or refine your design using MATLAB including a direct computation of the step response overshoot.

6.51. The open-loop transfer function of a unity feedback system is

$$G(s) = \frac{K}{s(1 + s/5)(1 + s/20)}.$$

(a) Sketch the system block diagram including input reference commands and sensor noise.

(b) Design a compensator for $G(s)$ using Bode plot sketches so that the closed-loop system satisfies the following specifications:

i. The steady-state error to a unit ramp input is less than 0.01.

ii. PM $\geq 45°$

iii. The steady-state error for sinusoidal inputs with $\omega < 0.2$ rad/sec is less than 1/250.

iv. Noise components introduced with the sensor signal at frequencies greater than 200 rad/sec are to be attenuated at the output by at least a factor of 100.

(c) Verify and/or refine your design using MATLAB including a computation of the closed-loop frequency response to verify (iv).

6.52. Consider a type I unity feedback system with

$$G(s) = \frac{K}{s(s + 1)}.$$

Design a lead compensator using Bode plot sketches so that $K_v = 20\,\text{sec}^{-1}$ and PM $> 40°$. Use MATLAB to verify and/or refine your design so that it meets the specifications.

6.53. Consider a satellite-attitude control system with the transfer function

$$G(s) = \frac{0.05(s + 25)}{s^2(s^2 + 0.1s + 4)}.$$

Amplitude-stabilize the system using lead compensation so that GM ≥ 2 (6 db), and PM $\geq 45°$, keeping the bandwidth as high as possible with a single lead.

6.54. In one mode of operation the autopilot of a jet transport is used to control altitude. For the purpose of designing the altitude portion of the autopilot loop, only the long-period airplane dynamics are important. The linearized relationship between altitude and elevator angle for the long-period dynamics is

$$G(s) = \frac{h(s)}{\delta(s)} = \frac{20(s + 0.01)}{s(s^2 + 0.01s + 0.0025)} \frac{\text{ft/sec}}{\text{deg}}.$$

Figure 6.105
Control system for
Problem 6.54

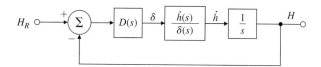

The autopilot receives from the altimeter an electrical signal proportional to altitude. This signal is compared with a command signal (proportional to the altitude selected by the pilot), and the difference provides an error signal. The error signal is processed through compensation, and the result is used to command the elevator actuators. A block diagram of this system is shown in Fig. 6.105. You have been given the task of designing the compensation. Begin by considering a proportional control law $D(s) = K$.

(a) Use MATLAB to draw a Bode plot of the open-loop system for $D(s) = K = 1$.

(b) What value of K would provide a crossover frequency (i.e., where $|G| = 1$) of 0.16 rad/sec?

(c) For this value of K, would the system be stable if the loop were closed?

(d) What is the PM for this value of K?

(e) Sketch the Nyquist plot of the system, and locate carefully any points where the phase angle is 180° or the magnitude is unity.

(f) Use MATLAB to plot the root locus with respect to K, and locate the roots for your value of K from part (b).

(g) What steady-state error would result if the command was a step change in altitude of 1000 ft?

For parts (h)and (i), assume a compensator of the form

$$D(s) = K \frac{Ts + 1}{\alpha Ts + 1}.$$

(h) Choose the parameters K, T, and α so that the crossover frequency is 0.16 rad/sec and the PM is greater that 50°. Verify your design by superimposing a Bode plot of $D(s)G(s)/K$ on top of the Bode plot you obtained for part (a), and measure the PM directly.

(i) Use MATLAB to plot the root locus with respect to K for the system including the compensator you designed in part (h). Locate the roots for your value of K from part (h).

(j) Altitude autopilots also have a mode where the rate of climb is sensed directly and commanded by the pilot.

   i. Sketch the block diagram for this mode.

   ii. Define the pertinent $G(s)$.

   iii. Design $D(s)$ so that the system has the same crossover frequency as the altitude hold mode and the PM is greater than 50°.

6.55. For a system with open-loop transfer function

$$G(s) = \frac{10}{s[(s/1.4) + 1][(s/3) + 1]},$$

design a lag compensator with unity DC gain so that PM $\geq 40°$. What is the approximate bandwidth of this system?

6.56. For the ship-steering system in Problem 6.38,

   (a) Design a compensator that meets the following specifications:

   i. Velocity constant $K_v = 2$.

   ii. PM $\geq 50°$.

   iii. Unconditional stability (PM > 0 for all $\omega \leq \omega_c$, the crossover frequency).

   (b) For your final design, draw a root locus with respect to K, and indicate the location of the closed-loop poles.

6.57. For a unity feedback system with

$$G(s) = \frac{1}{s\left(\dfrac{s}{20}+1\right)\left(\dfrac{s^2}{100^2}+0.5\dfrac{s}{100}+1\right)}. \tag{6.83}$$

   (a) A lead compensator is introduced with $\alpha = 1/5$ and a zero at $1/T = 20$. How must the gain be changed to obtain crossover at $\omega_c = 31.6$ rad/sec, and what is the resulting value of K_v?

   (b) With the lead compensator in place, what is the required value of K for a lag compensator that will readjust the gain to the original K_v value of 100?

   (c) Place the pole of the lag compensator at 3.16 rad/sec, and determine the zero location that will maintain the crossover frequency at $\omega_c = 31.6$ rad/sec. Plot the compensated frequency response on the same graph.

   (d) Determine the PM of the compensated design.

6.58. Golden Nugget Airlines had great success with their free bar near the tail of the airplane (see Problem 5.41). However, when they purchased a much larger airplane to handle the passenger demand, they discovered that there was some flexibility in the fuselage that caused a lot of unpleasant yawing motion at the rear of the airplane when in turbulence and was causing the revelers to spill their drinks. The approximate transfer function for the dutch roll mode (see Section 9.3.1) is

$$\frac{r(s)}{\delta_r(s)} = \frac{8.75(4s^2 + 0.4s + 1)}{(s/0.01 + 1)(s^2 + 0.24s + 1)},$$

where r is the airplane's yaw rate and δ_r is the rudder angle. In performing a finite element analysis (FEA) of the fuselage structure and adding those dynamics to the dutch roll motion, they found that the transfer function needed additional terms that reflected the fuselage lateral bending that occurred due to excitation from the rudder and turbulence. The revised transfer function is

$$\frac{r(s)}{\delta_r(s)} = \frac{8.75(4s^2 + 0.4s + 1)}{(s/0.01 + 1)(s^2 + 0.24s + 1)} \cdot \frac{1}{\left(\dfrac{s^2}{\omega_b^2} + 2\zeta\dfrac{s}{\omega_b} + 1\right)},$$

where ω_b is the frequency of the bending mode ($= 10$ rad/sec) and ζ is the bending mode damping ratio ($= 0.02$). Most swept wing airplanes have a "yaw damper" which essentially feeds back yaw rate measured by a rate gyro to the rudder with a simple proportional control law. For the new Golden Nugget airplane the proportional feedback gain, K, is 1, where

$$\delta_r(s) = -Kr(s). \tag{6.84}$$

(a) Make a Bode plot of the open-loop system, determine the PM and GM for the nominal design, and plot the step response and Bode magnitude of the closed-loop system. What is the frequency of the lightly damped mode that is causing the difficulty?

(b) Investigate remedies to quiet down the oscillations, but maintain the same low-frequency gain in order not to affect the quality of the dutch roll damping provided by the yaw rate feedback. Specifically, investigate one at a time:

   i. Increasing the damping of the bending mode from $\zeta = 0.02$ to $\zeta = 0.04$ (would require adding energy absorbing material in the fuselage structure).

   ii. Increasing the frequency of the bending mode from $\omega_b = 10$ rad/sec to $\omega_b = 20$ rad/sec (would require stronger and heavier structural elements).

   iii. Adding a low-pass filter in the feedback, that is, replace K in Eq. (6.84) with $KD(s)$ where

$$D(s) = \frac{1}{s/\tau_p + 1}. \tag{6.85}$$

   Pick τ_p so that the objectionable features of the bending mode are reduced while maintaing the PM $\geq 60°$.

   iv. Adding a notch filter as described in Section 5.5.3. Pick the frequency of the notch zero to be at ω_b with a damping of $\zeta = 0.04$ and pick the denominator poles to be $(s/100 + 1)^2$, keeping the DC gain of the filter equal to 1.

(c) Investigate the sensitivity of the two compensated designs above (iii and iv) by determining the effect of a reduction in the bending mode frequency of -10%. Specifically, reexamine the two designs by tabulating the GM, PM, closed-loop bending mode damping ratio, and resonant peak amplitude, and qualitatively describe the differences in the step response.

(d) What do you recommend to Golden Nugget to help their customers quit spilling their drinks? (Telling them to get back in their seats is not an acceptable answer for this problem! Make the recommendation in terms of improvements to the yaw damper.)

Problems for Section 6.8

6.59. A feedback control system is shown in Fig. 6.106. The closed-loop system is specified to have an overshoot of less than 30% to a step input.

(a) Determine the corresponding PM specification in the frequency domain and the corresponding closed-loop resonant peak value M_r (see Fig. 6.37).

(b) From Bode plots of the system, determine the maximum value of K that satisfies the PM specification.

Figure 6.106
Control system for
Problem 6.59

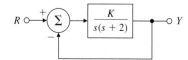

(c) Plot the data from the Bode plots [adjusted by the K obtained in part (b)] on a copy of the Nichols chart in Fig. 6.73 and determine the resonant peak magnitude M_r. Compare that with the approximate value obtained in part (a).

(d) Use the Nichols chart to determine the resonant peak frequency ω_r and the closed-loop bandwidth.

6.60. The Nichols plot of an uncompensated and a compensated system are shown in Fig. 6.107.

(a) What are the resonance peaks of each system?

(b) What are the PM and GM of each system?

(c) What are the bandwidths of each system?

(d) What type of compensation is used?

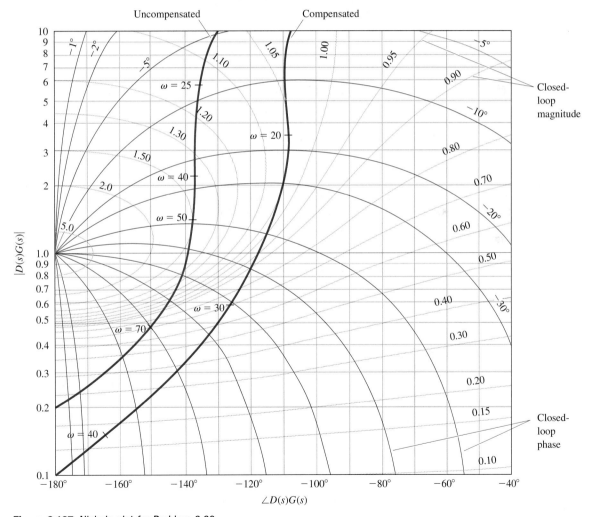

Figure 6.107 Nichols plot for Problem 6.60

6.61. Consider the system shown in Fig. 6.99.

(a) Construct an inverse Nyquist plot of $[Y(j\omega)/E(j\omega)]^{-1}$.

(b) Show how the value of K for neutral stability can be read directly from the inverse Nyquist plot.

(c) For $K = 4, 2$, and 1, determine the gain and phase margins.

(d) Construct a root-locus plot for the system, and identify corresponding points in the two plots. To what damping ratios ζ do the GM and PM of part (c) correspond?

6.62. An unstable plant has the transfer function

$$\frac{Y(s)}{F(s)} = \frac{s+1}{(s-1)^2}.$$

A simple control loop is to be closed around it, in the same manner as the block diagram in Fig. 6.99.

(a) Construct an inverse Nyquist plot of Y/F.

(b) Choose a value of K to provide a PM of $45°$. What is the corresponding GM?

(c) What can you infer from your plot about the stability of the system when $K < 0$?

(d) Construct a root-locus plot for the system, and identity corresponding points in the two plots. In this case, to what value of ζ does PM $= 45°$ correspond?

6.63. Consider the system shown in Fig. 6.108(a).

(a) Construct a Bode plot for the system.

(b) Use your Bode plot to sketch an inverse Nyquist plot.

(c) Consider closing a control loop around $G(s)$, as shown in Fig. 6.108(b). Using the inverse Nyquist plot as a guide, read from your Bode plot the values of GM and PM when $K = 0.7, 1.0, 1.4$, and 2. What value of K yields PM $= 30°$?

(d) Construct a root-locus plot, and label the same values of K on the locus. To what value of ζ does each pair of PM/GM values correspond? Compare the ζ versus PM with the rough approximation in Fig. 6.36.

Figure 6.108
Control system for
Problem 6.63

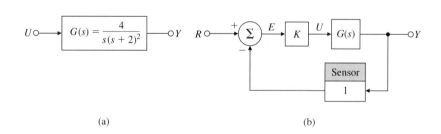

(a)   (b)

Problems for Section 6.9

6.64. Consider a system with the open-loop transfer function (loop gain)

$$G(s) = \frac{1}{s(s+1)(s/10+1)}.$$

(a) Create the Bode plot for the system, and find GM and PM.

(b) Compute the sensitivity sensitivity function and plot its magnitude frequency response.

(c) Compute the vector margin (VM).

6.65. Prove that the sensitivity function $S(s)$ has magnitude greater than 1 inside a circle with a radius of 1 centered at the -1 point. What does this imply about the shape of the Nyquist plot if closed-loop control is to outperform open-loop control at all frequencies?

6.66. Consider the system in Fig. 6.104 with the plant transfer function

$$G(s) = \frac{10}{s(s/10 + 1)}.$$

(a) We wish to design a compensator $D(s)$ that satisfies the following design specifications:

 i. $K_v = 100$.

 ii. PM $\geq 45°$.

 iii. Sinusoidal inputs of up to 1 rad/sec to be reproduced with $\leq 2\%$ error.

 iv. Sinusoidal inputs with a frequency of greater than 100 rad/sec to be attenuated at the output to $\leq 5\%$ of their input value.

(b) Create the Bode plot of $G(s)$, choosing the open-loop gain so that $K_v = 100$.

(c) Show that a *sufficient* condition for meeting the specification on sinusoidal inputs is that the magnitude plot lies outside the shaded regions in Fig. 6.109. Recall that

$$\frac{Y}{R} = \frac{KG}{1 + KG} \quad \text{and} \quad \frac{E}{R} = \frac{1}{1 + KG}.$$

(d) Explain why introducing a lead network alone cannot meet the design specifications.

(e) Explain why a lag network alone cannot meet the design specifications.

(f) Develop a full design using a lead–lag compensator that meets all the design specifications, without altering the previously chosen low frequency open-loop gain.

Figure 6.109
Control system constraints for Problem 6.66

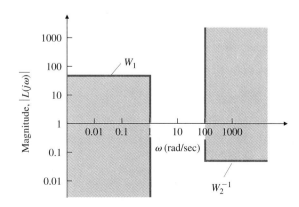

6.67. Assume that the system

$$G(s) = \frac{e^{-T_d s}}{s + 10},$$

has a 0.2-sec time delay ($T_d = 0.2$ sec). While maintaining a phase margin $\geq 40°$, find the maximum possible bandwidth using the following:

(a) One lead-compensator section

$$D(s) = K \frac{s + a}{s + b},$$

where $b/a = 100$;

(b) Two lead-compensator sections

$$D(s) = K \left(\frac{s + a}{s + b} \right)^2,$$

where $b/a = 10$.

(c) Comment on the statement in the text about the limitations on the bandwidth imposed by a delay.

6.68. Determine the range of K for which the following systems are stable:

(a) $G(s) = K \dfrac{e^{-4s}}{s}$.

(b) $G(s) = K \dfrac{e^{-s}}{s(s + 2)}$.

6.69. In Chapter 5, we used various approximations for the time delay, one of which is the first-order Padé

$$e^{-T_d s} \cong H_1(s) = \frac{1 - T_d s/2}{1 + T_d s/2}.$$

Using frequency response methods, the exact time delay

$$H_2(s) = e^{-T_d s}$$

can be used. Plot the phase of $H_1(s)$ and $H_2(s)$ and discuss the implications.

6.70. Consider the heat exchanger of Example 2.17 with the open-loop transfer function

$$G(s) = \frac{e^{-5s}}{(10s + 1)(60s + 1)}.$$

(a) Design a lead compensator that yields PM $\geq 45°$ and the maximum possible closed-loop bandwidth.

(b) Design a PI compensator that yields PM $\geq 45°$ and the maximum possible closed-loop bandwidth.

6.71. The Bode plot in Fig. 6.110 is for a transfer function of the form

$$G(s) = \frac{Ks}{(1 + T_1s)(1 + T_2s)^2},$$

where K, T_1, and T_2 are positive constants. Determine values for these three constants from the Bode plot.

Figure 6.110

Bode plot for Problem 6.71

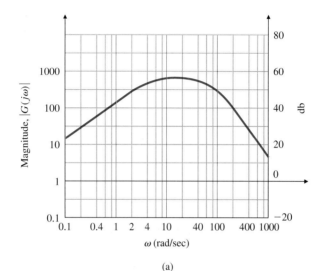

(a)

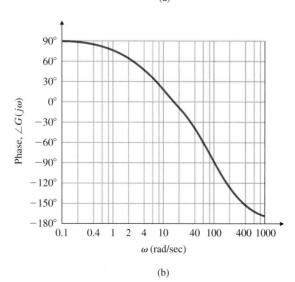

(b)

6.72. You are given the experimentally determined Bode plot shown in Fig. 6.111. Design a compensation that will yield a crossover frequency of $\omega_c = 10$ rad/sec with PM > 75°.

Figure 6.111
Bode plot for Problem 6.72

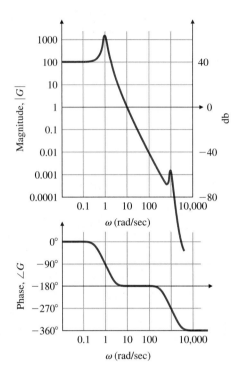

6.73. Consider the frequency-response plot in Fig. 6.112, which was obtained from a finite-element analysis. Determine a transfer function that approximately matches the measured frequency response up to 60 Hz.

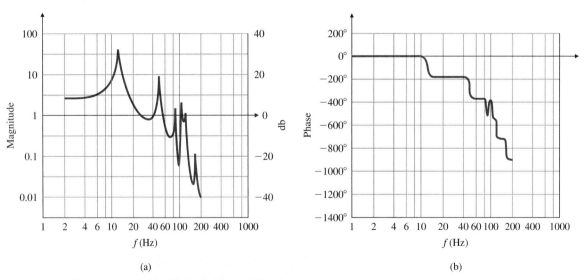

(a)   (b)

Figure 6.112 Frequency-response plot for Problem 6.73

6.74. The frequency response data shown in Table 6.1 were taken from a DC motor that is to be used in a position control system. Assume that the motor is linear and minimum-phase.

(a) Estimate $G(s)$, the transfer function of the system.

(b) Design a series compensator for the motor so that the closed-loop system meets the following specifications:

    i. The steady-state error to a unit ramp input is less than 0.01.

    ii. PM $\geq 45°$.

TABLE 6.1   **Frequency-Reponse Data for Problem 6.74**

ω (rad/sec)	$|G(s)|$ (db)	ω (rad/sec)	$|G(s)|$ (db)	ω (rad/sec)	$|G(s)|$ (db)
0.1	60.0	3.0	30.5	60.0	−20.0
0.2	54.0	4.0	27.0	65.0	−21.0
0.3	50.0	5.0	23.0	80.0	−24.0
0.5	46.0	7.0	19.5	100.0	−30.0
0.8	42.0	10.0	14.0	200.0	−48.0
1.0	40.0	20.0	2.0	300.0	−59.0
2.0	34.0	40.0	−10.0	500.0	−72.0

7 State-Space Design

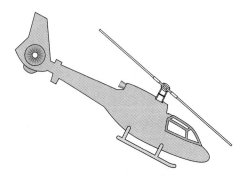

Chapter Overview

This chapter begins by considering the purposes and advantages of using state-space design. In Section 7.2 we review the development of state-variable equations from block diagrams. We then solve for the dynamic response, using state equations for both hand and computer analysis. Having covered these preliminary fundamentals, we next proceed to the major task of control system design via state space. The steps of the design method are as follows:

1. Select closed-loop pole (root as referred to in previous chapters) locations and develop the control law for the closed-loop system that corresponds to satisfactory dynamic response (Sections 7.3 and 7.4).

2. Design an estimator (Section 7.5).

3. Combine the control law and the estimator (Section 7.6).

4. Introduce the reference input (Sections 7.3.2 and 7.8).

After working through the central design steps, we briefly explore the use of integral control in state space (Section 7.9). The final three sections of this chapter consider briefly some additional concepts pertaining to the state-space method; because they are relatively advanced, they may be considered optional to some courses or readers.

A Perspective on State-Space Design

In addition to the transform techniques of root locus and frequency response, there is a third major method of designing feedback control systems: the state-space method. In Chapter 2 we introduced the state-variable method of describing differential equations. In state-space design, the control engineer designs a dynamic compensation by working directly with the state-variable description of the system. Like the transform techniques, the aim of the state-space method is to find a compensation $D(s)$, such as that shown in Fig. 7.1, that satisfies the design specifications. Because the state-space method of describing the plant and computing the compensation is so different from the transform techniques, it may seem at first to be solving an entirely different problem. We selected the examples and analysis given toward the end of this chapter to help convince you that, indeed, state-space design results in a compensator with a transfer function $D(s)$ that is equivalent to those $D(s)$ compensators obtained with the other two methods.

Because it is particularly well suited to the use of computer techniques, state-space design is increasingly studied and used today by control engineers.

Figure 7.1
A control system design definition

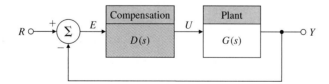

7.1 Advantages of State Space

The idea of **state space** comes from the state-variable method of describing differential equations, which we introduced in Chapter 2. In this method the differential equations describing a dynamic system are organized as a set of first-order differential equations in the vector-valued state of the system, and the solution is visualized as a trajectory of this state vector in space. **State-space control design** is the technique in which the control engineer designs a dynamic compensation by working directly with the state-variable description of the system. Thus far we have seen that the ordinary differential equations (ODEs) of physical dynamic systems can be manipulated into state-variable form. In the field of mathematics, where ODEs are studied, the state-variable form is called the **normal form** for the equations. There are several good reasons for studying equations in this form, three of which are listed here:

Normal form

- *To study more general models:* The ODEs do not have to be linear or stationary. Thus by studying the equations themselves, we can develop methods that are very general. Having them in state-variable form gives us a compact, standard form for study. Furthermore, the techniques of state-space analysis and design easily extend to systems with multiple inputs and/or multiple outputs. Of course, in this text we study mainly linear time-invariant models with single input and output (for the reasons given earlier).

- *To introduce the ideas of geometry into differential equations:* In physics the plane of position versus velocity of a particle or rigid body is called the **phase plane**, and the trajectory of the motion can be plotted as a curve in this plane. The state is a generalization of that idea to include more than two dimensions. While we cannot plot more than three dimensions, the concepts of distance and of orthogonal and parallel lines, as well as other concepts from geometry, can be useful in visualizing the solution of an ODE as a path in state space.

Phase plane

- *To connect internal and external descriptions:* The state of a dynamic system often directly describes the distribution of internal energy in the system. For example, it is common to select the following as state variables: position (potential energy), velocity (kinetic energy), capacitor voltage (electric energy), and inductor current (magnetic energy). The internal energy can always be computed from the state variables. By a system of analysis to be described shortly, we can relate the state to the system inputs and outputs and thus connect the internal variables to the external inputs and to the sensed outputs. In contrast, the transfer function relates only the input to the output and does not show the internal behavior. The state form keeps the latter information, which is sometimes important.

Use of the state-space approach has often been referred to as **modern control design**, and use of transfer-function-based methods such as root locus and frequency response has been referred to as **classical control design**. However, because the state-space method of description for ODEs has been in use for over 100 years and was introduced to control design in the late 1950s, it seems somewhat misleading to refer to it as modern. We prefer to refer to the two approaches to design as state-space methods and transform methods.

Advantages of state-space design are especially apparent when the system to be controlled has more than one control input or more than one sensed output. However, in this book we will examine the ideas of state-space design using the simpler single-input–single-output systems. The design approach used for systems described in state form is "divide and conquer." First we design the control as if all of the state were measured and available for use in the control law. This provides the possibility of assigning arbitrary dynamics for the system. Having a satisfactory control law based on full-state feedback, we introduce the concept of an observer and construct estimates of the state based on the sensed output. We then show that these estimates can be used in place of the actual state variables. Finally, we introduce the external reference-command inputs, and the structure is complete. Only at this point can we recognize that the resulting compensation has the same essential structure as that developed with transform methods.

Before we can begin the design using state descriptions, it is necessary to develop some analytical results and tools from matrix linear algebra for use throughout the chapter. We assume that you are familiar with elementary matrix concepts such as the identity matrix, triangular and diagonal matrices, and the transpose of a matrix. We also assume you have some familiarity with the mechanics of matrix algebra, including adding, multiplying, and inverting matrices. More advanced results will be developed in Section 7.2 in the context of the dynamic response of a linear system. All of the linear algebra results used in this chapter are repeated in Appendix C for your reference and review.

7.2   Analysis of the State Equations

In Chapter 2 we introduced and illustrated the process of selecting a state and organizing the equations in state form. In this section we review that process and describe how to analyze the dynamic response using the state description. In Section 7.2.1 we begin by relating the state description to block diagrams and the Laplace transform description and to consider the fact that for a given system the choice of state is not unique. We show how to use this nonuniqueness to select among several canonical forms for the one that will help solve the particular problem at hand; a control canonical form makes feedback gains of the state easy to design. After studying the structure of state equations in Section 7.2.2, we consider the dynamic response and show how transfer-function poles and zeros are related to the matrices of the state descriptions. To illustrate the results with hand calculations, we offer a simple example which represents the model of a thermal system. For more realistic examples, a computer-aided control systems design software package such as MATLAB is especially helpful; relevant MATLAB commands will be described from time to time.

7.2.1   Block Diagrams and Canonical Forms

We begin with a thermal system that has a simple transfer function

$$G = \frac{b(s)}{a(s)} = \frac{s+2}{s^2 + 7s + 12} = \frac{2}{s+4} + \frac{-1}{s+3}. \tag{7.1}$$

The roots of the numerator polynomial $b(s)$ are the zeros of the transfer function, and the roots of the denominator polynomial $a(s)$ are the poles. Notice that we have represented the transfer function in two forms, as a ratio of polynomials and as the result of a partial-fraction expansion. In order to develop a state description of this system (and this is a generally useful technique), we construct a block diagram that corresponds to the transfer function (and the differential equations) *using only isolated integrators as the dynamic elements*. One such block diagram, structured in **control canonical form**, is drawn in Fig. 7.2. The central feature of this structure is that each state variable is connected by the feedback to the control input.

Once we have drawn the block diagram in this form, we can identify the state description matrices simply by inspection; this is possible because when the output of an integrator is a state variable, the input of that integrator is the derivative of that variable. For example, in Fig. 7.2 the equation for the first state variable is

$$\dot{x}_1 = -7x_1 - 12x_2 + u.$$

Continuing in this fashion, we get

$$\dot{x}_2 = x_1,$$

$$y = x_1 + 2x_2.$$

These three equations can then be rewritten in the matrix form

$$\dot{\mathbf{x}} = \mathbf{A}_c\mathbf{x} + \mathbf{B}_c u, \tag{7.2}$$

$$y = \mathbf{C}_c\mathbf{x}, \tag{7.3}$$

Figure 7.2
A block diagram
representing Eq. (7.1) in
control form

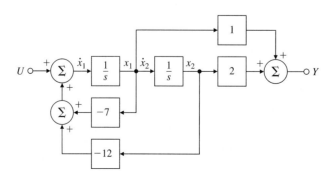

where

$$\mathbf{A}_c = \begin{bmatrix} -7 & -12 \\ 1 & 0 \end{bmatrix}, \qquad \mathbf{B}_c = \begin{bmatrix} 1 \\ 0 \end{bmatrix}, \tag{7.4a}$$

$$\mathbf{C}_c = \begin{bmatrix} 1 & 2 \end{bmatrix}, \qquad D_c = 0 \tag{7.4b}$$

and where the subscript c refers to control canonical form.

Two significant facts about this form are that the coefficients 1 and 2 of the numerator polynomial $b(s)$ appear in the $\mathbf{C}_c$ matrix, and (except for the leading term) the coefficients 7 and 12 of the denominator polynomial $a(s)$ appear (with opposite signs) as the first row of the $\mathbf{A}_c$ matrix. Armed with this knowledge, we can thus write down *by inspection* the state matrices in control canonical form for any system whose transfer function is known as a ratio of numerator and denominator polynomials. If $b(s) = b_1 s^{n-1} + b_2 s^{n-2} + \cdots + b_n$

and $a(s) = s^n + a_1 s^{n-1} + a_2 s^{n-2} + \cdots + a_n$, then the MATLAB steps are:

$$\mathsf{num} = \mathsf{b} = \begin{bmatrix} \mathsf{b_1} & \mathsf{b_2} & \cdots & \mathsf{b_n} \end{bmatrix}$$

$$\mathsf{den} = \mathsf{a} = \begin{bmatrix} 1 & \mathsf{a_1} & \mathsf{a_2} & \cdots & \mathsf{a_n} \end{bmatrix}$$

$$\begin{bmatrix} \mathsf{A_c}, & \mathsf{B_c}, & \mathsf{C_c}, & \mathsf{D_c} \end{bmatrix} = \mathsf{tf2ss(num, den)}.$$

We read tf2ss as "transfer function to state space." The result will be

$$\mathbf{A}_c = \begin{bmatrix} -a_1 & & -a_2 & \cdots & -a_n \\ 1 & & 0 & \cdots & 0 \\ \vdots & & & & \\ 0 & & & & \vdots \\ 0 & \cdots & 0 & 1 & 0 \end{bmatrix}, \qquad \mathbf{B}_c = \begin{bmatrix} 1 \\ 0 \\ \vdots \\ 0 \end{bmatrix}, \tag{7.5a}$$

$$\mathbf{C}_c = \begin{bmatrix} b_1 & b_2 & \cdots & b_n \end{bmatrix}, \qquad D_c = 0. \tag{7.5b}$$

The block diagram of Fig. 7.2 and the corresponding matrices of Eqs. (7.4a) and (7.4b) are not the only way to represent the transfer function $G(s)$. A block diagram corresponding to the partial-fraction expansion of $G(s)$ is given in Fig. 7.3. Using the same technique as before, with the state variables marked as shown on the figure, we can determine the matrices directly from the block diagram as being

$$\dot{\mathbf{z}} = \mathbf{A}_m \mathbf{z} + \mathbf{B}_m u,$$

$$y = \mathbf{C}_m \mathbf{z} + D_m u,$$

where

$$\mathbf{A}_m = \begin{bmatrix} -4 & 0 \\ 0 & -3 \end{bmatrix}, \qquad \mathbf{B}_m = \begin{bmatrix} 1 \\ 1 \end{bmatrix}, \tag{7.6a}$$

$$\mathbf{C}_m = \begin{bmatrix} 2 & -1 \end{bmatrix}, \qquad D_m = 0 \tag{7.6b}$$

Figure 7.3
Block diagram for Eq. (7.1) in modal canonical form

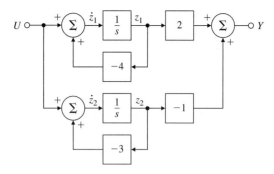

Modal form

and the subscript m refers to **modal canonical form**. The name for this form derives from the fact that the poles of the system transfer function are sometimes called the **normal modes** of the system. The important fact about the matrices in this form is that the system poles (here -4 and -3) appear as the elements along the diagonal of the $\mathbf{A}_m$ matrix, and the residues, the numerator terms in the partial-fraction expansion (here 2 and -1), appear in the $\mathbf{C}_m$ matrix.

Expressing a system in modal canonical form can be complicated by two factors: (1) The elements of the matrices will be complex when the poles of the system are complex, and (2) the system matrix *cannot* be diagonal when the partial-fraction expansion has repeated poles. To solve the first problem we express the complex poles of the partial-fraction expansion as conjugate pairs in second-order terms so that all the elements remain real. The corresponding $\mathbf{A}_m$ matrix will then have 2×2 blocks along the main diagonal representing the local coupling between the variables of the complex-pole set. To handle the second difficulty we also couple the corresponding state variables, so that the poles appear along the diagonal with off-diagonal terms indicating the coupling. A simple example of this latter case is the satellite system from Example 2.7, whose transfer function is $G(s) = 1/s^2$. The system matrices for this transfer function in a modal form are

$$\mathbf{F} = \begin{bmatrix} 0 & 1 \\ 0 & 0 \end{bmatrix}, \qquad \mathbf{G} = \begin{bmatrix} 0 \\ 1 \end{bmatrix}, \qquad \mathbf{H} = \begin{bmatrix} 1 & 0 \end{bmatrix}, \qquad J = 0. \qquad (7.7)$$

EXAMPLE 7.1   *State Equations in Modal Canonical Form*

A "quarter car model" [see Eq. 2.12] with one resonant mode has a transfer function given by

$$G(s) = \frac{2s + 4}{s^2(s^2 + 2s + 4)} = \frac{1}{s^2} - \frac{1}{s^2 + 2s + 4}. \qquad (7.8)$$

Find state matrices in modal form describing this system.

Solution.   The transfer function has been given in real partial-fraction form. To get state-description matrices, we draw a corresponding block diagram with integrators only, assign the state, and write down the corresponding matrices. This process is not

Figure 7.4
Block diagram for a
fourth-order system in
modal canonical form with
shading indicating portion in
control canonical form

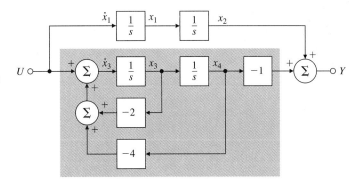

unique, so there are several acceptable solutions to the problem as stated, but they will differ in only trivial ways. A block diagram with a satisfactory assignment of variables is given in Fig. 7.4.

Notice that the second-order term to represent the complex poles has been realized in control canonical form. There are a number of other possibilities that can be used as alternatives for this part. This particular form allows us to write down the system matrices by inspection:

$$\mathbf{F} = \begin{bmatrix} 0 & 0 & 0 & 0 \\ 1 & 0 & 0 & 0 \\ 0 & 0 & -2 & -4 \\ 0 & 0 & 1 & 0 \end{bmatrix}, \qquad \mathbf{G} = \begin{bmatrix} 1 \\ 0 \\ 1 \\ 0 \end{bmatrix}, \qquad \mathbf{H} = \begin{bmatrix} 0 & 1 & 0 & -1 \end{bmatrix}, \qquad J = 0. \quad (7.9)$$

Thus far we have seen that we can obtain the state description from a transfer function in either control or modal form. Because these matrices represent the same dynamic system, we might ask: What is the relationship between the matrices in the two forms (and their corresponding state variables)? More generally, suppose we have a set of state equations that describe some physical system in no particular form and we are given a problem for which the control canonical form would be helpful. (We will see such a problem in Section 7.3.) Is it possible to calculate the desired canonical form without obtaining the transfer function first? To answer these questions requires a look at the topic of state transformations.

State description and output
equation

Consider a system described by the state equations

$$\dot{\mathbf{x}} = \mathbf{F}\mathbf{x} + \mathbf{G}u, \tag{7.10a}$$

$$y = \mathbf{H}\mathbf{x} + Ju. \tag{7.10b}$$

As we have seen, this is not a unique description of the dynamic system. We consider a change of state from $\mathbf{x}$ to a new state $\mathbf{z}$ that is a linear transformation of $\mathbf{x}$. For a nonsingular matrix $\mathbf{T}$ we let

$$\mathbf{x} = \mathbf{T}\mathbf{z}. \tag{7.11}$$

By substituting Eq. (7.11) into Eq. (7.10a), we have the equations of motion in terms of the new state $\mathbf{z}$:

$$\dot{\mathbf{x}} = \mathbf{T}\dot{\mathbf{z}} = \mathbf{F}\mathbf{T}\mathbf{z} + \mathbf{G}u, \tag{7.12a}$$

$$\dot{\mathbf{z}} = \mathbf{T}^{-1}\mathbf{F}\mathbf{T}\mathbf{z} + \mathbf{T}^{-1}\mathbf{G}u, \tag{7.12b}$$

$$\dot{\mathbf{z}} = \mathbf{A}\mathbf{z} + \mathbf{B}u, \tag{7.12c}$$

Transformation of state

where

$$\mathbf{A} = \mathbf{T}^{-1}\mathbf{F}\mathbf{T}, \tag{7.13a}$$

$$\mathbf{B} = \mathbf{T}^{-1}\mathbf{G}. \tag{7.13b}$$

Then we substitute Eq. (7.11) into Eq. (7.10b) to get the output in terms of the new state $\mathbf{z}$:

$$y = \mathbf{H}\mathbf{T}\mathbf{z} + Ju$$

$$= \mathbf{C}\mathbf{z} + Du,$$

where

$$\mathbf{C} = \mathbf{H}\mathbf{T}, \quad D = J. \tag{7.14}$$

Given the general matrices $\mathbf{F}$, $\mathbf{G}$, and $\mathbf{H}$ and scalar J, we would like to find the transformation matrix $\mathbf{T}$ such that $\mathbf{A}$, $\mathbf{B}$, $\mathbf{C}$, and D are in a particular form, for example, control canonical form. To find such a $\mathbf{T}$ we assume that $\mathbf{A}$, $\mathbf{B}$, $\mathbf{C}$, and D are already in the required form, further assume that the transformation $\mathbf{T}$ has a general form, and match terms. Here we will work out the third-order case; how to extend the analysis to the general case should be clear from the development. It goes like this.

First we rewrite Eq. (7.13a) as

$$\mathbf{A}\mathbf{T}^{-1} = \mathbf{T}^{-1}\mathbf{F}.$$

If $\mathbf{A}$ is in control canonical form, and we describe $\mathbf{T}^{-1}$ as a matrix with rows $\mathbf{t}_1$, $\mathbf{t}_2$, and $\mathbf{t}_3$, then

$$\begin{bmatrix} -a_1 & -a_2 & -a_3 \\ 1 & 0 & 0 \\ 0 & 1 & 0 \end{bmatrix} \begin{bmatrix} \mathbf{t}_1 \\ \mathbf{t}_2 \\ \mathbf{t}_3 \end{bmatrix} = \begin{bmatrix} \mathbf{t}_1\mathbf{F} \\ \mathbf{t}_2\mathbf{F} \\ \mathbf{t}_3\mathbf{F} \end{bmatrix}. \tag{7.15}$$

Working out the third and second rows gives the matrix equations

$$\mathbf{t}_2 = \mathbf{t}_3\mathbf{F}, \tag{7.16a}$$

$$\mathbf{t}_1 = \mathbf{t}_2\mathbf{F} = \mathbf{t}_3\mathbf{F}^2. \tag{7.16b}$$

From Eq. (7.13b), assuming that $\mathbf{B}$ is also in control canonical form, we have the relation

$$\mathbf{T}^{-1}\mathbf{G} = \mathbf{B},$$

or

$$\begin{bmatrix} \mathbf{t}_1\mathbf{G} \\ \mathbf{t}_2\mathbf{G} \\ \mathbf{t}_3\mathbf{G} \end{bmatrix} = \begin{bmatrix} 1 \\ 0 \\ 0 \end{bmatrix}. \tag{7.17}$$

Combining Eqs. (7.16) and (7.17), we get

$$\mathbf{t}_3\mathbf{G} = 0,$$

$$\mathbf{t}_2\mathbf{G} = \mathbf{t}_3\mathbf{F}\mathbf{G} = 0,$$

$$\mathbf{t}_1\mathbf{G} = \mathbf{t}_3\mathbf{F}^2\mathbf{G} = 1.$$

These equations can in turn be written in matrix form as

$$\mathbf{t}_3[\,\mathbf{G} \quad \mathbf{FG} \quad \mathbf{F}^2\mathbf{G}\,] = [\,0 \quad 0 \quad 1\,]$$

or

$$\mathbf{t}_3 = [\,0 \quad 0 \quad 1\,]\mathcal{C}^{-1}, \tag{7.18}$$

Controllability matrix transformation to control canonical form

where the **controllability matrix** $\mathcal{C} = [\,\mathbf{G} \quad \mathbf{FG} \quad \mathbf{F}^2\mathbf{G}\,]$. Having $\mathbf{t}_3$, we can now go back to Eq. (7.16) and construct all the rows of $\mathbf{T}^{-1}$.

To sum up, the recipe for converting a general state description of dimension n to control canonical form is as follows:

- From $\mathbf{F}$ and $\mathbf{G}$, form the controllability matrix $\mathcal{C}$ as

$$\mathcal{C} = [\,\mathbf{G} \quad \mathbf{FG} \quad \cdots \quad \mathbf{F}^{n-1}\mathbf{G}\,]. \tag{7.19}$$

- Compute the last row of the inverse of the transformation matrix as

$$\mathbf{t}_n = [\,0 \quad 0 \quad \cdots \quad 1\,]\mathcal{C}^{-1}. \tag{7.20}$$

- Construct the entire transformation matrix as

$$\mathbf{T}^{-1} = \begin{bmatrix} \mathbf{t}_n\mathbf{F}^{n-1} \\ \mathbf{t}_n\mathbf{F}^{n-2} \\ \vdots \\ \mathbf{t}_n \end{bmatrix}. \tag{7.21}$$

- Compute the new matrices from $\mathbf{T}^{-1}$ using Eqs. (7.13) and (7.14).

When the controllability matrix $\mathcal{C}$ is nonsingular, the corresponding **F** and **G** matrices are said to be **controllable**. This is a technical property that usually holds for physical systems and will be important when we consider feedback of the state in Section 7.3. We will also consider a few physical illustrations of loss of controllability at that time.

Because computing the transformation given by Eq. (7.21) is numerically difficult to do accurately, it is almost never done. The reason for developing this transformation in some detail is to show how such changes of state could be done in theory and to make the following important observation:

> One can *always* transform a given state description to control canonical form if (and only if) the controllability matrix $\mathcal{C}$ is nonsingular.

If we need to test for controllability in a real case with numbers, we use a numerically stable method that depends on converting the system matrices to "staircase" form rather than on trying to compute the controllability matrix. Problem 7.27 at the end of the chapter calls for consideration of this method.

An important question regarding controllability follows directly from our discussion so far: What is the effect of a state transformation on controllability? We can show the result by using Eqs. (7.19), (7.13a), and (7.13b). The controllability matrix of the system $(\mathbf{F}, \mathbf{G})$ is

$$\mathcal{C}_{\mathbf{x}} = [\mathbf{G} \quad \mathbf{FG} \quad \cdots \quad \mathbf{F}^{n-1}\mathbf{G}]. \tag{7.22}$$

After the state transformation, the new description matrices are given by Eqs. (7.13a) and (7.13b), and the controllability matrix changes to

$$\mathcal{C}_{\mathbf{z}} = [\mathbf{B} \quad \mathbf{AB} \quad \cdots \quad \mathbf{A}^{n-1}\mathbf{B}] \tag{7.23a}$$

$$= [\mathbf{T}^{-1}\mathbf{G} \quad \mathbf{T}^{-1}\mathbf{FTT}^{-1}\mathbf{G} \quad \cdots \quad \mathbf{T}^{-1}\mathbf{F}^{n-1}\mathbf{TT}^{-1}\mathbf{G}] \tag{7.23b}$$

$$= \mathbf{T}^{-1}\mathcal{C}_{\mathbf{x}}. \tag{7.23c}$$

Thus we see that $\mathcal{C}_{\mathbf{z}}$ is nonsingular if and only if $\mathcal{C}_{\mathbf{x}}$ is nonsingular, yielding the following observation:

> A change of state by a nonsingular linear transformation does *not* change controllability.

We return once again to the transfer function of Eq. (7.1), this time to represent it with the block diagram having the structure known as **observer canonical form** (Fig. 7.5). The corresponding matrices for this form are

$$\mathbf{A}_o = \begin{bmatrix} -7 & 1 \\ -12 & 0 \end{bmatrix}, \qquad \mathbf{B}_o = \begin{bmatrix} 1 \\ 2 \end{bmatrix}, \tag{7.24a}$$

$$\mathbf{C}_o = [1 \quad 0], \qquad D_o = 0. \tag{7.24b}$$

The significant fact about this canonical form is that all the feedback is from the output to the state variables.

Figure 7.5
Observer canonical form

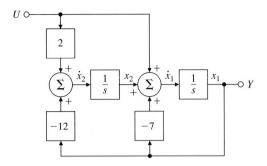

Let us now consider what happens to the controllability of this system as the zero at -2 is varied. For this purpose we replace the second element 2 of $\mathbf{B}_o$ with the variable zero location $-z_o$ and form the controllability matrix:

$$\mathcal{C}_\mathbf{x} = [\,\mathbf{B}_o \quad \mathbf{A}_o\mathbf{B}_o\,] \tag{7.25a}$$

$$= \begin{bmatrix} 1 & -7 - z_o \\ -z_o & -12 \end{bmatrix}. \tag{7.25b}$$

The determinant of this matrix is a function of z_o:

$$\det(\mathcal{C}_\mathbf{x}) = -12 + (z_o)(-7 - z_o)$$

$$= -(z_o^2 + 7z_o + 12).$$

This polynomial is zero for $z_o = -3$ or -4, implying that controllability is lost for these values. What does this mean? In terms of the parameter z_o, the transfer function is

$$G(s) = \frac{s - z_o}{(s + 3)(s + 4)}.$$

If $z_0 = -3$ or -4, there is a pole-zero cancellation and the transfer function reduces from a second-order system to a first-order one. When $z_o = -3$, for example, the mode at -3 is decoupled from the input and control of this mode is lost.

Notice that we have taken the transfer function given by Eq. (7.1) and have given it two realizations, one in control canonical form and one in observer canonical form. The control form is always controllable for any value of the zero, while the observer form loses controllability if the zero cancels either of the poles. Thus, these two forms may represent the same transfer function, but it may not be possible to transform the state of one to the state of the other (in this case, from observer to control canonical form). While a transformation of state cannot affect controllability, the particular state selected from a transfer function can:

Controllability is a function of the *state* of the system and cannot be decided from a transfer function.

To discuss controllability more at this point would take us too far afield. The closely related property of observability and the observer canonical form will be taken up in Section 7.5.1. A more detailed discussion of these properties of dynamic systems is given in the Appendix D, for those who would like to learn more.

We return now to the modal form for the equations, given by Eqs. (7.6a) and (7.6b) for the example transfer function. As mentioned before, it is not always possible to find a modal form for transfer functions that have repeated poles, so we assume our system has only distinct poles. Furthermore, we assume that the general state equations given by Eqs. (7.10a) and (7.10b) apply. We want to find a transformation matrix $\mathbf{T}$ defined by Eq. (7.11) such that the transformed Eqs. (7.13) and (7.14) will be in modal form. In this case, we assume that the $\mathbf{A}$ matrix is diagonal and that $\mathbf{T}$ is composed of the *columns* $\mathbf{t}_1$, $\mathbf{t}_2$, and $\mathbf{t}_3$. With this assumption, the state transformation Eq. (7.13a) becomes

$$\mathbf{TA} = \mathbf{FT}$$

$$[\,\mathbf{t}_1 \quad \mathbf{t}_2 \quad \mathbf{t}_3\,] \begin{bmatrix} p_1 & 0 & 0 \\ 0 & p_2 & 0 \\ 0 & 0 & p_3 \end{bmatrix} = \mathbf{F}[\,\mathbf{t}_1 \quad \mathbf{t}_2 \quad \mathbf{t}_3\,]. \tag{7.26}$$

Transformation to modal form

Equation (7.26) is equivalent to the three vector–matrix equations

$$p_i \mathbf{t}_i = \mathbf{Ft}_i, \qquad i = 1, 2, 3. \tag{7.27}$$

Eigenvectors
Eigenvalues

In matrix algebra, Eq. (7.27) is a famous equation whose solution is known as the **eigenvector/eigenvalue problem**. Recall that $\mathbf{t}_i$ is a vector, $\mathbf{F}$ is a matrix, and p_i is a scalar. The vector $\mathbf{t}_i$ is called an **eigenvector** of $\mathbf{F}$, and p_i is called the corresponding **eigenvalue**. Because we saw earlier that the modal form is equivalent to a partial-fraction-expansion representation with the system poles along the diagonal of the state matrix, it should be clear that these "eigenvalues" are precisely the poles of our system. The transformation matrix that will convert the state description matrices to modal form has as its columns the eigenvectors of $\mathbf{F}$ as shown in Eq. (7.26) for the third-order case. As it happens, there are robust, reliable computer algorithms to compute eigenvalues and the eigenvectors of quite large systems using the QR algorithm.[1] In MATLAB the command p = eig(F) is the way to compute the poles if the system equations are in state form.

MATLAB eig

Notice also that Eq. (7.27) is homogeneous in that, if $\mathbf{t}_i$ is an eigenvector, so is $\alpha \mathbf{t}_i$ for any scalar α. In most cases the scale factor is selected so that the length (square root of the sum of squares of the magnitudes of the elements) is unity. MATLAB will perform this operation. Another option is to select the scale factors so that the input matrix $\mathbf{B}$ is composed of all 1's. The latter choice

---

[1] This algorithm is part of MATLAB and all other well-known computer-aided design packages. It is carefully documented in the software package LAPACK (Anderson et al., 1999). See also Strang (1988).

is suggested by a partial-fraction expansion with each part realized in control canonical form. If the system is real, then each element of $\mathbf{F}$ is real, and if $p = \sigma + j\omega$ is a pole, so is the conjugate, $p^* = \sigma - j\omega$. For these eigenvalues the eigenvectors are also complex and conjugate. It is possible to compose the transformation matrix using the real and complex parts of the eigenvectors separately so the modal form is real but has 2×2 blocks for each pair of complex poles. Later we will see the result of the MATLAB function that does this, but first let us look at the simple real-poles case.

EXAMPLE 7.2

Transformation of Thermal System from Control to Modal Form

Find the matrix to transform the control form matrices in Eq. (7.4) into the modal form of Eq. (7.6).

Solution.   According to Eqs. (7.26) and (7.27), we need first to find the eigenvectors and eigenvalues of the $\mathbf{A}_c$ matrix. We take the eigenvectors to be

$$\begin{bmatrix} t_{11} \\ t_{21} \end{bmatrix} \quad \text{and} \quad \begin{bmatrix} t_{12} \\ t_{22} \end{bmatrix}.$$

The equations using the eigenvector on the left are:

$$\begin{bmatrix} -7 & -12 \\ 1 & 0 \end{bmatrix} \begin{bmatrix} t_{11} \\ t_{21} \end{bmatrix} = p \begin{bmatrix} t_{11} \\ t_{21} \end{bmatrix}, \tag{7.28a}$$

$$-7t_{11} - 12t_{21} = pt_{11}, \tag{7.28b}$$

$$t_{11} = pt_{21}. \tag{7.28c}$$

Substituting Eq. (7.28c) into Eq. (7.28b) results in

$$-7pt_{21} - 12t_{21} = p^2 t_{21}, \tag{7.29a}$$

$$p^2 t_{21} + 7pt_{21} + 12t_{21} = 0, \tag{7.29b}$$

$$p^2 + 7p + 12 = 0, \tag{7.29c}$$

$$p = -3, -4. \tag{7.29d}$$

We have found (again!) that the eigenvalues (poles) are -3 and -4; furthermore, Eq. (7.28c) tells us that the two eigenvectors are

$$\begin{bmatrix} -4t_{21} \\ t_{21} \end{bmatrix} \quad \text{and} \quad \begin{bmatrix} -3t_{22} \\ t_{22} \end{bmatrix},$$

where t_{21} and t_{22} are arbitrary nonzero scale factors. We want to select the two scale factors such that both elements of $\mathbf{B}_m$ in Eq. (7.6a) are unity. The equation for $\mathbf{B}_m$ in terms of $\mathbf{B}_c$ is $\mathbf{T}\mathbf{B}_m = \mathbf{B}_c$, and its solution is $t_{21} = -1$ and $t_{22} = 1$. Therefore, the transformation matrix and its inverse[2] are

$$\mathbf{T} = \begin{bmatrix} 4 & -3 \\ -1 & 1 \end{bmatrix}, \qquad \mathbf{T}^{-1} = \begin{bmatrix} 1 & 3 \\ 1 & 4 \end{bmatrix}. \tag{7.30}$$

Elementary matrix multiplication shows that, using $\mathbf{T}$ as defined by Eq. (7.30), the matrices of Eqs. (7.4) and (7.6) are related as follows:

$$\mathbf{A}_m = \mathbf{T}^{-1}\mathbf{A}_c\mathbf{T}, \qquad \mathbf{B}_m = \mathbf{T}^{-1}\mathbf{B}_c, \tag{7.31}$$

$$\mathbf{C}_m = \mathbf{C}_c\mathbf{T}, \qquad D_m = D_c,$$

These computations can be carried out using the MATLAB statements,

```
T = [4 −3;-1 1];
Am = inv(T)*Ac*T;
Bm = inv(T)*Bc;
Cm = Cc*T;
Dm = Dc;
```

The next example has five state variables and in state-variable form is too complicated for hand calculations. However, it is a good example for illustrating the use of computer software designed for the purpose. The model we will use is based on a physical state after amplitude and time scaling have been done.

EXAMPLE 7.3

Using MATLAB to Find Poles and Zeros of Tape-Drive System

Find the eigenvalues of the system matrix for the tape-drive control shown below. Also compute the transformation of the equations of the tape drive in their given form to modal canonical form. The system matrices are

$$\mathbf{F} = \begin{bmatrix} 0 & 2 & 0 & 0 & 0 \\ -0.1 & -0.35 & 0.1 & 0.1 & 0.75 \\ 0 & 0 & 0 & 2 & 0 \\ 0.4 & 0.4 & -0.4 & -1.4 & 0 \\ 0 & -0.03 & 0 & 0 & -1 \end{bmatrix}, \qquad \mathbf{G} = \begin{bmatrix} 0 \\ 0 \\ 0 \\ 0 \\ 1 \end{bmatrix},$$

$\mathbf{H}_2 = [\,0.0 \quad 0.0 \quad 1.0 \quad 0.0 \quad 0.0\,]$ Servomotor position output,

$\mathbf{H}_3 = [\,0.5 \quad 0.0 \quad 0.5 \quad 0.0 \quad 0.0\,]$ Position at read/write head as output,

$\mathbf{H}_T = [\,-0.2 \quad -0.2 \quad 0.2 \quad 0.2 \quad 0.0\,]$ tension output,

$J = 0.0.$

---

[2] To find the inverse of a 2×2 matrix, you need only interchange the elements subscripted "11" and "22", change the signs of the "12" and the "21" elements, and divide by the determinant [$= 1$ in Eq. (7.30)].

The state vector is defined as

$$\mathbf{x} = \begin{bmatrix} x_1 \text{ (tape position at capstan)} \\ \omega_1 \text{ (speed of the drive wheel)} \\ x_3 \text{ (position of the tape at the head)} \\ \omega_2 \text{ (output speed)} \\ i \text{ (current into capstan motor)} \end{bmatrix}.$$

The matrix $\mathbf{H}_3$ corresponds to making x_3 (the position of the tape over the read/write head) the output, and the matrix $\mathbf{H}_T$ corresponds to making tension the output.

Solution.   To compute the eigenvalues using MATLAB, we write

$$\mathsf{P} = \mathsf{eig}(\mathsf{F}),$$

which results in

$$\mathsf{P} = \begin{bmatrix} -0.6371 + 0.6669i \\ -0.6371 - 0.6669i \\ 0.0000 \\ -0.5075 \\ -0.9683 \end{bmatrix}.$$

Notice that the system has all poles in the left half-plane (LHP) except for one pole at the origin. This means that a step input will result in a ramp output, so we conclude the system has type 1 behavior.

MATLAB canon   To transform to modal form, we use the MATLAB function canon:

$$\mathsf{sysG} = \mathsf{ss}(\mathsf{F}, \mathsf{G}, \mathsf{H3}, \mathsf{J})$$
$$[\mathsf{Am}, \quad \mathsf{Bm}, \quad \mathsf{Cm}, \quad \mathsf{Dm}, \quad \mathsf{TI}] = \mathsf{canon}(\mathsf{sysG}, \,'\mathsf{modal}')$$

The result of this calculation is

$$\mathsf{Am} = \mathbf{A}_m = \begin{bmatrix} -0.6371 & 0.6669 & 0.0000 & 0.0000 & 0.0000 \\ -0.6669 & -0.6371 & 0.0000 & 0.0000 & 0.0000 \\ 0.0000 & 0.0000 & -0.0000 & 0.0000 & 0.0000 \\ 0.0000 & 0.0000 & 0.0000 & -0.5075 & 0.0000 \\ 0.0000 & 0.0000 & 0.0000 & 0.0000 & -0.9683 \end{bmatrix}.$$

Notice that the complex poles appear in the 2×2 block in the upper left corner of $\mathbf{A}_m$, and the real poles fall on the main diagonal of this matrix. The rest of the calculations from canon are

$$\mathsf{Bm} = \mathbf{B}_m = \begin{bmatrix} 3.0868 \\ 1.2388 \\ 4.0599 \\ 9.9016 \\ 3.9341 \end{bmatrix}, \qquad \mathsf{Cm} = \mathbf{C}_m = [\,0.3268 \quad 0.2190 \quad 0.7071 \quad -0.6625 \quad 0.6124\,],$$

$$\mathsf{Dm} = D_m = 0, \quad \mathsf{TI} = \mathbf{T}^{-1} = \begin{bmatrix} -1.1369 & 2.5951 & 1.1369 & 0.4764 & 3.0868 \\ -1.1895 & -2.1453 & 1.1895 & 3.2536 & 1.2388 \\ 0.7376 & 5.4133 & 0.6767 & 1.3533 & 4.0599 \\ -0.9227 & 6.5022 & 0.9227 & 2.7962 & 9.9016 \\ -0.0014 & 0.1663 & 0.0014 & 0.0449 & 3.9341 \end{bmatrix}.$$

It happens that canon was written to compute the *inverse* of the transformation we are working with (as you can see from TI above), so we need to invert our MATLAB results. The inverse is calculated from

$$T = inv(TI)$$

and results in

$$T = \mathbf{T} = \begin{bmatrix} -0.1748 & 0.1420 & 0.7071 & -0.4871 & 0.5887 \\ 0.0083 & -0.1035 & 0.0000 & 0.1236 & -0.2850 \\ 0.8284 & 0.2960 & 0.7071 & -0.8379 & 0.6360 \\ -0.3626 & 0.1820 & -0.0000 & 0.2126 & -0.3079 \\ 0.0034 & 0.0022 & -0.0000 & -0.0075 & 0.2697 \end{bmatrix}.$$

The eigenvectors computed with [V,P]=eig(F) are

$$V = \mathbf{V} = \begin{bmatrix} -0.1748 + 0.1420i & -0.1748 - 0.1420i & 0.7071 & -0.4871 & 0.5887 \\ 0.0083 - 0.1035i & 0.0083 + 0.1035i & 0.0000 & 0.1236 & -0.2850 \\ 0.8284 + 0.2960i & 0.8284 - 0.2960i & 0.7071 & -0.8379 & 0.6360 \\ -0.3626 + 0.1820i & -0.3626 - 0.1820i & -0.0000 & 0.2126 & -0.3079 \\ 0.0034 + 0.0022i & 0.0034 - 0.0022i & 0.0000 & -0.0075 & 0.2697 \end{bmatrix}.$$

Notice that the first two columns of the real transformation $\mathbf{T}$ are composed of the real and the imaginary parts of the first eigenvector in the first column of $\mathbf{V}$. It is this step that causes the complex roots to appear in the 2×2 block in the upper left of the $\mathbf{A}_m$ matrix. The vectors in $\mathbf{V}$ are normalized to unit length, which results in non-normalized values in $\mathbf{B}_m$ and $\mathbf{C}_m$. If we found it desirable to do so, we could readily find further transformations to make each element of $\mathbf{B}_m$ equal 1 or to interchange the order in which the poles appear.

7.2.2   Dynamic Response from the State Equations

Having considered the structure of the state-variable equations, we now turn to finding the dynamic response from the state description and to the relationships between the state description and our earlier discussion in Chapter 6 of the frequency response and poles and zeros. Let us begin with the general equations of state given by Eqs. (7.10a) and (7.10b), and consider the problem in the frequency domain. Taking the Laplace transform of

$$\dot{\mathbf{x}} = \mathbf{F}\mathbf{x} + \mathbf{G}u, \tag{7.32}$$

we obtain

$$s\mathbf{X}(s) - \mathbf{x}(0) = \mathbf{F}\mathbf{X}(s) + \mathbf{G}U(s), \tag{7.33}$$

which is now an algebraic equation. If we collect the terms involving $\mathbf{X}(s)$ on the left side of Eq. (7.33), keeping in mind that in matrix multiplication order is very important, we find that[3]

$$(s\mathbf{I} - \mathbf{F})\mathbf{X}(s) = \mathbf{G}U(s) + \mathbf{x}(0).$$

If we premultiply both sides by the inverse of $(s\mathbf{I} - \mathbf{F})$, then

$$\mathbf{X}(s) = (s\mathbf{I} - \mathbf{F})^{-1}\mathbf{G}U(s) + (s\mathbf{I} - \mathbf{F})^{-1}\mathbf{x}(0). \qquad (7.34)$$

The output of the system is

$$Y(s) = \mathbf{H}\mathbf{X}(s) + JU(s), \qquad (7.35a)$$

$$= \mathbf{H}(s\mathbf{I} - \mathbf{F})^{-1}\mathbf{G}U(s) + \mathbf{H}(s\mathbf{I} - \mathbf{F})^{-1}\mathbf{x}(0) + JU(s). \qquad (7.35b)$$

This equation expresses the output response to both an initial condition and an external forcing input. The coefficient of the external input is the transfer function of the system, which in this case is given by

Transfer function from state
equations

$$G(s) = \frac{Y(s)}{U(s)} = \mathbf{H}(s\mathbf{I} - \mathbf{F})^{-1}\mathbf{G} + J. \qquad (7.36)$$

EXAMPLE 7.4

Thermal System Transfer Function from the State Description

Use Eq. (7.36) to find the transfer function of the thermal system described by Eqs. (7.4a) and (7.4b).

Solution.   The state-variable description matrices of the system are

$$\mathbf{F} = \begin{bmatrix} -7 & -12 \\ 1 & 0 \end{bmatrix}, \qquad \mathbf{G} = \begin{bmatrix} 1 \\ 0 \end{bmatrix},$$

$$\mathbf{H} = \begin{bmatrix} 1 & 2 \end{bmatrix}, \qquad J = 0.$$

To compute the transfer function according to Eq. (7.36), we form

$$s\mathbf{I} - \mathbf{F} = \begin{bmatrix} s+7 & 12 \\ -1 & s \end{bmatrix}$$

and compute

$$(s\mathbf{I} - \mathbf{F})^{-1} = \frac{\begin{bmatrix} s & -12 \\ 1 & s+7 \end{bmatrix}}{s(s+7) + 12}. \qquad (7.37)$$

---

[3]   The identity matrix $\mathbf{I}$ is a matrix of 1s on the main diagonal and zeros everywhere else; therefore, $\mathbf{I}\mathbf{x} = \mathbf{x}$.

We then substitute Eq. (7.37) into Eq. (7.36) to get

$$G(s) = \frac{[1 \quad 2] \begin{bmatrix} s & -12 \\ 1 & s+7 \end{bmatrix} \begin{bmatrix} 1 \\ 0 \end{bmatrix}}{s(s+7) + 12} \tag{7.38}$$

$$= \frac{[1 \quad 2] \begin{bmatrix} s \\ 1 \end{bmatrix}}{s(s+7) + 12} \tag{7.39}$$

$$= \frac{(s+2)}{(s+3)(s+4)}. \tag{7.40}$$

The results can also be found using the MATLAB statements

[num,den] = ss2tf(F,G,H,J)

and yield num = [0 1 2] and den = [1 7 12], which agrees with hand calculations.

Because Eq. (7.36) expresses the transfer function in terms of the general state-space descriptor matrices $\mathbf{F}$, $\mathbf{G}$, $\mathbf{H}$, and J, we are able to express poles and zeros in terms of these matrices. We saw earlier that by transforming the state matrices to diagonal form, the poles appear as the eigenvalues on the main diagonal of the $\mathbf{F}$ matrix. We now take a systems theory point of view to look at the poles and zeros as they are involved in the transient response of a system.

As we saw in Chapter 3, a pole of the transfer function $G(s)$ is a value of generalized frequency s such that, if $s = p_i$, then the system can respond to an initial condition as $K_i e^{p_i t}$, *with no forcing function*. In this context, p_i is called a **natural frequency** or **natural mode** of the system. If we take the state-space equations (7.10a) and (7.10b) and set the forcing function u to zero, we have

$$\dot{\mathbf{x}} = \mathbf{F}\mathbf{x}. \tag{7.41}$$

If we assume some (as yet unknown) initial condition

$$\mathbf{x}(0) = \mathbf{x}_0 \tag{7.42}$$

and that the entire state motion behaves according to the same natural frequency, then the state can be written as $\mathbf{x}(t) = e^{p_i t}\mathbf{x}_0$. It follows from Eq. (7.41) that

$$\dot{\mathbf{x}}(t) = p_i e^{p_i t}\mathbf{x}_0 = \mathbf{F}\mathbf{x} = \mathbf{F}e^{p_i t}\mathbf{x}_0, \tag{7.43}$$

or

$$\mathbf{F}\mathbf{x}_0 = p_i \mathbf{x}_0. \tag{7.44}$$

We can rewrite Eq. (7.44) as

$$(p_i \mathbf{I} - \mathbf{F})\mathbf{x}_0 = 0. \tag{7.45}$$

Equations (7.44) and (7.45) comprise the eigenvector/eigenvalue problem we saw in Eq. (7.27) with eigenvalues p_i and in this case eigenvectors $\mathbf{x}_0$ of the matrix $\mathbf{F}$. If we are just interested in the eigenvalues, we can use the fact that for a nonzero $\mathbf{x}_0$, Eq. (7.45) has a solution if and only if

$$\det(p_i \mathbf{I} - \mathbf{F}) = 0. \tag{7.46}$$

These equations show again that the *poles* of the transfer function are the eigenvalues of the system matrix $\mathbf{F}$. The determinant equation (7.46) is a polynomial in the eigenvalues p_i known as the **characteristic equation**. In Example 7.2 we computed the eigenvalues and eigenvectors of a particular matrix in control canonical form. As an alternative computation for the poles of that system, we could solve the characteristic equation (7.46). For the system described by Eqs. (7.4a) and (7.4b), we can find the poles from Eq. (7.46) by solving

$$\det(s\mathbf{I} - \mathbf{F}) = 0, \tag{7.47a}$$

$$\det \begin{bmatrix} s+7 & 12 \\ -1 & s \end{bmatrix} = 0, \tag{7.47b}$$

$$s(s+7) + 12 = (s+3)(s+4) = 0. \tag{7.47c}$$

This confirms again that the poles of the system are the eigenvalues of $\mathbf{F}$.

We can also determine the zeros of a system from the state-variable description matrices $\mathbf{F}, \mathbf{G}, \mathbf{H}$, and J using a systems theory point of view. From this perspective a zero is a value of generalized frequency s such that the system can have a nonzero input and state and yet have an output of zero. If the input is exponential at the zero frequency z_i, given by

$$u(t) = u_0 e^{z_i t}, \tag{7.48}$$

then the output is identically zero:

$$y(t) \equiv 0. \tag{7.49}$$

The state-space description of Eqs. (7.48) and (7.49) would be

$$u = u_0 e^{z_i t}, \qquad \mathbf{x}(t) = \mathbf{x}_0 e^{z_i t}, \qquad y(t) \equiv 0. \tag{7.50}$$

Thus

$$\dot{\mathbf{x}} = z_i e^{z_i t} \mathbf{x}_0 = \mathbf{F} e^{z_i t} \mathbf{x}_0 + \mathbf{G} u_0 e^{z_i t}, \tag{7.51}$$

or

$$[\, z_i \mathbf{I} - \mathbf{F} \quad -\mathbf{G} \,] \begin{bmatrix} \mathbf{x}_0 \\ u_0 \end{bmatrix} = \mathbf{0} \tag{7.52}$$

and

$$y = \mathbf{H}\mathbf{x} + Ju = \mathbf{H} e^{z_i t} \mathbf{x}_0 + J u_0 e^{z_i t} \equiv 0. \tag{7.53}$$

Combining Eqs. (7.52) and (7.53), we get

$$
\begin{bmatrix} z_i\mathbf{I} - \mathbf{F} & -\mathbf{G} \\ \mathbf{H} & J \end{bmatrix}\begin{bmatrix} \mathbf{x}_0 \\ u_0 \end{bmatrix} = \begin{bmatrix} \mathbf{0} \\ 0 \end{bmatrix}.
\tag{7.54}
$$

Transfer function zeros from state equations

From Eq. (7.54) we can conclude that a zero of the state-space system is a value of z_i where Eq. (7.54) has a nontrivial solution. With one input and one output, the matrix is square, and a solution to Eq. (7.54) is equivalent to a solution to

$$
\det\begin{bmatrix} z_i\mathbf{I} - \mathbf{F} & -\mathbf{G} \\ \mathbf{H} & J \end{bmatrix} = 0.
\tag{7.55}
$$

EXAMPLE 7.5

Zeros for the Thermal System from a State Description

Compute the zero(s) of the thermal system described by Eq. (7.4).

Solution.   We use Eq. (7.55) to compute the zeros:

$$
\det\begin{bmatrix} s+7 & 12 & -1 \\ -1 & s & 0 \\ 1 & 2 & 0 \end{bmatrix} = 0,
$$

$$
-2 - s = 0,
$$

$$
s = -2.
$$

Note that this result agrees with the zero of the transfer function given by Eq. (7.1). The result can also be found using the MATLAB statements

sysG = ss(Ac,Bc,Cc,Dc);
z = tzero(sysG)

and yields z = −2.0.

Equation (7.46) for the characteristic equation and Eq. (7.55) for the zeros polynomial can be combined to express the transfer function in a compact form from state-description matrices as

$$
G(s) = \frac{\det\begin{bmatrix} s\mathbf{I} - \mathbf{F} & -\mathbf{G} \\ \mathbf{H} & J \end{bmatrix}}{\det(s\mathbf{I} - \mathbf{F})}.
\tag{7.56}
$$

(See Appendix C for more details.) While Eq. (7.56) is a compact formula for theoretical studies, it is very sensitive to numerical errors. A numerically stable algorithm for computing the transfer function is described in Emami-Naeini and Van Dooren (1982b). Given the transfer function, we can compute the frequency response as $G(j\omega)$; and as discussed earlier, we can use Eqs. (7.45) and (7.54) to find the poles and zeros, upon which the transient response depends, as we saw in Chapter 3.

EXAMPLE 7.6

Analysis of the State Equations of a Tape Drive

Compute the poles, zeros, and transfer function for the equations of the tape-drive servomechanism given in Example 7.3.

Solution.   The state equations are described by the system matrices:

$$
\mathbf{F} = \begin{bmatrix} 0 & 2 & 0 & 0 & 0 \\ -0.1 & -0.35 & 0.1 & 0.1 & 0.75 \\ 0 & 0 & 0 & 2 & 0 \\ 0.4 & 0.4 & -0.4 & -1.4 & 0 \\ 0 & -0.03 & 0 & 0 & -1 \end{bmatrix}, \quad \mathbf{G} = \begin{bmatrix} 0 \\ 0 \\ 0 \\ 0 \\ 1 \end{bmatrix}, \tag{7.57}
$$

$\mathbf{H}_2 = \begin{bmatrix} 0.0 & 0.0 & 1.0 & 0.0 & 0.0 \end{bmatrix}$ Servomotor position output,

$\mathbf{H}_3 = \begin{bmatrix} 0.5 & 0.0 & 0.5 & 0.0 & 0.0 \end{bmatrix}$ Position at read/write head as output,

$\mathbf{H}_T = \begin{bmatrix} -0.2 & -0.2 & 0.2 & 0.2 & 0.0 \end{bmatrix}$ Tension output,

$J = 0.0.$

MATLAB ss2tf

There are two different ways to compute the answer to this problem. The most direct is to use the MATLAB function ss2tf (state space to transfer function), which will give the numerator and denominator polynomials directly. This function permits multiple inputs and outputs; the fifth argument of the function tells which input is to be used. We have only one input here but must still provide the argument. The computation of the transfer function from motor-current input to the servomotor position output is

[N2, D2] = ss2tf(F, G, H2, J, 1)

which results in

$$N2 = \begin{bmatrix} 0 & 0.0000 & 0 & -0.0000 & 0.6000 & 1.2000 \end{bmatrix}$$

$$D2 = \begin{bmatrix} 1.0000 & 2.7500 & 3.2225 & 1.8815 & 0.4180 & 0.000 \end{bmatrix}.$$

Similarly, for the position at the read/write head, the transfer function polynomials are computed by

[N3, D3] = ss2tf(F, G, H3, J, 1),

which results in

$$N3 = \begin{bmatrix} 0 & 0.0000 & -0.0000 & 0.7500 & 1.3500 & 1.2000 \end{bmatrix},$$

$$D3 = \begin{bmatrix} 1.0000 & 2.7500 & 3.2225 & 1.8815 & 0.4180 & 0.0000 \end{bmatrix}.$$

Finally, the transfer function to tension is

[NT, DT] = ss2tf(F, G, HT, J, 1)

producing

$$NT = \begin{bmatrix} 0 & 0.0000 & -0.1500 & -0.4500 & -0.3000 & -0.0000 \end{bmatrix},$$

$$DT = \begin{bmatrix} 1.0000 & 2.7500 & 3.2225 & 1.8815 & 0.4180 & 0.0000 \end{bmatrix}.$$

It is interesting to check to see whether the poles and zeros determined this way agree with those found by other means. For a polynomial we use the function roots:

$$\text{roots(D3)} = \begin{bmatrix} -0.6371 + 0.6669i \\ -0.6371 - 0.6669i \\ -0.9683 \\ -0.5075 \\ -0.0000 \end{bmatrix}.$$

Checking with Example 7.3, we confirm that they agree.

How about the zeros? We can find these by finding the roots of the numerator polynomial. We compute the roots of the polynomial N3:

$$\text{roots(N3)} = \begin{bmatrix} 2.4 + 4.1096 \times 10^7 i \\ -2.4 - 4.1096 \times 10^7 i \\ -0.9000 + 0.8888i \\ -0.9000 - 0.8888i \end{bmatrix}.$$

Here we notice that roots are given with a magnitude of 10^7, which seems inconsistent with the values given for the polynomial. The problem is that MATLAB has used the very small leading terms in the polynomial as real values and thereby introduced extraneous roots that are for all practical purposes at infinity. The true zeros are found by truncating the polynomial to the significant values using the statement

$$\text{N3R} = \text{N3}(4:6)$$

to get

$$\text{N3R} = [\,0.7500 \quad 1.3500 \quad 1.200\,],$$

$$\text{roots(N3R)} = \begin{bmatrix} -0.9000 + 0.8888i \\ -0.9000 - 0.8888i \end{bmatrix}.$$

The other approach is to compute the poles and zeros separately and, if desired, combine these into a transfer function. The poles were computed with eig in Example 7.3 and are

$$P = \begin{bmatrix} -0.6371 + 0.6669i \\ -0.6371 - 0.6669i \\ 0.0000 \\ -0.5075 \\ -0.9683 \end{bmatrix}.$$

The zeros can be computed by the equivalent of Eq. (7.54) with the function tzero (transmission zeros). The zeros depend on which output is being used, of course, and are respectively given below. For the position of the tape at the servomotor as the output, the statement:

$$\text{sysG2} = \text{ss(F, G, H2, J)}$$
$$\text{ZER2} = \text{tzero(sysG2)}$$

yields

$$\text{ZER2} = -2.0000.$$

For the position of the tape over the read/write head as the output, the statement

$$sysG3 = ss(F, G, H3, J)$$

$$ZER3 = tzero(sysG3)$$

$$ZER3 = \begin{bmatrix} -0.9000 + 0.8888i \\ -0.9000 - 0.8888i \end{bmatrix}.$$

We note that these results agree with the values computed from the numerator polynomial N3 above. And, finally, for the tension as output we use

$$sysGT = ss(F, G, HT, J)$$
$$ZERT = \quad tzero(sysGT)$$

to get

$$ZERT = \begin{bmatrix} -1.0000 \\ -0.0000 \\ -2.0000 \end{bmatrix}.$$

From these results we can write down, for example, the transfer function to position x_3 as

$$G(s) = \frac{X_3(s)}{E_1(s)}$$

$$= \frac{0.75s^2 + 1.35s + 1.2}{s^5 + 2.75s^4 + 3.22s^3 + 1.88s^2 + 0.418s}$$

$$= \frac{0.75(s + 0.9 \pm 0.8888j)}{s(s + 0.507)(s + 0.968)(s + 0.637 \pm 0.667j)}. \tag{7.58}$$

7.3   Control-Law Design for Full-State Feedback

One of the attractive features of the state-space design method is that it consists of a sequence of independent steps, as mentioned in the chapter overview. The first step, discussed in Section 7.3.1, is to determine the control. The purpose of the control law is to allow us to assign a set of pole locations for the closed-loop system that will correspond to satisfactory dynamic response in terms of rise time and other measures of transient response. In Section 7.3.2 we will show how to introduce the reference input with full-state feedback, and in Section 7.4 we will describe the process of finding the poles for good design.

The second step—necessary if the full state is not available—is to design an **normal form** (sometimes called an **observer**), which computes an estimate of the entire state vector when provided with the measurements of the system indicated by Eq. (7.10b). We will examine estimator design in Section 7.5.

Estimator/observer

Figure 7.6
Schematic diagram of state-space design elements

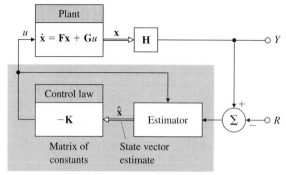

Compensation

The third step consists of combining the control law and the estimator. Figure 7.6 shows how the control law and the estimator fit together and how the combination takes the place of what we have been previously referring to as **compensation**. At this stage the control-law calculations are based on the estimated state rather than the actual state. In Section 7.6 we will show that this substitution is reasonable, and also that using the combined control law and estimator results in closed-loop pole locations that are the same as those determined when designing the control and estimator separately.

The control law and the estimator together form the compensation

The fourth and final step of state-space design is to introduce the reference input in such a way that the plant output will track external commands with acceptable rise-time, overshoot, and settling-time values. At this point in the design, all the closed-loop poles have been selected, and the designer is concerned with the zeros of the overall transfer function. Figure 7.6 shows the command input r introduced in the same relative position as was done with the transform design methods; however, in Section 7.8 we will show how to introduce the reference at another location, resulting in different zeros and (usually) superior control.

7.3.1 Finding the Control Law

Control law

The first step in the state-space design method, as mentioned earlier, is to find the control law as feedback of a linear combination of the state variables—that is,

$$u = -\mathbf{K}\mathbf{x} = -[\, K_1 \quad K_2 \quad \cdots \quad K_n \,] \begin{bmatrix} x_1 \\ x_2 \\ \vdots \\ x_n \end{bmatrix}. \tag{7.59}$$

We assume for feedback purposes that all the elements of the state vector are at our disposal. In practice, of course, this would usually be a ridiculous assumption; moreover, a well-trained control designer knows that other design methods do not require so many sensors. The assumption that all state variables are available merely allows us to proceed with this first step.

Figure 7.7
Assumed system for
control-law design

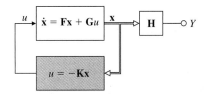

Equation (7.59) tells us that the system has a constant matrix in the state-vector feedback path, as shown in Fig. 7.7. For an nth-order system there will be n feedback gains, $K_1, \ldots, K_n$, and because there are n roots of the system, it is possible that there are enough degrees of freedom to select *arbitrarily* any desired root location by choosing the proper values of K_i. This freedom contrasts sharply with root-locus design, in which we have only one parameter and the closed-loop poles are restricted to the locus.

Substituting the feedback law given by Eq. (7.59) into the system described by Eq. (7.10) yields

$$\dot{\mathbf{x}} = \mathbf{Fx} - \mathbf{GKx}. \tag{7.60}$$

Control characteristic equation

The characteristic equation of this closed-loop system is

$$\det[s\mathbf{I} - (\mathbf{F} - \mathbf{GK})] = 0. \tag{7.61}$$

When evaluated, this yields an nth-order polynomial in s containing the gains $K_1, \ldots, K_n$. The control-law design then consists of picking the gains $\mathbf{K}$ so that the roots of Eq. (7.61) are in desirable locations. Selecting desirable root locations is an inexact science that may require some iteration by the designer. Issues in their selection are considered in Examples 7.7 to 7.9 as well as in Section 7.4. For now we assume that the desired locations are known, say,

$$s = s_1, s_2, \ldots, s_n.$$

Then the corresponding desired (control) characteristic equation is

$$\alpha_c(s) = (s - s_1)(s - s_2) \ldots (s - s_n) = 0. \tag{7.62}$$

Hence the required elements of $\mathbf{K}$ are obtained by matching coefficients in Eqs. (7.61) and (7.62). This forces the system's characteristic equation to be identical to the desired characteristic equation and the closed-loop poles to be placed at the desired locations.

EXAMPLE 7.7

Control Law for a Pendulum

Suppose you have a pendulum with frequency ω_0 and a state-space description given by

$$\begin{bmatrix} \dot{x}_1 \\ \dot{x}_2 \end{bmatrix} = \begin{bmatrix} 0 & 1 \\ -\omega_0^2 & 0 \end{bmatrix} \begin{bmatrix} x_1 \\ x_2 \end{bmatrix} + \begin{bmatrix} 0 \\ 1 \end{bmatrix} u. \tag{7.63}$$

Find the control law that places the closed-loop poles of the system so that they are both at $-2\omega_0$. In other words, you wish to double the natural frequency and increase the damping ratio ζ from 0 to 1.

Solution.   From Eq. (7.62) we find that

$$\alpha_c(s) = (s + 2\omega_0)^2 \tag{7.64a}$$

$$= s^2 + 4\omega_0 s + 4\omega_0^2. \tag{7.64b}$$

Equation (7.61) tells us that

$$\det[s\mathbf{I} - (\mathbf{F} - \mathbf{GK})] = \det\left\{ \begin{bmatrix} s & 0 \\ 0 & s \end{bmatrix} - \left(\begin{bmatrix} 0 & 1 \\ -\omega_0^2 & 0 \end{bmatrix} - \begin{bmatrix} 0 \\ 1 \end{bmatrix} [K_1 \quad K_2] \right) \right\},$$

or

$$s^2 + K_2 s + \omega_0^2 + K_1 = 0. \tag{7.65}$$

Equating the coefficients with like powers of s in Eqs. (7.64) and (7.65) yields the system of equations

$$K_2 = 4\omega_0,$$

$$\omega_0^2 + K_1 = 4\omega_0^2,$$

and therefore,

$$K_1 = 3\omega_0^2,$$

$$K_2 = 4\omega_0.$$

More concisely, the control law is

$$\mathbf{K} = [K_1 \quad K_2] = [3\omega_0^2 \quad 4\omega_0].$$

Figure 7.8 shows the response of the closed-loop system to the initial conditions $x_1 = 1.0$, $x_2 = 0.0$, and $\omega_0 = 1$. It shows a very well damped response, as would be expected from having two roots at $s = -2$. The MATLAB command impulse was used to generate the plot.

Figure 7.8
Impulse response of the undamped oscillator with full-state feedback $(\omega_0 = 1)$

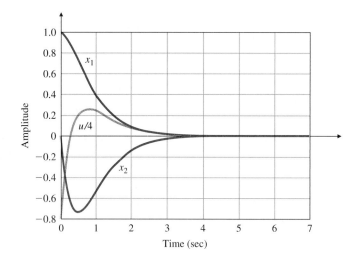

Calculating the gains using the technique illustrated in Example 7.7 becomes rather tedious when the order of the system is higher than 3. There are, however, special "canonical" forms of the state-variable equations where the algebra for finding the gains is especially simple. One such canonical form useful in control law design is the *control canonical form*. Consider the third-order system[4]

$$\dddot{y} + a_1\ddot{y} + a_2\dot{y} + a_3 y = b_1\ddot{u} + b_2\dot{u} + b_3 u, \tag{7.66}$$

which corresponds to the transfer function

$$G(s) = \frac{Y(s)}{U(s)} = \frac{b_1 s^2 + b_2 s + b_3}{s^3 + a_1 s^2 + a_2 s + a_3} = \frac{b(s)}{a(s)}. \tag{7.67}$$

Suppose we introduce an auxiliary variable (referred to as the *partial state*) ξ, which relates $a(s)$ and $b(s)$ as shown in Fig. 7.9(a). The transfer function from U to ξ is

$$\frac{\xi(s)}{U(s)} = \frac{1}{a(s)}, \tag{7.68}$$

or

$$\dddot{\xi} + a_1\ddot{\xi} + a_2\dot{\xi} + a_3\xi = u. \tag{7.69}$$

It is easy to draw a block diagram corresponding to Eq. (7.69) if we rearrange the equation as follows:

$$\dddot{\xi} = -a_1\ddot{\xi} - a_2\dot{\xi} - a_3\xi + u. \tag{7.70}$$

The summation indicates in Fig. 7.9(b), where each ξ on the right-hand side is obtained by sequential integration of $\dddot{\xi}$. To form the output, we go back to Fig. 7.9(a) and note that

$$Y(s) = b(s)\xi(s), \tag{7.71}$$

which means

$$y = b_1\ddot{\xi} + b_2\dot{\xi} + b_3\xi. \tag{7.72}$$

We again pick off the outputs of the integrators, multiply them by $\{b_i\}$'s, and form the right-hand side of Eq. (7.66) by using a summer to yield the output as shown in Fig. 7.9(c). In this case, all the feedback loops return to the point of the application of the input, or "control" variable, and hence the form is referred to as the *control canonical form*. Reduction of the structure by Mason's rule or by elementary block diagram operations verifies that this structure has the transfer function given by $G(s)$.

---

[4] This development is exactly the same for higher-order systems.

Figure 7.9
Derivation of control
canonical form.

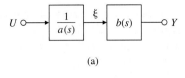

(a)

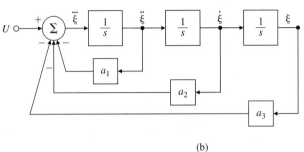

(b)

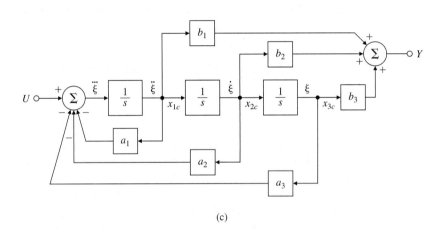

(c)

Taking the state as the outputs of the three integrators numbered, by convention, from the left,

$$x_1 = \ddot{\xi}_1, \quad x_2 = \dot{\xi}, \quad x_3 = \xi, \tag{7.73}$$

we obtain

$$\dot{x}_1 = \dddot{\xi} = -a_1 x_1 - a_2 x_2 - a_3 x_3 + u, \tag{7.74}$$

$$\dot{x}_2 = x_1,$$

$$\dot{x}_3 = x_2.$$

We may now write the matrices describing the control canonical form in general,

$$\mathbf{F}_c = \begin{bmatrix} -a_1 & -a_2 & \cdots & -a_n \\ 1 & 0 & \cdots & 0 \\ 0 & 1 & \cdots & 0 \end{bmatrix}, \quad \mathbf{G}_c = \begin{bmatrix} 1 \\ 0 \\ 0 \end{bmatrix}, \tag{7.75a}$$

$$\mathbf{H}_c = \begin{bmatrix} b_1 & b_2 & \cdots & b_n \end{bmatrix}, \quad J_c = 0. \tag{7.75b}$$

The special structure of this system matrix is referred to as the **upper companion form** because the characteristic equation is $a(s) = s^n + a_1 s^{n-1} + a_2 s^{n-2} \cdots + a_n$ and the coefficients of this monic "companion" polynomial are the elements in the first row of $\mathbf{F}_c$. If we now form the closed-loop system matrix $\mathbf{F}_c - \mathbf{G}_c \mathbf{K}_c$, we find that

$$\mathbf{F}_c - \mathbf{G}_c \mathbf{K}_c = \begin{bmatrix} -a_1 - K_1 & -a_2 - K_2 & \cdots & -a_n - K_n \\ 1 & 0 & \cdots & 0 \\ 0 & 1 & \cdots & 0 \end{bmatrix}. \tag{7.76}$$

By visually comparing Eqs. (7.75a) and (7.76) we see that the closed-loop characteristic equation is

$$s^n + (a_1 + K_1)s^{n-1} + (a_2 + K_2)s^{n-2} + \cdots + (a_n + K_n) = 0. \tag{7.77}$$

Therefore, if the desired pole locations result in the characteristic equation given by

$$\alpha_c(s) = s^n + \alpha_1 s^{n-1} + \alpha_2 s^{n-2} + \cdots + \alpha_n = 0, \tag{7.78}$$

then the necessary feedback gains can be found by equating the coefficients in Eqs. (7.77) and (7.78):

$$K_1 = -a_1 + \alpha_1, \qquad K_2 = -a_2 + \alpha_2, \ldots, \qquad K_n = -a_n + \alpha_n. \tag{7.79}$$

We now have the basis for a design procedure. Given a system of order n described by an arbitrary $(\mathbf{F}, \mathbf{G})$ and given a desired nth-order monic characteristic polynomial $\alpha_c(s)$, we (1) transform $(\mathbf{F}, \mathbf{G})$ to control canonical form $(\mathbf{F}_c, \mathbf{G}_c)$ by changing the state $\mathbf{x} = \mathbf{Tz}$ and (2) solve for the control gains by inspection using Eq. (7.79) to give the control law $u = -\mathbf{K}_c \mathbf{z}$. Because this gain is for the state in the *control* form, we must (3) transform the gain back to the original state to get $\mathbf{K} = \mathbf{K}_c \mathbf{T}^{-1}$.

An alternative to this transformation method is given by **Ackermann's formula** (Ackermann, 1972), which organizes the three-step process of converting to $(\mathbf{F}_c, \mathbf{G}_c)$, solving for the gains, and converting back again into the following very compact form:

$$\mathbf{K} = \begin{bmatrix} 0 & \cdots & 0 & 1 \end{bmatrix} \mathcal{C}^{-1} \alpha_c(\mathbf{F}) \tag{7.80}$$

such that

$$\mathcal{C} = \begin{bmatrix} \mathbf{G} & \mathbf{FG} & \mathbf{F}^2 \mathbf{G} & \cdots & \mathbf{F}^{n-1} \mathbf{G} \end{bmatrix}, \tag{7.81}$$

where $\mathcal{C}$ is the controllability matrix we saw in Section 7.2, n gives the order of the system and the number of state variables, and $\alpha_c(\mathbf{F})$ is a matrix defined as

$$\alpha_c(\mathbf{F}) = \mathbf{F}^n + \alpha_1 \mathbf{F}^{n-1} + \alpha_2 \mathbf{F}^{n-2} + \cdots + \alpha_n \mathbf{I}, \tag{7.82}$$

where the α_i are the coefficients of the desired characteristic polynomial Eq. (7.78). Note that Eq. (7.82) is a matrix equation.

EXAMPLE 7.8   *Ackermann's Formula for Undamped Oscillator*

(a) Use Ackermann's formula to solve for the gains for the undamped oscillator of Example 7.7.

(b) Verify the calculations with MATLAB for $\omega_0 = 1$.

Solution.

(a) The desired characteristic equation is $\alpha_c(s) = (s + 2\omega_0)^2$. Therefore, the desired characteristic polynomial coefficients,

$$\alpha_1 = 4\omega_0, \qquad \alpha_2 = 4\omega_0^2,$$

are substituted into Eq. (7.82) and result in

$$\alpha_c(\mathbf{F}) = \begin{bmatrix} -\omega_0^2 & 0 \\ 0 & -\omega_0^2 \end{bmatrix} + 4\omega_0 \begin{bmatrix} 0 & 1 \\ -\omega_0^2 & 0 \end{bmatrix} + 4\omega_0^2 \begin{bmatrix} 1 & 0 \\ 0 & 1 \end{bmatrix}, \tag{7.83a}$$

$$= \begin{bmatrix} 3\omega_0^2 & 4\omega_0 \\ -4\omega_0^3 & 3\omega_0^2 \end{bmatrix}. \tag{7.83b}$$

The controllability matrix is

$$\mathcal{C} = [\mathbf{G} \quad \mathbf{FG}] = \begin{bmatrix} 0 & 1 \\ 1 & 0 \end{bmatrix},$$

which yields

$$\mathcal{C}^{-1} = \begin{bmatrix} 0 & 1 \\ 1 & 0 \end{bmatrix}. \tag{7.84}$$

Finally, we substitute Eqs. (7.84) and (7.83b) into Eq. (7.80) to get

$$\mathbf{K} = [K_1 \quad K_2]$$

$$= [0 \quad 1] \begin{bmatrix} 0 & 1 \\ 1 & 0 \end{bmatrix} \begin{bmatrix} 3\omega_0^2 & 4\omega_0 \\ -4\omega_0^3 & 3\omega_0^2 \end{bmatrix}.$$

Therefore

$$\mathbf{K} = [3\omega_0^2 \quad 4\omega_0],$$

which is the same result we obtained previously.

(b) The MATLAB statements

```
wo = 1;
F = [0 1;-wo*wo 0];
G = [0;1];
pc = [-2*wo;-2*wo];
K = acker(F,G,pc)
```

yield $\mathbf{K} = [3 \quad 4]$, which agrees with hand calculations.

MATLAB acker, place

As was mentioned earlier, computation of the controllability matrix has very poor numerical accuracy, and this carries over to Ackermann's formula. Equation (7.80), implemented in MATLAB with the function acker, can be used for the design of single-input–single-output systems with a small (≤ 10) number of state variables. For more complex cases a more reliable formula is available, implemented in MATLAB with place. A modest limitation on place is that, because it is based on assigning closed-loop eigenvectors, none of the desired closed-loop poles may be repeated; that is, the poles must be distinct,[5] a requirement that does not apply to acker.

The fact that we can shift the poles of a system by state feedback to any desired location is a rather remarkable result. The development in this section reveals that this shift is possible if we can transform $(\mathbf{F}, \mathbf{G})$ to the control form $(\mathbf{F}_c, \mathbf{G}_c)$, which in turn is possible if the system is controllable. In rare instances the system may be uncontrollable, in which case no possible control will yield arbitrary pole locations. **Uncontrollable systems** have certain modes, or subsystems, that are unaffected by the control. This usually means that parts of the system are physically disconnected from the input. For example, in modal canonical form for a system with distinct poles, one of the modal state variables is not connected to the input if there is a zero entry in the $\mathbf{B}_m$ matrix. A good physical understanding of the system being controlled would prevent any attempt to design a controller for an uncontrollable system. As we saw earlier, there are algebraic tests for controllability; however, no mathematical test can replace the control engineer's understanding of the physical system. Often the physical situation is such that every mode is controllable to some degree, and, while the mathematical tests indicate the system is controllable, certain modes are so weakly controllable that designs to control them are virtually useless.

An example of weak controllability

Airplane control is a good example of weak controllability of certain modes. Pitch plane motion $\mathbf{x}_p$ is primarily affected by the elevator δ_e and weakly affected by rolling motion $\mathbf{x}_r$. Rolling motion is essentially only affected by the ailerons δ_a. The state-space description of these relationships is

$$
\begin{bmatrix} \dot{\mathbf{x}}_p \\ \dot{\mathbf{x}}_r \end{bmatrix} = \begin{bmatrix} \mathbf{F}_p & \varepsilon \\ 0 & \mathbf{F}_r \end{bmatrix} \begin{bmatrix} \mathbf{x}_p \\ \mathbf{x}_r \end{bmatrix} + \begin{bmatrix} \mathbf{G}_p & 0 \\ 0 & \mathbf{G}_r \end{bmatrix} \begin{bmatrix} \delta_e \\ \delta_a \end{bmatrix},
\tag{7.85}
$$

where the matrix of small numbers ε represents the weak coupling from rolling motion to pitching motion. A mathematical test of controllability for this system would conclude that pitch plane motion (and therefore altitude) is controllable by the ailerons as well as by the elevator! However, it is impractical to attempt to control an airplane's altitude by rolling the aircraft with the ailerons.

Another example will illustrate some of the properties of pole placement by state feedback and the effects of loss of controllability on the process.

---

[5] One may get around this restriction by moving the repeated poles by very small amounts to make them distinct.

EXAMPLE 7.9   ***How Zero Location Can Affect the Control Law***

A specific thermal system is described by Eq. (7.24) in observer canonical form with a zero at $s = z_0$.

(a) Find the state feedback gains necessary for placing the poles of this system at the roots of $s^2 + 2\zeta\omega_n s + \omega_n^2$—that is, at $-\zeta\omega_n \pm j\omega_n\sqrt{1 - \zeta^2}$.

(b) Repeat the computation with MATLAB using the parameter values $z_0 = 2$, $\zeta = 0.5$, and $\omega_n = 2$ rad/sec.

Solution.

(a) The state description matrices are

$$\mathbf{A}_o = \begin{bmatrix} -7 & 1 \\ -12 & 0 \end{bmatrix}, \qquad \mathbf{B}_o = \begin{bmatrix} 1 \\ -z_0 \end{bmatrix},$$

$$\mathbf{C}_o = \begin{bmatrix} 1 & 0 \end{bmatrix}, \qquad D_o = 0.$$

First we substitute these matrices into Eq. (7.61) to get the closed-loop characteristic equation in terms of the unknown gains and the zero position:

$$s^2 + (7 + K_1 - z_o K_2)s + 12 - K_2(7z_o + 12) - K_1 z_0 = 0.$$

Next we equate this equation to the desired characteristic equation to get the equations

$$K_1 - z_0 K_2 = 2\zeta\omega_n - 7,$$

$$-z_0 K_1 - (7z_0 + 12)K_2 = \omega_n^2 - 12.$$

The solutions to these equations are

$$K_1 = \frac{z_0(14\zeta\omega_n - 37 - \omega_n^2) + 12(2\zeta\omega_n - 7)}{(z_0 + 3)(z_0 + 4)},$$

$$K_2 = \frac{z_0(7 - 2\zeta\omega_n) + 12 - \omega_n^2}{(z_0 + 3)(z_0 + 4)}.$$

(b) The MATLAB statements

```
Ao = [-7 1;-12 0];
zo = 2;
Bo = [1;-zo];
pc = roots([1 2 4]);
K = place(Ao,Bo,pc)
```

can be used to find the solution, and yield K = [-3.80 0.60], which agrees with hand calculations. If the zero were close to one of the open-loop poles, say $z_0 = -2.99$, then we find K = [2052.5 −688.1].

Two important observations should be made from this example. The first is that the gains grow as the zero z_0 approaches either -3 or -4, the values where this system loses controllability. In other words, as controllability is almost lost, the control gains become very large.

The system has to work harder and harder to achieve control as controllability slips away.

The second important observation illustrated by the example is that both K_1 and K_2 grow as the desired closed-loop bandwidth given by ω_n is increased. From this, we conclude that

> To move the poles a long way requires large gains.

These observations lead us to a discussion of how we might go about selecting desired pole locations in general. Before we begin that topic we will complete the design with full-state feedback by showing how the reference input might be applied to such a system and what the resulting response characteristics are.

7.3.2   Introducing the Reference Input with Full-State Feedback

Thus far, the control has been given by Eq. (7.59), or $u = -\mathbf{K}\mathbf{x}$. In order to study the transient response of the pole-placement designs to input commands, it is necessary to introduce the reference input into the system. An obvious way to do this is to change the control to $u = -\mathbf{K}\mathbf{x} + r$. However, the system will now almost surely have a nonzero steady-state error to a step input. The way to correct this problem is to compute the steady-state values of the state and the control input that will result in zero output error and then force them to take these values. If the desired final values of the state and the control input are $\mathbf{x}_{ss}$ and u_{ss}, then the new control formula should be

$$u = u_{ss} - \mathbf{K}(\mathbf{x} - \mathbf{x}_{ss}), \tag{7.86}$$

so that when $\mathbf{x} = \mathbf{x}_{ss}$ (no error), $u = u_{ss}$. To pick the correct final values, we must solve the equations so that the system will have zero steady-state error to *any* constant input. The system differential equations are the standard ones:

$$\dot{\mathbf{x}} = \mathbf{F}\mathbf{x} + \mathbf{G}u, \tag{7.87a}$$

$$y = \mathbf{H}\mathbf{x} + Ju. \tag{7.87b}$$

In the constant steady-state, Eqs. (7.87a) and (7.87b) reduce to the pair

$$\mathbf{0} = \mathbf{F}\mathbf{x}_{ss} + \mathbf{G}u_{ss}, \tag{7.88a}$$

$$y_{ss} = \mathbf{H}\mathbf{x}_{ss} + Ju_{ss}. \tag{7.88b}$$

Gain calculation for reference input

We want to solve for the values for which $y_{ss} = r_{ss}$ for any value of r_{ss}. To do this we make $\mathbf{x}_{ss} = \mathbf{N}_x r_{ss}$ and $u_{ss} = N_u r_{ss}$. With these substitutions we can write Eq. (7.88) as a matrix equation; the common factor of r_{ss} cancels out to give the equation for the gains:

$$\begin{bmatrix} \mathbf{F} & \mathbf{G} \\ \mathbf{H} & J \end{bmatrix} \begin{bmatrix} \mathbf{N}_x \\ N_u \end{bmatrix} = \begin{bmatrix} \mathbf{0} \\ 1 \end{bmatrix}. \tag{7.89}$$

This equation can be solved for $\mathbf{N}_x$ and N_u to get

$$\begin{bmatrix} \mathbf{N}_x \\ N_u \end{bmatrix} = \begin{bmatrix} \mathbf{F} & \mathbf{G} \\ \mathbf{H} & J \end{bmatrix}^{-1} \begin{bmatrix} 0 \\ 1 \end{bmatrix}.$$

Control equation with reference input

With these values we finally have the basis for introducing the reference input so as to get zero steady-state error to a step input:

$$u = N_u r - \mathbf{K}(\mathbf{x} - \mathbf{N}_x r) \tag{7.90a}$$

$$= -\mathbf{K}\mathbf{x} + (N_u + \mathbf{K}\mathbf{N}_x)r. \tag{7.90b}$$

The coefficient of r in parentheses is a constant that can be computed beforehand. We give it the symbol $\bar{N}$, so

$$u = -\mathbf{K}\mathbf{x} + \bar{N}r. \tag{7.91}$$

The block diagram of the system is shown in Fig. 7.10.

Figure 7.10
Block diagram for introducing the reference input with full-state feedback: (a) with state and control gains; (b) with a single composite gain

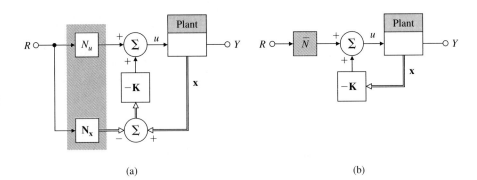

(a)   (b)

EXAMPLE 7.10   *Introducing the Reference Input*

Compute the necessary gains for zero steady-state error to a step command at x_1, and plot the resulting unit step response for the oscillator in Example 7.7 with $\omega_0 = 1$.

Solution.   We substitute the matrices of Eq. (7.63) (with $\omega_0 = 1$ and $\mathbf{H} = \begin{bmatrix} 1 & 0 \end{bmatrix}$ because $y = x_1$) into Eq. (7.89) to get

$$\begin{bmatrix} 0 & 1 & 0 \\ -1 & 0 & 1 \\ 1 & 0 & 0 \end{bmatrix} \begin{bmatrix} \mathbf{N}_x \\ N_u \end{bmatrix} = \begin{bmatrix} 0 \\ 0 \\ 1 \end{bmatrix}. \tag{7.92}$$

The solution is (x = a\b in MATLAB, where a and b are the left- and right-hand sides matrices respectively),

$$\mathbf{N}_x = \begin{bmatrix} 1 \\ 0 \end{bmatrix},$$

$$N_u = 1,$$

Figure 7.11
Step response of oscillator
to a reference input

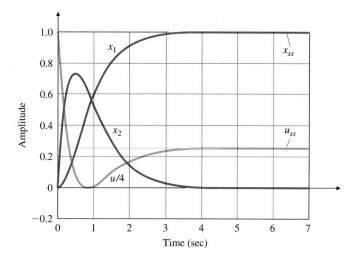

and, for the given control law, $\mathbf{K} = [\,3\omega_0^2 \;\; 4\omega_0\,] = [\,3 \;\; 4\,]$,

$$\bar{N} = N_u + \mathbf{K}\mathbf{N}_x = 4. \tag{7.93}$$

The corresponding step response (using the MATLAB step command) is plotted in Fig. 7.11.

Note that there are two equations for the control: Eqs. (7.90) and (7.91). While these expressions are equivalent in theory, they differ in practical implementation in that Eq. (7.90) is usually more robust to parameter errors than Eq. (7.91), particularly when the plant includes a pole at the origin and type 1 behavior is possible. The difference is most clearly illustrated by the following example.

EXAMPLE 7.11

Reference Input to a Type 1 System: DC Motor

Compute the input gains necessary to introduce a reference input with zero steady-state error to a step for the DC motor of Example 5.1, which in state-variable form is described by the matrices:

DC motor [Eq. (2.62)]

$$\mathbf{F} = \begin{bmatrix} 0 & 1 \\ 0 & -1 \end{bmatrix}, \qquad \mathbf{G} = \begin{bmatrix} 0 \\ 1 \end{bmatrix},$$
$$\mathbf{H} = [\,1 \;\; 0\,], \qquad J = 0.$$

Assume that the state feedback gain is $[\,K_1 \;\; K_2\,]$.

Solution.   If we substitute the system matrices of this example into the equation for the input gains, Eq. (7.89), we find that the solution is

$$\mathbf{N}_x = \begin{bmatrix} 1 \\ 0 \end{bmatrix},$$

$$N_u = 0,$$

$$\bar{N} = K_1.$$

Figure 7.12
Alternative structures for
introducing the reference
input. (a) Eq. (7.90);
(b) Eq. (7.91)

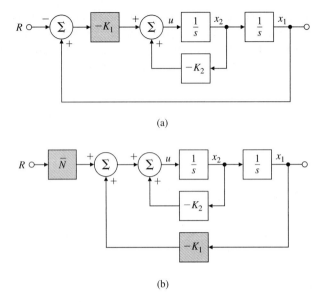

(a)

(b)

With these values the expression for the control using $\mathbf{N}_x$ and N_u [Eq. (7.90)] reduces
to

$$u = -K_1(x_1 - r) - K_2 x_2,$$

while the one using $\bar{N}$ [Eq. (7.91)] becomes

$$u = -K_1 x_1 - K_2 x_2 + K_1 r.$$

The block diagrams for the systems using each of the control equations are given in
Fig. 7.12. When using Eq. (7.91), as shown in Fig. 7.12(b), it is necessary to multiply the
input by a gain $K_1(= \bar{N})$ exactly equal to that used in the feedback. If these two gains
do not match exactly, there will be a steady-state error. On the other hand, if we use
Eq. (7.90) as shown in Fig. 7.12(a), there is only one gain to be used on the difference
between the reference input and the first state, and zero steady-state error will result
even if this gain is slightly in error. The system of Fig. 7.12(a) is more robust than the
system of Fig. 7.12(b).

With the reference input in place, the closed-loop system has input r and
output y. From the state description we know that the system poles are at the
eigenvalues of the closed-loop system matrix, $\mathbf{F} - \mathbf{GK}$. In order to compute the
closed-loop transient response, it is necessary to know where the closed-loop
zeros of the transfer function from r to y are. They are to be found by applying
Eq. (7.55) to the closed-loop description, which we assume has no direct path
from input u to output y so that $J = 0$. The zeros are values of s such that

$$\det \begin{bmatrix} s\mathbf{I} - (\mathbf{F} - \mathbf{GK}) & -\bar{N}\mathbf{G} \\ \mathbf{H} & 0 \end{bmatrix} = 0. \qquad (7.94)$$

We can use two elementary facts about determinants to simplify Eq. (7.94). In the first place, if we divide the last column by $\bar{N}$, which is a scalar, then the point where the determinant is zero remains unchanged. The determinant is also not changed if we multiply the last column by $\mathbf{K}$ and add it to the first (block) column, with the result that the $\mathbf{GK}$ term is cancelled out. Thus the matrix equation for the zeros reduces to

$$\det \begin{bmatrix} s\mathbf{I} - \mathbf{F} & -\mathbf{G} \\ \mathbf{H} & 0 \end{bmatrix} = 0. \tag{7.95}$$

Equation (7.95) is the same as Eq. (7.55) for the zeros of the plant *before* the feedback was applied. The important conclusion is as follows:

> When full-state feedback is used as in Eq. (7.90b) or (7.91), the zeros remain unchanged by the feedback.

7.4   Selection of Pole Locations for Good Design

The first step in the pole-placement design approach is to decide on the closed-loop pole locations. When selecting pole locations, it is always useful to keep in mind that the control effort required is related to how far the open-loop poles are moved by the feedback. Furthermore, when a zero is near a pole, the system may be nearly uncontrollable and as we saw in Section 7.3, moving such poles requires large control gains and thus large control effort; however, the designer is able to temper the choices to take control effort into account. Therefore, a pole-placement philosophy that aims to fix only the undesirable aspects of the open-loop response and avoids either large increases in bandwidth or efforts to move poles that are near zeros will typically allow smaller gains and thus smaller control actuators than a philosophy that arbitrarily picks all the poles without regard to the original open-loop pole and zero locations.

Two methods of pole selection

In this section we discuss two techniques to aid in the pole-selection process. The first approach—dominant second-order poles—deals with pole selection without explicit regard for their effect on control effort; however, the designer is able to temper the choices to take control effort into account. The second method (called optimal control, or symmetric root locus) does specifically address the issue of achieving a balance between good system response and control effort.

7.4.1   Dominant Second-Order Poles

The step response corresponding to the second-order transfer function with complex poles at radius ω_n and damping ratio ζ was discussed in Chapter 3. The rise time, overshoot, and settling time can be deduced directly from the pole locations. We can choose the closed-loop poles for a higher-order system as a desired pair of dominant second-order poles, and select the rest of the poles to have real parts corresponding to sufficiently damped modes, so that the system will mimic a second-order response with reasonable control effort. We also must make sure that the zeros are far enough into the left half-plane (LHP) to avoid having any appreciable effect on the second-order behavior. A system with several lightly damped high-frequency vibration modes plus two rigid-body low-frequency modes lends itself to this philosophy. Here we can pick the low-frequency modes to achieve desired values of ω_n and ζ and select the rest of the poles to increase the damping of the high-frequency modes, while holding their frequency constant in order to minimize control effort. To illustrate this design method we obviously need a system of higher than second order; we will use the tape drive servomotor described in Example 7.3.

EXAMPLE 7.12

Pole Placement as a Dominant Second-Order System

Design the tape servomotor by the dominant second-order poles method to have no more than 5% overshoot and a rise time of no more than 4 sec. Keep the peak tension as low as possible.

Solution.   From the plots of the second-order transients in Fig. 3.23, a damping ratio $\zeta = 0.7$ will meet the overshoot requirement and, for this damping ratio, a rise time of 4 sec suggests a natural frequency of about 1/1.5. There are five poles in all, so the other three need to be placed far to the left of the dominant pair. For our purposes, "far" means the transients due to the fast poles should be over well before the transients due to the dominant poles, and we assume a factor of 4 in the respective undamped natural frequencies to be adequate. From these considerations, the desired poles are given by

$$\text{pc} = [-0.707 + 0.707 * \text{j}; -0.707 - 0.707 * \text{j}; -4; -4; -4]/1.5; \qquad (7.96)$$

MATLAB acker

With these desired poles, we can use the function acker with **F** and **G** from Example 7.3, Eq. (7.59), to find the control gains

$$\mathbf{K}_2 = [\,8.5123 \quad 20.3457 \quad -1.4911 \quad -7.8821 \quad 6.1927\,]. \qquad (7.97)$$

These are found using the MATLAB statements

```
F = [0 2 0 0 0;-.1 -.35 .1 .1.75;0 0 0 2 0;.4 .4 -.4 -1.4 0;0 -.03 0 0 -1];
G = [0;0;0;0;1];
pc = [-.707+.707*j;-.707-.707*j;-4;-4;-4]/1.5;
K2 = acker(F,G,pc)
```

The step response and the corresponding tension plots for this and another design (to be discussed in Section 7.4.2) are given in Figs. 7.13 and Figs. 7.14. Notice that the rise time is approximately 4 sec and the overshoot is about 5%, as specified.

Because the design process is iterative, the poles we selected should be seen as only a first step, to be followed by further modifications to meet the specifications as accurately as necessary.

For this example we happened to select adequate pole locations on the first try.

Figure 7.13
Step responses of the tape servomotor designs

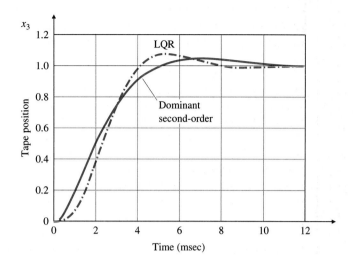

Figure 7.14
Tension plots for tape servomotor step responses

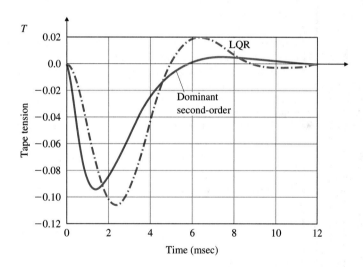

7.4.2   Symmetric Root Locus (SRL)

A most effective and widely used technique of linear control systems design is the optimal **linear quadratic regulator (LQR)**. The simplified version of the LQR problem is to find the control such that the performance index

LQR design

$$\mathcal{J} = \int_0^\infty [\rho z^2(t) + u^2(t)]\, dt \tag{7.98}$$

is minimized for the system

$$\dot{\mathbf{x}} = \mathbf{F}\mathbf{x} + \mathbf{G}u, \tag{7.99a}$$

$$z = \mathbf{H}_1\mathbf{x}, \tag{7.99b}$$

where ρ in Eq. (7.98) is a weighting factor of the designer's choice. A remarkable fact is that the control law that minimizes $\mathcal{J}$ is given by linear-state feedback

$$u = -\mathbf{K}\mathbf{x}. \tag{7.100}$$

Symmetric root locus

Here the optimal value of $\mathbf{K}$ is that which places the closed-loop poles at the stable roots (those in the LHP) of the symmetric root-locus (SRL) equation (Kailath, 1980)

$$1 + \rho G_0(-s)G_0(s) = 0, \tag{7.101}$$

where G_0 is the open-loop transfer function from u to z:

$$G_0(s) = \frac{Z(s)}{U(s)} = \mathbf{H}_1(s\mathbf{I} - \mathbf{F})^{-1}\mathbf{G} = \frac{N(s)}{D(s)}. \tag{7.102}$$

Note that this is a root-locus problem as discussed in Chapter 5 with respect to the parameter ρ, which weights the the relative cost of (tracking error) z^2 with respect to the control effort u^2 in the performance index Eq. (7.98). Note also that s and $-s$ affect Eq. (7.101) in an identical manner; therefore, for any root s_0 of Eq. (7.101), there will also be a root at $-s_0$. We call the resulting root locus a **symmetric root locus (SRL)**, since the locus in the LHP will have a mirror image in the right-half-plane (RHP); that is, there is symmetry with respect to the imaginary axis. We may thus choose the optimal closed-loop poles by first selecting the matrix $\mathbf{H}_1$, which defines the tracking error and which the designer wishes to keep small, and then choosing ρ, which balances the importance of this tracking error against the control effort. Notice that the output we select as tracking error does *not* need to be the plant sensor output. That is why we call the output in Eq. (7.99) z rather than y.

Selecting a set of stable poles from the solution of Eq. (7.101) results in desired closed-loop poles, which we can then use in a pole-placement calculation such as Ackermann's formula [Eq. (7.80)] to obtain $\mathbf{K}$. As with all root loci for real transfer functions G_0, the locus is also symmetric with respect to the real axis; thus there is symmetry with respect to both the real and imaginary axes.

SRL equation

We can write the SRL equation in the standard root-locus form

$$1 + \rho \frac{N(-s)N(s)}{D(-s)D(s)} = 0, \tag{7.103}$$

obtain the locus poles and zeros by reflecting the open-loop poles and zeros of the transfer function from U to Z across the imaginary axis (which doubles the number of poles and zeros), and then sketch the locus. Note that the locus could be either a $0°$ or $180°$ locus, depending on the sign of $G_0(-s)G_0(s)$ in

Eq. (7.101). A quick way to determine which type of locus to use ($0°$ or $180°$) is to pick the one that *has no part on the imaginary axis*. The real-axis rule of root locus plotting will reveal this right away. For the controllability assumptions we have made here, plus the assumption that all the system modes are present in the chosen output z, the optimal closed-loop system is guaranteed to be stable; thus no part of the locus can be on the imaginary axis. The following examples illustrate the use of the SRL.

EXAMPLE 7.13

Symmetric Root Locus for Servo Speed Control

Plot the SRL for the following servo speed control system with $z = y$:

$$\dot{y} = -ay + u, \tag{7.104a}$$

$$G_0(s) = \frac{1}{s + a}. \tag{7.104b}$$

Solution.   The SRL equation [Eq. (7.101)] for this example is

$$1 + \rho \frac{1}{(-s + a)(s + a)} = 0. \tag{7.105}$$

The SRL, shown in Fig. 7.15, is a $0°$ locus. The optimal (stable) pole can be determined explicitly in this case as

$$s = -\sqrt{a^2 + \rho}. \tag{7.106}$$

Thus the closed-loop root location that minimizes the performance index of Eq. (7.98) lies on the real axis at the distance given by Eq. (7.106) and is always to the left of the open-loop root.

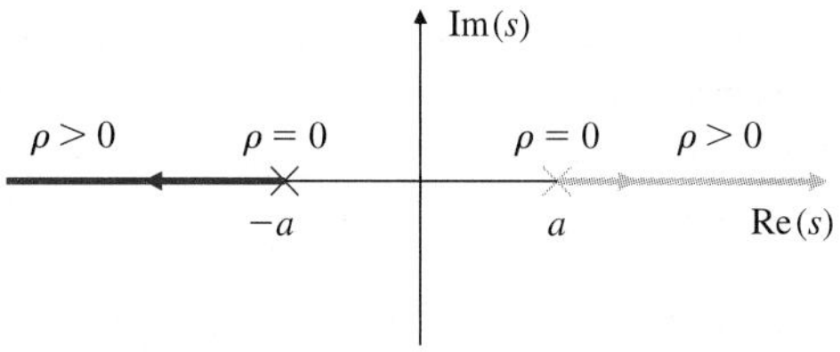

Figure 7.15
Symmetric root locus for a first-order system

EXAMPLE 7.14

SRL Design for Satellite Attitude Control

Draw the SRL for the satellite system with $z = y$.

Solution.   The equations of motion are

$$\dot{\mathbf{x}} = \begin{bmatrix} 0 & 1 \\ 0 & 0 \end{bmatrix} \mathbf{x} + \begin{bmatrix} 0 \\ 1 \end{bmatrix} u, \tag{7.107}$$

$$y = \begin{bmatrix} 1 & 0 \end{bmatrix} \mathbf{x}. \tag{7.108}$$

Figure 7.16
Symmetric root locus for
the satellite.

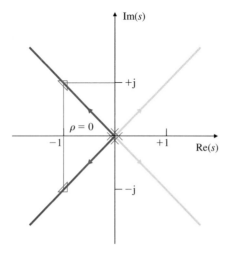

We then calculate from Eqs. (7.107) and (7.108) so that

$$G_0(s) = \frac{1}{s^2}. \tag{7.109}$$

The symmetric $180°$ loci are shown in Fig. 7.16. The MATLAB statements to generate the SRL are

```
numGG = [1];
denGG = conv([1 0 0],[1 0 0]);
sysGG = tf(numGG,denGG);
rlocus(sysGG);
```

It is interesting to note that the (stable) closed-loop poles have damping of $\zeta = 0.707$. We would choose two stable roots for a given value of ρ—for example $s = -1 \pm j1$ for $\rho = 4.07$—on the SRL and use them for pole-placement and control-law design.

Choosing different values of ρ can provide us with pole locations that achieve varying balances between a fast response (small values of $\int z^2 \, dt$ and a low control effort (small values of $\int u^2 \, dt$). Figure 7.17 shows the design-tradeoff curve for the satellite (double-integrator) plant [Eq. (7.7)] for various values of ρ ranging from .01 to 100. The curve has two asymptotes (dashed lines) corresponding to low (large ρ) and high (small ρ) penalty on the control usage. In practice, usually a value of ρ corresponding to a point close to the knee of the tradeoff curve is chosen. This is because it provides a reasonable compromise between the use of control and the speed of response. For the satellite plant, the value of $\rho = 1$ corresponds to the knee of the curve. In this case the closed-loop poles have a damping ratio of $\zeta = 0.707$! Figure 7.18 shows the associated Nyquist plot, which has a phase margin PM $= 65°$ and infinite gain margin. These excellent stability properties are a general feature of LQR designs.

Figure 7.17
Design trade-off curve for
satellite plant

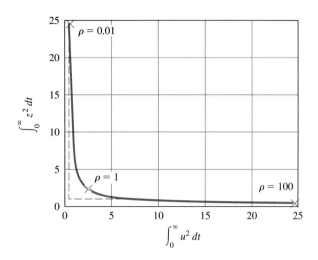

Figure 7.18
Nyquist plot for LQR design

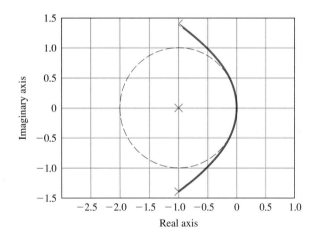

It is also possible to locate optimal pole locations for the design of an open-loop unstable system using the SRL and LQR method.

EXAMPLE 7.15   *SRL Design for an Inverted Pendulum*

Draw the SRL for the linearized equations of the simple inverted pendulum with $\omega_o = 1$. Take the output, z, to be the sum of twice the position plus the velocity (so as to weight or penalize *both* position and velocity).

Solution.   The equations of motion are

$$\dot{\mathbf{x}} = \begin{bmatrix} 0 & 1 \\ \omega_0^2 & 0 \end{bmatrix} \mathbf{x} + \begin{bmatrix} 0 \\ -1 \end{bmatrix} u. \tag{7.110}$$

For the specified output of $2\times$ position plus velocity we let the tracking error be

$$z = [\,2 \quad 1\,]\mathbf{x}. \tag{7.111}$$

We then calculate from Eqs. (7.110) and (7.111) so that

$$G_0(s) = -\frac{s+2}{s^2 - \omega_0^2}. \tag{7.112}$$

The symmetric $0°$ loci are shown in Fig. 7.19. The MATLAB statements to generate the SRL are (for $\omega_o = 1$),

```
numGG = conv(-[1 2],-[-1 2]);
denGG = conv([1 0 −1],[1 0 −1]);
sysGG = tf(numGG,denGG);
rlocus(sysGG);
```

For $\rho = 1$, we find that the closed-loop poles are at $-1.36 \pm j0.606$, corresponding to $\mathbf{K} = [\,-2.23 \quad -2.73\,]$. If we substitute the system matrices of this example into the equation for the input gains, Eq. (7.89), we find that the solution is

$$\mathbf{N}_x = \begin{bmatrix} 1 \\ 0 \end{bmatrix},$$
$$N_u = 1,$$
$$\bar{N} = -1.23.$$

With these values the expression for the control using $\mathbf{N}_x$ and N_u [Eq. (7.90)] the controller reduces to

$$u = -\mathbf{K}\mathbf{x} + \bar{N}r.$$

The corresponding step response for position is shown in Fig. 7.20.

Figure 7.19
Symmetric root locus for
the inverted pendulum

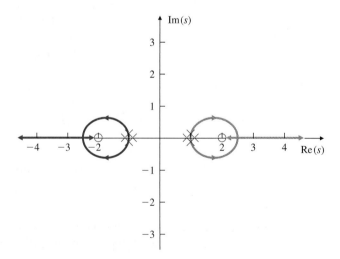

Figure 7.20
Step response for the
inverted pendulum

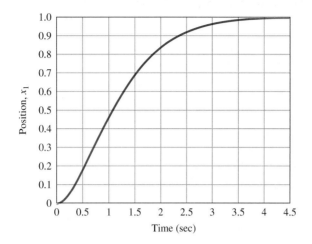

As a final example in this section, we consider again the tape servomotor and introduce LQR design using the computer directly to solve for the optimal control law. From Eqs. (7.98) and (7.100) we know the information needed to find the optimal control is given by the system matrices $\mathbf{F}$ and $\mathbf{G}$ and the output matrix $\mathbf{H}_1$. Most computer-aided software packages, including MATLAB, use a more general form of Eq. (7.98):

$$\mathcal{J} = \int_0^\infty (\mathbf{x}^T \mathbf{Q} \mathbf{x} + \mathbf{u}^T \mathbf{R} \mathbf{u}) \, dt. \tag{7.113}$$

MATLAB lqr

Equation (7.113) reduces to the simpler form of Eq. (7.98) if we take $\mathbf{Q} = \rho \mathbf{H}_1^T \mathbf{H}_1$ and $\mathbf{R} = 1$. The direct solution for the optimal control gain is the MATLAB statement

$$K = \mathsf{lqr}(F, G, Q, R). \tag{7.114}$$

Bryson's Rule

One reasonable method to start the LQR design iteration is suggested by Bryson's rule (Bryson and Ho, 1969). In practice, an appropriate choice to obtain acceptable values of $\mathbf{x}$ and $\mathbf{u}$ is to initially choose diagonal matrices $\mathbf{Q}$ and $\mathbf{R}$ such that

$$Q_{ii} = 1/\text{maximum acceptable value of } [x_i^2],$$
$$R_{ii} = 1/\text{maximum acceptable value of } [u_i^2].$$

The weighting matrices are then modified during subsequent iterations to achieve an acceptable tradeoff between performance and control effort.

EXAMPLE 7.16   *LQR Design for a Tape Drive*

(a) Find the optimal control for the tape drive of Example 7.6, using the position x_3 as the output for the performance index. Let $\rho = 1$. Compare the results with that of dominant second-order obtained before.

(b) Compare the LQR designs for $\rho = 0.1, 1, 10$.

Solution.

(a) All we need to do here is to substitute the matrices into Eq. (7.114), form the feedback system, and plot the response. The performance index matrix is the scalar $R = 1$; the most difficult part of the problem is finding the state-cost matrix **Q**. With the output-cost variable $z = x_3$, the output matrix from Example 7.3 is

$$\mathbf{H}_3 = [\,0.5 \quad 0 \quad 0.5 \quad 0 \quad 0\,],$$

and with $\rho = 1$, then the required matrix is

$$\mathbf{Q} = \mathbf{H}_3^T \mathbf{H}_3$$

$$= \begin{bmatrix} 0.25 & 0 & 0.25 & 0 & 0 \\ 0 & 0 & 0 & 0 & 0 \\ 0.25 & 0 & 0.25 & 0 & 0 \\ 0 & 0 & 0 & 0 & 0 \end{bmatrix}.$$

The gain is given by MATLAB, using the statements

```
F = [0 2 0 0 0;-.1 -.35 .1 .1 .75;0 0 0 2 0;.4 .4 -.4 −1.4;0 -.03 0 0 −1];
G = [0;0;0;0;1];
H3 = [.5 0 .5 0 0];
R = 1;
rho = 1;
Q = rho*H3'*H3;
K = lqr(F,G,Q,R)
```

as

$$\mathbf{K} = [\,0.6526 \quad 2.1667 \quad 0.3474 \quad 0.5976 \quad 1.0616\,]. \tag{7.115}$$

The results of a position step and the corresponding tension are plotted in Figs. 7.13 and 7.14 (using **step**) with the earlier responses for comparison. Obviously, there is a vast range of choice for the elements of **Q** and **R**, so substantial experience is needed in order to use the LQR method effectively.

(b) The LQR designs may be repeated as above with the same **Q** and **R**, but with $\rho = 0.1, 10$. Figure 7.21 shows a comparison of position step and the corresponding tension for the three designs. As seen from the results, the smaller values of ρ correspond to higher cost on the control and slower response, whereas the larger values of ρ correspond to lower cost on the control and relatively fast response.

Figure 7.21
(a) Step response of the
tape servomotor for LQR
designs, (b) Corresponding
tension for tape servomotor
step responses

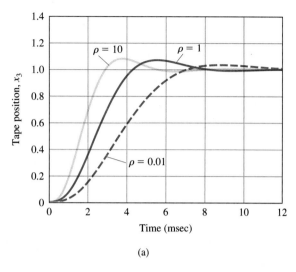

(a)

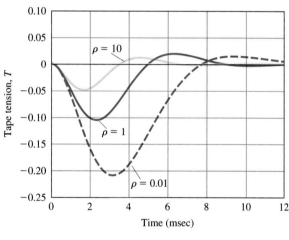

(b)

Limiting Behavior of LQR Regulator Poles

It is interesting to consider the limiting behavior of the optimal closed-loop
poles as a function of the root-locus parameter (i.e., ρ), although, in practice,
neither case would be used.

"*Expensive control*" case ($\rho \to 0$).   Equation (7.98) primarily penalizes
the use of control energy. If the control is expensive, then the optimal control
does not move any of the open-loop poles except for those which are in the
right half-plane (RHP). The poles in the RHP are simply moved to their mirror
images in the LHP. The optimal control does this to stabilize the system using

minimum control effort and makes no attempt to move any of the poles of the system in the LHP. The closed-loop pole locations are simply the starting points on the SRL in the LHP. The optimal control does not speed up the response of the system in this case. For the satellite plant, the vertical dashed line in Fig. 7.17 corresponds to the "expensive control" case and illustrates that the very low control usage results in a very large error in z.

"*Cheap control*" case ($\rho \to \infty$).   In this case control energy is no object and arbitrary control effort may be used by the optimal control law. The control law then moves some of the closed-loop pole locations right on top of the zeros in the LHP. The rest are moved to infinity along the SRL asymptotes. If the system is nonminimum phase, some of the closed-loop poles are moved to mirror images of these zeros in the LHP as shown in Example 7.15. The rest of the poles go to infinity and do so along a Butterworth filter pole pattern as shown in Example 7.14. The optimal control law provides the fastest possible response time consistent with the LQR cost function. The feedback gain matrix **K** becomes unbounded in this case. For the double-integrator plant, the horizontal dashed line in Fig. 7.17 corresponds to the "cheap control" case.

Robustness Properties of LQR Regulators

It has been proved (Anderson and Moore, 1990) that the Nyquist plot for LQR design avoids a circle of radius one centered at the -1 point as shown in Fig. 7.18. This leads to extraordinary phase and gain margin properties. It can be shown (see Problem 7.30) that the return difference must satisfy

$$|1 + \mathbf{K}(j\omega\mathbf{I} - \mathbf{F})^{-1}\mathbf{G}| \geq 1. \tag{7.116}$$

Let us rewrite the loop gain as sum of its real and imaginary parts:

$$L(j\omega) = \mathbf{K}(j\omega\mathbf{I} - \mathbf{F})^{-1}\mathbf{G} = \mathbf{Re}(L(j\omega)) + j\mathbf{Im}(L(j\omega)). \tag{7.117}$$

Equation (7.116) implies that

$$([\mathbf{Re}(L(j\omega)] + 1)^2 + [\mathbf{Im}(L(j\omega)]^2 \geq 1, \tag{7.118}$$

which means that the Nyquist plot must indeed avoid a circle centered at -1 with unit radius. This implies that $\frac{1}{2} <$ GM $< \infty$, which means that the "upward" gain margin is GM $= \infty$ and the "downward" gain margin is GM $= \frac{1}{2}$ (see also Problem 6.23). Hence the LQR gain matrix, **K**, can be multiplied by a large scalar or reduced by half with guaranteed closed-loop system stability. The phase margin, PM, is at least $\pm 60°$. These margins are remarkable and it is not realistic to assume that they can be achieved in practice because of the presence of modeling errors and lack of sensors!

LQR gain and phase margins

7.4.3   Comments on the Methods

The two methods of pole selection described in Sections 7.4.1 to 7.4.2 are alternatives the designer can use for an initial design by pole placement. Note that the first method (dominant second-order) suggests selecting closed-loop poles without regard to the effect on the control effort required to achieve that response. In some cases, therefore, the resulting control effort may be ridiculously high. The second method (SRL), on the other hand, selects poles that result in some balance between system errors and control effort. The designer can easily examine the relationship between shifts in that balance (by changing ρ) and system root locations, time response, and feedback gains. Whatever initial pole-selection method we use, some modification is almost always necessary to achieve the desired balance of bandwidth, overshoot, sensitivity, control effort, and other practical design requirements. Further insight into pole selection will be gained from the examples that illustrate compensation in Section 7.6 and from the case studies in Chapter 9.

7.5   Estimator Design

The control law designed in Section 7.3 assumed that all the state variables are available for feedback. In most cases, not all the state variables are measured. The cost of the required sensors may be prohibitive, or it may be physically impossible to measure all the state variables, as in, for example, a nuclear power plant. In this section we demonstrate how to reconstruct all the state variables of a system from a few measurements. If the estimate of the state is denoted by $\hat{\mathbf{x}}$, it would be convenient if we could replace the true state in the control law given by Eq. (7.91) with the estimates so that the control becomes $u = -\mathbf{K}\hat{\mathbf{x}} + \bar{N}r$. This is indeed possible, as we shall see in Section 7.6, so construction of a state estimate is a key part of state-space control design.

7.5.1   Full-Order Estimators

One method of estimating the state is to construct a full-order model of the plant dynamics,

$$\dot{\hat{\mathbf{x}}} = \mathbf{F}\hat{\mathbf{x}} + \mathbf{G}u, \tag{7.119}$$

where $\hat{\mathbf{x}}$ is the estimate of the actual state $\mathbf{x}$. We know $\mathbf{F}$, $\mathbf{G}$, and $u(t)$. Hence this estimator will be satisfactory if we can obtain the correct initial condition $\mathbf{x}(0)$ and set $\hat{\mathbf{x}}(0)$ equal to it. Figure 7.22 depicts this open-loop estimator. However, it is precisely the lack of information about $\mathbf{x}(0)$ that requires the construction of an estimator. Otherwise, the estimated state would track the true state exactly. Thus, if we made a poor estimate for the initial condition, the estimated state would have a continually growing error or an error that goes to zero too slowly to be of use. Furthermore, small errors in our knowledge of the system $(\mathbf{F}, \mathbf{G})$ would also cause the estimate to diverge from the true state.

Figure 7.22
Open-loop estimator

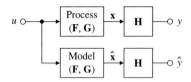

To study the dynamics of this estimator we define the error in the estimate to be

$$\tilde{\mathbf{x}} \overset{\triangle}{=} \mathbf{x} - \hat{\mathbf{x}}. \tag{7.120}$$

Then the dynamics of this error system are given by

$$\dot{\tilde{\mathbf{x}}} = \mathbf{F}\tilde{\mathbf{x}}, \quad \tilde{\mathbf{x}}(0) = \mathbf{x}(0) - \hat{\mathbf{x}}(0). \tag{7.121}$$

The error converges to zero for a stable system (**F** stable), but we have no ability to influence the rate at which the state estimate converges to the true state. Furthermore, the error is converging to zero at the same rate as the natural dynamics of **F**. If this convergence rate were satisfactory, no control or estimation would be required.

We now invoke the golden rule: When in trouble, use feedback. Consider feeding back the difference between the measured and estimated outputs and correcting the model continuously with this error signal. The equation for this scheme, shown in Fig. 7.23, is

Feed back the output error to correct the state estimate equation.

$$\dot{\hat{\mathbf{x}}} = \mathbf{F}\hat{\mathbf{x}} + \mathbf{G}u + \mathbf{L}(y - \mathbf{H}\hat{\mathbf{x}}). \tag{7.122}$$

Here **L** is a proportional gain defined as

$$\mathbf{L} = [l_1, l_2, \dots, l_n]^T \tag{7.123}$$

and is chosen to achieve satisfactory error characteristics. The dynamics of the error can be obtained by subtracting the estimate [Eq. (7.122)] from the state [Eq. (7.32)], to get the error equation

$$\dot{\tilde{\mathbf{x}}} = (\mathbf{F} - \mathbf{L}\mathbf{H})\tilde{\mathbf{x}}. \tag{7.124}$$

Figure 7.23
Closed-loop estimator

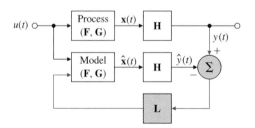

Estimate-error characteristic
equation

The characteristic equation of the error is now given by

$$\det[s\mathbf{I} - (\mathbf{F} - \mathbf{L}\mathbf{H})] = 0. \tag{7.125}$$

If we can choose $\mathbf{L}$ so that $\mathbf{F} - \mathbf{L}\mathbf{H}$ has stable and reasonably fast eigenvalues, then $\tilde{\mathbf{x}}$ will decay to zero and remain there—independent of the known forcing function $u(t)$ and its effect on the state $\mathbf{x}(t)$ and irrespective of the initial condition $\tilde{\mathbf{x}}(0)$. This means that $\hat{\mathbf{x}}(t)$ will converge to $\mathbf{x}(t)$, regardless of the value of $\hat{\mathbf{x}}(0)$; furthermore, we can choose the dynamics of the error to be stable as well as much faster than the open-loop dynamics determined by $\mathbf{F}$.

Note that in obtaining Eq. (7.124) we have assumed that $\mathbf{F}$, $\mathbf{G}$, and $\mathbf{H}$ are identical in the physical plant and in the computer implementation of the estimator. If we do not have an accurate model of the plant ($\mathbf{F}$, $\mathbf{G}$, $\mathbf{H}$), the dynamics of the error are no longer governed by Eq. (7.124). However, we can typically choose $\mathbf{L}$ so that the error system is still at least stable and the error remains acceptably small, even with (small) modeling errors and disturbing inputs. It is important to emphasize that the nature of the plant and the estimator are quite different. The plant is a physical system such as a chemical process or servomechanism, whereas the estimator is usually a digital processor computing the estimated state according to Eq. (7.122).

The selection of $\mathbf{L}$ can be approached in exactly the same fashion as $\mathbf{K}$ is selected in the control-law design. If we specify the desired location of the estimator error poles as

$$s_i = \beta_1, \beta_2, \ldots, \beta_n,$$

then the desired estimator characteristic equation is

$$\alpha_e(s) \stackrel{\triangle}{=} (s - \beta_1)(s - \beta_2) \cdots (s - \beta_n). \tag{7.126}$$

We can then solve for $\mathbf{L}$ by comparing coefficients in Eqs. (7.125) and (7.126).

EXAMPLE 7.17

An Estimator Design for a Simple Pendulum

Design an estimator for the simple pendulum. Compute the estimator gain matrix that will place both the estimator error poles at $-10\omega_0$ (five times as fast as the controller poles selected in Example 7.7). Verify the result using MATLAB for $\omega_0 = 1$. Evaluate the performance of the estimator.

Solution.   The equations of motion are

$$\dot{\mathbf{x}} = \begin{bmatrix} 0 & 1 \\ -\omega_0^2 & 0 \end{bmatrix} \mathbf{x} + \begin{bmatrix} 0 \\ 1 \end{bmatrix} u, \tag{7.127a}$$

$$y = [\,1 \quad 0\,]\mathbf{x}. \tag{7.127b}$$

We are asked to place the two estimator error poles at $-10\omega_0$. The corresponding characteristic equation is

$$\alpha_e(s) = (s + 10\omega_0)^2 = s^2 + 20\omega_0 s + 100\omega_0^2. \tag{7.128}$$

From Eq. (7.125) we get

$$\det[s\mathbf{I} - (\mathbf{F} - \mathbf{LH})] = s^2 + l_1 s + l_2 + \omega_0^2. \tag{7.129}$$

Comparing coefficients in Eqs. (7.128) and (7.129), we find that

$$\mathbf{L} = \begin{bmatrix} l_1 \\ l_2 \end{bmatrix} = \begin{bmatrix} 20\omega_0 \\ 99\omega_0^2 \end{bmatrix}. \tag{7.130}$$

The result can also be found from MATLAB for $\omega_0 = 1$, using the following MATLAB statements:

```
wo = 1;
F = [0 1;-wo*wo 0];
H = [1 0];
pe = [-10*wo;-10*wo];
Lt = acker(F',H',pe);
L = Lt'
```

This yields $\mathbf{L} = [20\ 99]^T$ and agrees with the above hand calculations.

Performance of the estimator can be tested by adding the actual state feedback to the plant and plotting the estimate errors. Note that this is not the way the system will ultimately be built, but this approach provides a means of to validate the estimator performance. Combining Eq. (7.60) of the plant with state feedback with Eq. (7.122) of the estimator with output feedback results in the following overall system equations:

$$\begin{bmatrix} \dot{\mathbf{x}} \\ \dot{\hat{\mathbf{x}}} \end{bmatrix} = \begin{bmatrix} \mathbf{F} - \mathbf{GK} & \mathbf{0} \\ \mathbf{LH} - \mathbf{GK} & \mathbf{F} - \mathbf{LH} \end{bmatrix} \begin{bmatrix} \mathbf{x} \\ \hat{\mathbf{x}} \end{bmatrix}, \tag{7.131}$$

$$y = [\mathbf{H} \quad \mathbf{0}] \begin{bmatrix} \mathbf{x} \\ \hat{\mathbf{x}} \end{bmatrix}, \tag{7.132}$$

$$\tilde{y} = [\mathbf{H} \quad -\mathbf{H}] \begin{bmatrix} \mathbf{x} \\ \hat{\mathbf{x}} \end{bmatrix}. \tag{7.133}$$

A block diagram of the setup is drawn in Fig. 7.24.

The response of this closed-loop system with $\omega_0 = 1$ to an initial condition $\mathbf{x}_0 = [1.0, 0.0]^T$ and $\hat{\mathbf{x}}_0 = [0, 0]^T$ is shown in Fig. 7.25, where $\mathbf{K}$ is obtained from Example 7.7 and $\mathbf{L}$ comes from Eq. (7.130). The response may be obtained using impulse or initial in MATLAB. Note that the state estimates converge to the actual state variables after an initial transient even though the initial value of $\hat{\mathbf{x}}$ had a large error. Also note that the estimate error decays approximately five times faster than the decay of the state itself, as we designed it to do.

MATLAB impulse, initial

Figure 7.24

Estimator connected to the plant

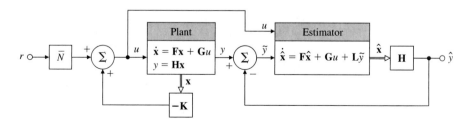

Figure 7.25
Initial-condition response of
oscillator showing x and $\hat{x}$

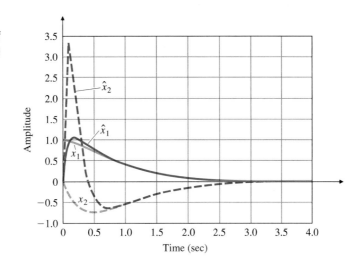

Observer Canonical Form

As was the case for control-law design, there is a canonical form for which the estimator-gain design equations are particularly simple and the existence of a solution is obvious. We introduced this form in Section 7.2.1: The equations are in the observer canonical form and have the following structure:

$$\dot{\mathbf{x}}_o = \mathbf{F}_o\mathbf{x}_o + \mathbf{G}_o u, \qquad (7.134a)$$

$$y = \mathbf{H}_o\mathbf{x}_o, \qquad (7.134b)$$

Observer canonical form

where

$$\mathbf{F}_o = \begin{bmatrix} -a_1 & 1 & 0 & 0 & \cdots & 0 \\ -a_2 & 0 & 1 & 0 & \cdots & \vdots \\ \vdots & \vdots & & \ddots & & 1 \\ -a_n & 0 & & 0 & & 0 \end{bmatrix}, \qquad \mathbf{G}_0 = \begin{bmatrix} b_1 \\ b_2 \\ \vdots \\ b_n \end{bmatrix},$$

$$\mathbf{H}_o = \begin{bmatrix} 1 & 0 & 0 & \cdots & 0 \end{bmatrix}$$

A block diagram for the third-order case is shown in Fig. 7.26. In observer canonical form, all the feedback loops come from the output, or observed signal. Like control canonical form, observer canonical form is a "direct" form because the values of the significant elements in the matrices are obtained directly from the coefficients of the numerator and denominator polynomials of the corresponding transfer function $G(s)$. The matrix $\mathbf{F}_o$ is called a **left companion matrix** to the characteristic equation because the coefficients of the equation appear on the left side of the matrix.

Figure 7.26
Block diagram for observer
canonical form of a
third-order system

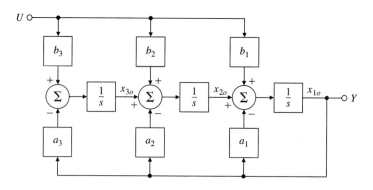

One of the advantages of the observer canonical form is that the estimator gains can be obtained from it by inspection. The estimator error closed-loop matrix for the third-order case is

$$\mathbf{F}_o - \mathbf{L}\mathbf{H}_o = \begin{bmatrix} -a_1 - l_1 & 1 & 0 \\ -a_2 - l_2 & 0 & 1 \\ -a_3 - l_3 & 0 & 0 \end{bmatrix}, \tag{7.135}$$

which has the characteristic equation

$$s^3 + (a_1 + l_1)s^2 + (a_2 + l_2)s + (a_3 + l_3) = 0, \tag{7.136}$$

and the estimator gain can be found by comparing the coefficients of Eq. (7.136) with $\alpha_e(s)$ from Eq. (7.126).

In a development exactly parallel with the control-law case, we can find a transformation to take a given system to observer canonical form if and only if the system has a structural property that in this case we call **observability**. Roughly speaking, observability refers to our ability to deduce information about all the modes of the system by monitoring only the sensed outputs. Unobservability results when some mode or subsystem is disconnected physically from the output and therefore no longer appears in the measurements. For example, if only derivatives of certain state variables are measured, and these state variables do not affect the dynamics, a constant of integration is obscured. This situation occurs with a plant having the transfer function $1/s^2$ if only velocity is measured, because then it is impossible to deduce the initial value of the position. On the other hand, for an oscillator a velocity measurement is sufficient to estimate position because the acceleration, and consequently the velocity observed, is affected by position. The mathematical test for determining observability is that the **observability matrix**

Observability

Observability matrix

$$\mathcal{O} = \begin{bmatrix} \mathbf{H} \\ \mathbf{H}\mathbf{F} \\ \vdots \\ \mathbf{H}\mathbf{F}^{n-1} \end{bmatrix} \tag{7.137}$$

must have independent columns. In the one output case we will study, $\mathcal{O}$ is square, so the requirement is that $\mathcal{O}$ be nonsingular or have a nonzero determinant. In general, we can find a transformation to observer canonical form if and only if the observability matrix is nonsingular. Note that this is analogous to our earlier conclusions for transforming system matrices to control canonical form.

As with control-law design, we could find the transformation to observer form, compute the gains from the equivalent of Eq. (7.136), and transform back. An alternative method of computing **L** is to use Ackermann's formula in estimator form, which is

Ackermann's estimator formula

$$\mathbf{L} = \alpha_e(\mathbf{F})\mathcal{O}^{-1}\begin{bmatrix} 0 \\ 0 \\ \vdots \\ 1 \end{bmatrix}, \tag{7.138}$$

where $\mathcal{O}$ is the observability matrix given in Eq. (7.137).

Duality

You may already have noticed from this discussion the considerable resemblance between estimation and control problems. In fact, the two problems are mathematically equivalent. This property is called **duality**. Table 7.1 shows the duality relationships between the estimation and control problems. For example, Ackermann's control formula [Eq. (7.80)] becomes the estimator formula Eq. (7.138) if we make the substitutions given in Table 7.1. We can demonstrate this directly using matrix algebra. The control problem is to select the row matrix **K** for satisfactory placement of the poles of the system matrix $\mathbf{F} - \mathbf{GK}$; the estimator problem is to select the column matrix **L** for satisfactory placement of the poles of $\mathbf{F} - \mathbf{LH}$. However, the poles of $\mathbf{F} - \mathbf{LH}$ equal those of $(\mathbf{F} - \mathbf{LH})^T = \mathbf{F}^T - \mathbf{H}^T\mathbf{L}^T$, and in this form, the algebra of the design for $\mathbf{L}^T$ is identical to that for **K**. Therefore, where we used Ackermann's formula or the place algorithm in the forms

Duality of estimation and control

MATLAB acker, place

$$K = \text{acker}(F, G, p_c),$$

$$K = \text{place}(F, G, p_c),$$

TABLE 7.1

Duality

Control	Estimation
F	$\mathbf{F}^T$
G	$\mathbf{H}^T$
H	$\mathbf{G}^T$

for the control problem, we use

$$\text{Lt} = \text{acker}(\text{F}', \text{H}', \text{p}_e),$$

$$\text{Lt} = \text{place}(\text{F}', \text{H}', \text{p}_e),$$

$$\text{L} = \text{Lt}',$$

where p_e is a vector containing the desired estimator error poles for the estimator problem.

Thus duality allows us to use the same design tools for estimator problems as for control problems with proper substitutions. The two canonical forms are also dual, as we can see from comparing the triples $(\mathbf{F}_c, \mathbf{G}_c, \mathbf{H}_c)$ and $(\mathbf{F}_o, \mathbf{G}_o, \mathbf{H}_o)$.

▲ 7.5.2   Reduced-Order Estimators

The estimator design method described in Section 7.5.1 reconstructs the entire state vector using measurements of some of the state variables. If the sensors have no noise, then a full-order estimator contains redundancies, and it seems reasonable to question the necessity for estimating state variables that are measured directly. Can we reduce the complexity of the estimator by using the state variables that are measured directly and exactly? Yes. However, it is better to implement a full-order estimator if there is significant noise on the measurements because, in addition to estimating unmeasured state variables, the estimator filters the measurements.

The **reduced-order estimator** reduces the order of the estimator by the number (1 in this text) of sensed outputs. To derive this estimator we start with the assumption that the output equals the first state as, for example, $y = x_a$. If this is not so, a preliminary step is required. Transforming to observer form is possible but overkill; any nonsingular transformation with $\mathbf{H}$ as the first row will do. We now partition the state vector into two parts: x_a, which is directly measured, and $\mathbf{x}_b$, which represents the remaining state variables that need to be estimated. If we partition the system matrices accordingly, the complete description of the system is given by

$$\begin{bmatrix} \dot{x}_a \\ \dot{\mathbf{x}}_b \end{bmatrix} = \begin{bmatrix} F_{aa} & F_{ab} \\ \mathbf{F}_{ba} & \mathbf{F}_{bb} \end{bmatrix} \begin{bmatrix} x_a \\ \mathbf{x}_b \end{bmatrix} + \begin{bmatrix} G_a \\ \mathbf{G}_b \end{bmatrix} u \qquad (7.139a)$$

$$y = \begin{bmatrix} 1 & \mathbf{0} \end{bmatrix} \begin{bmatrix} x_a \\ \mathbf{x}_b \end{bmatrix}. \qquad (7.139b)$$

The dynamics of the unmeasured state variables are given by

$$\dot{\mathbf{x}}_b = \mathbf{F}_{bb}\mathbf{x}_b + \underbrace{\mathbf{F}_{ba}x_a + \mathbf{G}_b u}_{\text{known input}}, \qquad (7.140)$$

where the rightmost two terms are known and can be considered as an input into the $\mathbf{x}_b$ dynamics. Because $x_a = y$, the measured dynamics are given by the scalar equation

$$\dot{x}_a = \dot{y} = F_{aa}y + \mathbf{F}_{ab}\mathbf{x}_b + G_a u. \tag{7.141}$$

If we collect the known terms of Eq. (7.141) on one side,

$$\underbrace{\dot{y} - F_{aa}y - G_a u}_{\text{known measurement}} = \mathbf{F}_{ab}\mathbf{x}_b, \tag{7.142}$$

we obtain a relationship between known quantities on the left side, which we consider measurements, and unknown state variables on the right. Therefore, Eqs. (7.141) and (7.142) have the same relationship to the state $\mathbf{x}_b$ that the original equation [Eq. (7.140)] had to the entire state $\mathbf{x}$. Following this line of reasoning, we can establish the following substitutions in the original estimator equations to obtain a (reduced-order) estimator of $\mathbf{x}_b$:

$$\mathbf{x} \leftarrow \mathbf{x}_b, \tag{7.143a}$$

$$\mathbf{F} \leftarrow \mathbf{F}_{bb}, \tag{7.143b}$$

$$Gu \leftarrow \mathbf{F}_{ba}y + \mathbf{G}_b u, \tag{7.143c}$$

$$y \leftarrow \dot{y} - F_{aa}y - G_a u, \tag{7.143d}$$

$$\mathbf{H} \leftarrow \mathbf{F}_{ab}. \tag{7.143e}$$

Therefore, the reduced-order estimator equations are obtained by substituting Eq. (7.143) into the full-order estimator [Eq. (7.122)]:

$$\dot{\hat{\mathbf{x}}}_b = \mathbf{F}_{bb}\hat{\mathbf{x}}_b + \underbrace{\mathbf{F}_{ba}y + \mathbf{G}_b u}_{\text{input}} + \mathbf{L}\underbrace{(\dot{y} - F_{aa}y - G_a u}_{\text{measurement}} - \mathbf{F}_{ab}\hat{\mathbf{x}}_b). \tag{7.144}$$

If we define the estimator error to be

$$\tilde{\mathbf{x}}_b \overset{\triangle}{=} \mathbf{x}_b - \hat{\mathbf{x}}_b, \tag{7.145}$$

then the dynamics of the error are given by subtracting Eq. (7.140) from Eq. (7.144) to get

$$\dot{\tilde{\mathbf{x}}}_b = (\mathbf{F}_{bb} - \mathbf{LF}_{ab})\tilde{\mathbf{x}}_b, \tag{7.146}$$

and its characteristic equation is given by

$$\det[s\mathbf{I} - (\mathbf{F}_{bb} - \mathbf{LF}_{ab})] = 0. \tag{7.147}$$

We design the dynamics of this estimator by selecting $\mathbf{L}$ so that Eq. (7.147) matches a reduced order $\alpha_e(s)$. Now Eq. (7.144) can be rewritten as

$$\dot{\hat{\mathbf{x}}}_b = (\mathbf{F}_{bb} - \mathbf{LF}_{ab})\hat{\mathbf{x}}_b + (\mathbf{F}_{ba} - \mathbf{L}F_{aa})y + (\mathbf{G}_b - \mathbf{L}G_a)u + \mathbf{L}\dot{y}. \tag{7.148}$$

Figure 7.27
Reduced-order estimator
structure

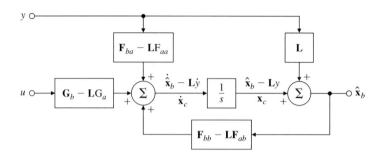

The fact that we must form the derivative of the measurements in Eq. (7.148) appears to present a practical difficulty. It is known that differentiation amplifies noise, so if y is noisy, the use of $\dot{y}$ is unacceptable. To get around this difficulty, we define the new controller state to be

$$\mathbf{x}_c \overset{\triangle}{=} \hat{\mathbf{x}}_b - \mathbf{L}y. \tag{7.149}$$

In terms of this new state the implementation of the reduced-order estimator is given by

$$\dot{\mathbf{x}}_c = (\mathbf{F}_{bb} - \mathbf{L}\mathbf{F}_{ab})\hat{\mathbf{x}}_b + (\mathbf{F}_{ba} - \mathbf{L}\mathbf{F}_{aa})y + (\mathbf{G}_b - \mathbf{L}\mathbf{G}_a)u, \tag{7.150}$$

Reduced-order estimator

and $\dot{y}$ no longer appears directly. A block-diagram representation of the reduced-order estimator is shown in Fig. 7.27.

EXAMPLE 7.18   *A Reduced-Order Estimator Design for Pendulum*

Design a reduced-order estimator for the pendulum that has the error pole at $-10\omega_0$.

Solution.   We are given the system equations

$$\begin{bmatrix} \dot{x}_1 \\ \dot{x}_2 \end{bmatrix} = \begin{bmatrix} 0 & 1 \\ -\omega_0^2 & 0 \end{bmatrix} \begin{bmatrix} x_1 \\ x_2 \end{bmatrix} + \begin{bmatrix} 0 \\ 1 \end{bmatrix} u,$$

$$y = \begin{bmatrix} 1 & 0 \end{bmatrix} \begin{bmatrix} x_1 \\ x_2 \end{bmatrix}.$$

The partitioned matrices are

$$\begin{bmatrix} F_{aa} & F_{ab} \\ F_{ba} & F_{bb} \end{bmatrix} = \begin{bmatrix} 0 & 1 \\ -\omega_0^2 & 0 \end{bmatrix},$$

$$\begin{bmatrix} G_a \\ G_b \end{bmatrix} = \begin{bmatrix} 0 \\ 1 \end{bmatrix}.$$

From Eq. (7.147) we find the characteristic equation in terms of L:

$$s - (0 - L) = 0.$$

Figure 7.28
Initial-condition response of
the reduced-order estimator

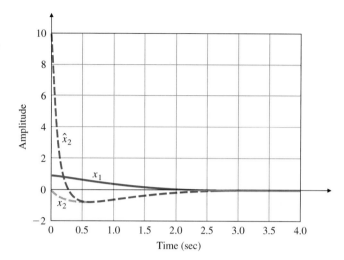

We compare it with the desired equation,

$$\alpha_e(s) = s + 10\omega_0 = 0,$$

which yields

$$L = 10\omega_0.$$

The estimator equation, from Eq. (7.150), is

$$\dot{x}_c = -10\omega_0\hat{x}_2 - \omega_0^2 y + u,$$

and the state estimate, from Eq. (7.149), is

$$\hat{x}_2 = x_c + 10\omega_0 y.$$

We use the control law given in the earlier examples. The response of the estimator to a plant initial condition $\mathbf{x}_0 = [1.0, 0.0]^T$ and an estimator initial condition $x_{c0} = 0$ is shown in Fig. 7.28 for $\omega_0 = 1$. The response may be obtained using impulse or initial in MATLAB. Note the similarity of the initial-condition response to that of the full-order estimator plotted in Fig. 7.25.

MATLAB impulse, initial

The reduced-order estimator gain can also be found from MATLAB by using

$$\text{Lt} = \text{acker}(\text{F}'_{bb}, \text{F}'_{ab}, \text{p}_e),$$

$$\text{Lt} = \text{place}(\text{F}'_{bb}, \text{F}'_{ab}, \text{p}_e),$$

$$\text{L} = \text{Lt}'.$$

The conditions for the existence of the reduced-order estimator are the same as for the full-order estimator—namely, observability of $(\mathbf{F}, \mathbf{H})$.

7.5.3   Estimator Pole Selection

We can base our selection of estimator-pole locations on the techniques discussed in Section 7.4 for the case of controller poles. As a rule of thumb, the estimator poles can be chosen to be faster than the controller poles by a factor of 2 to 6. This ensures a faster decay of the estimator errors compared with the desired dynamics, thus causing the controller poles to dominate the total response. If sensor noise is large enough to be a major concern, we may choose the estimator poles to be slower than two times the controller poles, which would yield a system with lower bandwidth and more noise smoothing. However, we would expect the total system response in this case to be strongly influenced by the location of the estimator poles. If the estimator poles are slower than the controller poles, we would expect the system response to disturbances to be dominated by the dynamic characteristics of the estimator rather than by those selected by the control law.

> **Design rules of thumb for selecting estimator poles**

In comparison with the selection of controller poles, estimator pole selection requires us to be concerned with a much different relationship than with control effort. As in the controller, there is a feedback term in the estimator that grows in magnitude as the requested speed of response increases. However, this feedback is in the form of an electronic signal or a digital word in a computer, so its growth causes no special difficulty. In the controller, increasing the speed of response increases the control effort; this implies the use of a larger actuator, which in turn increases size, weight, and cost. The important consequence of increasing the speed of response of an estimator is that the bandwidth of the estimator becomes higher, thus causing more sensor noise to pass on to the control actuator. Of course, if $(\mathbf{F}, \mathbf{H})$ are not observable, for then no amount of estimator gain can produce a reasonable state estimate. Thus, as with controller design, the best estimator design is a balance between good transient response and low-enough bandwidth that sensor noise does not significantly impair actuator activity. Dominant second-order characteristic equation ideas can be used to meet the requirements.

There is also a result for estimator gain design based on the SRL. In optimal estimation theory, the best choice for estimator gain is dependent on the ratio of sensor noise intensity v to process (disturbance) noise intensity [w in Eq. (7.152) below]. This is best understood by reexamining the estimator equation

$$\dot{\hat{\mathbf{x}}} = \mathbf{F}\hat{\mathbf{x}} + \mathbf{G}u + \mathbf{L}(y - \mathbf{H}\hat{\mathbf{x}}) \tag{7.151}$$

> **Process noise**

to see how it interacts with the system when process noise w is present. The plant with process noise is described by

$$\dot{\mathbf{x}} = \mathbf{F}\mathbf{x} + \mathbf{G}u + \mathbf{G}_1 w, \tag{7.152}$$

> **Sensor noise**

and the measurement equation with sensor noise v is described by

$$y = \mathbf{H}\mathbf{x} + v. \tag{7.153}$$

The estimator error equation with these additional inputs is found directly by subtracting Eq. (7.151) from Eq. (7.152) and substituting Eq. (7.153) for y:

$$\dot{\tilde{\mathbf{x}}} = (\mathbf{F} - \mathbf{LH})\tilde{\mathbf{x}} + \mathbf{G}_1 w - \mathbf{L}\nu. \qquad (7.154)$$

In Eq. (7.154) the sensor noise is multiplied by $\mathbf{L}$ and the process noise is not. If $\mathbf{L}$ is very small, then the effect of sensor noise is removed but the estimator's dynamic response will be "slow," so the error will not reject effects of w very well. The state of a low-gain estimator will not track uncertain plant inputs very well. These results can, with some success, also be applied to model errors in, for example, $\mathbf{F}$ or $\mathbf{G}$. Such model errors will add terms to Eq. (7.154) and act like additional process noise. On the other hand, if $\mathbf{L}$ is large, then the estimator response will be fast and the disturbance or process noise will be rejected, but the sensor noise, multiplied by $\mathbf{L}$, results in large errors. Clearly, a balance between these two effects is required. It turns out that the optimal solution to this balance can be found under very reasonable assumptions by solving an **SRL** equation for the estimator that is very similar to the one for the optimal control formulation [Eq. (7.101)]. The estimator **SRL** equation is

Estimator SRL equation

$$1 + q G_e(-s)G_e(s) = 0, \qquad (7.155)$$

where q is the ratio of input disturbance noise intensity to sensor noise intensity and G_e is the transfer function from the process noise to the sensor output and is given by

$$G_e(s) = \mathbf{H}(s\mathbf{I} - \mathbf{F})^{-1}\mathbf{G}_1. \qquad (7.156)$$

Note from Eqs. (7.101) and (7.155) that $G_e(s)$ is similar to $G_0(s)$. However, a comparison of Eqs. (7.102) and (7.156) shows that $G_e(s)$ has the input matrix $\mathbf{G}_1$ instead of $\mathbf{G}$, and that G_0 is the transfer function from the control input u to *cost* output z and has output matrix $\mathbf{H}_1$ instead of $\mathbf{H}$.

The use of the estimator **SRL** [Eq. (7.155)] is identical to the use of the controller **SRL**. A root locus with respect to q is generated, thus yielding sets of optimal estimator poles corresponding more or less to the ratio of process noise intensity to sensor noise intensity. The designer then picks the set of (stable) poles that seems best considering all aspects of the problem. An important advantage of using the **SRL** technique is that after the process noise input matrix $\mathbf{G}_1$ has been selected, the arbitrariness is reduced to one degree of freedom, the selection q, instead of the many degrees of freedom required to select the poles directly in a higher-order system.

A final comment concerns the reduced-order estimator. Because of the presence of a direct transmission term from y through $\mathbf{L}$ to $\mathbf{x}_b$ (see Fig. 7.27), the reduced-order estimator has a much higher bandwidth from sensor to control when compared with the full-order estimator. Therefore, if sensor noise is a significant factor, the reduced-order estimator is less attractive because the potential savings in complexity is more than offset by the increased sensitivity to noise.

EXAMPLE 7.19   *SRL Estimator Design for a Simple Pendulum*

Draw the estimator SRL for the linearized equations of the simple inverted pendulum with $\omega_o = 1$. Take the output to be a noisy measurement of position with noise intensity ratio q.

Solution.   We are given the system equations

$$\begin{bmatrix} \dot{x}_1 \\ \dot{x}_2 \end{bmatrix} = \begin{bmatrix} 0 & 1 \\ -\omega_0^2 & 0 \end{bmatrix} \begin{bmatrix} x_1 \\ x_2 \end{bmatrix} + \begin{bmatrix} 0 \\ 1 \end{bmatrix} w,$$

$$y = \begin{bmatrix} 1 & 0 \end{bmatrix} \begin{bmatrix} x_1 \\ x_2 \end{bmatrix} + v.$$

We then calculate from Eq. (7.156) that

$$G_e(s) = \frac{1}{s^2 + \omega_0^2}.$$

The symmetric 180° loci are shown in Fig. 7.29. The MATLAB statements to generate the SRL are (for $\omega_o = 1$),

```
numGG = 1;
denGG = conv([1 0 1],[1 0 1]);
sysGG = tf(numGG,denGG);
rlocus(sysGG);
```

We would choose two stable roots for a given value of q—for example, $s = -3 \pm j3.18$ for $q = 365$—and use them for estimator pole placement.

Figure 7.29
Symmetric root locus for the inverted pendulum estimator design

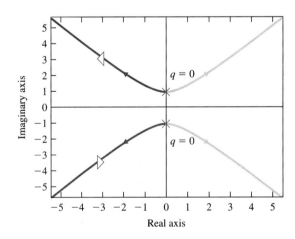

7.6 Compensator Design: Combined Control Law and Estimator

Regulator

If we take the control-law design described in Section 7.3, combine it with the estimator design described in Section 7.5, and implement the control law using the estimated state variables, the design is complete for a **regulator** that is able to reject disturbances but has no reference input to be tracked. However, because the control law was designed for feedback of the actual (not the estimated) state, you may wonder what effect using $\hat{\mathbf{x}}$ in place of $\mathbf{x}$ has on the system dynamics. In this section we compute this effect. In doing so we will compute the closed-loop characteristic equation and the open-loop compensator transfer function. We will use these results to compare the state-space designs with root-locus and frequency-response designs.

The plant equation with feedback is now

$$\dot{\mathbf{x}} = \mathbf{F}\mathbf{x} - \mathbf{G}\mathbf{K}\hat{\mathbf{x}}, \tag{7.157}$$

which can be rewritten in terms of the state error $\tilde{\mathbf{x}}$ as

$$\dot{\mathbf{x}} = \mathbf{F}\mathbf{x} - \mathbf{G}\mathbf{K}(\mathbf{x} - \tilde{\mathbf{x}}). \tag{7.158}$$

The overall system dynamics in state form are obtained by combining Eq. (7.158) with the estimator error [Eq. (7.124)] to get

$$\begin{bmatrix} \dot{\mathbf{x}} \\ \dot{\tilde{\mathbf{x}}} \end{bmatrix} = \begin{bmatrix} \mathbf{F} - \mathbf{G}\mathbf{K} & \mathbf{G}\mathbf{K} \\ \mathbf{0} & \mathbf{F} - \mathbf{L}\mathbf{H} \end{bmatrix} \begin{bmatrix} \mathbf{x} \\ \tilde{\mathbf{x}} \end{bmatrix}. \tag{7.159}$$

The characteristic equation of this closed-loop system is

$$\det \begin{bmatrix} s\mathbf{I} - \mathbf{F} + \mathbf{G}\mathbf{K} & -\mathbf{G}\mathbf{K} \\ \mathbf{0} & s\mathbf{I} - \mathbf{F} + \mathbf{L}\mathbf{H} \end{bmatrix} = 0. \tag{7.160}$$

Because the matrix is block triangular (see Appendix C), we can rewrite Eq. (7.160) as

$$\det(s\mathbf{I} - \mathbf{F} + \mathbf{G}\mathbf{K}) \cdot \det(s\mathbf{I} - \mathbf{F} + \mathbf{L}\mathbf{H}) = \alpha_c(s)\alpha_e(s) = 0. \tag{7.161}$$

Poles of the combined control law and estimator

In other words, the set of poles of the combined system consists of the union of the control poles and the estimator poles. This means that the designs of the control law and the estimator can be carried out independently, yet when they are used together in this way, the poles remain unchanged.[6]

To compare the state-variable method of design with the transform methods discussed in Chapters 5 and 6, we note from Fig. 7.30 that the blue shaded portion corresponds to a compensator. The state equation for this compensator

---

[6] This is a special case of the **separation principle** (Gunckel and Franklin, 1963), which holds in much more general contexts and allows us to obtain an overall optimal design by combining the separate designs of control law and estimator in certain stochastic cases.

Figure 7.30
Estimator and controller
mechanization

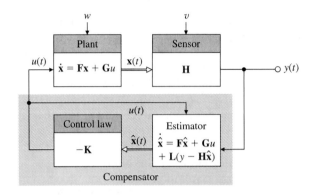

is obtained by including the feedback law $u = -\mathbf{K}\hat{\mathbf{x}}$ (because it is part of the compensator) in the estimator Eq. (7.122) to get

$$\dot{\hat{\mathbf{x}}} = (\mathbf{F} - \mathbf{GK} - \mathbf{LH})\hat{\mathbf{x}} + \mathbf{L}y, \qquad (7.162a)$$

$$u = -\mathbf{K}\hat{\mathbf{x}}, \qquad (7.162b)$$

Note that Eq. (7.162) has the same structure as Eq. (7.10), which we repeat here:

$$\dot{\mathbf{x}} = \mathbf{F}\mathbf{x} + \mathbf{G}u. \qquad (7.163)$$

Because the characteristic equation of Eq. (7.163) is

$$\det(s\mathbf{I} - \mathbf{F}) = 0, \qquad (7.164)$$

the characteristic equation of the compensator is found by comparing Eqs. (7.162a) and (7.163) and substituting the equivalent matrices into Eq. (7.164) to get

$$\det(s\mathbf{I} - \mathbf{F} + \mathbf{GK} + \mathbf{LH}) = 0. \qquad (7.165)$$

Note that we never specified the roots of Eq. (7.165) nor used them in our discussion of the state-space design technique. [Note also that the compensator is not guaranteed to be stable; the roots of Eq. (7.165) can be in the RHP.] The transfer function from y to u representing the dynamic compensator is obtained by inspecting Eq. (7.36) and substituting in the corresponding matrices from Eq. (7.162):

Compensator transfer function

$$D_c(s) = \frac{U(s)}{Y(s)} = -\mathbf{K}(s\mathbf{I} - \mathbf{F} + \mathbf{GK} + \mathbf{LH})^{-1}\mathbf{L}. \qquad (7.166)$$

The same development can be carried out for the reduced-order estimator. Here the control law is

$$u = -[\, K_a \quad K_b \,] \begin{bmatrix} x_a \\ \hat{x}_b \end{bmatrix} = -K_a y - K_b \hat{x}_b. \tag{7.167}$$

Substituting Eq. (7.167) into Eq. (7.163) and using Eq. (7.150) and some algebra, we obtain

$$\dot{x}_c = A_r x_c + B_r y, \tag{7.168a}$$

$$u = C_r x_c + D_r y, \tag{7.168b}$$

where

$$A_r = F_{bb} - LF_{ab} - (G_b - LG_a)K_b, \tag{7.169a}$$

$$B_r = A_r L + F_{ba} - LF_{aa} - (G_b - LG_a)K_a, \tag{7.169b}$$

$$C_r = -K_b, \tag{7.169c}$$

$$D_r = -K_a - K_b L. \tag{7.169d}$$

Reduced-order compensator
transfer function

The dynamic compensator now has the transfer function

$$D_{cr}(s) = \frac{U(s)}{Y(s)} = C_r(sI - A_r)^{-1}B_r + D_r. \tag{7.170}$$

When we compute $D_c(s)$ or $D_{cr}(s)$ for a specific case, we will find that they are very similar to the classical compensators given in Chapters 5 and 6, in spite of the fact that they are arrived at by entirely different means.

EXAMPLE 7.20   *Full-Order Compensator Design for Satellite Attitude Control*

Design a compensator using pole placement for the satellite plant with transfer function $1/s^2$. Place the control poles at $s = -0.707 \pm 0.707j$ ($\omega_n = 1$ rad/sec, $\zeta = 0.707$) and place the estimator poles at $\omega_n = 5$ rad/sec, $\zeta = 0.5$.

Solution.   A state-variable description for the given transfer function $G(s) = 1/s^2$ is

$$\dot{x} = \begin{bmatrix} 0 & 1 \\ 0 & 0 \end{bmatrix} x + \begin{bmatrix} 0 \\ 1 \end{bmatrix} u,$$

$$y = [\, 1 \quad 0 \,] x.$$

If we place the control roots at $s = -0.707 \pm 0.707j$ ($\omega_n = 1$ rad/sec, $\zeta = 0.7$), then

$$\alpha_c(s) = s^2 + s\sqrt{2} + 1. \tag{7.171}$$

The state feedback gain is found using K = place(F,G,pc) to be

$$\mathbf{K} = [1 \quad \sqrt{2}].$$

If the estimator-error roots are at $\omega_n = 5$ rad/sec and $\zeta = 0.5$, then the desired estimator characteristic polynomial is

$$\alpha_e(s) = s^2 + 5s + 25 = s + 2.5 \pm 4.3j, \tag{7.172}$$

and the estimator feedback-gain matrix is found using Lt = place(F',H',pe) to be

$$\mathbf{L} = \begin{bmatrix} 5 \\ 25 \end{bmatrix}.$$

The compensator transfer function given by Eq. (7.166) is

$$D_c(s) = -40.4 \frac{(s+0.619)}{s+3.21 \pm 4.77j}, \tag{7.173}$$

which looks very much like a lead compensator in that it has a zero on the real axis to the right of its poles; however, rather than one real pole, Eq. (7.173) has two complex poles. The zero provides the derivative feedback with phase lead, and the two poles provide some smoothing of sensor noise.

The effect of the compensation on this system's closed-loop poles can be evaluated in exactly the same way we evaluated compensation in Chapter 5 and 6 using root-locus or frequency-response tools. The gain of 40.4 in Eq. (7.173) is a result of the pole selection inherent in Eqs. (7.171) and (7.172). If we replace this specific value of compensator gain with a variable gain K, then the characteristic equation for the closed-loop system of plant plus compensator becomes

$$1 + K \frac{(s+0.619)}{(s+3.21 \pm 4.77j)s^2} = 0. \tag{7.174}$$

The root-locus technique allows us to evaluate the roots of this equation with respect to K, as drawn in Fig. 7.31. Note that the locus goes through the roots selected for Eqs. (7.171) and (7.172), and, when $K = 40.4$, the four roots of the closed-loop system are equal to those specified.

Figure 7.31
Root locus for the combined control and estimator, with process gain as the parameter

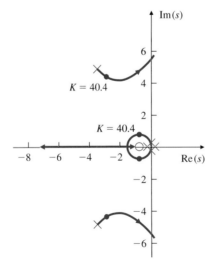

Figure 7.32
Frequency response for
$G(s) = 1/s^2$

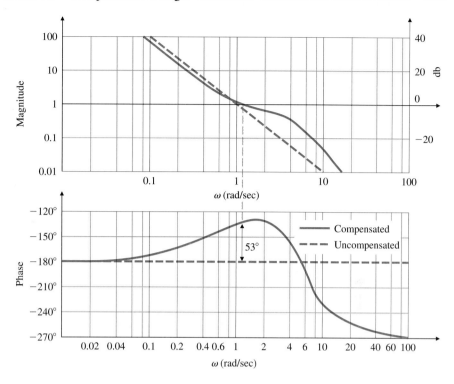

Identical results of state space
and frequency response
design methods

The frequency-response plots given in Fig. 7.32 show that the compensation de-signed using state-space accomplishes the same results that one would strive for using frequency-response design. Specifically, the uncompensated phase margin of $0°$ increases to $53°$ in the compensated case, and the gain $K = 40.4$ produces a crossover frequency $\omega_c = 1.35$ rad/sec. Both these values are roughly consistent with the controller closed-loop roots, with $\omega_n = 1$ rad/sec and $\zeta = 0.7$, as we would expect because these slow controller poles are dominant in the system response over the fast estimator poles.

Now we consider a reduced-order estimator for the same system.

EXAMPLE 7.21

Reduced-Order Compensator Design for a Satellite Attitude Control

Repeat the design for the $1/s^2$ satellite plant but use a reduced-order estimator. Place the one estimator pole at -5 rad/sec.

Solution.   From Eq. (7.147) we know that the estimator gain is

$$L = 5$$

Figure 7.33
Simplified block diagram of a reduced-order controller that is a lead network

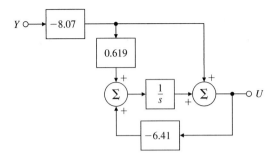

and the scalar compensator equations from Eq. (7.169) are

$$\dot{x}_c = -6.41x_c - 33.1y,$$

$$u = -1.41x_c - 8.07y,$$

where, from Eq. (7.149),

$$x_c = \hat{x}_2 - 5y.$$

The compensator has the transfer function calculated from Eq. (7.170) to be

$$D_{cr}(s) = -\frac{8.07(s + 0.619)}{s + 6.41}$$

and is shown in Fig. 7.33.

The reduced-order compensator here is precisely a lead network. This is a pleasant discovery, because it shows that transform and state-variable techniques can result in exactly the same type of compensation. The root locus of Fig. 7.34 shows that the closed-loop poles occur at the assigned locations. The frequency response of the compensated system seen in Fig. 7.35 shows a phase margin of about 55°. As with the full-order estimator, analysis by other methods confirms the selected root locations.

Figure 7.34
Root locus of a reduced-order controller and $1/s^2$ process, root locations at $K = 8.07$ shown by the dots

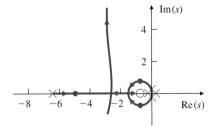

Figure 7.35
Frequency response for
$G(s) = 1/s^2$ with a
reduced-order estimator

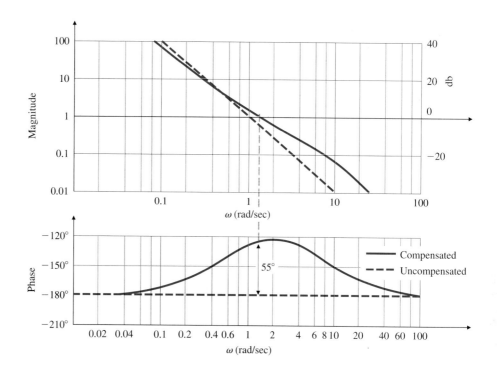

More subtle properties of the pole-placement method can be illustrated by a third-order system.

EXAMPLE 7.22

DC Servo

Use the state-space pole-placement method to design a compensator for the DC servo system with the transfer function

$$G(s) = \frac{10}{s(s+2)(s+8)}.$$

Using a state description in observer canonical form, place the control poles at $pc = [-1.42; \quad -1.04 \pm 2.14j]$ locations and the full-order estimator poles at $pe = [-4.25; \quad -3.13 \pm 6.41j]$.

Solution.   A block diagram of this system in observer canonical form is shown in Fig. 7.36. The corresponding state-space matrices are

$$\mathbf{F} = \begin{bmatrix} -10 & 1 & 0 \\ -16 & 0 & 1 \\ 0 & 0 & 0 \end{bmatrix}, \qquad \mathbf{G} = \begin{bmatrix} 0 \\ 0 \\ 10 \end{bmatrix},$$

$$\mathbf{H} = \begin{bmatrix} 1 & 0 & 0 \end{bmatrix}, \qquad J = 0.$$

Figure 7.36
DC Servo in observer
canonical form

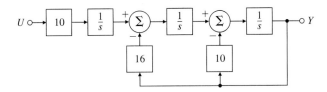

The desired poles are

$$pc = [-1.42; -1.04 + 2.14 * j; -1.04 - 2.14 * j].$$

We compute the state feedback gain to be K = place(F,G,pc),

$$\mathbf{K} = [-46.4 \quad 5.76 \quad -0.65].$$

The estimator error poles are at

$$pe = [-4.25; -3.13 + j * 6.41; -3.13 - j * 6.41].$$

We compute the estimator gain to be Lt = place(F',H',pe), L = Lt',

$$\mathbf{L} = \begin{bmatrix} 0.5 \\ 61.4 \\ 216 \end{bmatrix}.$$

The compensator transfer function, as given by substituting into Eq. (7.166), is

$$D_c(s) = -190 \frac{(s + 0.432)(s + 2.10)}{(s - 1.88)(s + 2.94 \pm 8.32 j)}$$

Figure 7.37 shows the root locus of the system of compensator and plant in series plotted with the compensator gain as the parameter. It verifies that the roots are in the

Figure 7.37
Root locus for DC Servo
pole assignment

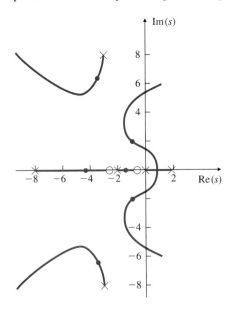

desired locations specified when the gain $K = 190$ in spite of the peculiar (unstable) compensation that has resulted. Even though this compensator has an unstable root at $s = +1.88$, all system closed-loop poles (controller and estimator) are stable.

An unstable compensator is typically not acceptable because of the difficulty in testing either the compensator by itself or the system in open loop during a bench checkout. In some cases, however, better control can be achieved with an unstable compensator; then its inconvenience in checkout may be worthwhile.[7]

Figure 7.37 shows that a direct consequence of the unstable compensator is that the system becomes unstable as the gain is reduced from its nominal value. Such a system is called **conditionally stable** and should be avoided if possible. As we saw in Chapter 5, actuator saturation in response to large signals has the effect of lowering the effective gain, and in a conditionally stable system instability can result. Also, if the electronics are such that the control amplifier gain rises continuously from zero to the nominal value during startup, such a system would be initially unstable. These considerations lead us to consider alternative designs for this system.

Conditionally stable
compensator

Redesign of the DC Servo System with a Reduced-Order Estimator

EXAMPLE 7.23

Design a compensator for the DC servo system of Example 7.22 using the same control poles but with a reduced-order estimator. Place the estimator poles at $-4.24 \pm 4.24j$ positions with $\omega_n = 6$ and $\zeta = 0.707$.

Solution.   The reduced-order estimator correspond to

$$\text{pc} = [-4.24 + 4.24 * j; \ -4.24 - 4.24 * j].$$

After partitioning we have

$$\begin{bmatrix} \mathbf{F}_{aa} & \mathbf{F}_{ab} \\ \mathbf{F}_{ba} & \mathbf{F}_{bb} \end{bmatrix} = \begin{bmatrix} -10 & 1 & 0 \\ -16 & 0 & 1 \\ 0 & 0 & 0 \end{bmatrix}, \qquad \begin{bmatrix} G_a \\ \mathbf{G}_b \end{bmatrix} = \begin{bmatrix} 0 \\ 0 \\ 10 \end{bmatrix}.$$

Solving for the estimator error characteristic polynomial,

$$\det(s\mathbf{I} - \mathbf{F}_{bb} + \mathbf{LF}_{ab}) = \alpha_e(s),$$

we find (using place) that

$$\mathbf{L} = \begin{bmatrix} 8.5 \\ 36 \end{bmatrix}.$$

---

[7] There are even systems that cannot be stabilized with a stable compensator.

Figure 7.38

Root locus for DC Servo
reduced-order controller

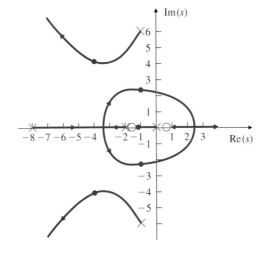

The compensator transfer function, given by Eq. (7.170), is computed to be

$$D_{cr}(s) = 20.93 \frac{(s - 0.735)(s + 1.871)}{s + 0.990 \pm 6.120j}.$$

A nonminimum-phase
compensator

The associated root locus for this system is shown in Fig. 7.38. Note that this time we have a stable but nonminimum-phase compensator and a zero-degree root locus. The RHP portion of the locus will not cause difficulties because the gain has to be selected to keep all closed-loop poles in LHP.

As a next pass at the design for this system, we attempt a design with the SRL.

EXAMPLE 7.24

Redesign of the DC Servo Compensator Using the SRL

Design a compensator for the DC servo system of Example 7.22 using pole placement based on the SRL. For the control law, let the cost output z be the same as the plant output; for the estimator design, assume that the process noise enters at the same place as the system control signal. Select roots for a control bandwidth of about 2.5 rad/sec, and choose the estimator roots for a bandwidth of about 2.5 times faster than the control bandwidth (6.3 rad/sec). Derive an equivalent discrete controller with a sampling period of $T_s = 0.1$ sec (10 times the fastest pole) and compare the continuous and digital control outputs and control efforts.

Solution.   Because the problem has specified that $\mathbf{G}_1 = \mathbf{G}$ and $\mathbf{H}_1 = \mathbf{H}$, then the SRL is the same for the control as for the estimator, so we need to generate only one locus based on the plant transfer function. The SRL for the system is shown in Fig. 7.39. From the locus we select $-2 \pm 1.56j$ and -8.04 as the desired control poles (pc = [-2+1.56*j;-2-1.56*j;-8.04]) and $-4 \pm 4.9j$ and -9.169 (pe = [-4+4.9*j;-4-4.9*j;-9.169]) as the desired estimator poles. The state feedback gain is K = place(F,G,pc),

$$\mathbf{K} = [\,-0.285 \quad 0.219 \quad 0.204\,],$$

Figure 7.39

Symmetric root locus

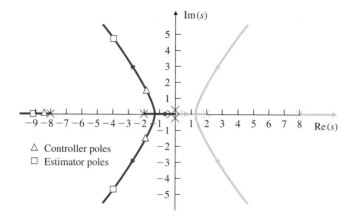

and the estimator gain is Lt = place(F',H',pe), L = Lt',

$$L = \begin{bmatrix} 7.17 \\ 97.4 \\ 367 \end{bmatrix}.$$

Notice that the feedback gains are much smaller than before. The resulting compensator transfer function is computed from Eq. (7.166) to be

$$D_c(s) = -\frac{94.5(s + 7.98)(s + 2.52)}{(s + 4.28 \pm 6.42\,j)(s + 10.6)}.$$

We now take this compensator, put it in series with the plant, and use the compensator gain as the parameter. The resulting ordinary root locus of the closed-loop system is shown in Fig. 7.40. When the root-locus gain equals the nominal gain of 94.5, the roots are at the closed-loop locations selected from the SRL, as they should be.

Figure 7.40

Root locus for pole assignment from the SRL

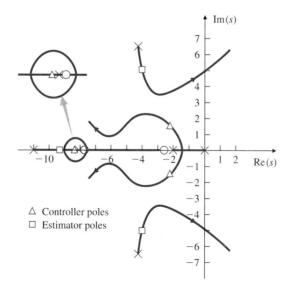

Note that the compensator is now stable and minimum-phase. This improved design comes about in large part because the plant pole at $s = -8$ is virtually unchanged by either controller or estimator. It does not need to be changed for good performance; in fact, the only feature in need of repair in the original $G(s)$ is the pole at $s = 0$. Using the SRL technique, we essentially discovered that the best use of control effort is to shift the two low-frequency poles at $s = 0$ and -2 and to leave the pole at $s = -8$ virtually unchanged. As a result, the control gains are much lower and the compensator design is less radical. This example illustrates why LQR design is typically preferable over pole placement.

The discrete equivalent for the controller is obtained from MATLAB using the c2d command as shown below

```
nc = 94.5*conv([1 7.98],[1 2.52]);        % form controller numerator
dc = conv([1 8.56 59.5348],[1 10.6]);     % form controller denominator
sysDc = tf(nc,dc);                        % form controller system description
ts = 0.1;                                 % sampling time of 0.1 sec
sysDd = c2d(sysDc,ts,'zoh');              % convert controller to discrete time
```

Discrete controller

and has the discrete transfer function

$$D_d(z) = \frac{5.9157(z + 0.766)(z + 0.4586)}{(z - 0.522 \pm 0.3903j)(z + 0.3465)}.$$

The equation for the control law is (with the sample period suppressed for clarity)

$$u(k+1) = 1.3905u(k) - 0.7866u(k-1) + 0.1472u(k-2) + e(k) - 7.2445e(k-2) + 2.0782e(k-2).$$

A Simulink diagram for simulating both the continuous and discrete systems is shown in Fig. 7.41. A comparison of the continuous and discrete step responses and control signals are shown in Fig.7.42. Better agreement between the two responses can be obtained if the sampling period were reduced.

Simulink simulation

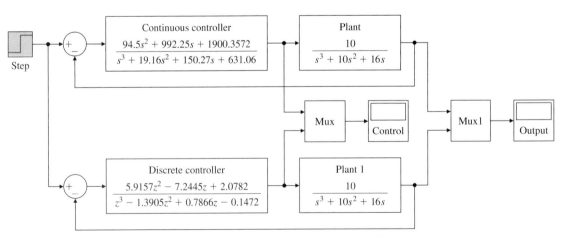

Figure 7.41 Simulink block diagram to compare continuous and discrete controllers

Figure 7.42

Comparison of step responses and control signals for continuous and discrete controllers: (a) step responses, (b) control signals

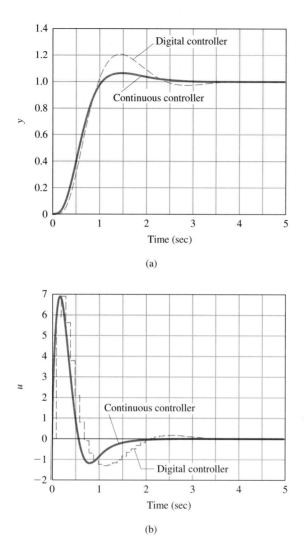

(a)

(b)

Armed with the knowledge gained from Example 7.24, let us go back, with a better selection of poles, to investigate the use of pole placement for this example. Initially we used the third-order locations, which produced three poles with a natural frequency of about 2 rad/sec. This design moved the pole at $s = -8$ to $s = -1.4$, thus violating the principle that open-loop poles should not be moved unless they are a problem. Now let us try it again, this time using dominant second-order locations to shift the slow poles, and leaving the fast pole alone at $s = -8$.

EXAMPLE 7.25

DC Servo System Redesign with Modified Dominant Second-Order Pole Locations

Design a compensator for the DC servo system of Example 7.22 using pole placement with control poles given by

$$pc = [-1.41 \pm 1.41j; -8]$$

and the estimator poles given by

$$pe = [-4.24 \pm 4.24j; -8].$$

Solution.   With these pole locations, we find that the required feedback gain is (using K = place(F,G,pc))

$$\mathbf{K} = [\, -0.469 \quad 0.234 \quad 0.0828\,].$$

which has a smaller magnitude than the case where the pole at $s = -8$ was moved.
We find the estimator gain to be (using Lt = place(F',H',pe), L = Lt')

$$\mathbf{L} = \begin{bmatrix} 6.48 \\ 87.8 \\ 288 \end{bmatrix}.$$

The compensator transfer function is

$$D_c(s) = -\frac{414(s + 2.78)(s + 8)}{(s + 4.13 \pm 5.29j)(s + 9.05)},$$

which is stable and minimum phase. This example illustrates the value of judicious pole selection and of the SRL technique.

The poor pole selection inherent in the initial use of the poles results in higher control effort and produces an unstable compensator. Both these undesirable features are eliminated by using the SRL (or LQR) or by improved pole selection.

But we really need to use SRL to guide the proper selection of poles. The bottom line is that *SRL (or LQR) is the method of choice*!

___

As seen from some of the above examples, we have shown use of optimal design via the SRL. However, it is more common in practice to skip that step and use LQR directly.

▲ 7.7 Loop Transfer Recovery (LTR)

The introduction of an estimator in a state feedback controller loop may adversely affect the stability robustness properties of the system, that is, the phase margin (PM) and gain margin (GM) properties may become arbitrarily poor as shown by Doyle's famous example (Doyle, 1978). However, it is possible to modify the estimator design so as to try to "recover" the LQR stability robustness properties to some extent. This process, called loop transfer recovery (LTR), is especially effective for minimum-phase systems. To achieve the recovery, some of the estimator poles are placed at (or near) the zeros of the plant and the remaining poles are moved (sufficiently far) into the LHP. The idea behind LTR is to redesign the estimator in such a way to shape the loop gain properties to approximate those of LQR.

The use of loop transfer recovery means that feedback controllers can be designed to achieve desired sensitivity ($\mathcal{S}(s)$) and complementary sensitivity functions ($\mathcal{T}(s)$) at critical (loop-breaking) points in the feedback system—for example, at either the input or output of the plant. Of course, there is a price to be paid for this improvement in stability robustness! The newly designed control system may have worse sensor noise sensitivity properties. Intuitively, one can think of making (some of) the estimator poles arbitrarily fast so that the loop gain is approximately that of LQR. Alternatively, one can think of essentially "inverting" the plant transfer function so that all the LHP poles of the plant are cancelled by the dynamic compensator to achieve the desired loop shape. There are obvious tradeoffs and the designer needs to be careful to make the correct choice for the given problem depending on the control system specifications.

LTR is a well-known technique now and specific practical design procedures have been identified (Athans, 1986; Stein and Athans, 1987; Saberi et al., 1993). The same procedures may also be applied to nonminimum phase systems, but there is no guarantee on the extent of possible recovery. The LTR technique may be viewed as a systematic procedure to study design trade-offs for Linear Quadratic-based compensator design (Doyle and Stein, 1981). We will now formulate the LTR problem.

Consider the linear system

$$\dot{\mathbf{x}} = \mathbf{F}\mathbf{x} + \mathbf{G}u + w, \tag{7.175a}$$

$$y = \mathbf{H}\mathbf{x} + v, \tag{7.175b}$$

where w and v are uncorrelated zero-mean white gaussian process and sensor noise with covariance matrices $\mathbf{R}_w \geq 0$ and $\mathbf{R}_v \geq 0$. The estimator design yields

$$\dot{\hat{\mathbf{x}}} = \mathbf{F}\hat{\mathbf{x}} + \mathbf{G}u + \mathbf{L}(y - \hat{y}), \tag{7.176a}$$

$$\hat{y} = \mathbf{H}\hat{\mathbf{x}}, \tag{7.176b}$$

LTR

resulting in the usual dynamic compensator

$$D_c(s) = -\mathbf{K}(s\mathbf{I} - \mathbf{F} + \mathbf{G}\mathbf{K} + \mathbf{L}\mathbf{H})^{-1}\mathbf{L} \tag{7.177}$$

We will now treat the noise parameters $\mathbf{R}_w$, and $\mathbf{R}_v$, as design "knobs" in the dynamic compensator design. Without loss of generality let us choose $\mathbf{R}_w = \mathbf{\Gamma}\mathbf{\Gamma}^T$ and $\mathbf{R}_v = 1$. For loop transfer recovery, assume $\mathbf{\Gamma} = q\mathbf{G}$ where q is a scalar design parameter. The estimator design is then based on the specific design parameters $\mathbf{R}_w$ and $\mathbf{R}_v$. It can be shown that for a minimum-phase system as q becomes large (Doyle and Stein, 1979)

$$\lim_{q\to\infty} D_c(s)G(s) = \mathbf{K}(s\mathbf{I} - \mathbf{F})^{-1}\mathbf{G} \tag{7.178}$$

and the convergence is pointwise in s and the degree of recovery can be arbitrarily good. This design procedure in effect "inverts" the plant transfer function in the limit as $q \to \infty$,

Plant inversion

$$\lim_{q\to\infty} D_c(s) = \mathbf{K}(s\mathbf{I} - \mathbf{F})^{-1}\mathbf{G}G^{-1}(s), \tag{7.179}$$

and this is precisely the reason that full loop transfer recovery is not possible for a nonminimum-phase system. This limiting behavior may be explained using the symmetric root loci. As $q \to \infty$, some of the estimator poles approach the zeros of

$$G_e(s) = \mathbf{H}(s\mathbf{I} - \mathbf{F})^{-1}\mathbf{\Gamma}, \tag{7.180}$$

and the rest tend to infinity[8] [see Eqs. (7.155) and (7.156)]. In practice, the LTR design procedure can still be applied to a nonminimum-phase plant. The degree of recovery will depend on the specific locations of the nonminimum-phase zeros. Sufficient recovery should be possible at many frequencies if the RHP zeros are located outside the specified closed-loop bandwidth. Limits on achievable performance of feedback systems due to RHP zeros are discussed in (Freudenberg and Looze, 1985). We will next illustrate the LTR procedure by a simple example.

LTR for nonminimum-phase systems

EXAMPLE 7.26   *LTR Design for Satellite Attitude Control*

Consider the satellite system with state-space description

$$\mathbf{F} = \begin{bmatrix} 0 & 1 \\ 0 & 0 \end{bmatrix}, \qquad \mathbf{G} = \begin{bmatrix} 0 \\ 1 \end{bmatrix},$$
$$\mathbf{H} = \begin{bmatrix} 1 & 0 \end{bmatrix}, \qquad J = 0.$$

---

[8] In a Butterworth configuration.

(a) Design an LQR controller with $\mathbf{Q} = \rho\mathbf{H}^T\mathbf{H}$ and $R = 1$, $\rho = 1$, and determine the loop gain.

(b) Then design a compensator which recovers the LQR loop gain of part (a) using the loop transfer recovery (LTR) technique for $q = 1, 10, 100$.

(c) Compare the different candidate designs in part (b) with respect to the actuator activity due to additive white gaussian sensor noise.

Solution.   Using lqr, the selected LQR weights result in the feedback gain $\mathbf{K} = \begin{bmatrix} 1 & 1.414 \end{bmatrix}$. The loop transfer function is

$$\mathbf{K}(s\mathbf{I} - \mathbf{F})^{-1}\mathbf{G} = \frac{1.414(s + 0.707)}{s^2}.$$

A magnitude frequency response plot of this LQR loop gain is shown in Fig. 7.43. For the estimator design using lqe, let $\mathbf{\Gamma} = q\mathbf{G}$, $\mathbf{R}_w = \mathbf{\Gamma}^T\mathbf{\Gamma}$, and $\mathbf{R}_v = 1$ and choose $q = 10$, resulting in the estimator gain

$$\mathbf{L} = \begin{bmatrix} 14.142 \\ 100 \end{bmatrix}.$$

The compensator transfer function is

$$\begin{aligned} D_c(s) &= \mathbf{K}(s\mathbf{I} - \mathbf{F} + \mathbf{GK} + \mathbf{LH})^{-1}\mathbf{L} \\ &= \frac{155.56(s + 0.6428)}{(s^2 + 15.556s + 121)} = \frac{155.56(s + 0.6428)}{(s + 7.77 + j7.77)(s + 7.77 - j7.77)} \end{aligned}$$

and the loop transfer function is

$$D_c(s)G(s) = \frac{155.56(s + 0.6428)}{s^2(s + 7.77 + j7.77)(s + 7.77 - j7.77)}.$$

Figure 7.43

Frequency response plots for LTR design

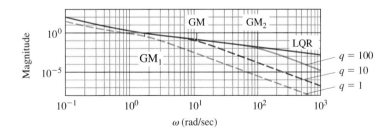

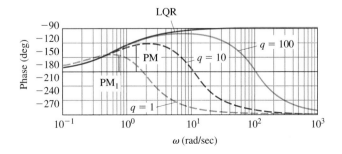

Figure 7.43 shows the frequency response of the loop transfer function for several values of q ($q = 1, 10, 100$) along with the ideal LQR loop transfer function frequency response. As seen from this figure, the loop gain tends to approach that of LQR as the value of q increases. As seen in Fig. 7.43, for $q = 10$ the "recovered" gain margin is GM = 11.1 = 20.9 db and PM = 55.06°. Sample MATLAB statements to carry out the above LTR design procedure are shown below.

```
F = [0 1; 0 0];
G = [0;1];
H = [1 0];
J = [0];
sys0 = ss(F,G,H,J);
H1 = [1 0];
sys = ss(F,G,H1,J);
w = logspace(-1,3,1000);
rho = 1.0;
Q = rho*H1'*H1;
r = 1;
```

MATLAB lqr
```
[K] = lqr(F,G,Q,r)
sys1 = ss(F,G,K,0);
[maggk1,phasgk1,w] = bode(sys1,w);
```

```
q = 10;
gam = q*G;
Q1 = gam'*gam;
rv = 1;
```

MATLAB lqe
```
[L] = lqe(F,gam,H,Q1,rv)
```

```
aa = F-G*K-L*H;
bb = L;
cc = K;
dd = 0;
sysk = ss(aa,bb,cc,dd);
sysgk = series(sys0,sysk);
```

MATLAB bode
MATLAB margin
```
[maggk,phsgk,w] = bode(sysgk,w);
[gm,phm,wcg,wcp] = margin(maggk,phsgk,w)
loglog(w,[maggk1(:) maggk(:)]);
semilogx(w,[phasgk1(:) phsgk(:)]);
```

(c) To determine the effect of sensor noise, v, on the actuator activity, we determine the transfer function from v to u as shown in Fig. 7.44. For the selected value of LTR design parameter, $q = 10$, we have

$$\frac{U(s)}{V(s)} = H(s) = \frac{-D_c(s)}{1 + D_c(s)G(s)} = \frac{-155.56s^2(s + 0.6428)}{s^4 + 15.556s^3 + 121s^2 + 155.56s + 99.994}.$$

One reasonable measure of the effect of the sensor noise on the actuator activity is the root-mean-square (RMS) value of the control, u, due to the additive noise, v. The RMS value of the control may be computed as follows:

RMS value

$$\| u \|_{rms} = \left(\frac{1}{T_0} \int_0^{T_0} u(t)^2 \, dt \right)^{1/2} \tag{7.181}$$

Figure 7.44
Closed-loop system for LTR

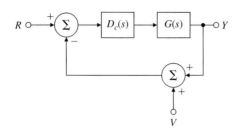

where T_0 is the signal duration. Assuming white gaussian noise v, the RMS value of the control can also be determined analytically (see Boyd and Barratt, 1991). The closed-loop Simulink diagram with band-limited white sensor noise excitation is shown in Fig. 7.45. The values of the RMS control were computed for different values of the LTR design parameter, q, using the Simulink simulations and are tabulated in Table 7.2. The results suggest increased vulnerability due to actuator wear as q is increased.

Simulink simulation

TABLE 7.2

Computed RMS Control for Various Values of LTR Tuning Parameter q.

q	$\| u \|_{rms}$
1	0.1454
10	2.8054
100	70.5216

Figure 7.45
Simulink block diagram for LTR

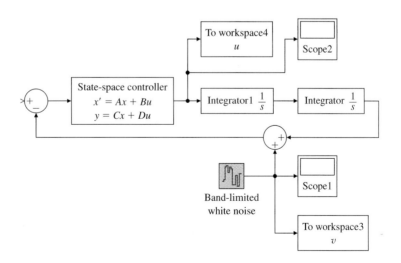

7.8 Introduction of the Reference Input with the Estimator

The controller obtained by combining the control law studied in Section 7.3 with the estimator discussed in Section 7.6 is essentially a **regulator design**. This means that the characteristic equations of the control and the estimator are chosen for good disturbance rejection—that is, to give satisfactory transients to disturbances such as $w(t)$. However, this design approach does not consider a reference input, nor does it provide for **command following**, which is evidenced by a good transient response of the combined system to command changes. In general, good disturbance rejection and good command following both need to be taken into account in designing a control system. Good command following is done by properly introducing the reference input into the system equations.

Let us repeat the plant and controller equations for the full-order estimator; the reduced-order case is the same in concept, differing only in detail:

$$\text{Plant:} \quad \dot{\mathbf{x}} = \mathbf{Fx} + \mathbf{Gu}, \tag{7.182a}$$

$$y = \mathbf{Hx}; \tag{7.182b}$$

$$\text{Controller:} \quad \dot{\hat{\mathbf{x}}} = (\mathbf{F} - \mathbf{GK} - \mathbf{LH})\hat{\mathbf{x}} + \mathbf{L}y, \tag{7.183a}$$

$$u = -\mathbf{K}\hat{\mathbf{x}}. \tag{7.183b}$$

Figure 7.46 shows two possibilities for introducing the command input r into the system. This figure illustrates the general issue of whether the compensation

Figure 7.46
Possible locations for introducing the command input: (a) compensation in the feedback path; (b) compensation in the feedforward path

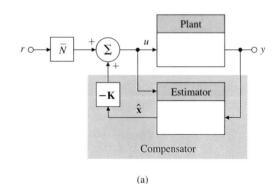

(a)

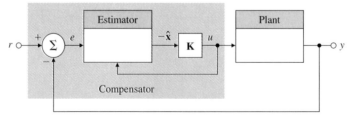

(b)

should be put in the feedback or feedforward path. The response of the system to command inputs is different, depending on the configuration, because the zeros of the transfer functions are different. The closed-loop poles are identical, however, as can be easily verified by letting $r = 0$ and noting that the systems are then identical.

The difference in the responses of the two configurations can be seen quite easily. Consider the effect of a step input in r. In Fig. 7.46(a) the step will excite the estimator in precisely the same way that it excites the plant; thus the estimator error will remain zero during and after the step. This means that the estimator dynamics are not excited by the command input, so the transfer function from r to y must have zeros at the estimator pole locations that cancel those poles. As a result, a step command will excite system behavior that is consistent with the control poles alone—that is, with the roots of $\det(s\mathbf{I} - \mathbf{F} + \mathbf{GK}) = 0$.

In Fig. 7.46(b) a step command in r enters directly only into the estimator, thus causing an estimation error that decays with the estimator dynamic characteristics in addition to the response corresponding to the control poles. Therefore, a step command will excite system behavior consistent with both control roots and estimator roots—that is, the roots of

$$\det(s\mathbf{I} - \mathbf{F} + \mathbf{GK}) \cdot \det(s\mathbf{I} - \mathbf{F} + \mathbf{LH}) = 0.$$

For this reason the configuration shown in Fig. 7.46(a) is typically the superior way to command the system where $\bar{N}$ is found using Eqs. (7.89)–(7.91).

In Section 7.8.1, we will show a general structure for introducing the reference input with three choices of parameters that implement either the feedforward or the feedback case. We will analyze the three choices from the point of view of the system zeros and the implications the zeros have for the system transient response. Finally, in Section 7.8.2 we will show how to select the remaining parameter to eliminate constant errors.

▲ 7.8.1   A General Structure for the Reference Input

Given a reference input $r(t)$, the most general linear way to introduce r into the system equations is to add terms proportional to it in the controller equations. We can do this by adding $\bar{N}r$ to Eq. (7.183b) and $\mathbf{M}r$ to Eq. (7.183a). Note that in this case, $\bar{N}$ is a scalar and $\mathbf{M}$ is an $n \times 1$ vector. With these additions the **controller equations** become

Controller equations

$$\dot{\hat{\mathbf{x}}} = (\mathbf{F} - \mathbf{GK} - \mathbf{LH})\hat{\mathbf{x}} + \mathbf{L}y + \mathbf{M}r, \qquad (7.184a)$$

$$u = -\mathbf{K}\hat{\mathbf{x}} + \bar{N}r. \qquad (7.184b)$$

The block diagram is shown in Fig. 7.47(a). The alternatives shown in Fig. 7.46 correspond to different choices of $\mathbf{M}$ and $\bar{N}$. Because $r(t)$ is an external signal, it is clear that neither $\mathbf{M}$ nor $\bar{N}$ affects the characteristic equation of the combined controller–estimator system. In transfer-function terms, the selection of

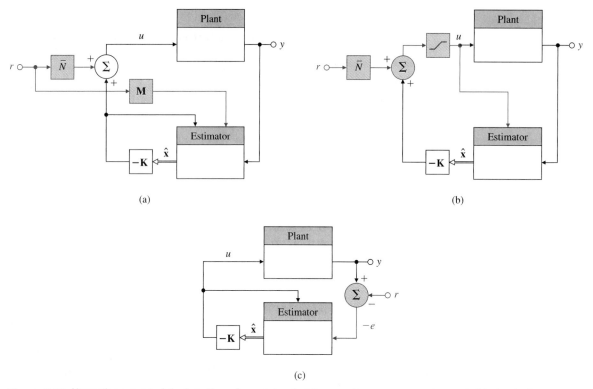

Figure 7.47 Alternative ways to introduce the reference input: (a) general case—zero assignment; (b) standard case—estimator not excited, zeros $= \alpha_e(s)$; (c) error-control case—classical compensation

$\mathbf{M}$ and $\bar{N}$ will only affect the zeros of transmission from r to y and as a consequence can significantly affect the transient response but not the stability. How can we choose $\mathbf{M}$ and $\bar{N}$ to obtain satisfactory transient response? We should point out that we assigned the poles of the system by feedback gains $\mathbf{K}$ and $\mathbf{L}$ and we are now going to assign zeros by feedforward gains $\mathbf{M}$ and $\bar{N}$.

There are three strategies for choosing $\mathbf{M}$ and $\bar{N}$:

Three methods for selecting $\mathbf{M}$ and N

1. *Autonomous estimator:* Select $\mathbf{M}$ and $\bar{N}$ so that the state estimator error equation is independent of r [Fig. 7.47(b)].

2. *Tracking-error estimator:* Select $\mathbf{M}$ and $\bar{N}$ so that only the tracking error, $e = (r - y)$, is used in the control [Fig. 7.47(c)].

3. *Zero-assignment estimator:* Select $\mathbf{M}$ and $\bar{N}$ so that n of the zeros of the overall transfer function are assigned at places of the designer's choice [Fig. 7.47(a)].

CASE 1. From the viewpoint of estimator performance, the first method is quite attractive and the most widely used of the alternatives. If $\hat{\mathbf{x}}$ is to generate a good estimate of $\mathbf{x}$, then surely $\tilde{\mathbf{x}}$ should be as free of external excitation as possible; that is, $\tilde{\mathbf{x}}$ should be uncontrollable from r. The computation of $\mathbf{M}$ and $\bar{N}$ to bring this about is quite easy. The estimator error equation is found by subtracting Eq. (7.184a) from Eq. (7.182a), with the plant output (Eq. 7.182b) substituted into the estimator [Eq. (7.183a)] and the control [Eq. (7.183b)] substituted into the plant [Eq. (7.182a)]:

$$\dot{\mathbf{x}} - \dot{\hat{\mathbf{x}}} = \mathbf{Fx} + \mathbf{G}(-\mathbf{K}\hat{\mathbf{x}} + \bar{N}r) - [(\mathbf{F} - \mathbf{GK} - \mathbf{LH})\hat{\mathbf{x}} + \mathbf{L}y + \mathbf{M}r], \quad (7.185a)$$

$$\dot{\tilde{\mathbf{x}}} = (\mathbf{F} - \mathbf{LH})\tilde{\mathbf{x}} + \mathbf{G}\bar{N}r - \mathbf{M}r. \qquad (7.185b)$$

If r is not to appear in Eq. (7.185b), then we should choose

$$\mathbf{M} = \mathbf{G}\bar{N}. \qquad (7.186)$$

Because $\bar{N}$ is a scalar, $\mathbf{M}$ is fixed to within a constant factor. Note that with this choice of $\mathbf{M}$ we can write the controller equations as

$$u = -\mathbf{K}\hat{\mathbf{x}} + \bar{N}r, \qquad (7.187a)$$

$$\dot{\hat{\mathbf{x}}} = (\mathbf{F} - \mathbf{LH})\hat{\mathbf{x}} + \mathbf{G}u + \mathbf{L}y, \qquad (7.187b)$$

which matches the configuration in Fig. 7.47(b). The net effect of this choice is that the control is computed from the feedback gain and the reference input *before* it is applied, and then the same control is input to both the plant and the estimator. In this form, if the plant control is subject to saturation, the same control limits can be applied in Eq. (7.187) to the control entering the equation for the estimate $\hat{\mathbf{x}}$, and the nonlinearity cancels out of the $\tilde{\mathbf{x}}$ equation. This behavior is essential for proper estimator performance. The block diagram corresponding to this technique is shown in Fig. 7.47(b). We will return to the selection of the gain factor on the reference input, $\bar{N}$, in Section 7.8.2 after discussing the other two methods of selecting $\mathbf{M}$.

CASE 2. The second approach suggested earlier is to use the tracking error. This solution is sometimes forced on the control designer when the sensor measures only the output error. For example, in many thermostats the output is the difference between the temperature to be controlled and the setpoint temperature, and there is no absolute indication of the reference temperature available to the controller. Also, some radar tracking systems have a reading that is proportional to the pointing error, and this error signal alone must be used for feedback control. In these situations, we must select $\mathbf{M}$ and $\bar{N}$ so that Eq. (7.184) is driven by the error only. This requirement is satisfied if we select

$$\bar{N} = 0 \quad \text{and} \quad \mathbf{M} = -\mathbf{L}. \qquad (7.188)$$

Then the estimator equation is

$$\dot{\hat{\mathbf{x}}} = (\mathbf{F} - \mathbf{GK} - \mathbf{LH})\hat{\mathbf{x}} + \mathbf{L}(y - r). \qquad (7.189)$$

The compensator in this case, for low-order designs, is a standard lead compensator in the forward path. As we have seen in earlier chapters, this design can have a considerable amount of overshoot because of the zero of the compensator. This design corresponds exactly to the compensators designed by the transform methods given in Chapters 5 and 6.

CASE 3. The third method of selecting $\mathbf{M}$ and $\bar{N}$ is to choose the values so as to assign the system's zeros to arbitrary locations of the designers choice. This method provides the designer with the maximum flexibility in satisfying transient-response and steady-state gain constraints. The other two methods are special cases of this third method. All three methods depend on the zeros. As we saw in Section 7.3.2, when there is no estimator and the reference input is added to the control, the closed-loop system zeros remain fixed as the zeros of the open-loop plant. We now examine what happens to the zeros when an estimator is present. To do so, we reconsider the controller of Eq. (7.184b). If there is a zero of transmission from r to u, then there is necessarily a zero of transmission from r to y unless there is a pole at the same location as the zero. It is therefore sufficient to treat the controller alone to determine what effect the choices of $\mathbf{M}$ and $\bar{N}$ will have on the system zeros. The equations for a zero from r to u from Eqs. (7.184a) and (7.184b) are given by

$$\det \begin{bmatrix} s\mathbf{I} - \mathbf{F} + \mathbf{GK} + \mathbf{LH} & -\mathbf{M} \\ -\mathbf{K} & \bar{N} \end{bmatrix} = 0. \tag{7.190}$$

(We let $y = 0$ because we care only about the effect of r.) If we divide the last column by the (nonzero) scalar $\bar{N}$ and then add to the rest the product of $\mathbf{K}$ times the last column, we find the feedforward zeros are at the values of s such that

$$\det \begin{bmatrix} s\mathbf{I} - \mathbf{F} + \mathbf{GK} + \mathbf{LH} - \frac{\mathbf{M}}{\bar{N}}\mathbf{K} & -\frac{\mathbf{M}}{\bar{N}} \\ 0 & 1 \end{bmatrix} = 0,$$

or

$$\det \left(s\mathbf{I} - \mathbf{F} + \mathbf{GK} + \mathbf{LH} - \frac{\mathbf{M}}{\bar{N}}\mathbf{K} \right) = \gamma(s) = 0. \tag{7.191}$$

Now Eq. (7.191) is exactly in the form of Eq. (7.125) for selecting $\mathbf{L}$ to yield desired locations for the estimator poles. Here we have to select $\mathbf{M}/\bar{N}$ for a desired zero polynomial $\gamma(s)$ in the transfer function from the reference input to the control. Thus the selection of $\mathbf{M}$ provides a substantial amount of freedom to influence the transient response. We can add an arbitrary nth-order polynomial to the transfer function from r to u and hence from r to y; that is, we can assign n zeros in addition to all the poles that we assigned previously. If the roots of $\gamma(s)$ are not canceled by the poles of the system, then they will be included in zeros of transmission from r to y.

Two considerations can guide us in the choice of $\mathbf{M}/\bar{N}$—that is, in the location of the zeros. The first is dynamic response. We have seen in Chapter 3 that the zeros influence the transient response significantly, and the heuristic guidelines given there may suggest useful locations for the available zeros. The second consideration, which will connect state-space design to another result from transform techniques, is steady-state error or velocity-constant control. In Chapter 4 we derived the relationship between the steady-state accuracy of a type 1 system and the closed-loop poles and zeros. If the system is type 1, then the steady-state error to a step input will be zero and to a unit ramp input will be

$$e_\infty = \frac{1}{K_v},$$

(7.192)

Truxal's formula

where K_v is the velocity constant. Furthermore, it was shown that if the *closed-loop* poles are at $\{p_i\}$ and the *closed-loop* zeros are at $\{z_i\}$, then (for a type 1 system) **Truxal's formula** gives

$$\frac{1}{K_v} = \sum \frac{1}{z_i} - \sum \frac{1}{p_i}$$

(7.193)

Equation (7.193) forms the basis for a partial selection of $\gamma(s)$, and hence of **M** and $\bar{N}$. The choice is based on two observations:

1. If $|z_i - p_i| \ll 1$, then the effect of this pole-zero pair on the dynamic response will be small, because the pole is almost canceled by the zero, and in any transient the residue of the pole at p_i will be very small.

2. Even though $z_i - p_i$ is small, it is possible for $1/z_i - 1/p_i$ to be substantial and thus to have a significant influence on K_v according to Eq. (7.193).

Application of these two guidelines to the selection of $\gamma(s)$, and hence of **M** and $\bar{N}$, results in a lag-network design. We illustrate this with an example.

EXAMPLE 7.27

Servomechanism: Increasing the Velocity Constant through Zero Assignment

Lag compensation by a state-space method

Consider the second-order servomechanism system described by

$$G(s) = \frac{1}{s(s+1)}$$

and with state description

$$\dot{x}_1 = x_2,$$

$$\dot{x}_2 = -x_2 + u.$$

Design a controller using pole placement so that both poles are at $s = -2$ and the system has a velocity constant $K_v = 10$. Derive an equivalent discrete controller with a sampling period of $T_s = 0.1$ sec ($20 \times \omega_n = 20 \times 0.05 = 0.1$ sec) and compare the continuous and digital control outputs as well as the control efforts.

Solution.   For this problem the state feedback gain

$$\mathbf{K} = [8 \quad 3]$$

results in the desired control poles. However, with this gain, $K_v = 2$, and we need $K_v = 10$. What effect will using estimators designed according to the three methods for $\mathbf{M}$ and $\bar{N}$ selection have on our design? Using the first strategy (the autonomous estimator), we find that the value of K_v does not change. If we use the second method (error control), we introduce a zero at a location unknown beforehand, and the effect on K_v will not be under direct design control. However, if we use the third option (zero placement) along with Truxal's formula [Eq. (7.193)], we can satisfy both the dynamic response and the steady-state requirements.

First we must select the estimator pole p_3 and the zero z_3 to satisfy Eq. (7.193) for $K_v = 10$. We want to keep $z_3 - p_3$ small so that there is little effect on the dynamic response and yet have $1/z_3 - 1/p_3$ be large enough to increase the value of K_v. To do this, we arbitrarily set p_3 small compared with the control dynamics. For example, we let

$$p_3 = -0.1.$$

Notice that this approach is opposite to the usual philosophy of estimation design, where fast response is the requirement. Now we use Eq. (7.193) to get

$$\frac{1}{K_v} = \frac{1}{z_3} - \frac{1}{p_1} - \frac{1}{p_2} - \frac{1}{p_3},$$

where $p_1 = -2 + 2j$, $p_2 = -2 - 2j$, and $p_3 = -0.1$. We solve for z_3 such that $K_v = 10$:

$$\frac{1}{K_v} = \frac{4}{8} + \frac{1}{0.1} + \frac{1}{z_3} = \frac{1}{10}$$

or

$$z_3 = -\frac{1}{10.4} = -0.096.$$

We thus design a reduced-order estimator to have a pole at -0.1 and choose $\mathbf{M}/\bar{N}$ such that $\gamma(s)$ has a zero at -0.096. A block diagram of the resulting system is shown in Fig. 7.48(a). You can readily verify that this system has the overall transfer function

$$\frac{Y(s)}{R(s)} = \frac{8.32(s + 0.096)}{(s^2 + 4s + 8)(s + 0.1)}, \tag{7.194}$$

for which $K_v = 10$, as specified.

The compensation shown in Fig. 7.48(a) is nonclassical in the sense that it has two inputs (e and y) and one output. If we resolve the equations to provide pure error compensation by finding the transfer function from e and u, which would give Eq. (7.194), we obtain the system shown in Fig. 7.48(b). This can be seen as follows. The relevant controller equations are

$$\dot{x}_c = 0.8\, e - 3.1\, u,$$
$$u = 8.32\, e + 3.02\, y + x_c,$$

Figure 7.48
Servomechanism with assigned zeros (a lag network): (a) the two-input compensator; (b) equivalent unity feedback system

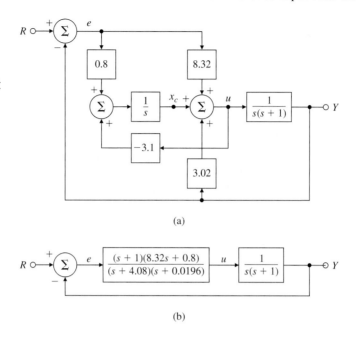

(a)

(b)

where x_c is the controller state. Taking the Laplace transform of these equations, eliminating $X_c(s)$, and substituting for the output ($Y(s) = G(s)U(s)$), we find that the compensator is described by

$$\frac{U(s)}{E(s)} = D_c(s) = \frac{(s+1)(8.32s+0.8)}{(s+4.08)(s+0.0196)}.$$

This compensation is a classical lag–lead network. The root locus of the system in Fig. 7.48(b) is shown in Fig. 7.49. Note the pole-zero pattern near the origin that is characteristic of a lag network. The Bode plot in Fig. 7.50 shows the phase lag at low frequencies and phase lead at high frequencies. The step response of the system is shown in Fig. 7.51 and shows the presence of a "tail" on the response due to the slow pole at -0.1. Of course, the system is type 1 and the system will have zero tracking error eventually.

The discrete equivalent for the controller is obtained from MATLAB using the c2d command

```
nc = conv([1 1],[8.32 0.8]);        % controller numerator
dc = conv([1 4.08],[1 0.0196]);     % controller denominator
sysDc = tf(nc,dc);                  % form controller system description
ts = 0.1;                           % sampling time of 0.1 sec
sysDd = c2d(sysDc,ts,'zoh');        % convert to discrete time controller
```

MATLAB c2d

and has the discrete transfer function

$$D_d(z) = \frac{8.32z^2 - 15.8855z + 7.5721}{z^2 - 1.6630z + 0.6637} = \frac{8.32(z-0.9903)(z-0.9191)}{(z-0.998)(z-0.6665)}.$$

Figure 7.49
Root locus of lag-lead
compensation

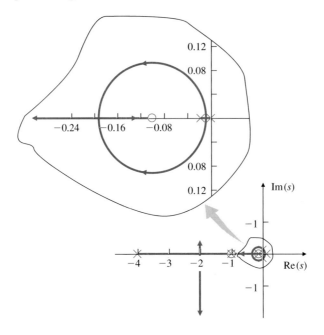

Figure 7.50
Frequency response of
lag-lead compensation

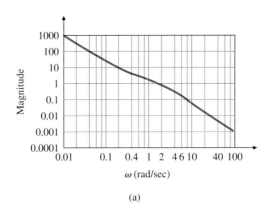

(a)

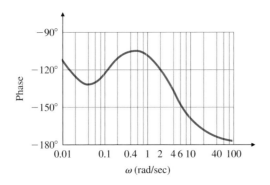

(b)

Figure 7.51
Step response of
the system with lag
compensation

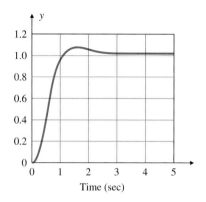

The equation for the control law is (with sample period suppressed for clarity)

$$u(k+1) = 1.6630u(k) + 0.6637u(k-1) + 8.32e(k+1) - 15.8855e(k) + 7.5721e(k-1).$$

A Simulink diagram for simulating both the continuous and discrete systems is shown in Fig. 7.52. A comparison of the continuous and discrete step responses and control signals are shown in Fig. 7.53. Better agreement between the two responses can be achieved if the sampling period were to be reduced.

Simulink simulation

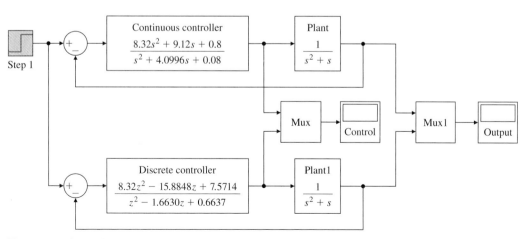

Figure 7.52 Simulink block diagram to compare continuous and discrete controllers

Figure 7.53

Comparison of step responses and control signals for continuous and discrete controllers: (a) step responses, (b) control signals

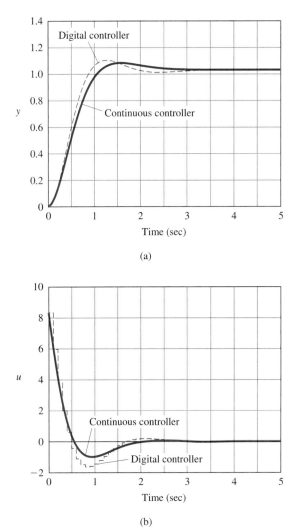

(a)

(b)

We now reconsider the first two methods for choosing $\mathbf{M}$ and $\bar{N}$, this time to examine their implications in terms of zeros. Under the first rule (for the autonomous estimator), we let $\mathbf{M} = \mathbf{G}\bar{N}$. Substituting this into Eq. (7.191) yields, for the controller feedforward zeros, we obtain

$$\det(s\mathbf{I} - \mathbf{F} + \mathbf{L}\mathbf{H}) = 0. \tag{7.195}$$

This is exactly the equation from which $\mathbf{L}$ was selected to make the characteristic polynomial of the estimator equation equal to $\alpha_e(s)$. Thus we have created n

zeros in exactly the same locations as the n poles of the estimator. Because of this pole-zero cancellation (which causes "uncontrollability" of the estimator modes), the overall transfer function poles consist only of the state feedback controller poles.

The second rule (for a tracking-error estimator) selects $\mathbf{M} = -\mathbf{L}$ and $\bar{N} = 0$. If these are substituted into Eq. (7.190), then the feedforward zeros are given by

$$\det\begin{bmatrix} s\mathbf{I} - \mathbf{F} + \mathbf{GK} + \mathbf{LH} & \mathbf{L} \\ -\mathbf{K} & 0 \end{bmatrix} = 0. \qquad (7.196)$$

If we postmultiply the last column by $\mathbf{H}$ and subtract the result from the first n columns, and then premultiply the last row by $\mathbf{G}$ and add it to the first n rows, Eq. (7.196) then reduces to

$$\det\begin{bmatrix} s\mathbf{I} - \mathbf{F} & \mathbf{L} \\ -\mathbf{K} & 0 \end{bmatrix} = 0. \qquad (7.197)$$

If we compare Eq. (7.197) with the equations for the zeros of a system in a state description, Eq. (7.54), we see that the added zeros are those obtained by replacing the input matrix with $\mathbf{L}$ and the output with $\mathbf{K}$. Thus, if we wish to use error control, we have to accept the presence of these compensator zeros that depend on the choice of $\mathbf{K}$ and $\mathbf{L}$ and over which we have no direct control. For low-order cases this results, as we said before, in a lead compensator as part of a unity feedback topology.

Let us now summarize our findings on the effect of introducing the reference input. When the reference input signal is included in the controller, the overall transfer function of the closed-loop system is

Transfer function for the closed-loop system when reference input is included in controller

$$\mathcal{T}(s) = \frac{Y(s)}{R(s)} = \frac{K_s \gamma(s) b(s)}{\alpha_e(s) \alpha_c(s)}, \qquad (7.198)$$

where K_s is the total system gain and $\gamma(s)$ and $b(s)$ are monic polynomials. The polynomial $\alpha_c(s)$ results in a control gain $\mathbf{K}$ such that $\det[s\mathbf{I} - \mathbf{F} + \mathbf{GK}] = \alpha_c(s)$. The polynomial $\alpha_e(s)$ results in estimator gains $\mathbf{L}$ such that $\det[s\mathbf{I} - \mathbf{F} + \mathbf{LH}] = \alpha_e(s)$. Because as designers we get to choose $\alpha_c(s)$ and $\alpha_e(s)$, we have complete freedom in assigning the poles of the closed-loop system. There are three ways to handle the polynomial $\gamma(s)$: We can select it so that $\gamma(s) = \alpha_e(s)$ by using the implementation of Fig. 7.47(b), in which case $\mathbf{M}/\bar{N}$ is given by Eq. (7.186); we may accept $\gamma(s)$ as given by Eq. (7.197), so that error control is used; or we may give $\gamma(s)$ arbitrary coefficients by selecting $\mathbf{M}/\bar{N}$ from Eq. (7.191). It is important to point out that the plant zeros represented by $b(s)$ are not moved by this technique and remain as part of the closed-loop transfer function unless α_c or α_e are selected to cancel some of these zeros.

▲   7.8.2   Selecting the Gain

We now turn to the process of determining the gain $\bar{N}$ for the three methods of selecting $\mathbf{M}$. If we choose method 1, the control is given by Eq. (7.187a) and $\hat{x}_{ss} = x_{ss}$. Therefore, we can use either $\bar{N} = N_u + \mathbf{K}N_x$, as in Eq. (7.91), or $u = N_u r - \mathbf{K}(\hat{\mathbf{x}} - \mathbf{N}_x r)$. *This is the most common choice.* If we use the second method, the result is trivial; recall that $\bar{N} = 0$ for error control. If we use the third method, we pick $\bar{N}$ such that the overall closed-loop DC gain is unity.[9]

The overall system equations then are

$$\begin{bmatrix} \dot{\mathbf{x}} \\ \dot{\tilde{\mathbf{x}}} \end{bmatrix} = \begin{bmatrix} \mathbf{F} - \mathbf{GK} & \mathbf{GK} \\ \mathbf{0} & \mathbf{F} - \mathbf{LH} \end{bmatrix} \begin{bmatrix} \mathbf{x} \\ \tilde{\mathbf{x}} \end{bmatrix} + \begin{bmatrix} \mathbf{G} \\ \mathbf{G} - \bar{\mathbf{M}} \end{bmatrix} \bar{N}r, \qquad (7.199a)$$

$$y = [\mathbf{H} \quad \mathbf{0}] \begin{bmatrix} \mathbf{x} \\ \tilde{\mathbf{x}} \end{bmatrix}, \qquad (7.199b)$$

where $\bar{\mathbf{M}}$ is the outcome of selecting zero locations with either Eq. (7.191) or Eq. (7.186). The closed-loop system has unity DC gain if

$$-[\mathbf{H} \quad \mathbf{0}] \begin{bmatrix} \mathbf{F} - \mathbf{GK} & \mathbf{GK} \\ \mathbf{0} & \mathbf{F} - \mathbf{LH} \end{bmatrix}^{-1} \begin{bmatrix} \mathbf{G} \\ \mathbf{G} - \bar{\mathbf{M}} \end{bmatrix} \bar{N} = 1. \qquad (7.200)$$

If we solve Eq. (7.200) for $\bar{N}$, we get[10]

$$\bar{N} = -\frac{1}{\mathbf{H}(\mathbf{F} - \mathbf{GK})^{-1}\mathbf{G}[1 - \mathbf{K}(\mathbf{F} - \mathbf{LH})^{-1}(\mathbf{G} - \bar{\mathbf{M}})]}. \qquad (7.201)$$

The techniques in this section can be readily extended to reduced-order estimators.

7.9   Integral Control and Robust Tracking

The choices of $\bar{N}$ gain in Section 7.8 will result in zero steady-state error to a step command, but the result is not robust because any change in the plant

---

[9]   A reasonable alternative is to select $\bar{N}$ such that, when r and y are both unchanging, the DC gain from r to u is the *negative* of the DC gain from y to u. The consequences of this choice are that our controller can be structured as a combination of error control and generalized derivative control, and if the system is capable of type 1 behavior, that capability will be realized.

[10] We have used the fact that

$$\begin{bmatrix} \mathbf{A} & \mathbf{C} \\ \mathbf{0} & \mathbf{B} \end{bmatrix}^{-1} = \begin{bmatrix} \mathbf{A}^{-1} & -\mathbf{A}^{-1}\mathbf{CB}^{-1} \\ \mathbf{0} & \mathbf{B}^{-1} \end{bmatrix}.$$

parameters will cause the error to be nonzero. We need to use integral control to obtain robust tracking.

In the state-space design methods discussed so far, no mention has been made of integral control, and no design examples have produced a compensation containing an integral term. In Section 7.9.1 we show how integral control can be introduced by a direct method of adding the integral of the system error to the equations of motion. Integral control is a special case of tracking a signal that does not go to zero in the steady-state. We introduce (in Section 7.9.2) a general method for robust tracking that will present the internal model principle, which solves an entire class of tracking problems and disturbance-rejection controls. Finally, in Section 7.9.3 we show that, if the system has an estimator and also needs to reject a disturbance of known structure, we can include a model of the disturbance in the estimator equations and use the computer estimate of the disturbance to cancel the effects of the real plant disturbance on the output.

7.9.1   Integral Control

We start with an ad hoc solution to integral control by augmenting the state vector with the desired dynamics. For the system

$$\dot{\mathbf{x}} = \mathbf{F}\mathbf{x} + \mathbf{G}u + \mathbf{G}_1 w, \tag{7.202a}$$

$$y = \mathbf{H}\mathbf{x}, \tag{7.202b}$$

we can feed back the integral of the error,[11] $e = y - r$, as well as the state of the plant, $\mathbf{x}$, by augmenting the plant state with the extra (integral) state x_I, which obeys the differential equation

$$\dot{x}_I = \mathbf{H}\mathbf{x} - r \ (= e).$$

Thus

$$x_I = \int^t e\,dt.$$

Augmented state equations with integral control

The augmented state equations become

$$\begin{bmatrix} \dot{x}_I \\ \dot{\mathbf{x}} \end{bmatrix} = \begin{bmatrix} 0 & \mathbf{H} \\ \mathbf{0} & \mathbf{F} \end{bmatrix} \begin{bmatrix} x_I \\ \mathbf{x} \end{bmatrix} + \begin{bmatrix} 0 \\ \mathbf{G} \end{bmatrix} u - \begin{bmatrix} 1 \\ \mathbf{0} \end{bmatrix} r + \begin{bmatrix} 0 \\ \mathbf{G}_1 \end{bmatrix} w, \tag{7.203}$$

and the feedback law is

$$u = -\begin{bmatrix} K_1 & \mathbf{K}_0 \end{bmatrix} \begin{bmatrix} x_I \\ \mathbf{x} \end{bmatrix},$$

---

[11] Watch out for the sign here; we are using the negative of the usual convention.

Figure 7.54
Integral control structure

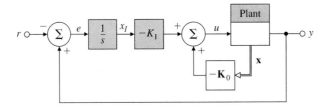

Feedback law with integral control

or simply

$$u = -\mathbf{K}\begin{bmatrix} x_I \\ \mathbf{x} \end{bmatrix}.$$

With this revised definition of the system, we can apply the design techniques from Section 7.3 in a similar fashion; they will result in the control structure shown in Fig. 7.54.

EXAMPLE 7.28   *Integral Control of a Motor Speed System*

Consider the motor speed system described by

$$\frac{Y(s)}{U(s)} = \frac{1}{s+3};$$

that is, $F = -3$, $G = 1$, and $H = 1$. Design the system to have integral control and two poles at $s = -5$. Design an estimator with pole at $s = -10$. The disturbance enters at the same place as the control. Evaluate the tracking and disturbance rejection responses.

Solution.   The pole-placement requirement is equivalent to

$$\mathsf{p}_c = [-5;\ -5].$$

The augmented system description including the disturbance w is

$$\begin{bmatrix} \dot{x}_I \\ \dot{x} \end{bmatrix} = \begin{bmatrix} 0 & 1 \\ 0 & -3 \end{bmatrix}\begin{bmatrix} x_I \\ x \end{bmatrix} + \begin{bmatrix} 0 \\ 1 \end{bmatrix}(u + w) - \begin{bmatrix} 1 \\ 0 \end{bmatrix}r.$$

Therefore, we can find $\mathbf{K}$ from

$$\det\left(s\mathbf{I} - \begin{bmatrix} 0 & 1 \\ 0 & -3 \end{bmatrix} + \begin{bmatrix} 0 \\ 1 \end{bmatrix}\mathbf{K} \right) = s^2 + 10s + 25,$$

Figure 7.55

Integral control example

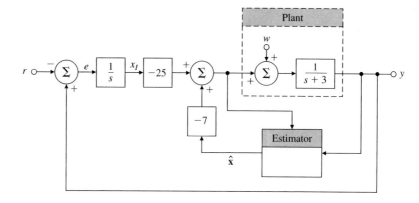

or

$$s^2 + (3 + K_0)s + K_1 = s^2 + 10s + 25.$$

Therefore,

$$\mathbf{K} = [\, K_1 \quad K_0 \,] = [\, 25 \quad 7 \,].$$

We may verify this result using acker. The system is shown with feedbacks in Fig. 7.55 along with a disturbance input w.

The estimator gain $L = 7$ is obtained from

$$\alpha_e(s) = s + 10 = s + 3 + L.$$

The estimator equation is of the form

$$\hat{x} = (F - LH)\hat{x} + Gu + Ly$$
$$= -10\hat{x} + u + 7y$$

and

$$u = -K_0\hat{x} = -7\hat{x}.$$

The step response y_1 due to a step reference input r, and the output disturbance response y_2 due to a step disturbance input w are shown in Fig. 7.56(a) and the associated control efforts (u_1 and u_2) are shown in Fig. 7.56(b). As expected, the system is type 1 and tracks the step reference input and rejects the step disturbance asymptotically.

Figure 7.56

Transient response for motor speed system:(a) step responses, (b) control efforts

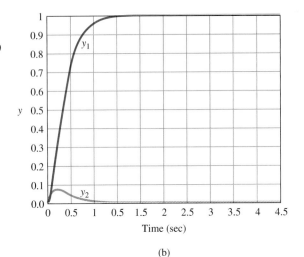

(b)

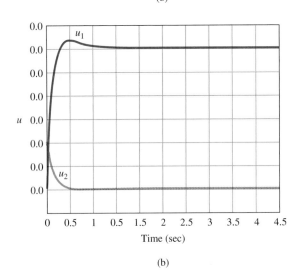

(b)

▲ 7.9.2　Robust Tracking Control: The Error-Space Approach

In Section 7.9.1 we introduced integral control in a direct way and selected the structure of the implementation so as to achieve integral action with respect to reference and disturbance inputs. We now present a more analytical approach to giving a control system the ability to track (with zero steady-state error) a nondecaying input and to reject (with zero steady-state error) a nondecaying disturbance such as a step, ramp, or sinusoidal input. The method is based on including the equations satisfied by these external signals as part of the problem formulation and solving the problem of control in an **error space** so we are assured that the error approaches zero even if the output is following a nondecaying, or even a growing, command (such as a ramp signal) and even if some parameters change (the robustness property). The method is illustrated in detail for signals that satisfy differential equations of order 2, but the extension to more complex signals is not difficult.

Suppose we have the system state equations

$$\dot{\mathbf{x}} = \mathbf{F}\mathbf{x} + \mathbf{G}u + \mathbf{G}_1 w,$$

$$y = \mathbf{H}\mathbf{x} \qquad\qquad (7.204a)$$

and a reference signal that is known to satisfy a specific differential equation. The initial conditions on the equation generating the input are unknown. For example, the input could be a ramp whose slope and initial value are unknown. Plant disturbances of the same class may also be present. We wish to design a controller for this system so that the closed-loop system will have specified poles and will have the ability to track input command signals and to reject disturbances of the type described without steady-state error. We will develop the results only for second-order differential equations. We define the reference input to satisfy the relation

$$\ddot{r} + \alpha_1 \dot{r} + \alpha_2 r = 0 \qquad\qquad (7.205)$$

and the disturbance to satisfy exactly the same equation:

$$\ddot{w} + \alpha_1 \dot{w} + \alpha_2 w = 0. \qquad\qquad (7.206)$$

The (tracking) error is defined as

$$e = y - r. \qquad\qquad (7.207)$$

The meaning of robust control   The problem of tracking r and rejecting w can be seen as an exercise in designing a control law to provide *regulation of the error*, which is to say that the error e tends to zero as time gets large. The control must also be **structurally stable** or **robust**, in the sense that regulation of e to zero in the steady-state occurs even in the presence of "small" perturbations of the original system parameters. Note that in practice we never have a perfect model of the plant and the values of parameters are virtually always subject to some change, so robustness is always very important.

We know that the command input satisfies Eq. (7.205), and we would like to eliminate the reference from the equations in favor of the error. We begin by replacing r in Eq. (7.205) with the error of Eq. (7.207). When we do this, the reference cancels because of Eq. (7.205), and we have the formula for the error in terms of the state

$$\ddot{e} + \alpha_1 \dot{e} + \alpha_2 e = \ddot{y} + \alpha_1 \dot{y} + \alpha_2 y \qquad\qquad (7.208a)$$

$$= \mathbf{H}\ddot{\mathbf{x}} + \alpha_1 \mathbf{H}\dot{\mathbf{x}} + \alpha_2 \mathbf{H}\mathbf{x}. \qquad\qquad (7.208b)$$

We now replace the plant state vector with the error-space state defined by

$$\xi \overset{\triangle}{=} \ddot{\mathbf{x}} + \alpha_1 \dot{\mathbf{x}} + \alpha_2 \mathbf{x}. \tag{7.209}$$

Similarly, we replace the control with the control in error space, defined as

$$\mu \overset{\triangle}{=} \ddot{u} + \alpha_1 \dot{u} + \alpha_2 u. \tag{7.210}$$

With these definitions we can replace Eq. (7.208b) with

$$\ddot{e} + \alpha_1 \dot{e} + \alpha_2 e = \mathbf{H}\xi. \tag{7.211}$$

Robust control equations in the error space

The state equation for ξ is given by[12]

$$\dot{\xi} = \ddot{\mathbf{x}} + \alpha_1 \ddot{\mathbf{x}} + \alpha_2 \dot{\mathbf{x}} = \mathbf{F}\xi + \mathbf{G}\mu. \tag{7.212}$$

Notice that the disturbance as well as the reference cancels from Eq. (7.212). Equations (7.211) and (7.212) now describe the overall system in an error space. In standard state-variable form, the equations are

$$\dot{\mathbf{z}} = \mathbf{A}\mathbf{z} + \mathbf{B}\mu, \tag{7.213}$$

where $\mathbf{z} = [\, e \quad \dot{e} \quad \xi^T \,]^T$ and

$$\mathbf{A} = \begin{bmatrix} 0 & 1 & \mathbf{0} \\ -\alpha_2 & -\alpha_1 & \mathbf{H} \\ \mathbf{0} & \mathbf{0} & \mathbf{F} \end{bmatrix}, \quad \mathbf{B} = \begin{bmatrix} 0 \\ 0 \\ \mathbf{G} \end{bmatrix}. \tag{7.214}$$

The error system $(\mathbf{A}, \mathbf{B})$ can be given arbitrary dynamics by state feedback if it is controllable. If the plant $(\mathbf{F}, \mathbf{G})$ is controllable and does not have a zero at any of the roots of the reference-signal characteristic equation

$$\alpha_r(s) = s^2 + \alpha_1 s + \alpha_2,$$

then the error system $(\mathbf{A}, \mathbf{B})$ is controllable.[13]   We assume these conditions hold; therefore, there exists a control law of the form

$$\mu = -[\, K_2 \quad K_1 \quad \mathbf{K_0} \,] \begin{bmatrix} e \\ \dot{e} \\ \xi \end{bmatrix} = -\mathbf{K}\mathbf{z}, \tag{7.215}$$

such that the error system has arbitrary dynamics by pole placement. We now need to express this control law in terms of the actual process state $\mathbf{x}$ and the

---

[12] Notice that this concept can be extended to more complex equations in r and to multivariable systems.

[13] For example, it is not possible to add integral control to a plant that has a zero at the origin.

Figure 7.57
Integral control using the
internal model approach

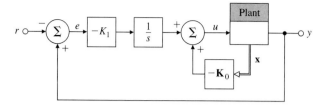

actual control. We combine Eqs. (7.215), (7.209), and (7.210) to get the control law in terms of u and $\mathbf{x}$ (we write $u^{(2)}$ to mean $d^2 u/dt^2$):

$$(u + \mathbf{K}_0 \mathbf{x})^{(2)} + \sum_{i=1}^{2} \alpha_i (u + \mathbf{K}_0 \mathbf{x})^{(2-i)} = -\sum_{i=1}^{2} K_i e^{(2-i)}. \tag{7.216}$$

The structure for implementing Eq. (7.216) is very simple for tracking constant inputs. In that case the equation for the reference input is $\dot{r} = 0$. In terms of u and $\mathbf{x}$ the control law [Eq. (7.216)] reduces to

$$\dot{u} + \mathbf{K}_0 \dot{\mathbf{x}} = -K_1 e. \tag{7.217}$$

Here we only need to integrate to reveal the control law and the action of integral control:

$$u = -K_1 \int^{t} e d\tau - \mathbf{K}_0 \mathbf{x}. \tag{7.218}$$

A block diagram of the system, shown in Fig. 7.57, clearly shows the presence of a pure integrator in the controller. In this case the only difference between the internal model method of Fig. 7.57 and the ad hoc method of Fig. 7.54 is the relative location of the integrator and the gain.

A more complex problem that clearly shows the power of the error-space approach to robust tracking is posed by requiring that a sinusoid be tracked with zero steady-state error. The problem arises, for instance, in the control of a mass-storage disk-head assembly.

Disk-Drive Servomechanism: Robust Control to Follow a Sinusoid

EXAMPLE 7.29

A simple normalized model of a computer disk-drive servomechanism is given by the equations

$$\mathbf{F} = \begin{bmatrix} 0 & 1 \\ 0 & -1 \end{bmatrix}, \qquad \mathbf{G} = \begin{bmatrix} 0 \\ 1 \end{bmatrix};$$

$$\mathbf{G}_1 = \begin{bmatrix} 0 \\ 1 \end{bmatrix}; \qquad \mathbf{H} = [1 \quad 0]; \qquad J = 0.$$

Because the data on the disk is not exactly on a centered circle, the servo must follow a sinusoid of radian frequency ω_0 determined by the spindle speed.

(a) Give the structure of a controller for this system that will follow the given reference input with zero steady-state error.

(b) Assume $\omega_0 = 1$ and that the desired closed-loop poles are at $-1 \pm j\sqrt{3}$ and $-\sqrt{3} \pm j1$. (c) Demonstrate the tracking and disturbance rejection properties of the system using MATLAB or Simulink.

Solution.

(a) The reference input satisfies the differential equation $\ddot{r} = -\omega_0^2 r$ so that $\alpha_1 = 0$ and $\alpha_2 = \omega_0^2$. With these values the error-state matrices according to Eq. (7.214) are

$$\mathbf{A} = \begin{bmatrix} 0 & 1 & 0 & 0 \\ -\omega_0^2 & 0 & 1 & 0 \\ 0 & 0 & 0 & 1 \\ 0 & 0 & 0 & -1 \end{bmatrix}, \quad \mathbf{B} = \begin{bmatrix} 0 \\ 0 \\ 0 \\ 1 \end{bmatrix}.$$

The characteristic equation of $\mathbf{A} - \mathbf{BK}$ is

$$s^4 + (1 + K_{02})s^3 + (\omega_0^2 + K_{01})s^2 + [K_1 + \omega_0^2(1 + K_{02})]s + K_{01}\omega_0^2 K_2 = 0,$$

from which the gain may be selected by pole assignment. The compensator implementation from Eq. (7.216) has the structure shown in Fig. 7.58, which clearly shows the presence of the oscillator with frequency ω_0 (known as the **internal model of the input generator**) in the controller.[14]

Internal model principle

(b) Now assume that $\omega_0 = 1$ rad/sec and the desired closed-loop poles are as given above. If

$$pc = [-1 + j * \sqrt{3}; -1 - j * \sqrt{3}; -\sqrt{3} + j; -\sqrt{3} - j],$$

then the feedback gain is

$$\mathbf{K} = [K_2 \ K_1 : \mathbf{K}_o] = [2.0718 \ 16.3923 : 13.9282 \ 4.4641],$$

Figure 7.58
Structure of the compensator for the servomechanism to track exactly the sinusoid of frequency ω_0

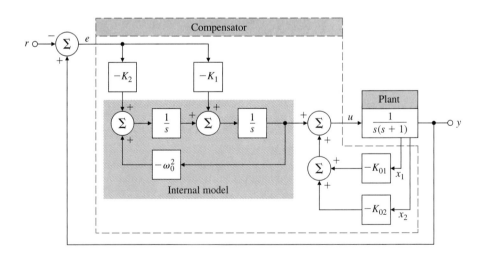

---

[14] This is a particular case of the **internal model principle**, which requires that a model of the external or exogenous signal be in the controller for robust tracking and disturbance rejection.

which results in the controller,

$$\dot{\mathbf{x}}_c = \mathbf{A}_c \mathbf{x}_c + \mathbf{B}_c e$$
$$u = \mathbf{C}_c \mathbf{x}_c$$

with

$$\mathbf{A}_c = \begin{bmatrix} 0 & 1 \\ -1 & 0 \end{bmatrix}, \qquad \mathbf{B}_c = \begin{bmatrix} -16.3923 \\ -2.0718 \end{bmatrix}$$
$$\mathbf{C}_c = \begin{bmatrix} 1 & 0 \end{bmatrix}$$

The relevant MATLAB statements are

```
% plant matrices
F = [0 1; 0 −1];
G = [0;1];
H = [1 0];
J = [0];
% form error space matrices
omega = 1;
A = [0 1 0 0;-omega*omega 0 1 0;0 0 0 1;0 0 0 −1];
B = [0;0;G];
% desired closed-loop poles
pc = [-1+sqrt(3)*j;-1-sqrt(3)*j;-sqrt(3)+j;-sqrt(3)-j];
K = place(A,B,pc);
% form controller matrices
K1 = K(:,1:2);
Ko = K(:,3:4);
Ac = [0 1;-omega*omega 0];
Bc = -[K(2);K(1)];
Cc = [1 0];
Dc = [0];
```

The controller frequency response is shown in Fig. 7.59 and shows a gain of infinity at the rotation frequency of $\omega_0 = 1$ rad/sec. The frequency response from

Figure 7.59
Controller frequency response

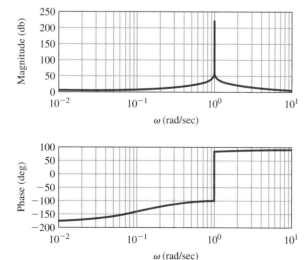

Figure 7.60

Sensitivity function frequency response

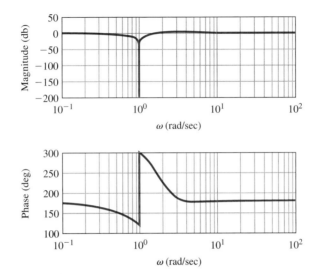

r to e, i.e. the sensitivity function $\mathcal{S}(s)$, is shown in Fig. 7.60 and shows a sharp notch at the rotation frequency $\omega_0 = 1$ rad/sec. The same notch is also present in the frequency response of the transfer function from w to y.

(c) Figure 7.61 shows the Simulink simulation diagram for the system. Although the simulations can also be done in MATLAB, it is more instructive to use the interactive

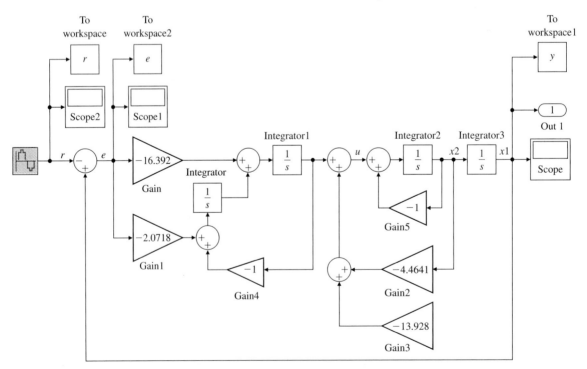

Figure 7.61 Simulink block diagram for robust servomechanism

Figure 7.62
Tracking properties for
robust servomechanism

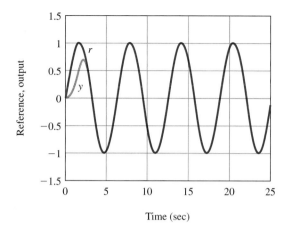

graphical environment of Simulink. Simulink also provides the capability to add
nonlinearities and carry out robustness studies efficiently.[15]   The tracking properties of the system are shown in Fig. 7.62 showing asymptotic tracking property
of the system. The disturbance rejection properties of the system are illustrated
in Fig. 7.63 displaying asymptotic disturbance rejection of sinusoidal disturbance
input. The closed-loop frequency response—that is, the complementary transfer
function $\mathcal{T}(s)$—for the robust servomechanism is shown in Fig. 7.64. As seen from
the figure, the frequency response from r to y is unity at $\omega_0 = 1$ rad/sec as expected.

Simulink simulation

The zeros of the system from r to e are located at $\pm j, -2.7321 \pm j2.5425$. The
robust tracking properties are due to the presence of the blocking zeros at $\pm j$. The
zeros from w to y, both blocking zeros, are located at $\pm j$. The robust disturbance
rejection properties are due to the presence of these blocking zeros.

Blocking zeros

Figure 7.63
Disturbance rejection
properties for robust
servomechanism

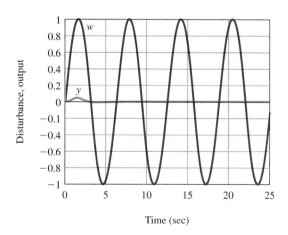

---

[15] In general, the design can be done in MATLAB and (nonlinear) simulations can be carried out
in Simulink.

Figure 7.64
Closed-loop frequency
response for robust
servomechanism

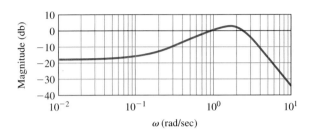

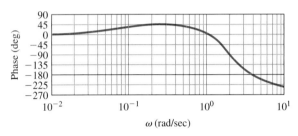

From the nature of the pole-placement problem, the state $\mathbf{z}$ in Eq. (7.213) will tend toward zero for all perturbations in the system parameters as long as $\mathbf{A} - \mathbf{BK}$ remains stable. Notice that the signals that are rejected are those that satisfy the equations with the values of α_i actually implemented in the model of the external signals. The method assumes that these are known and implemented exactly. If the implemented values are in error, then a steady-state error will result.

Now let us repeat the example of Section 7.9.1 for integral control.

EXAMPLE 7.30   *Integral Control Using the Error-Space Design*

For the system

$$H(s) = \frac{1}{s+3}$$

with the state-variable description

$$F = -3, \qquad G = 1, \qquad H = 1,$$

construct a controller with poles at $s = -5$ to track an input that satisfies $\dot{r} = 0$.

Solution. The error system is

$$\begin{bmatrix} \dot{e} \\ \dot{z} \end{bmatrix} = \begin{bmatrix} 0 & 1 \\ 0 & -3 \end{bmatrix} \begin{bmatrix} e \\ z \end{bmatrix} + \begin{bmatrix} 0 \\ 1 \end{bmatrix} \mu.$$

Figure 7.65
Example of internal model with feedforward

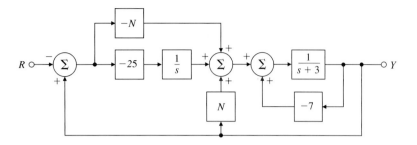

If we take the desired characteristic equation to be

$$\alpha_c(s) = s^2 + 10s + 25,$$

then the pole-placement equation for **K** is

$$\det[s\mathbf{I} - \mathbf{A} + \mathbf{BK}] = \alpha_c(s). \tag{7.219}$$

In detail Eq. (7.219) is

$$s^2 + (3 + K_0)s + K_1 = s^2 + 10s + 25,$$

which gives

$$\mathbf{K} = [\,25 \quad 7\,] = [\,K_1 \quad K_0\,],$$

and the system is implemented as shown in Fig. 7.65. The transfer function from r to e for this system, the sensitivity function,

$$\frac{E(s)}{R(s)} = \mathcal{S}(s) = -\frac{s(s+10)}{s^2 + 10s + 25},$$

shows a blocking zero at $s = 0$, which prevents the constant input from affecting the error. The closed-loop transfer function, that is, the complementary sensitivity function is

$$\frac{Y(s)}{R(s)} = \mathcal{T}(s) = \frac{25}{s^2 + 10s + 25}.$$

The structure of Fig. 7.65 permits us to add a feedforward of the reference input, which provides one extra degree of freedom in zero assignment. If we add a term proportional to r in Eq. (7.218), then

$$u = -K_1 \int^t e(\tau)\,d\tau - \mathbf{K}_0\mathbf{x} + Nr. \tag{7.220}$$

This relationship has the effect of creating a zero at $-K_1/N$. The location of this zero can be chosen to improve the transient response of the system. For

Figure 7.66
Internal model as integral
control with feedforward

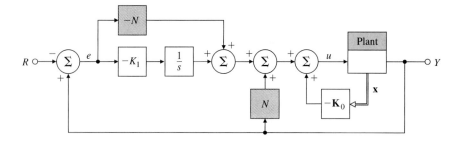

actual implementation we can rewrite Eq. (7.220) in terms of e to get

$$u = -K_1 \int^t e(\tau)\,d\tau - \mathbf{K}_0\mathbf{x} + N(y - e). \qquad (7.221)$$

The block diagram for the system is shown in Fig. 7.66. For our example the overall transfer function now becomes

$$\frac{Y(s)}{R(s)} = \frac{Ns + 25}{s^2 + 10s + 25}.$$

Notice that the DC gain is unity for any value of N and that through our choice of N we can place the zero at any real value to improve the dynamic response. A natural strategy for locating the zero is to have it cancel one of the system poles, in this case at $s = -5$. The step response of the system is shown in Fig. 7.67 for $N = 5$, as well as for $N = 0$ and 8. With the understanding that one pole can be canceled in integral control designs, we make sure to choose one of the desired control poles such that it is both real and able to be canceled through the proper choice of N.

Figure 7.67
Step responses with
integral control and
feedforward

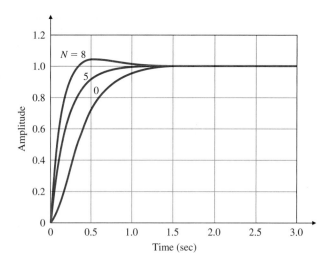

▲ 7.9.3   The Extended Estimator

Our discussion of robust control so far has used a control based on full-state feedback. If the state is not available, then as in the regular case, the full-state feedback, $\mathbf{Kx}$, can be replaced by the estimates, $\mathbf{K\hat{x}}$, where the estimator is built as before. As a final look at ways to design control with external inputs, in this section we develop a method for tracking a reference input and rejecting disturbances. The method is based on augmenting the estimator to include estimates from external signals in a way that permits us to cancel out their effects on the system error.

Suppose the plant is described by the equations

$$\dot{\mathbf{x}} = \mathbf{Fx} + \mathbf{G}u + \mathbf{G}w, \tag{7.222a}$$

$$y = \mathbf{Hx}, \tag{7.222b}$$

$$e = \mathbf{Hx} - r. \tag{7.222c}$$

Furthermore, assume that both the reference r and the disturbance w are known to satisfy the equations[16]

$$\alpha_w(s)w = \alpha_\rho(s)w = 0, \tag{7.223}$$

$$\alpha_r(s)r = \alpha_\rho(s)r = 0, \tag{7.224}$$

where

$$\alpha_\rho(s) = s^2 + \alpha_1 s + \alpha_2$$

corresponding to polynomials $\alpha_w(s)$ and $\alpha_r(s)$ in Figure 7.68(a). In general, we would select the equivalent disturbance polynomial $\alpha_\rho(s)$ in Figure 7.68(b) to be the *least common multiple* of $\alpha_w(s)$ and $\alpha_r(s)$. The first step is to recognize that, as far as the steady-state response of the output is concerned, there is an input-equivalent signal ρ that satisfies the same equation as r and w and enters the system at the same place as the control signal as shown in Figure 7.68(b). As before, we must assume that the plant does not have a zero at any of the roots of Eq. (7.223). For our purposes here, we can replace Eq. (7.222) with

$$\dot{\mathbf{x}} = \mathbf{Fx} + \mathbf{G}(u + \rho), \tag{7.225a}$$

$$e = \mathbf{Hx}. \tag{7.225b}$$

If we can estimate this equivalent input, we can add to the control a term $-\hat{\rho}$ that will cancel out the effects of the real disturbance and reference and cause

---

[16] Again we develop the results for a second-order equation in the external signals; the discussion can be extended to higher-order equations.

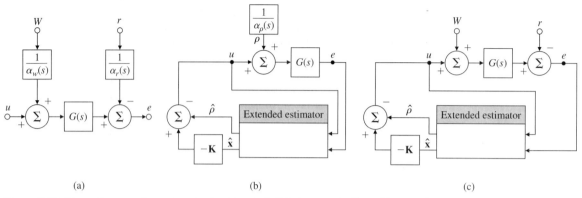

(a)   (b)   (c)

Figure 7.68 Block diagram of a system for tracking and disturbance rejection with extended estimator: (a) equivalent disturbance; (b) block diagram for *design*; (c) block diagram for *implementation*

the output to track r in the steady-state. To do this we combine Eqs. (7.223) and (7.225) into a state description to get

$$\dot{\mathbf{z}} = \mathbf{A}\mathbf{z} + \mathbf{B}u, \tag{7.226a}$$

$$e = \mathbf{C}\mathbf{z}, \tag{7.226b}$$

where $\mathbf{z} = [\,\rho \quad \dot{\rho} \quad \mathbf{x}^T\,]^T$. The matrices are

$$\mathbf{A} = \begin{bmatrix} 0 & 1 & \mathbf{0} \\ -\alpha_2 & -\alpha_1 & \mathbf{0} \\ \mathbf{G} & 0 & \mathbf{F} \end{bmatrix}, \qquad \mathbf{B} = \begin{bmatrix} 0 \\ 0 \\ \mathbf{G} \end{bmatrix} \tag{7.227a}$$

$$\mathbf{C} = [\,0 \quad 0 \quad \mathbf{H}\,] \tag{7.227b}$$

The system given by Eqs. (7.227) is not controllable since we cannot influence ρ from u. However, if $\mathbf{F}$ and $\mathbf{H}$ are observable and if the system $(\mathbf{F}, \mathbf{G}, \mathbf{H})$ does not have a zero that is also a root of Eq. (7.223), then the system of Eq. (7.227) will be observable, and we can construct an observer that will compute estimates of both the state of the plant and of ρ. The estimator equations are standard, but the control is not:

$$\dot{\hat{\mathbf{z}}} = \mathbf{A}\hat{\mathbf{z}} + \mathbf{B}u + \mathbf{L}(e - \mathbf{C}\hat{\mathbf{z}}) \tag{7.228a}$$

$$u = -\mathbf{K}\hat{\mathbf{x}} - \hat{\rho} \tag{7.228b}$$

In terms of the original variables, the estimator equations are

$$\begin{bmatrix} \dot{\hat{\rho}} \\ \ddot{\hat{\rho}} \\ \dot{\hat{\mathbf{x}}} \end{bmatrix} = \begin{bmatrix} 0 & 1 & \mathbf{0} \\ -\alpha_2 & -\alpha_1 & \mathbf{0} \\ \mathbf{G} & 0 & \mathbf{F} \end{bmatrix} \begin{bmatrix} \hat{\rho} \\ \dot{\hat{\rho}} \\ \hat{\mathbf{x}} \end{bmatrix} + \begin{bmatrix} 0 \\ 0 \\ \mathbf{G} \end{bmatrix} u + \begin{bmatrix} l_1 \\ l_2 \\ \mathbf{L}_3 \end{bmatrix} [e - \mathbf{H}\hat{\mathbf{x}}]. \tag{7.229}$$

The overall block diagram of the system for *design* is shown in Fig. 7.68(b). If we write out the last equation for $\hat{\mathbf{x}}$ in Eq. (7.229) and substitute Eq. (7.228b), a simplification of sorts results because a term in $\hat{\rho}$ cancels out:

$$\dot{\hat{\mathbf{x}}} = \mathbf{G}\hat{\rho} + \mathbf{F}\hat{\mathbf{x}} + \mathbf{G}(-\mathbf{K}\hat{\mathbf{x}} - \hat{\rho}) + \mathbf{L}_3(e - \mathbf{H}\hat{\mathbf{x}})$$

$$= \mathbf{F}\hat{\mathbf{x}} + \mathbf{G}(-\mathbf{K}\hat{\mathbf{x}}) + \mathbf{L}_3(e - \mathbf{H}\hat{\mathbf{x}})$$

$$= \mathbf{F}\hat{\mathbf{x}} + \mathbf{G}\bar{u} + \mathbf{L}_3(e - \mathbf{H}\hat{\mathbf{x}}).$$

With the estimator of Eq. (7.229) and the control of Eq. (228b), the state equation is

$$\dot{\mathbf{x}} = \mathbf{F}\mathbf{x} + \mathbf{G}(-\mathbf{K}\hat{\mathbf{x}} - \hat{\rho}) + \mathbf{G}\rho. \tag{7.230}$$

In terms of the estimate errors, Eq. (7.230) can be rewritten as

$$\dot{\mathbf{x}} = (\mathbf{F} - \mathbf{G}\mathbf{K})\mathbf{x} + \mathbf{G}\mathbf{K}\tilde{\mathbf{x}} + \mathbf{G}\tilde{\rho}. \tag{7.231}$$

Because we designed the estimator to be stable, the values of $\tilde{\rho}$ and $\tilde{\mathbf{x}}$ go to zero in the steady-state, and the final value of the state is not affected by the external input. The block diagram of the system for *implementation* is drawn in Fig. 7.68(c). A very simple example will illustrate the steps in this process.

EXAMPLE 7.31

Steady-State Tracking and Disturbance Rejection of Motor Speed by Extended Estimator

Construct an estimator to control the state and cancel a constant bias at the output and track a constant reference in the motor speed system described by

$$\dot{x} = -3x + u, \tag{7.232a}$$

$$y = x + w, \tag{7.232b}$$

$$\dot{w} = 0, \tag{7.232c}$$

$$\dot{r} = 0. \tag{7.232d}$$

Place the control pole at $s = -5$ and the two extended estimator poles at $s = -15$.

Solution.   To begin, we design the control law by ignoring the equivalent disturbance. Rather, we notice by inspection that a gain of -2 will move the single pole from -3 to the desired -5. Therefore, $K = 2$. The system augmented with equivalent external input ρ, which replaces the actual disturbance w and the reference r, is given by

$$\dot{\rho} = 0,$$

$$\dot{x} = -3x + u + \rho,$$

$$e = x.$$

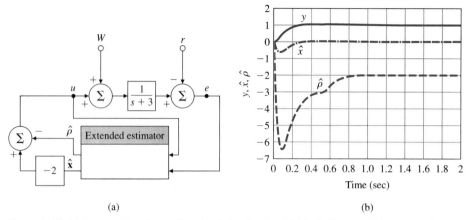

Figure 7.69 Motor speed system with extended estimator (a) block diagram; (b) command step response and disturbance step response

The extended estimator equations are

$$\dot{\hat{\rho}} = l_1(e - \hat{x}),$$

$$\dot{\hat{x}} = -3\hat{x} + u + \hat{\rho} + l_2(e - \hat{x}).$$

The estimator error gain is found to be $\mathbf{L} = [\,225 \quad 27\,]^T$ from the characteristic equation

$$\det \begin{bmatrix} s & l_1 \\ 1 & s+3+l_2 \end{bmatrix} = s^2 + 30s + 225.$$

A block diagram of the system is given in Fig. 7.69(a), and the step responses to input at the command r (applied at $t = 0$ sec) and at the disturbance w (applied at $t = 0.5$ sec) are shown in Fig. 7.69(b).

▲ 7.10   Direct Design with Rational Transfer Functions

An alternative to the state-space methods discussed so far is to postulate a general-structure dynamic controller with two inputs (r and y) and one output (u) and to solve for the transfer function of the controller to give a specified overall r-to-y transfer function. A block diagram of the situation is shown in Fig. 7.70. We model the plant as the transfer function

$$\frac{Y(s)}{U(s)} = \frac{b(s)}{a(s)}, \tag{7.233}$$

Figure 7.70
Direct transfer-function
formulation

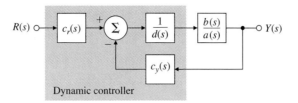

Dynamic controller

General controller in
polynomial form

rather than by state equations. The controller is also modeled by its transfer
function, in this case a transfer function with two inputs and one output:

$$U(s) = -\frac{c_y(s)}{d(s)} Y(s) + \frac{c_r(s)}{d(s)} R(s), \tag{7.234}$$

where $d(s)$, $c_y(s)$, and $c_r(s)$ are polynomials. In order for the controller of
Fig. 7.70 and Eq. (7.234) to be implemented, the orders of the numerator polynomials $c_y(s)$ and $c_r(s)$ must not be higher than the order of the denominator
polynomial $d(s)$.

To carry out the design, we require that the closed-loop transfer function
defined by Eqs. (7.233) and (7.234) be matched to the desired transfer function

$$\frac{Y(s)}{R(s)} = \frac{c_r(s)b(s)}{\alpha_c(s)\alpha_e(s)}. \tag{7.235}$$

Equation (7.235) tells us that the zeros of the plant must be zeros of the overall
system. The only way to change this is to have factors of $b(s)$ appear in either
α_c or α_e. We continue Eqs. (7.233) and (7.234) to get

$$a(s)Y(s) = b(s)\left[-\frac{c_y(s)}{d(s)} Y(s) + \frac{c_r(s)}{d(s)} R(s)\right], \tag{7.236}$$

which can be rewritten as

$$[a(s)d(s) + b(s)c_y(s)]Y(s) = b(s)c_r(s)R(s). \tag{7.237}$$

Diophantine equation

Comparing Eq. (7.235) with Eq. (7.236), we immediately see that the design
can be accomplished if we can solve the **Diophantine equation**

$$a(s)d(s) + b(s)c_y(s) = \alpha_c(s)\alpha_e(s) \tag{7.238}$$

for given arbitrary a, b, α_c, and α_e. Because each transfer function is a ratio
of polynomials, we can assume that $a(s)$ and $d(s)$ are **monic polynomials**; that
is, the coefficient of the highest power of s in each polynomial is unity. The
question is, How many equations and how many unknowns are there if we
match coefficients of equal powers of s in Eq. (7.238)? If $a(s)$ is of degree
n (given) and $d(s)$ is of degree m (to be selected), then a direct count yields

Dimension of the controller $2m + 1$ unknowns in $d(s)$ and $c_y(s)$ and $n + m$ equations from the coefficients of powers of s. Thus the requirement is that

$$2m + 1 \geq n + m$$

or

$$m \geq n - 1.$$

One possibility for a solution is to choose $d(s)$ of degree n and $c_y(s)$ of degree $n - 1$. In that case, which corresponds to the state-space design for a full-order estimator, there are $2n$ equations and $2n$ unknowns with $\alpha_c \alpha_e$ of degree $2n$. The resulting equations will then have a solution for arbitrary α_i if and only if $a(s)$ and $b(s)$ have no common factors.[17]

EXAMPLE 7.32   ***Pole Placement for Polynomial Transfer Functions***

Using the polynomial method, design a controller of order n for the third-order plant in Example 7.22. Note that if the polynomials $\alpha_c(s)$ and $\alpha_e(s)$ from Example 7.22 are multiplied, the result is the desired closed-loop characteristic equation:

$$\alpha_c(s)\alpha_e(s) = s^6 + 14s^5 + 122.75s^4 + 585.2s^3 + 1505.64s^2 + 2476.8s + 1728. \quad (7.239)$$

Solution.   Using Eq. (7.238) with $b(s) = 10$, we find that

$$(d_0 s^3 + d_1 s^2 + d_2 s + d_3)(s^3 + 10s^2 + 16s) + 10(c_0 s^2 + c_1 s + c_2) \equiv \alpha_c(s)\alpha_e(s). \quad (7.240)$$

We have expanded the polynomial $d(s)$ with coefficients d_i and the polynomial $c_y(s)$ with coefficients c_i.

Now we equate the coefficients of the like powers of s in Eq. (7.240) to find that the parameters must satisfy[18]

$$
\begin{bmatrix}
1 & 0 & 0 & 0 & 0 & 0 & 0 \\
10 & 1 & 0 & 0 & 0 & 0 & 0 \\
16 & 10 & 1 & 0 & 0 & 0 & 0 \\
0 & 16 & 10 & 1 & 0 & 0 & 0 \\
0 & 0 & 16 & 10 & 10 & 0 & 0 \\
0 & 0 & 0 & 16 & 0 & 10 & 0 \\
0 & 0 & 0 & 0 & 0 & 0 & 10
\end{bmatrix}
\begin{bmatrix}
d_0 \\
d_1 \\
d_2 \\
d_3 \\
c_0 \\
c_1 \\
c_2
\end{bmatrix}
=
\begin{bmatrix}
1 \\
14 \\
122.75 \\
585.2 \\
1505.64 \\
2476.8 \\
1728
\end{bmatrix}. \quad (7.241)
$$

---

[17] If they do have a common factor, it will show up on the left side of Eq. (7.238); for there to be a solution, the same factor must be on the right side of Eq. (7.238) and thus must be a factor of either α_c or α_e.

[18] The matrix on the left side of Eq. (7.241) is called a **Sylvester matrix** and is nonsingular if and only if $a(s)$ and $b(s)$ have no common factor.

The solution to Eq. (7.241) is

$$
\begin{aligned}
d_0 &= 1, & c_0 &= 190.1, \\
d_1 &= 4, & c_1 &= 481.8, \\
d_2 &= 66.75, & c_2 &= 172.8. \\
d_3 &= -146.3,
\end{aligned}
$$

MATLAB a\b

[The solution can be found using x = a\b command in MATLAB where a is the Sylvester matrix and b is the right hand side in Eq. (7.241).] Hence the controller transfer function is

$$
\frac{c_y(s)}{d(s)} = \frac{190.1s^2 + 481.8s + 172.8}{s^3 + 4s^2 + 66.75s - 146.3}. \tag{7.242}
$$

Note that the coefficients of Eq. (7.242) are the same as those of the controller $D_c(s)$ (which we obtained using the state-variable techniques), once the factors in $D_c(s)$ are multiplied out.

The reduced-order compensator can also be derived using a polynomial solution.

Reduced-Order Design for a Polynomial Transfer Function Model

EXAMPLE 7.33

Design a reduced-order controller for the third-order system in Example 7.22. The desired characteristic equation is

$$
\alpha_c(s)\alpha_e(s) = s^5 + 12s^4 + 74s^3 + 207s^2 + 378s + 288.
$$

Solution.   The equations needed to solve this problem are the same as those used to obtain Eq. (7.240), except that we take both $d(s)$ and $c_y(s)$ to be of degree $n - 1$. We need to solve

$$
(d_0s^2 + d_1s + d_2)(s^3 + 10s^2 + 16s) + 10(c_0s^2 + c_1s + c_2) \equiv \alpha_c(s)\alpha_e(s). \tag{7.243}
$$

Equating coefficients of like powers of s in Eq. (7.243), we obtain

$$
\begin{bmatrix}
1 & 0 & 0 & 0 & 0 & 0 \\
10 & 1 & 0 & 0 & 0 & 0 \\
16 & 10 & 1 & 0 & 0 & 0 \\
0 & 16 & 10 & 10 & 0 & 0 \\
0 & 0 & 16 & 0 & 10 & 0 \\
0 & 0 & 0 & 0 & 0 & 10
\end{bmatrix}
\begin{bmatrix}
d_0 \\ d_1 \\ d_2 \\ c_0 \\ c_1 \\ c_2
\end{bmatrix}
=
\begin{bmatrix}
1 \\ 12 \\ 74 \\ 207 \\ 378 \\ 288
\end{bmatrix}. \tag{7.244}
$$

MATLAB a\b

The solution is (again using the x = a\b command in MATLAB)

$$
\begin{aligned}
d_0 &= 1, & c_0 &= -20.8, \\
d_1 &= 2.0, & c_1 &= -23.6, \\
d_2 &= 38, & c_2 &= 28.8,
\end{aligned}
$$

and the resulting controller is

$$\frac{c_y(s)}{d(s)} = \frac{-20.8s^2 - 23.6s + 28.8}{s^2 + 2.0s + 38}. \tag{7.245}$$

Again, Eq. (7.245) is exactly the same as $D_{cr}(s)$ derived using the state-variable techniques in Example 7.23, once the polynomials of $D_{cr}(s)$ are multiplied out and minor numerical differences are considered.

Notice that the reference input polynomial $c_r(s)$ does not enter into the analysis of Examples 7.32 and 7.33. We can select $c_r(s)$ so that it will assign zeros in the transfer function from $R(s)$ to $Y(s)$. This is the same role played by $\gamma(s)$ in Section 7.8. One choice is to select $c_r(s)$ to cancel $\alpha_e(s)$ so that the overall transfer function is

$$\frac{Y(s)}{R(s)} = \frac{K_s b(s)}{\alpha_c(s)}.$$

This corresponds to the first and most common choice of $\mathbf{M}$ and $\bar{N}$ for introducing the reference input described in Section 7.8.

Adding integral control to the polynomial solution

It is also possible to introduce integral control and, indeed, internal-model-based robust tracking control into the polynomial design method. What is required is that we have error control and that the controller have poles at the internal model locations. To get error control with the structure of Fig. 7.70, we need only let $c_r = c_y$. To get desired poles into the controller, we need to require that a specific factor be part of $d(s)$. For integral control—the most common case—this is almost trivial. The polynomial $d(s)$ will have a root at zero if we set the last term, d_m, to zero. The resulting equations can be solved if $m = n$. For a more general internal model we define $d(s)$ to be the product of a reduced-degree polynomial and a specified polynomial such as Eq. (7.223), and we match coefficients in the Diophantine equation as before. The process is straightforward but tedious. Again we caution that, while the polynomial design method can be effective, the numerical problems of this method are often much worse than those associated with methods based on state equations. For higher-order systems the state-space methods are preferable.

▲ 7.11   Design for Systems with Pure Time Delay

In any linear system consisting of lumped elements, the response of the system appears immediately after an excitation of the system. In some feedback systems—for example, process control systems, whether controlled by a human operator in the loop or by computer—there is a **pure time delay** (also called **transportation lag**) in the system. As a result of the distributed nature of these systems, the response remains identically zero until after a delay of λ seconds. A typical step response is shown in Fig. 7.71(a). The transfer function of a pure

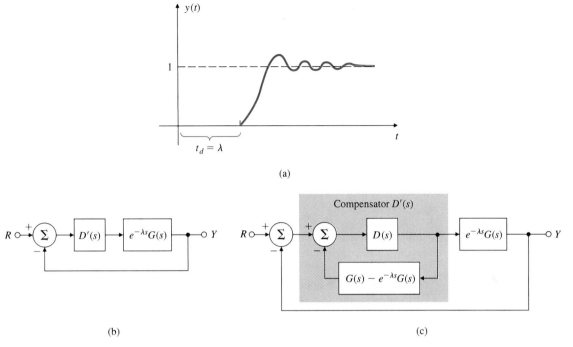

Figure 7.71 A Smith regulator for systems with time delay

Overall transfer function for a time-delayed system

transportation lag is $e^{-\lambda s}$. We can represent an overall transfer function of a single-input–single-output (SISO) system with time delay as

$$G_I(s) = G(s)e^{-\lambda s}, \tag{7.246}$$

where $G(s)$ has no pure time delay. Because $G_I(s)$ does not have a finite state description, standard use of state-variable methods is impossible. However, O. J. M. Smith (1958) showed how to construct a feedback structure that effectively takes the delay outside the loop and allows a feedback design based on $G(s)$ alone, which can be done with standard methods. The result of this method is a design having closed-loop transfer function with delay λ but otherwise showing the same response as the closed-loop design based on no delay. To see how the method works, let us consider the feedback structure shown in Fig. 7.71(b). The overall transfer function is

$$\frac{Y(s)}{R(s)} = \mathcal{T}(s) = \frac{D'(s)G(s)e^{-\lambda s}}{1 + D'(s)G(s)e^{-\lambda s}}. \tag{7.247}$$

Smith suggested that we solve for $D'(s)$ by setting up a dummy overall transfer function in which the controller transfer function $D(s)$ is in a loop with $G(s)$ with *no loop* delay but with an overall delay of λ:

$$\frac{Y(s)}{R(s)} = \mathcal{T}(s) = \frac{D(s)G(s)}{1 + D(s)G(s)}e^{-\lambda s}. \tag{7.248}$$

We then equate Eqs. (7.247) and (7.248) to solve for $D'(s)$:

$$D'(s) = \frac{D(s)}{1 + D(s)[G(s) - G(s)e^{-\lambda s}]}. \qquad (7.249)$$

If the plant transfer function and the delay are known, $D'(s)$ can be realized with real components by means of the block diagram shown in Fig. 7.71(c). With this knowledge we can design the compensator $D(s)$ in the usual way based on Eq. (7.248), as if there were no delay, and then implement it as shown in Fig. 7.71(c). The resulting closed-loop system would exhibit the behavior of a finite closed-loop system except for the time delay λ. This design approach is particularly suitable when the pure delay, λ, is significant as compared to the process time constant, for example, in pulp and paper process applications.

Notice that, conceptually, the Smith compensator is feeding back a simulated plant output to cancel the true plant output and then adding in a simulated plant output without the delay. It can be demonstrated that $D'(s)$ in Fig. 7.71(c) is equivalent to an ordinary regulator in line with a compensator that provides significant phase lead. To implement such compensators in analog systems, it is usually necessary to approximate the delay required in $D'(s)$ by a Padé approximant; with digital compensators the delay can be implemented exactly (see Chapter 8). It is also a fact that the compensator $D'(s)$ is a strong function of $G(s)$, and a small error in the model of the plant used in the controller could lead to large errors in the closed-loop, perhaps even to instability. This design is very sensitive. If $D(s)$ is implemented as a PI controller, then one could detune (i.e., reduce the gain) to try to ensure stability and reasonable performance. For automatic-tuning of Smith regulator and a recent application to Stanford's quiet hydraulic precision lathe fluid temperature control, the reader is referred to J-J. Huang and D. B. DeBra (2000).

EXAMPLE 7.34   *Heat Exchanger: Design with Pure Time Delay*

Figure 7.72 shows the heat exchanger from Example 2.17. The temperature of the product is controlled by controlling the flow rate of steam in the exchanger jacket. The temperature sensor is several meters downstream from the steam control valve, which introduces a transportation lag into the model. A suitable model is given by

$$G(s) = \frac{e^{-5s}}{(10s + 1)(60s + 1)}.$$

Design a controller for the heat exchanger using the Smith compensator and pole placement. The control poles are to be at

$$p_c = -0.05 \pm 0.087j,$$

and the estimator poles are to be at three times the control poles' natural frequency:

$$p_e = -0.15 \pm 0.26j.$$

Simulate the response of the system using Simulink.

Figure 7.72
A heat exchanger

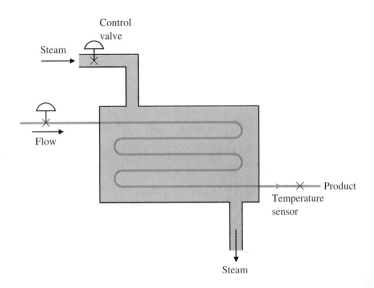

Solution.   A suitable set of state-space equations is

$$\dot{\mathbf{x}}(t) = \begin{bmatrix} -0.017 & 0.017 \\ 0 & -0.1 \end{bmatrix} \mathbf{x}(t) + \begin{bmatrix} 0 \\ 0.1 \end{bmatrix} u(t-5),$$
$$y = [1 \quad 0]\mathbf{x},$$
$$\lambda = 5.$$

For the specified control pole locations and for the moment ignoring the time delay we find that the state feedback gain is

$$\mathbf{K} = [5.2 \quad -0.17].$$

For the given estimator poles, the estimator gain matrix for a full-order estimator is

$$\mathbf{L} = \begin{bmatrix} 0.18 \\ 4.2 \end{bmatrix}.$$

The resulting controller transfer function is

$$D(s) = \frac{U(s)}{Y(s)} = \frac{-0.25(s+1.8)}{s+0.14 \pm 0.27j}.$$

If we choose to adjust for unity closed-loop DC gain, then

$$\bar{N} = 1.2055.$$

The Simulink diagram for the system is shown in Fig. 7.73. The open-loop and closed-loop step responses of the system and the control effort are shown in Figs. 7.74 and 7.75, and the root locus of the system (without the delay) is shown in Fig. 7.76. Note that the time delay of 5 sec in Figs. 7.74 and 7.75 is quite small compared with the response of the system, and it is barely noticeable in this case.

Simulink simulation

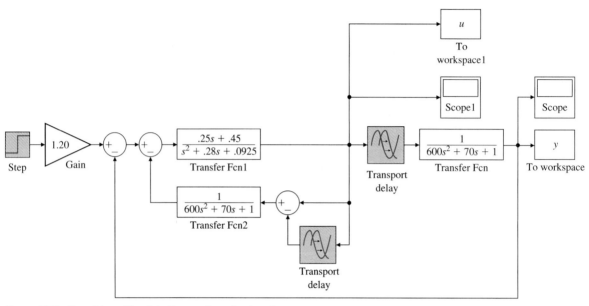

Figure 7.73 Closed-loop Simulink diagram for a heat exchanger

Figure 7.74
Step response for a heat exchanger

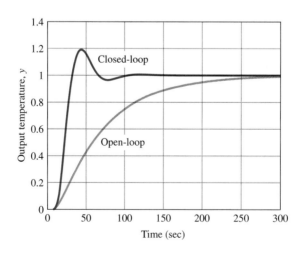

Figure 7.75
Control effort for a heat exchanger

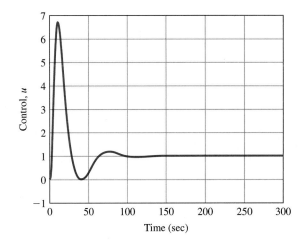

Figure 7.76
Root locus for a heat exchanger

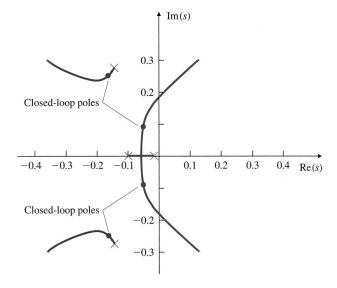

▲ 7.12 Lyapunov Stability

It should be obvious by now that stability plays a major role in control systems design. We have showed that for linear constant systems, bounded-input–bounded-output (BIBO) stability requires that all system poles be in the LHP; also, we have given Routh's criterion based on the coefficients of the characteristic equation and Nyquist's criterion based on the frequency response to test for unstable poles. In this section we consider another important technique for analyzing the stability of nonlinear as well as linear dynamic systems based on the state-variable form of ordinary differential equations (ODEs).

A. M. Lyapunov[19] considered the stability of general systems described by ODEs in state-variable form. To illustrate a few points of his very useful theory, we consider first the general set of state equations

$$\dot{\mathbf{x}} = \mathbf{f}(\mathbf{x}). \tag{7.250}$$

If $\mathbf{f}(\mathbf{x}) = \mathbf{F}\mathbf{x}$, then we have the linear case considered throughout this chapter. We assume that the equations have been written so that $\mathbf{x} = \mathbf{0}$ is an equilibrium point, which is to say that $\mathbf{f}(\mathbf{0}) = \mathbf{0}$. This equilibrium point is said to be **Lyapunov-stable** if we are able to select a bound on initial conditions that will result in trajectories that remain within a chosen finite limit. The equilibrium is said to be **asymptotically stable** if it is Lyapunov-stable and if the state approaches zero as time approaches infinity. More formally, the system described by Eq. (7.250) has a stable equilibrium at $\mathbf{x} = \mathbf{0}$ if, for every ε there is a δ such that if $\|\mathbf{x}(0)\| < \delta$, then $\|\mathbf{x}(t)\| < \varepsilon$ for all t.[20] Pictures of Lyapunov-stable and -unstable responses are shown in Fig. 7.77.

> Stability in the sense of Lyapunov

For constant linear systems we see at once that $\dot{\mathbf{x}} = \mathbf{F}\mathbf{x}$ is Lyapunov-stable if none of the eigenvalues of $\mathbf{F}$ are in the **RHP** and if any eigenvalues on the imaginary axis are simple. (A multiple root on the imaginary axis would have a

Figure 7.77
Definition of Lyapunov stability

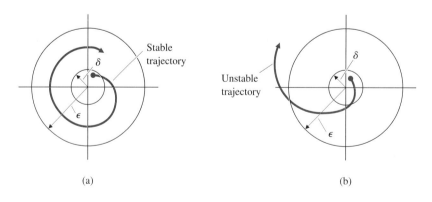

(a)   (b)

---

[19] A. M. Lyapunov studied dynamic systems in Russia, and his fundamental work, *On the General Problem of Stability of Motion*, was published in 1892 by the Kharkov Mathematical Society.

[20] $\|\mathbf{x}\|2 = \mathbf{x}^T\mathbf{x} = \sum_{i=1}^{n} x_i^2$ is a measure of the length of the vector $\mathbf{x}$.

response that grows in time and could not be stable.) Furthermore, the response of a linear constant system is asymptotically stable if all the eigenvalues of $\mathbf{F}$ are inside the **LHP**.

Let us consider a nonlinear system such as Eq. (7.250) and expand it into the form

$$\dot{\mathbf{x}} = \mathbf{F}\mathbf{x} + \mathbf{g}(\mathbf{x}), \qquad (7.251)$$

where $\mathbf{g}(\mathbf{x})$ contains all the higher powers of $\mathbf{x}$ to the extent that

$$\lim_{\|x\| \to 0} \frac{\|\mathbf{g}(\mathbf{x})\|}{\|\mathbf{x}\|} = 0$$

[that is, $\mathbf{g}(\mathbf{x})$ goes to zero faster than $\mathbf{x}$ does]. Lyapunov showed that such a system will be stable if all the eigenvalues of $\mathbf{F}$ are strictly inside the **LHP** and unstable if at least one eigenvalue is in the RHP. For systems with eigenvalues in the **LHP** and on the imaginary axis, the stability depends on the terms in $\mathbf{g}$.

In order to prove these results, Lyapunov introduced a function that has many of the properties of energy. He then proved that if the "energy" never increases, then the state must be confined to a volume bounded by a surface of constant energy. This argument goes as follows: For a state $\mathbf{x}$ satisfying Eq. (7.251), we consider the scalar function $V(\mathbf{x})$ with the following properties:

Properties of the Lyapunov function

1. $V(\mathbf{0}) = 0$.

2. $V(\mathbf{x}) > 0$, $\|\mathbf{x}\| \neq 0$.

3. V is continuous and has continuous derivatives with respect to all components of $\mathbf{x}$.

4. $\dot{V}(\mathbf{x}) \leq 0$ along trajectories of the equation.

We call V having these properties a **Lyapunov function** for the system. Properties 1 and 2 mean that, like energy, $V > 0$ if any state is different from zero, but $V = 0$ when the state is zero. Property 3 ensures that V is a smooth function and generally has the shape of a bowl near the equilibrium. Property 4 guarantees that any trajectory moves so as never to climb higher on the bowl than where it started out. If property 4 is made stronger so that $\dot{V} < 0$ for $\|\mathbf{x}\| \neq 0$, then the trajectory must be drawn to the origin. The **Lyapunov stability theorem** states that, given the system of equations $\dot{\mathbf{x}} = \mathbf{f}(\mathbf{x})$ with $\mathbf{f}(\mathbf{0}) = \mathbf{0}$, if there exists a Lyapunov function for this equation, then the origin is a stable equilibrium point; in addition, if $\dot{V} < 0$, then the stability is asymptotic.

Lyapunov stability theorem

For the linear constant systems of main concern in this text, the quadratic function is adequate for demonstrating Lyapunov stability. Consider the function $V = \mathbf{x}^T \mathbf{P}\mathbf{x}$, where $\mathbf{P}$ is a symmetric positive matrix. For example, if $\mathbf{P} = \mathbf{I}$, then V is the sum of squares of the components of $\mathbf{x}$; that is, $V = \sum_{i=1}^{n} x_i^2$. In general, if $\mathbf{P} > \mathbf{0}$, it means we can find a matrix $\mathbf{T}$ such that $\mathbf{P} = \mathbf{T}^T \mathbf{T}$ and $V = \sum_{i=1}^{n} z_i^2$ where $\mathbf{z} = \mathbf{T}\mathbf{x}$. For such a $\mathbf{P}$ the quadratic function satisfies properties 1, 2, and 3 of a Lyapunov function. A well-known proof that a symmetric matrix is positive is if the determinants of the n principal minors of the matrix are all positive. We need to consider property 4, the derivative condition. To compute the derivative of V, we use the chain rule to get

$$\dot{V} = \frac{d}{dt} \mathbf{x}^T \mathbf{P} \mathbf{x}$$

$$= \dot{\mathbf{x}}^T \mathbf{P} \mathbf{x} + \mathbf{x}^T \mathbf{P} \dot{\mathbf{x}}$$

$$= \mathbf{x}^T (\mathbf{F}^T \mathbf{P} + \mathbf{P} \mathbf{F}) \mathbf{x}$$

$$= -\mathbf{x}^T \mathbf{Q} \mathbf{x},$$

Lyapunov equation

where $\mathbf{Q}$ is defined by

$$\mathbf{F}^T \mathbf{P} + \mathbf{P} \mathbf{F} = -\mathbf{Q}. \tag{7.252}$$

Lyapunov showed that, for any positive $\mathbf{Q}$, the solution $\mathbf{P}$ of the Lyapunov equation (7.252) is positive if and only if all the eigenvalues of $\mathbf{F}$ have negative real parts. In other words, given the system matrix $\mathbf{F}$, we can select a positive $\mathbf{Q}$ (such as the identity matrix $\mathbf{I}$), solve the Lyapunov equation (which is simply a system of linear equations in $n(n-1)/2$ unknowns), and test to see whether $\mathbf{P}$ is positive by looking at the determinants of the n principal minors. From this process we can determine the stability of the equations without either solving them or finding the eigenvalues, both much harder problems.

EXAMPLE 7.35

Lyapunov Stability for a Second-Order System

Use Lyapunov's method to find conditions for the stability of a linear system described by the state matrix

$$\mathbf{F} = \begin{bmatrix} -\alpha & \beta \\ -\beta & -\alpha \end{bmatrix}. \tag{7.253}$$

Solution.   For the linear case we can take any positive definite $\mathbf{Q}$ we like; the simplest is $\mathbf{Q} = \mathbf{I}$. The corresponding Lyapunov equation is

$$\begin{bmatrix} -\alpha & -\beta \\ \beta & -\alpha \end{bmatrix} \begin{bmatrix} p & q \\ q & r \end{bmatrix} + \begin{bmatrix} p & q \\ q & r \end{bmatrix} \begin{bmatrix} -\alpha & \beta \\ -\beta & -\alpha \end{bmatrix} = \begin{bmatrix} -1 & 0 \\ 0 & -1 \end{bmatrix}. \tag{7.254}$$

Multiplying out Eq. (7.254) and equating coefficients, we get

$$-\alpha p - \beta q - \alpha p - \beta q = -1, \tag{7.255}$$

$$-\alpha q - \beta r + \beta p - \alpha q = 0, \tag{7.256}$$

$$\beta q - \alpha r + \beta q - \alpha r = -1. \tag{7.257}$$

Equations (7.255)–(7.257) are readily solved to get $p = r = 1/2\alpha, q = 0$, so that

$$\mathbf{P} = \begin{bmatrix} \frac{1}{2\alpha} & 0 \\ 0 & \frac{1}{2\alpha} \end{bmatrix},$$

and the determinants are $1/2\alpha > 0$ and $1/4\alpha^2 > 0$.
   Thus $\mathbf{P} > \mathbf{0}$, so we conclude that the system is stable if $\alpha > 0$.

For systems with many state variables, solution of the Lyapunov equation can be burdensome, but the result is an alternative to, for example, Routh's method for computing the conditions for stability in a system with literal parameters.

Lyapunov stability of a nonlinear system

Testing for stability by considering the linear approximation to a differential equation is referred to as **Lyapunov's first, (or indirect) method**; using the idea of the Lyapunov function for a direct attack on the stability question is **Lyapunov's second (or direct) method**.

EXAMPLE 7.36

Lyapunov's Direct Method for Position Feedback System

Consider the position feedback system modeled in Fig. 7.78. Illustrate the use of the direct method on this nonlinear system. Simulate the system using Simulink assuming $T = 1$ and evaluate the step response of the system.

Solution.   We assume that the actuator, which is perhaps only an amplifier in this case, has a significant nonlinearity, which in the figure is shown as a saturation but is possibly more complex. We will assume only that $u = f(e)$ where the function lies in the first and third quadrants so that $\int_0^e f(e)\, de > 0$. We also assume that $f(e) = 0$ implies that $e = 0$ and we will assume $T > 0$, so the system is open-loop stable. The equations of motion are

$$\dot{e} = -x_2, \tag{7.258a}$$

$$\dot{x}_2 = -\frac{1}{T}x_2 + \frac{f(e)}{T}. \tag{7.258b}$$

For a Lyapunov function, consider something like kinetic plus potential energy:

$$V = \frac{T}{2}x_2^2 + \int_0^e f(\sigma)\, d\sigma. \tag{7.259}$$

Clearly, $V = 0$ if $x_2 = e = 0$ and, because of the assumptions about f, $V > 0$ if $x_2^2 + e^2 \neq 0$. To see whether the V in Eq. (7.259) is a Lyapunov function, we compute $\dot{V}$ as follows:

$$\dot{V} = Tx_2\dot{x}_2 + f(e)\dot{e}$$

$$= Tx_2\left[-\frac{1}{T}x_2 + \frac{f(e)}{T}\right] + f(e)(-x_2)$$

$$= -x_2^2.$$

Figure 7.78

An elementary position feedback system with a nonlinear actuator

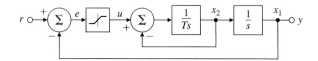

Figure 7.79

Simulink diagram for
position feedback system

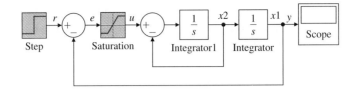

Hence $\dot{V} \leq 0$ and the origin is Lyapunov-stable. Moreover, $\dot{V}$ is always decreasing if $x_2 \neq 0$, and Eq. (7.258b) indicates that the system has no trajectory with $x_2 \equiv 0$, except $x_2 = 0$. Thus we can conclude that the system is asymptotically stable for every f that satisfies two conditions: (1) $\int f \, d\sigma > 0$ and (2) $f(e) = 0$ implies that $e = 0$. The Simulink diagram for the system is shown in Fig. 7.79 for $T = 1$. The step response of the system is shown in Fig. 7.80.

Simulink simulation

Figure 7.80

Step response for position
control system

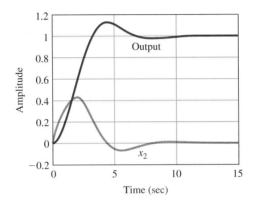

As we mentioned earlier, the study of the stability of nonlinear systems is vast, so we have only touched here on some important points and methods. Further material for study can be found in LaSalle and Lefschetz (1961), Kalman and Bertram (1960), Vidyasagar (1993), Khalil (1996), and Sastry (1999).

SUMMARY

- To every transfer function that has no more zeros than poles, there corresponds a differential equation in state-space form.

- State-space descriptions can be in several canonical forms. Among these are **control, observer** and **modal canonical forms.**

- Open-loop poles and zeros can be computed from the state description matrices $(\mathbf{F}, \mathbf{G}, \mathbf{H}, J)$.

$$\text{Poles}: \quad p = \text{eig}(\mathbf{F}), \qquad \det(p\mathbf{I} - \mathbf{F}) = 0,$$

$$\text{Zeros}: \qquad \det\begin{bmatrix} z\mathbf{I} - \mathbf{F} & -\mathbf{G} \\ \mathbf{H} & J \end{bmatrix} = 0.$$

- For any controllable system of order n, there exists a state feedback control law that will place the closed-loop poles at the roots of an arbitrary **control characteristic equation** of order n.

- The reference input can be introduced so as to result in zero steady-state error to a step command. This property is not expected to be robust to parameter changes.

- Good closed-loop pole locations depend on the desired transient response, the robustness to parameter changes, and a balance between dynamic performance and control effort.

- Closed-loop pole locations can be selected to result in a dominant second-order response or to minimize a quadratic performance measure.

- For any observable system of order n, an estimator (or observer) can be constructed with only sensor inputs and a state that estimates the plant state. The n poles of the estimator error system can be placed arbitrarily.

- The control law and the estimator can be combined into a controller such that the poles of the closed-loop system are the sum of the control-law-only poles and the estimator-only poles.

- With the estimator-based controller the **reference input** can be introduced in such a way as to permit n arbitrary zeros to be assigned. The most common choice is to assign the zeros to cancel the estimator poles, thus not exciting an estimator error.

- **Integral control** can be introduced to obtain robust steady-state tracking of a step by augmenting the plant state. The design is also robust with respect to rejecting constant disturbances.

- General **robust control** can be realized by combining the equations of the plant and the reference model into an **error space** and designing a control law for the extended system. Implementation of the robust design demonstrates the **internal model principle**. An estimator of the plant state can be added while retaining the robustness properties.

- The estimator can be extended to include estimates of the equivalent control disturbance and so result in robust tracking and disturbance rejection.

- Pole-placement designs, including integral control, can be computed using the polynomials of the plant transfer function in place of the state descriptions. Designs using polynomials frequently have problems with numerical accuracy.

- Controllers for plants that include a pure time delay can be designed as if there were no delay and then a controller implemented for the plant with the delay. The design can be expected to be sensitive to parameter changes.

- The stability of a nonlinear system in state-space description form can be studied by the methods of **Lyapunov**.

- Table 7.3 gives the important equations discussed in this chapter. The blue triangles indicate equations taken from optional sections in the text.

TABLE 7.3   **Important Equations in Chapter 7**

Name	Equation	Page
Control canonical form	$$\mathbf{A}_c = \begin{bmatrix} -a_1 & -a_2 & \cdots & & -a_n \\ 1 & 0 & \cdots & & 0 \\ \vdots & & & & \\ 0 & & & & \vdots \\ 0 & \cdots & 0 & 1 & 0 \end{bmatrix},$$	497
	$$\mathbf{B}_c = \begin{bmatrix} 1 \\ 0 \\ \vdots \\ 0 \end{bmatrix}, \quad \mathbf{C}_c = [\, b_1 \quad b_2 \quad \cdots \quad b_n \,], \quad D_c = 0.$$	
State description	$\dot{\mathbf{x}} = \mathbf{F}\mathbf{x} + \mathbf{G}u$	499
Output equation	$y = \mathbf{H}\mathbf{x} + Ju$	499
Transformation of state	$\mathbf{A} = \mathbf{T}^{-1}\mathbf{F}\mathbf{T}$	500
	$\mathbf{B} = \mathbf{T}^{-1}\mathbf{G}$	
	$y = \mathbf{H}\mathbf{T}\mathbf{z} + Ju = \mathbf{C}\mathbf{z} + Du,$	
	where $\mathbf{C} = \mathbf{H}\mathbf{T}$, $D = J$	
Controllability matrix	$\mathcal{C} = [\, \mathbf{G} \quad \mathbf{F}\mathbf{G} \quad \cdots \quad \mathbf{F}^{n-1}\mathbf{G} \,]$	501
Transfer function from state equations	$G(s) = \dfrac{Y(s)}{U(s)} = \mathbf{H}(s\mathbf{I} - \mathbf{F})^{-1}\mathbf{G} + J$	509
Transfer-function poles	$\det(p_i\mathbf{I} - \mathbf{F}) = 0$	511
Transfer-function zeros	$\alpha_z(s) = \det\begin{bmatrix} z_i\mathbf{I} - \mathbf{F} & -\mathbf{G} \\ \mathbf{H} & J \end{bmatrix} = 0$	512
Control characteristic equation	$\det[s\mathbf{I} - (\mathbf{F} - \mathbf{G}\mathbf{K})] = 0$	517
Ackermann's control formula for pole placement	$\mathbf{K} = [\, 0 \quad \cdots \quad 0 \quad 1 \,]\mathcal{C}^{-1}\alpha_c(\mathbf{F})$	521
Reference input gains	$\begin{bmatrix} \mathbf{F} & \mathbf{G} \\ \mathbf{H} & J \end{bmatrix}\begin{bmatrix} \mathbf{N}_x \\ N_u \end{bmatrix} = \begin{bmatrix} \mathbf{0} \\ 1 \end{bmatrix}$	525
Control equation with reference input	$u = N_u r - \mathbf{K}(\mathbf{x} - \mathbf{N}_x r)$	526
	$= -\mathbf{K}\mathbf{x} + (N_u + \mathbf{K}\mathbf{N}_x)r$	
	$= -\mathbf{K}\mathbf{x} + \bar{N}r$	
Symmetric root locus	$1 + \rho G_0(-s)G_0(s) = 0$	532

TABLE 7.3 **Important Equations in Chapter 7** (*continued*)

Name	Equation	Page
Estimator-error characteristic equation	$\alpha_e(s) = \det[s\mathbf{I} - (\mathbf{F} - \mathbf{LH})] = 0$	543
Observer canonical form	$\dot{\mathbf{x}}_0 = \mathbf{F}_0\mathbf{x}_0 + \mathbf{G}_0 u,$ $y = \mathbf{H}_0\mathbf{x}_0,$ where $$\mathbf{F}_o = \begin{bmatrix} -a_1 & 1 & 0 & 0 & \cdots & 0 \\ -a_2 & 0 & 1 & 0 & \cdots & \vdots \\ \vdots & \vdots & & \ddots & & 1 \\ -a_n & 0 & & 0 & & 0 \end{bmatrix}$$ $$\mathbf{G}_o = \begin{bmatrix} b_1 \\ b_2 \\ \vdots \\ b_n \end{bmatrix} \qquad \mathbf{H}_o = \begin{bmatrix} 1 & 0 & 0 & \cdots & 0 \end{bmatrix}$$	545
Observability matrix	$$\mathcal{O} = \begin{bmatrix} \mathbf{H} \\ \mathbf{HF} \\ \vdots \\ \mathbf{HF}^{n-1} \end{bmatrix}$$	546
Ackermann's estimator formula	$$\mathbf{L} = \alpha_e(\mathbf{F})\mathcal{O}^{-1} \begin{bmatrix} 0 \\ 0 \\ \vdots \\ 1 \end{bmatrix}$$	547
Compensator transfer function	$D_c(s) = \dfrac{U(s)}{Y(s)} = -\mathbf{K}(s\mathbf{I} - \mathbf{F} + \mathbf{GK} + \mathbf{LH})^{-1}\mathbf{L}$	556
Reduced-order compensator transfer function	$D_{cr}(s) = \dfrac{U(s)}{Y(s)} = \mathbf{C}_r(s\mathbf{I} - \mathbf{A}_r)^{-1}\mathbf{B}_r + D_r$	557
Controller equations	$\dot{\hat{\mathbf{x}}} = (\mathbf{F} - \mathbf{GK} - \mathbf{LH})\hat{\mathbf{x}} + \mathbf{L}y + \mathbf{M}r$ $u = -\mathbf{K}\hat{\mathbf{x}} + \bar{N}r$	574
Augmented state equations for integral control	$\begin{bmatrix} \dot{x}_I \\ \dot{\mathbf{x}} \end{bmatrix} = \begin{bmatrix} 0 & \mathbf{H} \\ \mathbf{0} & \mathbf{F} \end{bmatrix}\begin{bmatrix} x_I \\ \mathbf{x} \end{bmatrix} + \begin{bmatrix} 0 \\ \mathbf{G} \end{bmatrix}u - \begin{bmatrix} 1 \\ \mathbf{0} \end{bmatrix}r + \begin{bmatrix} 0 \\ \mathbf{G}_1 \end{bmatrix}w$	587
▲ General controller in polynomial form	$U(s) = -\dfrac{C_y(s)}{d(s)}Y(s) + \dfrac{c_r(s)}{d(s)}R(s)$	605
▲ Diophantine equation for closed-loop characteristic equation	$a(s)d(s) + b(s)c_y(s) = \alpha_c(s)\alpha_e(s)$	605
▲ Lyapunov's equation	$\mathbf{F}^T\mathbf{P} + \mathbf{PF} = -\mathbf{Q}$	616

- Determining a model from experimental data, or verifying an analytically based model by experiment, is an important step in system design by state-space analysis, a step that is not necessarily needed for compensator design via frequency response methods.

Review Questions

The following questions are based on a system in state variable form with matrices $\mathbf{F}, \mathbf{G}, \mathbf{H}, J$, input u, output y, and state $\mathbf{x}$.

1. Give an expression for the transfer function of this system.

2. Give two expressions for the poles of the transfer function of the system.

3. Give an expression for the zeros of the system transfer function.

4. Under what condition will the state of the system be controllable?

5. Under what conditions will the system be observable from the output y?

6. Give an expression for the *closed-loop* poles if state feedback of the form $u = -\mathbf{Kx}$ is used.

7. Under what conditions can the feedback matrix, $\mathbf{K}$, be selected so that the roots of $\alpha_c(s)$ are arbitrary?

8. What is the advantage of using the LQR or symmetrical root locus in designing the feedback matrix $\mathbf{K}$?

9. What is the main reason for using an estimator in feedback control?

10. If the estimator gain, $\mathbf{L}$, is used, give an expression for the closed-loop poles due to the estimator.

11. Under what conditions can the estimator gain, $\mathbf{L}$, be selected so that the roots of $\alpha_e(s) = 0$ are arbitrary?

12. If the reference input is arranged so that the input to the estimator is identical to the input to the process, what will be the over all closed-loop transfer function?

13. If the reference input is introduced in such a way as to permit the zeros to be assigned as the roots of $\gamma(s)$, what will the over all closed-loop transfer function be?

14. What are the three standard techniques for introducing integral control in the state feedback design method?

Problems

Problems for Section 7.2

7.1. Give the state description matrices in control-canonical form for the following transfer functions:

(a) $\dfrac{1}{4s+1}$

(b) $\dfrac{5(s/2+1)}{(s/10+1)}$

(c) $\dfrac{2s+1}{s^2+3s+2}$

(d) $\dfrac{s+3}{s(s^2+2s+2)}$

(e) $\dfrac{(s+10)(s^2+s+25)}{s^2(s+3)(s^2+s+36)}$

7.2. Use the MATLAB function tf2ss to obtain the state matrices called for Problem 7.1.

7.3. Give the state description matrices in normal-mode form for the transfer functions of Problem 7.1. Make sure that all entries in the state matrices are real-valued by keeping any pairs of complex conjugate poles together, and realize them as a separate subblock in control canonical form.

7.4. A certain system with state $\mathbf{x}$ is described by the state matrices

$$\mathbf{F} = \begin{bmatrix} -2 & 1 \\ -2 & 0 \end{bmatrix}, \qquad \mathbf{G} = \begin{bmatrix} 1 \\ 3 \end{bmatrix},$$
$$\mathbf{H} = \begin{bmatrix} 1 & 0 \end{bmatrix}, \qquad J = 0.$$

Find the transformation $\mathbf{T}$ so that if $\mathbf{x} = \mathbf{Tz}$, the state matrices describing the dynamics of $\mathbf{z}$ are in control canonical form. Compute the new matrices $\mathbf{A}$, $\mathbf{B}$, $\mathbf{C}$, and D.

7.5. Show that the transfer function is not changed by a linear transformation of state.

7.6. Use block-diagram reduction or Mason's rule to find the transfer function for the system in observer canonical form depicted by Fig. 7.26.

7.7. Suppose we are given a system with state matrices $\mathbf{F}, \mathbf{G}, \mathbf{H}$ ($J = 0$ in this case). Find the transformation $\mathbf{T}$ so that, under Eqs. (7.13) and (7.14), the new state description matrices will be in observer canonical form.

7.8. Use the transformation matrix in Eq. (7.30) to explicitly multiply out the equations at the end of Example 7.2.

7.9. Find the state transformation that takes the observer canonical form of Eq. (7.24) to the modal canonical form.

7.10. (a) Find the transformation $\mathbf{T}$ that will keep the description of the tape-drive system of Example 7.3 in modal canonical form but will convert each element of the input matrix $\mathbf{B}_m$ to unity.

   (b) Use MATLAB to verify that your transformation does the job.

7.11. (a) Find the state transformation that will keep the description of the tape-drive system of Example 7.3 in modal canonical form but will cause the poles to be displayed in $\mathbf{A}_m$ in order of increasing magnitude.

   (b) Use MATLAB to verify your result in part (a), and give the complete new set of state matrices as $\mathbf{A}, \mathbf{B}, \mathbf{C}$, and D.

7.12. Find the characteristic equation for the modal-form matrix $\mathbf{A}_m$ of Eq. (7.6) using Eq. (7.46).

7.13. Given the system

$$\dot{\mathbf{x}} = \begin{bmatrix} -4 & 1 \\ -2 & -1 \end{bmatrix} \mathbf{x} + \begin{bmatrix} 0 \\ 1 \end{bmatrix} u$$

with zero initial conditions, find the steady-state value of $\mathbf{x}$ for a step input u.

7.14. Consider the system shown in Fig. 7.81:

Figure 7.81
A block diagram for
Problem 7.14

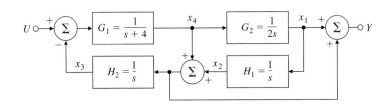

(a) Find the transfer function from U to Y.

(b) Write state equations for the system using the state variables indicated.

7.15. Using the indicated state variables, write the state equations for each of the systems shown in Fig. 7.82. Find the transfer function for each system using both block-diagram manipulation and matrix algebra [as in Eq. (7.36)].

Figure 7.82
Block diagrams for
Problem 7.15

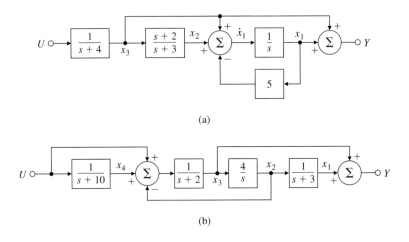

(a)

(b)

7.16. For each of the transfer functions below, write the state equations in both control and observer canonical form. In each case draw a block diagram and give the appropriate expressions for $\mathbf{F}$, $\mathbf{G}$, and $\mathbf{H}$.

(a) $\dfrac{s^2 - 2}{s^2(s^2 - 1)}$ (control of an inverted pendulum by a force on the cart)

(b) $\dfrac{3s + 4}{s^2 + 2s + 2}$

7.17. Consider the transfer function

$$G(s) = \frac{Y(s)}{U(s)} = \frac{s+1}{s^2 + 5s + 6}.$$
(7.260)

(a) By rewriting Eq. (7.260) in the form

$$G(s) = \frac{1}{s+3}\left(\frac{s+1}{s+2}\right),$$

find a **series realization** of $G(s)$ as a cascade of two first-order systems.

(b) Using a partial-fraction expansion of $G(s)$, find a **parallel realization** of $G(s)$.

(c) Realize $G(s)$ in control canonical form.

Problems for Section 7.3

7.18. Consider the plant described by

$$\dot{\mathbf{x}} = \begin{bmatrix} 0 & 1 \\ 7 & -4 \end{bmatrix}\mathbf{x} + \begin{bmatrix} 1 \\ 2 \end{bmatrix}u,$$

$$y = [\,1 \quad 3\,]\mathbf{x}.$$

(a) Draw a block diagram for the plant with one integrator for each state variable.

(b) Find the transfer function using matrix algebra.

(c) Find the closed-loop characteristic equation if the feedback is:

$$(1)\ u = -[\,K_1 \quad K_2\,]\mathbf{x}; \qquad (2)\ u = -Ky.$$

7.19. For the system

$$\dot{\mathbf{x}} = \begin{bmatrix} 0 & 1 \\ -6 & -5 \end{bmatrix}\mathbf{x} + \begin{bmatrix} 0 \\ 1 \end{bmatrix}u,$$

$$y = [\,1 \quad 0\,]\mathbf{x}.$$

design a state feedback controller that satisfies the following specifications:

• Closed-loop poles have a damping coefficient $\zeta = 0.707$.

• Step-response peak time is under 3.14 sec.

Verify your design with MATLAB.

7.20. **(a)** Design a state feedback controller for the following system so that the closed-loop step response has an overshoot of less than 25% and a 1% settling time under 0.115 sec:

$$\dot{\mathbf{x}} = \begin{bmatrix} 0 & 1 \\ 0 & -10 \end{bmatrix} \mathbf{x} + \begin{bmatrix} 0 \\ 1 \end{bmatrix} u,$$

$$y = [1 \quad 0]\mathbf{x}.$$

(b) Use the step command in MATLAB to verify that your design meets the specifications. If it does not, modify your feedback gains accordingly.

7.21. Consider the system

$$\dot{\mathbf{x}} = \begin{bmatrix} -1 & -2 & -2 \\ 0 & -1 & 1 \\ 1 & 0 & -1 \end{bmatrix} \mathbf{x} + \begin{bmatrix} 2 \\ 0 \\ 1 \end{bmatrix} u,$$

$$y = [1 \quad 0 \quad 0]\mathbf{x}.$$

(a) Design a state feedback controller for the following system so that the closed-loop step response has an overshoot of less than 5% and a 1% settling time under 4.6 sec.

(b) Use the step command in MATLAB to verify that your design meets the specifications. If it does not, modify your feedback gains accordingly.

7.22. Consider the system in Fig. 7.83.

(a) Write a set of equations that describes this system in the control canonical form as $\dot{\mathbf{x}} = \mathbf{F}\mathbf{x} + \mathbf{G}u$ and $y = \mathbf{H}\mathbf{x}$.

(b) Design a control law of the form

$$u = -[K_1 \quad K_2]\begin{bmatrix} x_1 \\ x_2 \end{bmatrix}$$

that will place the closed-loop poles at $s = -2 \pm 2j$.

Figure 7.83
System for Problem 7.22

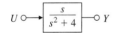

$U \circ \!\!\longrightarrow\!\! \boxed{\dfrac{s}{s^2 + 4}} \!\!\longrightarrow\!\! \circ Y$

7.23. *Output Controllability*: In many situations a control engineer may be interested in controlling the output y rather than the state $\mathbf{x}$. A system is said to be **output controllable** if at any time you are able to transfer the output from zero to any desired output y^* in a finite time using an appropriate control signal u^*. Derive necessary and sufficient conditions for a continuous system $(\mathbf{F}, \mathbf{G}, \mathbf{H})$ to be output controllable. Are output and state controllability related? If so, how?

7.24. Consider the system

$$\dot{\mathbf{x}} = \begin{bmatrix} 0 & 4 & 0 & 0 \\ -1 & -4 & 0 & 0 \\ 5 & 7 & 1 & 15 \\ 0 & 0 & 3 & -3 \end{bmatrix} \mathbf{x} + \begin{bmatrix} 0 \\ 0 \\ 1 \\ 0 \end{bmatrix} u.$$

(a) Find the eigenvalues of this system. (*Hint:* Note the block-triangular structure.)

(b) Find the controllable and uncontrollable modes of this system.

(c) For each of the uncontrollable modes, find a vector **v** such that

$$\mathbf{v}^T\mathbf{G} = 0, \quad \mathbf{v}^T\mathbf{F} = \lambda\mathbf{v}^T.$$

(d) Show that there are an infinite number of feedback gains **K** that will relocate the modes of the system to -5, -3, -2, and -2.

(e) Find the unique matrix **K** that achieves these pole locations and prevents initial conditions on the uncontrollable part of the system from ever affecting the controllable part.

7.25. Two pendulums, coupled by a spring, are to be controlled by two equal and opposite forces u, which are applied to the pendulum bobs as shown in Fig. 7.84. The equations of motion are

$$ml^2\ddot{\theta}_1 = -ka^2(\theta_1 - \theta_2) - mgl\theta_1 - lu,$$

$$ml^2\ddot{\theta}_2 = -ka^2(\theta_2 - \theta_1) - mgl\theta_2 + lu.$$

(a) Show that the system is uncontrollable. Can you associate a physical meaning with the controllable and uncontrollable modes?

(b) Is there any way that the system can be made controllable?

Figure 7.84

Coupled pendulums for Problem 7.25

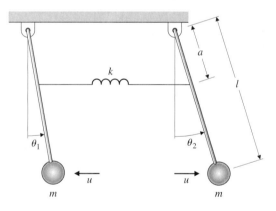

7.26. The state-space model for a certain application has been given to us with the following state description matrices:

$$\mathbf{F} = \begin{bmatrix} 0.174 & 0 & 0 & 0 & 0 \\ 0.157 & 0.645 & 0 & 0 & 0 \\ 0 & 1 & 0 & 0 & 0 \\ 0 & 0 & 1 & 0 & 0 \\ 0 & 0 & 0 & 1 & 0 \end{bmatrix}, \quad \mathbf{G} = \begin{bmatrix} -0.207 \\ -0.005 \\ 0 \\ 0 \\ 0 \end{bmatrix}, \quad \mathbf{H} = [1 \ 0 \ 0 \ 0 \ 0].$$

(a) Draw a block diagram of the realization with an integrator for each state variable.

(b) A student has computed $\det \mathcal{C} = 2.3 \times 10^{-7}$ and claims that the system is uncontrollable. Is the student right or wrong? Why?

(c) Is the realization observable?

7.27. *Staircase Algorithm (Van Dooren et al., 1978):* Any realization $(\mathbf{F}, \mathbf{G}, \mathbf{H})$ can be transformed by an **orthogonal similarity transformation** to $(\bar{\mathbf{F}}, \bar{\mathbf{G}}, \bar{\mathbf{H}})$, where $\bar{\mathbf{F}}$ is an **upper Hessenberg matrix** (having one nonzero diagonal above the main diagonal):

$$
\bar{\mathbf{F}} = \mathbf{T}^T \mathbf{F} \mathbf{T} = \begin{bmatrix} * & \alpha_1 & 0 \dots & & 0 \\ * & * & \ddots & & 0 \\ * & * & & \ddots & \alpha_{n-1} \\ * & * & & \cdots & * \end{bmatrix}, \quad \bar{\mathbf{G}} = \mathbf{T}^T \mathbf{G} = \begin{bmatrix} 0 \\ 0 \\ \vdots \\ 0 \\ g_1 \end{bmatrix},
$$

where $g_1 \neq 0$, and

$$
\bar{\mathbf{H}} = \mathbf{H}\mathbf{T} = [h_1 \cdots h_n], \qquad \mathbf{T}^{-1} = \mathbf{T}^T.
$$

Orthogonal transformations correspond to a **rotation** of the vectors (represented by the matrix columns) being transformed with no change in length.

(a) Prove that if $\alpha_i = 0$ and $\alpha_{i+1}, \dots, \alpha_{n-1} \neq 0$ for some i, then the controllable and uncontrollable modes of the system can be identified after this transformation has been done.

(b) How would you use this technique to identify the observable and unobservable modes of $(\mathbf{F}, \mathbf{G}, \mathbf{H})$?

(c) What advantage does this approach for determining the controllable and uncontrollable modes have over transforming the system to any other form?

(d) How can we use this approach to determine a basis for the controllable and uncontrollable subspaces, as in Problem 7.41?

This algorithm can be used to design a numerically stable algorithm for pole placement (see Minimis and Paige, 1982). The name of the algorithm comes from the multi-input version in which the α_i are the blocks that make $\bar{\mathbf{F}}$ resemble a staircase.

Problems for Section 7.4

7.28. The normalized equations of motion for an inverted pendulum at angle θ on a cart are

$$
\ddot{\theta} = \theta + u, \qquad \ddot{x} = -\beta \theta - u,
$$

where x is the cart position, and the control input u is a force acting on the cart.

(a) With the state defined as $\mathbf{x} = [\theta, \dot{\theta}, x, \dot{x}]^T$, find the feedback gain $\mathbf{K}$ that places the closed-loop poles at $s = -1, -1, -1 \pm 1j$.
For parts (b) through (d), assume that $\beta = 0.5$.

(b) Use the symmetric root locus to select poles with a bandwidth as close as possible to those of part (a), and find the control law that will place the closed-loop poles at the points you selected.

(c) Compare the responses of the closed-loop systems in parts (a) and (b) to an initial condition of $\theta = 10°$. You may wish to use the initial command in MATLAB.

(d) Compute $\mathbf{N}_x$ and N_u for zero steady-state error to a constant command input on the cart position, and compare the step responses of each of the two closed-loop systems.

7.29. Consider the feedback system in Fig. 7.85. Find the relationship between K, T, and ξ such that the closed-loop transfer function minimizes the integral of the time multiplied by the absolute value of the error (ITAE) criterion,

$$\mathcal{J} = \int_0^\infty t|e|\,dt,$$

for a step input. Assume $\omega_0 = 1$.

Figure 7.85

Control system for Problem 7.29

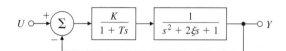

7.30. Prove that the Nyquist plot for LQR design avoids a circle of radius one centered at the -1 point as shown in Fig. 7.86. Show that this implies that $\frac{1}{2} < \text{GM} < \infty$ the "upward" gain margin is $\text{GM} = \infty$, and there is a "downward" $\text{GM} = \frac{1}{2}$, and the phase margin is at least $\text{PM} = \pm 60°$. Hence the LQR gain matrix, $\mathbf{K}$, can be multiplied by a large scalar or reduced by half with guaranteed closed-loop system stability.

Figure 7.86

Nyquist plot for an optimal regulator

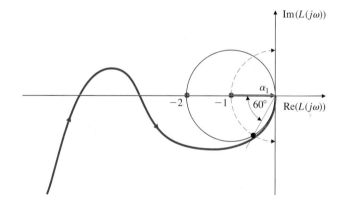

Problems for Section 7.5

7.31. Consider the system

$$\mathbf{F} = \begin{bmatrix} -2 & 1 \\ 1 & 0 \end{bmatrix}, \qquad \mathbf{G} = \begin{bmatrix} 1 \\ 0 \end{bmatrix}, \qquad \mathbf{H} = [1 \ 2],$$

and assume you are using feedback of the form $u = -\mathbf{K}\mathbf{x} + r$, where r is a reference input signal.

(a) Show that $(\mathbf{F}, \mathbf{H})$ is observable.

(b) Show that there exists a $\mathbf{K}$ such that $(\mathbf{F} - \mathbf{GK}, \mathbf{H})$ is unobservable.

(c) Compute a $\mathbf{K}$ of the form $\mathbf{K} = [1, K_2]$ that will make the system unobservable as in part (b); that is, find K_2 so that the closed-loop system is not observable.

(d) Compare the open-loop transfer function with the transfer function of the closed-loop system of part (c). What is the unobservability due to?

7.32. Consider a system with the transfer function

$$G(s) = \frac{9}{s^2 - 9}.$$

(a) Find $(\mathbf{F}_0, \mathbf{G}_0, \mathbf{H}_0)$ for this system in observer canonical form.
(b) Is $(\mathbf{F}_0, \mathbf{G}_0)$ controllable?
(c) Compute $\mathbf{K}$ so that the closed-loop poles are assigned to $s = -3 \pm 3j$.
(d) Is the closed-loop system of part (c) observable?
(e) Design a full-order estimator with estimator-error poles at $s = -12 \pm 12j$.
(f) Suppose the system is modified to have a zero:

$$G_1(s) = \frac{9(s + 1)}{s^2 - 9}.$$

Prove that if $u = -\mathbf{K}\mathbf{x} + r$, there is a feedback gain $\mathbf{K}$ that makes the closed-loop system unobservable. [Again assume an observer canonical realization for $G_1(s)$.]

7.33. Explain how the controllability, observability, and stability properties of a linear system are related.

7.34. Consider the electric circuit shown in Fig. 7.87.

(a) Write the internal (state) equations for the circuit. The input $u(t)$ is a current, and the output y is a voltage. Let $x_1 = i_L$ and $x_2 = v_c$.
(b) What condition(s) on R, L, and C will guarantee that the system is controllable?
(c) What condition(s) on R, L, and C will guarantee that the system is observable?

Figure 7.87
Electric circuit for
Problem 7.34

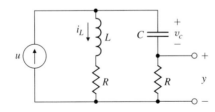

7.35. The block diagram of a feedback system is shown in Fig. 7.88. The system state is

$$\mathbf{x} = \begin{bmatrix} \mathbf{x}_p \\ \mathbf{x}_f \end{bmatrix},$$

and the dimensions of the matrices are as follows:

$$\mathbf{F} = n \times n, \qquad \mathbf{L} = n \times 1,$$
$$\mathbf{G} = n \times 1, \qquad \mathbf{x} = 2n \times 1,$$
$$\mathbf{H} = 1 \times n, \qquad r = 1 \times 1,$$
$$\mathbf{K} = 1 \times n, \qquad y = 1 \times 1,$$

Figure 7.88
Block diagram for
Problem 7.35

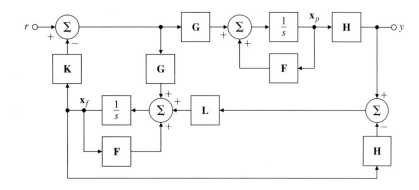

(a) Write state equations for the system.

(b) Let $\mathbf{x} = \mathbf{Tz}$, where

$$\mathbf{T} = \begin{bmatrix} \mathbf{I} & \mathbf{0} \\ \mathbf{I} & -\mathbf{I} \end{bmatrix}.$$

Show that the system is not controllable.

(c) Find the transfer function of the system from r to y.

7.36. This problem is intended to give you more insight into controllability and observability. Consider the circuit in Fig. 7.89, with an input voltage source $u(t)$ and an output current $y(t)$.

(a) Using the capacitor voltage and inductor current as state variables, write state and output equations for the system.

(b) Find the conditions relating R_1, R_2, C, and L that render the system uncontrollable. Find a similar set of conditions that result in an unobservable system.

(c) Interpret the conditions found in part (b) physically in terms of the time constants of the system.

(d) Find the transfer function of the system. Show that there is a pole-zero cancellation for the conditions derived in part (b) (that is, when the system is uncontrollable or unobservable).

Figure 7.89
Electric circuit for
Problem 7.36

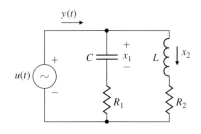

7.37. The linearized equations of motion for a satellite are

$$\dot{\mathbf{x}} = \mathbf{Fx} + \mathbf{Gu},$$

$$\mathbf{y} = \mathbf{Hx},$$

where

$$
\mathbf{F} = \begin{bmatrix} 0 & 1 & 0 & 0 \\ 3\omega^2 & 0 & 0 & 2\omega \\ 0 & 0 & 0 & 1 \\ 0 & -2\omega & 0 & 0 \end{bmatrix}, \qquad \mathbf{G} = \begin{bmatrix} 0 & 0 \\ 1 & 0 \\ 0 & 0 \\ 0 & 1 \end{bmatrix}, \qquad \mathbf{H} = \begin{bmatrix} 1 & 0 & 0 & 0 \\ 0 & 0 & 1 & 0 \end{bmatrix},
$$

$$
\mathbf{u} = \begin{bmatrix} u_1 \\ u_2 \end{bmatrix}, \qquad \mathbf{y} = \begin{bmatrix} y_1 \\ y_2 \end{bmatrix}.
$$

The inputs u_1 and u_2 are the radial and tangential thrusts, the state variables x_1 and x_3 are the radial and angular deviations from the reference (circular) orbit, and the outputs y_1 and y_2 are the radial and angular measurements, respectively.

(a) Show that the system is controllable using both control inputs.

(b) Show that the system is controllable using only a single input. Which one is it?

(c) Show that the system is observable using both measurements.

(d) Show that the system is observable using only one measurement. Which one is it?

7.38.   Consider the system in Fig. 7.90.

(a) Write the state-variable equations for the system, using $[\,\theta_1 \quad \theta_2 \quad \dot\theta_1 \quad \dot\theta_2\,]^T$ as the state vector and F as the single input.

(b) Show that all the state variables are observable using measurements of θ_1 alone.

(c) Show that the characteristic polynomial for the system is the product of the polynomials for two oscillators. Do so by first writing a new set of system equations involving the state variables

$$
\begin{bmatrix} y_1 \\ y_2 \\ \dot y_1 \\ \dot y_2 \end{bmatrix} = \begin{bmatrix} \theta_1 + \theta_2 \\ \theta_1 - \theta_2 \\ \dot\theta_1 + \dot\theta_2 \\ \dot\theta_1 - \dot\theta_2 \end{bmatrix}
$$

Figure 7.90
Coupled pendulums for
Problem 7.38

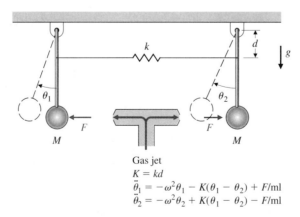

Gas jet
$K = kd$
$\ddot\theta_1 = -\omega^2\theta_1 - K(\theta_1 - \theta_2) + F/ml$
$\ddot\theta_2 = -\omega^2\theta_2 + K(\theta_1 - \theta_2) - F/ml$

Hint: If **A** and **D** are invertible matrices, then

$$\begin{bmatrix} \mathbf{A} & \mathbf{0} \\ \mathbf{0} & \mathbf{D} \end{bmatrix}^{-1} = \begin{bmatrix} \mathbf{A}^{-1} & \mathbf{0} \\ \mathbf{0} & \mathbf{D}^{-1} \end{bmatrix}.$$

(d) Deduce that the spring mode is controllable with F but the pendulum mode is not.

7.39. A certain fifth-order system is found to have a characteristic equation with roots at $0, -1, -2$, and $-1 \pm 1j$. A decomposition into controllable and uncontrollable parts discloses that the controllable part has a characteristic equation with roots 0, and $-1 \pm 1j$. A decomposition into observable and nonobservable parts discloses that the observable modes are at $0, -1$, and -2.

(a) Where are the zeros of $b(s) = \mathbf{H} \mathrm{adj}(s\mathbf{I} - \mathbf{F})\mathbf{G}$ for this system?

(b) What are the poles of the reduced-order transfer function that includes only controllable and observable modes?

7.40. Consider the systems shown in Fig. 7.91, employing series, parallel, and feedback configurations.

(a) Suppose we have controllable-observable realizations for each subsystem:

$$\dot{\mathbf{x}}_i = \mathbf{F}_i \mathbf{x}_i + \mathbf{G}_i u_i,$$

$$y_i = \mathbf{H}_i \mathbf{x}_i, \qquad \text{where } i = 1, 2.$$

Give a set of state equations for the combined systems in Fig. 7.91.

(b) For each case, determine what condition(s) on the roots of the polynomials N_i and D_i is necessary for each system to be controllable and observable. Give a brief reason for your answer in terms of pole-zero cancellations.

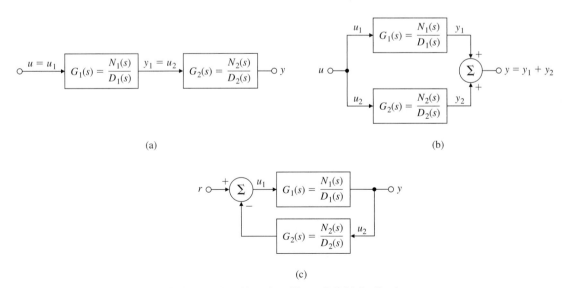

(a)

(b)

(c)

Figure 7.91 Block diagrams for Problem 7.40. (a) series; (b) parallel; (c) feedback

7.41. Consider the system $\ddot{y} + 3\dot{y} + 2y = \dot{u} + u$.

(a) Find the state matrices $\mathbf{F}_c$, $\mathbf{G}_c$, and $\mathbf{H}_c$ in control canonical form that correspond to the given differential equation.

(b) Sketch the eigenvectors of $\mathbf{F}_c$ in the (x_1, x_2) plane, and draw vectors that correspond to the completely observable $(\mathbf{x}_0)$ and the completely unobservable $(\mathbf{x}_{\bar{0}})$ state variables.

(c) Express $\mathbf{x}_0$ and $\mathbf{x}_{\bar{0}}$ in terms of the observability matrix $\mathcal{O}$.

(d) Give the state matrices in observer canonical form and repeat parts (b) and (c) in terms of controllability instead of observability.

7.42. The equations of motion for a station-keeping satellite (such as a weather satellite) are

$$\ddot{x} - 2\omega\dot{y} - 3\omega^2 x = 0, \qquad \ddot{y} + 2\omega\dot{x} = u,$$

where

$$x = \text{radial perturbation,}$$

$$y = \text{longitudinal position perturbation,}$$

$$u = \text{engine thrust in the y} - \text{direction}$$

as depicted in Fig. 7.92. If the orbit is synchronous with the Earth's rotation, then $\omega = 2\pi/(3600 \times 24)$ rad/sec.

(a) Is the state $[x \quad \dot{x} \quad y \quad \dot{y}]^T$ observable?

(b) Choose $\mathbf{x} = [x \quad \dot{x} \quad y \quad \dot{y}]^T$ as the state vector and y as the measurement, and design a full-order observer with poles placed at $s = -2\omega$, -3ω, and $-3\omega \pm 3\omega j$.

Figure 7.92
Diagram of a
station-keeping satellite in
orbit

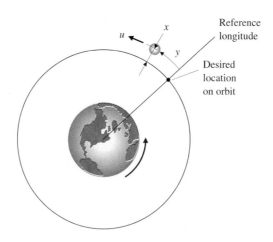

7.43. The linearized equations of motion of the simple pendulum in Fig. 7.93 are

$$\ddot{\theta} + \omega^2\theta = u.$$

(a) Write the equations of motion in state-space form.

(b) Design an estimator (observer) that reconstructs the state of the pendulum given measurements of $\dot{\theta}$. Assume $\omega = 5$ rad/sec, and pick the estimator roots to be at $s = -10 \pm 10j$.

(c) Write the transfer function of the estimator between the measured value of $\dot{\theta}$ and the estimated value of θ.

(d) Design a controller (that is, determine the state feedback gain $\mathbf{K}$) so that the roots of the closed-loop characteristic equation are at $s = -4 \pm 4j$.

Figure 7.93
Pendulum diagram for
Problem 7.43

7.44. An error analysis of an inertial navigator leads to the following set of normalized state equations:

$$\begin{bmatrix} \dot{x}_1 \\ \dot{x}_2 \\ \dot{x}_3 \end{bmatrix} = \begin{bmatrix} 0 & -1 & 0 \\ 1 & 0 & 1 \\ 0 & 0 & 0 \end{bmatrix} \begin{bmatrix} x_1 \\ x_2 \\ x_3 \end{bmatrix} + \begin{bmatrix} 0 \\ 0 \\ 1 \end{bmatrix} u,$$

where

$$x_1 = \text{east—velocity error,}$$

$$x_2 = \text{platform tilt about the north axis,}$$

$$x_3 = \text{north—gyro drift,}$$

$$u = \text{gyro drift rate of change.}$$

Design a reduced-order estimator with $y = x_1$ as the measurement, and place the observer error poles at -0.1 and -0.1. Be sure to provide all the relevant estimator equations.

7.45. A certain process has the transfer function $G(s) = 4/(s^2 - 4)$.

(a) Find $\mathbf{F}$, $\mathbf{G}$, and $\mathbf{H}$ for this system in observer canonical form.

(b) If $u = -\mathbf{Kx}$, compute $\mathbf{K}$ so that the closed-loop control poles are located at $s = -2 \pm 2j$.

(c) Compute $\mathbf{L}$ so the estimator error poles are located at $s = -10 \pm 10j$.

(d) Give the transfer function of the resulting controller [for example, using Eq. (7.166)].

(e) What are the gain and phase margins of the controller and the given open-loop system?

7.46. The linearized longitudinal motion of a helicopter near hover (Fig. 7.94) can be modeled by the normalized third-order system

$$\begin{bmatrix} \dot{q} \\ \dot{\theta} \\ \dot{u} \end{bmatrix} = \begin{bmatrix} -0.4 & 0 & -0.01 \\ 1 & 0 & 0 \\ -1.4 & 9.8 & -0.02 \end{bmatrix} \begin{bmatrix} q \\ \theta \\ u \end{bmatrix} + \begin{bmatrix} 6.3 \\ 0 \\ 9.8 \end{bmatrix} \delta,$$

where

q = pitch rate,

θ = pitch angle of fuselage,

u = horizontal velocity (standard aircraft notation),

δ = rotor tilt angle (control variable).

Suppose our sensor measures the horizontal velocity u as the output; that is, $y = u$.

(a) Find the open-loop pole locations.

(b) Is the system controllable?

(c) Find the feedback gain that places the poles of the system at $s = -1 \pm 1j$ and $s = -2$.

(d) Design a full-order estimator for the system, and place the estimator poles at -8 and $-4 \pm 4\sqrt{3}j$.

Figure 7.94
Helicopter

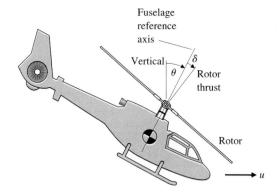

Fuselage
reference
axis

Vertical

δ

θ

Rotor
thrust

Rotor

u

(e) Design a reduced-order estimator with both poles at -4. What are the advantages and disadvantages of the reduced-order estimator compared with the full-order case?

(f) Compute the compensator transfer function using the control gain and the full-order estimator designed in part (d), and plot its frequency response using MATLAB. Draw a Bode plot for the compensated design, and indicate the corresponding gain and phase margins.

(g) Repeat part (f) with the reduced-order estimator.

(h) Draw the symmetrical root locus (SRL) and select roots for a control law that will give a control bandwidth matching the design of part (c), and select roots for a full-order estimator that will result in an estimator error bandwidth comparable to the design of part (d). Draw the corresponding Bode plot and compare the pole placement and SRL designs with respect to bandwidth, stability margins, step response, and control effort for a unit-step rotor-angle input. Use MATLAB for the computations.

7.47. Suppose a DC drive motor with motor current u is connected to the wheels of a cart in order to control the movement of an inverted pendulum mounted on the cart. The linearized and normalized equations of motion corresponding to this system can be put in the form

$$\ddot{\theta} = \theta + v + u,$$

$$\dot{v} = \theta - v - u,$$

where

$$\theta = \text{angle of the pendulum,}$$

$$v = \text{velocity of the cart.}$$

(a) We wish to control θ by feedback to u of the form

$$u = -K_1\theta - K_2\dot{\theta} - K_3 v.$$

Find the feedback gains so that the resulting closed-loop poles are located at $-1, -1 \pm j\sqrt{3}$.

(b) Assume that θ and v are measured. Construct an estimator for θ and $\dot{\theta}$ of the form

$$\dot{\hat{x}} = F\hat{x} + L(y - \hat{y}),$$

where $x = [\theta \quad \dot{\theta}]^T$ and $y = \theta$. Treat both v and u as known. Select L so that the estimator poles are at -2 and -2.

(c) Give the transfer function of the controller, and draw the Bode plot of the compensated system, indicating the corresponding gain and phase margins.

(d) Using MATLAB, plot the response of the system to an initial condition on θ, and give a physical explanation for the initial motion of the cart.

7.48. Consider the control of

$$G(s) = \frac{Y(s)}{U(s)} = \frac{10}{s(s+1)}.$$

(a) Let $y = x_1$ and $\dot{x}_1 = x_2$, and write state equations for the system.

(b) Find K_1 and K_2 so that $u = -K_1 x_1 - K_2 x_2$ yields closed-loop poles with a natural frequency $\omega_n = 3$ and a damping ratio $\zeta = 0.5$.

(c) Design a state estimator for the system that yields estimator error poles with $\omega_{n1} = 15$ and $\zeta_1 = 0.5$.

(d) What is the transfer function of the controller obtained by combining parts (a) through (c)?

(e) Sketch the root locus of the resulting closed-loop system as plant gain (nominally 10) is varied.

7.49. Unstable equations of motion of the form

$$\ddot{x} = x + u$$

arise in situations where the motion of an upside-down pendulum (such as a rocket) must be controlled.

(a) Let $u = -Kx$ (position feedback alone), and sketch the root locus with respect to the scalar gain K.

(b) Consider a lead compensator of the form

$$U(s) = K\frac{s+a}{s+10}X(s).$$

Select a and K so that the system will display a rise time of about 2 sec and no more than 25% overshoot. Sketch the root locus with respect to K.

(c) Sketch the Bode plot (both magnitude and phase) of the uncompensated plant.

(d) Sketch the Bode plot of the compensated design, and estimate the phase margin.

(e) Design state feedback so that the closed-loop poles are at the same locations as those of the design in part (b).

(f) Design an estimator for x and $\dot{x}$ using the measurement of $x = y$, and select the observer gain $\mathbf{L}$ so that the equation for $\tilde{x}$ has characteristic roots with a damping ratio $\zeta = 0.5$ and a natural frequency $\omega_n = 8$.

(g) Draw a block diagram of your combined estimator and control law, and indicate where $\hat{x}$ and $\dot{x}$ appear. Draw a Bode plot for the closed-loop system, and compare the resulting bandwidth and stability margins with those obtained using the design of part (b). Determine the step response of the closed-loop system.

7.50. A simplified model for the control of a flexible robotic arm is shown in Fig. 7.95, where

$$k/M = 900 \text{ rad/sec}^2,$$

$$y = \text{output, the mass position};$$

$$u = \text{input, the position of the end of the spring}.$$

Figure 7.95
Simple robotic arm

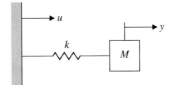

(a) Write the equations of motion in state-space form.

(b) Design an estimator with roots as $s = -100 \pm 100j$.

(c) Could both state-variables of the system be estimated if only a measurement of $\dot{y}$ was available?

(d) Design a full-state-feedback controller with roots at $s = -20 \pm 20j$.

(e) Would it be reasonable to design a control law for the system with roots at $s = -200 \pm 200j$? State your reasons.

(f) Write equations for the compensator, including a command input for y. Draw a Bode plot for the compensated system, and give the gain and phase margins for the design.

7.51. The linearized differential equations governing the fluid-flow dynamics for the two cascaded tanks in Fig. 7.96 are as follows:

$$\delta \dot{h}_1 + \sigma \delta h_1 = \delta u,$$

$$\delta \dot{h}_2 + \sigma \delta h_2 = \sigma \delta h_1,$$

where

$$\delta h_1 = \text{deviation of depth in tank 1 from the nominal level,}$$

$$\delta h_2 = \text{deviation of depth in tank 2 from the nominal level,}$$

$$\delta u = \text{deviation in fluid inflow rate to tank 1 (control).}$$

Figure 7.96
Coupled tanks for
Problem 7.51

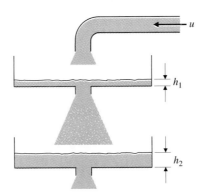

Figure 7.97

View of ship from above

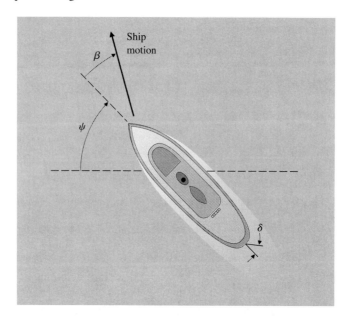

(a) *Level Controller for Two Cascaded Tanks:* Using state feedback of the form

$$\delta u = -K_1 \delta h_1 - K_2 \delta h_2,$$

choose values of K_1 and K_2 that will place the closed-loop eigenvalues at

$$s = -2\sigma(1 \pm j).$$

(b) *Level Estimator for Two Cascaded Tanks:* Suppose that only the deviation in the level of tank 2 is measured (that is, $y = \delta h_2$). Using this measurement, design an estimator that will give continuous, smooth estimates of the deviation in levels of tank 1 and tank 2, with estimator error poles at $-8\sigma(1 \pm j)$.

(c) *Estimator/Controller for Two Cascaded Tanks:* Sketch a block diagram (showing individual integrators) of the closed-loop system obtained by combining the estimator of part (b) with the controller of part (a).

(d) Using MATLAB, compute and plot the response at y to an initial offset in δh_1. Assume $\sigma = 1$ for the plot.

7.52. The lateral motions of a ship that is 100 m long, moving at a constant velocity of 10 m/sec, are described by

$$\begin{bmatrix} \dot{\beta} \\ \dot{r} \\ \dot{\psi} \end{bmatrix} = \begin{bmatrix} -0.0895 & -0.286 & 0 \\ -0.0439 & -0.272 & 0 \\ 0 & 1 & 0 \end{bmatrix} \begin{bmatrix} \beta \\ r \\ \psi \end{bmatrix} + \begin{bmatrix} 0.0145 \\ -0.0122 \\ 0 \end{bmatrix} \delta,$$

where

$$\beta = \text{sideslip angle (deg)},$$

$$\psi = \text{heading angle (deg)}$$

$$\delta = \text{rudder angle (deg)},$$

$$r = \text{yaw rate (see Fig. 7.97)}.$$

(a) Determine the transfer function from δ to ψ and the characteristic roots of the uncontrolled ship.

(b) Using complete state feedback of the form

$$\delta = -K_1\beta - K_2 r - K_3(\psi - \psi_d),$$

where ψ_d is the desired heading, determine values of K_1, K_2, and K_3 that will place the closed-loop roots at $s = -0.2, -0.2 \pm 0.2j$.

(c) Design a state estimator based on the measurement of ψ (obtained from a gyrocompass, for example). Place the roots of the estimator error equation at $s = -0.8$ and $-0.8 \pm 0.8j$.

(d) Give the state equations and transfer function for the compensator $D_c(s)$ in Fig. 7.98, and plot its frequency response.

(e) Draw the Bode plot for the closed-loop system, and compute the corresponding gain and phase margins.

(f) Compute the feedforward gains for a reference input, and plot the step response of the system to a change in heading of $5°$.

Figure 7.98
Ship control block diagram

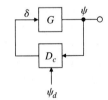

Problem for Section 7.8

▲ **7.53.** As mentioned in footnote 9 in Section 7.8.2, a reasonable approach for selecting the feedforward gain in Eq. (7.201) is to choose $\bar{N}$ such that when r and y are both unchanging, the DC gain from r to u is the negative of the DC gain from y to u. Derive a formula for $\bar{N}$ based on this selection rule. Show that if the plant is type 1, this choice is the same as that given by Eq. (7.201).

Problems for Section 7.9

7.54. Assume the linearized and time-scaled equation of motion for the ball-bearing levitation device is $\ddot{x} - x = u + w$. Here w is a constant bias due to the power amplifier. Introduce integral error control, and select three control gains $\mathbf{K} = [K_1 \quad K_2 \quad K_3]$ so that the closed-loop poles are at -1 and $-1 \pm j$ and the steady-state error to w and to a (step) position command will be zero. Let $y = x$ and the reference input $r \overset{\triangle}{=} y_{\text{ref}}$ be a constant. Draw a block diagram of your design showing the locations of the feedback gains K_i. Assume that both $\dot{x}$ and x can be measured. Plot the response of the closed-loop system to a step command input and the response to a step change in the bias input. Verify that the system is type 1. Use MATLAB (Simulink) software to simulate the system responses.

7.55. Consider a system with state matrices

$$\mathbf{F} = \begin{bmatrix} -2 & 1 \\ 0 & -3 \end{bmatrix}, \qquad \mathbf{G} = \begin{bmatrix} 1 \\ 1 \end{bmatrix}, \qquad \mathbf{H} = [1 \quad 3].$$

(a) Use feedback of the form $u(t) = -\mathbf{Kx}(t) + \bar{N}r(t)$, where $\bar{N}$ is a nonzero scalar, to move the poles to $-3 \pm 3j$.

(b) Choose $\bar{N}$ so that if r is a constant, the system has zero steady-state error; that is $y(\infty) = r$.

(c) Show that if $\mathbf{F}$ changes to $\mathbf{F} + \delta\mathbf{F}$, where $\delta\mathbf{F}$ is an arbitrary 2×2 matrix, then your choice of $\bar{N}$ in part (b) will no longer make $y(\infty) = r$. Therefore, the system is not robust under changes to the system parameters in $\mathbf{F}$.

(d) The system steady-state error performance can be made robust by augmenting the system with an integrator and using unity feedback; that is, by setting $\dot{x}_I = r - y$, where x_I is the state of the integrator. To see this, first use state feedback of the form $u = -\mathbf{Kx} - K_1 x_I$ so that the poles of the augmented system are at $-3, -2 \pm j\sqrt{3}$.

(e) Show that the resulting system will yield $y(\infty) = r$ no matter how the matrices $\mathbf{F}$ and $\mathbf{G}$ are changed, as long as the closed-loop system remains stable.

(f) For part (d), use MATLAB (Simulink) software to plot the time response of the system to a constant input. Draw Bode plots of the controller as well as the sensitivity function ($\mathcal{S}$) and the complementary sensitivity function ($\mathcal{T}$).

▲ **7.56.** Consider a servomechanism for following the data track on a computer-disk memory system. Because of various unavoidable mechanical imperfections, the data track is not exactly a centered circle, and thus the radial servo must follow a sinusoidal input of radian frequency ω_0 (the spin rate of the disk). The state matrices for a linearized model of such a system are

$$\mathbf{F} = \begin{bmatrix} 0 & 1 \\ 0 & -1 \end{bmatrix}, \qquad \mathbf{G} = \begin{bmatrix} 0 \\ 1 \end{bmatrix}, \qquad \mathbf{H} = [1 \quad 0].$$

The sinusoidal reference input satisfies $\ddot{r} = -\omega_0^2 r$.

(a) Let $\omega_0 = 1$, and place the poles of the error system for an internal model design at

$$\alpha_c(s) = (s + 2 \pm j2)(s + 1 \pm 1j)$$

and place the pole of the reduced-order estimator at

$$\alpha_e(s) = (s + 6).$$

(b) Draw a block diagram of the system, and clearly show the presence of the oscillator with frequency ω_0 (the internal model) in the controller. Also verify the presence of the blocking zeros at $\pm j\omega_0$.

(c) Use MATLAB (Simulink) software to plot the time response of the system to a sinusoidal input at frequency $\omega_0=1$.

(d) Draw a Bode plot to show how this system will respond to sinusoidal inputs at frequencies different from but near ω_0.

▲ **7.57.** Compute the controller transfer function (from $Y(s)$ to $U(s)$) in Example 7.31. What is the prominent feature of the controller that allows tracking and disturbance rejection?

▲ **7.58.** Consider the pendulum problem with control torque T_c and disturbance torque T_d:

$$\ddot{\theta} + 4\theta = T_c + T_d$$

(here $g/l = 4$). Assume there is a potentiometer at the pin that measures the output angle θ, but with a constant unknown bias b. Thus the measurement equation is $y = \theta + b$.

(a) Take the "augmented" state vector to be

$$\begin{bmatrix} \theta \\ \dot{\theta} \\ w \end{bmatrix}.$$

where w is the input-equivalent bias. Write the system equations in state-space form. Give values for the matrices $\mathbf{F}$, $\mathbf{G}$, and $\mathbf{H}$.

(b) Using state-variable methods, show that the characteristic equation of the model is $s(s^2 + 4) = 0$.

(c) Show that w is observable if we assume $y = \theta$, and write the estimator equations for

$$\begin{bmatrix} \hat{\theta} \\ \dot{\hat{\theta}} \\ \hat{w} \end{bmatrix}.$$

Pick estimator gains $[l_1 \quad l_2 \quad l_3]^T$ to place all the roots of the estimator-error characteristic equation at -10.

(d) Using full state feedback of the estimated (controllable) state-variables, derive a control law to place the closed-loop poles at $-2 \pm 2j$.

(e) Draw a block diagram of the complete closed-loop system (estimator, plant, and controller) using integrator blocks.

(f) Introduce the estimated bias into the control so as to yield zero steady-state error to the output bias b. Demonstrate the performance of your design by plotting the response of the system to a step change in b; that is, b changes from 0 to some constant value.

Problems for Section 7.11

▲ **7.59.** Consider the system with the transfer function $e^{-Ts}G(s)$, where

$$G(s) = \frac{1}{s(s+1)(s+2)}.$$

The Smith compensator for this system is given by

$$D'_c(s) = \frac{D_c}{1 + (1 - e^{-sT})G(s)D_c}.$$

Plot the frequency response of the compensator for $T = 5$ and $D_c = 1$, and draw a Bode plot that shows the gain and phase margins of the system.[21]

Problems for Section 7.12

▲ **7.60.** A first-order nonlinear system is described by the equation $\dot{x} = -f(x)$, where $f(x)$ is a continuous and differentiable nonlinear function that satisfies the following:

$$f(0) = 0,$$

$$f(x) > 0 \quad \text{for} \quad x > 0,$$

$$f(x) < 0 \quad \text{for} \quad x < 0.$$

Use the Lyapunov function $V(x) = x^2/2$ to show that the system is stable near the origin $(x = 0)$.

▲ **7.61.** Use the Lyapunov equation

$$\mathbf{F}^T\mathbf{P} + \mathbf{P}\mathbf{F} = -\mathbf{Q} = -\mathbf{I}$$

to find the range of K for which the system in Fig. 7.99 will be stable. Compare your answer with the stable values for K obtained using Routh's stability criterion.

Figure 7.99
Control system for
Problem 7.61

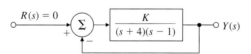

▲ **7.62.** Consider the system

$$\frac{d}{dt}\begin{bmatrix} x_1 \\ x_2 \end{bmatrix} = \begin{bmatrix} x_1 + x_2 u \\ x_2(x_2 + u) \end{bmatrix}, \qquad y = x_1.$$

Find all values of α and β for which the input $u(t) = \alpha y(t) + \beta$ will achieve the goal of maintaining the output $y(t)$ near 1.

---

[21] This problem was given by Åström (1977).

▲ **7.63.** Consider the nonlinear autonomous system

$$\frac{d}{dt}\begin{bmatrix} x_1 \\ x_2 \\ x_3 \end{bmatrix} = \begin{bmatrix} x_2(x_3 - x_1) \\ x_1^2 - 1 \\ -x_1 x_3 \end{bmatrix}.$$

(a) Find the equilibrium point(s).

(b) Find the linearized system about each equilibrium point.

(c) For each case in part (b), what does Lyapunov theory tell us about the stability of the nonlinear system near the equilibrium point?

8
Digital Control

Chapter Overview

In Section 8.1 we describe the basic structure of digital control systems and introduce the issues that arise due to the sampling. The digital implementation described in Section 4.4 is sufficient for implementing a feedback control law in a digital control system, which you can then evaluate via Simulink to determine the degradation with respect to the continuous case. However, to fully understand the effect of sampling, it is useful to learn about discrete linear analysis tools. This requires an understanding of the z-transform, which we discuss in Section 8.2. Section 8.3 builds on this understanding to provide a better foundation for the "emulation" method used in Section 4.4.

In contrast with emulation, which is an approximate method, Section 8.4 explores direct digital design (also called discrete design), which provides an exact method independent of whether the sample rate is fast or not. Design using the state-variable form is introduced in Section 8.5. We end the chapter with a discussion of several topics related to digital control: hardware characteristics, word size, and sample rates, respectively, in Sections 8.6 to 8.8.

A Perspective on Digital Control

Most of the controllers we have studied so far were described by the Laplace transform or differential equations which, strictly speaking, are assumed to be built using analog electronics, such as those in Figs. 5.33 and 5.35. However, as discussed in Section 4.4, most control systems today use digital computers (usually microprocessors) to implement the controllers. The intent of this chapter is to expand on the design of control systems that will be implemented in a digital computer. The implementation leads to an average delay of half the sample period and to a phenomenon called aliasing, which need to be addressed in the controller design

Analog electronics can integrate and differentiate signals. In order for a digital computer to accomplish these tasks, the differential equations describing compensation must be approximated by reducing them to algebraic equations involving addition, division, and multiplication, as developed in Sections 3.7 and 4.4. This chapter expands on various ways to make these approximations. The resulting design can then be tuned up, if needed, using direct digital analysis and design.

You should be able to design, analyze, and implement a digital control system from the material in this chapter. However, our treatment here is a limited version of a complex subject covered in more detail in *Digital Control of Dynamic Systems* by Franklin et al. (3rd ed., 1998).

8.1 Digitization

Figure 8.1(a) shows the topology of the typical continuous system that we have been considering in previous chapters. The computation of the error signal e and the dynamic compensation $D(s)$ can all be accomplished in a digital computer as shown in Fig. 8.1(b). The fundamental differences between the two implementations are that the digital system operates on **samples** of the sensed plant output rather than on the continuous signal and that the control provided by $D(s)$ must be generated by algebraic recursive equations.

We consider first the action of the **analog-to-digital (A/D) converter** on a signal. This device samples a physical variable, most commonly an electrical voltage, and converts it into a binary number that usually consists of 10 to 16 bits. Conversion from the analog signal $y(t)$ to the samples, $y(kT)$, occurs repeatedly at instants of time T seconds apart. T is the **sample period**, and $1/T$ is the **sample rate** in hertz. The sampled signal is $y(kT)$, where k can take on any integer value. It is often written simply as $y(k)$. We call this type of variable a **discrete signal** to distinguish it from a continuous signal like $y(t)$, which changes continuously in time. A system having both discrete and continuous signals is called a **sampled data system**.

We make the assumption that the sample period is fixed. In practice, digital control systems sometimes have varying sample periods and/or different

Sample period

Figure 8.1

Block diagrams for a basic control system: (a) continuous system; (b) with a digital computer

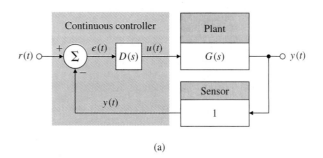

(a)

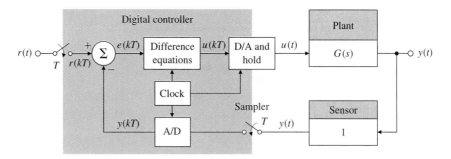

(b)

periods in different feedback paths. Usually, the computer logic includes a **clock** that supplies a pulse, or **interrupt**, every T seconds, and the A/D converter sends a number to the computer each time the interrupt arrives. An alternative implementation, often referred to as **free-running**, is to access the A/D converter after each cycle of code execution has been completed. In the former case the sample period is precisely fixed; in the latter case the sample period is fixed essentially by the length of the code, provided that no logic branches are present, which could vary the amount of code executed.

There also may be a sampler and an A/D converter for the input command $r(t)$, which produces the discrete $r(kT)$ from which the sensed output $y(kT)$ will be subtracted to arrive at the discrete error signal $e(kT)$. As we saw in Sections 4.4 and 5.5 and in Examples 6.14, 7.26, and 7.29, the continuous compensation is approximated by difference equations which are the discrete version of differential equations and can be made to duplicate the dynamic behavior of $D(s)$ if the sample period is short enough. The result of the difference equations is a discrete $u(kT)$ at each sample instant. This signal is converted to a continuous $u(t)$ by the **digital-to-analog (D/A) converter** and the hold: The D/A converter changes the binary number to an analog voltage, and a **zero-order hold** (ZOH) maintains that same voltage throughout the sample period. The resulting $u(t)$ is then applied to the actuator in precisely the same manner as the continuous implementation. There are two basic techniques for finding the difference equations for the digital controller. One technique, called **emulation**, consists of designing a continuous compensation $D(s)$ using methods described in the previous chapters, then approximating that $D(s)$ using the method of Section 4.4 (Tustin's Method) or one of the other methods described in Section 8.3. The other technique is **discrete design**, described in Sections 8.4 and 8.5. Here the difference equations are found directly without designing $D(s)$ first.

The sample rate required depends on the closed-loop bandwidth of the system. Generally, sample rates should be about 20 times the bandwidth or faster in order to assure that the digital controller will match the performance of the continuous controller. Slower sample rates can be used if some adjustments are made in the digital controller or some performance degradation is acceptable. Use of the discrete design method described in Section 8.4 allows for a much slower sample rate if that is desirable to mimimize hardware costs; however, best performance of a digital controller is obtained when the sample rate is greater than 20 times the bandwidth.

It is worth noting that the single most important impact of implementing a control system digitally is the delay associated with the hold. Because each value of $u(kT)$ in Fig. 8.1(b) is held constant until the next value is available from the computer, the continuous value of $u(t)$ consists of steps (see Fig. 8.2) that, on average, are delayed from $u(kT)$ by $T/2$ as shown in the figure. If we simply incorporate this $T/2$ delay into a continuous analysis of the system, an excellent prediction of the effects of sampling results for sample rates much slower than 20 times bandwidth. We will discuss this further in Section 8.3.3.

Zero-order hold (ZOH)

Emulation

Sample rate selection

Figure 8.2
The delay due to the hold operation

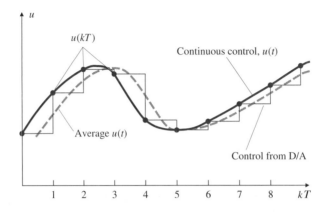

8.2   Dynamic Analysis of Discrete Systems

The z-transform is the mathematical tool for the analysis of linear discrete systems. It plays the same role for discrete systems that the Laplace transform does for continuous systems. This section will give a short description of the z-transform, describe its use in analyzing discrete systems, and show how it relates to the Laplace transform.

8.2.1   z-Transform

In the analysis of continuous systems, we use the Laplace transform, which is defined by

$$\mathcal{L}\{f(t)\} = F(s) = \int_0^\infty f(t)e^{-st}\, dt,$$

which leads directly to the important property that (with zero initial conditions)

$$\mathcal{L}\{\dot{f}(t)\} = sF(s). \tag{8.1}$$

Relation (8.1) enables us to find easily the transfer function of a linear continuous system, given the differential equation of that system.

For discrete systems a very similar procedure is available. The **z-transform**
z-Transform   is defined by

$$\mathcal{Z}\{f(k)\} = F(z) = \sum_{k=0}^{\infty} f(k)z^{-k}, \tag{8.2}$$

where $f(k)$ is the sampled version of $f(t)$, as shown in Fig. 8.3, and $k = 0, 1, 2, 3, \ldots$ refers to discrete sample times $t_0, t_1, t_2, t_3, \ldots$. This leads directly to a property analogous to Eq. (8.1), specifically, that

$$\mathcal{Z}\{f(k-1)\} = z^{-1}F(z). \tag{8.3}$$

Figure 8.3
A continuous, sampled version of signal f

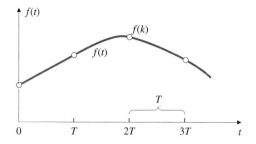

This relation allows us to easily find the transfer function of a discrete system, given the difference equations of that system. For example, the general second-order difference equation

$$y(k) = -a_1 y(k-1) - a_2 y(k-2) + b_0 u(k) + b_1 u(k-1) + b_2 u(k-2)$$

can be converted from this form to the z-transform of the variables $y(k), u(k), \ldots$ by invoking Eq. (8.3) once or twice to arrive at

$$Y(z) = (-a_1 z^{-1} - a_2 z^{-2})Y(z) + (b_0 + b_1 z^{-1} + b_2 z^{-2})U(z). \qquad (8.4)$$

Discrete transfer function

Equation (8.4) then results in the discrete transfer function

$$\frac{Y(z)}{U(z)} = \frac{b_0 + b_1 z^{-1} + b_2 z^{-2}}{1 + a_1 z^{-1} + a_2 z^{-2}}.$$

8.2.2 z-Transform Inversion

Table 8.1 relates simple discrete-time functions to their z-transforms and gives the Laplace transforms for the same time functions.

Given a general z-transform, we could expand it into a sum of elementary terms using partial-fraction expansion (see Appendix A) and find the resulting time series from the table. These procedures are exactly the same as those used for continuous systems; like the continuous case, most designers would use a numerical evaluation of the discrete equations to obtain a time history rather than inverting the z-transform.

A z-transform inversion technique that has no continuous counterpart is called **long division**. Given the z-transform

$$Y(z) = \frac{N(z)}{D(z)}, \qquad (8.5)$$

z-Transform inversion: long division

we simply divide the denominator into the numerator using long division. The result is a series (perhaps with an infinite number of terms) in z^{-1}, from which the time series can be found by using Eq. (8.2).

TABLE 8.1   Laplace Transforms and z-Transforms of Simple Discrete Time Functions

$F(s)$ is the Laplace transform of $f(t)$, and $F(z)$ is the z-transform of $f(kT)$. *Note:* $f(t) = 0$ for $t = 0$.

No.	$F(s)$	$f(kT)$	$F(z)$
1		$1, k = 0; 0, k \neq 0$	1
2		$1, k = k_o; 0, k \neq k_o$	z^{-k_o}
3	$\dfrac{1}{s}$	$1(kT)$	$\dfrac{z}{z-1}$
4	$\dfrac{1}{s^2}$	kT	$\dfrac{Tz}{(z-1)^2}$
5	$\dfrac{1}{s^3}$	$\dfrac{1}{2!}(kT)^2$	$\dfrac{T^2}{2}\left[\dfrac{z(z+1)}{(z-1)^3}\right]$
6	$\dfrac{1}{s^4}$	$\dfrac{1}{3!}(kT)^3$	$\dfrac{T^3}{6}\left[\dfrac{z(z^2+4z+1)}{(z-1)^4}\right]$
7	$\dfrac{1}{s^m}$	$\lim_{a\to 0}\dfrac{(-1)^{m-1}}{(m-1)!}\left(\dfrac{\partial^{m-1}}{\partial a^{m-1}}e^{-akT}\right)$	$\lim_{a\to 0}\dfrac{(-1)^{m-1}}{(m-1)!}\left(\dfrac{\partial^{m-1}}{\partial a^{m-1}}\dfrac{z}{z-e^{-aT}}\right)$
8	$\dfrac{1}{s+a}$	e^{-akT}	$\dfrac{z}{z-e^{-aT}}$
9	$\dfrac{1}{(s+a)^2}$	kTe^{-akT}	$\dfrac{Tze^{-aT}}{(z-e^{-aT})^2}$
10	$\dfrac{1}{(s+a)^3}$	$\dfrac{1}{2}(kT)^2e^{-akT}$	$\dfrac{T^2}{2}e^{-aT}z\dfrac{(z+e^{-aT})}{(z-e^{-aT})^3}$
11	$\dfrac{1}{(s+a)^m}$	$\dfrac{(-1)^{m-1}}{(m-1)!}\left(\dfrac{\partial^{m-1}}{\partial a^{m-1}}e^{-akT}\right)$	$\dfrac{(-1)^{m-1}}{(m-1)!}\left(\dfrac{\partial^{m-1}}{\partial a^{m-1}}\dfrac{z}{z-e^{-aT}}\right)$
12	$\dfrac{a}{s(s+a)}$	$1-e^{-akT}$	$\dfrac{z(1-e^{-aT})}{(z-1)(z-e^{-aT})}$
13	$\dfrac{a}{s^2(s+a)}$	$\dfrac{1}{a}(akT-1+e^{-akT})$	$\dfrac{z[(aT-1+e^{-aT})z+(1-e^{-aT}-aTe^{-aT})]}{a(z-1)^2(z-e^{-aT})}$
14	$\dfrac{b-a}{(s+a)(s+b)}$	$e^{-akT}-e^{-bkT}$	$\dfrac{(e^{-aT}-e^{-bT})z}{(z-e^{-aT})(z-e^{-bT})}$
15	$\dfrac{s}{(s+a)^2}$	$(1-akT)e^{-akT}$	$\dfrac{z[z-e^{-aT}(1+aT)]}{(z-e^{-aT})^2}$
16	$\dfrac{a^2}{s(s+a)^2}$	$1-e^{-akT}(1+akT)$	$\dfrac{z[z(1-e^{-aT}-aTe^{-aT})+e^{-2aT}-e^{-aT}+aTe^{-aT}]}{(z-1)(z-e^{-aT})^2}$
17	$\dfrac{(b-a)s}{(s+a)(s+b)}$	$be^{-bkT}-ae^{-akT}$	$\dfrac{z[z(b-a)-(be^{-aT}-ae^{-bT})]}{(z-e^{-aT})(z-e^{-bT})}$
18	$\dfrac{a}{s^2+a^2}$	$\sin akT$	$\dfrac{z\sin aT}{z^2-(2\cos aT)z+1}$
19	$\dfrac{s}{s^2+a^2}$	$\cos akT$	$\dfrac{z(z-\cos aT)}{z^2-(2\cos aT)z+1}$

TABLE 8.1   **Laplace Transforms and z-Transforms of Simple Discrete Time Functions** (*continued*)

$F(s)$ is the Laplace transform of $f(t)$, and $F(z)$ is the z-transform of $f(kT)$. *Note:* $f(t) = 0$ for $t = 0$.

No.	$F(s)$	$f(kT)$	$F(z)$
20	$\dfrac{s+a}{(s+a)^2+b^2}$	$e^{-akT}\cos bkT$	$\dfrac{z(z-e^{-aT}\cos bT)}{z^2-2e^{-aT}(\cos bT)z+e^{-2aT}}$
21	$\dfrac{b}{(s+a)^2+b^2}$	$e^{-akT}\sin bkT$	$\dfrac{ze^{-aT}\sin bT}{z^2-2e^{-aT}(\cos bT)z+e^{-2aT}}$
22	$\dfrac{a^2+b^2}{s[(s+a)^2+b^2]}$	$1-e^{-akT}\left(\cos bkT+\dfrac{a}{b}\sin bkT\right)$	$\dfrac{z(Az+B)}{(z-1)[z^2-2e^{-aT}(\cos bT)z+e^{-2aT}]}$

$$A = 1 - e^{-aT}\cos bT - \frac{a}{b}e^{-aT}\sin bT$$

$$B = e^{-2aT} + \frac{a}{b}e^{-aT}\sin bT - e^{-aT}\cos bT$$

For example, a first-order system described by the difference equation

$$y(k) = \alpha y(k-1) + u(k)$$

yields the discrete transfer function

$$\frac{Y(z)}{U(z)} = \frac{1}{1-\alpha z^{-1}}.$$

For a unit pulse input defined by

$$u(0) = 1,$$
$$u(k) = 0, \qquad k \neq 0,$$

the z-transform is then

$$U(z) = 1, \tag{8.6}$$

so

$$Y(z) = \frac{1}{1-\alpha z^{-1}}. \tag{8.7}$$

Therefore, to find the time series, we divide the numerator of Eq. (8.7) by its denominator using long division,

$$
1 - \alpha z^{-1} \overline{\smash{\big)}\,
\begin{aligned}
& 1 + \alpha z^{-1} + \alpha^2 z^{-2} + \alpha^3 z^{-3} + \cdots \\
& \underline{1} \\
& \underline{1 - \alpha z^{-1}} \\
& \quad \alpha z^{-1} + 0 \\
& \quad \underline{\alpha z^{-1} - \alpha^2 z^{-2}} \\
& \qquad \alpha^2 z^{-2} + 0 \\
& \qquad \underline{\alpha^2 z^{-2} - \alpha 3 z^{-3}} \\
& \qquad\qquad \alpha^3 z^{-3} \\
& \qquad\qquad\qquad \ddots
\end{aligned}}
$$

to yield the infinite series

$$
Y(z) = 1 + \alpha z^{-1} + \alpha^2 z^{-2} + \alpha^3 z^{-3} + \cdots. \tag{8.8}
$$

From Eqs. (8.8) and (8.2) we see that the sampled time history of y is

$$
y(0) = 1,
$$

$$
y(1) = \alpha,
$$

$$
y(2) = \alpha^2,
$$

$$
\vdots \quad \vdots
$$

$$
y(k) = \alpha^k.
$$

8.2.3   Relationship Between s and z

For continuous systems, we saw in Chapter 3 that certain behaviors result from different pole locations in the s-plane: oscillatory behavior for poles near the imaginary axis, exponential decay for poles on the negative real axis, and unstable behavior for poles with a positive real part. A similar kind of association would also be useful to know when designing discrete systems. Consider the continuous signal

$$
f(t) = e^{-at}, \qquad t > 0,
$$

which has the Laplace transform

$$
F(s) = \frac{1}{s + a}
$$

and corresponds to a pole at $s = -a$. The z-transform of $f(kT)$ is

$$F(z) = \mathcal{Z}\{e^{-akT}\}. \tag{8.9}$$

From Table 8.1 we can see Eq. (8.9) is equivalent to

$$F(z) = \frac{z}{z - e^{-aT}},$$

which corresponds to a pole at $z = e^{-aT}$. This means that a pole at $s = -a$ in the s-plane corresponds to a pole at $z = e^{-aT}$ in the discrete domain. This is true in general:

Relationship between z-plane and s-plane characteristics

> The equivalent characteristics in the z-plane are related to those in the s-plane by the expression
>
> $$z = e^{sT}, \tag{8.10}$$
>
> where T is the sample period.

Table 8.1 also includes the Laplace transforms, which demonstrates the $z = e^{sT}$ relationship for the roots of the denominators of the table entries for $F(s)$ and $F(z)$.

Figure 8.4 shows the mapping of lines of constant damping ζ and natural frequency ω_n from the s-plane (Fig. 3.20) to the upper half of the z-plane, using Eq. (8.10). The mapping has several important features (see Problem 8.20):

1. The stability boundary is the unit circle $|z| = 1$.

2. The small vicinity around $z = +1$ in the z-plane is essentially identical to the vicinity around $s = 0$ in the s-plane.

3. The z-plane locations give response information normalized to the sample rate, rather than to time as in the s-plane.

4. The negative real z-axis always represents a frequency of $\omega_s/2$, where $\omega_s = 2\pi/T$ = sample rate in radians per second.

5. Vertical lines in the left half of the s-plane (the constant real part or time constant) map into circles within the unit circle of the z-plane.

6. Horizontal lines in the s-plane (the constant imaginary part of the frequency) map into radial lines in the z-plane.

Nyquist frequency $= \omega_s/2$

7. Frequencies greater than $\omega_s/2$, called the **Nyquist frequency**, appear in the z-plane on top of corresponding lower frequencies because of the circular character of the trigonometric functions imbedded in Eq. (8.10). This overlap is called **aliasing** or **folding**. As a result it is necessary to sample at least twice as fast as a signal's highest frequency component in order to represent that signal with the samples. (We will discuss aliasing in greater detail in Section 8.6.3.)

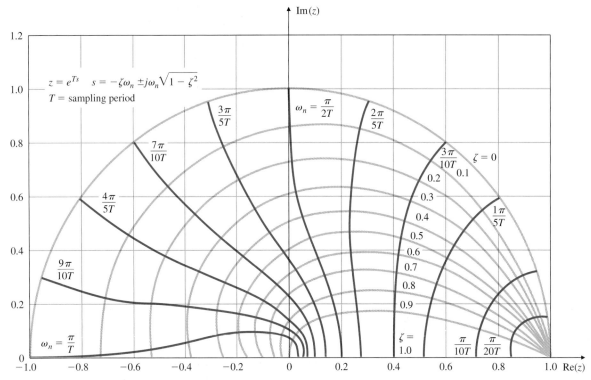

Figure 8.4 Natural frequency (solid color) and damping loci (light color) in the z-plane; the portion below the Re(z)-axis (not shown) is the mirror image of the upper half shown.

To provide insight into the correspondence between z-plane locations and the resulting time sequence, Fig. 8.5 sketches time responses that would result from poles at the indicated locations. This figure is the discrete companion of Fig. 3.20.

8.2.4   Final Value Theorem

The Final Value Theorem for continuous systems, which we discussed in Section 3.1.6, states that

$$\lim_{t \to \infty} x(t) = x_{ss} = \lim_{s \to 0} s X(s), \tag{8.11}$$

as long as all the poles of $sX(s)$ are in the left half-plane (LHP). It is often used to find steady-state system errors and/or steady-state gains of portions of a control system. We can obtain a similar relationship for discrete systems by noting that a constant continuous steady-state response is denoted by $X(s) = A/s$ and leads to the multiplication by s in Eq. (8.11). Therefore, since the constant

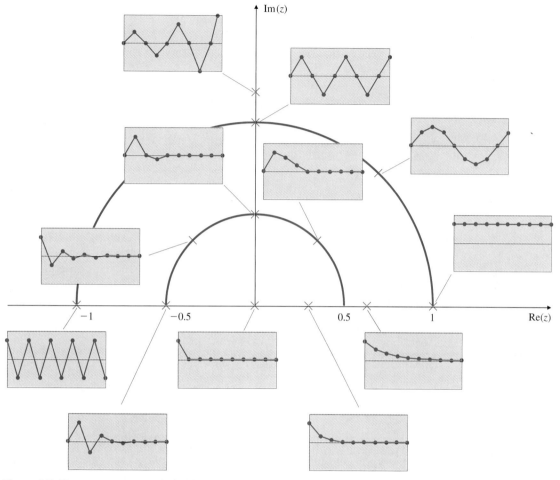

Figure 8.5 Time sequences associated with points in the z-plane

steady-state response for discrete systems is

$$X(z) = \frac{A}{1 - z^{-1}},$$

Final Value Theorem for
discrete systems

the discrete Final Value Theorem is

$$\lim_{k \to \infty} x(k) = x_{ss} = \lim_{z \to 1}(1 - z^{-1})X(z), \tag{8.12}$$

if all the poles of $(1 - z^{-1})X(z)$ are inside the unit circle.
For example, to find the DC gain of the transfer function

$$G(z) = \frac{X(z)}{U(z)} = \frac{0.58(1 + z)}{z + 0.16},$$

we let $u(k) = 1$ for $k \geq 0$, so that

$$U(z) = \frac{1}{1 - z^{-1}}$$

and

$$X(z) = \frac{0.58(1 + z)}{(1 - z^{-1})(z + 0.16)}.$$

Applying the Final Value Theorem yields

$$x_{ss} = \lim_{z \to 1} \left[\frac{0.58(1 + z)}{z + 0.16} \right] = 1,$$

DC gain

so the DC gain of $G(z)$ is unity. To find the DC gain of any stable transfer function, we simply substitute $z = 1$ and compute the resulting gain. Because the DC gain of a system should not change whether represented continuously or discretely, this calculation is an excellent aid to check that an equivalent discrete controller matches a continuous controller. It is also a good check on the calculations associated with determining the discrete model of a system.

8.3   Design by Emulation

Design by emulation, partially described in Section 4.4, proceeds through the following stages:

Stages in emulation design

1. Design a continuous compensation as described in Chapters 1 through 7.

2. Digitize the continuous compensation.

3. Use discrete analysis, simulation, or experimentation to verify the design.

In Section 4.4 we discussed Tustin's method for performing the digitization. Armed with an understanding of the z-transform from Section 8.2, we now develop more digitization procedures and analyze the performance of the digitally controlled system.

   Assume we are given a continuous compensation $D(s)$ as shown in Fig. 8.1(a). We wish to find a set of difference equations or $D(z)$ for the digital implementation of that compensation in Fig. 8.1(b). First we rephrase the problem as one of finding the best $D(z)$ in the digital implementation shown in Fig. 8.6(a) to match the continuous system represented by $D(s)$ in Fig. 8.6(b). In this section we examine and compare three methods for solving this problem.

   It is important to remember, as stated earlier, that these methods are approximations; there is no exact solution for all possible inputs because $D(s)$ responds to the complete time history of $e(t)$, whereas $D(z)$ only has access to the samples $e(kT)$. In a sense the various digitization techniques simply make different assumptions about what happens to $e(t)$ between the sample points.

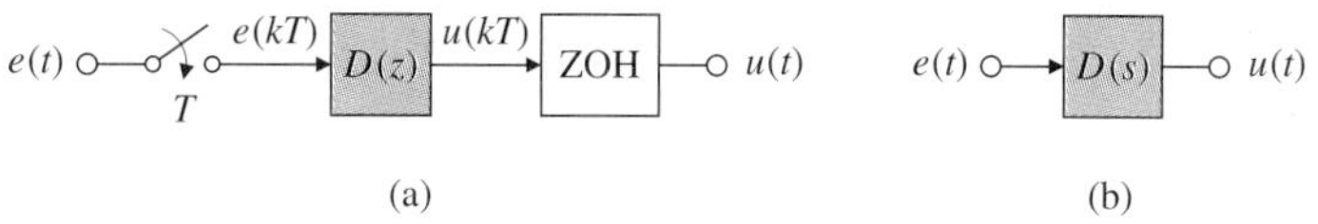

Figure 8.6
Comparison of (a) digital and (b) continuous implementation

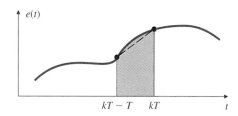

(a)

(b)

Tustin's Method

As discussed in Section 4.4, one digitization technique is to approach the problem as one of numerical integration. Suppose

$$\frac{U(s)}{E(s)} = D(s) = \frac{1}{s},$$

which is integration. Therefore,

$$u(kT) = \int_{0}^{kT-T} e(t)\,dt + \int_{kT-T}^{kT} e(t)\,dt, \tag{8.13}$$

which can be rewritten as

$$u(kT) = u(kT - T) + \text{area under } e(t) \text{ over last } T, \tag{8.14}$$

where T is the sample period.

For Tustin's method, the task at each step is to use trapezoidal integration—that is, to approximate $e(t)$ by a straight line between the two samples (Fig. 8.7). Writing $u(kT)$ as $u(k)$ and $u(kT-T)$ as $u(k-1)$ for short, we convert Eq. (8.14) to

$$u(k) = u(k - 1) + \frac{T}{2}[e(k - 1) + e(k)], \tag{8.15}$$

or, taking the z-transform,

$$\frac{U(z)}{E(z)} = \frac{T}{2}\left(\frac{1 + z^{-1}}{1 - z^{-1}}\right) = \frac{1}{\frac{2}{T}\left(\frac{1 - z^{-1}}{1 + z^{-1}}\right)}. \tag{8.16}$$

Figure 8.7
Trapezoidal integration

For $D(s) = a/(s+a)$, applying the same integration approximation yields

$$D(z) = \frac{a}{\dfrac{2}{T}\left(\dfrac{1-z^{-1}}{1+z^{-1}}\right) + a}.$$

In fact, substituting

$$s = \frac{2}{T}\left(\frac{1-z^{-1}}{1+z^{-1}}\right)$$

Tustin's or bilinear approximation

for every occurrence of s in any $D(s)$ yields a $D(z)$ based on the trapezoidal integration formula. This is called **Tustin's Method** or the **bilinear approximation**. Finding Tustin's approximation by hand for even a simple transfer function requires fairly extensive algebraic manipulations. MATLAB's c2d function expedites the process, as shown in the following example.

EXAMPLE 8.1

Digital Controller for Example 6.14 Using Tustin's Approximation

Determine the difference equations to implement the compensation from Example 6.14,

$$D(s) = 10\frac{s/2 + 1}{s/10 + 1},$$

at a sample rate of 25 times bandwidth using Tustin's approximation. Compare the performance against the continuous system.

Solution.   The bandwidth (ω_{BW}) for Example 6.14 is approximately 10 rad/sec as can be deduced by observing that the crossover frequency (ω_c) is approximately 5 rad/sec and noting the relationship between ω_c and ω_{BW} in Fig. 6.50. Therefore, the sample frequency should be

$$\omega_s = 25 \times \omega_{BW} = (25)(10) = 250 \text{ rad/sec.}$$

Normally, when a frequency is indicated with the units of cycles per second, or Hz, it is given the symbol f, so with this convention we have

$$f_s = \omega_s/(2\pi) \simeq 40 \text{ Hz} \tag{8.17}$$

and the sample period is then

$$T = 1/f_s = 1/40 = 0.025 \text{ sec.}$$

The discrete compensation is computed by the MATLAB statement

```
sysDs = tf(10*[0.5 1],[0.1 1]);
sysDd = c2d(sysDs,0.025,'tustin');
```

which produces

$$D(z) = \frac{45.56 - 43.33\,z^{-1}}{1 - 0.7778\,z^{-1}}. \tag{8.18}$$

We can then write the difference equation by inspecting Eq. (8.18) to get

$$u(k) = 0.7778u(k-1) + 45.56e(k) - 43.33e(k-1)$$

or indexing all time variables by 1, the equivalent is

$$u(k+1) = 0.7778u(k) + 45.56[e(k+1) - 0.9510e(k)]. \tag{8.19}$$

Equation (8.19) computes the new value of the control, $u(k+1)$, given the past value of the control, $u(k)$, and the new and past values of the error signal, $e(k+1)$ and $e(k)$.

In principle the difference equation is evaluated initially with $k = 0$, then $k = 1, 2, 3, \ldots$. However, there is usually no requirement that values for all times be saved in memory. Therefore, the computer need only have variables defined for the current and past values. The instructions to the computer to implement the feedback loop in Fig. 8.1(b) with the difference equation from Eq. (8.19) would call for a continual looping through the following code:

READ y, r
  $e = r - y$
  $u = 0.7778u_p + 45.56[e - 0.9510e_p]$
OUTPUT u
  $u_p = u$ (where u_p will be the past value for the next loop through)
  $e_p = e$

go back to READ when T seconds have elapsed since last READ

Use of Simulink to compare the two implementations, in a similiar manner as was done for Example 6.14, yields the step responses shown in Fig. 8.8. Note that sampling at 25 times bandwidth causes the digital implementation to match the continuous one quite well.

Figure 8.8

Comparison between the digital and the continuous controller step response with a sample rate 25 times bandwidth. (a) Position, (b) Control

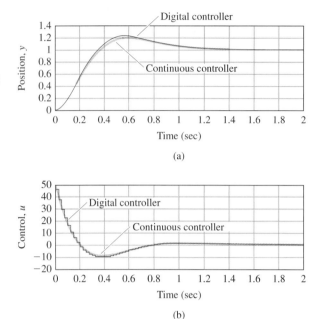

8.3.1   Matched Pole-Zero (MPZ) Method

Another digitization method, called the **matched pole-zero** method, is found by extrapolating from the relationship between the s- and z-planes stated in Eq. (8.10). If we take the z-transform of a sampled function $x(k)$, then the poles of $X(z)$ are related to the poles of $X(s)$ according to the relation $z = e^{sT}$. The MPZ technique applies the relation $z = e^{sT}$ to the poles and zeros of a transfer function, even though, strictly speaking, this relation applies neither to transfer functions nor even to the zeros of a time sequence. Like all transfer-function digitization methods, the MPZ method is an approximation; here the approximation is motivated partly by the fact that $z = e^{sT}$ is the correct s-to-z transformation for the poles of the transform of a time sequence, and partly by the minimal amount of algebra required to determine the digitized transfer function by hand so as to facilitate checking the computer calculations.

Because physical systems often have more poles than zeros, it is useful to arbitrarily add zeros at $z = -1$, resulting in a $1 + z^{-1}$ term in $D(z)$. This causes an averaging of the current and past input values, as in Tustin's method. We select the low-frequency gain of $D(z)$ so that it equals that of $D(s)$.

MPZ Method Summary

1. Map poles and zeros according to the relation $z = e^{sT}$.

2. If the numerator is of lower order than the denominator, add powers of $(z + 1)$ to the numerator until numerator and denominator are of equal order.

3. Set the DC or low-frequency gain of $D(z)$ equal to that of $D(s)$.

The MPZ approximation of

$$D(s) = K_c \frac{s + a}{s + b} \tag{8.20}$$

is

$$D(z) = K_d \frac{z - e^{-aT}}{z - e^{-bT}}, \tag{8.21}$$

where K_d is found by causing the DC gain of $D(z)$ to equal the DC gain of $D(s)$ using the continuous and discrete versions of the Final Value Theorem. The result is

$$K_c \frac{a}{b} = K_d \frac{1 - e^{-aT}}{1 - e^{-bT}},$$

or

$$K_d = K_c \frac{a}{b} \left(\frac{1 - e^{-bT}}{1 - e^{-aT}} \right). \tag{8.22}$$

For a $D(s)$ with a higher-order denominator, Step 2 in the method calls for adding the $(z + 1)$ term. For example,

$$D(s) = K_c \frac{s + a}{s(s + b)} \Rightarrow D(z) = K_d \frac{(z + 1)(z - e^{-aT})}{(z - 1)(z - e^{-bT})}, \tag{8.23}$$

where, after dropping the poles at $s = 0$ and $z = 1$,

$$K_d = K_c \frac{a}{2b} \left(\frac{1 - e^{-bT}}{1 - e^{-aT}} \right). \tag{8.24}$$

In the digitization methods described so far, the same power of z appears in the numerator and denominator of $D(z)$. This implies that the difference equation output at time k will require a sample of the input at time k. For example, the $D(z)$ in Eq. (8.21) can be written as

$$\frac{U(z)}{E(z)} = D(z) = K_d \frac{1 - \alpha z^{-1}}{1 - \beta z^{-1}}, \tag{8.25}$$

where $\alpha = e^{-aT}$ and $\beta = e^{-bT}$. By inspection we can see that Eq. (8.25) results in the difference equation

$$u(k + 1) = \beta u(k) + K_d[e(k + 1) - \alpha e(k)]. \tag{8.26}$$

EXAMPLE 8.2

Emulation Design of a Space Station Attitude Digital Controller

A very simplified model of the space station attitude control dynamics has the plant transfer function

$$G(s) = \frac{1}{s^2}.$$

Design a digital controller to have a closed-loop natural frequency $\omega_n \cong 0.3$ rad/sec and a damping ratio $\zeta = 0.7$.

Solution.   The first step is to find the proper $D(s)$ for the system defined in Fig. 8.9. After some trial and error, we find that the specifications can be met by the lead compensation

$$D(s) = 0.81 \frac{s + 0.2}{s + 2}, \tag{8.27}$$

The root locus in Fig. 8.10 verifies the appropriateness of using Eq. (8.27).

To digitize this $D(s)$, we first need to select a sample rate. For a system with $\omega_n = 0.3$ rad/sec, the bandwidth will also be about 0.3 rad/sec, and a safe sample rate would be about 20 times ω_n. Thus

$$\omega_s = 0.3 \times 20 = 6 \text{ rad/sec.}$$

A sample rate of 6 rad/sec is about 1 Hz; therefore, the sample period should be $T = 1$ sec. The MPZ digitization of Eq. (8.27), given by Eqs. (8.21) and (8.22), yields

$$D(z) = 0.389 \frac{z - 0.82}{z - 0.135}$$

$$= \frac{0.389 - 0.319z^{-1}}{1 - 0.135z^{-1}}. \tag{8.28}$$

Inspection of Eq. (8.28) gives us the difference equation

$$u(k+1) = 0.135u(k) + 0.389e(k+1) - 0.319e(k), \tag{8.29}$$

where

$$e(k) = r(k) - y(k),$$

and this completes the digital algorithm design. The complete digital system is shown in Fig. 8.11.

Figure 8.9

Continuous-design definition for Example 8.2

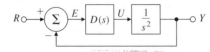

Figure 8.10
s-plane locus with respect to K

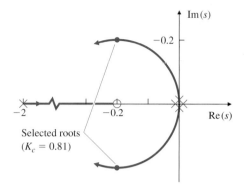

Figure 8.11
A digital control system for emulating Fig. 8.9

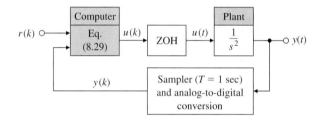

Figure 8.12
Step responses of the continuous and digital implementations

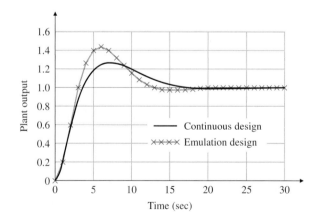

The last step in emulation design is to verify the design by implementing it on the computer. Figure 8.12 compares the step response of the digital system with the step response of the continuous compensation. Note that there is greater overshoot and a longer settling time in the digital system, which suggests a decrease in the damping. The average $T/2$ delay shown in Fig. 8.2 is the cause of the reduced damping.

It is impossible to sample $e(k + 1)$, compute $u(k + 1)$, and then output $u(k+1)$ all in zero elapsed time; therefore, Eqs. (8.26) and (8.29) are impossible to implement precisely. However, if the equation is simple enough and/or the computer is fast enough, a slight computational delay between the $e(k + 1)$ sample and the $u(k + 1)$ output will have a negligible effect on the actual response of the system compared with that expected from the original design. A rule of thumb would be to keep the computational delay on the order of 1/10 of T. The real-time code and hardware can be structured so that this delay is minimized by making sure that computations between read A/D and write D/A are minimized and that $u(k + 1)$ is sent to the ZOH immediately after its calculation.

8.3.2   Modified Matched Pole-Zero (MMPZ) Method

The $D(z)$ in Eq. (8.23) results in $u(k+1)$ being dependent on $e(k+1)$, the input at the same time point. If the structure of the computer hardware prohibits this relation or if the computations are particularly lengthy, it may be desirable to derive a $D(z)$ that has one less power of z in the numerator than in the denominator; hence the computer output $u(k + 1)$ would only require input from the previous time, that is, $e(k)$. To do this, we simply modify Step 2 in the matched pole-zero procedure so that the numerator is of lower order than the denominator by 1. For example, if

$$D(s) = K_c \frac{s + a}{s(s + b)},$$

we skip Step 2 to get

$$D(z) = K_d \frac{z - e^{-aT}}{(z - 1)(z - e^{-bT})}, \tag{8.30}$$

$$K_d = K_c \frac{a}{b} \left(\frac{1 - e^{-bT}}{1 - e^{-aT}} \right).$$

To find the difference equation, we multiply the top and bottom of Eq. (8.30) by z^{-2} to find

$$D(z) = K_d \frac{z^{-1}(1 - e^{-aT} z^{-1})}{1 - z^{-1}(1 + e^{-bT}) + z^{-2} e^{-bT}}. \tag{8.31}$$

By inspecting Eq. (8.31) we can see that the difference equation is

$$u(k + 1) = (1 + e^{-bT})u(k) - e^{-bT} u(k - 1) + K_d[e(k) - e^{-aT} e(k - 1)].$$

In this equation an entire sample period is available to perform the calculation and to output $u(k+1)$ because it depends only on $e(k)$. A discrete analysis of this controller would therefore more accurately explain the behavior of the actual system. However, because this controller is using data that are one cycle old, it will typically not perform as well as the MPZ controller in terms of the deviations of the desired system output in the presence of random disturbances.

8.3.3   Comparison of Digital Approximation Methods

A numerical comparison of the magnitude of the frequency response for a first-order lag,

$$D(s) = \frac{5}{s+5},$$

is made in Fig. 8.13 for the three approximation techniques at two different sample rates. The results of the $D(z)$ computations used in Fig. 8.13 are shown in Table 8.2.

Figure 8.13 shows that all the approximations are quite good at frequencies below about 1/4 the sample rate, or $\omega_s/4$. If $\omega_s/4$ is sufficiently larger than the filter break-point frequency—that is, if the sampling is fast enough—the break-point characteristics of the lag will be accurately reproduced. Tustin's technique and the two MPZ methods show a notch at $\omega_s/2$ because of their zero at $z = -1$ from the $z+1$ term. Other than the large difference at $\omega_s/2$, which is typically outside the range of interest, the three methods have similar accuracies.

TABLE 8.2	Comparing Digital Approximations of $D(z)$ for $D(s) = 5/(s+5)$	
		ω_s
Method	100 rad/sec	20 rad/sec
Matched pole-zero (MPZ)	$0.143\dfrac{z+1}{z-0.715}$	$0.405\dfrac{z+1}{z-0.189}$
Modified MPZ (MMPZ)	$0.285\dfrac{1}{z-0.715}$	$0.811\dfrac{1}{z-0.189}$
Tustin's	$0.143\dfrac{z+1}{z-0.713}$	$0.454\dfrac{z+1}{z-0.0914}$

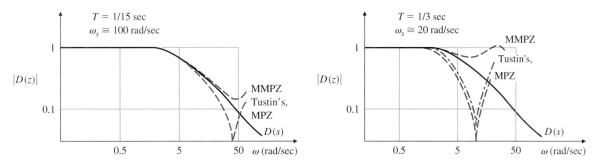

Figure 8.13 A comparison of the frequency response of three discrete approximations

8.3.4   Applicability Limits of the Emulation Design Method

If we performed an exact discrete analysis or a simulation of a system and determined the digitization for a wide range of sample rates, the system would often be unstable for rates slower than approximately $5\omega_n$, and the damping would be degraded significantly for rates slower than about $10\omega_n$. At sample rates $\stackrel{>}{\approx} 20\omega_n$ (or $\stackrel{>}{\approx} 20$ times the bandwidth for more complex systems), design by emulation yields reasonable results, and at sample rates of 30 times the bandwidth or higher, emulation can be used with confidence.

ZOH transfer function

As shown by Fig. 8.2, the errors come about because the technique ignores the lagging effect of the ZOH which, on the average, is $T/2$. A method to account for this is to approximate the $T/2$ delay with Eq. (5.119) by including a transfer function approximation for the ZOH:[1]

$$G_{\text{ZOH}}(s) = \frac{2/T}{s + 2/T}. \tag{8.32}$$

Once an initial design is carried out and the sampling rate has been selected, we could improve on our discrete design by inserting Eq. (8.32) into the original plant model and adjusting the $D(s)$ so that a satisfactory response in the presence of the sampling delay is achieved. Therefore, we see that use of Eq. (8.32) partly alleviates the approximate nature of the emulation method.

For sample rates slower than about $10\omega_n$ it is advisable to analyze the entire system using an exact discrete analysis. If a discrete analysis shows an unacceptable degradation of performance due to the sampling, the design can then be refined using exact discrete methods. We cover this approach next.

8.4   Discrete Design

It is possible to obtain an exact discrete model that relates the samples of the continuous plant $y(k)$ to the input control sequence $u(k)$. This plant model can be used as part of a discrete model of the feedback system including the compensation $D(z)$. Analysis and design using this discrete model is called **discrete design** or, alternatively, **direct digital design**. The following sections will describe how to find the discrete plant model (Section 8.4.1), what the feedback compensation looks like when designing with a discrete model (Sections 8.4.2 and 8.4.3), and how the design process is carried out (Section 8.4.4).

8.4.1   Analysis Tools

The first step in performing a discrete analysis of a system with some discrete elements is to find the discrete transfer function of the continuous portion. For a system similar to that shown in Fig. 8.1(b), we wish to find the transfer

---

[1]   Or other Padé approximate as discussed in Section 5.7.2.

function between $u(kT)$ and $y(kT)$. Unlike the cases discussed in the previous sections, there is an *exact* discrete equivalent for this system because the ZOH precisely describes what happens between samples of $u(kT)$, and the output $y(kT)$ is dependent only on the input at the sample times $u(kT)$.

The exact discrete equivalent   For a plant described by $G(s)$ and preceded by a ZOH, the discrete transfer function is

$$G(z) = (1 - z^{-1})\mathcal{Z}\left\{\frac{G(s)}{s}\right\}, \tag{8.33}$$

where $\mathcal{Z}\{F(s)\}$ is the z-transform of the sampled time series whose Laplace transform is the expression for $F(s)$, given on the same line in Table 8.1. Equation (8.33) has the term $G(s)/s$ because the control comes in as a step input from the ZOH during each sample period. The term $1 - z^{-1}$ reflects the fact that a one-sample duration step can be thought of as an infinite duration step followed by a negative step one cycle delayed. For a more complete derivation, see Franklin et al. (1998). Equation (8.33) allows us to replace the mixed (continuous and discrete) system shown in Fig. 8.14(a) with the equivalent pure discrete system shown in Fig. 8.14(b).

The analysis and design of discrete systems is very similar to the analysis and design of continuous systems; in fact, all the same rules apply. The closed-loop transfer function of Fig. 8.14(b) is obtained using the same rules of block-diagram reduction; that is,

$$\frac{Y(z)}{R(z)} = \frac{D(z)G(z)}{1 + D(z)G(z)}. \tag{8.34}$$

To find the characteristic behavior of the closed-loop system, we need to find the factors in the denominator of Eq. (8.34), that is, the roots of the discrete characteristic equation

$$1 + D(z)G(z) = 0.$$

The root-locus techniques used in continuous systems to find roots of a polynomial in s apply equally well and without modification to the polynomial in z; however, the interpretation of the results is quite different, as we saw in Fig. 8.4. A major difference is that the stability boundary is now the unit circle instead of the imaginary axis.

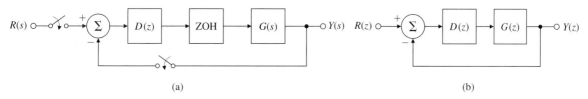

(a)   (b)

Figure 8.14 Comparison of a (a) mixed control system and (b) its pure discrete equivalent

EXAMPLE 8.3 *Discrete Root Locus*

For the case where $G(s)$ in Fig. 8.14(a) is

$$G(s) = \frac{a}{s+a}$$

and $D(z) = K$, draw the root locus with respect to K, and compare your results with a root locus of a continuous version of the system. Discuss the implications of your loci.

Solution.   It follows from Eq. (8.33) that

$$G(z) = (1 - z^{-1})\mathcal{Z}\left[\frac{a}{s(s+a)}\right]$$

$$= (1 - z^{-1})\left[\frac{(1 - e^{-aT})z^{-1}}{(1 - z^{-1})(1 - e^{-aT}z^{-1})}\right]$$

$$= \frac{1 - \alpha}{z - \alpha},$$

where

$$\alpha = e^{-aT}.$$

To analyze the performance of the closed-loop system, standard root-locus rules apply. The result is shown in Fig. 8.15(a) for the discrete case and in Fig. 8.15(b) for the continuous case. In contrast to the continuous case, in which the system remains stable for all values of K, in the discrete case the system becomes oscillatory with decreasing damping ratio as z goes from 0 to -1 and eventually becomes unstable. This instability is due to the lagging effect of the ZOH, which is properly accounted for in the discrete analysis.

Figure 8.15
Root loci for (a) the z-plane and (b) the s-plane

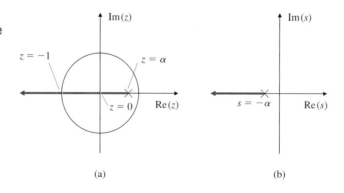

(a)   (b)

8.4.2   Feedback Properties

Discrete control laws

In continuous systems we typically start the design process by using the following basic design elements: proportional, derivative, or integral control laws or some combination of these, sometimes with a lag included. The same ideas can be used in discrete design. Alternatively, the $D(z)$ resulting from the digitization of a continuously designed $D(s)$ will produce these basic design elements, which will then be used as a starting point in a discrete design. The discrete control laws are as follows.

Proportional

$$u(k) = Ke(k) \Rightarrow D(z) = K. \tag{8.35}$$

Derivative

$$u(k) = KT_D[e(k) - e(k-1)], \tag{8.36}$$

for which the transfer function is

$$D(z) = KT_D(1 - z^{-1}) = KT_D\frac{z-1}{z} = k_D\frac{z-1}{z}. \tag{8.37}$$

Integral

$$u(k) = u(k-1) + \frac{K_p}{T_I}e(k), \tag{8.38}$$

for which the transfer function is

$$D(z) = \frac{K}{T_I}\left(\frac{1}{1-z^{-1}}\right) = \frac{K}{T_I}\left(\frac{z}{z-1}\right) = k_I\left(\frac{z}{z-1}\right). \tag{8.39}$$

Lead Compensation

The examples in Section 8.3 showed that a continuous lead compensation leads to difference equations of the form

$$u(k+1) = \beta u(k) + K[e(k+1) - \alpha e(k)], \tag{8.40}$$

for which the transfer function is

$$D(z) = K\frac{1 - \alpha z^{-1}}{1 - \beta z^{-1}}. \tag{8.41}$$

8.4.3   Discrete Design Example

Digital control design consists of using the basic feedback elements of Eqs. (8.35) to (8.41) and iterating on the design parameters until all specifications are met.

EXAMPLE 8.4   *Direct Discrete Design of the Space Station Digital Controller*

Design a digital controller to meet the same specifications as in Example 8.2 using discrete design.

Solution.   The discrete model of the $1/s^2$ plant, preceded by a ZOH, is found through Eq. (8.33) to be

$$G(z) = \frac{T^2}{2}\left[\frac{z+1}{(z-1)^2}\right],$$

which, with $T = 1$ sec, becomes

$$G(z) = \frac{1}{2}\left[\frac{z+1}{(z-1)^2}\right].$$

Proportional feedback in the continuous case yields pure oscillatory motion, so in the discrete case we should expect even worse results. The root locus in Fig. 8.16 verifies this. For very low values of K (where the locus represents roots at very low frequencies compared to the sample rate) the locus is tangent to the unit circle ($\zeta \cong 0$ indicating pure oscillatory motion), thus matching the proportional continuous design.

For higher values of K, Fig. 8.16 shows that the locus diverges into the unstable region because of the effect of the ZOH and sampling. To compensate for this, we will add a derivative term to the proportional term so that the control law is

$$U(z) = K[1 + T_D(1 - z^{-1})]E(z), \tag{8.42}$$

which yields compensation of the form

$$D(z) = K\frac{z-\alpha}{z} \tag{8.43}$$

where the new K and α replace the K and T_D in Eq. (8.42). Now the task is to find the values of α and K that yield good performance. The specifications for the design are that $\omega_n = 0.3$ rad/sec and $\zeta = 0.7$. Figure 8.4 indicates that this s-plane root location maps into a desired z-plane location of

$$z = 0.78 \pm 0.18j.$$

Figure 8.16
z-plane root locus
for a $1/s^2$ plant with
proportional feedback

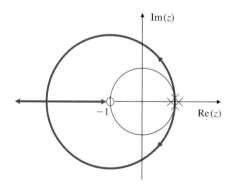

Figure 8.17
z-plane locus for
the $1/s^2$ plant with
$D(z) = K(z - 0.85)/z$

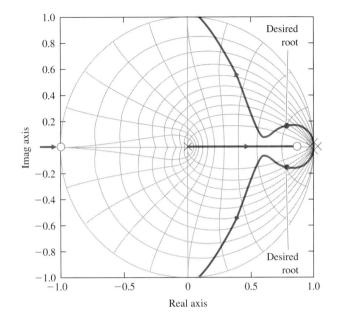

Figure 8.17 is the locus with respect to K for $\alpha = 0.85$. The location of the zero (at $z = 0.85$) was determined by trial and error until the locus passed through the desired z-plane location. The value of the gain when the locus passes through $z = 0.78 \pm 0.18j$ is $K = 0.374$. Equation (8.43) now becomes

$$D(z) = 0.374\frac{z - 0.85}{z}. \tag{8.44}$$

Normally, it is not particularly advantageous to match specific z-plane root locations; rather it is only necessary to pick K and α (or T_D) to obtain acceptable z-plane roots, a much easier task. In this example, we want to match a specific location only so that we can compare the result with the design in Example 8.2.

The control law that results is

$$U(z) = 0.374(1 - 0.85z^{-1})E(z),$$

or

$$u(k + 1) = 0.374e(k + 1) - 0.318e(k), \tag{8.45}$$

which is similar to the control equation [Eq. (8.29)] obtained previously.

The controller in Eq. (8.45) basically differs from the continuously designed controller [Eq. (8.29)] only in the absence of the $u(k)$ term. The $u(k)$ term in Eq. (8.29) results from the lag term $(s + b)$ in the compensation [Eq. (8.27)]. The lag term is typically included in analog controllers both because it supplies noise attenuation and because pure analog differentiators are difficult to build. Some equivalent lag in discrete design naturally appears as a pole at $z = 0$ (see

Fig. 8.17) and represents the one sample delay in computing the derivative by a first difference. For more noise attenuation we could move the pole to the right of $z = 0$, thus resulting in less derivative action and more smoothing, the same tradeoff that exists in continuous control design.

8.4.4   Discrete Analysis of Designs

Any digital controller, whether designed by emulation or directly in the z-plane, can be analyzed using discrete analysis, which consists of the following steps:

1. Find the discrete model of the plant and ZOH using Eq. (8.33).

2. Form the feedback system including $D(z)$.

3. Analyze the resulting discrete system.

We can determine the roots of the system using a root locus as described in Section 8.4.3, or we can determine the time history (at the sample instants) of the discrete system.

EXAMPLE 8.5

Damping and Step Response in Digital versus Continuous Design

Use discrete analysis to determine the equivalent s-plane damping and the step responses of the digital designs in Examples 8.2 and 8.4, and compare your results with the damping and step response of the continuous case in Example 8.2.

Solution.   The MATLAB statements to evaluate the damping and step response of the continuous case in Example 8.2 are

```
sysGs = tf(1,[1   0   0]);
sysDs = tf(0.81 * [1   0.2],[1   2]);
sysGDs = series(sysGs,sysDs);
sysCLs = feedback(sysGDs,1,1);
step(sysCLs)
damp(sysCLs)
```

To analyze the digital control cases, the model of the plant preceded by the ZOH is found using the statements

```
T = 1;
sysGz = c2d(sysGs,T,'zoh')
```

Figure 8.18

Step response of the continuous and digital systems in Examples 8.2 and 8.4

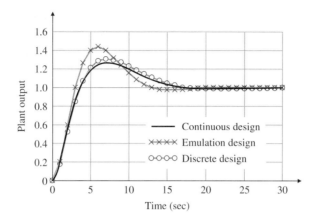

Analysis of the digital control designed by emulation [Eq. (8.29)] in Example 8.2 is performed by the statements

```
sysDz = tf( [.389 -.319],[1 -.135])
sysDGz = series(sysGz,sysDz)
sysCLz = feedback(sysDGz,1)
step(sysCLz,T)
damp(sysCLz,T)
```

Likewise, the discrete design of $D(z)$ from Eq. (8.44) can be analyzed by the same sequence.

The resulting step responses are shown in Fig. 8.18. The calculated damping ζ and complex root natural frequencies ω_n of the closed-loop systems are as follows:

$$\text{Continuous case :} \qquad \zeta = 0.705, \qquad \omega_n = 0.324;$$

$$\text{Emulation design :} \qquad \zeta = 0.645, \qquad \omega_n = 0.441;$$

$$\text{Discrete design :} \qquad \zeta = 0.733, \qquad \omega_n = 0.306.$$

The figure shows increased overshoot for the emulation method that occurred because of the decreased damping of that case. Very little increased overshoot occurred in the discrete design because that compensation was adjusted specifically so that the equivalent s-plane damping of the discrete system was approximately at the desired damping value of $\zeta = 0.7$.

Although the analysis showed some differences between the performance of the digital controllers designed by the two methods, neither the performance nor the control equations [Eqs. (8.29) and (8.45)] are very different. This similarity results because the sample rate is fairly fast compared with ω_n; that is, $\omega_s \cong 20 \times \omega_n$. If we were to decrease the sample rate, the numerical values in the compensations would become increasingly different and the performance would degrade considerably for the emulation case.

As a general rule, discrete design should be used if the sampling frequency is slower than $20 \times \omega_n$. At the very least, an emulation design with slow sampling ($\omega_s < 20 \times \omega_n$) should be verified by a discrete analysis or by simulation as described in Section 4.4, and the compensation should be adjusted if needed. A simulation of a digital control system is a good idea in any case. If it properly accounts for all delays and possibly asynchronous behavior of different modules, it may expose instabilities that are impossible to detect using continuous or discrete linear analysis. A more complete discussion regarding the effects of sample rate on the design is contained in Section 8.8.

8.5 State-Space Design Methods

We have seen in previous chapters that a linear, constant-coefficient continuous system can be represented by a set of first-order matrix differential equations of the form

$$\dot{\mathbf{x}} = \mathbf{F}\mathbf{x} + \mathbf{G}u, \tag{8.46}$$

where u is the control input to the system. The output equation can be expressed as

$$y = \mathbf{H}\mathbf{x} + Ju. \tag{8.47}$$

The solution to these equations, discussed in Section 3.7.2, resulted in Eq. (3.79), which we repeat here:

$$\mathbf{x}(t) = e^{\mathbf{F}(t-t_0)}\mathbf{x}(t_0) + \int_{t_0}^{t} e^{\mathbf{F}(t-\tau)}\mathbf{G}u(\tau)\,d\tau. \tag{8.48}$$

It is possible to use Eq. (8.48) to obtain a discrete state-space representation of the system. Because the solution over one sample period results in a difference equation, we can alter the notation a bit (letting $t = kT + T$ and $t_0 = kT$) to arrive at a particularly useful version of Eq. (8.48):

$$\mathbf{x}(kT + T) = e^{\mathbf{F}T}\mathbf{x}(kT) + \int_{kT}^{kT+T} e^{\mathbf{F}(kT+T-\tau)}\mathbf{G}u(\tau)\,d\tau. \tag{8.49}$$

This result is not dependent on the type of hold, since u is specified in terms of its continuous time history $u(\tau)$ over the sample interval. To find the discrete model of a continuous system where the input $u(t)$ is the output of a ZOH, we let $u(\tau)$ be a constant throughout the sample interval; that is,

$$u(\tau) = u(kT), \qquad kT \leq \tau < kT + T.$$

To facilitate the solution of Eq. (8.49) for a ZOH, we let

$$\eta = kT + T - \tau,$$

which converts Eq. (8.49) to

$$\mathbf{x}(kT + T) = e^{\mathbf{F}T}\mathbf{x}(kT) + \left(\int_{0}^{T} e^{\mathbf{F}\eta}d\eta\right)\mathbf{G}u(kT).$$

If we let

$$\mathbf{\Phi} = e^{\mathbf{F}T}$$

and

$$\mathbf{\Gamma} = \left(\int_0^T e^{\mathbf{F}\eta} d\eta \right) \mathbf{G}, \tag{8.50}$$

Difference equations in standard form

Eqs. (8.49) and (8.47) reduce to difference equations in standard form:

$$\mathbf{x}(k+1) = \mathbf{\Phi}\mathbf{x}(k) + \mathbf{\Gamma}u(k), \tag{8.51}$$

$$y(k) = \mathbf{H}\mathbf{x}(k) + Ju(k), \tag{8.52}$$

where $\mathbf{x}(k+1)$ is a shorthand notation for $\mathbf{x}(kT+T)$, $\mathbf{x}(k)$ for $\mathbf{x}(kT)$, and $u(k)$ for $u(kT)$. The series expansion

$$\mathbf{\Phi} = e^{\mathbf{F}T} = \mathbf{I} + \mathbf{F}T + \frac{\mathbf{F}^2 T^2}{2!} + \frac{\mathbf{F}^3 T^3}{3!} + \cdots$$

can also be written as

$$\mathbf{\Phi} = \mathbf{I} + \mathbf{F}T\mathbf{\Psi}, \tag{8.53}$$

where

$$\mathbf{\Psi} = \mathbf{I} + \frac{\mathbf{F}T}{2!} + \frac{\mathbf{F}^2 T^2}{3!} + \cdots.$$

The $\mathbf{\Gamma}$ integral in Eq. (8.50) can be evaluated term by term to give

$$\mathbf{\Gamma} = \sum_{k=0}^{\infty} \frac{\mathbf{F}^k T^{k+1}}{(k+1)!} \mathbf{G}$$

$$= \sum_{k=0}^{\infty} \frac{\mathbf{F}^k T^k}{(k+1)!} T\mathbf{G}$$

$$= \mathbf{\Psi}T\mathbf{G}. \tag{8.54}$$

We evaluate $\mathbf{\Psi}$ by a series in the form

$$\mathbf{\Psi} \cong \mathbf{I} + \frac{\mathbf{F}T}{2} \left\{ \mathbf{I} + \frac{\mathbf{F}T}{3} \left[\mathbf{I} + \cdots \frac{\mathbf{F}T}{N-1} \left(\mathbf{I} + \frac{\mathbf{F}T}{N} \right) \right] \right\},$$

MATLAB c2d

which has better numerical properties than the direct series. We then find $\mathbf{\Gamma}$ from Eq. (8.54) and $\mathbf{\Phi}$ from Eq. (8.53). For a discussion of various methods of numerical determination of $\mathbf{\Phi}$ and $\mathbf{\Gamma}$, see Franklin et al. (1998) and Moler and van Loan (1978). The evaluation of the $\mathbf{\Phi}$ and $\mathbf{\Gamma}$ matrices in practice is carried out by the c2d function in MATLAB.

To compare this method of representing the plant with the discrete transfer function, we can take the z-transform of Eqs. (8.51) and (8.52) with $J = 0$ to obtain

$$(z\mathbf{I} - \boldsymbol{\Phi})\mathbf{X}(z) = \boldsymbol{\Gamma}U(z), \tag{8.55}$$

$$Y(z) = \mathbf{H}\mathbf{X}(z). \tag{8.56}$$

Therefore,

$$\frac{Y(z)}{U(z)} = G(z) = \mathbf{H}(z\mathbf{I} - \boldsymbol{\Phi})^{-1}\boldsymbol{\Gamma}. \tag{8.57}$$

EXAMPLE 8.6

Discrete State-Space Representation of $1/s^2$ *Plant*

Use the relation in this section to verify that the discrete model of the $1/s^2$ plant preceded by a ZOH is that given in the solution to Example 8.4.

Solution.   The $\boldsymbol{\Phi}$ and $\boldsymbol{\Gamma}$ matrices can be calculated using Eqs. (8.53) and (8.54). Example 2.7 (with $I = 1$) showed that the values for $\mathbf{F}$ and $\mathbf{G}$ are

$$\mathbf{F} = \begin{bmatrix} 0 & 1 \\ 0 & 0 \end{bmatrix}, \qquad \mathbf{G} = \begin{bmatrix} 0 \\ 1 \end{bmatrix}.$$

Because $\mathbf{F}^2 = \mathbf{0}$ in this case, we have

$$\boldsymbol{\Phi} = \mathbf{I} + \mathbf{F}T + \frac{{}^2T^2}{2!} + \cdots$$

$$= \begin{bmatrix} 1 & 0 \\ 0 & 1 \end{bmatrix} + \begin{bmatrix} 0 & 1 \\ 0 & 0 \end{bmatrix} T = \begin{bmatrix} 1 & T \\ 0 & 1 \end{bmatrix},$$

$$\boldsymbol{\Gamma} = \left(\mathbf{I} + \mathbf{F}\frac{T}{2!}\right)T\mathbf{G}$$

$$= \left(\begin{bmatrix} T & 0 \\ 0 & T \end{bmatrix} + \begin{bmatrix} 0 & 1 \\ 0 & 0 \end{bmatrix}\frac{T^2}{2}\right)\begin{bmatrix} 0 \\ 1 \end{bmatrix} = \begin{bmatrix} T^2/2 \\ T \end{bmatrix}.$$

Hence, using Eq. (8.57), we obtain

$$G(z) = \frac{Y(z)}{U(z)} = \begin{bmatrix} 1 & 0 \end{bmatrix}\left(z\begin{bmatrix} 1 & 0 \\ 0 & 1 \end{bmatrix} - \begin{bmatrix} 1 & T \\ 0 & 1 \end{bmatrix}\right)^{-1}\begin{bmatrix} T^2/2 \\ T \end{bmatrix}$$

$$= \frac{T^2}{2}\left[\frac{z+1}{(z-1)^2}\right] \tag{8.58}$$

This is the same result we obtained using Eq. (8.33) and the z-transform tables in Example 8.4.

Note that to compute Y/U we find that the denominator of Eq. (8.58) is $\det(z\mathbf{I} - \boldsymbol{\Phi})$, which was created by the matrix inverse in Eq. (8.57). This determinant is the characteristic polynomial of the transfer function, and the zeros of the determinant are the poles of the plant. We have two poles at $z = 1$ in this case, corresponding to two integrations in this plant's equations of motion.

We can further explore the question of poles and zeros and the state-space description by considering again the transform formulas [Eqs. (8.55) and (8.56)]. One way to interpret transfer-function poles from the perspective of the corresponding difference equation is that a pole is a value of z such that the equation has a nontrivial solution when the forcing input is zero. From Eq. (8.55), this interpretation implies that the linear equations

$$(z\mathbf{I} - \boldsymbol{\Phi})\mathbf{X}(z) = \mathbf{0}$$

have a nontrivial solution. From matrix algebra the well-known requirement for a nontrivial solution is that $\det(z\mathbf{I} - \boldsymbol{\Phi}) = 0$. Using the system in Example 8.6, we get

$$\det(z\mathbf{I} - \boldsymbol{\Phi}) = \det\left(\begin{bmatrix} z & 0 \\ 0 & z \end{bmatrix} - \begin{bmatrix} 1 & T \\ 0 & 1 \end{bmatrix}\right)$$

$$= \det\begin{bmatrix} z - 1 & -T \\ 0 & z - 1 \end{bmatrix}$$

$$= (z - 1)^2 = 0,$$

which is the characteristic equation, as we have seen. In MATLAB, the poles of the system are found by P = eig(Phi).

Along the same line of reasoning, a system zero is a value of z such that the system output is zero even with a nonzero state-and-input combination. Thus, if we are able to find a nontrivial solution for $\mathbf{X}(z_0)$ and $U(z_0)$ such that $Y(z_0)$ is identically zero, then z_0 is a zero of the system. In combining Eqs. (8.55) and (8.56), we must satisfy the requirement that

$$\begin{bmatrix} z\mathbf{I} - \boldsymbol{\Phi} & -\boldsymbol{\Gamma} \\ \mathbf{H} & 0 \end{bmatrix}\begin{bmatrix} \mathbf{X}(z) \\ U(z) \end{bmatrix} = \mathbf{0}.$$

Once more the condition for the existence of nontrivial solutions is that the determinant of the square coefficient system matrix be zero. For Example 8.6, the calculation is

$$\det\begin{bmatrix} z - 1 & -T & -T^2/2 \\ 0 & z - 1 & -T \\ 1 & 0 & 0 \end{bmatrix} = \det\begin{bmatrix} -T & -T^2/2 \\ z - 1 & -T \end{bmatrix}$$

$$= T^2 + \frac{T^2}{2}(z - 1)$$

$$= \frac{T^2}{2}z + \frac{T^2}{2}$$

$$= \frac{T^2}{2}(z + 1).$$

Thus we have a single zero at $z = -1$, as we have seen from the transfer function. In MATLAB, the zeros are found by Z=tzero(Phi,Gam,H,J).

Much of the algebra for discrete state-space control design is the same as for the continuous time case discussed in Chapter 7. The poles of a discrete system can be moved to desirable locations by linear state-variable feedback:

$$u = -\mathbf{K}\mathbf{x}$$

such that

$$\det(z\mathbf{I} - \mathbf{\Phi} + \mathbf{\Gamma}\mathbf{K}) = \alpha_c(z), \tag{8.59}$$

provided the system is controllable. The system is controllable if the controllability matrix

$$C = [\mathbf{\Gamma} \quad \mathbf{\Phi}\mathbf{\Gamma} \quad \mathbf{\Phi}^2\mathbf{\Gamma} \quad \ldots \quad \mathbf{\Phi}^{n-1}\mathbf{\Gamma}]$$

is full-rank.

A discrete full-order estimator has the form

$$\bar{\mathbf{x}}(k + 1) = \mathbf{\Phi}\bar{\mathbf{x}}(k) + \mathbf{\Gamma}u(k) + \mathbf{L}[y(k) - \mathbf{H}\bar{\mathbf{x}}(k)],$$

where $\bar{\mathbf{x}}$ is the state estimate. The error equation

$$\tilde{\mathbf{x}}(k + 1) = (\mathbf{\Phi} - \mathbf{L}\mathbf{H})\tilde{\mathbf{x}}(k)$$

can be given arbitrary dynamics $\alpha_e(z)$, provided that the system is observable, which requires that the observability matrix

$$\mathcal{O} = \begin{bmatrix} \mathbf{H} \\ \mathbf{H}\mathbf{\Phi} \\ \mathbf{H}\mathbf{\Phi}^2 \\ \vdots \\ \mathbf{H}\mathbf{\Phi}^{n-1} \end{bmatrix}$$

be full-rank.

As was true for the continuous-time case, if the open-loop transfer function is

$$G(z) = \frac{Y(z)}{U(z)} = \frac{b(z)}{a(z)},$$

then a state-space compensator can be designed such that

$$\frac{Y(z)}{R(z)} = \frac{K_s\gamma(z)b(z)}{\alpha_c(z)\alpha_e(z)},$$

where r is the reference input. The polynomials $\alpha_c(z)$ and $\alpha_e(z)$ are selected by the designer using exactly the same methods discussed in Chapter 7 for continuous systems. $\alpha_c(z)$ results in a control gain $\mathbf{K}$ such that $\det(z\mathbf{I} - \mathbf{\Phi} + \mathbf{\Gamma}\mathbf{K}) = \alpha_c(z)$, and $\alpha_e(z)$ results in an estimator gain $\mathbf{L}$ such that $\det(z\mathbf{I} - \mathbf{\Phi} + \mathbf{L}\mathbf{H}) = \alpha_e(z)$. If the estimator is structured according to Fig. 7.46(a), the system zeros $\gamma(z)$ will be identical to the estimator poles $\alpha_e(z)$, thus removing the estimator response from the closed-loop system response. However, if desired, we can arbitrarily select the polynomial $\gamma(z)$ by providing suitable feedforward from the reference input. Refer to Franklin et al. (1998) for details.

EXAMPLE 8.7   *State-Space Design of a Digital Controller*

Design a digital controller for a $1/s^2$ plant to meet the specifications given in Example 8.2. Use state-space design methods including use of an estimator, and structure the reference input in two ways:

(a) Use the error command shown in Fig. 7.46(b), and

(b) use the state command shown in Fig. 7.46(a).

Solution.   We find the state-space model of the $1/s^2$ plant preceded by a ZOH using the MATLAB statements

sysSSc = ss([0 1;0 0], [0; 1], [1 0], 0);
T = 1;
sysSSd = c2d(sysSSc, T);
[Phi,Gam,H] = ssdata(sysSSd);

Using discrete analysis for Example 8.4, we find that the desired z-plane roots are at $z = 0.78 \pm 0.18j$. Solving the discrete pole-placement problem involves placing the eigenvalues of $\boldsymbol{\Phi} - \boldsymbol{\Gamma}\mathbf{K}$ as indicated by Eq. (8.59). Likewise, the solution of the continuous pole-placement problem involves placing the eigenvalues of $\mathbf{F} - \mathbf{G}\mathbf{K}$ as indicated by Eq. (7.61). Because these two tasks are identical, we use the same function in MATLAB for the continuous and discrete cases. Therefore, the control feedback matrix $\mathbf{K}$ is found by

pc = [0.78 + 0.18*j; 0.78 - 0.18*j];
K = acker(Phi,Gam,pc);

which yields

$$\mathbf{K} = [\,0.0808 \quad 0.3996\,].$$

To ensure that the estimator roots are substantially faster than the control roots (so that the estimator roots will have little effect on the output), we choose them to be at $z = 0.2 \pm 0.2j$. Therefore, the estimator feedback matrix $\mathbf{L}$ is found by

pe = [0.2 + 0.2*j; 0.2 - 0.2*j];
L = acker(Phi,' H,' pe)';

which yields

$$\mathbf{L} = \begin{bmatrix} 1.6 \\ 0.68 \end{bmatrix}.$$

The equations of the compensation for $r = 0$ (regulation to $\mathbf{x}^T = [0 \quad 0]$) are then

$$\bar{\mathbf{x}}(k+1) = \boldsymbol{\Phi}\bar{\mathbf{x}}(k) + \boldsymbol{\Gamma}u(k) + \mathbf{L}[y(k) - \mathbf{H}\bar{\mathbf{x}}(k)], \qquad (8.60)$$

$$u(k) = -\mathbf{K}\bar{\mathbf{x}}(k). \qquad (8.61)$$

(a) For the error command structure where the compensator is placed in the feedforward path, as shown in Fig. 7.46(b), $y(k)$ from Eq. (8.60) is replaced with $y(k) - r$, so the state description of the plant plus the estimator (a fourth-order system whose state vector is $[\mathbf{x}\,\bar{\mathbf{x}}]^T$) is

```
A = [Phi    - Gam*K; L*H   Phi - Gam*K - L*H];
B = [0; 0; -L];
C = [1 0 0 0];
D = 0;
step(A,B,C,D).
```

The resulting step response in Fig. 8.19 shows a response similar to that of the previous step responses in Fig. 8.18.

(b) For the state command structure described in Section 7.3.2, we wish to command the position element of the state vector so that

$$\mathbf{N}_x = \begin{bmatrix} 1 \\ 0 \end{bmatrix},$$

and the $1/s^2$ plant requires no steady control input for a constant output y, therefore $N_u = 0$. To analyze a system with this command structure, we need to modify matrix B from the MATLAB statement above to properly introduce the reference input r according to Fig. 7.46(a). The MATLAB statement

```
B = [Gam*K*Nx; Gam*K*Nx];
```

channels r into both the plant and estimator equally, thus not exciting the estimator dynamics. The resulting step response in Fig. 8.19 shows a substantial reduction in the overshoot with this structure. In fact, the overshoot is now about 5%, which is expected for a second-order system with $\zeta \cong 0.7$. The previous designs all had considerably greater overshoot because of the effect of the extra zero and pole.

Figure 8.19

Step response of Example 8.7

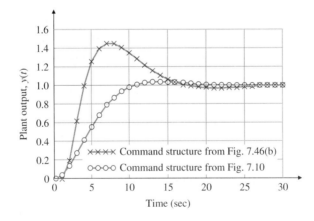

8.6   Hardware Characteristics

A digital control system includes several unique components not found in continuous control systems: An **analog-to-digital converter** is a device to sample the continuous signal voltage from the sensor and to convert that signal to a digital word, a **digital-to-analog converter** is a device to convert the digital word from the computer to an analog voltage, an **anti-alias prefilter** is an analog device designed to reduce the effects of aliasing, and the **computer** is the device where the compensation $D(z)$ is programmed and the calculations are carried out. This section provides a brief description of each of these.

8.6.1   Analog-to-Digital (A/D) Converters

As discussed in Section 8.1, A/D converters are devices that convert a voltage level from a sensor to a digital word usable by the computer. At the most basic level, all digital words are binary numbers consisting of many bits that are set to either 1 or 0. Therefore, the task of the A/D converter at each sample time is to convert a voltage level to the correct bit pattern and often to hold that pattern until the next sample time.

Of the many A/D conversion techniques that exist, the most common are based on counting schemes or a successive-approximation technique. In counting methods the input voltage may be converted to a train of pulses whose frequency is proportional to the voltage level. The pulses are then counted over a fixed period using a binary counter, thus resulting in a binary representation of the voltage level. A variation on this scheme is to start the count simultaneously with a voltage that is linear in time and to stop the count when the voltage reaches the magnitude of the input voltage to be converted.

The successive-approximation technique tends to be much faster than the counting methods. It is based on successively comparing the input voltage to reference levels representing the various bits in the digital word. The input voltage is first compared with a reference value that is half the maximum. If the input voltage is greater, the most significant bit is set, and the signal is then compared with a reference level that is three-fourths the maximum to determine the next bit, and so on. One clock cycle is required to set each bit, so an n-bit converter would require n cycles. At the same clock rate a counter-based converter might require as many as 2^n cycles, which would usually be much slower.

With either technique, the greater the number of bits, the longer it will take to perform the conversion. The price of A/D converters generally goes up with both speed and bit size. In 2001, a 12-bit (resolution of 0.025%) converter with a good performance capability of a 150-n sec conversion time sold for ~$20 while an 8 bit (0.4% resolution) with a 1-μsec conversion time sold for ~$2.

If more than one channel of data needs to be sampled and converted to digital words, it is usually accomplished by use of a multiplexer rather than by multiple A/D converters. The multiplexer sequentially connects the converter into the channel being sampled.

8.6.2   Digital-to-Analog (D/A) Converters

D/A converters, as mentioned in Section 8.1, are used to convert the digital words from the computer to a voltage level and are somtimes referred to as **Sample and Hold** devices. They provide analog outputs from a computer for driving actuators or perhaps a recording device such as an oscilloscope or strip-chart recorder. The basic idea behind their operation is that the binary bits cause switches (electronic gates) to open or close, thus routing the electric current through an appropriate network of resistors to generate the correct voltage level. Because no counting or iteration is required for such converters, they tend to be much faster than A/D converters. In fact, A/D converters that use the successive-approximation method of conversion include D/A converters as components.

8.6.3   Anti-Alias Prefilters

An analog **anti-alias prefilter** is often placed between the sensor and the A/D converter. Its function is to reduce the higher-frequency noise components in the analog signal in order to prevent aliasing—that is, having the noise be modulated to a lower frequency by the sampling process.

Analog prefilters reduce aliasing

An example of aliasing is shown in Fig. 8.20, where a 60-Hz oscillatory signal is being sampled at 50 Hz. The figure shows the result from the samples as a 10-Hz signal and also shows the mechanism by which the frequency of the signal is aliased from 60 to 10 Hz. Aliasing will occur any time the sample rate is not at least twice as fast as any of the frequencies in the signal being sampled. Therefore, to prevent aliasing of a 60-Hz signal, the sample rate would have to be faster than 120 Hz, clearly much higher than the 50-Hz rate in the figure.

Nyquist–Shannon sampling theorem

Aliasing is one of the consequences of the **sampling theorem of Nyquist and Shannon**. Their theorem basically states that, for the signal to be accurately reconstructed from the samples, it must have no frequency component greater than half the sample rate $(\omega_s/2)$. Another consequence of their theorem is that the highest frequency that can be unambiguously represented by discrete samples is the Nyquist rate of $\omega_s/2$, an idea we discussed in Section 8.2.3.

Figure 8.20

An example of aliasing

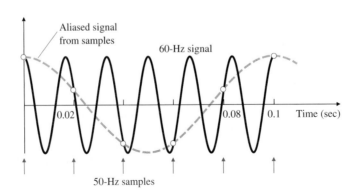

The consequence of aliasing on a digital control system can be substantial. In a continuous system, noise components with a frequency much higher than the control-system bandwidth normally have a small effect because the system will not respond at the high frequency. However, in a digital system, the frequency of the noise can potentially be aliased down in frequency to the vicinity of the system bandwidth so that the closed-loop system would respond to the noise. Thus, the noise in a poorly designed digitally controlled system could have a substantially greater effect than if the control had been implemented using analog electronics.

The solution is to place an analog prefilter before the sampler. In many cases a simple first-order low-pass filter will do; that is,

$$H_p(s) = \frac{a}{s + a},$$

where the break point a is selected to be lower than $\omega_s/2$ so that any noise present with frequencies greater than $\omega_s/2$ is attenuated by the prefilter. The lower the break-point frequency selected, the more the noise above $\omega_s/2$ is attenuated. However, too low a break point may force the designer to reduce the control system's bandwidth. The prefilter does not completely eliminate the aliasing; however, through judicious choice of the prefilter break point and the sample rate, the designer has the ability to reduce the magnitude of the aliased noise to some acceptable level.

8.6.4   The Computer

The computer is the unit that does all the computations. Most digital controllers used today are built around a microcontroller that contains both a microprocessor and most of the other functions needed, including the A/D and D/A conversion. For development purposes in a laboratory, a digital controller could be a desktop-sized workstation or a PC. The relatively low cost of microprocessor technology has accounted for the large increase in the use of digital control systems, which started in the 1980s and continues into the 2000s.

The computer consists of a central processor unit (CPU), which does the computations and provides the system logic; a clock to synchronize the system; memory modules for data and instruction storage; and a power supply to provide the various required voltages. The memory modules come in three basic varieties:

1. **Read-only memory (ROM)** is the least expensive, but after its manufacture its contents cannot be changed. Most of the memory in products manufactured in quantity is ROM. It retains its stored values when power is removed.

2. **Random-access memory (RAM)** is the most expensive, but its values can be changed by the CPU. It is only required to store the values that will be changed during the control process and typically represents only a small fraction of the total memory of a developed product. It loses the values in memory when power is removed.

3. **Programmable read-only memory (EPROM)** is a ROM whose values can be changed by a technician using a special device. It is typically used during product development to enable the designer to try different algorithms and parameter values. It retains its stored values when power is removed. In some products, it is useful to have a few of the stored quantities in EPROMs so that individual calibrations can be carried out for each unit.

Microprocessors for control applications generally come with a digital word size of 8, 16, or 32 bits, although some have been available with 12 bits. Larger word sizes give better accuracy, but at an increase in cost. The most economical solution is often to use an 8-bit microprocessor, but to use two digital words to store one value (**double precision**) in the areas of the controller that are critical to the system accuracy. Many digital control systems use computers originally designed for digital signal-processing applications, so-called DSP chips.

8.7   Word-Size Effects

A numerical value can be represented with only limited precision in a digital computer. For **fixed-point arithmetic**, the resolution is 0.4% of full range for 8 bits and 0.1% for 10 bits; resolution drops by a factor of 2 for each additional bit. The effect of this quantization shows up in A/D conversion, multiplication truncation, and parameter storage errors. If the computer uses **floating-point arithmetic**, the resolution of the multiplication and parameter storage changes with the magnitude of the number being stored; thus the resolution affects only the mantissa, while the exponent continually adjusts scale factors relating the bit size to the physical variable.

8.7.1   Random Effects

As long as a system has varying inputs or disturbances, A/D errors and multiplication errors act in a random manner on the system producing noise at the output of the system. The output noise due to a particular noise source (multiplication or A/D conversion) has a mean value of

$$\bar{n}_o = H_{DC}\bar{n}_I,$$

where

$H_{DC} =$ DC gain of transfer function between noise source and output,

$\bar{n}_I =$ mean value of noise source.

The mean value of the noise will be zero for a roundoff process, but otherwise it has the value

$$\bar{n}_I = \frac{q}{2},$$

where q is the resolution level for a truncation process. Although most A/D converters round off, producing no mean error, some truncate and thus do produce an error. The total noise effect is the sum of all noise sources.

Lyapunov equation

The variance of the output noise, σ_o, is always nonzero, irrespective of whether the process truncates or rounds off. The value of the output variance is most easily found by first solving the discrete **Lyapunov equation** for the covariance of the state,

$$\mathbf{R}_x = \mathbf{\Phi}\mathbf{R}_x\mathbf{\Phi}^T + \mathbf{\Gamma}\mathbf{\Gamma}^T\sigma_I^2, \tag{8.62}$$

and then solving

$$\sigma_o^2 = \mathbf{H}\mathbf{R}_x\mathbf{H}^T, \tag{8.63}$$

where

$$\sigma_o = \text{output noise root-mean-square (rms)},$$

$$\mathbf{R}_x = \text{state covariance matrix},$$

$$\sigma_I^2 = \text{input noise variance},$$

$$\mathbf{\Phi}, \mathbf{H} = \text{system description matrices from Section 8.5},$$

$$\mathbf{\Gamma} = \text{noise input matrix}.$$

The magnitude of the input noise variance is

$$\sigma_I^2 = \frac{q^2}{12}$$

for either roundoff or truncation.

The evaluation of Eq. (8.62) is usually done using computer tools, as described in Franklin et al. (1998). Carrying out the calculations for a system with a small-word-size computer (8 bits or less) will generally show that the multi-plication noise response of the system becomes more sensitive as the sampling rate increases and can be a design issue for very fast sample rates. However, for 16- or 32-bit computers, the onset of significant noise errors occurs at much higher sample rates, and quantization noise is typically not a factor in design.

Noise response vs. sampling rate

The sensitivity of a system to A/D errors can be alleviated by sampling faster or by adding more bits to the A/D converter. Different structures of the digital controller have no effect.

In contrast, multiplication errors can be reduced substantially for high-order (third-order and above) controllers by properly structuring a given con-trol transfer function. For example, a second-order transfer function with real roots,

$$D(z) = \frac{U(z)}{E(z)} = \frac{z - 0.8}{(z - 0.2)(z - 0.3)}, \tag{8.64}$$

Figure 8.21
An example of parallel implementation

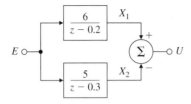

Direct implementation

can be implemented in a direct manner, yielding

$$u(k) = 0.5u(k-1) - 0.06u(k-2) + e(k) - 0.8e(k-1). \tag{8.65}$$

Equation (8.64) can also be implemented in a parallel manner by performing a partial-fraction expansion on the equation. The result, shown in Fig. 8.21, yields

Parallel implementation

$$x_1(k) = 0.2x_1(k-1) + 6e(k),$$

$$x_2(k) = 0.3x_2(k-1) + 5e(k), \tag{8.66}$$

$$u(k) = x_1(k) - x_2(k).$$

Cascade implementation

Note that the transfer functions to the output from the multiplications in Eq. (8.65) are substantially different from those in Eq. (8.66). It is also possible to implement $D(z)$ with a cascade factorization, which, for this example, would consist of two first-order blocks arranged serially.

For complex digital controllers with a small word size, either cascade or parallel implementation is preferable to the direct implementation because of the substantially reduced response of such controllers to quantization errors.

8.7.2 Systematic Effects

Parameters such as the numerical values in Eqs. (8.65) and (8.66), if in error, will change the dynamic behavior of a system. In a high-order controller with a small word size and a direct implementation, a very small percentage error in a stored parameter can result in substantial root-location changes and can sometimes cause instability. In the Apollo command module, a 14-bit word size for parameter storage would have resulted in instability if a direct implementation had been used in the sixth-order compensator. These effects are amplified when there are two compensator poles close together (or repeated), or when there are fast sample rates, in which case all poles tend to clump around $z = +1$ and are close to one another. These effects can be reduced by using larger word sizes, parallel or cascade implementations, double-precision parameter storage, and/or slower sample rates.

Under conditions of constant disturbances and input commands, multiplication errors can also cause systematic errors. Typically, the result is a steady-state error or possibly a limit cycle. The steady-state error results from the dead band created by the quantization. These effects are also reduced by larger word sizes, parallel or cascade implementations, and slower sample rates. For more detail, see Franklin et al. (1998).

8.8   Sample-Rate Selection

The selection of the best sample rate for a digital control system is the result of a compromise of many factors. Sampling too fast can cause a loss of accuracy, while the basic motivation for lowering the sample rate ω_s is cost. A decrease in sample rate means more time is available for the control calculations; hence slower computers can be used for a given control function or more control capability can be achieved from a given computer. Either way, the cost per function is lowered. For systems with A/D converters, less demand on conversion speed will also lower cost. These economic arguments indicate that the best engineering choice is the slowest-possible sample rate that still meets all performance specifications.

There are several factors that could provide a lower limit on the acceptable sample rate:

1. Tracking effectiveness as measured by closed-loop bandwidth or by time-response requirements, such as rise time and settling time.

2. Regulation effectiveness as measured by the error response to random plant disturbances.

3. Error due to measurement noise and the associated prefilter design methods.

A fictitious limit occurs when using emulation design techniques. The inherent approximation in the method may give rise to system instabilities as the sample rate is lowered. This can lead the designer to conclude that a lower limit on ω_s has been reached when, in fact, the proper conclusion is that the approximations are invalid; the solution is not to sample faster but to refine the design with a direct digital-design method.

The ease of designing digital control systems with fast sample rates and the low cost of very capable computers often drives the designer to select a sample rate that is $40 \times \omega_{BW}$ or higher. For computers with fixed-point arithmetic, this can lead to multiplication errors that can produce significant offsets or limit cycles in the control, as discussed in Section 8.7.2.

8.8.1   Tracking Effectiveness

An absolute lower bound on the sample rate is set by a specification to track a command input with a certain frequency (the system bandwidth). The sampling theorem (see Section 8.6.3 and Franklin et al., 1998) states that in order to reconstruct an unknown, band-limited, continuous signal from samples of that signal, we must sample at least twice as fast as the highest frequency contained in the signal. Therefore, in order for a closed-loop system to track an input at a certain frequency, it must have a sample rate twice as fast; that is, ω_s must be at least twice the system bandwidth ($\omega_s \Rightarrow 2 \times \omega_{BW}$). We also saw from the results of mapping the s-plane into the z-plane ($z = e^{sT}$) that the highest frequency that can be represented by a discrete system is $\omega_s/2$, which supports the conclusion of the theorem.

It is important to note the distinction between the closed-loop bandwidth ω_{BW} and the highest frequency in the open-loop plant dynamics, because the two frequencies can be quite different. For example, closed-loop bandwidths can be an order of magnitude *less* than open-loop modes of resonances for some control problems. Information concerning the state of the plant resonances for purposes of control can be extracted from sampling the output without satisfying the sampling theorem because some a priori knowledge concerning these dynamics (albeit imprecise) is available, and the system is not required to track these frequencies. Thus a priori knowledge of the dynamic model of the plant can be included in the compensation in the form of a notch filter.

The closed-loop-bandwidth limitation provides the fundamental lower bound on the sample rate. In practice, however, the theoretical lower bound of sampling at twice the bandwidth of the reference input signal would not be judged sufficient in terms of the quality of the desired time responses. For a system with a rise time on the order of 1 sec (thus yielding a closed-loop bandwidth on the order of 0.5 Hz), it is reasonable to insist on a sampling rate of 10 to 20 Hz, which is a factor of 20 to 40 times ω_{BW}. The purposes of choosing a sample rate much greater than the bandwidth are to reduce the delay between a command and the system response to the command and also to smooth the system output to the control steps coming out of the ZOH.

8.8.2   Disturbance Rejection

Disturbance rejection is an important—if not the most important—aspect of any control system. Disturbances enter a system with various frequency characteristics ranging from steps to white noise. For the purpose of sample-rate selection, the higher-frequency random disturbances are the most influential.

The ability of the control system to reject disturbances with a good continuous controller represents the lower bound on the error response that we can hope for when implementing the controller digitally. In fact, some degradation relative to the continuous design must occur because the sampled values are slightly out of date at all times except precisely at the sampling instants. However, if the sample rate is very fast compared with the frequencies contained in the noisy disturbance, we should expect no appreciable loss from the digital system as compared with the continuous controller. At the other extreme, if the sample rate is very slow compared with the characteristic frequencies of the noise, the response of the system because of noise is essentially the same as the response we would get if the system had no control at all. The selection of a sample rate will place the response somewhere in between these two extremes. Thus the impact of the sample rate on the ability of the system to reject disturbances may be very important to consider when choosing the sample rate.

Although the best choice of sample rate in terms of the ω_{BW} multiple is dependent on the frequency characteristics of the noise and the degree to which random disturbance rejection is important to the quality of the controller, sample rates on the order of 20 times ω_{BW} or higher are typical.

8.8.3   Effect of Anti-alias Prefilter

Digital control systems with analog sensors typically include an analog anti-alias prefilter between the sensor and the sampler as described in Section 8.6.3. The prefilters are low-pass, and the simplest transfer function is

$$H_p(s) = \frac{a}{s+a}$$

so that the noise above the prefilter break point a is attenuated. The goal is to provide enough attenuation at half the sample rate $(\omega_s/2)$ so that the noise above $\omega_s/2$, when aliased into lower frequencies by the sampler will not be detrimental to control system performance.

A conservative design procedure is to select ω_s and the break point to be sufficiently higher than the system bandwidth so that the phase lag from the prefilter does not significantly alter the system stability. This would allow the prefilter to be ignored in the basic control system design. Furthermore, for a good reduction in the high-frequency noise at $\omega_s/2$, we choose a sample rate that is about 5 or 10 times higher than the prefilter break point. The implication of this prefilter design procedure is that sample rates need to be on the order of 30 to 100 times faster than the system bandwidth. Using this conservative design procedure, the prefilter influence will likely provide the lower bound on the selection of the sample rate.

An alternative design procedure is to allow significant phase lag from the prefilter at the system bandwidth. This requires us to include the analog prefilter characteristics in the plant model when carrying out the control design. It allows the use of lower sample rates, but at the possible expense of increased complexity in the compensation because additional phase lead must be provided to counteract the prefilter's phase lag. If this procedure is used and low prefilter break points are allowed, the effect of sample rate on sensor noise is small and the prefilter essentially has no effect on the sample rate.

It may seem counterintuitive that placing a lag (the analog prefilter) in one portion of the controller and a counteracting lead (extra lead in $D(z)$) in another portion of the controller provides a net positive effect on the overall system. The net gain is a result of the fact that the lag is in the analog part of the system where high frequencies can exist. The counteracting lead is in the digital part of the system where frequencies above the Nyquist rate do not exist. The result is a reduction in the high frequencies before the sampling which are not reamplified by the counteracting digital lead, thus producing net reduction in high frequencies. Furthermore, these high frequencies are particularly insiduous with a digital controller because of the aliasing that would result from the sampling.

8.8.4 Asynchronous Sampling

As noted in the previous paragraphs, divorcing the prefilter design from the control-law design may require using a faster sample rate than otherwise. This same result may show up in other types of architecture. For example, a smart sensor with its own computer running asynchronously relative to the primary control computer will not be amenable to direct digital design because the overall system transfer function depends on the phasing between the smart sensor and the primary digital controller. This situation is similar to that of the digitization errors discussed in Section 8.4. Therefore, if asynchronous digital subsystems are present, sample rates on the order of $20 \times \omega_{BW}$ or slower in any module should be used with caution and the system performance checked through simulation or experiment.

SUMMARY

- There are two basic design techniques for finding compensation equations for implementation in a digital computer: emulation and discrete design.

- **Emulation design** entails (a) finding the continuous compensation $D(s)$ using the ideas in Chapters 1 to 7, and (b) approximating $D(s)$ with difference equations using Euler's method, Tustin's method, or the matched-pole-zero method.

- **Discrete design** entails (a) finding the discrete model of the plant $G(s)$ and (b) using the discrete model to design the compensation directly in its discrete form.

- The z-**transform** is the primary tool used to determine the behavior of discrete linear systems. The z-transform of a time sequence $f(k)$ is given by

$$\mathcal{Z}\{f(k)\} = F(z) = \sum_{k=0}^{\infty} f(k) z^{-k}$$

and has the key property that

$$\mathcal{Z}\{f(k-1)\} = z^{-1} F(z).$$

This property allows us to find the discrete transfer function of a difference equation. Analysis using z-transforms closely parallels that using Laplace transforms.

- Normally z-transforms are found using Table 8.1 or by computer.

- The discrete Final Value Theorem is

$$\lim_{k \to \infty} x(k) = \lim_{z \to 1} (1 - z^{-1}) X(z),$$

provided that all poles of $(1 - z^{-1}) X(z)$ are inside the unit circle.

- For a continuous signal $f(t)$ whose samples are $f(k)$, the poles of $F(s)$ are related to the poles of $F(z)$ by

$$z = e^{sT}.$$

- The most common emulation methods are:

  1. **Tustin's approximation**:

  $$D(z) = D(s)\big|_{s = \frac{2}{T} \left(\frac{z-1}{z+1} \right)}$$

  2. The **matched-pole-zero approximation**:
     (a) Map poles and zeros by $z = e^{sT}$.
     (b) Add powers of $z + 1$ to the numerator until numerator and denominator are of equal order or numerator is one order less than the denominator.
     (c) Set the low-frequency gain of $D(z)$ equal to that of $D(s)$.

- The discrete model of the continuous plant $G(s)$ preceded by a **ZOH** is

$$G(z) = (1 - z^{-1}) \mathcal{Z} \left\{ \frac{G(s)}{s} \right\}.$$

- Discrete design using $G(z)$ closely parallels continuous design, but the stability boundary and interpretation of z-plane root locations is different. Figure 8.5 summarizes the response characteristics.

- The continuous state-space form of a differential equation,

$$\dot{\mathbf{x}} = \mathbf{F}\mathbf{x} + \mathbf{G}u,$$

$$y = \mathbf{H}\mathbf{x} + Ju,$$

has a discrete counterpart in the difference equations

$$\mathbf{x}(k + 1) = \boldsymbol{\Phi}\mathbf{x}(k) + \boldsymbol{\Gamma}u(k),$$

$$y(k) = \mathbf{H}x(k) + Ju(k),$$

where

$$\boldsymbol{\Phi} = e^{\mathbf{F}T},$$

$$\boldsymbol{\Gamma} = \left(\int_0^T e^{\mathbf{F}\eta} \, d\eta \right) \mathbf{G}.$$

These matrices can be computed in MATLAB by c2d.

- The pole placement and estimation ideas are identical in the continuous and discrete domains.

- If designing by emulation, a minimum sample rate of 20 times the bandwidth is recommended. If using discrete design, system stability can be assured when sampling at a rate as slow as two times the bandwidth. However, to reject random disturbances, best results are obtained by sampling at 10 times the closed-loop bandwidth or faster.

- Analog **prefilters** are commonly placed before the **sampler** in order to attenuate the effects of high-frequency measurement noise. A sampler **aliases** all frequencies in the signal that are greater than half the sample frequency to lower frequencies; therefore, prefilter break points should be selected so that no significant frequency content remains above half the sample rate.

Review Questions

1. What is the Nyquist rate? What are its characteristics?

2. Describe the emulation process.

3. Describe how to arrive at a $D(z)$ if the sample rate is $30 \times \omega_{BW}$.

4. Describe how to arrive at a $D(z)$ if the sample rate is $5 \times \omega_{BW}$.

5. For a system with a 1 rad/sec bandwidth, describe the consequences of various sample rates.

6. How can the effects of quantization in an A/D converter be studied?

7. How can the effects of parameter quantization in the controller be minimized?

8. Give two advantages for selecting a digital processor rather than analog circuitry to implement a controller.

9. Give two disadvantages for selecting a digital processor rather than analog circuitry to implement a controller.

Problems

8.1.   The z-transform of a discrete-time filter $h(k)$ at a $1Hz$ sample rate is

$$H(z) = \frac{1 + (1/2)z^{-1}}{[1 - (1/2)z^{-1}][1 + (1/3)z^{-1}]}.$$

   (a) Let $u(k)$ and $y(k)$ be the discrete input and output of this filter. Find a difference equation relating $u(k)$ and $y(k)$.

   (b) Find the natural frequency and damping coefficient of the filter's poles

   (c) Is the filter stable?

8.2.   Use the z-transform to solve the difference equation

$$y(k) - 3y(k-1) + 2y(k-2) = 2u(k-1) - 2u(k-2),$$

where

$$u(k) = \begin{cases} k, & k \geq 0, \\ 0, & k < 0, \end{cases}$$

$$y(k) = 0, \qquad k > 0.$$

8.3.   The one-sided z-transform is defined as

$$F(z) = \sum_{0}^{\infty} f(k)z^{-k}.$$

(a) Show that the one-sided transform of $f(k+1)$ is $\mathcal{Z}\{f(k+1)\} = zF(z) - zf(0)$.

(b) Use the one-sided transform to solve for the transforms of the Fibonacci numbers generated by the difference equation $u(k+2) = u(k+1) + u(k)$. Let $u(0) = u(1) = 1$. [*Hint:* You will need to find a general expression for the transform of $f(k+2)$ in terms of the transform of $f(k)$].

(c) Compute the pole locations of the transform of the Fibonacci numbers.

(d) Compute the inverse transform of the Fibonacci numbers.

(e) Show that if $u(k)$ represents the kth Fibonacci number, then the ratio $u(k+1)/u(k)$ will approach $(1 + \sqrt{5})/2$. This is the golden ratio valued so highly by the Greeks.

8.4.   A unity feedback system has an open-loop transfer function given by

$$G(s) = \frac{250}{s[(s/10) + 1]}.$$

The following lag compensator added in series with the plant yields a phase margin of $50°$:

$$D(s) = \frac{s/1.25 + 1}{50s + 1}.$$

Using the matched pole-zero approximation, determine an equivalent digital realization of this compensator.

8.5.   The following transfer function is a lead network designed to add about $60°$ of phase at $\omega_1 = 3$ rad/sec:

$$H(s) = \frac{s + 1}{0.1s + 1}.$$

(a) Assume a sampling period of $T = 0.25$ sec, and compute and plot in the z-plane the pole and zero locations of the digital implementations of $H(s)$ obtained using (1) Tustin's method and (2) pole-zero mapping. For each case, compute the amount of phase lead provided by the network at $z_1 = e^{j\omega_1 T}$.

(b) Using a log–log scale for the frequency range $\omega = 0.1$ to $\omega = 100$ rad/sec, plot the magnitude Bode plots for each of the equivalent digital systems you found in part (a), and compare with $H(s)$. (*Hint:* Magnitude Bode plots are given by $|H(z)| = |H(e^{j\omega T})|$.)

8.6. The following transfer function is a lag network designed to introduce a gain attenuation of $10(-20 \text{ db})$ at $\omega = 3$ rad/sec:

$$H(s) = \frac{10s + 1}{100s + 1}.$$

(a) Assume a sampling period of $T = 0.25$ sec, and compute and plot in the z-plane the pole and zero locations of the digital implementations of $H(s)$ obtained using (1) Tustin's method and (2) pole-zero mapping. For each case, compute the amount of gain attenuation provided by the network at $z_1 = e^{j\omega_1 T}$.

(b) For each of the equivalent digital systems in part (a), plot the Bode magnitude curves over the frequency range $\omega = 0.01$ to 10 rad/sec.

8.7. Consider the linear equation $\mathbf{Ax} = \mathbf{b}$, where $\mathbf{A}$ is an $n \times n$ matrix. When $\mathbf{b}$ is given, one way of solving for $\mathbf{x}$ is to use the discrete-time recursion

$$\mathbf{x}(k + 1) = (\mathbf{I} + c\mathbf{A})\mathbf{x}(k) - c\mathbf{b},$$

where c is a scalar to be chosen.

(a) Show that the solution of $\mathbf{Ax} = \mathbf{b}$ is the equilibrium point $\mathbf{x}^*$ of the discrete-time system. An equilibrium point $\mathbf{x}^*$ of a discrete-time system $\mathbf{x}(k + 1) = \mathbf{f}(\mathbf{x}(k))$ satisfies the relation $\mathbf{x}^* = \mathbf{f}(\mathbf{x}^*)$.

(b) Consider the error $\mathbf{e}(k) = \mathbf{x}(k) - \mathbf{x}^*$. Write the linear equation that relates the error $\mathbf{e}(k + 1)$ to $\mathbf{e}(k)$.

(c) Suppose $|1 + c\lambda_i(\mathbf{A})| < 1, i = 1, \ldots, n$, where $\lambda_i(\mathbf{A})$ denotes the ith eigenvalue of $\mathbf{A}$. Show that starting from any initial guess $\mathbf{x}_0$, the algorithm converges to $\mathbf{x}^*$. [*Hint:* For any matrix $\mathbf{B}$, $\lambda_i(\mathbf{I} + \mathbf{B}) = 1 + \lambda_i(\mathbf{B})$.]

8.8. The open-loop plant of a unity feedback system has the transfer function

$$G(s) = \frac{1}{s(s + 2)}.$$

Determine the transfer function of the equivalent digital plant using a sampling period of $T = 1$ sec, and design a proportional controller for the discrete-time system that yields dominant closed-loop poles with a damping ratio ζ of 0.7.

8.9. Consider the system configuration shown in Fig. 8.22, where

$$G(s) = \frac{40(s + 2)}{(s + 10)(s^2 - 1.4)}.$$

Figure 8.22
Control system for
Problem 8.9

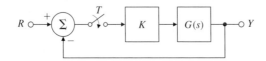

(a) Find the transfer function $G(z)$ for $T = 1$ assuming the system is preceded by a ZOH.

(b) Use MATLAB to draw the root locus of the system with respect to K.

(c) What is the range of K for which the closed-loop system is stable?

(d) Compare your results of part (c) to the case where an analog controller is used (that is, where the sampling switch is always closed). Which system has a larger allowable value of K?

(e) Use MATLAB to compute the step response of both the continuous and discrete systems with K chosen to yield a damping factor of $\zeta = 0.5$ for the continuous case.

8.10. Write a computer program to compute $\boldsymbol{\Phi}$ and $\boldsymbol{\Gamma}$ from $\mathbf{F}, \mathbf{G}$, and the sample period T. OK to use MATLAB, but don't use C2D; write code in MATLAB to compute the discrete matrices using the relations developed in this chapter. Use your program to compute $\boldsymbol{\Phi}$ and $\boldsymbol{\Gamma}$ when

(a) $\mathbf{F} = \begin{bmatrix} -1 & 0 \\ 0 & -2 \end{bmatrix}, \quad \mathbf{G} = \begin{bmatrix} 1 \\ 1 \end{bmatrix}, \quad T = 0.2 \text{ sec},$

(b) $\mathbf{F} = \begin{bmatrix} -3 & -2 \\ 1 & 0 \end{bmatrix}, \quad \mathbf{G} = \begin{bmatrix} 1 \\ 0 \end{bmatrix}, \quad T = 0.2 \text{ sec}.$

8.11. Consider the following discrete-time system in state-space form:

$$\begin{bmatrix} x_1(k+1) \\ x_2(k+1) \end{bmatrix} = \begin{bmatrix} 0 & 1 \\ 0 & -1 \end{bmatrix} \begin{bmatrix} x_1(k) \\ x_2(k) \end{bmatrix} + \begin{bmatrix} 0 \\ 10 \end{bmatrix} u(k).$$

Use state feedback to relocate all of the system's poles to 0.5.

8.12. For

$$\boldsymbol{\Phi} = \begin{bmatrix} 1 & T \\ 0 & 1 \end{bmatrix} \quad \text{and} \quad \boldsymbol{\Gamma} = \begin{bmatrix} T^2/2 \\ T \end{bmatrix},$$

(a) Find a transformation matrix $\mathbf{T}$ so that, if $\mathbf{x} = \mathbf{Tw}$, the state equations for $\mathbf{w}$ will be in control canonical form.

(b) Compute the gain $\mathbf{K}_w$ so that if $u = -\mathbf{K}_w\mathbf{w}$, the characteristic equation will be $\alpha_c(z) = z^2 - 1.6z + 0.7$.

(c) Use $\mathbf{T}$ from part (a) to compute $\mathbf{K}_x$, the feedback gain required by the state equations in $\mathbf{x}$ to achieve the desired characteristic polynomial.

8.13. Consider a system whose plant transfer function is $1/s^2$ and which has a piecewise constant input of the form

$$u(t) = u(kT), \qquad kT \leq t < (k+1)T.$$

(a) Show that if we restrict attention to the time instants kT, $k = 0, 1, 2, \ldots$, the resulting sampled-data system can be described by the equations

$$\begin{bmatrix} x_1(k+1) \\ x_2(k+1) \end{bmatrix} = \begin{bmatrix} 0 & 1 \\ 0 & -1 \end{bmatrix} \begin{bmatrix} x_1(k) \\ x_2(k) \end{bmatrix} + \begin{bmatrix} 0 \\ 10 \end{bmatrix} u(k).$$

$$y(k) = [0 \quad 1][x_1(k) \, x_2(k)]^T.$$

(b) Design a second-order estimator that will always drive the error in the estimate of the initial state vector to zero in time $2T$ or less.

(c) Is it possible to estimate the initial state exactly with a first-order estimator? Justify your answer.

8.14. *Single-Axis Satellite Attitude Control:* Satellites often require attitude control for proper orientation of antennas and sensors with respect to Earth. Figure 2.6 shows a communication satellite with a three-axis attitude-control system. To gain insight into the three-axis problem we often consider one axis at a time. Figure 8.23 depicts this case where motion is only allowed about an axis perpendicular to the page. The equations of motion of the system are given by

$$I\ddot{\theta} = M_C + M_D,$$

where

I = moment of inertia of the satellite about its mass center,

M_C = control torque applied by the thrusters,

M_D = disturbance torques,

θ = angle of the satellite axis with respect to an inertial reference with no angular acceleration.

We normalize the equations of motion by defining

$$u = \frac{M_C}{I}, \qquad w_d = \frac{M_D}{I},$$

and obtain

$$\ddot{\theta} = u + w_d.$$

Taking the Laplace transform yields

$$\theta(s) = \frac{1}{s^2}[u(s) + w_d(s)],$$

which with no disturbance becomes

$$\frac{\theta(s)}{u(s)} = \frac{1}{s^2} = G_1(s).$$

In the discrete case where u is applied through a ZOH, we can use the methods described in this chapter to obtain the discrete transfer function

$$G_1(z) = \frac{\theta(z)}{u(z)} = \frac{T^2}{2}\left[\frac{z+1}{(z-1)^2}\right].$$

Figure 8.23

Satellite control schematic for Problem 8.14

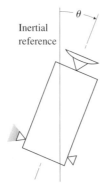

Inertial reference θ

(a) Sketch the root locus of this system by hand assuming proportional control.

(b) Draw the root locus using MATLAB to verify the hand sketch.

(c) Add a lead network to your controller so that the dominant poles correspond to $\zeta = 0.5$ and $\omega_n = 3\pi/(10T)$.

(d) What is the feedback gain if $T = 1\,\text{sec}$? If $T = 2\,\text{sec}$?

(e) Plot the closed-loop step response and the associated control time history for $T = 1\,\text{sec}$.

8.15. In this problem you will show how to compute Φ by changing states so that the system matrix is diagonal.

(a) Using an infinite series expansion, compute $e^{\mathbf{A}T}$ for

$$\mathbf{A} = \begin{bmatrix} -1 & 0 \\ 0 & -2 \end{bmatrix}.$$

(b) Show that if $\mathbf{F} = \mathbf{T}\mathbf{A}\mathbf{T}^{-1}$ for some nonsingular transformation matrix $\mathbf{T}$, then

$$e^{\mathbf{F}T} = \mathbf{T}e^{\mathbf{A}T}\mathbf{T}^{-1}.$$

(c) Show that if

$$\mathbf{F} = \begin{bmatrix} -3 & 1 \\ -2 & 0 \end{bmatrix},$$

there exists a $\mathbf{T}$ such that $\mathbf{T}\mathbf{A}\mathbf{T}^{-1} = \mathbf{F}$. (*Hint*: Write $\mathbf{T}\mathbf{A} = \mathbf{F}\mathbf{T}$, assume four unknowns for the elements of $\mathbf{T}$, and solve. Next show that the columns of $\mathbf{T}$ are the eigenvectors of $\mathbf{F}$.)

(d) Compute $e^{\mathbf{F}T}$.

8.16. It is possible to suspend a mass of magnetic material by means of an electromagnet whose current is controlled by the position of the mass (Woodson and Melcher, 1968). The schematic of a possible setup is shown in Fig. 8.24, and a photo of a working system at Stanford University is shown in Fig. 2.35. The equations of motion are

$$m\ddot{x} = -mg + f(x, I),$$

where the force on the ball due to the electromagnet is given by $f(x, I)$. At equilibrium the magnet force balances the gravity force. Suppose we let I_0 represent the current at equilibrium. If we write $I = I_0 + i$, expand f about $x = 0$ and $I = I_0$, and neglect higher-order terms, we obtain the linearized equation

$$m\ddot{x} = k_1 x + k_2 i. \tag{8.67}$$

Reasonable values for the constants in Eq. (8.67) are $m = 0.02\,\text{kg}$, $k_1 = 20\,\text{N/m}$, and $k_2 = 0.4\,\text{N/A}$.

Figure 8.24

Schematic of magnetic
levitation device for
Problems 8.16 and 8.17

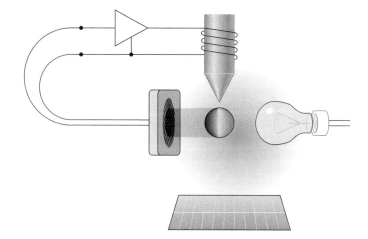

(a) Compute the transfer function from I to x, and draw the (continuous) root locus for the simple feedback $i = -Kx$.

(b) Assume the input is passed through a ZOH, and let the sampling period be 0.02 sec. Compute the transfer function of the equivalent discrete-time plant.

(c) Design a digital control for the magnetic levitation device so that the closed-loop system meets the following specifications: $t_r \leq 0.1$ sec, $t_s \leq 0.4$ sec, and overshoot $\leq 20\%$.

(d) Plot a root locus with respect to k_1 for your design, and discuss the possibility of using your closed-loop system to balance balls of various masses.

(e) Plot the step response of your design to an initial disturbance displacement on the ball, and show both x and the control current i. If the sensor can measure x only over a range of $\pm 1/4$cm and the amplifier can only provide a current of 1 A, what is the *maximum* displacement possible for control, neglecting the nonlinear terms in $f(x, I)$?

8.17. In Problem 8.16 we described an experiment in magnetic levitation described by Eq. (8.67), which reduces to

$$\ddot{x} = 1000x + 20i.$$

Let the sampling time be 0.01 sec.

(a) Use pole placement to design a controller for the magnetic levitator so that the closed-loop system meets the following specifications: settling time, $t_s \leq$ 0.25 sec, and overshoot to an initial offset in x that is less than 20%.

(b) Plot the step response of $x, \tilde{x}$, and i to an initial displacement in x.

(c) Plot the root locus for changes in the plant gain, and mark the pole locations of your design.

(d) Introduce a command reference input r (as discussed in Section 7.3.2) that does not excite the estimate of x. Measure or compute the frequency response from r to the system error $r - x$ and give the highest frequency for which the error amplitude is less than 20% of the command amplitude.

8.18. *Servomechanism for Antenna Elevation Control:* Suppose it is desired to control the elevation of an antenna designed to track a satellite. A photo of such a system is shown in Fig. 8.25, and a schematic diagram is depicted in Fig. 8.26. The antenna and drive parts have a moment of inertia J and damping B, arising to some extent from bearing and aerodynamic friction, but mostly from the back emf of the DC drive motor. The equation of motion is

$$J\ddot{\theta} + B\dot{\theta} = T_c + T_d,$$

where

$$T_c = \text{net torque from the drive motor,} \tag{8.68}$$

$$T_d = \text{disturbance torque due to wind.} \tag{8.69}$$

Figure 8.25
Satellite-tracking antenna *(Courtesy Space Systems/Loral)*

Figure 8.26
Schematic diagram of satellite-tracking antenna

If we define

$$\frac{B}{J} = a, \qquad u = \frac{T_c}{B}, \qquad w_d = \frac{T_d}{B},$$

the equation simplifies to

$$\frac{1}{a}\ddot{\theta} + \dot{\theta} = u + w_d.$$

After Laplace transformation, we obtain

$$\theta(s) = \frac{1}{s(s/a + 1)}[u(s) + w_d(s)],$$

or, with no disturbance,

$$\frac{\theta(s)}{u(s)} = \frac{1}{s(s/a + 1)} = G_2(s).$$

With $u(k)$ applied through a ZOH, the transfer function for an equivalent discrete-time system is

$$G_2(z) = \frac{\theta(z)}{u(z)} = K\frac{z + b}{(z - 1)(z - e^{-aT})},$$

where

$$K = \frac{aT - 1 + e^{-aT}}{a}, \qquad b = \frac{1 - e^{-aT} - aTe^{-aT}}{aT - 1 + e^{-aT}}.$$

(a) Let $a = 0.1$ and $x_1 = \dot{\theta}$, and write the continuous-time state equations for the system.

(b) Let $T = 1$ sec, and find a state feedback gain $\mathbf{K}$ for the equivalent discrete-time system that yields closed-loop poles corresponding to the following points in the s-plane:

$$s = -1/2 \pm j\left(\sqrt{\frac{3}{2}}\right).$$

Plot the step response of the resulting design.

(c) Design an estimator: Select $\mathbf{L}$ so that $\alpha_e(z) = z^2$.

(d) Using the values for $\mathbf{K}$ and $\mathbf{L}$ computed in parts (b) and (c) as the gains for a combined estimator/controller, introduce a reference input that will leave the state estimate undisturbed. Plot the response of the closed-loop system due to a step change in the reference input. Also plot the system response to a step wind-gust disturbance.

(e) Plot the root locus of the closed-loop system with respect to the plant gain, and mark the locations of the closed-loop poles.

8.19. *Tank Fluid Temperature Control*: The temperature of a tank of fluid with a constant inflow and outflow rate is to be controlled by adjusting the temperature of the incoming fluid. The temperature of the incoming fluid is controlled by a mixing valve that adjusts the relative amounts of hot and cold supplies of the fluid (see Fig. 8.27). The distance between the valve and the point of discharge into the tank creates a time delay between the application of a temperature change at the mixing valve and the discharge of the flow with the changed temperature into the tank. The differential equation governing the tank temperature is

$$\dot{T}_e = \frac{1}{cM}(q_{in} - q_{out}),$$

where

$$T_e = \text{tank temperature,}$$

$$c = \text{specific heat of the fluid,}$$

$$M = \text{fluid mass contained in the tank,}$$

$$q_{in} = c\dot{m}_{in}T_{ei},$$

$$q_{out} = c\dot{m}_{out}T_e,$$

$$\dot{m} = \text{mass flow rate}(\dot{m}_{in} = \dot{m}_{out}),$$

$$T_{ei} = \text{temperature of fluid entering tank.}$$

However, the temperature at the input to the tank at time t is equal to the control temperature τ_d seconds in the past. This relationship may be expressed as

$$T_{ei}(t) = T_{ec}(t - \tau_d),$$

where

$$\tau_d = \text{delay time,}$$

$$T_{ec} = \text{temperature of fluid immediately after the control valve and directly controllable by the valve.}$$

Combining constants, we obtain

$$\dot{T}_e(t) + aT_e(t) = aT_{ec}(t - \tau_d),$$

Figure 8.27
Tank temperature control

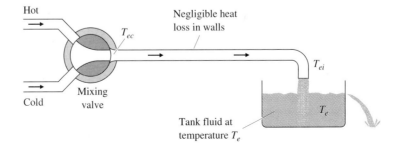

Hot

Negligible heat
loss in walls

T_{ec}

T_{ei}

Cold

Mixing
valve

Tank fluid at
temperature T_e

T_e

where

$$a = \frac{\dot{m}}{M}.$$

The transfer function of the system is thus

$$\frac{T_e(s)}{T_{ec}(s)} = \frac{e^{-\tau_d s}}{s/a + 1} = G_3(s).$$

To form a discrete transfer function equivalent to G_3 preceded by a ZOH, we must compute

$$G_3(z) = \mathcal{Z}\left\{\left(\frac{1 - e^{-sT}}{s}\right)\left(\frac{e^{-\tau_d s}}{s/a + 1}\right)\right\}.$$

We assume that for some integer l, $\tau_d = lT - mT$, where $0 \le m < 1$. Then

$$G_3(z) = \mathcal{Z}\left\{\left(\frac{1 - e^{-sT}}{s}\right)\left(\frac{e^{-lsT} e^{msT}}{s/a + 1}\right)\right\}$$

$$= (1 - z^{-1})z^{-l}\mathcal{Z}\left\{\frac{e^{msT}}{s(s/a + 1)}\right\}$$

$$= (1 - z^{-1})z^{-l}\mathcal{Z}\left\{\frac{e^{msT}}{s} - \frac{e^{msT}}{s + a}\right\}$$

$$= \frac{z - 1}{z}\left(\frac{1}{z^l}\right)\mathcal{Z}\{1(t + mT) - e^{-a(t + mT)}1(t + mT)\}$$

$$= \frac{z - 1}{z}\left(\frac{1}{z^l}\right)\left(\frac{z}{z - 1} - \frac{e^{-amT}z}{z - e^{-aT}}\right)$$

$$= \frac{1}{z^l}\left[\frac{(1 - e^{-amT})z + e^{-amT} - e^{-aT}}{z - e^{-aT}}\right]$$

$$= \left(\frac{1 - e^{-amT}}{z^l}\right)\left(\frac{z + \alpha}{z - e^{-aT}}\right);$$

and

$$\alpha = \frac{e^{-amT} - e^{-aT}}{1 - e^{-amT}}.$$

The zero location $-\alpha$ varies from $\alpha = \infty$ at $m = 0$ to $\alpha = 0$ as $m \to 1$. Note also that $G_3(1) = 1.0$ for all a, m, and l. For the specific values $\tau_d = 1.5$, $T = 1$, $a = 1$, $l = 2$, and $m = \frac{1}{2}$, the transfer function reduces to

$$G_3(z) = 0.3935\frac{z + 0.6065}{z^2(z - 0.3679)}.$$

(a) Write the discrete-time system equations in state-space form.
(b) Design a state feedback gain that yields $\alpha_c(z) = z^3$.
(c) Design a state estimator with $\alpha_e(z) = z^3$.
(d) Plot the root locus of the system with respect to the plant gain.
(e) Plot the step response of the system.

8.20. Prove the seven properties of the s-plane-to-z-plane mapping listed in Section 8.2.3.

8.21. For the system shown in Fig. 8.28, find values for K, T_D, and T_I so that the closed-loop poles satisfy $\zeta > 0.5$ and $\omega_n > 1$ rad/sec. Discretize the PID controller using

(a) Tustin's method

(b) Matched pole-zero method

Use MATLAB to simulate the step response of each of these digital implementations for sample times of $T = 1, 0.1$, and 0.01 sec.

Figure 8.28
Control system for
Problem 8.21

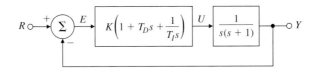

8.22. Repeat Problem 5.29 by constructing discrete root loci and performing the designs directly in the z-plane. Assume that the output y is sampled, the input u is passed through a ZOH as it enters the plant, and the sample rate is 15 Hz.

8.23. Design a digital controller for the antenna servo system shown in Figs. 3.63 and 3.64 and described in Problem 3.30. The design should provide a step response with an overshoot of less than 10% and a rise time of less than 80 sec.

(a) What should the sample rate be?

(b) Use emulation design with the matched pole-zero method.

(c) Use discrete design and the z-plane root locus.

8.24. The system

$$G(s) = \frac{1}{(s + 0.1)(s + 3)}$$

is to be controlled with a digital controller having a sampling period of $T = 0.1$ sec. Using a z-plane root locus, design compensation that will respond to a step with a rise time $t_r \le 1$ sec and an overshoot $M_p \le 5\%$. What can be done to reduce the steady-state error?

8.25. The transfer function for pure derivative control is

$$D(z) = K T_D \frac{z - 1}{T z},$$

where the pole at $z = 0$ adds some destabilizing phase lag. Can this phase lag be removed by using derivative control of the form

$$D(z) = K T_D \frac{(z - 1)}{T}?$$

Support your answer with the difference equation that would be required, and discuss the requirements to implement it.

9 Control-System Design: Principles and Case Studies

Chapter Overview

Section 9.1 opens the chapter with a step-by-step design process that is sufficiently general to apply to any control design process, but which also provides useful definitions and directions. We then apply the design process to six practical, complex applications: design of an attitude control system for a satellite (Section 9.2); lateral and longitudinal control of a Boeing 747 (Section 9.3); control of the fuel–air ratio in an automotive engine (Section 9.4); control of a digital tape transport (Section 9.5); control of a disk drive (Section 9.6); and control of a rapid thermal processing system (Section 9.7).

A Perspective on Design Principles

In Chapters 5, 6, and 7 we presented techniques for analyzing and designing feedback systems based on the root-locus, frequency-response, and state-variable methods. Thus far we have had to consider somewhat isolated, idealized aspects of larger systems and to focus on applying one analysis method at a time. In this chapter we return to the theme of Chapter 4 — the advantages of feedback control — to reconsider the overall problem of control systems design with the sophisticated tools developed in Chapters 5 to 7 in hand. We will apply these tools to several complex, real-world applications in a case-study-type format.

Having an overarching, step-by-step design approach serves two purposes: It provides a useful starting point for any real-world controls problem, and it provides meaningful checkpoints once the design process is underway. This chapter develops just such a general approach, which will be applied in the case studies.

9.1   An Outline of Control Systems Design

Control engineering is an important part of the design process of many dynamic systems. As suggested in Chapter 4, the deliberate use of feedback can stabilize an otherwise unstable system, reduce the error due to disturbance inputs, reduce the tracking error while following a command input, and reduce the sensitivity of a closed-loop transfer function to small variations in internal system parameters. In those situations where feedback control is required, it is possible to outline an approach to control systems design that often leads to a satisfactory solution.

Before describing this approach, we wish to emphasize that the purpose of control is to aid the product or process—the mechanism, the robot, the chemical plant, the aircraft, or whatever—to do its job. Engineers engaged in other areas of the design process are increasingly taking the contribution of control into account early in their plans. As a result, more and more systems have been designed so that they will not work at all without feedback. This is especially significant in the design of high-performance aircraft, where control has taken its place along with structures and aerodynamics as essential to assuring that the craft will even fly at all. It is impossible to give a description of such overall design in this book, but recognizing the existence of such cases places in perspective not only the specific task of control system design but also the central role this task can play in an enterprise.

Control system design begins with a proposed product or process whose satisfactory dynamic performance depends on feedback for stability, disturbance regulation, tracking accuracy, or reduction of the effects of parameter variations. We will give an outline of the design process that is general enough to be useful whether the product is an electronic amplifier or a large structure to be placed in earth orbit. Obviously, to be so widely applicable, our outline has to be vague with respect to physical details and specific only with respect to the feedback-control problem. To present our results, we will divide the control design problem into a sequence of characteristic steps.

Specifications

STEP 1. *Understand the process and translate dynamic performance requirements into time, frequency, or pole-zero specifications.* The importance of understanding the process, what it is intended to do, how much system error is permissible, how to describe the class of command and disturbance signals to be expected, and what the physical capabilities and limitations are can hardly be overemphasized. Regrettably, in a book such as this, it is easy to view the process as a linear, time-invariant transfer function capable of responding to inputs of arbitrary size, and we tend to overlook the fact that the linear model is a *very* limited representation of the real system, valid only for small signals, short times, and particular environmental conditions. Do not confuse the approximation with the reality. You must be able to (a) use the simplified model for its intended purpose and (b) to return to an accurate model or the actual physical system to really verify the design performance.

Typical results of this step are specifications that the system have a step response inside some constraint boundaries [as shown in Fig. 9.1(a)], an open-loop frequency response satisfying certain constraints [Fig. 9.1(b)], or closed-loop poles to the left of some constraint boundary [Fig. 9.1(c)].

Sensors

STEP 2. *Select sensors.* In **sensor** selection, consider which variables are important to control and which variables can physically be measured. For example, in a jet engine there are critical internal temperatures that must be controlled but that cannot be measured directly in an operational engine. Select sensors that indirectly allow a good estimate to be made of these critical variables. It is important to consider sensors for the disturbance. Sometimes, especially in chemical processes, it is beneficial to sense a load disturbance directly because improved performance can be obtained if this information is fed forward to the controller.

Factors that influence sensor selection are as follows:

Number of sensors and locations:	Select minimum required number of sensors and their optimal locations
Technology:	Electric or magnetic, mechanical, electromechanical, electrooptical, piezoelectric
Functional performance:	Linearity, bias, accuracy, bandwidth, resolution, dynamic range, noise
Physical properties:	Weight, size, strength
Quality factors:	Reliability, durability, maintainability
Cost:	Expense, availability, facilities for testing, and maintenance

Actuators

STEP 3. *Select actuators.* In order to control a dynamic system, obviously you must be able to influence the response. The device that does this is the **actuator**. Before choosing a specific actuator, consider which variables can be influenced. For example, in a flight vehicle many configurations of movable surfaces are possible, and the influence these have on the performance and controllability

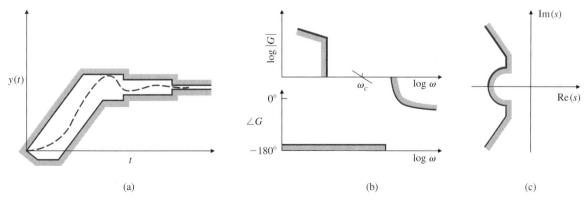

(a)   (b)   (c)

Figure 9.1 Examples of (a) time-response, (b) frequency-response, and (c) pole-zero specifications resulting from Step 1

of the craft can be profound. The locations of jets or other torque devices are also a major part of the control design of spacecraft.

Having selected a particular variable to control, you may need to consider other factors:

Number of actuators and locations:	Select required actuators and their optimal locations
Technology:	Electric, hydraulic, pneumatic, thermal, other
Functional performance:	Maximum force possible, extent of the linear range, maximum speed possible, power, efficiency
Physical properties:	Weight, size, strength
Quality factors:	Reliability, durability, maintainability
Cost:	Expense, availability, facilities for testing, and maintenance

Linearization

STEP 4. *Make a linear model.* Here you take the best choice for process, actuator, and sensor; identify the equilibrium point of interest; and construct a small-signal dynamic model valid over the range of frequencies included in the specifications of Step 1. You should also validate the model with experimental data where possible. To be able to make use of all the available tools, express the model in state-variable and pole-zero form as well as in frequency-response form. As we have seen, MATLAB and other computer-aided control systems design software packages have the means to perform the transformations among these forms. Simplify and reduce the order of the model if necessary. Quantify model uncertainty.

Simple compensation
PID/lead–lag design

STEP 5. *Try a simple proportional-integral-derivative (PID) or lead–lag design.* To form an initial estimate of the complexity of the design problem, sketch a frequency response (Bode plot) and a root locus with respect to plant gain. If the plant–actuator–sensor model is stable and minimum-phase, the Bode plot will probably be the most useful; otherwise, the root locus shows very important information with respect to behavior in the right half-plane (RHP). In any case, try to meet the specifications with a simple controller of the lead–lag variety, including integral control if steady-state error response requires it. Do not overlook feedforward of the disturbances if the necessary sensor information is available. Consider the effect of sensor noise, and compare a lead network to a direct sensor of velocity to see which gives a better design.

STEP 6. *Evaluate/modify plant.* Based on the simple control design, evaluate the source of the undesirable characteristics of the system performance. Reevaluate the specifications, the physical configuration of the process, and the actuator and sensor selections in the light of the preliminary design, and return to Step 1 if improvement seems necessary or feasible. For example, in many motion-control problems, after testing the first-pass design, you might find vibrational modes that prevent the design from meeting the initial specifications of the problem. It may be much easier to meet the specifications by altering the structure of the plant through the addition of stiffening members or by passive damping than to meet them by control strategies alone. An alternative solution may be to move a sensor so it is at a node of a vibration mode, thus providing no feedback of the motion. Also, some actuator technologies (such as hydraulic) have many more low-frequency vibrations than others (such as electric) do and changing the actuator technology may be indicated. In a digital implementation, it may be possible to revise the sensor–controller–actuator system structure so as to reduce time delay, which is always a destabilizing element. In thermal systems, it is often possible to change heat capacity or conductivities by material substitution that will enhance the control design. It is important to consider all parts of the design, not only the control logic, to meet the specifications in the most cost-effective way. If the plant is modified, go back to Step 1. If the design now seems satisfactory, go to Step 8; otherwise try Step 7.

Optimal design

STEP 7. *Try an optimal design.* If the trial-and-error compensators do not give entirely satisfactory performance, consider a design based on optimal control. The symmetric root locus will show possible root locations from which to select locations for the control poles that meet the response specifications; you can select locations for the estimator poles that represent a compromise between sensor and process noise. Plot the corresponding open-loop frequency response and the root locus to evaluate the stability margins of this design and its robustness to parameter changes. You can modify the pole locations until a best compromise results. Returning to the symmetric root locus with different cost measures is often a part of this step, or computations via the direct functions lqr and lqe can be used. Another variation on optimal control is to propose a fixed structure controller with unknown parameters, formulate a performance cost function, and use parameter optimization to find a good set of parameter values.

Compare the optimal design yielding the most satisfactory frequency response with the transform-method design you derived in Step 5. Select the better of the two before proceeding to Step 8.

Simulation

STEP 8. *Build a computer model, and compute (simulate) the performance of the design.*[1] After reaching the best compromise among process modification, actuator and sensor selection, and controller design choice, run a computer model of the system. This model should include important nonlinearities such as actuator saturation, realistic noise sources, and parameter variations you expect to find during operation of the system. The simulation will often identify sensitivities that may lead to going back to Step 5 or even Step 2. Design iterations should continue until the simulation confirms acceptable stability

---

[1] Extensive computer software is available to assist in this difficult but critical step. Simulink (from The Mathworks), SystemBuild (Wind River Systems, Inc.), 20-sim (Controllab Products, BV), Dymola (Dynasim, AB), SIMNON (SSPA), ACSL (AEgis Research), DSL (IBM), and EASY5 (Boeing Computer Services) are just some of these packages.

and robustness. As part of this simulation you can often include parameter optimization, in which the computer tunes the free parameters for best performance. In the early stages of design, the model you simulate will be relatively simple; as the design progresses, you will study more complete and detailed models. At this step it is also possible to compute a digital equivalent of the analog controller as described in Chapters 4 and 8. Some refinement of the controller parameters may be required to account for the effects of digitization. This allows the final design to be implemented with digital processor logic.

   If the results of the simulation prove the design satisfactory, go to Step 9; otherwise return to Step 1.

STEP 9. *Build a prototype*. As the final test before production, it is common to build and test a prototype. At this point you verify the quality of the model, discover unsuspected vibration and other modes, and consider ways to improve the design. Implement the controller using an embedded software/hardware. Tune the controller if necessary. After these tests, you may want to reconsider the sensor, actuator, and process and return to Step 1—unless time, money, or ideas have run out.

This outline is an approximation of good practice; other engineers will have variations on these themes. In some cases you may wish to carry out the steps in a different order, to omit a step, or to add one. The stages of simulation and prototype construction vary widely, depending on the nature of the system. For systems where a prototype is difficult to test and rework (for example, a satellite) or where a failure is dangerous (for example, active stabilization of a high-speed centrifuge or landing a human on the moon), most of the design verification is done through simulation of some sort. It may take the form of a digital numerical simulation, a laboratory-scale model, or a full-size laboratory model with a simulated environment. For systems that are easy to build and modify (for example, feedback control for an automotive fuel system), the simulation step is often skipped entirely; design verification and refinement are instead accomplished by working with prototypes.

   One of the issues raised above (Step 6) was the important consideration for **changing the plant itself**. In many cases, proper plant modifications can provide additional damping or increase in stiffness, change in mode shapes, reduction of system response to disturbances, reduction of Coulomb friction, change in thermal capacity or conductivity, and so on. It is worth elaborating on this by way of specific examples from the authors' experiences. In a semiconductor wafer processing example, the edge ring holding the wafer was identified as a limiting factor in closed-loop control. Modifying the thickness of the edge ring and using a different coating material reduced the heat losses and, together with relocating one of the temperature sensors closer to the edge ring, resulted in significant improvement in control performance. In another application, thin film processing, simply changing the order of the two incoming flows resulted in significant improvement in the mixing of the precursor and oxidizer materials and led to improvement in uniformity of the film. In an application on physical vapor deposition using RF plasma, the shape of the target was modified to be curved to counter the geometry effects of the chamber and yielded substantial improvements in deposition uniformity. As the last example, in a hydraulic spindle control problem adding oil temperature control with ceramic insulation and a temperature sink for the bell housing resulted in several orders of magnitude reduction in disturbances not achievable by feedback control alone.[2] One can also mention aerospace applications where the control was an afterthought, and the feedback control problem became exceedingly difficult and resulted in poor closed-loop performance. The moral of this discussion is that one must not forget the option of modifying the plant itself to make the control problem easier and provide maximum closed-loop performance.

The usual approach of designing the system and "throwing it over the fence" to the control group has proved to be inefficient and flawed. A better approach that is gaining momentum is to get the control engineer involved from the onset of a project to provide early feedback on how hard it is to control the system. The control engineer can provide valuable feedback on choice of actuators and sensors and even suggest modifications to the plant. It is often much more efficient to change the plant design while it is on the drawing board before "any metal has been bent." Closed-loop performance studies can then be performed on a simple model of the system early on.

Implicit in the process of design is the well-known fact that designs within a given category often draw on experience gained from earlier models. Thus good designs evolve rather than appear in their best form after the first pass. We will illustrate the method with several cases (Sections 9.2 to 9.7). For easy reference, we summarize the steps here.

Summary of Control Design Steps

1. Understand the process and its performance requirements.

2. Select the types and number of sensors considering location, technology, and noise.

3. Select the types and number of actuators considering location, technology, noise and power.

4. Make a linear model of the process, actuator, and sensor.

5. Make a simple trial design based on the concepts of lead-lag compensation or PID control. If satisfied, go to Step 8.

6. Consider modifying the plant itself for improved closed-loop control.

7. Make a trial pole-placement design based on optimal control or other criteria.

---

[2] Our colleague Professor Daniel DeBra strongly believes in considering modyfying the plant itself an an option for improved control. He cites this particular application to make the point. Of course, we agree with him!

8. Simulate the design, including the effects of nonlinearities, noise, and parameter variations. If the performance is not satisfactory, return to Step 1 and repeat. Consider modifying the plant itself for improved closed-loop control.

9. Build a prototype and test it. If not satisfied, return to Step 1 and repeat.

9.2   Design of a Satellite's Attitude Control

Our first example, taken from the space program, is suggested by the need to control the pointing direction, or attitude, of a satellite in orbit about the Earth. We will go through each step in our design outline and touch on some of the factors that might be considered for the control of such a system.

STEP 1. *Understand the process and its performance specifications.* A satellite is sketched in Fig. 9.2. We imagine that the vehicle has an astronomical survey mission requiring accurate pointing of a scientific sensor package. This package must be maintained in the quietest possible environment, which entails isolating it from the vibrations and electrical noise of the main service body and from its power supplies, thrusters, and communication gear. We model the resulting structure as two masses connected by a flexible boom. In Fig. 9.2(a), the satellite attitude, θ_2, is the angle between the star sensor and the instrument package, and θ_1 is the angle of the main satellite with respect to the star. Figure 9.2(b) shows the equivalent mechanical system diagram for the satellite where the sensor is mounted to the disk associated with θ_2. Disturbance torques due to solar pressure, micrometeorites, and orbit perturbations are computed to be negligible. The pointing requirement arises when it is necessary to point the unit in another direction, it can be met by dynamics with a transient settling time of 20 sec and an overshoot of no more than 15%. The dynamics of the

Figure 9.2
Diagram of a satellite and its two-body model

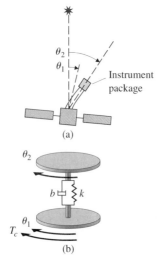

satellite include parameters that can vary. The control must be satisfactory for any parameter values in a prespecified range to be given when the equations are written.

STEP 2. *Select sensors.* In order to orient the scientific package, it is necessary to measure the attitude angles of the package. For this purpose we propose to use a **star tracker**, a system based on gathering an image of a specific star and keeping it centered on the focal plane of a telescope. This sensor gives a relatively noisy but very accurate (on the average) reading proportional to θ_2, the angle of deviation of the instrument package from the desired angle. To stabilize the control, we include a rate gyro to give a clean reading of $\dot{\theta}_2$, because a lead network on the star-tracker signal would amplify the noise too much. Furthermore, the rate gyro can stabilize large motions before the star tracker has acquired the target star image.

STEP 3. *Select actuators.* Major considerations in selecting the actuator are precision, reliability, weight, power requirements, and lifetime. Alternatives for applying torque are cold-gas jets, reaction wheels or gyros, magnetic torquers, and a gravity gradient. The jets have the most power and are the least accurate. Reaction wheels are precise but can only transfer momentum, so jets or magnetic torquers are required to "dump" momentum from time to time. Magnetic torquers provide relatively low levels of torque and are only suitable for some low-altitude satellite missions. A gravity gradient also provides a very small torque that limits the speed of response and places severe restrictions on the shape of the satellite. For purposes of this mission, we select cold-gas jets as being fast and adequately accurate.

STEP 4. *Make a linear model.* For the satellite we assume two masses connected by a spring with torque constant k and viscous-damping constant b as shown in Fig. 9.2. The equations of motion are

$$J_1\ddot{\theta}_1 + b(\dot{\theta}_1 - \dot{\theta}_2) + k(\theta_1 - \theta_2) = T_c, \tag{9.1a}$$

$$J_2\ddot{\theta}_2 + b(\dot{\theta}_2 - \dot{\theta}_1) + k(\theta_2 - \theta_1) = 0, \tag{9.1b}$$

where T_c is the control torque on the main body. With inertias $J_1 = 1$ and $J_2 = 0.1$, the transfer function is

$$G(s) = \frac{10bs + 10k}{s^2(s^2 + 11bs + 11k)}. \tag{9.2}$$

If we choose

$$\mathbf{x} = [\,\theta_2 \quad \dot{\theta}_2 \quad \theta_1 \quad \dot{\theta}_1\,]^T$$

as the state vector, then, using Eqs. (9.1a) and (9.1b) and assuming $T_c \equiv u$, the equations of motion in state-variable form are

$$\dot{\mathbf{x}} = \begin{bmatrix} 0 & 1 & 0 & 0 \\ -\dfrac{k}{J_2} & -\dfrac{b}{J_2} & \dfrac{k}{J_2} & \dfrac{b}{J_2} \\ 0 & 0 & 0 & 1 \\ \dfrac{k}{J_1} & \dfrac{b}{J_1} & -\dfrac{k}{J_1} & -\dfrac{b}{J_1} \end{bmatrix}\mathbf{x} + \begin{bmatrix} 0 \\ 0 \\ 0 \\ \dfrac{1}{J_1} \end{bmatrix}u, \tag{9.3a}$$

$$y = [\,1 \quad 0 \quad 0 \quad 0\,]\mathbf{x}. \tag{9.3b}$$

Physical analysis of the boom leads us to assume that the parameters k and b vary as a result of temperature fluctuations but are bounded by

$$0.09 \leq k \leq 0.4, \tag{9.4a}$$

$$0.038\sqrt{\frac{k}{10}} \leq b \leq 0.2\sqrt{\frac{k}{10}}. \tag{9.4b}$$

As a result, the vehicle's natural resonance frequency ω_n can vary between 1 and 2 rad/sec, and the damping ratio ζ varies between 0.02 and 0.1.

Selecting nominal values for
varying parameters

One approach to control design when parameters are subject to variation is to select nominal values for the parameters, construct the design for this model, and then test the controller performance with other parameter values. In the present case we choose nominal values of $\omega_n = 1$ and $\zeta = 0.02$. The choice is somewhat arbitrary, being based on experience and heuristic analysis. However, note that these are the lowest values in their respective ranges and thus correspond to the plant that is probably the most difficult to control so as to meet the specifications. We assume that a design for this model has a good chance to meet the specifications for other parameter values as well. (Another choice would be to select a model with average values for each parameter). The selected parameter values are $k = 0.091$ and $b = 0.0036$; with $J_1 = 1$ and $J_2 = 0.1$, the nominal equations become

$$\dot{\mathbf{x}} = \begin{bmatrix} 0 & 1 & 0 & 0 \\ -0.91 & -0.036 & 0.91 & 0.036 \\ 0 & 0 & 0 & 1 \\ 0.091 & 0.0036 & -0.091 & -0.0036 \end{bmatrix} \mathbf{x} + \begin{bmatrix} 0 \\ 0 \\ 0 \\ 1 \end{bmatrix} u, \tag{9.5a}$$

$$y = \begin{bmatrix} 1 & 0 & 0 & 0 \end{bmatrix} \mathbf{x}. \tag{9.5b}$$

The corresponding transfer function using the MATLAB ss2tf function is then

$$G(s) = \frac{0.036(s + 25)}{s^2(s^2 + 0.04s + 1)}. \tag{9.6}$$

When a trial design is completed, the computer simulation should be run with a range of possible parameter values to ensure that the design has adequate robustness to tolerate these changes. Equations (3.52)–(3.54) tell us that the dynamic performance specifications will be met if the closed-loop poles have a natural frequency of 0.5 rad/sec and a closed-loop damping ratio of 0.5; these correspond to an open-loop crossover frequency of $\omega_c \cong 0.5$ rad/sec and a phase margin of about PM $= 50°$. We will try to meet these design criteria.

STEP 5. *Try a lead–lag or PID controller.* The proportional-gain root locus for the nominal plant is drawn in Fig. 9.3, and the Bode plot is given in Fig. 9.4. We can see from Fig. 9.4 that this may be a difficult design problem because the frequency of the lightly damped resonance is greater than the crossover-frequency design point by only a factor of 2. This situation will require that the compensation can correct for the phase lag of the plant at the resonance frequency. Such correction is very dependent on knowing the resonance frequency, which is subject to change in this case. There may be trouble ahead.

Figure 9.3
Root locus of $KG(s)$

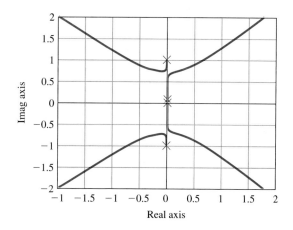

Figure 9.4
Open-loop Bode plot of $KG(s)$ for $K = 0.5$

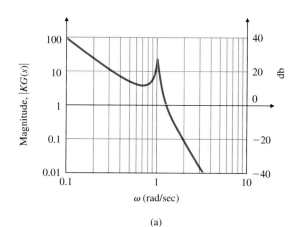

(a)

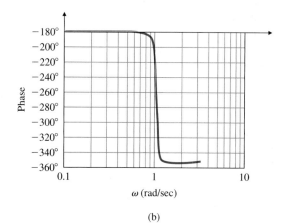

(b)

In order to illustrate some important aspects of compensation design, we will at first ignore the resonance and generate a design that would be acceptable for the rigid body alone. We take the process transfer function to be $1/s^2$, the feedback to be position plus derivative (star tracker plus rate gyro) or PD control with the transfer function $D(s) = K(sT_D + 1)$, and the response objective to be $\omega_n = 0.5$ rad/sec and $\zeta = 0.5$. A suitable controller would be

$$D_1(s) = 0.25(2s + 1). \tag{9.7}$$

The root locus for the actual plant with D_1 is shown in Fig. 9.5 and the Bode plot in Fig. 9.6. From these plots we can see that the low-frequency poles are reasonable but that the system will be unstable because of the resonance.[3] At this point we take the simple actions of reducing our expectation with respect to bandwidth and of slowing the system down by lowering the gain until the system is stable. With so little damping, we must really go slowly. A bit of experimentation leads to

$$D_2(s) = 0.001(30s + 1), \tag{9.8}$$

for which the root locus is drawn in Fig. 9.7 and the Bode plot given in Fig. 9.8. The Bode plot shows we have a phase margin of 50° but a crossover frequency of only $\omega_c = 0.04$ rad/sec. While this is too low to meet the settling-time specification, a low crossover frequency is unavoidable if we expect to keep the gain at the resonance frequency below unity so that it is gain stabilized.

Figure 9.5
Root locus of
$K D_1(s)G(s)$

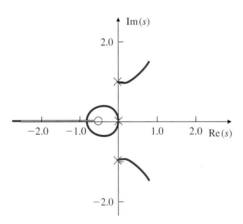

_____

[3] If this system were built, the actuator jets would saturate as the response grew. We could analyze the response using the method described in Section 5.7.3 for nonlinear systems. From the analysis we would expect the signal to grow and the equivalent gain of the actuator to fall until the roots return to the imaginary axis near ω_n. The resulting limit cycle would rapidly deplete the control gas supply.

Figure 9.6
Bode plot of $KD_1(s)G(s)$

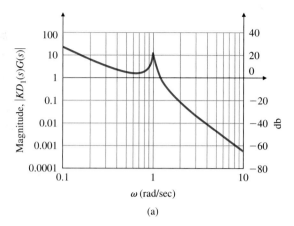

(a)

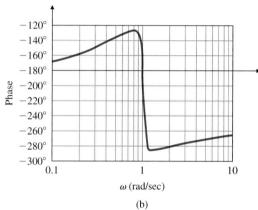

(b)

Figure 9.7
Root locus of
$KD_2(s)G(s)$

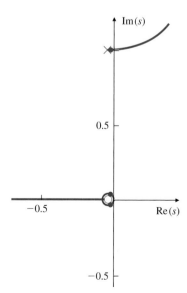

Figure 9.8

Bode plot of $D_2(s)G(s)$

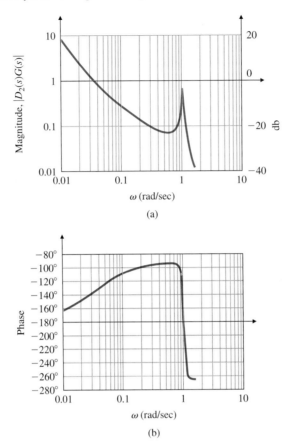

(a)

(b)

An alternative approach to the problem is to place zeros near the lightly damped poles and use them to hold these poles back from the RHP. Such a compensation has a frequency response with a very low gain near the frequency of the offending poles and a reasonable gain elsewhere. Because the frequency response seems to have a dent or notch in it, the device is called a **notch filter**. (It is also called a **band reject filter** in electric network theory.) An RC circuit with a notch characteristic is shown in Fig. 9.9, its pole-zero pattern in Fig. 9.10,

Notch filter

Figure 9.9

Twin-tee realization of a notch filter

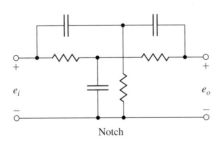

Notch

Figure 9.10

Notch filter pole-zero pattern

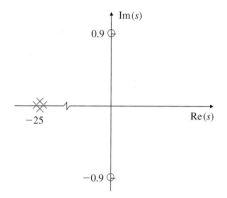

and its frequency response in Fig. 9.11. The $+180°$ phase lead of the notch can be used to correct for the $180°$ phase lag of the resonance; if the notch frequency is *lower* than the plant's resonance frequency, the system phase is kept above $180°$ near resonance.

Figure 9.11

Bode plot of a notch filter

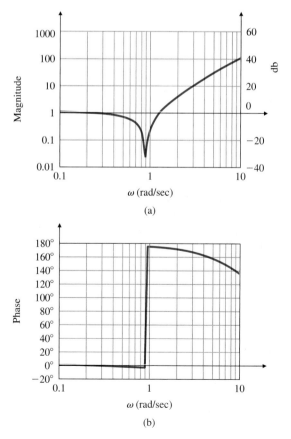

With this idea we return to the compensation given by Eq. (9.7) and add the notch, producing the revised compensator transfer function,

$$D_3(s) = 0.25(2s + 1)\frac{(s/0.9)^2 + 1}{[(s/25) + 1]^2}. \tag{9.9}$$

The Bode plot for this case is shown in Fig. 9.12, the root locus in Fig. 9.13, and the unit step response in Fig. 9.14. The settling time of the design is too long for the specification and the overshoot is too large, but this design approach seems promising; with iteration it could lead to a satisfactory compensator.

We now recall that the compensator is expected to provide adequate performance as the parameters vary over the ranges given by Eqs. (9.4). An examination of the robustness of the design can be made by looking at the root locus shown in Fig. 9.15, which is drawn using the compensator of Eq. (9.9) and the plant with $\omega_n = 2$ rather than 1, such that

$$\hat{G}(s) = \frac{(s/50 + 1)}{s^2(s^2/4 + 0.02s + 1)}. \tag{9.10}$$

This assumes that the boom is as stiff as possible. Notice that now the low-frequency poles have a damping ratio of only 0.02. Combining the various parameter values we get the frequency response and transient response shown in Figs. 9.16 and 9.17. We could make a few more trial-and-error iterations with the notch filter and rate feedback, but the system is complex enough that a look at state-space designs now seems reasonable. We go to Step 7.

STEP 6. *Evaluate/modify plant.* Refer to the collocated control discussion after STEP 8.

STEP 7. *Try an optimal design using pole placement.* Using the state-variable formulation of the equations of motion in Eq. (9.5), we devise a controller that will place the closed-loop poles in arbitrary locations. Of course, used without thought, the method of pole placement can also result in a design that requires unreasonable levels of control effort or is very sensitive to changes in the plant transfer function. Guidelines for pole placement are given in Chapter 7; an often successful approach is to derive optimal pole locations using the symmetric root locus (SRL). Figure 9.18 shows the SRL for the problem at hand. To obtain a bandwidth of about 0.5 rad/sec, we select closed-loop control poles from this locus at $-0.45 \pm 0.34j$ and $-0.15 \pm 1.05j$.

If we select $\alpha_c(s)$, as discussed earlier, from the SRL, the control gain using the MATLAB function place is

$$\mathbf{K} = [-0.2788 \quad 0.0546 \quad 0.6814 \quad 1.1655]. \tag{9.11}$$

Figure 9.19 shows the step responses for the nominal-plant-parameters and stiff-spring-plant models. The Bode plot of the SRL controller design with the nominal plant parameters can be computed from the loop transfer function (by breaking the loop at u)

$$\frac{\mathbf{K}\mathbf{X}(s)}{U(s)} = \mathbf{K}(s\mathbf{I} - \mathbf{F})^{-1}\mathbf{G}$$

Figure 9.12
Bode plot of $KD_3(s)G(s)$

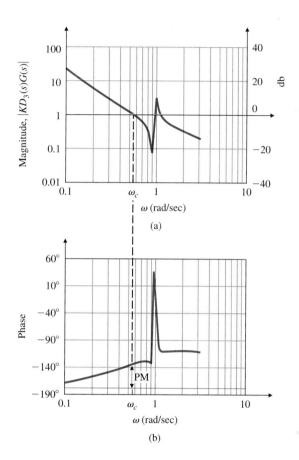

(a)

(b)

Figure 9.13
Root locus of
$KD_3(s)G(s)$

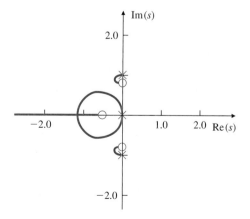

Figure 9.14

Closed-loop step response of $D_3(s)G(s)$ where $\theta_2(0) = 0.2$ rad

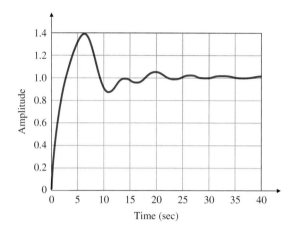

Figure 9.15

Root locus of $K D_3(s)\hat{G}(s)$

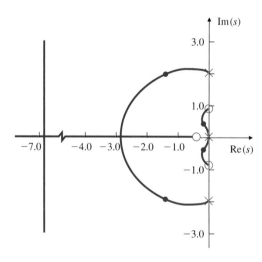

Figure 9.16
Bode plot of $KD_3(s)\hat{G}(s)$

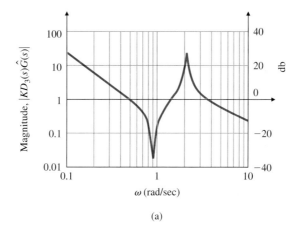

(a)

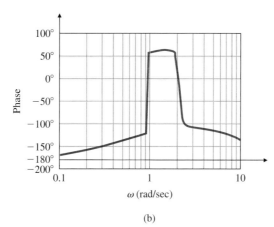

(b)

Figure 9.17
Closed-loop step response
of $D_3(s)\hat{G}(s)$

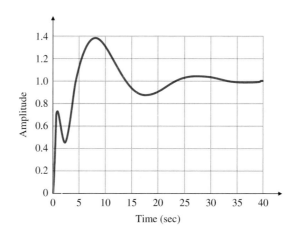

Figure 9.18

Symmetric root locus of the satellite

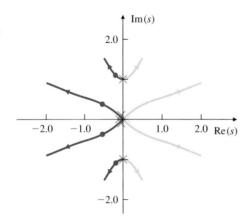

Figure 9.19

Closed-loop step response of the SRL design

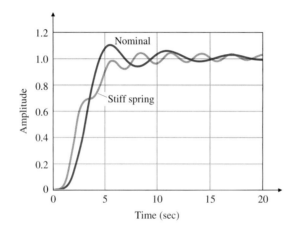

and results in a phase margin of about 60° as shown in Fig. 9.20. While the speed of response of the design meets the specifications with the nominal plant, the settling time when the plant has the stiff spring is a bit longer than the specifications call for. We might be able to get a better compromise between the nominal and the stiff-spring cases by selecting another point on the SRL; at this point we do not know. The designer must face alternatives such as these and select the best compromise for the problem at hand.

The design of Fig. 9.19 is based on full-state feedback. To complete the optimal design we need an estimator. We select the closed-loop estimator error poles to be about eight times faster than the control poles. The reason for this is to keep the error poles from reducing the robustness of the design; a fast estimator will have almost the same effect on the response as no estimator at all. We choose the error poles from the SRL at $-7.7 \pm 3.12j$ and $-3.32 \pm 7.85j$.

Figure 9.20
Frequency response of the
SRL design from u to $\mathbf{Kx}$

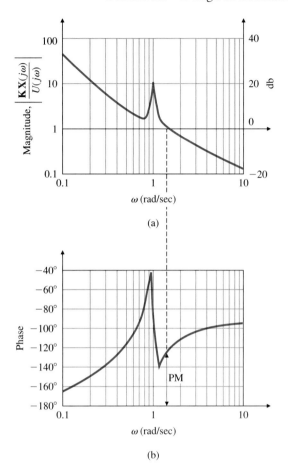

(a)

(b)

Pole placement with these values leads to an estimator (filter) gain, using the MATLAB function place,

$$\mathbf{L} = \begin{bmatrix} 22 \\ 242.3 \\ 1515.4 \\ 5503.9 \end{bmatrix}. \tag{9.12}$$

After we combine the control gain and estimator as described in Section 7.6, the compensator transfer function that results from Eq. (7.166) is

$$D_4(s) = \frac{-745(s + 0.3217)(s + 0.0996 \pm 0.9137j)}{(s + 3.1195 \pm 8.3438j)(s + 8.4905 \pm 3.6333j)}. \tag{9.13}$$

The frequency response of this compensator (Fig. 9.21) shows that pole placement has introduced a notch directly. The frequency response and the root locus of the combined system $D_4(s)G(s)$ are given in Figs. 9.22 and 9.23, while Fig. 9.24 shows the step response for both the nominal and the stiff-spring plants. Notice that the design almost meets the specifications.

Figure 9.21
Bode plot of optimal
compensator $D_4(s)$

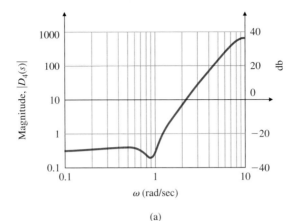

(a)

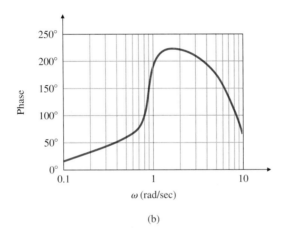

(b)

STEP 8. *Simulate the design, and compare the alternatives.* At this point we
have two designs, with differing complexities and different robustness properties. The notch-filter design might be improved with further iterations or by
starting with a different nominal case. The SRL design meets the specifications for the nominal plant but is too slow for the stiff-spring case, although
alternative selections for the pole locations might lead to a better design. In
either case, much more extensive studies need to be made to explore the robustness and noise-response properties. Rather than follow any of these paths,
we consider some aspects of the physical system.

Both designs are strongly influenced by the presence of the lightly damped
resonant mode caused by the coupled masses. However, the transfer function
of this system is strongly dependent on the fact that the actuator is on one body
and the sensor is on the other (that is, not collocated). Suppose that, rather than
considering pointing the star tracker on the small mass, we have the mission
of pointing the main mass, perhaps toward a Earth station for communications
purposes. For this purpose we can put the sensor on the *same* mass that holds

Figure 9.22
Bode plot of the compensated system $D_4(s)G(s)$

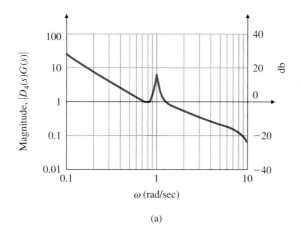

(a)

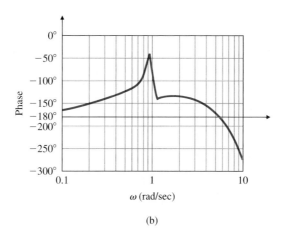

(b)

Figure 9.23
Root locus of $D_4(s)G(s)$

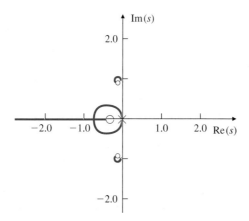

Figure 9.24
Closed-loop step response
of $D_4(s)G(s)$

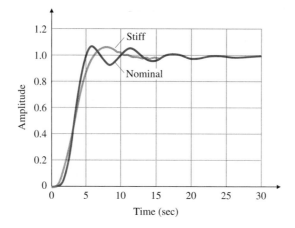

Collocated actuator and
sensor

the actuator—to give control with a collocated actuator and sensor. Due to the
physics of the situation, the system's transfer function now has zeros close to the
flexible modes, so control can be achieved by using PD feedback alone because
the plant already has the effect of a notch compensator. Consider the transfer
function of the satellite with collocated actuator and sensor (to measure θ_1) for
which the state matrices are

$$\mathbf{F} = \begin{bmatrix} 0 & 1 & 0 & 0 \\ -0.91 & -0.036 & 0.91 & 0.036 \\ 0 & 0 & 0 & 1 \\ 0.091 & 0.0036 & -0.091 & -0.0036 \end{bmatrix}, \qquad \mathbf{G} = \begin{bmatrix} 0 \\ 0 \\ 0 \\ 1 \end{bmatrix},$$

$$\mathbf{H} = \begin{bmatrix} 0 & 0 & 1 & 0 \end{bmatrix}.$$

The transfer function of the system using the MATLAB ss2tf function is

$$G_{co}(s) = \mathbf{H}(s\mathbf{I} - \mathbf{F})^{-1}\mathbf{G} = \frac{(s + 0.018 \pm 0.954j)}{s^2(s + 0.02 \pm j)}. \tag{9.14}$$

Notice the presence of the zeros in the vicinity of the complex conjugate poles.
If we now use the same PD feedback as before,

$$D_5(s) = 0.25(2s + 1), \tag{9.15}$$

then the system will not only be stabilized, but will also have a satisfactory
response (if we consider θ_1 as the output), because the resonant poles tend
to be canceled by the complex conjugate zeros. Figures 9.25 to 9.27 show the
frequency response, the root locus, and the step response, respectively, for this
system. Note from Fig. 9.27 that the step response has the excess overshoot
associated with the zero of the compensator in the forward path of the transfer
function.

Figure 9.25
Bode plot of $D_5(s)G_{co}(s)$

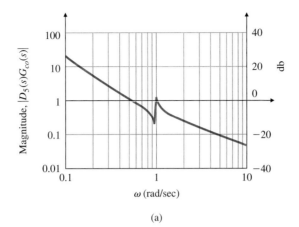

(a)

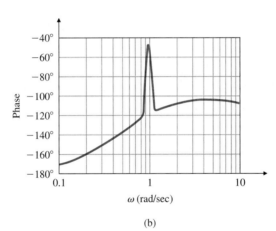

(b)

Figure 9.26
Root locus for
$D_5(s)G_{co}(s)$

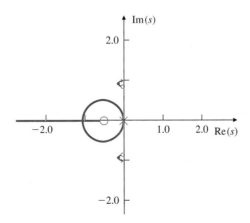

Figure 9.27
Closed-loop step response
of the system with
collocated control,
$D_5(s)G_{co}(s)$ and
$D_5(s)\hat{G}_{co}(s)$

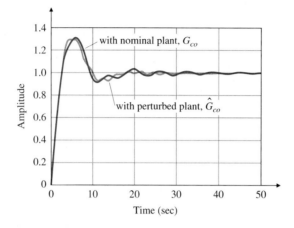

The result is a very simple robust design achieved by moving the sensor from a noncollocated position to one collocated with the actuator. The result illustrates that, to achieve good feedback control, it is very important to consider sensor location and other features of the physical problem. However, this last control design will *not* do for pointing the star tracker. This is evident from plotting the output θ_2 corresponding to the nice step response of Fig. 9.27. The result is shown in Fig. 9.28.

An architecture suggested by the results is to place a coarse star tracker on the satellite body to be used for search and initial settling. Then switch to a star tracker on the instrument package with longer settling time for fine control.

Figure 9.28
Response at θ_2 of the
collocated design

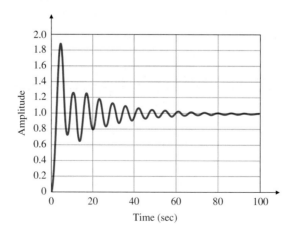

9.3   Lateral and Longitudinal Control of a Boeing 747

The Boeing 747 (Fig. 9.29) is a large wide-body transport jet. A schematic with the relevant coordinates that move with the airplane is shown in Fig. 9.30. The linearized equations of (rigid-body) motion[4] for the Boeing 747 are of eighth order but are separated into two fourth-order sets representing the perturbations in longitudinal (U, W, θ, and q in Fig. 9.30) and lateral motion (ϕ, β, r, and p). The longitudinal motion consists of axial (X), vertical (Z), and pitching (θ, q) motion, while the lateral motion consists of rolling (ϕ, p), yawing (r, β) movement. The side-slip angle β is a measure of the direction of forward velocity relative to the direction of the nose of the airplane. The elevator control surfaces and the throttle affect the longitudinal motion, whereas the aileron and rudder primarily affect lateral motion. Although there is a small amount of coupling of lateral motion into longitudinal motion, this is usually ignored, so the equations of motion are treated as two decoupled fourth-order sets for designing the control, or **stability augmentation**, for the aircraft.

The nonlinear rigid body equations of motion in body-axis coordinates, under proper assumptions,[5] can be derived as (Bryson, 1994)

$$m\,(\dot{U} + qW - rV) = X - mg\sin\theta + \kappa T\cos\theta, \qquad (9.16)$$

$$m\,(\dot{V} + rU - pW) = Y + mg\cos\theta\sin\phi,$$

$$m\,(\dot{W} + pV - qU) = Z + mg\cos\theta\cos\phi - \kappa T\sin\theta,$$

$$I_x\dot{p} + I_{xz}\dot{r} + (I_z - I_y)\,qr + I_{xz}\,qp = L, \qquad (9.17)$$

$$I_y\dot{q} + (I_x - I_z)\,pr + I_{xz}\,(r^2 - p^2) = M,$$

$$I_z\dot{r} + I_{xz}\dot{p} + (I_y - I_x)\,qp - I_{xz}\,qr = N,$$

Figure 9.29
Boeing 747 (*Courtesy Boeing Commercial Airplane Co.*)

---

[4] For derivation of equations of motion for an aircarft, the reader is referred to Bryson (1994), Etkin and Reid (1996), and McRuer et al. (1973).

[5] x–z is the body axis plane of mass symmetry.

Figure 9.30
Definition of aircraft coordinates

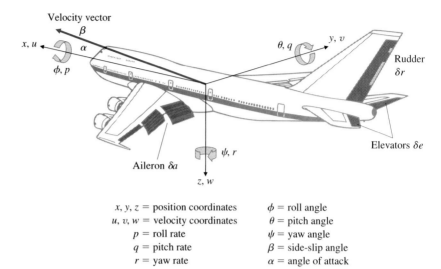

Velocity vector

x, y, z = position coordinates	ϕ = roll angle
u, v, w = velocity coordinates	θ = pitch angle
p = roll rate	ψ = yaw angle
q = pitch rate	β = side-slip angle
r = yaw rate	α = angle of attack

where

$$m = \text{mass of the aircraft,}$$

$[U, V, W]$ = body-axis components of the velocity of the center of mass (c.m.),

$$\beta = \tan^{-1}\left(\frac{V}{U}\right),$$

$[U_0, V_0, W_0]$ = reference velocities,

$[p, q, r]$ = the body-axis components of the angular velocity of the aircraft, (roll, pitch, and yaw respectively)

$[X, Y, Z]$ = the body-axis aerodynamic forces about the c.m.,

$[L, M, N]$ = the body-axis aerodynamic torques about the c.m.,

g_o = the gravitational force per unit mass,

I_i = the inertias in body axes,

(θ, ϕ) = the Euler pitch and roll angles of the aircraft body axes with respect to horizontal,

V_{ref} = reference flight speed,

T = the propulsive thrust resultant, and

κ = the angle between thrust and body x-axis.

The linearization of these equations can be carried out as follows. In the steady-state straight, level, and constant speed flight condition, $\dot{U} = \dot{V} = \dot{W} = p = \dot{q} = \dot{r} = 0$. Furthermore, there is no turning in any axis so that $p_o = q_o = r_o = 0$, and the wings will be level so that $\phi = 0$. However, there will be an angle of attack in order to provide some lift from the wings to counteract the aircraft's weight, so θ_o and $W_o \neq 0$, where

$$U = U_o + u, \tag{9.18}$$

$$V = V_o + v,$$

$$W = W_o + w.$$

The steady-state velocity body axis components will be

$$U_o = V_{ref} \cos(\theta_o), \tag{9.19}$$

$$V_o = 0 \; (\beta_o = 0),$$

$$W_o = V_{ref} \sin(\theta_o),$$

as depicted in Fig. 9.31. With these conditions, the *equilibrium* equations are

$$0 = X_0 - mg_o \sin\theta_0 + \kappa T \cos\theta_0, \tag{9.20}$$

$$0 = Y_0,$$

$$0 = Z_0 + mg_o \cos\theta_0 - \kappa T \sin\theta_0,$$

$$0 = L_0,$$

$$0 = M_0,$$

$$0 = N_0.$$

With the assumptions (Bryson, 1994)

$$(v^2, w^2) \ll u^2, \tag{9.21}$$

$$(\phi^2, \theta^2) \ll 1,$$

$$(p^2, q^2, r^2) \ll \frac{u^2}{b^2},$$

where b denotes the wing span, many of the nonlinear terms in Eqs. (9.16) and (9.17) can be neglected. Substitution of Eq. (9.20) in the nonlinear equations of motion leads to a set of linear perturbational equations that describe small deviations from constant speed, and straight and level flight. The equations of motion then divide into two uncoupled sets of *longitudinal* and *lateral* equations of motion as seen below.

Figure 9.31
Steady-state flight condition

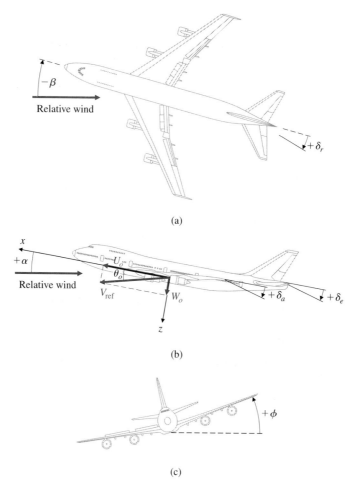

(a)

(b)

(c)

For linearized longitudinal motion, the results are

$$
\begin{bmatrix} \dot{u} \\ \dot{w} \\ \dot{q} \\ \dot{\theta} \end{bmatrix} = \begin{bmatrix} X_u & X_w & -W_o & -g_o\cos\theta_o \\ Z_u & Z_w & U_o & -g_o\sin\theta_o \\ M_u & M_w & M_q & 0 \\ 0 & 0 & 1 & 0 \end{bmatrix} \begin{bmatrix} u \\ w \\ q \\ \theta \end{bmatrix} + \begin{bmatrix} X_{\delta e} \\ Z_{\delta e} \\ M_{\delta e} \\ 0 \end{bmatrix} \delta e \qquad (9.22)
$$

where

u = forward velocity perturbation in the aircraft in x direction (see Fig. 9.30),

w = velocity perturbation in the z direction (also proportional to perturbations in the angle of attack, $\alpha = \frac{w}{U_0}$),

q = angular rate about positive y axis, or pitch rate,

θ = pitch-angle perturbation from the reference θ_o value,

$X_{u,w,\delta e}$ = partial derivative of the aerodynamic force in x direction with respect to perturbations in u, w, and δe,[6]

$Z_{u,w,\delta e}$ = partial derivative of the aerodynamic force in z direction with respect to perturbations in u, w, and δe,

$M_{u,w,q,\delta e}$ = partial derivative of the aerodynamic (pitching) moment with respect to perturbations in u, w, q, and δe,

δe = movable tail-section, or "elevator," angle for pitch control,

$W_o q$, $U_o q$ terms in the equations are due to the angular velocity of the body fixed (rotating) reference frame and arise directly from the left-hand side of Eq. (9.16).

To determine altitude changes, we need to add the following equation to the longitudinal equations of motion

$$\dot{h} = V_{ref} \sin\theta - w\cos\theta, \tag{9.23}$$

which would result in the linearized altitude equation,

$$\dot{h} = V_{ref}\,\theta - w, \tag{9.24}$$

which is to be augmented with Eq. (9.22).

For linearized lateral motion, the results are

$$
\begin{bmatrix} \dot{\beta} \\ \dot{r} \\ \dot{p} \\ \dot{\phi} \end{bmatrix}
=
\begin{bmatrix}
Y_v & -U_o & V_o & g_o\cos\theta_o \\
N_v & N_r & N_p & 0 \\
L_v & L_r & L_p & 0 \\
0 & \tan\theta_o & 1 & 0
\end{bmatrix}
\begin{bmatrix} \beta \\ r \\ p \\ \phi \end{bmatrix}
+
\begin{bmatrix}
Y_{\delta r} & Y_{\delta a} \\
N_{\delta r} & N_{\delta a} \\
L_{\delta r} & L_{\delta a} \\
0 & 0
\end{bmatrix}
\begin{bmatrix} \delta r \\ \delta a \end{bmatrix}
\tag{9.25}
$$

where

β = side-slip angle, defined to be $\dfrac{v}{U_o}$,

r = yaw rate,

p = roll rate,

ϕ = roll angle,

$Y_{v,\delta r,\delta a}$ = partial derivative of the aerodynamic force in y direction with respect to perturbations in β, δr and δa,

$N_{v,r,p,\delta r,\delta a}$ = aerodynamic (yawing) moment stability derivatives,

$L_{v,r,p,\delta r,\delta a}$ = aerodynamic (rolling) moment stability derivatives,

δr = rudder deflection,

δa = aileron deflection.

---

[6] X, Z, and M are stability derivatives and are identified from wind tunnel and flight tests.

We will next discuss the design of a stability-augmentation system for the lateral dynamics, called a **yaw damper**, and the autopilot affecting the longitudinal behavior.

9.3.1 Yaw Damper

STEP 1. *Understand the process and its performance specifications.* Swept-wing aircraft have a natural tendency to be lightly damped in the lateral modes of motion. At typical commercial-aircraft cruising speeds and altitudes, this dynamic mode is sufficiently difficult to control that virtually every swept wing aircraft has a feedback system to help the pilot. Therefore, the goal of our control system is to modify the natural dynamics so that the plane is acceptable for the pilot to fly.[7] Studies have shown that pilots like natural frequencies $\omega_n \lesssim 0.5$ and damping ratio of $\zeta \geq 0.5$. Aircraft with dynamics that violate these guidelines are generally considered fatiguing to fly and highly undesirable. Thus our system specifications are to achieve lateral dynamics that meet these constraints.

STEP 2. *Select sensors.* The easiest measurement of aircraft motion to take is the angular rate. The side-slip angle can be measured with a wind-vane device, but it is noisier and less reliable for stabilization. Two angular rates—roll and yaw—partake in the lateral motion. Study of the lightly damped lateral mode indicates that it is primarily a yawing phenomenon, so measurement of the yaw rate is a logical starting point for the design. Until the early 1980s the measurement was made with a **gyroscope** with a small, fast-spinning rotor that can yield an electric output proportional to the angular yaw rate of the aircraft. Since the early 1980s most new aircraft systems have relied on a laser device (called a **ring-laser gyroscope**) for the measurement. Here two laser beams traverse a closed path (often a triangle) in opposite directions. As the triangular device rotates, the detected frequencies of the two beams appear to shift, and this frequency shift is measured, producing a measure of rotational rate. These devices have fewer moving parts and are more reliable at less cost than the spinning-rotor variety of gyroscope.

STEP 3. *Select actuators.* Two aerodynamic surfaces typically influence the lateral aircraft motion: the rudder and the ailerons (see Fig. 9.30). The lightly damped yaw mode that will be stabilized by the yaw damper is most affected by the rudder. Therefore, use of that single control input is a logical starting point for the design. Hence it is best to choose the rudder as our actuator. Hydraulic devices are universally employed in large aircraft to provide the force that moves the aerodynamic surfaces. No other kind of device has been developed to provide the combination of high force, high speed, and light weight desirable for the actuation of the controlling aerodynamic surfaces. On the other hand,

---

[7] The mode is sufficiently difficult to control manually that, if the yaw damper fails in cruise, the pilot is instructed to descend and slow down where the mode is more manageable.

the low-speed flaps, which are extended slowly prior to landing, are typically actuated by an electric motor with a worm gear. For small aircraft with no autopilot, no actuator is required at all; the pilot yoke is directly connected to the aerodynamic surface by means of wire cables, and all the force required to move the surfaces is provided by the pilot.

STEP 4. *Make a linear model.* The lateral-perturbation equations of motion for Boeing 747 in horizontal flight at 40,000 ft and nominal forward speed $U_0 = 774$ ft/sec (Mach 0.8) (Heffley and Jewell, 1972), with the rudder chosen as the actuator (STEP 3), are

$$
\begin{bmatrix} \dot{\beta} \\ \dot{r} \\ \dot{p} \\ \dot{\phi} \end{bmatrix} = \begin{bmatrix} -0.0558 & -0.9968 & 0.0802 & 0.0415 \\ 0.598 & -0.115 & -0.0318 & 0 \\ -3.05 & 0.388 & -0.4650 & 0 \\ 0 & 0.0805 & 1 & 0 \end{bmatrix} \begin{bmatrix} \beta \\ r \\ p \\ \phi \end{bmatrix} + \begin{bmatrix} 0.00729 \\ -0.475 \\ 0.153 \\ 0 \end{bmatrix} \delta r,
$$

$$
y = \begin{bmatrix} 0 & 1 & 0 & 0 \end{bmatrix} \begin{bmatrix} \beta \\ r \\ p \\ \phi \end{bmatrix},
$$

where β and ϕ are in radians and r and p are in radians per second. The transfer function using the MATLAB ss2tf function is

$$
G(s) = \frac{r(s)}{\delta r(s)} = \frac{-0.475(s + 0.498)(s + 0.012 \pm 0.488j)}{(s + 0.0073)(s + 0.563)(s + 0.033 \pm 0.947j)} \tag{9.26}
$$

so that the system has two stable real poles and a pair of stable complex poles. Notice first that the low-frequency gain is negative, corresponding to the simple physical fact that a positive or clockwise rudder motion causes a negative or counterclockwise yaw rate. In other words, turning the rudder left (clockwise) causes the front of the aircraft to rotate left (counterclockwise). The natural motion corresponding to the complex poles is referred to as the **Dutch roll**; the name comes from the motions of a person skating on the frozen canals of Holland. The motion corresponding to the stable real poles is referred to as the **spiral mode** ($s_1 = -0.0073$) and the **roll mode** ($s_2 = -0.563$). From looking at the system poles, we see that the offending mode that needs repair for good pilot handling is the Dutch roll, with the poles at $s = -0.033 \pm 0.95j$. The roots have an acceptable frequency, but their damping ratio $\zeta \cong 0.03$ is far short of the desired value $\zeta \cong 0.5$.

STEP 5. *Try a lead–lag or PID design.* As a first try at the design, we will consider proportional feedback of the yaw rate to the rudder. The root locus with respect to the gain of this feedback is shown in Fig. 9.32, and its frequency response is shown in Fig. 9.33. The figures show that $\zeta \cong 0.45$ is achievable and can be computed to occur at a gain of about 3.0.

Dutch roll
Spiral mode
Roll mode

Figure 9.32
Root locus for yaw damper
with proportional feedback

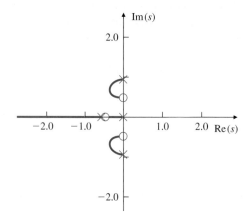

Figure 9.33
Bode plot of yaw damper
with proportional feedback

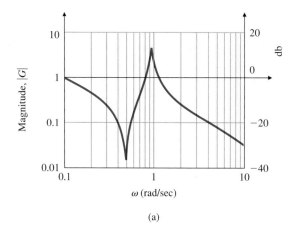

(a)

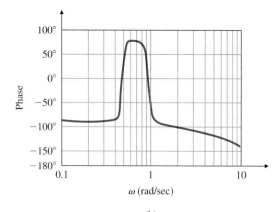

(b)

This feedback, however, creates an objectionable situation during a steady turn when the yaw rate is constant: Because the feedback produces a steady rudder input opposite the yaw rate, the pilot must introduce a much larger steady command for the same yaw rate than is necessary in the open-loop case. This dilemma is solved by attenuating the feedback at DC—that is, "washing out" the feedback. This is accomplished by inserting

Washout

$$H(s) = \frac{s}{s + 1/\tau},$$

in the feedback which passes the yaw rate feedback at frequencies above $1/\tau$, and provides no feedback at DC. Therefore, in a steady turn, the damper will provide no correction. Figure 9.34 shows a block diagram of the yaw damper with the washout.

For a more complete model, we include the rudder servo, which represents the actuator dynamics and has the transfer function

$$A(s) = \frac{\delta r(s)}{e_{\delta r}(s)} = \frac{10}{s + 10},$$

which is fast compared with the dynamics of the rest of the system and is not expected to change the response very much. The root locus including actuator dynamics and a washout circuit with $\tau = 3$ is shown in Fig. 9.35. As seen from the root locus, the addition of the yaw rate feedback including the washout allows the damping ratio to be increased from 0.03 to about 0.35. The associated frequency response of the system is shown in Fig. 9.36. The response of the closed-loop system to an initial condition of $\beta_0 = 1°$ is shown in Fig. 9.37 for a

Figure 9.34
Yaw damper: (a) Functional block diagram; (b) block diagram for analysis

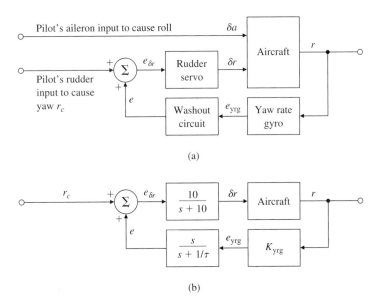

Figure 9.35
Root locus with washout
circuit, $\tau = 3$

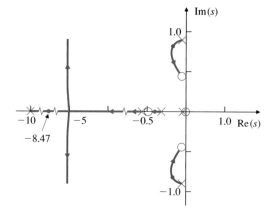

Figure 9.36
Bode plot of yaw damper,
including washout and
actuator

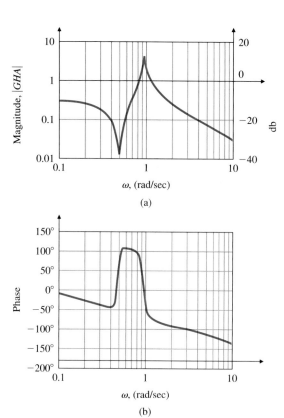

root-locus gain of 2.6. For reference the response of yaw rate with no feedback
is also given. Although feedback of yaw rate through the washout circuit results
in a considerable improvement over the original aircraft control, the response
is not as good as originally specified. Further iterations, not included here,
could include other gain values or more complex compensations.

Figure 9.37
Initial condition response
with yaw damper and
washout, and SRL design,
for $\beta_0 = 1°$

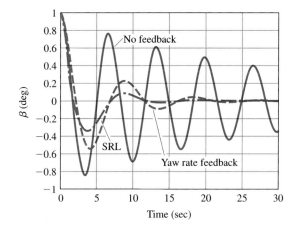

STEP 6. *Evaluate/modify plant.* The solution would be to unsweep the wings, which would cause a large drag penalty.

STEP 7. *Try an optimal design using pole placement.* If we augment the dynamic model of the system by adding the actuator and washout, we obtain the following state-variable model:

$$
\begin{bmatrix} \dot{x}_A \\ \dot{\beta} \\ \dot{r} \\ \dot{p} \\ \dot{\phi} \\ \dot{x}_{wo} \end{bmatrix} = \begin{bmatrix} -10 & 0 & 0 & 0 & 0 & 0 \\ 0.0729 & -0.0558 & -0.997 & 0.0802 & 0.0415 & 0 \\ -4.75 & 0.598 & -0.1150 & -0.0318 & 0 & 0 \\ 1.53 & -3.05 & 0.388 & -0.465 & 0 & 0 \\ 0 & 0 & 0.0805 & 1 & 0 & 0 \\ 0 & 0 & 1 & 0 & 0 & -0.333 \end{bmatrix} \begin{bmatrix} x_A \\ \beta \\ r \\ p \\ \phi \\ x_{wo} \end{bmatrix} + \begin{bmatrix} 10 \\ 0 \\ 0 \\ 0 \\ 0 \\ 0 \end{bmatrix} e_{\delta r},
$$

$$
e = \begin{bmatrix} 0 & 0 & 1 & 0 & 0 & -0.333 \end{bmatrix} \begin{bmatrix} x_A \\ \beta \\ r \\ p \\ \phi \\ x_{wo} \end{bmatrix},
$$

where $e_{\delta r}$ is the input to the actuator and e is the output of the washout. The SRL for the augmented system is as shown in Fig. 9.38. If we select the state-feedback poles from the SRL so that the complex roots have maximum damping ($\zeta = 0.4$), we find that

pc = [−0.0051; −0.468; 0.279 + 0.628 ∗ j; 0.279 − 0.628 ∗ j; −1.106; −9.89].

Then we can compute the state-feedback gain, using the MATLAB function place, to be

$$\mathbf{K} = \begin{bmatrix} 1.059 & -0.191 & -2.32 & 0.0992 & 0.0370 & 0.486 \end{bmatrix}.$$

Note that the third entry in **K** is larger than the others, so the feedback of all six state variables is essentially the same as proportional feedback of r. This is also evident from the similarity of the root locus in Fig. 9.32 and the SRL of Fig. 9.38. If we select the estimator poles to be five times faster than the controller poles, then

$$pe = [-0.0253; -2.34; -1.39 + 3.14 * j; -1.39 - 3.14 * j; -5.53; -49.5],$$

and the estimator gain, again using the MATLAB function place, is found to be

$$\mathbf{L} = \begin{bmatrix} 25.0 \\ -2{,}044 \\ -5{,}158 \\ -24{,}843 \\ -40{,}113 \\ -15{,}624 \end{bmatrix}.$$

The compensator transfer function from Eq. (7.166) is

$$D_c(s) = \frac{-844(s + 10.0)(s - 1.04)(s + 0.974 \pm 0.559 j)(s + 0.0230)}{(s + 0.0272)(s + 0.837 \pm 0.671 j)(s + 4.07 \pm 10.1 j)(s + 51.3)} \quad (9.27)$$

Figure 9.37 also shows the response of the yaw rate to an initial condition of $\beta_0 = 1°$. It is clear from the root locus that the damping can be improved by the SRL approach, and this is borne out by the reduced oscillatory behavior in the transient response of the system. However, this improvement has come at a considerable price. Note that the order of the compensator has increased from 1 in the original design (Fig. 9.34) to 6 and washout in the design obtained using the controller-estimator-SRL approach.

Figure 9.38
SRL of lateral dynamics, including washout filter and actuator

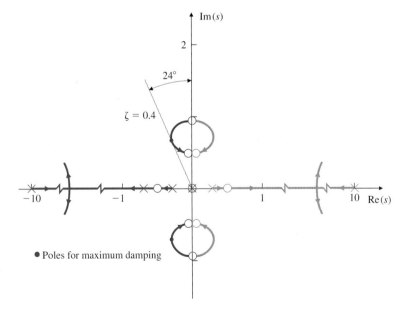

Design tradeoff: system
response vs. system
complexity

Aircraft yaw dampers in use today generally employ a proportional feedback of yaw rate to rudder through a washout or through minor modifications to this design. The improved performance achievable with an optimal design approach utilizing full-state feedback and estimation is not judged to be worth the increase in complexity.

Perhaps a more fruitful approach to improving the design would be to add the aileron surface as a control variable along with the rudder.

STEPS 8 and 9. *Verify the design.* Linear models of aircraft motion are reasonably accurate as long as the motion is small enough that the actuators and surfaces do not saturate. Because actuators are sized for safety in order to handle large transients, such saturation is very rare. Therefore, the linear-analysis-based design is reasonably accurate, and we will not pursue a nonlinear simulation or further design verification. However, aircraft manufacturers do carry out extensive nonlinear simulations and flight testing under all possible flight conditions before obtaining FAA certification to carry passengers.

9.3.2   Altitude-Hold Autopilot

STEP 1. *Understand the process and its performance specifications.* One of the pilot's many tasks is to hold a specific altitude. As an aid to keeping aircraft from colliding, those craft on an easterly path are required to be on an odd multiple of 1000 ft and those on a westerly path on an even multiple of 1000 ft. Therefore, the pilot needs to be able to hold the altitude to less than a hundred feet. A well-trained, attentive pilot can easily accomplish this task manually to within ± 50 ft, and air-traffic controllers expect pilots to maintain this kind of tolerance. However, because this task requires the pilot to be fairly diligent, sophisticated aircraft often have an altitude-hold autopilot to lessen the pilot's work. This system differs fundamentally from the yaw damper because its role is to replace the pilot for certain periods of time, while the yaw damper's role is to help the pilot fly. Dynamic specifications, therefore, need not require that pilots like the craft's "feel" (how it responds to their handling of the controls); instead, the design should provide the kind of ride that pilots and passengers like. The damping ratio should still be in the vicinity of $\zeta \cong 0.5$, but for a smooth ride the natural frequency should be much slower than $\omega_n = 1$ rad/sec.

STEP 2. *Select sensors.* Clearly needed is a device to measure altitude, a task most easily done by measuring the atmospheric pressure. Almost from the time of the first Wright brothers' flight, this basic idea has been used in a device called a **barometric altimeter**. Before autopilots the device consisted of a bellows whose free end was connected to a needle that directly indicated altitude on a dial. The same bellows concept is used today for the altitude display, but the pressure is sensed electrically for the autopilot.

Because the transfer function from the controlling elevator input to the altitude control consists of five poles [see Eq. (9.30)], stabilization of the feedback loop cannot be accomplished by simple proportional feedback. Therefore pitch rate q is also used as a stabilizing feedback; it is measured by a gyroscope or ring-laser gyro identical to that used for yaw-rate measurement. Further stabilization using pitch-angle feedback is also helpful. It is obtained either from an inertial reference system based on a ring-laser gyro or from a rate-integrating gyro. The latter is a device similar to the rate gyro but structured differently so that its outputs are proportional to the angles of the aircraft's pitch θ and roll ϕ.

STEP 3. *Select actuators.* The only aerodynamic surface typically used for pitch control on most aircraft is the elevator δe. It is located on the horizontal tail, well removed from the aircraft's center of gravity so that its force produces an angular pitch rate and thus a pitch angle, which acts to change the lift from the wing. In some high-performance aircraft there are direct-lift control devices on the wing or perhaps small canard surfaces, which are like tiny wings forward of the main wing, which produce vertical forces on the aircraft that are much faster than elevators on the tail are able to generate. However, for purposes of our altitude hold, we will only consider the typical case of an elevator surface on the tail.

As for the rudder, hydraulic actuators are the preferred devices to move the elevator surface mainly because of their favorable force-to-weight ratio.

STEP 4. *Make a linear model.* The longitudinal perturbation equations of motion for the Boeing 747 in horizontal flight at a nominal speed $U_0 = 830$ ft/sec at 20,000 ft (Mach 0.8) with a weight of 637,000 lb are

$$\dot{\mathbf{x}} = \mathbf{F}\mathbf{x} + \mathbf{G}\delta e$$

$$\begin{bmatrix} \dot{u} \\ \dot{w} \\ \dot{q} \\ \dot{\theta} \\ \dot{h} \end{bmatrix} = \begin{bmatrix} -0.00643 & 0.0263 & 0 & -32.2 & 0 \\ -0.0941 & -0.624 & 820 & 0 & 0 \\ -0.000222 & -0.00153 & -0.668 & 0 & 0 \\ 0 & 0 & 1 & 0 & 0 \\ 0 & -1 & 0 & 830 & 0 \end{bmatrix} \begin{bmatrix} u \\ w \\ q \\ \theta \\ h \end{bmatrix} + \begin{bmatrix} 0 \\ -32.7 \\ -2.08 \\ 0 \\ 0 \end{bmatrix} \delta e \quad (9.28)$$

where the desired output for an altitude-hold autopilot is

$$h = \mathbf{H}\mathbf{x}$$

$$h = \begin{bmatrix} 0 & 0 & 0 & 0 & 1 \end{bmatrix} \begin{bmatrix} u \\ w \\ q \\ \theta \\ h \end{bmatrix} \quad (9.29)$$

and

$$\frac{h(s)}{\delta e(s)} = \frac{32.7(s + 0.0045)(s + 5.645)(s - 5.61)}{s(s + 0.003 \pm 0.0098j)(s + 0.6463 \pm 1.1211j)}. \quad (9.30)$$

The system has two pairs of stable complex poles and a pole at $s = 0$. The complex pair at $-0.003 \pm 0.0098j$ are referred to as the **phugoid mode**,[8] and

---

[8] The name was adopted by F. W. Lanchester (1908), who was the first to study the dynamic stability of aircraft analytically. It is apparently an incorrect version of a Greek word.

Phugoid mode
Short-period modes

Inner-loop design

the poles at $-0.6463 \pm 1.1211j$ are the **short-period modes** as computed using the MATLAB eig command.

STEP 5. *Try a lead–lag PID controller.* As a first step in the design, it is typically helpful to use an inner-loop feedback of pitch rate q to δe so as to improve the damping of the short-period mode of the aircraft (see Fig. 9.39). The transfer function from δe to q using the MATLAB ss2tf function is

$$\frac{q(s)}{\delta e(s)} = -\frac{2.08s(s + 0.0105)(s + 0.596)}{(s + 0.003 \pm 0.0098j)(s + 0.646 \pm 1.21j)}. \tag{9.31}$$

The inner loop root locus for q feedback using Eq. (9.31) is as shown in Fig. 9.40. Because k_q is the root-locus parameter, the system matrix [Eq. (9.29)] is now modified as follows:

$$\mathbf{F}_q = \mathbf{F} + k_q\mathbf{GH}_q, \tag{9.32}$$

where $\mathbf{F}$ and $\mathbf{G}$ are defined in Eq. (9.29) and $\mathbf{H}_q = [0 \quad 0 \quad 1 \quad 0 \quad 0]$. The process of picking a suitable gain k_q is an iterative one. The selection procedure is the same one discussed in Chapter 5 (recall the tachometer feedback example in Section 5.7.1). If we choose $k_q = 1$, then the closed-loop poles will be located at $-0.0039 \pm 0.0067j$, $-1.683 \pm 0.277j$ on the root locus and

$$\mathbf{F}_q = \begin{bmatrix} -0.00643 & 0.0263 & 0 & -32.2 & 0 \\ -0.0941 & -0.624 & 787.3 & 0 & 0 \\ -0.000222 & -0.00153 & -2.75 & 0 & 0 \\ 0 & 0 & 1 & 0 & 0 \\ 0 & -1 & 0 & 830 & 0 \end{bmatrix}. \tag{9.33}$$

Note that only the third column of $\mathbf{F}_q$ is different from $\mathbf{F}$. To further improve the damping, it is useful to feedback the pitch angle of the aircraft. By trial and error, we select

$$\mathbf{K}_{\theta q} = [0 \quad 0 \quad -0.8 \quad -6 \quad 0]$$

Figure 9.39
Altitude-hold feedback system

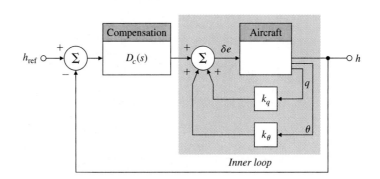

Inner loop

Figure 9.40
Inner-loop root locus for altitude-hold dynamics with q feedback

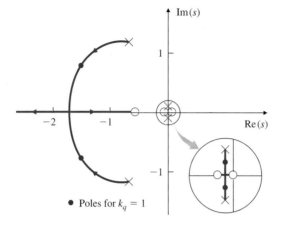

● Poles for $k_q = 1$

in order to feed back θ and q, and the system matrix becomes

$$\mathbf{F}_{\theta q} = \mathbf{F}_q - \mathbf{GK}_{\theta q}$$

$$= \begin{bmatrix} -0.0064 & 0.0263 & 0 & -32.2 & 0 \\ -0.0941 & -0.624 & 761 & -196.2 & 0 \\ -0.0002 & -0.0015 & -4.41 & -12.48 & 0 \\ 0 & 0 & 1 & 0 & 0 \\ 0 & -1 & 0 & 830 & 0 \end{bmatrix},$$

with poles at $s = 0,\ -2.25 \pm 2.99j,\ -0.531,\ -0.0105$.

So far, the inner loop of the aircraft has been stabilized significantly. The uncontrolled aircraft has a natural tendency to return to equilibrium in level flight as evidenced by the open-loop roots in the LHP. The inner-loop stabilization is necessary to enable an outer-loop feedback of h and $\dot{h}$ to be successful; furthermore, the feedbacks of θ and q can be used by themselves in an attitude-hold mode of the autopilot, where a pilot wishes to control θ directly through input command. Fig. 9.41 shows the response of the inner loop to a 2° (0.035-rad) step command in θ. With the inner loop in place the transfer function of the system from elevator angle to altitude is now

$$\frac{h(s)}{\delta e(s)} = \frac{32.7(s + 0.0045)(s + 5.645)(s - 5.61)}{s(s + 2.25 \pm 2.99j)(s + 0.0105)(s + 0.0531)}. \tag{9.34}$$

The root locus for this system, given in Fig. 9.42, shows that proportional feedback of altitude by itself does not yield an acceptable design. For stabilization we may also feed back the rate of change in the altitude in a PD controller. The root locus of the system with feedback of both h and $\dot{h}$ is shown in Fig. 9.43. After some iteration we find that the best ratio of $\dot{h}$ to h is 10:1, that is,

$$D_e(s) = K_h(s + 0.1).$$

Figure 9.41
Response of
altitude-hold autopilot
to a command
in θ

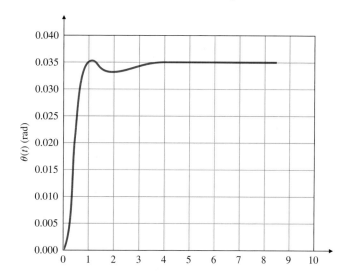

Figure 9.42
$0°$ root locus with
feedback of h only

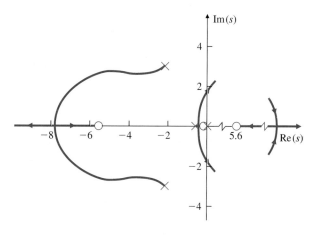

Figure 9.43
$0°$ root locus with
feedback of h and $\dot{h}$

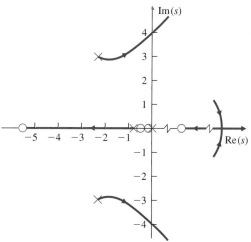

The final design is the result of iterations between the q, θ, $\dot{h}$, and h feedback gains, obviously a lengthy process. Although this trial design was successful, use of the SRL approach promises to expedite the process.

STEP 6. *Evaluate/modify plant.* Not applicable here.

STEP 7. *Do an optimal design.* The symmetric root locus of the system is shown in Fig. 9.44. If we choose the closed-loop poles at

$$pc = [-0.0045; -0.145; -0.513; -2.25 + 2.98 * j; -2.25 - 2.98 * j]$$

then the required feedback gain using the MATLAB function place, is

$$\mathbf{K} = [\,-0.0009 \quad 0.0016 \quad -1.883 \quad -7.603 \quad -0.001\,].$$

The step response of the system to a 100-ft step command in h is shown in Fig. 9.45, and the associated control effort is shown in Fig. 9.46.

This design has been carried out with the assumption that the linear model is valid for the altitude changes under consideration. We should perform simulations to verify this or to determine the range of validity of the linear model.

Figure 9.44

SRL for altitude-hold design

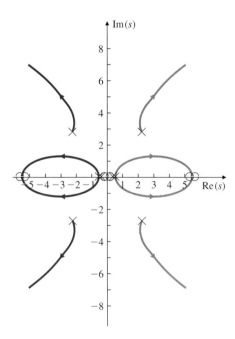

Figure 9.45

Step response of altitude-hold autopilot to a 100-ft step command

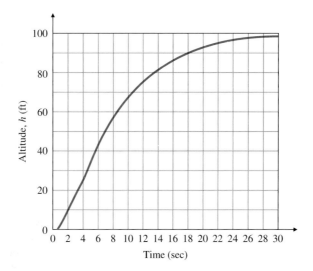

STEPS 8 and 9. *Verify the design.* The comments in Steps 7 and 8 of Section 9.3.1 apply to this design as well.

For small airplane autopilots now in production, such as the one described in Chapter 5, it is interesting to note that, for the inner loop, some manufacturers employ only θ feedback while others use q feedback. The use of θ enables faster response, but use of q is less costly. Both, of course, use the altimeter for h feedback.

Figure 9.46

Control effort for 100-ft step command in altitude

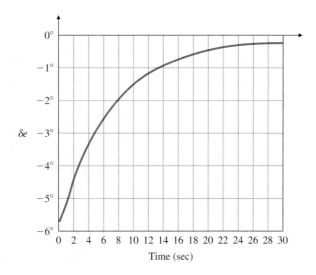

9.4   Control of the Fuel–Air Ratio in an Automotive Engine

Until the 1980s most automobile engines had a carburetor to meter the fuel so that the ratio of the gasoline-mass flow to air-mass flow, or fuel-to-air ratio (F/A), remained in the vicinity of 1:15. This device metered the fuel by relying on a pressure drop produced by the air flowing through a venturi. The device performed adequately in terms of keeping the engine running satisfactorily, but it historically allowed excursions of up to 20% in the F/A. After the implementation of federal exhaust-pollution regulations, this level of inaccuracy in the F/A was unacceptable because neither excess hydrocarbons (HCs) nor excess oxygen could be accepted. During the 1970s, automobile companies improved the design and manufacturing process of the carburetors so that they became more accurate and delivered an F/A accuracy in the vicinity of 3% to 5%. Through a combination of factors, this improved F/A accuracy helped lower the exhaust pollution levels. However, the carburetors were still open-loop devices because the system did not measure the F/A of the mixture entering the engine for subsequent feedback into the carburetor. During the 1980s almost all manufacturers have turned to feedback control systems to provide a much-improved level of F/A accuracy, an action made necessary by the decreasing levels of allowable exhaust pollutants.

We now turn to the design of a typical feedback system for engine control, again using the step-by-step design outline given in Section 9.1.

STEP 1. *Understand the process and its performance.* The method chosen to meet the exhaust-pollution standards has been to use a catalytic converter that simultaneously oxidizes excess levels of exhaust carbon monoxide (CO) and unburned HCs and reduces excess levels of the oxides of nitrogen (NO and NO_2, or NO_x). This device is usually referred to as a three-way catalyst because of its effect on all three pollutants. This catalyst is ineffective when the F/A is more than 1% different from the stoichiometric level of 1:14.7; therefore, a feedback control system is required to maintain the F/A within ±1% of that desired level. The system is depicted in Fig. 9.47.

The dynamic phenomena that affect the relationship between the sensed F/A output from the exhaust and the fuel-metering command in the intake manifold are (1) intake fuel and air mixing, (2) cycle delays due to the piston strokes in the engine, and (3) the time required for the exhaust to travel from

Figure 9.47
F/A feedback control system

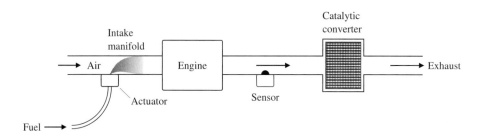

the engine to the sensor. All these effects are strongly dependent on the speed and load of the engine. For example, engine speeds typically vary from 600 to 6000 rpm. The result of these variations is that the time delays in the system that will affect the feedback control-system behavior will also vary by at least 10:1, depending on the operating condition. The system undergoes transients as the driver demands more or less power through changes in the accelerator pedal, with the changes taking place over fractions of a second. Ideally, the feedback control system should be able to keep up with these transients.

STEP 2. *Select sensors.* The discovery and development of the exhaust sensor was the key technological step that made possible this concept of exhaust-emission reduction by feedback control. The active element in the device, zirconium oxide, is placed in the exhaust stream where it yields a voltage that is a monotonic function of the oxygen content of the exhaust gas. The F/A is uniquely related to the oxygen level. The voltage of the sensor is highly non-linear with respect to F/A (Fig. 9.48); almost all the change in voltage occurs precisely at the F/A value at which the feedback system must operate for effec-tive performance of the catalyst. Therefore, the gain of the sensor will be very high when the F/A is at the desired point (1:14.7) but will fall off considerably for F/A excursions away from 1:14.7.

Nonlinear sensor

Although other sensors have been under development for possible use in F/A feedback control, no other cost-effective sensor has so far demonstrated the capability to perform adequately. All manufacturers of production-line automobiles currently use zirconium oxide sensors in their feedback control systems.

STEP 3. *Select actuators.* Fuel metering can be accomplished by a carburetor or by fuel injection. Implementing a feedback F/A system requires the capability of adjusting the fuel metering electrically, because the sensor used provides an electric output. Initially, carburetors were designed to provide this capability by including adjustable orifices that modify the primary fuel flow in response to the electric error signal. However, today manufacturers accomplish the metering by use of fuel injection. Fuel-injection systems are typically electrical by nature, so they can be used to perform the fuel adjustment for F/A feedback

Figure 9.48
Exhaust sensor output

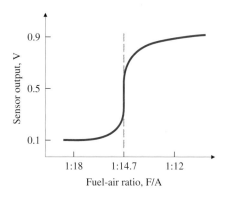

simply by including the capability of using the feedback signal from the sensor. In some cases, fuel injectors are placed at the inlet to every cylinder (called **multipoint injection**); in others there is one large injector upstream from all the cylinders (called **single-point** or **throttle body injection**). Multipoint injection offers improved performance because the fuel is introduced much closer to the engine, with better distribution to the cylinders. Being closer reduces the time delays and thus yields better engine response. It is used universally today.

STEP 4. *Make a linear model.* The sensor nonlinearity shown in Fig. 9.48 is severe enough that any design effort based on a linearized model of it should be used with caution. Figure 9.49 shows a block diagram of the system, with the sensor shown to have a gain K_s. The time constants τ_1 and τ_2 indicated for the inlet manifold dynamics represent, respectively, fast fuel flow in the form of vapor or droplets and slow fuel flow in the form of a liquid film on the manifold walls. The time delay is the sum of (1) the time it takes the pistons to move through the four strokes from the intake process until the exhaust process and (2) the time required for the exhaust to travel from the engine to the sensor located roughly 1 ft away. A sensor lag with time constant τ is also included in the process to account for the mixing that occurs in the exhaust manifold. Although the time constants and the delay time change considerably, primarily as a function of engine load and speed, we will examine the design at a specific point where the values are

$$\tau_1 = 0.02 \text{ sec}, \qquad T_d = 0.2 \text{ sec},$$
$$\tau_2 = 1 \text{ sec}, \qquad \tau = 0.1 \text{ sec}.$$

In an actual engine, designs would be carried out for all speed loads.

STEP 5. *Try a lead–lag or PID controller.* Given the tight error specifications and the wide variations in the required fuel command u_f due to varying

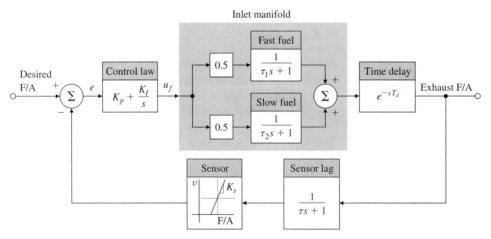

Figure 9.49 Block diagram of an F/A control system

engine-operating conditions, an integral control term is mandatory. With integral control, any required steady-state u_f can be provided when the error signal $e = 0$. The addition of a proportional term, although not often used, allows for an increase (doubling) in the bandwidth without degrading steady-state characteristics. In this example we use a control law that is proportional plus integral (PI). The output from the control law is a voltage that drives the injector's pulse former to give a fuel pulse whose duration is proportional to the voltage. The controller transfer function can be written as

$$D_c(s) = K_p + \frac{K_I}{s} = \frac{K_p}{s}(s + z), \qquad (9.35)$$

where

$$z = \frac{K_I}{K_p}$$

and z can be chosen as desired.

First, let us assume that the sensor is linear and can be represented by a gain K_s. Then we can choose z for good stability and good response of the system. Figure 9.50 shows the frequency response of the system for $K_s K_p = 1.0$ and $z = 0.3$, while Fig. 9.51 shows a root locus of the system with respect to $K_s K_p$ with $z = 0.3$. Both analyses show that the system becomes unstable for $K_s K_p \cong 2.8$. Figure 9.50 shows that to achieve a phase margin of approximately $60°$ the gain $K_s K_p$ should be ~ 2.2. Figure 9.50 also shows that this produces a crossover frequency of 6.0 rad/sec (~ 1 Hz). The root locus in Fig. 9.51 verifies that this candidate design will achieve acceptable damping ($\zeta \cong 0.5$).

Complications of nonlinearity

Although this linear analysis shows that acceptable stability at a reasonable bandwidth (~ 1 Hz) can be achieved with a proportional-integral (PI) controller, a look at the nonlinear sensor characteristics (Fig. 9.48) shows that this indeed may not be achievable. Note that the slope of the sensor output is extremely high near the desired setpoint, thus producing a very high value of K_s. Therefore, lower values of the controller gain K_p need to be used to maintain the overall $K_s K_p$ value of 2.2 when including the effect of the high sensor gain. On the other hand, a value of K_p low enough to yield a stable system at F/A $= 1 : 14.7$ ($= 0.068$) will yield a very sluggish response to transient errors that deviate much from the set point, because the effective sensor gain will be reduced substantially. It is therefore necessary to account for the sensor nonlinearity in order to obtain satisfactory response characteristics of the system for anything other than minute disturbances about the setpoint. A first approximation to the sensor is shown in Fig. 9.52. Because the actual sensor gain at the setpoint is still quite different from its approximation, this approximation will yield erroneous conclusions regarding stability about the set point; however, it will be useful in a simulation to determine the response to initial conditions.

STEP 6. *Evaluate/modify plant.* The nonlinear sensor is undesirable; however, no suitable linear sensor has been found.

Figure 9.50

Bode plot of a PI F/A
controller

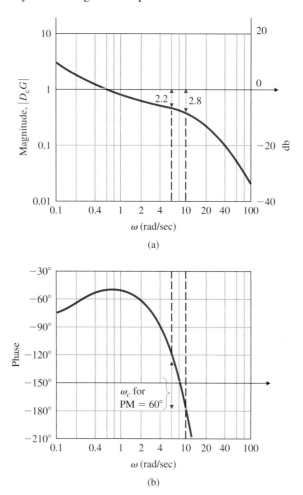

(a)

(b)

Figure 9.51

Root locus of a PI F/A
controller

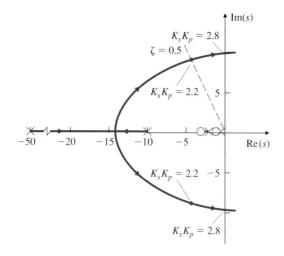

Figure 9.52
Sensor approximation

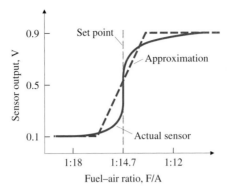

STEP 7. *Try an optimal controller.* The response of this system is dominated by the sensor nonlinearity, and any fine tuning of the control needs to account for that feature. Furthermore, the system dynamics are relatively simple, and it is unlikely that an optimal design approach will yield any improvement over the PI controller used. We will thus omit this step.

STEP 8. *Simulate design with nonlinearities.* The nonlinear closed-loop simulation of the system implemented in Simulink is shown in Figure 9.53. The MATLAB function (fas) implements the approximate nonlinear sensor characteristics of Fig. 9.52:

```
function y = fas(u)
if u < 0.0606,
    y = 0.1 ;
elseif u < 0.0741,
    y = 0.1 + (u - 0.0606) * 20;
else y = 0.9;
end
```

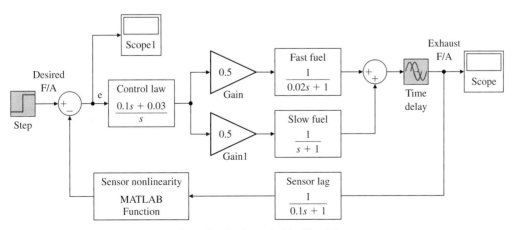

Figure 9.53 Closed-loop nonlinear simulation implemented in Simulink

Figure 9.54(a) is a plot of the system error using the approximate sensor of Fig. 9.52 and $K_p K_s = 2.0$. The slow response is apparent with 12.5 sec before the error comes out of saturation and a time constant of almost 5 sec once the linear region is reached. In real automobiles these systems are operated with much higher gains. To show these effects, a simulation with $K_p K_s = 6.0$ is plotted in Fig. 9.54(b, c). At this gain the linear system is unstable and up until about 5 sec the signals grow. The growth halts after 5 sec due to the fact that, as the input to the sensor nonlinearity gets large, the *effective* gain of the sensor decreases due to the saturation, and eventually, a limit cycle is reached. The frequency of this limit cycle corresponds to the point where the root locus crosses the imaginary axis and has an amplitude such that the total effective

Figure 9.54

System response
with nonlinear sensor
approximation

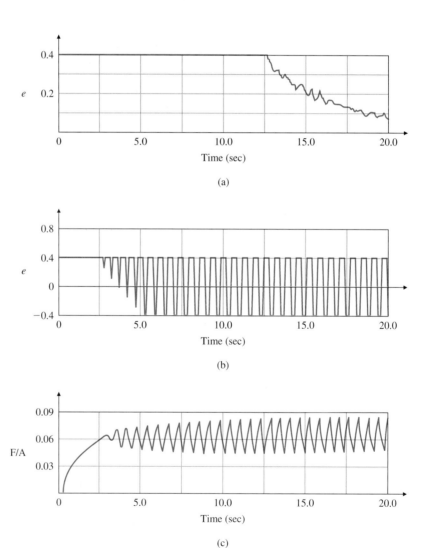

gain of $K_p K_{s,eq} = 2.8$. As described in Section 5.7.3, the effective gain of a saturation for moderately large inputs can be computed and is approximately $4N/\pi A$, where N is the saturation level and A is the amplitude of the input signal. Here $N = 0.4$, and, if $K_p = 0.1$, then $K_{s,eq} = 28$. Thus we predict an input signal amplitude of $A = 4(0.4)/28\pi = 0.018$. This value is closely verified by the plot of Fig. 9.54(c), the input to the nonlinearity in this case. The frequency of oscillation is also nearly 10.1 rad/sec, as predicted by the root locus in Fig. 9.51.

Simulink nonlinear simulation

In the actual implementation of F/A feedback controllers in automobile engines, sensor degradation over thousands of miles of use is of primary concern because the federal government mandates that the engines meet the exhaust-pollution standards for the first 50,000 mi. In order to reduce the sensitivity of the average set point to changes in the sensor output characteristics, manufacturers typically modify the design discussed here. One approach is to feed the sensor output into a relay function [see Fig. 5.48(b)], thus completely eliminating any dependency on the sensor gain at the set point. The frequency of the limit cycle is then solely determined by controller constants and engine characteristics. Average steady-state F/A accuracy is also improved. The oscillations in the F/A are acceptable because they are not noticeable to the car's occupants. In fact, the F/A excursions are beneficial to the catalyst operation in reducing pollutants.

9.5   Control of a Digital Tape Transport

The control of tension and velocity of a moving tape is a generic control problem in industry. Applications vary widely from digital tape transport to thin film manufacturing.

STEP 1. *Understand the process and its performance specifications.* Some high-performance tape transports are designed with a small capstan to pull the tape past the read/write heads with the take-up reels turned by DC motors. The capstan motion is isolated from the reels by vacuum columns that provide nearly constant tension to the tape. This structure permits separate design of capstan, vacuum, and reel controls. A sketch of the system is shown in Fig. 9.55, and a schematic of the capstan control is drawn in Fig. 9.56. The objective of the capstan control system is to control the speed and the tension of the tape at the read/write head. The tape is to be controlled at speeds of up to 200 in./sec, and startup is to be as fast as possible but the tension must remain below 12 N at all times to prevent permanent distortion or breakage of the tape. For the purposes of this design exercise, we will consider the speed of response to be satisfactory if the settling time of the response to a step input is less than 12 msec and the overshoot of the response to a step input is less than 10%.

STEP 2. *Select sensors.* Selecting the sensors is easy in this case: We will use a DC tachometer to measure idler speed and use a shaft encoder to measure idler position. An alternative to the tachometer would be an AC generator, which

Figure 9.55
Front view of a typical
tape-drive mechanism with
reference track 1 nearest
the front (observer) edge of
the tape

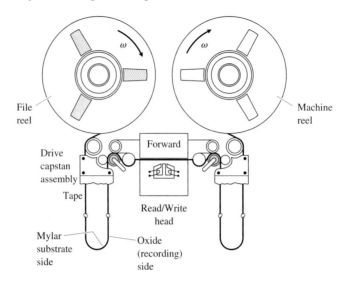

gives a sinusoidal voltage signal whose amplitude is proportional to speed. We
chose not to use it because the output voltage needs additional electronic signal
processing (such as rectifying and filtering) to provide a suitable signal for the
controller. As for the shaft recorder, models that provide a digital reading
of shaft position have low inertia and low friction and thus are often used in
digital control, which is increasingly attractive for motion-control problems.

Figure 9.56
Model of a tape drive

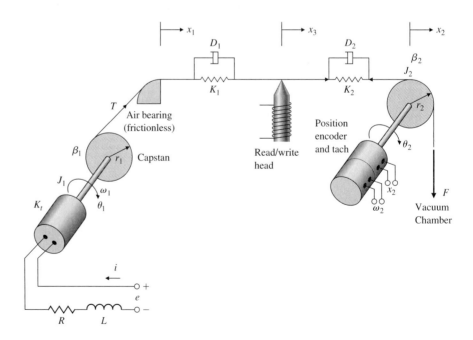

In the present case, however, the output of the encoder will be processed to provide an analog measure of position, which can be used in the feedback to the drive motor.

STEP 3. *Select actuators.* There are many choices for motion-control actuators, including hydraulic, pneumatic, AC electric, DC electric, and stepping motors. Hydraulic and pneumatic actuators require an auxiliary pump but can provide high forces and torques in a lightweight package. They are common in aircraft control-surface servomechanisms and chemical-process valve controls. However, for our purpose the electric motors are the best choice, because they are reliable, inexpensive, and easy to maintain. Of these, the DC motor is the machine of choice here because it has the best low-speed acceleration and low-frequency noise characteristics.

STEP 4. *Make a linear model.* The system is in static equilibrium when the tape tension equals the vacuum forces, $T_0 = F$, and when the torque from the motor equals the torque on the capstan, $K_t i_0 = r_1 T_0$, where

$$T_0 = \text{tape tension at the read/write head at equilibrium, } N,$$

$$F = \text{constant force (tape tension from vacuum column), 6 N,}$$

$$K = \text{motor torque constant,}$$

$$i_0 = \text{equilibrium motor current,}$$

$$r_1 = \text{radius of the capstan take-up wheel.}$$

We define the variables as deviations from this equilibrium. Thus the linearized system has torque deviations limited to 6 N. The equations of motion of the system are given by the laws of dynamics as follows:

$$J_1 \frac{d\omega_1}{dt} + \beta_1 \omega_1 - r_1 T = K_t i, \tag{9.36}$$

$$\dot{x}_1 = r_1 \omega_1, \tag{9.37}$$

$$L\frac{di}{dt} + Ri + K_e \, \omega_1 = e, \tag{9.38}$$

$$\dot{x}_2 = r_2 \omega_2, \tag{9.39}$$

$$J_2 \frac{d\omega_2}{dt} + \beta_2 \omega_2 + r_2 T = 0, \tag{9.40}$$

$$T = K_1(x_3 - x_1) + D_1(\dot{x}_3 - \dot{x}_1) \tag{9.41}$$

$$T = K_2(x_2 - x_3) + D_2(\dot{x}_2 - \dot{x}_3), \tag{9.42}$$

$$x_1 = r_1 \theta_1, \tag{9.43}$$

$$x_2 = r_2 \theta_2, \tag{9.44}$$

$$x_3 = \frac{x_1 + x_2}{2}, \tag{9.45}$$

where

$$D_{1,2} = \text{damping in the tape-stretch motion} = 20 \text{ N/m} \cdot \text{sec},$$

$$e = \text{applied voltage, } V,$$

$$i = \text{current into capstan motor}$$

$$J_1 = \text{combined inertia of the wheel and take-up motor, } 4 \times 10^{-5} \text{ kg} \cdot \text{m}^2,$$

$$J_2 = \text{inertia of the idler} = 10^{-5} \text{ kg} \cdot \text{m}^2,$$

$$K_{1,2} = \text{spring constant in the tape} - \text{stretch motion, } 4 \times 10^4 \text{ N/m},$$

$$K_e = \text{electric constant of the motor} = 0.03 \text{ V} \cdot \text{sec},$$

$$K_t = \text{torque constant of the motor} = 0.03 \text{ V} \cdot \text{sec},$$

$$L = \text{armature inductance} = 10^{-3} \text{ H},$$

$$R = \text{armature resistance} = 1\Omega,$$

$$r_1 = \text{radius of take} - \text{up wheel, } 0.02 \text{ m},$$

$$r_2 = \text{radius of the tape on the idler} = 0.02 \text{ m},$$

$$T = \text{tape tension at the read/write head N},$$

$$x_3 = \text{position of the tape at the head, m},$$

$$\dot{x}_3 = \text{velocity of the tape at the head, m/sec}.$$

$$\beta_1 = \text{viscous friction at the take} - \text{up wheel, } 0.01 \text{ N} \cdot \text{m} \cdot \text{sec},$$

$$\beta_2 = \text{viscous friction at the wheel} = 0.01 \text{ N} \cdot \text{m} \cdot \text{sec},$$

$$\theta_1 = \text{angular displacement of the capstan},$$

$$\theta_2 = \text{tachometer shaft angle},$$

$$\omega_1 = \text{speed of the drive wheel} = \dot{\theta}_1, \text{ rad/sec},$$

$$\omega_2 = \text{output speed measured by the tachometer output } \theta_2, \text{ rad/sec},$$

The equations of the system are fifth-order and are already almost in the state form

$$\dot{\mathbf{x}} = \mathbf{F}\mathbf{x} + \mathbf{G}u, \tag{9.46}$$

$$y = \mathbf{H}\mathbf{x}, \tag{9.47}$$

where

$$\mathbf{x} = [\, x_1 \quad \omega_1 \quad x_2 \quad \omega_2 \quad i \,]^T,$$

$$u = e.$$

There are two outputs of interest, tape position at the head, x_3, and tape tension T. For the tape position we have Eq. (9.45). Thus, for the position at the head,

$$\mathbf{H}_3 = [\, 0.5 \quad 0 \quad 0.5 \quad 0 \quad 0 \,].$$

The tension is given by Eq. (9.41), which we can write as

$$T = K_1 \left(\frac{x_1 + x_2}{2} - x_1 \right) + D_1 \left(\frac{r_1\omega_1 + r_2\omega_2}{2} - r_1\omega_1 \right)$$

$$= \frac{K_1}{2}(x_2 - x_1) + \frac{D_1}{2}(r_2\omega_2 - r_1\omega_1). \tag{9.48}$$

Using time and amplitude scaling

If we substitute the numbers into these equations, we find very large and very small values, indicating the need for time and amplitude scaling. Ideally, we would like the elements in $\mathbf{F}$, $\mathbf{G}$, and $\mathbf{H}$ to be in the range 0.1 to 10 if possible. We write the equations in state form with the numbers and use x'_1 and x'_2 for the unscaled positions:

$$\dot{x}'_1 = 0.02\omega_1,$$

$$4 \times 10^{-5}\dot{\omega}_1 = -400x'_1 - 0.014\omega_1 + 400x'_2 + 0.004\omega_2 + 0.03i,$$

$$\dot{x}'_2 = 0.02\omega_2,$$

$$10^{-5}\dot{\omega}_2 = 400x'_1 + 0.004\omega_1 - 400x'_2 - 0.014\omega_2,$$

$$10^{-3}i = -0.03\omega_1 - i + e. \tag{9.49}$$

After some experimentation, we decide to measure time in milliseconds, so $\tau = 10^3 t$, and to measure position in units of 10^{-5} m, so we take $x_1 = 10^5 x'_1$ and $x_2 = 10^5 x'_2$. With these conversions, Eqs. (9.49) become

$$\frac{dx_1}{d\tau} = 2\omega_1,$$

$$\frac{d\omega_1}{d\tau} = -0.1x_1 - 0.35\omega_1 + 0.1x_2 + 0.1\omega_2 + 0.75i,$$

$$\frac{dx_2}{d\tau} = 2\omega_2,$$

$$\frac{d\omega_2}{d\tau} = 0.4x_1 + 0.4\omega_1 - 0.4x_2 - 1.4\omega_2,$$

$$\frac{di}{d\tau} = -0.03\omega_1 - i + e. \tag{9.50}$$

These equations are described by the matrices:

$$\mathbf{F} = \begin{bmatrix} 0 & 2 & 0 & 0 & 0 \\ -0.1 & -0.35 & 0.1 & 0.1 & 0.75 \\ 0 & 0 & 0 & 2 & 0 \\ 0.4 & 0.4 & -0.4 & -1.4 & 0 \\ 0 & -0.03 & 0 & 0 & -1 \end{bmatrix}, \qquad \mathbf{G} = \begin{bmatrix} 0 \\ 0 \\ 0 \\ 0 \\ 1 \end{bmatrix},$$

$$\mathbf{H_2} = [0 \ \ 0 \ \ 1 \ \ 0 \ \ 0], \qquad x_2 \text{ output},$$

$$\mathbf{H_3} = [0.5 \ \ 0 \ \ 0.5 \ \ 0 \ \ 0], \qquad x_3 \text{ output},$$

$$\mathbf{H_T} = [-0.2 \ \ -0.2 \ \ 0.2 \ \ 0.2 \ \ 0], \qquad \text{tension output}.$$

The transfer function to measured position x_2 using the MATLAB ss2tf function is

$$G(s) = \frac{X_2(s)}{E(s)} = \frac{0.6(s+2)}{s^5 + 2.75s^4 + 3.22s^3 + 1.88s^2 + 0.418s}$$

$$= \frac{0.6(s+2)}{s(s+0.507)(s+0.968)(s+0.637 \pm 0.667j)}. \tag{9.51}$$

STEP 5. *Try a lead–lag or PID controller.* The root locus with respect to K for proportional feedback only and the corresponding open-loop frequency response of the system are shown in Figs. 9.57 and 9.58. The problem states that we are to achieve a settling time of about 12 msec and low overshoot in response to a step command.

As a start on this design, we will aim for a natural frequency ω_n of about 1 rad/msec at the "dominant" second-order poles, and check the response of the higher-order system at that time to see which direction the design modifications should go. With a pole-zero excess (that is, the difference between number of poles and zeros) of 4, we expect the compensator to be required to

Figure 9.57
Uncompensated root locus
of a tape drive

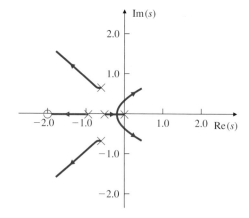

Figure 9.58
Uncompensated Bode plot
of tape transport, $K_v = 1$

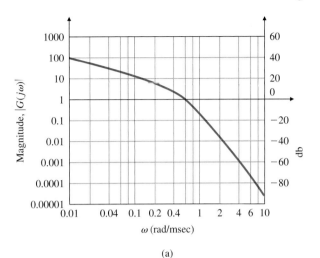

(a)

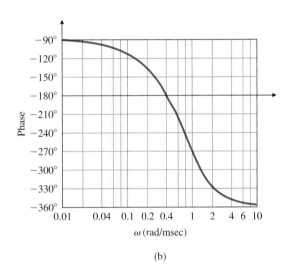

(b)

provide substantial lead if the crossover frequency is $\gtrsim 0.2$ rad/msec. One way to obtain this is to design a compensator with *both* derivative feedback (from the tachometer) and lead-network feedback. If we place both the lead-network zero and the zero due to the tachometer[9] at -0.968 and place the lead network pole a factor of 10 away at -9.68, then the compensator is

$$D_c(s) = K_c \frac{(s + 0.968)^2}{s + 9.68}. \tag{9.52}$$

---

[9] Pole-zero cancellation like this is usually avoided because of sensitivity problems, but it may be acceptable if, as in this case, the canceled pole is sufficiently stable. Cancellation of poles in the unstable region is, of course, absolutely forbidden.

Figure 9.59
Compensated root locus of the tape drive

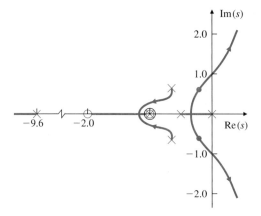

with $K_c = 3$. The root locus of the system is as shown in Fig. 9.59. If we place this controller in the forward loop, in a unity feedback configuration, the step response (not shown) has an overshoot of 48%, which does not meet the specification. The high overshoot is due to the presence of the zeros. If we implement the controller in the feedback loop, the overshoot reduces to 28% with a settling time of about 40 msec as shown in Fig. 9.60. This can be done in this application as both the reference input as well as the output are available to the controller. A disadvantage of implementing the controller in the feedback loop is that the system is no longer type 1 (nonunity feedback) and the reference input needs to be multiplied by the inverse of the closed-loop DC gain, $N = 0.2904$. The ramp response of the system and the associated tension are shown in Fig. 9.61. The frequency response of the compensated system is given in Fig. 9.62. Clearly more design iterations are necessary to meet the specifications.

STEP 6. *Evaluate/modify plant.* Not applicable in this case.

Figure 9.60
Step response with $D_c(s)$ compensation implemented in the feedback loop

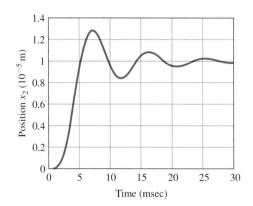

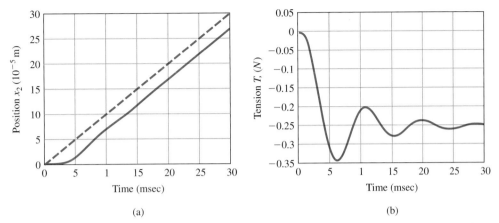

(a)

(b)

Figure 9.61 Ramp response with $D_c(s)$ compensation implemented in the feedback loop: (a) position; (b) tension

Figure 9.62
Bode plot of the
compensated system

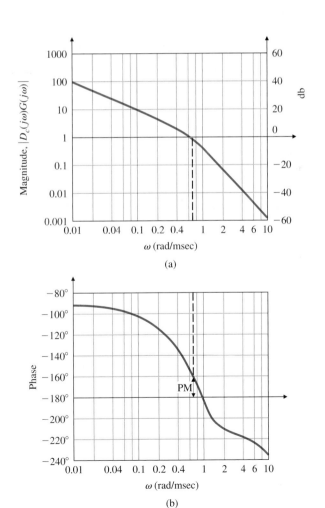

(a)

(b)

Figure 9.63

SRL of the tape drive

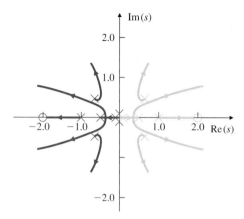

STEP 7. *Try an optimal design using pole placement.* We now consider designing a compensator using the state-variable approach. The SRL of the system
is shown in Fig. 9.63. If we select the closed-loop poles of the system to be on
the SRL at a bandwidth of about 1 rad/msec,

$$pc = [-0.451 + 0.937 * j - 0.451 - 0.937 * j; -0.947 + 0.581 * j;$$

$$-0.947 - 0.581 * j; -1.16],$$

then the required state feedback gain, using the MATLAB function place, is

$$\mathbf{K} = [\, 0.802 \quad 2.58 \quad 0.489 \quad 0.964 \quad 1.21 \,].$$

For the step response of the system shown in Fig. 9.64, the reference input has
again been multiplied by the inverse of the closed-loop DC gain ($N = 1.29$),

Figure 9.64

Step response of full state
feedback

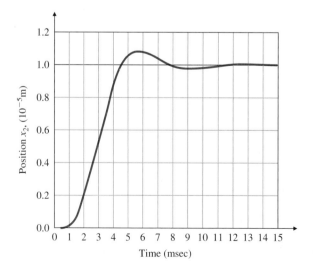

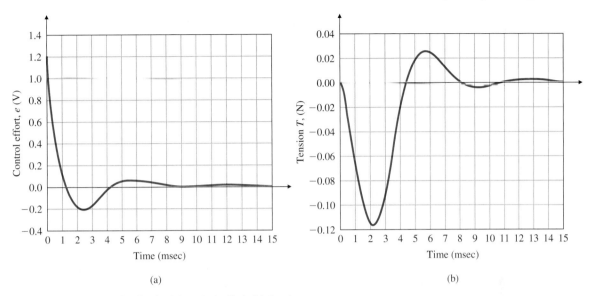

Figure 9.65 Full state feedback: (a) control effort; (b) tension

to get

$$u = -\mathbf{K}\mathbf{x} + Nr,$$

which results in zero steady-state error to a unit step input. The step response shows an overshoot of 8.8% which does meet and exceeds the specification. The associated control effort and tension are given in Fig. 9.65, while Fig. 9.66

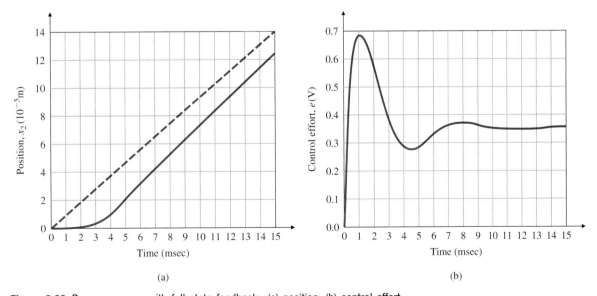

Figure 9.66 Ramp response with full-state feedback: (a) position; (b) control effort

shows the ramp response of the system and its associated control effort. If measurement of all the state is not available and we are to design an estimator based on position measurement alone, such that its poles are five times faster than the control, then

$$pe = [-1.80 + 3.75 * j;\ -1.80 - 3.75 * j;\ -3.79 - 2.32 * j;$$

$$-3.79 + 2.32 * j;\ -4.64],$$

and the estimator gain using the MATLAB place function is

$$\mathbf{L} = \begin{bmatrix} 403.9 \\ 50.6 \\ 13.1 \\ 38.6 \\ 1166.2 \end{bmatrix}.$$

The compensator transfer function is then given by Eq. (7.166)

$$D_c(s) = -\frac{1905(s + 0.640 \pm 0.797j)(s + 0.984 \pm 0.247j)}{(s + 2.38)(s + 1.60 \pm 4.73j)(s + 5.73 \pm 2.64j)}. \tag{9.53}$$

It has the frequency response shown in Fig. 9.67, which is essentially a lead network.

For the complete compensated system, Fig. 9.68 shows the frequency response, and Fig. 9.69 shows the root locus. The step response has an overshoot of about 9.8% as shown in Fig. 9.70, which again meets the specification. The reference input has been again multiplied by the inverse of the closed-loop DC gain (namely, $N = 0.868$). The ramp response of the system and the associated tension are shown in Fig. 9.71.

STEPS 8 and 9.   Verify the design. At this point the system should be simulated with all known nonlinear effects such as friction, and the best design should be selected. A prototype system is typically built and tested with the proposed control hardware.

9.6   Control of the Read/Write Head Assembly of a Hard Disk

The first mass storage device based on recording data on hard disks was introduced by IBM in 1956 as the model 350 RAMAC.[10] The stack of 50 24-in. diameter aluminum disks were coated with a magnetic material and the data was recorded in concentric tracks at 100 bytes per inch with 20 tracks per inch.

---

[10] RAMAC is an acronym for random access method of accounting and control.

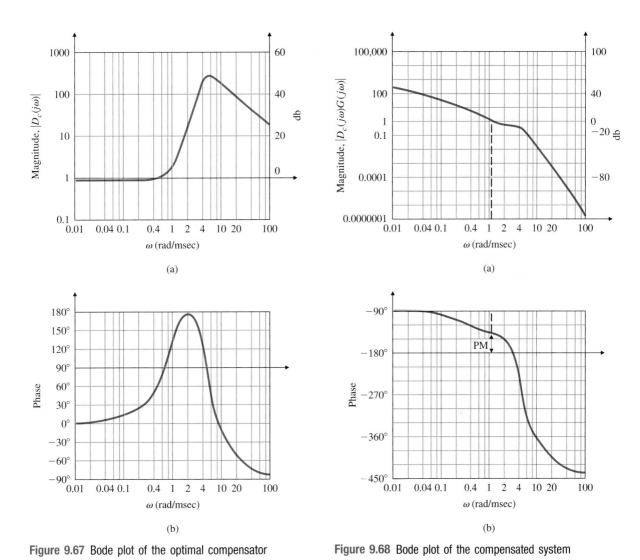

Figure 9.67 Bode plot of the optimal compensator

Figure 9.68 Bode plot of the compensated system

The disks were rotated at 1200 rpm. There was a single read/write head assembly mounted on an arm that could be moved vertically from disk to disk and horizontally across the chosen disk to reach a desired data track. The heads were held above the disk surface by an air bearing generated by blowing air through holes in the fixture holding the heads. The assembly was held on a particular disk by a detent on the elevator mechanism and held on a particular track by an arm detent. The entire head assembly was driven by a single electric motor. The system held 5 megabytes of data and consideration had to be given

Figure 9.69
Root locus of the
compensated tape drive

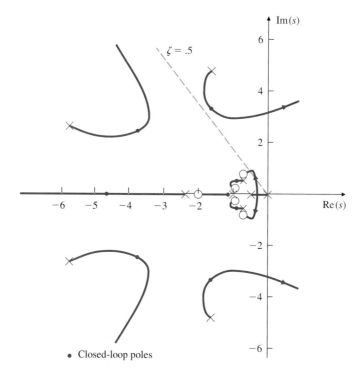

• Closed-loop poles

to be sure that the final device could be passed through a door 36 in. wide. The
technical advances in this field have been such that in the year 2000, Seagate
introduced a hard drive magnetic memory consisting of three disks each 2.5 in.
in diameter rotating at 15,000 rpm designed to be included in a portable laptop
computer. This device could hold 18,350 megabytes of data. The read/write
assembly consisted of a single arm moving a comb of heads, one per surface, in a
rotary motion to move the heads from track to track. The heads are mountable
on a gimbal at the end of the arm and fly above the surfaces of the disks. To

Figure 9.70
Step response for SRL
design

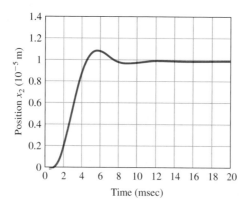

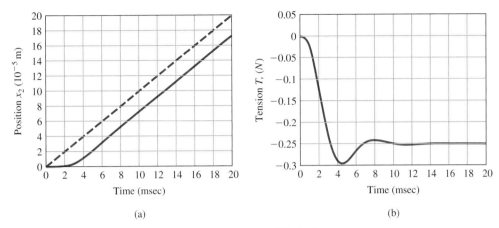

Figure 9.71 Ramp response and associated tension for SRL design

follow a track, the assembly is under active feedback control using samples of position data recorded between the sectors of user data around each track. An economic measure of the progress in the field is that while the cost of the RAMAC data was about $10,000 per megabyte, that of a modern drive is less than 1 cent per megabyte. A brief summary of this remarkable history with many references is given in Abramovitch and Franklin (2001), and a table of a few disk parameters over time is presented in Table 9.1. A large number of people from both industrial and academic institutions have contributed to the many technologies involved in the advances in hard-disk memory devices made over the past 45 years, and one of the enabling technologies has been feedback control. A picture of a Maxtor 100GB disk drive is shown in Fig. 9.72. In this brief case study we will point out a number of issues involving control, but the design example will be concerned only with the issue of track following. We will follow the outline given in Section 9.1 in presenting the case.

STEP 1. *Understand the process.* An exploded view of the track following servo problem is given in Fig. 9.73. The mechanism consists of a rotary voice-coil motor moving an assembly of a light arm supporting gimbal-mounted sliders that include the magnetoresistive read heads and the light, thin-film inductive write heads. The slider flies above the disk surface on an air bearing produced by the disk rotation. The power amplifier is usually connected as a current amplifier so the basic motion can be modeled as simple inertia described by

$$G_o(s) = \frac{A}{Js^2},\qquad(9.54)$$

where J is the total inertia and A includes both the motor torque constant and the amplifier gain. The structure is flexible, however, and the detailed motion is very complex with many lightly damped modes. It is also subject to buffeting from the air flow and from vibration caused by housing motion. For purposes of

TABLE 9.1 Disk Drive Parameters Over Time

No.	Year	Unit	Capacity	Size (N/d)	tpi	bpi	rpm	Fly Height	Head Type	Sensor Type	Actuator Type	Seek Time	Comment
1	1956	IBM RAMAC	5 MB	50/24"	20	100	1200	20 μ	Air bearing	Detent	dc motor		The first hard disk
2	1962	IBM 1301	28 MB	25/24"	50	520	1800		Flying head	Detent	Hydraulic piston	165 ms	
3	1971	IBM 3330	100 MB	11/14"	192	4040		1.2 μ	Ferrite, flying	Dedicated surface	Linear voice coil	30 ms	The first feedback
4	1973	3340 Winchester	70 MB	4/14"	270	5600		0.5 μ	Ferrite, flying	Dedicated surface	Linear voice coil		Low-mass heads
5	1979	IBM 3370	571 MB	7/14"	635	12,134	2964	0.324 μ	Thin film	Dedicated surface	Linear voice coil		
6	1979	IBM 3310	64.5 MB	6/8"	450	8530				Hybrid, sector servo	Rotary voice coil	27 ms	
7	1980	SeagateST506	5 MB	4/5.25"	255	7690				Open loop	Stepper motor	170 ms	5.25" disk for PCs
8	1983	MaxtorXT1140	126 MB	8/5.25"						Sector servo	Rotary voice coil		In-hub spindle motor
9	1991	IBM Corsair	1 GB	8/3.5"	2238	58,874			MR head	Sector servo	Rotary voice coil		
10	1993	Seagate 12550	2.19 GB	10/3.5"			7200			Sector servo	Rotary voice coil		
11	1997	IBMTravelstar	4 GB	3/2.5"	12,500	211,000				Sector servo	Rotary voice coil		
12	2000	Seagate318451	18.3 GB	3/2.5"			15,000			Sector servo	Rotary voice coil		

control design, a single resonant mode will be included according to the model

$$G(s) = \frac{A}{Js^2} \frac{\left(2\zeta \dfrac{s}{\omega_1} + 1\right)}{\left(\dfrac{s^2}{\omega_1} + 2\zeta \dfrac{s}{\omega_1} + 1\right)}, \tag{9.55}$$

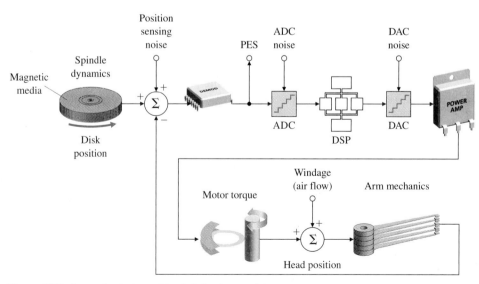

Figure 9.73 Generalized view of track-following model

where the vibration frequency, ω_1, and the damping ratio, ζ, are known only within bounds.

The motion control of the head assembly is in two modes, which are the seek motion to move the head from track to track and the track-follow motion to maintain the heads over the center of the selected track. In the seek mode the criterion is minimum time and theory would call for "on–off" or "bang–bang"[11] control. In order to use the same controller for many units, which differ in the maximum torque available and other critical parameters, the method used in disk drives is a bang–curve–follow technique in which the assembly is accelerated under full torque until the velocity reaches a torque reversal curve based on the distance to the desired track, and deceleration is under feedback control to follow this curve to reach the desired track with zero velocity. The curve approximates the optimal minimum time switching curve with torque discounted to the extent that the weakest motor will have a reserve of torque adequate to follow the curve. When the selected track is reached, the control transfers to track-following mode. A scheme to avoid mode switching when the selected track is approached and cause the servo to move seamlessly into track-follow mode has been called the proximate time optimal servo (PTOS).[12]

As a mature technology, many trends have influenced the nature of the control problem over the years. For example, as Table 9.1 shows, disks have become smaller and thus stiffer and smoother. As the arm assembly has become smaller, it has less inertia to the extent that for very small motions as in a one- or two-track transfer, friction is more important than inertia. For recent drives, the width of a track is on the order of 0.2 micron, a value comparable to the feature dimensions on a modern integrated circuit chip! To counter this trend, research is exploring ways to add a second actuator either on the arm or on the gimbal to make small moves much as the wrist acts on the end of a robot arm. Because of the difficulty of controlling a very lightly damped flexibility, consideration is also given to adding a coating to the arm to increase the damping of the principal modes of vibration. Other proposals include adding sensors on the arm to allow extra feedback to control the flexibility. In this case study, we will assume a single voice-coil actuator and that the flexibility is described as in Eq. (9.55) where $\omega_1 \geq 2\pi \times 2.500$ and $\zeta \geq 0.05$. Because the details of the actual resonance are not well known, the resonance will need to be gain stabilized.

STEP 2. *Select sensors.* The earliest drives were controlled open-loop with one mechanical detent to hold the assembly on a disk and another detent to hold the heads on a track. Feedback control was introduced in 1971 using position information recorded on a special disk surface dedicated to this data. The entire comb of heads was positioned by the servo surface information. If the comb were to tilt or otherwise be misaligned, the data would be that much more

---

[11] Common names for the case where the control is saturated with one polarity for half the time, then reversed for the remaining half.

[12] Workman (1987), Franklin, Powell, and Workman (1998).

difficult to read. Such issues limited the number of disks and the track density possible with this arrangement. The track position information in modern disks is recorded on each track in a gap between the sectors of user data. Controls based on this information are called sector servos, and the data are of necessity sampled. There is a conflict between the desire to record large amounts of data, which calls for fewer and larger sectors, and the control requirement to have a high sample rate which calls for smaller sectors. Each case is a compromise between these conflicting demands. Because the position data are sampled, the controllers are digital devices to make the best possible use of the position data. Theoretical study has been given to using a multi-rate control to apply more than one control correction for each sensor reading, but the method has not been found to be cost effective yet. For the case study here, we will design an analog controller.

The position information extracted from data recorded on the disk is subject to errors caused by run-out in the track path, which means that the radius of the track is not constant. In general, there is a repeatable component in each trip around the track; this element can be estimated, often harmonic by harmonic, and a signal can be used as feedforward to the motor to cancel it out. The position error signal (PES) also contains random noise from many sources. These include the buffeting by the air flow over the slider, wobble and vibration of the disks, noise in the signal processing electronics used to decode the position information, noise from the power amplifier used to provide torque to the motor, and errors caused by the analog to digital converters needed in the process.

STEP 3. *Select actuators.* The RAMAC used a DC motor as actuator, and later drives used hydraulic actuators. When the 5.25-in. drive was introduced by Seagate in 1980, the actuator was a stepping motor. Each of these were used in open loop. The first feedback control of the head position was on the IBM 3330 in 1971, and the actuator was a linear-motion voice-coil motor. In 1979 a rotary voice-coil motor was introduced and today almost all hard disk drives use a rotary motion actuator. The power amplifier is usually connected as a current amplifier to simplify the dynamics. The feedback from the current sensing resistor to the amplifier constitutes a "torque loop" that is designed separately and carefully so the dynamics of the motor can be ignored most of the time in considering the outer loop position control in track following.

STEP 4. *Make a linear model.* As mentioned in the discussion of the process, the linear model has one flexible mode $G(s)$ as

$$G(s) = \frac{1}{s^2} \frac{(2\zeta s/\omega_1 + 1)}{\left(\dfrac{s^2}{\omega_1} + 2\zeta \dfrac{s}{\omega_1} + 1\right)}, \tag{9.56}$$

where we take $\zeta = 0.05$ and $\omega_1 = 2.5$, corresponding to measuring time in milliseconds rather than seconds. The gain, A and the inertia, J will be absorbed in the gain of the compensator. The power amplifier is thus assumed to be an ideal current amplifier. Also we are considering only track following, and not seek.

STEP 5. *Try a PID or lead–lag design.* Because the nominal model is so simple, the first design will be a lead compensation with the objective to achieve the greatest possible bandwidth subject to having a phase margin of 50° and such that it will gain stabilize the resonance with a gain margin of at least 4. This approach was already published by R. K. Oswald (1974). We will try two designs and compare them for bandwidth and the quality of their step responses. In the first case, we will use a simple lead compensation, selected to give 50° phase margin and a factor of 4 gain margin. To get the phase margin, the lead will be designed with an α of 0.1 and the crossover frequency will be placed as high as possible while keeping a gain margin of 4 at the resonance, which rises by a factor of $1/2\zeta = 10$ above the Bode asymptote. Thus the crossover must be located so that the asymptote is a factor of $10 \times 4 = 40$ below 1 at $\omega_1 = 5\pi$. The resulting lead transfer function is

$$D(s) = 0.617 \frac{(2.22s + 1)}{(0.222s + 1)}, \qquad (9.57)$$

and the Bode plot of the lead design is shown in Fig. 9.74.

The gain crossover frequency for this design is $\omega_c = 1.39$ rad/msec and the step response is plotted in Fig. 9.75, which shows a rise time of about $t_r = 0.8$ msec with an overshoot of about 25%. We have shown before that a phase margin of 50° should correspond to a damping of 0.5 and thus an overshoot of about 17%. However, because the zero of the lead is in the forward path, we get the extra overshoot that goes with such a zero.

As a second design, a roll-off filter is to be added to try to suppress the resonance peak in order to gain a bit in speed of response and bandwidth. The idea is to put the filter cutoff frequency between the crossover frequency and the resonance frequency and to give it a damping ratio low enough that it does

Figure 9.74
The Bode plot of the design with a single lead

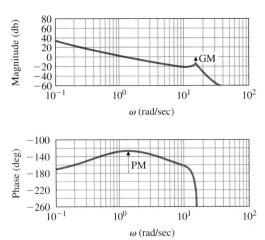

Figure 9.75

Step response of disk drive control with PM=50°

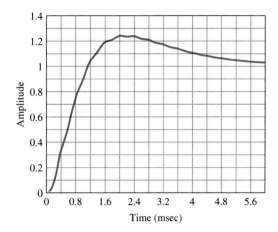

not reduce the phase margin too much but high enough that it does not interfere with the gain margin. After some experimentation, the following trial design is tested:

$$D(s) = 1.44 \frac{(1.48s + 1)}{(0.148s + 1)}, \tag{9.58}$$

with a filter of

$$F(s) = \frac{1}{\dfrac{s^2}{(10.3)^2} + 0.6\dfrac{s}{10.3} + 1}. \tag{9.59}$$

For this case the Bode plot is given in Fig. 9.76 and the step response in Fig. 9.77.

Figure 9.76

Bode plot of system with lead plus roll-off filter

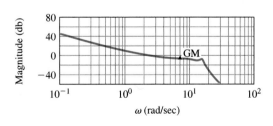

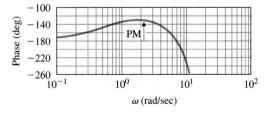

Figure 9.77

Step response of system
with lead plus roll-off filter

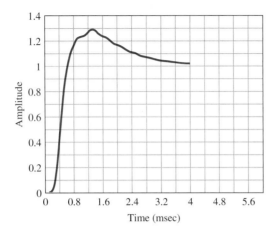

In this case the crossover frequency is 2.13, a 35% increase and the rise
time is 0.3 msec, a 60% reduction from the case without the roll-off filter. The
overshoot is a bit higher in this case. Although not presented here, further
possibilities for the control compensation might include a notch filter rather
than the low-pass filter designed here. A notch might be able to further suppress the resonance and permit further increase in the bandwidth. A great deal
depends on the degree of understanding of the resonance and how much uncertainty surrounds its behavior. In some cases, it is possible to phase stabilize the
resonance and to raise the crossover to be higher than the resonance frequency.

STEP 6. *Evaluate/modify plant.* Possible changes to the process that involve
major design changes were introduced in the discussion concerning understanding the process in Step 1 above. Once the major parameters of the design have
been selected, the remaining possibilities for improvement might include a
change in the fabrication of the arm to add stiffness which will raise the frequency of the vibration and to add a damping coating to the arm to increase the
damping ratio of the flexibility. Other possibilities for improvement concern
changes in the PES decoding methodology to reduce the noise.

STEP 7. *Try an optimal controller or adaptive control.* A design was done with
the linear quadratic performance measure with the performance index (loss
function) selected to obtain a rise time of about 0.3 msec to match the classical
design. The result is shown in Fig. 9.78. Although further effort might produce
an acceptable design, the clearly oscillatory response tolerated by this particular
technique does not look promising. In particular, a design that includes a cost
on $\dot{y}$ as well as y should be considered. Such extensions are considered in more
advanced courses.

STEP 8. *Simulate the design, and compare the alternatives.* Usually done in
parallel with the design.

STEP 9. *Build a prototype.* Done early in the design process as a bench model
so trial schemes can be tested on hardware as designed.

Figure 9.78
Step response for LQR
design

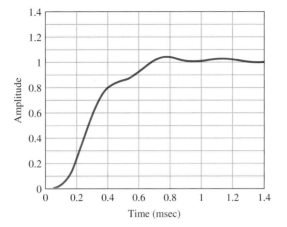

For digital control design and implementation of disk drive servos, the reader is referred to Franklin, Powell, Workman (1998).

9.7 Control of Rapid Thermal Processing (RTP) Systems in Semiconductor Wafer Manufacturing

Figure 9.79 diagrams the major steps in the manufacture of an ultra large scale integrated circuit such as a microprocessor and some of the associated control aspects. Many of the steps described in this process, such as chemical vapor deposition or etching must be done at closely controlled and timed temperature sequences (Sze, 1988). The standard practice for many years has been to perform these steps in batches on many wafers at a time to produce large numbers of identical chips. In response to the demand for ever-smaller critical dimensions of the devices on the chip, and to give more flexibility in the variety and number of chips to be produced, the makers of the tools for fabrication of integrated circuits are asked to provide more and more precise control of temperature and time profiles during thermal processing. In response to these demands, an important trend is to perform the thermal steps on one wafer at a time in a chamber with cold walls and a flexible heat source called a rapid thermal processor (RTP).

RTP

The demands on an RTP system are illustrated by the requirement that it cause the wafer temperature to follow a profile such as that shown in Fig. 9.80, where the ramp-up speeds are at rates of 25–150°C/sec, and the soak temperatures range from 600°C to 1100°C and last from a few seconds to up to 120 sec. The ramp-up rates are limited by the danger of causing damage to the crystal structure if the temperature gradients become too large. The ability of the RTP to change temperature rapidly permits fabrication of devices with very small critical lengths by being able to stop the processes such as deposition or etching quickly and accurately.

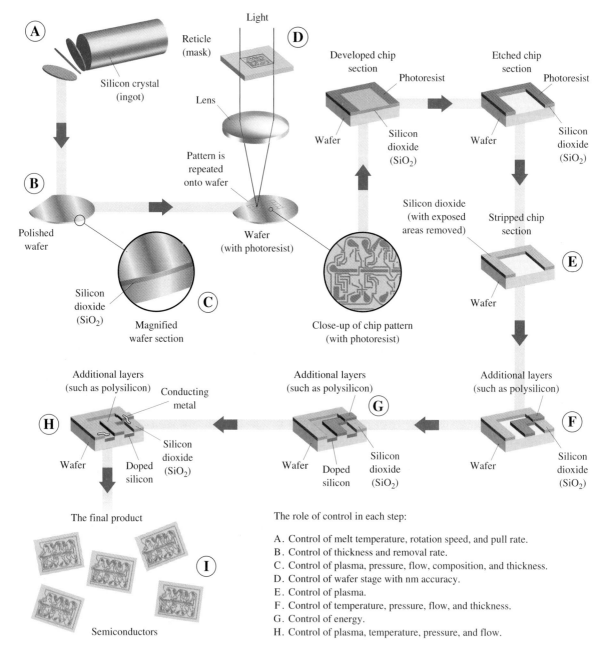

The role of control in each step:

A. Control of melt temperature, rotation speed, and pull rate.
B. Control of thickness and removal rate.
C. Control of plasma, pressure, flow, composition, and thickness.
D. Control of wafer stage with nm accuracy.
E. Control of plasma.
F. Control of temperature, pressure, flow, and thickness.
G. Control of energy.
H. Control of plasma, temperature, pressure, and flow.

Figure 9.79 Steps in making an integrated circuit *(Courtesy International Sematech)*

Figure 9.81 shows a generic RTP reactor with tungsten halogen lamps, stainless steel walls that are water-cooled, and quartz windows. Temperature measurement can be done by a variety of methods including thermocouples,

Figure 9.80
Typical RTP temperature trajectory

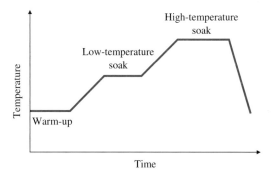

RTDs, and pyrometers. For various reasons (particle generation, minimal disturbance, etc.), it is desirable to use noncontact temperature sensing; therefore, pyrometric techniques are the most commonly employed. A pyrometer is a non-contact temperature sensor and measures the infrared (IR) radiation, which is directly a function of the temperature. It is known that objects emit radiant energy proportional to T^4, where T is the temperature of the object. Among the advantages of pyrometers are that they have very fast response time, can be used to measure the temperature of moving objects (e.g., a rotating semiconductor wafer), and can be used in vacuum for semiconductor manufacturing.

Pyrometer

The selection of the actuator depends on choice of techniques for supplying power (tungsten halogen lamps, arc lamps, hot susceptor, etc.) to heat the wafer. Tungsten halogen lamps are now commonly used in rapid thermal processing in semiconductor manufacturing (Ebert et al., 1995a). Figure 9.82(a) shows a system with two-sided heating by linear tungsten halogen lamps (typical of systems produced by Steag RTP systems). The lamp arrays on the top and bottom are at right angles to provide more of an axisymmetry. Figure 9.82(b) shows a one-sided heating with lamps in a honeycomb configuration (typical of the

Figure 9.81
Generic RTP system

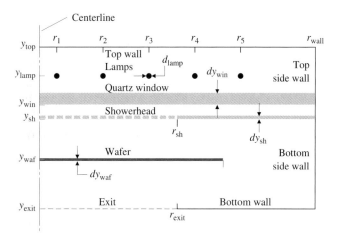

Figure 9.82
Various lamp geometries for RTP (Norman, 1992)

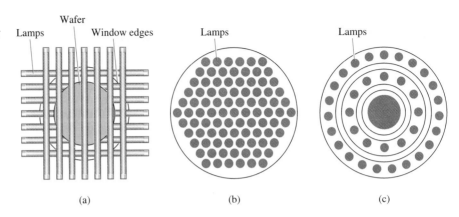

(a)　　　　　(b)　　　　　(c)

Tungsten halogen lamp

Applied Materials systems). Finally Fig. 9.82(c) shows a configuration of lamps arranged in concentric rings (typical of the Stanford-TI MMST chamber, Gyugyi et al., 1993). The lamps do saturate and, for practical reasons, it is desired to operate them within 5% to 95% of power settings.

To illustrate the design of an RTP system, we give the results of a specific design carried out at SC Solutions as a laboratory model constructed to study problems associated with RTP design and operation. The laboratory model is shown schematically in Fig. 9.83. It is made of aluminum. It consists of three standard 35-watt, 12-volt tungsten halogen lamps heating a rectangular plate, which simulates the wafer. The plate measures 4 in. $\times$ $1\frac{3}{4}$ in. and is blackened to increase its radiation absorption. The plate is mounted parallel to the lamps. The lamps are mounted in the lamp housing. The lamp assembly is mounted on a railing so that the distance from the lamps to the plate is adjustable. As the lamps are moved out, the gain of the system decreases but the radiation cross-talk (coupling) increases. On the other hand, as the lamps are moved closer to the plate, the gain of the system increases and the coupling is reduced. The nominal distance from the lamps to the plate is 1 in., but it is adjustable to several inches. The lamps are driven by a pulse-width modulated (PWM) amplifier driver. There is a separate power supply unit. There are three dials mounted on the side for open-loop and manual system operations. There are 14 resistive temperature detector (RTD) strips mounted vertically behind the back of the plate: 12 on the plate and 2 on each support on either side. There is a noise source filter which generates periodic sensor noise at 1.5 Hz so as to represent noise seen in real RTP systems. All electronics—that is, sensor signal processing and PWM amplifier—reside in the enclosure at the bottom of the unit. Because there is exposure to the outside, the surrounding environment

RTP laboratory model

provides sources of disturbance.

STEP 1. *Understand the process and its performance specifications.* RTP is an inherently dynamic and nonlinear process. Among interesting properties of the system are multiple time scales (time constants for lamps, wafer, showerhead, and quartz window are different), nonlinear (radiation dominant) behavior,

Figure 9.83

Block diagram of the RTP laboratory model

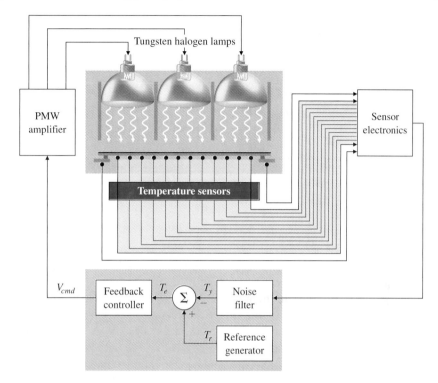

nonlinear lamps, effects of power supplies, number and placement of sensors, number, placement, grouping of lamps, large temperature variations, and so on. The DC gain in the system (δ temperature/δ power) decreases with increasing temperature due to the nonlinear increase in radiative losses. Various types of physical models are needed. Detailed physical models are required for equipment design, but reduced-order models are needed for fast evaluation of geometry changes, recipe development, and feedback control design. Smooth transition between manual and automatic control is also required.

STEP 2. *Select sensors.* This was discussed earlier. For the laboratory model, the sensors were a set of 14 RTDs, but 3 (located at the center and the support edges of the plate) can be used for feedback and the rest can be used for temperature monitoring purposes. In our case, we will use *only the center temperature* for feedback control (another alternative would be to sum up the three temperatures into one signal and control the *average* temperature).

STEP 3. *Select actuators.* This was also discussed earlier. For the laboratory model, the actuators were composed of three standard tungsten halogen lamps described earlier. In our case, we will tie up all three lamps *into a single actuator* by applying the same input command to each lamp.

STEP 4. *Make a linear model.* The laboratory model was built (see Step 9). The nonlinear system equations involve both conduction (recall Chapter 2) and radiation terms (see Ebert et al., 1995a). Nonlinear system identification approaches were used to derive a model for the system. Specifically, the three lamps were stepped up, held constant, and then stepped down sequentially and the three output temperatures were recorded. System identification studies[13] resulted in the following nonlinear model for the system that contains the radiation and conduction terms ($\mathbf{A}_r$ and $\mathbf{A}_{con}$ respectively),

Nonlinear radiation

Heat transfer

$$\mathbf{M}\,\dot{\mathbf{T}} = \mathbf{A}_r \left[\begin{array}{c} \mathbf{T} \\ T_\infty \end{array}\right]^4 + \mathbf{A}_{con} \left[\begin{array}{c} \mathbf{T} \\ T_\infty \end{array}\right] + \mathbf{B}\mathbf{u}, \tag{9.60}$$

where $\mathbf{T} = [T_1\ T_2\ T_3]^T$ denote the temperatures, $T_\infty =$ constant ambient temperature ($\dot{T}_\infty = 0$), $\mathbf{u} = [v_{cmd1}\ v_{cmd2}\ v_{cmd3}]^T$ are the voltage commands, and the system matrices are

$$\mathbf{M}^{-1} = \left[\begin{array}{ccc} 1.000040 & 0 & 0 \\ 0 & 5.557443 & 0 \\ 0 & 0 & 13.638218 \end{array}\right],$$

$$\mathbf{A}_r = \left[\begin{array}{ccc} 5.4762e-2 & -8.5706e-3 & -8.2961e-4 & -4.5361e-2 \\ -8.5706e-3 & 8.5709e-3 & -1.6213e-7 & -8.9134e-8 \\ -8.2961e-4 & -1.6213e-7 & 8.2998e-4 & 2.0976e-7 \end{array}\right],$$

$$\mathbf{A}_{con} = \left[\begin{array}{ccc} 3.5599e-7 & -1.1136e-7 & -1.1976e-7 & -4.7011e-8 \\ -1.1136e-7 & 1.1602e-2 & -2.5027e-3 & -9.0992e-3 \\ -1.9761e-7 & -2.5027e-3 & 6.3736e-3 & -3.8707e-3 \end{array}\right],$$

$$\mathbf{B} = \left[\begin{array}{ccc} 3.4600e-1 & 1.1772e-1 & 2.8380e-2 \\ 3.8803e-11 & 8.0249e-2 & 1.8072e-2 \\ 8.0041e-9 & 2.7216e-3 & 3.1713e-2 \end{array}\right].$$

A linear model for the system was derived as

$$\dot{\mathbf{T}} = \mathbf{F}_3\,\mathbf{T} + \mathbf{G}_3\mathbf{u}, \tag{9.61}$$

$$\mathbf{y} = \mathbf{H}_3\mathbf{T} + \mathbf{J}_3\mathbf{u},$$

RTP linear model

where $\mathbf{y} = [T_{y1}\ T_{y2}\ T_{y3}]^T$ and

$$\mathbf{F}_3 = \left[\begin{array}{ccc} -0.0682 & 0.0149 & 0.0000 \\ 0.0458 & -0.1181 & 0.0218 \\ 0.0000 & 0.04683 & -0.1008 \end{array}\right], \quad \mathbf{G}_3 = \left[\begin{array}{ccc} 0.3787 & 0.1105 & 0.0229 \\ 0.0000 & 0.4490 & 0.0735 \\ 0.0000 & 0.0007 & 0.4177 \end{array}\right],$$

$$\mathbf{H}_3 = \left[\begin{array}{ccc} 1 & 0 & 0 \\ 0 & 1 & 0 \\ 0 & 0 & 1 \end{array}\right], \quad \mathbf{J}_3 = \left[\begin{array}{ccc} 0 & 0 & 0 \\ 0 & 0 & 0 \\ 0 & 0 & 0 \end{array}\right].$$

The three open-loop poles are computed from MATLAB and are located at -0.0527, -0.0863, and -0.1482.

---

[13] Performed by Dr. G. van der Linden.

For our case, because we tied the three lamps into one actuator and are only using the center temperature for feedback, the linear model is then

$$\mathbf{F} = \begin{bmatrix} -0.0682 & 0.0149 & 0.0000 \\ 0.0458 & -0.1181 & 0.0218 \\ 0.0000 & 0.04683 & -0.1008 \end{bmatrix}, \qquad \mathbf{G} = \begin{bmatrix} 0.5122 \\ 0.5226 \\ 0.4185 \end{bmatrix},$$

$$\mathbf{H} = \begin{bmatrix} 0 & 1 & 0 \end{bmatrix}, \qquad\qquad J = [0],$$

resulting in the transfer function

$$G(s) = \frac{T_{y2}(s)}{V_{cmd}(s)} = \frac{0.5226(s + 0.0876)(s + 0.1438)}{(s + 0.1482)(s + 0.0527)(s + 0.0863)}.$$

STEP 5. *Try a lead–lag or PID controller.* We may try a simple PI controller of the form

$$D_c(s) = \frac{(s + 0.0527)}{s},$$

so as to cancel the effect of one of the slower poles. The linear closed-loop response is shown in Fig. 9.84(a) and the associated control effort is shown in Fig. 9.84(b). The system response follows the commanded trajectory with a time delay of approximately 2 sec and no overshoot. The lamp has its normal response until 75 sec and goes negative (shown by dashed line) to try to follow the sharp drop in commanded temperature. This behavior is not possible in the system as there is no means of active cooling and the lamps do saturate low. Note that there is no explicit means of controlling the temperature nonuniformity here.

STEP 6. *Evaluate/modify plant.* This was discussed already in connection with actuator and sensor selection.

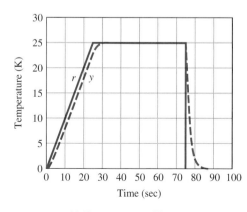

(a) Temperature tracking response

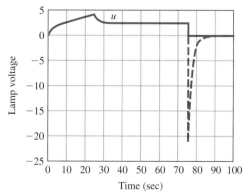

(b) Control effort

Figure 9.84 Linear closed-loop RTP response for PI controller

STEP 7. *Try an optimal design.* We use the error space approach for inclusion of integral control and employ the linear quadratic gaussian technique of Chapter 7. The error system is

$$\begin{bmatrix} \dot{e} \\ \dot{\xi} \end{bmatrix} = \begin{bmatrix} 0 & \mathbf{H} \\ 0 & \mathbf{F} \end{bmatrix} \begin{bmatrix} e \\ \xi \end{bmatrix} + \begin{bmatrix} J \\ \mathbf{G} \end{bmatrix} \mu, \tag{9.62}$$

where

$$\mathbf{A} = \begin{bmatrix} 0 & \mathbf{H} \\ 0 & \mathbf{F} \end{bmatrix}, \mathbf{B} = \begin{bmatrix} J \\ \mathbf{G} \end{bmatrix}$$

and $e = y - r$, $\xi = \dot{\mathbf{T}}$ with $\mu = \dot{u}$. For state feedback design, the LQR formulation of Chapter 7 is used:

$$\mathcal{J} = \int_0^\infty \{\mathbf{z}^T \mathbf{Q}\, \mathbf{z} + \rho \mu^2\}\, dt,$$

where $\mathbf{z} = [e\ \ \xi^T]^T$. Note that $\mathcal{J}$ needs to be chosen in such a way as to penalize the tracking error, e, the control, u, and the differences in the three temperatures; therefore, the performance index should include a term of the form

Temperature uniformity

$$10\left\{(T_1 - T_2)^2 + (T_1 - T_3)^2 + (T_2 - T_3)^2\right\}$$

and hence minimizes the *temperature nonuniformity*. The factor of 10 was determined by trial and error as the relative weighting between the error state and the plant state. The state and control weighting matrices, $\mathbf{Q}$ and R, are then

$$\mathbf{Q} = \begin{bmatrix} 1 & 0 & 0 & 0 \\ 0 & 20 & -10 & -10 \\ 0 & -10 & 20 & -10 \\ 0 & -10 & -10 & 20 \end{bmatrix}, \qquad R = \rho = 1.$$

The following MATLAB command is used to design the feedback gain,

[K] = lqr(A,B,Q,R).

The resulting feedback gain matrix computed from MATLAB is

$$\mathbf{K} = [K_1 : \mathbf{K}_0],$$

where

$$K_1 = 1, \qquad \mathbf{K}_0 = [\,0.1221\quad 2.0788\quad -0.2140\,],$$

which results in the internal model controller of the form

$$\dot{x}_c = B_c\, e \tag{9.63}$$

$$u = C_c\, x_c - \mathbf{K}_0 \mathbf{T}$$

with x_c denoting the controller state and

$$B_c = -K_1 = -1, \qquad C_c = 1.$$

The resulting state-feedback closed-loop poles computed from MATLAB are at $-0.5574 \pm 0.4584 j$, -0.1442, and -0.0877. The full-order estimator was designed with the process and sensor noise intensities selected as the estimator design knobs:

$$R_w = 1, \qquad R_v = 0.001.$$

The following MATLAB command is used to design the estimator:

[L] = lqe(F,G,H,Rw,Rv).

The resulting estimator gain matrix is

$$\mathbf{L} = \begin{bmatrix} 16.1461 \\ 16.4710 \\ 13.2001 \end{bmatrix}$$

with estimator error poles at -16.5268, -0.1438, and -0.0876. The estimator equation is

$$\dot{\hat{\mathbf{T}}} = \mathbf{F}\hat{\mathbf{T}} + \mathbf{G}u + \mathbf{L}(y - \mathbf{H}\hat{\mathbf{T}}). \qquad (9.64)$$

With the estimator, the internal model controller equation is modified as

$$\dot{x}_c = B_c e, \qquad (9.65)$$
$$u = C_c\, x_c - \mathbf{K}_0\hat{\mathbf{T}}.$$

The closed-loop system equations are given by

$$\dot{\mathbf{x}}_{cl} = \mathbf{A}_{cl}\, \mathbf{x}_{cl} + \mathbf{B}_{cl}\, r, \qquad (9.66)$$
$$y = \mathbf{C}_{cl}\, \mathbf{x}_{cl} + \mathbf{D}_{cl}\, r,$$

where r is the reference input temperature trajectory, the closed-loop state vector is $\mathbf{x}_{cl} = [\mathbf{T}^T\ x_c^T\ \hat{\mathbf{T}}^T]^T$, and the system matrices are

$$\mathbf{A}_{cl} = \begin{bmatrix} \mathbf{F} & \mathbf{G}C_c & -\mathbf{G}\mathbf{K}_0 \\ B_c\mathbf{H} & 0 & 0 \\ \mathbf{L}\mathbf{H} & \mathbf{G}C_c & \mathbf{F} - \mathbf{G}\mathbf{K}_0 - \mathbf{L}\mathbf{H} \end{bmatrix}, \qquad \mathbf{B}_{cl} = \begin{bmatrix} 0 \\ -B_c \\ 0 \end{bmatrix},$$
$$\mathbf{C}_{cl} = [\mathbf{H}\ \ 0\ \ 0], \qquad\qquad\qquad\qquad \mathbf{D}_{cl} = [0]$$

with closed-loop poles (computed with MATLAB) located at $-0.5574 \pm 0.4584 j$, -0.1442, -0.0877, -16.5268, -0.1438, and -0.0876 as expected. The closed-loop control structure is shown in Fig. 9.85.

Figure 9.85
Closed-loop control
structure diagram

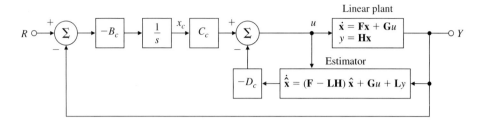

The closed-loop control system diagram implemented in Simulink is shown in Fig. 9.86. The linear closed-loop response is shown in Fig. 9.87(a) and the associated control effort is shown in Fig. 9.87(b). The commanded temperature trajectory, r, is a ramp from $0°C$ to $25°C$ with a $1°C/sec$ slope followed by 50-sec soak time and drop back to $0°C$ (note that the ramp rate is very slow here because we only have three lamps for our RTP laboratory model, whereas a real RTP system would have hundreds of lamps, and the much faster ramp rates mentioned earlier would be relevant). The system tracks the commanded temperature trajectory—albeit with a time delay of approximately 2 sec for the ramp and a maximum of $0.089°C$ overshoot. As expected, the system tracks a constant input asymptotically with zero steady-state error. The lamp command increases as expected to allow for tracking the ramp input, reaches a maximum value at 25 sec, and then drops to a steady-state value around 35 sec. The normal response of the lamp is seen from 0 to 75 sec followed by negative commanded voltage for a few seconds corresponding to fast cooling. Again, the negative control effort voltage (shown by dashed lines) is physically impossible because there is no active cooling in the system. Hence in the nonlinear simulations, commanded lamp power must be constrained to be strictly non-negative (see

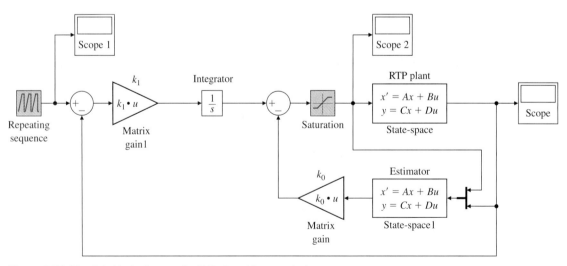

Figure 9.86 Simulink block diagram for RTP closed-loop control

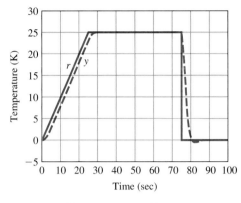

(a) Temperature tracking response

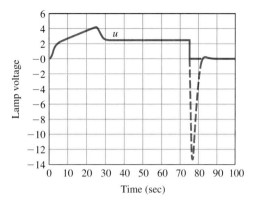

(b) Control effort

Figure 9.87 Linear closed-loop RTP response

Temperature trajectory

Step 8). Note that the response from 75 to 100 sec is that of the (negative) step response of the system.

Simulink nonlinear simulation

STEP 8. *Simulate the design with nonlinearities.* The nonlinear closed-loop system was simulated in Simulink as shown in Fig. 9.88. The model was implemented in temperature units of degrees kelvin and the ambient temperature is 301 K.[14] The nonlinear plant model is the implementation of Eq. (9.60). There is a prefilter following the reference temperature trajectory (to smoothen the sharp corners) with the transfer function,

Pre-filter

$$G_{pf}(s) = \frac{0.2}{s + 0.2}.$$ (9.67)

Lamp nonlinearity

Note that conversion from voltage to power was determined experimentally to be given by

$$P = V^{1.6}$$ (9.68)

and is implemented as a nonlinear block (named VtoPower) in the Simulink diagram accordingly. The inverse of the static nonlinear lamp model is also included as a block (named InvLamp):

$$V = P^{0.625}$$ (9.69)

so as to cancel the lamp nonlinearity. The voltage range for system operation is between 1 to 4 volts as seen from the diagram. A saturation nonlinearity is included for the lamp as well as integrator anti-windup logic to deal with lamp saturation. The nonlinear dynamic response is shown in Fig. 9.89(a) and the control effort is shown in Fig. 9.89(b). Note that the nonlinear response is in general agreement with the linear response.

---

[14] [°K]=[°C]+273.

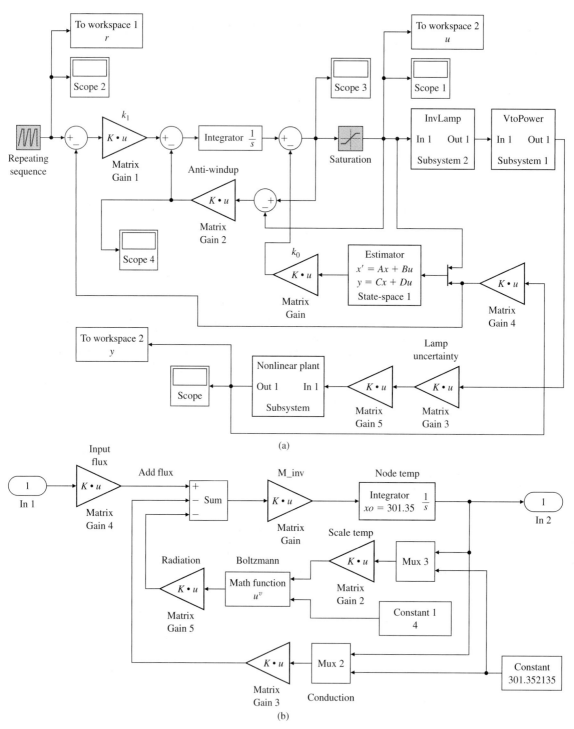

Figure 9.88 Simulink diagram for nonlinear closed-loop RTP system (a) nonlinear closed-loop, (b) nonlinear plant

Figure 9.88
(Continued) Simulink
diagram for nonlinear
closed-loop RTP system
(c) subsystem to convert
voltage-to-power, and
(d) subsystem for lamp
model inversion

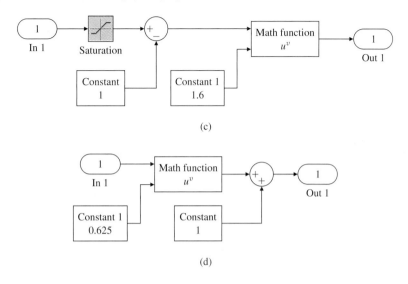

STEP 9. *Build a prototype.* A prototype of the RTP laboratory model was designed, built,[15] and demonstrated at the SEMATECH AEC/APC'98 Conference, in Vail, Colorado. Figure 9.90 shows a photograph of the operational system. This system is really multivariable in nature. The three-input, three-output multivariable controller used on the prototype system was designed using the same approach discussed in Step 7, and it was implemented on an embedded controller platform that uses a real-time operating system.

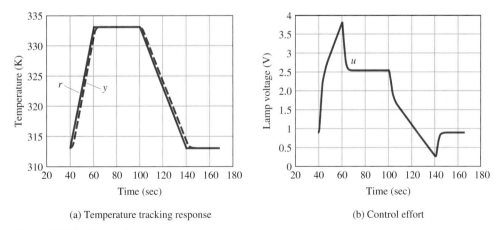

(a) Temperature tracking response

(b) Control effort

Figure 9.89 Nonlinear closed-loop response

---

[15] By Dr. J. L. Ebert.

Figure 9.90
RTP temperature control
laboratory model

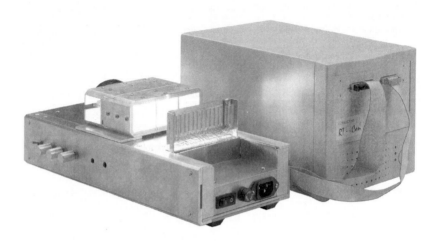

The continuous controller—that is, the combined internal model controller and the estimator—is of the form

$$\dot{\mathbf{x}}^c = \mathbf{A}^c\mathbf{x}^c + \mathbf{B}^c\mathbf{e}, \tag{9.70}$$

$$\mathbf{u} = \mathbf{C}^c\mathbf{x}^c,$$

where $\mathbf{x}^c = [\mathbf{x}_c^T \ \hat{\mathbf{T}}^T]^T$ and

$$\mathbf{A}^c = \begin{bmatrix} \mathbf{0} & \mathbf{0} \\ \mathbf{GC}_c & \mathbf{F} - \mathbf{GK}_0 - \mathbf{LH} \end{bmatrix}, \qquad \mathbf{B}^c = \begin{bmatrix} \mathbf{B}_c \\ \mathbf{L} \end{bmatrix}, \tag{9.71}$$

$$\mathbf{C}^c = [\,\mathbf{C}_c \quad -\mathbf{K}_0\,].$$

The controller was discretized (see Chapter 8) with a sampling period of $T_s = 0.1$ sec and implemented digitally (with appropriate anti-windup logic) as

$$\mathbf{x}^c_{k+1} = \mathbf{\Phi}^c\mathbf{x}_k^c + \mathbf{\Gamma}^c\mathbf{e}_k, \tag{9.72}$$

$$\mathbf{u}_k = \mathbf{C}^c\mathbf{x}_k^c.$$

The response of the actual system to the reference temperature trajectory along with the three lamp voltages is shown in Fig. 9.91. It is in good agreement with the nonlinear closed-loop simulation of the system (once noise is accounted for).

For further information on modeling and control of RTP systems, the reader is referred to Ebert et al. (1995a,b), de Roover et al. (1998), and Gyugyi et al. (1993).

Figure 9.91
Response of the RTP
temperature control
laboratory model

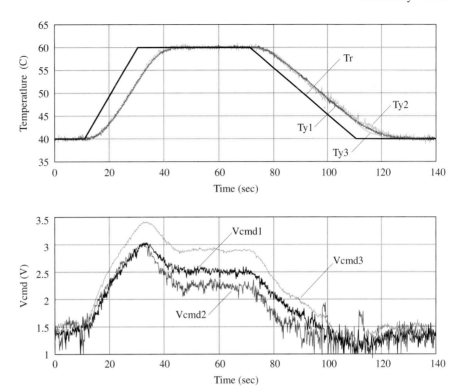

SUMMARY

- In this chapter we have laid out a basic outline of control systems design and applied it to six typical case studies. The design outline calls for a number of explicit steps.

    1. *Make a system model and determine the required performance specifications*. The purpose of this step is to answer the question, What is the system, and what is it supposed to do?

    2. *Select sensors*. A basic rule of control is that if you can't observe it, you can't control it. Factors to consider in the selection of sensors are:

        (a) Number and location of sensors.

        (b) Technology to be used.

        (c) Performance of the sensor, such as its accuracy.

        (d) Physical size and weight.

        (e) Quality of the sensor such as life time and robustness to environment changes.

        (f) Cost.

3. *Select actuators.* The actuators must be capable of driving the system so as to meet the required performance specifications. The selection is governed by the same factors that apply to sensor selection.

4. *Make a linear model.* All our design methods are based on linear models. Both small-signal perturbation models and feedback-linearization methods can be used.

5. *Try a simple PID controller.* An effort to meet the specifications with a PID or its cousin, the lead–lag compensator, may succeed; in any case such an effort will expose the nature of the control problem.

6. *Evaluate/modify plant.* Evaluate whether plant modifications enhance closed-loop performance; if so, return to Step 1 or Step 4.

7. *Try an optimal design.* The SRL method for control-law selection and estimator design based on state equations is guaranteed to produce a stable control system and can be structured to show a tradeoff between error reduction and control effort. A related alternative is arbitrary pole placement, which gives the designer direct control over the dynamic response. Both the SRL and the pole-placement methods may result in designs that are not robust to parameter changes.

8. *Simulate the design, and verify its performance.* All the tools of analysis should be used here, including the root locus, the frequency response, GM and PM measurements, and transient responses. Also, the performance of the design can be tested in simulation against changes in model parameters and the effects of approximating the compensator with a discrete model if digital control is to be used.

9. *Build a prototype, and measure the performance with typical input signals.* The proof of the pudding is in the eating, and no control design is acceptable until it has been tested. No model can include all the features of a real physical device, so the final step before fixing the design is to try it out on a physical prototype if time and budget permit.

- The satellite case study illustrated particularly the use of a notch compensation for a system with lightly damped resonance. Also shown was that collocated actuator and sensor systems are much easier to control than the noncollocated case.

- The Boeing 747 lateral-stabilization case study illustrated the use of feedback as an inner-loop designed to aid the pilot, who provides the primary outer-loop control.

- The Boeing 747 altitude control showed how to combine inner-loop feedback with outer-loop compensation to design a complete control system.

- The automobile fuel–air ratio control illustrated the use of the Bode plot to design a system that includes time delay. Simulation of the design with the nonlinear sensor verified our heuristic analysis of limit cycles using the concept of equivalent gain with a root locus.

- The magnetic-tape case study illustrated a design where two output responses are important: position and tension.

- The disk-drive case study illustrated control in an uncertain environment where bandwidth is very important.

- The rapid thermal processing case study illustrated modeling and control of a nonlinear thermal system.

- In all cases the designer needs to be able to use multiple tools including the root locus, the frequency response, pole placement by state feedback, and simulation of time responses to get a good design. We promised an understanding of these tools at the beginning of the text, and we trust you are now ready to practice the art of control engineering.

Review Questions

1. Why is a collocated actuator and sensor arrangement for a lightly damped structure such as a robot arm easier to design than a noncollocated setup?

2. Why should the control engineer be involved in the design of the process to be controlled?

3. Give examples of an actuator and a sensor for the following control problems:

    (a) Attitude control of a geosynchronous communication satellite.

    (b) Pitch control of a Boeing 747 airliner.

    (c) Track-following control of a CD player.

    (d) Fuel–air ratio control of a spark-ignited automobile engine.

    (e) Position control for an arm of a robot used to paint automobiles.

    (f) Heading control of a ship.

    (g) Attitude control of a helicopter.

Problems

9.1. Of the three types of PID control (proportional, integral, or derivative), which one is the most effective in reducing the error resulting from a constant disturbance? Explain.

9.2. Is there a greater chance of instability when the sensor in a feedback control system for a mechanical structure is not collocated with the actuator? Explain.

9.3. Consider the plant $G(s) = 1/s^3$. Determine whether or not it is possible to stabilize this plant by adding the lead compensator

$$D_c(s) = K\frac{s+a}{s+b}, \qquad a < b.$$

    (a) What is the maximum phase margin of the resulting feedback system?

    (b) Can a system with this plant, together with any number of lead compensators, be made unconditionally stable? Explain why or why not.

9.4. Consider the closed-loop system shown in Fig. 9.92.

   (a) What is the phase margin if $K = 70,000$?

   (b) What is the gain margin if $K = 70,000$?

   (c) What value of K will yield a phase margin of $\sim 70°$?

   (d) What value of K will yield a phase margin of $\sim 0°$?

   (e) Sketch the root locus with respect to K for the system, and determine what value of K causes the system to be on the verge of instability.

   (f) If the disturbance w is a constant and $K = 10,000$, what is the maximum allowable value for w if $y(\infty)$ is to remain less than 0.1? (Assume $r = 0$.)

   (g) Suppose the specifications require you to allow larger values of w than the value you obtained in part (f) but with the same error constraint $[|y(\infty)| < 0.1]$. Discuss what steps you could take to alleviate the problem.

Figure 9.92
Control system for
Problem 9.4

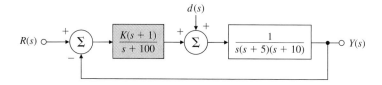

9.5. Consider the system shown in Fig. 9.93, which represents the attitude rate control for a certain aircraft.

   (a) Design a compensator so that the dominant poles are at $-2 \pm 2j$.

   (b) Sketch the Bode plot for your design, and select the compensation so that the crossover frequency is at least $2\sqrt{2}$ rad/sec and PM$\geq 50°$.

   (c) Sketch the root locus for your design, and find the velocity constant when $\omega_n > 2\sqrt{2}$ and $\zeta \geq 0.5$.

Figure 9.93
Block diagram for
aircraft-attitude rate control

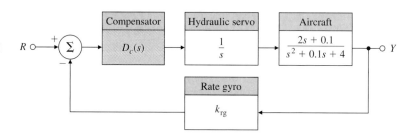

9.6. Consider the block diagram for the servomechanism drawn in Fig. 9.94. Which of the following claims are true?

   (a) The actuator dynamics (the pole at 1000 rad/sec) must be included in an analysis to evaluate a usable maximum gain for which the control system is stable.

   (b) The gain K must be negative for the system to be stable.

   (c) There exists a value of K for which the control system will oscillate at a frequency between 4 and 6 rad/sec.

   (d) The system is unstable if $|K| > 10$.

Figure 9.94
Servomechanism for
Problem 9.6

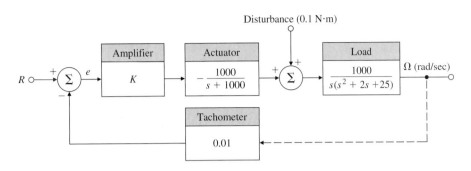

(e) If K must be negative for stability, the control system cannot counteract a positive disturbance.

(f) A positive constant disturbance will speed up the load, thereby making the final value of e negative.

(g) With only a positive constant command input r, the error signal e must have a final value greater than zero.

(h) For $K = -1$ the closed-loop system is stable, and the disturbance results in a speed error whose steady-state magnitude is less than 5 rad/sec.

9.7. A stick balancer and its corresponding control block diagram are shown in Fig. 9.95. The control is a torque applied about the pivot.

(a) Using root-locus techniques, design a compensator $D(s)$ that will place the dominant roots at $s = -5 \pm 5j$ (corresponding to $\omega_n = 7$ rad/sec, $\zeta = 0.707$).

(b) Use Bode plotting techniques to design a compensator $D(s)$ to meet the following specifications:

• Steady-state θ-displacement of less than 0.001 for a constant input torque $T_d = 1$.

• Phase margin $\geq 50°$.

• Closed-loop bandwidth $\cong 7$ rad/sec.

Figure 9.95
Stick balancer

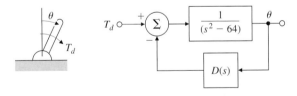

9.8. Consider the standard feedback system drawn in Fig. 9.96.

(a) Suppose

$$G(s) = \frac{2500 K}{s(s + 25)}.$$

Design a lead compensator so that the phase margin of the system is more than $45°$; the steady-state error due to a ramp should be less than or equal to 0.01.

Figure 9.96

Block diagram of a standard feedback control system

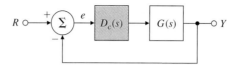

(b) Using the plant transfer function from part (a), design a lead compensator so that the overshoot is less than 25% and the 1% settling time is less than 0.1 sec.

(c) Suppose

$$G(s) = \frac{K}{s(1+0.1s)(1+0.2s)}$$

and the performance specifications are now $K_v = 100$, and PM $\geq 40°$. Is the lead compensation effective for this system? Find a lag compensator, and plot the root locus of the compensated system.

(d) Using $G(s)$ from part (c), design a lag compensator such that the peak overshoot is less than 20% and $K_v = 100$.

(e) Repeat part (c) using a lead–lag compensator.

(f) Find the root locus of the compensated system in part (e), and compare your findings with those from part (c).

9.9. Consider the system in Fig. 9.96, where

$$G(s) = \frac{300}{s(s+0.225)(s+4)(s+180)}.$$

The compensator $D_c(s)$ is to be designed so that the closed-loop system satisfies the following specifications:

- Zero steady-state error for step inputs.
- PM $= 55°$, GM ≥ 6 db.
- Gain crossover frequency is not smaller than that of the uncompensated plant.

(a) What kind of compensation should be used and why?

(b) Design a suitable compensator $D_c(s)$ to meet the specifications.

9.10. We have discussed three design methods: the root-locus method of Evans, the frequency-response method of Bode, and the state-variable pole-assignment method. Explain which of these methods is *best* described by the following statements. If you feel more than one method fits a given statement equally well, say so and explain why.

(a) This method is the one most commonly used when the plant description must be obtained from experimental data.

(b) This method provides the most direct control over dynamic response characteristics such as rise time, percent overshoot, and settling time.

(c) This method lends itself most easily to an automated (computer) implementation.

(d) This method provides the most direct control over the steady-state error constants K_p and K_v.

(e) This method is most likely to lead to the *least complex* controller capable of meeting the dynamic and static accuracy specifications.

(f) This method allows the designer to guarantee that the final design will be unconditionally stable.

(g) This method can be used without modification for plants that include transportation lag terms, for example,

$$G(s) = \frac{e^{-2s}}{(s+3)^2}.$$

9.11. Lead and lag networks are typically employed in designs based on frequency response (Bode) methods. Assuming a type 1 system, indicate the effect of these compensation networks on each of the following performance specifications. In each case, indicate the effect as "an increase," "substantially unchanged," or "a decrease." Use the second-order plant $G(s) = K/[s(s+1)]$ to illustrate your conclusions.

(a) K_v

(b) Phase margin

(c) Closed-loop bandwidth

(d) Percent overshoot

(e) Settling time .

9.12. *Altitude Control of a Hot-Air Balloon*: The equations of vertical motion for a hot-air balloon (Fig. 9.97) linearized about vertical equilibrium are

$$\delta\dot{T} + \frac{1}{\tau_1}\delta T = \delta q,$$

$$\tau_2\ddot{z} + \dot{z} = a\delta T + w,$$

where

$$\delta T = \text{deviation of the hot-air temperature from the equilibrium}$$
$$\text{temperature where buoyant force equals weight,}$$

$$z = \text{altitude of the balloon,}$$

$$\delta q = \text{deviation in the burner heating rate from the equilibrium}$$
$$\text{rate (normalized by the thermal capacity of the hot air),}$$

$$w = \text{vertical component of wind-velocity}$$

$$\tau_1, \tau_2, a = \text{parameters of the equations}$$

An altitude-hold autopilot is to be designed for a balloon whose parameters are

$$\tau_1 = 250 \text{ sec}, \qquad \tau_2 = 25 \text{ sec}, \qquad a = 0.3 \text{ m/(sec} \cdot {}^\circ\text{C)}.$$

Only altitude is sensed, so a control law of the form

$$\delta q(s) = D(s)[z_d(s) - z(s)]$$

will be used, where z_d is the desired (commanded) altitude.

Figure 9.97

Hot-air balloon

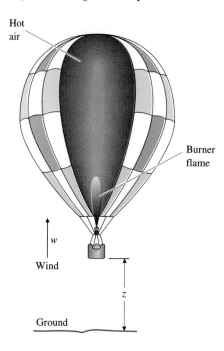

Hot
air

Burner
flame

w

Wind

z

Ground

(a) Sketch a root locus of the closed-loop eigenvalues with respect to the gain K for a proportional feedback controller, $\delta q = -K(z - z_d)$. Use Routh's criterion (or let $s = j\omega$ and find the roots of the characteristic polynomial) to determine the value of the gain and the associated frequency at which the system is marginally stable.

(b) Our intuition and the results of part (a) indicate that a relatively large amount of lead compensation is required to produce a satisfactory autopilot. Sketch a root locus of the closed-loop eigenvalues with respect to the gain K for a double-lead compensator, $\delta q = D(s)(z_d - z)$, where

$$D(s) = K \left(\frac{s + 0.03}{s + 0.12} \right)^2 .$$

(c) Select a gain K for the lead-compensated system to give a crossover frequency of 0.06 rad/sec.

(d) Sketch the magnitude portions of the Bode plots (straight-line asymptotes only) for the open-loop transfer functions of the proportional feedback and lead-compensated systems.

(e) With the gain selected in part (d), what is the steady-state error in altitude for a steady vertical wind of 1 m/sec? (Be careful: First find the closed-loop transfer function from w to the error.)

(f) If the error in part (e) is too large, how would you modify the compensation to give higher low-frequency gain? (Give a qualitative answer only.)

9.13. Satellite-attitude control systems often use a reaction wheel to provide angular motion. The equations of motion for such a system are

$$\text{Satellite}: \quad I\ddot{\phi} = T_c + T_{ex},$$

$$\text{Wheel}: \quad J\dot{r} = -T_c,$$

$$\text{Measurement}: \quad \dot{Z} = \dot{\phi} - aZ,$$

$$\text{Control}: \quad T_c = -D(s)(Z - Z_d).$$

where

$$J = \text{moment of inertia of the wheel},$$

$$r = \text{wheel speed},$$

$$T_c = \text{control torque},$$

$$T_{ex} = \text{disturbance torque},$$

$$\phi = \text{angle to be controlled},$$

$$Z = \text{measurement from the sensor},$$

$$Z_d = \text{reference angle},$$

$$I = \text{satellite inertia } (1000 \text{ kg/m}^2),$$

$$a = \text{sensor constant } (1 \text{ rad/sec}),$$

$$D(s) = \text{compensation}.$$

(a) Suppose $D(s) = K_0$, a constant. Draw the root locus with respect to K_0 for the resulting closed-loop system.

(b) For what range of K_0 is the closed-loop system stable?

(c) Add a lead network with a pole at $s = -1$ so that the closed-loop system has a bandwidth $\omega_{BW} = 0.04$ rad/sec and a damping ratio $\zeta = 0.5$ and the compensation is given by

$$D(s) = K_1 \frac{s+z}{s+1}.$$

Where should the zero of the lead network be located? Draw the root locus of the compensated system, and give the value of K_1 that allows the specifications to be met.

(d) For what range of K_1 is the system stable?

(e) What is the steady-state error (the difference between Z and some reference input Z_d) to a constant disturbance torque T_{ex} for the design of part (c)?

(f) What is the type of this system with respect to rejection of T_{ex}?

(g) Draw the Bode plot asymptotes of the *open-loop* system, with the gain adjusted for the value of K_1 computed in part (c). Add the compensation of part (c), and compute the phase margin of the closed-loop system.

(h) Write state equations for the open-loop system, using the state variables ϕ, $\dot{\phi}$, and Z. Select the gains of a state-feedback controller $T_c = -K_{\dot{\phi}}\dot{\phi} - K_{\phi}\phi$ to locate the closed-loop poles at $s = -0.02 \pm 0.02j\sqrt{3}$.

9.14. Three alternative designs are sketched in Fig. 9.98 for the closed-loop control of a system with the plant transfer function $G(s) = 1/s(s + 1)$. The signal w is the plant noise and may be analyzed as if it were a step; the signal v is the sensor noise and may be analyzed as if it contained power to very high frequencies.

(a) Compute values for the parameters K_1, a, K_2, K_T, K_3, d, and K_D so that in each case (assuming $w = 0$ and $v = 0$),

$$\frac{Y}{R} = \frac{16}{s^2 + 4s + 16}.$$

Note that in system III, a pole is to be placed at $s = -4$.

(b) Complete the following table. Express the last entries as A/s^k to show how fast noise from v is attenuated at high frequencies.

System	K_v	$\left.\dfrac{y}{w}\right|_{s=0}$	$\left.\dfrac{y}{v}\right|_{s\to\infty}$
I			
II			
III			

Figure 9.98
Alternative feedback
structures for Problem 9.14

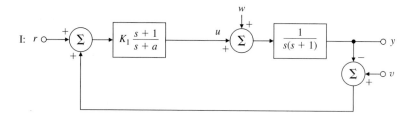

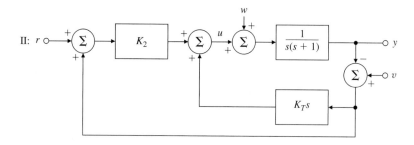

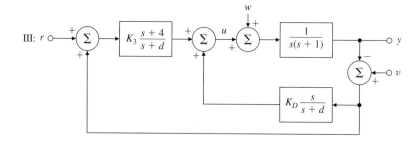

(c) Rank the three designs according to the following characteristics (the best as "1," the poorest as "3"):

	I	II	III
Tracking			
Plant-noise rejection			
Sensor-noise rejection			

9.15. The equations of motion for a cart-stick balancer with state variables of stick angle, stick angular velocity, and cart velocity are

$$\dot{\mathbf{x}} = \begin{bmatrix} 0 & 1 & 0 \\ 31.33 & 0 & 0.016 \\ -31.33 & 0 & -0.216 \end{bmatrix} \mathbf{x} + \begin{bmatrix} 0 \\ -0.649 \\ 8.649 \end{bmatrix} u,$$

$$y = [\,10 \quad 0 \quad 0\,]\mathbf{x},$$

where the output is stick angle, and the control input is voltage on the motor that drives the cart wheels.

(a) Compute the transfer function from u to y, and determine the poles and zeros.

(b) Determine the feedback gain $\mathbf{K}$ necessary to move the poles of the system to the locations -2.832 and $-0.521 \pm 1.068j$, with $\omega_n = 4$ rad/sec.

(c) Determine the estimator gain $\mathbf{L}$ needed to place the three estimator poles at -10.

(d) Determine the transfer function of the estimated-state-feedback compensator defined by the gains computed in parts (b) and (c).

(e) Suppose we use a reduced-order estimator with poles at -10, and -10. What is the required estimator gain?

(f) Repeat part (d) using the reduced-order estimator.

(g) Compute the frequency response of the two compensators.

9.16. A 282-ton Boeing 747 is on landing approach at sea level. If we use the state given in the case study (Section 9.3) and assume a velocity of 221 ft/sec (Mach 0.198), then the lateral-direction perturbation equations are

$$\begin{bmatrix} \dot{\beta} \\ \dot{r} \\ \dot{p} \\ \dot{\phi} \end{bmatrix} = \begin{bmatrix} -0.0890 & -0.989 & 0.1478 & 0.1441 \\ 0.168 & -0.217 & -0.166 & 0 \\ -1.33 & 0.327 & -0.975 & 0 \\ 0 & 0.149 & 1 & 0 \end{bmatrix} \begin{bmatrix} \beta \\ r \\ p \\ \phi \end{bmatrix} + \begin{bmatrix} 0.0148 \\ -0.151 \\ 0.0636 \\ 0 \end{bmatrix} \delta r,$$

$$y = [\,0 \quad 1 \quad 0 \quad 0\,] \begin{bmatrix} \beta \\ r \\ p \\ \phi \end{bmatrix}.$$

The corresponding transfer function is

$$G(s) = \frac{r(s)}{\delta r(s)} = \frac{-0.151(s + 1.05)(s + 0.0328 \pm 0.414j)}{(s + 1.109)(s + 0.0425)(s + 0.0646 \pm 0.731j)}.$$

(a) Draw the uncompensated root locus [for $1 + KG(s)$] and the frequency response of the system. What type of classical controller could be used for this system?

(b) Try a state-variable design approach by drawing a symmetric root locus for the system. Choose the closed-loop poles of the system on the SRL to be

$$\alpha_c(s) = (s + 1.12)(s + 0.165)(s + 0.162 \pm 0.681 j),$$

and choose the estimator poles to be five times faster at

$$\alpha_e(s) = (s + 5.58)(s + 0.825)(s + 0.812 \pm 3.40 j).$$

(c) Compute the transfer function of the SRL compensator.

(d) Discuss the robustness properties of the system with respect to parameter variations and unmodeled dynamics.

(e) Note the similarity of this design with the one developed for different flight conditions earlier in the chapter. What does this suggest about providing a continuous (nonlinear) control throughout the operating envelope?

9.17. (Contributed by Prof. L. Swindlehurst) The feedback control system shown in Fig. 9.99 is proposed as a position control system. A key component of this system is an armature-controlled DC motor. The input potentiometer produces a voltage E_i that is proportional to the desired shaft position: $E_i = K_p \theta_i$. Similarly, the output potentiometer produces a voltage E_0 that is proportional to the actual shaft position: $E_0 = K_p \theta_0$. Note we have assumed that both potentiometers have the same proportionality constant. The error signal $E_i - E_0$ drives a compensator, which in turn produces an armature voltage that drives the motor. The motor has an armature resistance R_a, an armature inductance L_a, a torque constant K_t, and a back-emf constant K_e. The moment of inertia of the motor shaft is J_m, and the rotational damping due to bearing friction is B_m. Finally, the gear ratio is $N : 1$, the moment of inertia of the load is J_L, and the load damping is B_L.

Figure 9.99
A servomechanism with gears on the motor shaft and potentiometer sensors

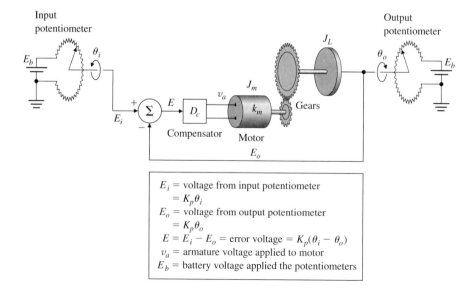

E_i = voltage from input potentiometer
  = $K_p \theta_i$
E_o = voltage from output potentiometer
  = $K_p \theta_o$
$E = E_i - E_o$ = error voltage = $K_p(\theta_i - \theta_o)$
v_a = armature voltage applied to motor
E_b = battery voltage applied the potentiometers

(a) Write the differential equations that describe the operation of this feedback system.

(b) Find the transfer function relating $\theta_0(s)$ and $\theta_i(s)$ for a general compensator $D_c(s)$.

(c) The open-loop frequency-response data shown in Table 9.2 was taken using the armature voltage v_a of the motor as an input and the output potentiometer voltage E_0 as the output. Assuming the motor is linear and minimum-phase, make an estimate of the transfer function of the motor:

$$G(s) = \frac{\theta_m(s)}{V_a(s)},$$

where θ_m is the angular position of the motor shaft.

(d) Determine a set of performance specifications that are appropriate for a position control system and will yield good performance. Design $D_c(s)$ to meet these specifications.

(e) Verify your design through analysis and simulation using MATLAB.

TABLE 9.2

Frequency-Response Data for Problem 9.17

Frequency (rad/sec)	$\left|\frac{E_0(s)}{V_a(s)}\right|$ (db)	Frequency (rad/sec)	$\left|\frac{E_0(s)}{V_a(s)}\right|$ (db)
0.1	60.0	10.0	14.0
0.2	54.0	20.0	2.0
0.3	50.0	40.0	-10.0
0.5	46.0	60.0	-20.0
0.8	42.0	65.0	-21.0
1.0	40.0	80.0	-24.0
2.0	34.0	100.0	-30.0
3.0	30.5	200.0	-48.0
4.0	27.0	300.0	-59.0
5.0	23.0	500.0	-72.0
7.0	19.5		

9.18. Design and construct a device to keep a ball centered on a freely swinging beam. An example of such a device is shown in Fig. 9.100. It uses coils surrounding permanent magnets as the actuator to move the beam, solar cells to sense the ball position, and a hall-effect device to sense the beam position. Research other possible actuators and sensors as part of your design effort. Compare the quality of the control achievable for ball-position-feedback only with that of multiple-loop feedback of both ball and beam position.

Figure 9.100
Ball-balancer design
example

9.19. *Run-to-Run Control*: Consider the Rapid Thermal Processing (RTP) system
shown in Fig. 9.101. We wish to heat up a semiconductor wafer, and control
the wafer surface temperature accurately using rings of tungsten halogen lamps.
The output of the system is temperature, T, as a function of time, $y = T(t)$.
The system reference input, R, is a desired step in temperature ($700°$ C) and the
control input is lamp power. A pyrometer is used to measure the wafer center
temperature. The model of the system is first-order and an integral controller is
used as shown in Fig. 9.101. Normally, there is not a sensor bias ($b = 0$).

(a) Suppose, the system suddenly develops a sensor bias , $b \neq 0$, where b is
known. What can be done to ensure zero steady-state tracking of temperature
command, R, despite the presence of the sensor bias?

(b) Now assume $b = 0$. In reality, we are trying to control the thickness of the
oxide film grown (Ox) on the wafer and not the temperature. At present,
there is not a sensor which measures Ox in real-time. The semiconductor
process engineer must use an off-line equipment (called metrology) to measure the thickness of the oxide film grown on the wafer. There is a nonlinear
relationship between the system output temperature and Ox as follows:

$$\text{Oxide thickness} = \int_0^{t_f} pe^{-\frac{c}{T(t)}} \, dt,$$

Figure 9.101
RTP system

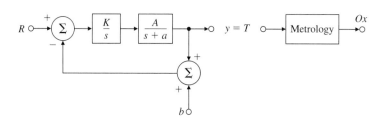

where t_f is the process duration, and p and c are known constants. Suggest a scheme where the center wafer oxide thickness, Ox, can be controlled to a desired value (say, $Ox = 5000$ Å) by employing the temperature controller and the output of the metrology.

9.20. Develop a nonlinear model for a tungsten halogen lamp and simulate it in Simulink.

9.21. Develop a nonlinear model for a pyrometer. Show how temperature can be deduced from the model.

9.22. Repeat the RTP case study design by summing the three sensors to form a single signal to control the average temperature. Demonstrate the performance of the linear design, and validate the performance on the nonlinear Simulink simulation.

A Laplace Transforms

A.1 The $\mathcal{L}_-$ Laplace Transform

Laplace transforms can be used to study the complete response characteristics of feedback systems, including the transient response. This is in contrast to Fourier transforms, in which the steady-state response is the main concern. In many applications it is useful to define the Laplace transform of $f(t)$, denoted by $\mathcal{L}_-\{f(t)\} = F(s)$, as a function of the complex variable $s = \sigma + j\omega$, where

$$F(s) \triangleq \int_{0^-}^{\infty} f(t)e^{-st}\,dt, \qquad (A.1)$$

which uses 0^- (that is, a value just before $t = 0$) as the lower limit of integration and is referred to as the **unilateral** (or **one-sided**) **Laplace transform.**[1] A function $f(t)$ will have a Laplace transform if it is of **exponential order**, which means there exists a real number σ such that

$$\lim_{t \to \infty} |f(t)e^{-\sigma t}| = 0. \qquad (A.2)$$

---

[1] **Bilateral** (or **two-sided**) **Laplace transforms** and the so-called $\mathcal{L}_+$ transforms, in which the lower value of integral is 0^+, also arise elsewhere.

The decaying exponential term in the integrand in effect provides a built-in convergence factor. This means that even if $f(t)$ does not vanish as $t \to \infty$, the integrand will vanish for sufficiently large values of σ if f does not grow at a faster than exponential rate. For example, ae^{bt} is of exponential order, whereas e^{t^2} is not. If $F(s)$ exists for some $s_0 = \sigma_0 + j\omega_0$, then it exists for all values of s such that

$$\mathrm{Re}(s) \geq \sigma_0. \tag{A.3}$$

The smallest value of σ_0 for which $F(s)$ exists is called the **abscissa of convergence**, and the region to the right of $\mathrm{Re}(s) \geq \sigma_0$ is called the **region of convergence**. Typically, two-sided Laplace transforms exist for a specified range,

$$\alpha < \mathrm{Re}(s) < \beta, \tag{A.4}$$

which defines the strip of convergence. Table A.2 gives some Laplace transform pairs. Each entry in the table follows from direct application of the transform definition.

A.1.1   Properties of Laplace Transforms

In this section we will address and prove each of the significant properties of the Laplace transform as discussed in Chapter 3 and listed in Table A.1. In addition we show how these properties can be used through examples.

TABLE A.1

Properties of Laplace Transforms

Number	Laplace Transform	Time Function	Comment
—	$F(s)$	$f(t)$	Transform pair
1	$\alpha F_1(s) + \beta F_2(s)$	$\alpha f_1(t) + \beta f_2(t)$	Superposition
2	$F(s)e^{-s\lambda}$	$f(t - \lambda)$	Time delay ($\lambda \geq 0$)
3	$\frac{1}{|a|} F\left(\frac{s}{a}\right)$	$f(at)$	Time scaling
4	$F(s + a)$	$e^{-at} f(t)$	Shift in frequency
5	$s^m F(s) - s^{m-1} f(0)$ $-s^{m-2} \dot{f}(0) - \cdots - f^{(m-1)}(0)$	$f^{(m)}(t)$	Differentiation
6	$\frac{1}{s} F(s)$	$\int f(\zeta)\, d\zeta$	Integration
7	$F_1(s) F_2(s)$	$f_1(t) * f_2(t)$	Convolution
8	$\lim\limits_{s \to \infty} s F(s)$	$f(0^+)$	Initial Value Theorem
9	$\lim\limits_{s \to 0} s F(s)$	$\lim\limits_{t \to \infty} f(t)$	Final Value Theorem
10	$\frac{1}{2\pi j} \int_{\sigma_c - j\infty}^{\sigma_c + j\infty} F_1(\zeta) F_2(s - \zeta)\, d\zeta$	$f_1(t) f_2(t)$	Time product
11	$-\frac{d}{ds} F(s)$	$t f(t)$	Multiplication by time

TABLE A.2   **Table of Laplace Transforms**

Number	$F(s)$	$f(t),\ t \geq 0$
1	1	$\delta(t)$
2	$1/s$	$1(t)$
3	$1/s^2$	t
4	$2!/s^3$	t^2
5	$3!/s^4$	t^3
6	$m!/s^{m+1}$	t^m
7	$\dfrac{1}{s+a}$	e^{-at}
8	$\dfrac{1}{(s+a)^2}$	te^{-at}
9	$\dfrac{1}{(s+a)^3}$	$\frac{1}{2!}t^2 e^{-at}$
10	$\dfrac{1}{(s+a)^m}$	$\frac{1}{(m-1)!}t^{m-1}e^{-at}$
11	$\dfrac{a}{s(s+a)}$	$1 - e^{-at}$
12	$\dfrac{a}{s^2(s+a)}$	$\frac{1}{a}(at - 1 + e^{-at})$
13	$\dfrac{b-a}{(s+a)(s+b)}$	$e^{-at} - e^{-bt}$
14	$\dfrac{s}{(s+a)^2}$	$(1 - at)e^{-at}$
15	$\dfrac{a^2}{s(s+a)^2}$	$1 - e^{-at}(1 + at)$
16	$\dfrac{(b-a)s}{(s+a)(s+b)}$	$be^{-bt} - ae^{-at}$
17	$\dfrac{a}{s^2 + a^2}$	$\sin at$
18	$\dfrac{s}{s^2 + a^2}$	$\cos at$
19	$\dfrac{s+a}{(s+a)^2 + b^2}$	$e^{-at}\cos bt$
20	$\dfrac{b}{(s+a)^2 + b^2}$	$e^{-at}\sin bt$
21	$\dfrac{a^2 + b^2}{s[(s+a)^2 + b^2]}$	$1 - e^{-at}\left(\cos bt + \frac{a}{b}\sin bt\right)$

1. Superposition

One of the more important properties of the Laplace transform is that it is linear. We can prove this as follows:

$$\mathcal{L}\{\alpha f_1(t) + \beta f_2(t)\} = \int_0^\infty [\alpha f_1(t) + \beta f_2(t)]e^{-st}\, dt \qquad (A.5)$$

$$= \alpha \int_0^\infty f_1(t)e^{-st}\, dt + \beta \int_0^\infty f_2(t)e^{-st}\, dt$$

$$= \alpha F_1(s) + \beta F_2(s).$$

The scaling property is a special case of this; that is,

$$\mathcal{L}\{\alpha f(t)\} = \alpha F(s). \qquad (A.6)$$

EXAMPLE A.1

Sinusoidal Signal

Find the Laplace transform of $f(t) = 1 + 2\sin(\omega t)$.

Solution.   The Laplace transform of $\sin(\omega t)$ is

$$\mathcal{L}\{\sin(\omega t)\} = \frac{\omega}{s^2 + \omega^2}.$$

Therefore, using Eq. (A.5) we obtain

$$F(s) = \frac{1}{s} + \frac{2\omega}{s^2 + \omega^2} = \frac{s^2 + 2\omega s + \omega^2}{s^3 + \omega^2 s}.$$

---

2. Time Delay

Suppose a function $f(t)$ is delayed by $\lambda > 0$ units of time. Its Laplace transform is

$$F_1(s) = \int_0^\infty f(t - \lambda)e^{-st}\, dt.$$

Let us define $t' = t - \lambda$. Then $dt' = dt$ because λ is a constant and $f(t) = 0$ for $t < 0$. Thus

$$F_1(s) = \int_{-\lambda}^\infty f(t')e^{-s(t'+\lambda)}\, dt' = \int_0^\infty f(t')e^{-s(t'+\lambda)}\, dt'.$$

Because $e^{-s\lambda}$ is independent of time, it can be taken out of the integrand, so

$$F_1(s) = e^{-s\lambda} \int_0^\infty f(t')e^{-st'}\, dt' = e^{-s\lambda} F(s). \qquad (A.7)$$

From this result we see that a time delay of λ corresponds to multiplication of the transform by $e^{-s\lambda}$.

EXAMPLE A.2   *Delayed Sinusoidal Signal*

Find the Laplace transform of $f(t) = A \sin(t - t_d)$.

Solution.   The Laplace transform of $\sin(t)$ is

$$\mathcal{L}\{\sin(t)\} = \frac{1}{s^2 + 1}.$$

Therefore, using Eq. (A.7) we obtain

$$F(s) = \frac{A}{s^2 + 1} e^{-st_d}.$$

3. Time Scaling

If the time t is scaled by a factor a, then the Laplace transform of the time-scaled signal is

$$F_1(s) = \int_0^\infty f(at) e^{-st} dt.$$

Again we define $t' = at$. As before, $dt' = a\, dt$ and

$$F_1(s) = \int_0^\infty f(t') \frac{e^{-st'/a}}{|a|} dt' = \frac{1}{|a|} F\left(\frac{s}{a}\right). \tag{A.8}$$

EXAMPLE A.3   *Sinusoid with Frequency ω*

Find the Laplace transform of $f(t) = A \sin(\omega t)$.

Solution.   The Laplace transform of $\sin(t)$ is

$$\mathcal{L}\{\sin(t)\} = \frac{1}{s^2 + 1}.$$

Therefore, using Eq. (A.8) we obtain

$$F(s) = \frac{1}{|\omega|} \frac{1}{(\frac{s}{\omega})^2 + 1}$$
$$= \frac{\omega}{s^2 + \omega^2}$$

as expected.

4. Shift in Frequency

Multiplication (modulation) of $f(t)$ by an exponential expression in the time domain corresponds to a shift in frequency:

$$F_1(s) = \int_0^\infty e^{-at} f(t) e^{-st} dt = \int_0^\infty f(t) e^{-(s+a)t} dt = F(s+a). \qquad (A.9)$$

EXAMPLE A.4   *Exponentially Decaying Sinusoid*

Find the Laplace transform of $f(t) = A\ \sin(\omega t)e^{-at}$.

Solution.   The Laplace transform of $\sin(\omega t)$ is

$$\mathcal{L}\{\sin(\omega t)\} = \frac{\omega}{s^2 + \omega^2}.$$

Therefore, using Eq. (A.9) we obtain

$$F(s) = \frac{A\omega}{(s+a)^2 + \omega^2}.$$

---

5. Differentiation

The transform of the derivative of a signal is related to its Laplace transform and its initial condition as follows:

$$\mathcal{L}\left\{\frac{df}{dt}\right\} = \int_{0^-}^\infty \left(\frac{df}{dt}\right) e^{-st} dt = e^{-st} f(t)|_{0^-}^\infty + s \int_{0^-}^\infty f(t) e^{-st} dt. \qquad (A.10)$$

Because $f(t)$ is assumed to have a Laplace transform, $e^{-st} f(t) \to 0$ as $t \to \infty$. Thus

$$\mathcal{L}\{\dot{f}\} = -f(0^-) + sF(s). \qquad (A.11)$$

Another application of Eq. (A.11) leads to

$$\mathcal{L}\{\ddot{f}\} = s^2 F(s) - sf(0^-) - \dot{f}(0^-). \qquad (A.12)$$

Repeated application of Eq. (A.11) leads to

$$\mathcal{L}\{f^m(t)\} = s^m F(s) - s^{m-1} f(0^-) - s^{m-2} \dot{f}(0^-) - \cdots - f^{(m-1)}(0^-), \qquad (A.13)$$

where $f^m(t)$ denotes the mth derivative of $f(t)$ with respect to time.

EXAMPLE A.5   *Derivative of Cosine Signal*

Find the Laplace transform of $g(t) = \frac{d}{dt} f(t)$, where $f(t) = \cos(\omega t)$.

Solution.   The Laplace transform of $\cos(\omega t)$ is

$$F(s) = \mathcal{L}\{\cos(\omega t)\} = \frac{s}{s^2 + \omega^2}.$$

Using Eq. (A.11) with $f(0^-) = 1$, we have

$$G(s) = \mathcal{L}\{g(t)\} = s \cdot \frac{s}{s^2 + \omega^2} - 1 = -\frac{\omega^2}{s^2 + \omega^2}.$$

---

6. Integration

Let us assume we wish to determine the Laplace transform of the integral of a time function—that is, to find

$$F_1(s) = \mathcal{L}\left\{\int_0^t f(\xi)\, d\xi\right\} = \int_0^\infty \left[\int_0^t f(\xi)\, d\xi\right] e^{-st}\, dt.$$

Employing integration by parts, where

$$u = \int_0^t f(\xi)\, d\xi \quad \text{and} \quad dv = e^{-st}\, dt,$$

we get

$$F_1(s) = \left[-\frac{1}{s} e^{-st} \left(\int_0^t f(\xi)\, d\xi\right)\right]_0^\infty - \int_0^\infty -\frac{1}{s} e^{-st} f(t)\, dt = \frac{1}{s} F(s). \quad \text{(A.14)}$$

EXAMPLE A.6

Time Integral of Sinusoidal Signal

Find the Laplace transform of $f(t) = \int_0^t \sin \omega \tau \, d\tau$.

Solution.   The Laplace transform of $\sin(\omega t)$ is

$$\mathcal{L}\{\sin(\omega t)\} = \frac{\omega}{s^2 + \omega^2}.$$

Therefore, using Eq. (A.14) then

$$F(s) = \frac{\omega}{s^3 + \omega^2 s}.$$

7. Convolution

Convolution in the time domain corresponds to multiplication in the frequency domain. Assume that $\mathcal{L}\{f_1(t)\} = F_1(s)$ and $\mathcal{L}\{f_2(t)\} = F_2(s)$. Then

$$\mathcal{L}\{f_1(t) * f_2(t)\} = \int_0^\infty f_1(t) * f_2(t)e^{-st}dt = \int_0^\infty \left[\int_0^t f_1(\tau)f_2(t-\tau)d\tau\right]e^{-st}dt.$$

Reversing the order of integration and changing the limits of integration yield

$$\mathcal{L}\{f_1(t) * f_2(t)\} = \int_0^\infty \int_\tau^\infty f_1(\tau)f_2(t-\tau)e^{-st}dtd\tau.$$

Multiplying by $e^{-s\tau}e^{s\tau}$ results in

$$\mathcal{L}\{f_1(t) * f_2(t)\} = \int_0^\infty f_1(\tau)e^{-s\tau}\left[\int_\tau^\infty f_2(t-\tau)e^{-s(t-\tau)}dt\right]d\tau.$$

If $t' \stackrel{\triangle}{=} t - \tau$, then

$$\mathcal{L}\{f_1(t) * f_2(t)\} = \int_0^\infty f_1(\tau)e^{-s\tau}d\tau \int_0^\infty f_2(t')e^{-st'}dt'$$

$$\mathcal{L}\{f_1(t) * f_2(t)\} = F_1(s)F_2(s).$$

This implies that

$$\mathcal{L}^{-1}\{F_1(s)F_2(s)\} = f_1(t) * f_2(t). \tag{A.15}$$

EXAMPLE A.7

Ramp Response of a First-Order System

Find the ramp response of a first order system with a pole at $+a$.

Solution.   Let $f_1(t) = t$ be the ramp input and $f_2(t) = e^{at}$ be the impulse response of the first order system. Then, using Eq. (A.15) we find

$$\mathcal{L}^{-1}\left\{\frac{1}{s^2}\frac{1}{s-a}\right\} = f_1(t) * f_2(t)$$

$$= \int_0^t f_1(\tau)f_2(t-\tau)\,d\tau$$

$$= \int_0^t \tau e^{a(t-\tau)}d\tau$$

$$= \frac{1}{a^2}(e^{at} - at - 1).$$

8. Time Product

Multiplication in the time domain corresponds to convolution in the frequency domain:

$$\mathcal{L}\{f_1(t) f_2(t)\} = \frac{1}{2\pi j} \int_{\sigma_c - j\infty}^{\sigma_c + j\infty} F_1(\xi) F_2(s - \xi) \, d\xi.$$

To see this, consider the relation

$$\mathcal{L}\{f_1(t) f_2(t)\} = \int_0^\infty f_1(t) f_2(t) e^{-st} \, dt.$$

Substituting the expression for $f_1(t)$ given by Eq. (3.14) yields

$$\mathcal{L}\{f_1(t) f_2(t)\} = \int_0^\infty \left[\frac{1}{2\pi j} \int_{\sigma_c - j\infty}^{\sigma_c + j\infty} F_1(\xi) e^{\xi t} \, d\xi \right] f_2(t) e^{-st} \, dt.$$

Changing the order of integration results in

$$\mathcal{L}\{f_1(t) f_2(t)\} = \frac{1}{2\pi j} \int_{\sigma_c - j\infty}^{\sigma_c + j\infty} F_1(\xi) \int_0^\infty f_2(t) e^{-(s - \xi)t} \, dt \, d\xi.$$

Using Eq. (A.9), we get

$$\mathcal{L}\{f_1(t) f_2(t)\} = \frac{1}{2\pi j} \int_{\sigma_c - j\infty}^{\sigma_c + j\infty} F_1(\xi) F_2(s - \xi) \, d\xi = \frac{1}{2\pi j} F_1(s) * F_2(s). \quad \text{(A.16)}$$

9. Multiplication by Time

Multiplication by time corresponds to differentiation in the frequency domain. Let us consider

$$\frac{d}{ds} F(s) = \frac{d}{ds} \int_0^\infty e^{-st} f(t) \, dt$$

$$= \int_0^\infty -t e^{-st} f(t) \, dt$$

$$= -\int_0^\infty e^{-st} [t f(t)] \, dt$$

$$= -\mathcal{L}\{t f(t)\}.$$

Then

$$\mathcal{L}\{t f(t)\} = -\frac{d}{ds} F(s), \quad \text{(A.17)}$$

which is the desired result.

EXAMPLE A.8   *Time Product of Sinusoidal Signal*

Find the Laplace transform of $f(t) = t \sin \omega t$.

Solution.   The Laplace transform of $\sin \omega t$ is

$$\mathcal{L}\{\sin(\omega t)\} = \frac{\omega}{s^2 + \omega^2}.$$

Hence using Eq. (A.17), we obtain

$$F(s) = -\frac{d}{ds}\left[\frac{\omega}{s^2 + \omega^2}\right] = \frac{2\omega s}{(s^2 + \omega^2)^2}.$$

A.1.2   Inverse Laplace Transform by Partial-Fraction Expansion

As we saw in Chapter 3, the easiest way to find $f(t)$ from its Laplace transform $F(s)$, if $F(s)$ is rational, is to expand $F(s)$ as a sum of simpler terms that can be found in the tables via partial fraction expansion. We have already discussed this method in connection with simple roots in Section 3.1.5. In this section, we discuss partial-fraction expansion for cases of complex and repeated roots.

Complex Poles In the case of quadratic factors in the denominator, the numerator of the quadratic factor is chosen to be first-order as shown in Example 3.8. Whenever there exists a complex conjugate pair of poles such as

$$F(s) = \frac{C_1}{s - p_1} + \frac{C_2}{s - p_1^*},$$

we can show that

$$C_2 = C_1^*$$

(see Problem 3.1), and that

$$f(t) = C_1 e^{p_1 t} + C_1^* e^{p_1^* t} = 2\mathrm{Re}(C_1 e^{p_1 t}).$$

Assuming that $p_1 = \alpha + j\beta$, we may rewrite $f(t)$ in a more compact form as

$$f(t) = 2\mathrm{Re}\{C_1 e^{p_1 t}\} = 2\mathrm{Re}\{|C_1| e^{j \arg(C_1)} e^{(\alpha + j\beta)t}\} \qquad \text{(A.18)}$$

$$= 2|C_1| e^{\alpha t} \cos[\beta t + \arg(C_1)].$$

EXAMPLE A.9   *Partial-fraction Expansion: Distinct Complex Roots*

Find the function $f(t)$ for which the Laplace transform is

$$F(s) = \frac{1}{s(s^2 + s + 1)}.$$

Solution.   We rewrite $F(s)$ as

$$F(s) = \frac{C_1}{s} + \frac{C_2 s + C_3}{s^2 + s + 1}.$$

Using the cover-up method, we find C_1 to be

$$C_1 = s F(s)|_{s=0} = 1.$$

Setting $C_1 = 1$ and then equating the numerators in the partial fraction expansion relation, we obtain

$$(s^2 + s + 1) + (C_2 s + C_3)s = 1.$$

After solving for C_2 and C_3, we find that $C_2 = -1$ and $C_3 = -1$. To make it more suitable for using the Laplace transform tables, we rewrite the partial fraction as

$$F(s) = \frac{1}{s} - \frac{s + \dfrac{1}{2} + \dfrac{1}{2}}{\left(s + \dfrac{1}{2}\right)^2 + \dfrac{3}{4}}.$$

From the tables we have

$$f(t) = \left(1 - e^{-t/2} \cos \sqrt{\frac{3}{4}} t - \frac{1}{\sqrt{3}} e^{-t/2} \sin \sqrt{\frac{3}{4}} t\right) 1(t)$$

$$= \left(1 - \frac{2}{\sqrt{3}} e^{-t/2} \cos\left(\frac{\sqrt{3}}{2} t - \frac{\pi}{6}\right)\right) 1(t).$$

Alternatively, we may write $F(s)$ as

$$F(s) = \frac{C_1}{s} + \frac{C_2}{s - p_1} + \frac{C_2^*}{s - p_1^*}, \tag{A.19}$$

where $p_1 = -\frac{1}{2} + j\frac{\sqrt{3}}{2}$. $C_1 = 1$ as before, and now

$$C_2 = (s - p_1)F(s)|_{s=p_1} = -\frac{1}{2} + j\frac{1}{2\sqrt{3}}$$

$$C_2^* = -\frac{1}{2} - j\frac{1}{2\sqrt{3}}$$

and

$$f(t) = \quad (1 + 2|C_2|e^{\alpha t}\cos[\beta t + \arg(C_2)])1(t)$$

$$= \left(1 + \frac{2}{\sqrt{3}} e^{-t/2} \cos\left[\frac{\sqrt{3}}{2} t + \frac{5\pi}{6}\right]\right) 1(t)$$

$$= \left(1 - \frac{2}{\sqrt{3}} e^{-t/2} \cos\left[\frac{\sqrt{3}}{2} t - \frac{\pi}{6}\right]\right) 1(t).$$

The latter partial fraction expansion can be readily computed using MATLAB,

```
num = 1;                        % form numerator
den = conv([1 0],[1 1 1]);      % form denominator
[r,p,k] = residue(num,den)      % compute residues
```

which yields the following result:

$$r = [-0.5000 + 0.2887i - 0.5000 - 0.2887i \ 1.0000]';$$

$$p = [-0.5000 + 0.8660i - 0.5000 - 0.8660i \ 0]'; \ k = [];$$

and agrees with the above hand calculations. Note that if we are using the tables, the first method is preferable, while the second method is preferable for checking MATLAB results.

Repeated Poles For the case where $F(s)$ has repeated roots, the procedure to compute the partial-fraction expansion must be modified. If p_1 is repeated three times, we write the partial fraction as

$$F(s) = \frac{C_1}{s - p_1} + \frac{C_2}{(s - p_1)^2} + \frac{C_3}{(s - p_1)^3} + \frac{C_4}{s - p_4} + \cdots + \frac{C_n}{s - p_n}.$$

We determine the constants C_4 through C_n as discussed previously. If we multiply both sides of the above equation by $(s - p_1)^3$,

$$(s - p_1)^3 F(s) = C_1(s - p_1)^2 + C_2(s - p_1) + C_3 + \cdots + \frac{C_n(s - p_1)^3}{s - p_n}, \quad (A.20)$$

and set $s = p_1$, then all the factors on the right side of Eq. (A.20) will go to zero except C_3, which is

$$C_3 = (s - p_1)^3 F(s)|_{s=p_1}$$

as before. To determine the other factors, we differentiate Eq. (A.20) with respect to the Laplace variable s:

$$\frac{d}{ds}[(s - p_1)^3 F(s)] = 2C_1(s - p_1) + C_2 + \ldots + \frac{d}{ds}\left[\frac{C_n(s - p_1)^3}{s - p_n}\right]. \quad (A.21)$$

Again, if we set $s = p_1$, we have

$$C_2 = \frac{d}{ds}[(s - p_1)^3 F(s)]_{s=p_1}.$$

Similarly, if we differentiate Eq. (A.21) again and set $s = p_1$ a second time, we get

$$C_1 = \frac{1}{2}\frac{d^2}{ds^2}[(s - p_1)^3 F(s)]_{s=p_1}.$$

In general, we may compute C_i for a factor with multiplicity k as

$$C_{k-i} = \frac{1}{i!}\left[\frac{d^i}{ds^i}[(s - p_1)^k F(s)]\right]_{s=p_1}, \qquad i = 0, \ldots, k - 1.$$

EXAMPLE A.10   ***Partial-Fraction Expansion: Repeated Real Roots***

Find the function $f(t)$ that has the Laplace transform

$$F(s) = \frac{s+3}{(s+1)(s+2)^2}.$$

Solution.   We write the partial fraction as

$$F(s) = \frac{C_1}{s+1} + \frac{C_2}{s+2} + \frac{C_3}{(s+2)^2}.$$

Then

$$C_1 = (s+1)F(s)\Big|_{s=-1} = \frac{s+3}{(s+2)^2}\Big|_{s=-1} = 2,$$

$$C_2 = \frac{d}{ds}\left[(s+2)^2 F(s)\right]|_{s=-2} = -2,$$

$$C_3 = (s+2)^2 F(s)\Big|_{s=-2} = \frac{s+3}{s+1}\Big|_{s=-2} = -1.$$

The function $f(t)$ is

$$f(t) = (2e^{-t} - 2e^{-2t} - te^{-2t})1(t).$$

The partial fraction computation can also be carried out using MATLAB's residue function,

```
num = [1 3];              % form numerator
den = conv([1 1],[1 4 4]);  % form denominator
[r,p,k] = residue(num,den)  % compute residues
```

which yields the following result

r = [−2 −1 2]'; p = [−2 −2 −1]'; and k = [];

and agrees with the hand calculations.

A.1.3   The Initial Value Theorem

We discussed the Final Value Theorem in Chapter 3. A second valuable Laplace transform theorem is the **Initial Value Theorem,** which states that it is always possible to determine the initial value of the time function $f(t)$ from its Laplace transform. We may also state the theorem in this way:

The Initial Value Theorem

For any Laplace transform pair,

$$\lim_{s \to \infty} sF(s) = f(0^+) \tag{A.22}$$

We may show this as follows. Using Eq. (A.11), we obtain

$$\mathcal{L}\left\{\frac{df}{dt}\right\} = sF(s) - f(0^-) = \int_{0^-}^{\infty} e^{-st}\frac{df}{dt}\,dt. \qquad (A.23)$$

Let us consider the case where $s \to \infty$ and rewrite the integral as

$$\int_{0^-}^{\infty} e^{-st}\frac{df(t)}{dt}\,dt = \int_{0^+}^{\infty} e^{-st}\frac{df(t)}{dt}\,dt + \int_{0^-}^{0^+} e^{-st}\frac{df(t)}{dt}\,dt.$$

Taking the limit of Eq. (A.23) as $s \to \infty$, we get

$$\lim_{s\to\infty}[sF(s) - f(0^-)] = \lim_{s\to\infty}\left[\int_{0^-}^{0^+} e^{0}\frac{df(t)}{dt}\,dt + \int_{0^+}^{\infty} e^{-st}\frac{df(t)}{dt}\,dt\right]$$

The second term on the right side of the above equation approaches zero as $s \to \infty$ because $e^{-st} \to 0$. Hence

$$\lim_{s\to\infty}[sF(s) - f(0^-)] = \lim_{s\to\infty}[f(0^+) - f(0^-)] = f(0^+) - f(0^-),$$

or

$$\lim_{s\to\infty} sF(s) = f(0^+).$$

In contrast with the Final Value Theorem, the Initial Value Theorem can be applied to any function $F(s)$.

EXAMPLE A.11   *Initial Value Theorem*

Find the initial value of the signal in Example 3.10.

Solution.   From the Initial Value Theorem, we get

$$y(0^+) = \lim_{s\to\infty} sY(s) = \lim_{s\to\infty} s\frac{3}{s(s-2)} = 0,$$

which checks with the expression for $y(t)$ computed in Example 3.10.

For a thorough study of Laplace transforms and extensive tables, see Churchill (1972) and Campbell and Foster (1948); for the two-sided transform, see Van der Pol and Bremmer (1955).

B A Review of Complex Variables

This appendix is a brief summary of some results on complex variables theory with emphasis on the facts needed in control theory. For a comprehensive study of basic complex variables theory, see standard textbooks such as Brown and Churchill (1996) or Marsden and Hoffman (1998).

B.1 Definition of a Complex Number

The complex numbers are distinguished from purely real numbers in that they also contain the **imaginary operator**, which we shall denote j. By definition,

$$j^2 = -1 \quad \text{or} \quad j = \sqrt{-1}. \tag{B.1}$$

A **complex number** may be defined as

$$A = \sigma + j\omega, \tag{B.2}$$

where σ is the real part and ω is the imaginary part, denoted respectively as

$$\sigma = \text{Re}(A), \qquad \omega = \text{Im}(A). \tag{B.3}$$

Note that the imaginary part of A is itself a real number.

Graphically, we may represent the complex number A in two ways: in the Cartesian coordinate system [Fig. B.1(a)], A is represented by a single point in the complex plane. In the polar coordinate system, A is represented by a vector with length r and an angle θ; the angle is measured in radians counterclockwise from the positive real axis [Fig. B.1(b)]. In polar form the complex number A is denoted by

$$A = |A| \cdot \angle \arg A = r \cdot \angle \theta = re^{j\theta}, \qquad 0 \leq \theta \leq 2\pi, \qquad \text{(B.4)}$$

where r—called the **magnitude, modulus**, or **absolute value** of A—is the length of the vector representing A,

$$r = |A| = \sqrt{\sigma^2 + \omega^2}, \qquad \text{(B.5)}$$

and where θ is given by

$$\tan \theta = \frac{\omega}{\sigma} \qquad \text{(B.6)}$$

or

$$\theta = \arg(A) = \tan^{-1}\left(\frac{\omega}{\sigma}\right). \qquad \text{(B.7)}$$

Care must be taken to compute the correct value of the angle depending on the sign of the real and imaginary parts; that is, one must find the quadrant in which the complex number lies.

The **conjugate** of A is defined as

$$A^* = \sigma - j\omega. \qquad \text{(B.8)}$$

Figure B.1
The complex number A represented in (a) Cartesian and (b) polar coordinates

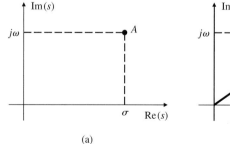

(a)

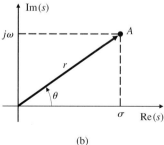

(b)

Therefore,

$$(A^*)^* = A, \tag{B.9}$$

$$(A_1 \pm A_2)^* = A_1^* \pm A_2^*, \tag{B.10}$$

$$\left(\frac{A_1}{A_2}\right)^* = \frac{A_1^*}{A_2^*}, \tag{B.11}$$

$$(A_1 A_2)^* = A_1^* A_2^* \tag{B.12}$$

$$\mathrm{Re}(A) = \frac{A + A^*}{2}, \qquad \mathrm{Im}(A) = \frac{A - A^*}{2j}, \tag{B.13}$$

$$AA^* = (|A|)^2. \tag{B.14}$$

B.2   Algebraic Manipulations

The rules for addition, multiplication, and division are as would be expected.

B.2.1   Complex Addition

If we let

$$A_1 = \sigma_1 + j\omega_1 \quad \text{and} \quad A_2 = \sigma_2 + j\omega_2, \tag{B.15}$$

then

$$A_1 + A_2 = (\sigma_1 + j\omega_1) + (\sigma_2 + j\omega_2) = (\sigma_1 + \sigma_2) + j(\omega_1 + \omega_2). \tag{B.16}$$

Because each complex number is represented by a vector extending from the origin, we can add or subtract complex numbers graphically. The sum is obtained by adding the two vectors. This we do by constructing a parallelogram and finding its diagonal, as shown in Fig. B.2(a). Alternatively, we could start at the tail of one vector, draw a vector parallel to the other vector, and then connect the origin to the new arrowhead.

Complex subtraction is very similar to complex addition.

B.2.2   Complex Multiplication

For two complex numbers defined according to Eq. (B.15),

$$A_1 A_2 = (\sigma_1 + j\omega_1)(\sigma_2 + j\omega_2)$$

$$= (\sigma_1\sigma_2 - \omega_1\omega_2) + j(\omega_1\sigma_2 + \sigma_1\omega_2). \tag{B.17}$$

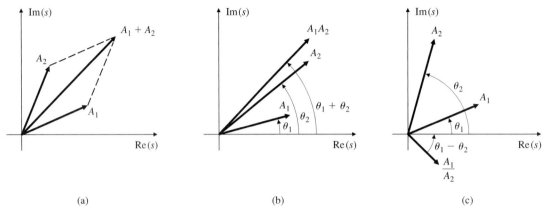

(a)   (b)   (c)

Figure B.2 Arithmetic of complex numbers: (a) addition; (b) multiplication; (c) division

The product of two complex numbers may be done graphically using polar representations as shown in Fig. B.2(b).

B.2.3   Complex Division

The division of two complex numbers is carried out by **rationalization**. This means that both the numerator and denominator in the ratio are multiplied by the conjugate of the denominator:

$$\frac{A_1}{A_2} = \frac{A_1 A_2^*}{A_2 A_2^*}$$

$$= \frac{(\sigma_1\sigma_2 + \omega_1\omega_2) + j(\omega_1\sigma_2 - \sigma_1\omega_2)}{\sigma_2^2 + \omega_2^2}. \tag{B.18}$$

From Eq. (B.4) it follows that

$$A^{-1} = \frac{1}{r}e^{-j\theta}, \qquad r \neq 0. \tag{B.19}$$

Also, if $A_1 = r_1 e^{j\theta_1}$ and $A_2 = r_2 e^{j\theta_2}$, then

$$A_1 A_2 = r_1 r_2 e^{j(\theta_1 + \theta_2)}, \tag{B.20}$$

where $|A_1 A_2| = r_1 r_2$ and $\arg(A_1 A_2) = \theta_1 + \theta_2$ and

$$\frac{A_1}{A_2} = \frac{r_1}{r_2} e^{j(\theta_1 - \theta_2)}, \qquad r_2 \neq 0. \tag{B.21}$$

where $\left|\frac{A_1}{A_2}\right| = \frac{r_1}{r_2}$ and $\arg\left(\frac{A_1}{A_2}\right) = \theta_1 - \theta_2$. The division of complex numbers may be carried out graphically in polar coordinates as shown in Fig. B.2(c).

EXAMPLE B.1

Frequency Response of First-Order System

Find the magnitude and phase of the transfer function $G(s) = \frac{1}{s+1}$ where $s = j\omega$.

Solution.   Substituting $s = j\omega$, and rationalizing, we obtain

$$G(j\omega) = \frac{1}{\sigma + 1 + j\omega} \frac{\sigma + 1 - j\omega}{\sigma + 1 - j\omega}$$
$$= \frac{\sigma + 1 - j\omega}{(\sigma + 1)^2 + \omega^2}.$$

Therefore, the magnitude and phase are

$$|G(j\omega)| = \frac{\sqrt{(\sigma + 1)^2 + \omega^2}}{(\sigma + 1)^2 + \omega^2} = \frac{1}{\sqrt{(\sigma + 1)^2 + \omega^2}},$$
$$\arg(G(j\omega)) = \tan^{-1}\left(\frac{\text{Im}(G(j\omega))}{\text{Re}(G(j\omega))}\right) = \tan^{-1}\left(\frac{-\omega}{\sigma + 1}\right).$$

B.3 Graphical Evaluation of Magnitude and Phase

Consider the transfer function

$$G(s) = \frac{\prod_{i=1}^{m}(s + z_i)}{\prod_{i=1}^{n}(s + p_i)}. \tag{B.22}$$

The value of the transfer function for sinusoidal inputs is found by replacing s with $j\omega$. The gain and phase are given by $G(j\omega)$ and may be determined analytically or by a graphical procedure. Consider the pole-zero configuration for such a $G(s)$ and a point $s_0 = j\omega_0$ on the imaginary axis, as shown in Fig. B.3. Also consider the vectors drawn from the poles and the zero to s_0. The magnitude of the transfer function evaluated at $s_0 = j\omega_0$ is simply the ratio of the distance from the zero to the product of all the distances from the poles:

$$|G(j\omega_0)| = \frac{r_1}{r_2 r_3 r_4}. \tag{B.23}$$

The phase is given by the sum of the angles from the zero minus the sum of the angles from the poles:

$$\arg G(j\omega_0) = \angle G(j\omega_0) = \theta_1 - (\theta_2 + \theta_3 + \theta_4). \tag{B.24}$$

Figure B.3
Graphical determination of
magnitude and phase

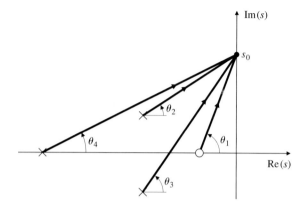

Figure B.4
Illustration of graphical
computation of $s + z_1$

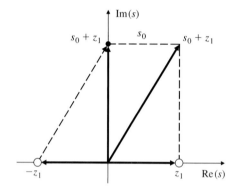

This may be explained as follows. The term $s + z_1$ is a vector addition of its two
components. We may determine this equivalently as $s - (-z_1)$, which amounts
to translation of the vector $s + z_1$ starting at $-z_1$, as shown in Fig. B.4. This
means that a vector drawn from the zero location to s_0 is equivalent to $s + z_1$.
The same reasoning applies to the poles. We reflect p_1, p_2 and p_3 about the
origin to obtain the pole locations. Then the vectors drawn from $-p_1$, $-p_2$ and
$-p_3$ to s_0 are the same as the vectors in the denominator represented in polar
coordinates. Note that this method may also be used to evaluate s_0 at places
in the complex plane besides the imaginary axis.

B.4   Differentiation and Integration

The usual rules apply to complex differentiation. Let $G(s)$ be differentiable
with respect to s. Then the derivative at s_0 is defined as

$$G'(s_0) = \lim_{s \to s_0} \frac{G(s) - G(s_0)}{s - s_0},$$

(B.25)

provided that the limit exists. For conditions on the existence of the derivative see Brown and Churchill (1996).

The standard rules also apply to integration, except that the constant of integration c is a complex constant:

$$\int G(s)\,ds = \int \text{Re}[G(s)]\,ds + j\int \text{Im}[G(s)]\,ds + c. \qquad \text{(B.26)}$$

B.5 Euler's Relations

Let us now derive an important relationship involving the complex exponential. If we define

$$A = \cos\theta + j\sin\theta, \qquad \text{(B.27)}$$

where θ is in radians, then

$$\frac{dA}{d\theta} = -\sin\theta + j\cos\theta = j^2\sin\theta + j\cos\theta$$

$$= j(\cos\theta + j\sin\theta) = jA. \qquad \text{(B.28)}$$

We collect the terms involving A to obtain

$$\frac{dA}{A} = j\,d\theta. \qquad \text{(B.29)}$$

Integrating both sides of Eq. (B.29) yields

$$\ln A = j\theta + c, \qquad \text{(B.30)}$$

where c is a constant of integration. If we let $\theta = 0$ in Eq. (B.30), we find that $c = 0$ or

$$A = e^{j\theta} = \cos\theta + j\sin\theta. \qquad \text{(B.31)}$$

Similarly,

$$A^* = e^{-j\theta} = \cos\theta - j\sin\theta. \qquad \text{(B.32)}$$

Euler's relations

From Eqs. (B.31) and (B.32) it follows that

$$\cos\theta = \frac{e^{j\theta} + e^{-j\theta}}{2}, \qquad \text{(B.33)}$$

$$\sin\theta = \frac{e^{j\theta} - e^{-j\theta}}{2j}. \qquad \text{(B.34)}$$

B.6 Analytic Functions

Let us assume that G is a complex-valued function defined in the complex plane. Let s_0 be in the domain of G, which is assumed to be finite within some disk centered at s_0. Thus $G(s)$ is defined not only at s_0 but also at all points in the disk centered at s_0. The function G is said to be **analytic** if its derivative exists at s_0 and at each point in the neighborhood of s_0.

B.7 Cauchy's Theorem

A **contour** is a piecewise-smooth arc that consists of a number of smooth arcs joined together. **A simple closed contour** is a contour that does not intersect itself and ends on itself. Let C be a closed contour as shown in Fig. B.5(a) and let G be analytic inside and on C. Cauchy's theorem states that

$$\oint_C G(s)\,ds = 0. \tag{B.35}$$

Figure B.5
Contours in the s-plane:
(a) A closed contour;
(b) two different paths
between A_1 and A_2

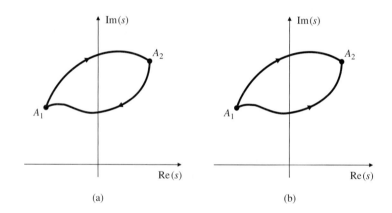

(a)   (b)

A corollary to this theorem is the following. Let C_1 and C_2 be two paths connecting the points A_1 and A_2 as in Fig. B.5(b), then

$$\int_{C_1} G(s)\,ds = \int_{C_2} G(s)\,ds. \tag{B.36}$$

B.8 Singularities and Residues

If a function $G(s)$ is not analytic at s_0 but is analytic at some point in every neighborhood of s_0, it is said to be a **singularity**. A singular point is said to be an **isolated singularity** if $G(s)$ is analytic everywhere else in the neighborhood of s_0 except at s_0. Let $G(s)$ be a **rational function** (that is, a ratio of polynomials).

If the numerator and denominator are both analytic, then $G(s)$ will be analytic except at the locations of the poles (that is, at roots of the denominator). All singularities of rational algebraic functions are the pole locations.

Let $G(s)$ be analytic except at s_0. Then we may write $G(s)$ in its Laurent series-expansion form

$$G(s) = \frac{A_{-n}}{(s - s_0)^n} + \cdots + \frac{A_{-1}}{(s - s_0)} + B_0 + B_1(s - s_0) + \cdots. \tag{B.37}$$

The coefficient A_{-1} is called the **residue** of $G(s)$ at s_0 and may be evaluated as

$$A_{-1} = \text{Res}[G(s); s_0] = \frac{1}{2\pi j} \oint_C G(s)\, ds, \tag{B.38}$$

where C denotes a closed arc within an analytic region centered at s_0 that contains no other singularity as shown in Fig. B.6. When s_0 is not repeated with $n = 1$, then

$$A_{-1} = \text{Res}[G(s); s_0] = (s - s_0)G(s)|_{s=s_0}. \tag{B.39}$$

This is the familiar cover-up method of computing residues.

Figure B.6
Contour around an isolated singularity

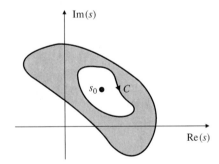

B.9   Residue Theorem

If the contour C contains l singularities, then Eq. (B.39) may be generalized to yield **Cauchy's residue theorem**:

$$\frac{1}{2\pi j} \oint G(s)ds = \sum_{i=1}^{l} \text{Res}[G(s); s_i]. \tag{B.40}$$

B.10 The Argument Principle

Before stating the argument principle, we need a preliminary result from which the principle follows readily.

B.10.1   **Number of Poles and Zeros**

Let $G(s)$ be an analytic function inside and on a closed contour C except for a finite number of poles inside C. Then, for C described in the positive sense (clockwise direction),

$$\frac{1}{2\pi j} \oint \frac{G'(s)}{G(s)} ds = N - P \tag{B.41}$$

or

$$\frac{1}{2\pi j} \oint d(\ln G) = N - P, \tag{B.42}$$

where N and P are the total number of zeros and poles of G inside C, respectively. A pole or zero of multiplicity k is counted k times.

Proof Let s_0 be a zero of G with multiplicity k. Then in some neighborhood of that point we may write $G(s)$ as

$$G(s) = (s - s_0)^k f(s), \tag{B.43}$$

where $f(s)$ is analytic and $f(s_0) \neq 0$. If we differentiate Eq. (B.43) we obtain

$$G'(s) = k(s - s_0)^{k-1} f(s) + (s - s_0)^k f'(s). \tag{B.44}$$

Equation (B.41) may be rewritten as

$$\frac{G'(s)}{G(s)} = \frac{K}{s - s_0} + \frac{f'(s)}{f(s)}. \tag{B.45}$$

Therefore, $G'(s)/G(s)$ has a pole at $s = s_0$ with residue K. This analysis may be repeated for every zero. Hence, the sum of the residues of $G'(s)/G(s)$ is the number of zeros of $G(s)$ inside C. If s_0 is a pole with multiplicity l, we may write

$$h(s) = (s - s_0)^l G(s), \tag{B.46}$$

where $h(s)$ is analytic and $h(s_0) \neq 0$. Then Eq. (B.46) may be rewritten as

$$G(s) = \frac{h(s)}{(s - s_0)^l}. \tag{B.47}$$

Differentiating Eq. (B.47), we obtain

$$G'(s) = \frac{h'(s)}{(s - s_0)^l} - \frac{lh(s)}{(s - s_0)^{l+1}} \tag{B.48}$$

so that

$$\frac{G'(s)}{G(s)} = \frac{-l}{s - s_0} + \frac{h'(s)}{h(s)}. \tag{B.49}$$

This analysis may be repeated for every pole. The result is that the sum of the residues of $G'(s)/G(s)$ at all the poles of $G(s)$ is $-P$.

B.10.2   The Argument Principle

Using Eq. (B.38) we get

$$\frac{1}{2\pi j} \oint_C d[\ln G(s)] = N - P,\tag{B.50}$$

where $d[\ln G(s)]$ was substituted for $G'(s)/G(s)$. If we write $G(s)$ in polar form, then

$$\oint_\Gamma d[\ln G(s)] = \oint_\Gamma d\{\ln |G(s)| + j \arg[\ln G(s)]\}$$

$$= \ln |G(s)| \Big|_{s=s_1}^{s=s_2} + j \arg G(s) \Big|_{s=s_1}^{s=s_2}.\tag{B.51}$$

Because Γ is a closed contour, the first term is zero, but the second term is 2π times the net encirclements of the origin:

$$\frac{1}{2\pi j} \oint_\Gamma d[\ln G(s)] = N - P.\tag{B.52}$$

Intuitively, the argument principle may be stated as follows. We let $G(s)$ be a rational function that is analytic except at possibly a finite number of points. We select an arbitrary contour in the s-plane so that $G(s)$ is analytic at every point on the contour (the contour does not pass through any of the singularities). The corresponding mapping into the $G(s)$-plane may encircle the origin. The number of times it does so is determined by the difference between the number of zeros and the number of poles of $G(s)$ encircled by the s-plane contour. The direction of this encirclement is determined by which is greater, N (clockwise) or P (counterclockwise). For example, if the contour encircles a single zero, the mapping will encircle the origin once in the clockwise direction. Similarly, if the contour encloses only a single pole, the mapping will encircle the origin, this time in the counterclockwise direction. If the contour encircles no singularities, or if the contour encloses an equal number of poles and zeros, there will be no encirclement of the origin. A contour evaluation of $G(s)$ will encircle the origin if there is a nonzero net difference between the encircled singularities. The mapping is **conformal** as well, which means that the magnitude and sense of the angles between smooth arcs is preserved. Chapter 6 provides a more detailed intuitive treatment of the argument principle and its application to feedback control in the form of the Nyquist stability theorem.

C | Summary of Matrix Theory

In the text, we assume you are already somewhat familiar with matrix theory and with the solution of linear systems of equations. However, for the purposes of review we present here a brief summary of matrix theory with an emphasis on the results needed in control theory. For further study, see Strang (1988) and Gantmacher (1959).

C.1 Matrix Definitions

An array of numbers arranged in rows and columns is referred to as a **matrix**. If $\mathbf{A}$ is a matrix with m rows and n columns, an $m \times n$ (read "m by n") matrix, it is denoted by

$$\mathbf{A} = \begin{bmatrix} a_{11} & a_{12} & \cdots & a_{1n} \\ a_{21} & a_{22} & \cdots & a_{2n} \\ \vdots & \vdots & & \vdots \\ a_{m1} & a_{m2} & \cdots & a_{mn} \end{bmatrix}, \tag{C.1}$$

where the entries a_{ij} are its elements. If $m = n$, then the matrix is **square**; otherwise it is **rectangular**. Sometimes a matrix is simply denoted by $\mathbf{A} = [a_{ij}]$. If $m = 1$ or $n = 1$, then the matrix reduces to a **row vector** or a **column vector**, respectively. A **submatrix** of $\mathbf{A}$ is the matrix with certain rows and columns removed.

C.2 Elementary Operations on Matrices

If **A** and **B** are matrices of the same dimension, then their sum is defined by

$$\mathbf{C} = \mathbf{A} + \mathbf{B}, \tag{C.2}$$

where

$$c_{ij} = a_{ij} + b_{ij}. \tag{C.3}$$

Commutative law for addition
Associative law for addition

That is, the addition is done element by element. It is easy to verify the following properties of matrices:

$$\mathbf{A} + \mathbf{B} = \mathbf{B} + \mathbf{A}, \tag{C.4}$$

$$(\mathbf{A} + \mathbf{B}) + \mathbf{C} = \mathbf{A} + (\mathbf{B} + \mathbf{C}). \tag{C.5}$$

Two matrices can be multiplied if they are compatible. Let $\mathbf{A} = m \times n$ and $\mathbf{B} = n \times p$. Then the $m \times p$ matrix

$$\mathbf{C} = \mathbf{AB} \tag{C.6}$$

is the product of the two matrices, where

$$c_{ij} = \sum_{k=1}^{n} a_{ik} b_{kj}. \tag{C.7}$$

Associative law for
multiplication

Matrix multiplication satisfies the associative law

$$\mathbf{A}(\mathbf{BC}) = (\mathbf{AB})\mathbf{C}, \tag{C.8}$$

but not the commutative law; that is, in general,

$$\mathbf{AB} \neq \mathbf{BA}. \tag{C.9}$$

C.3 Trace

The **trace** of a square matrix is the sum of its diagonal elements:

$$\text{trace } \mathbf{A} = \sum_{i=1}^{n} a_{ii}. \tag{C.10}$$

C.4 Transpose

The $n \times m$ matrix obtained by interchanging the rows and columns of $\mathbf{A}$ is called the **transpose of matrix A**:

$$\mathbf{A}^T = \begin{bmatrix} a_{11} & a_{21} & \cdots & a_{m1} \\ a_{12} & a_{22} & \cdots & a_{m2} \\ \vdots & \vdots & & \vdots \\ a_{1n} & a_{2n} & \cdots & a_{mn} \end{bmatrix}.$$

A matrix is said to be **symmetric** if

$$\mathbf{A}^T = \mathbf{A}. \tag{C.11}$$

Transposition

It is easy to show that

$$(\mathbf{AB})^T = \mathbf{B}^T \mathbf{A}^T, \tag{C.12}$$

$$(\mathbf{ABC})^T = \mathbf{C}^T \mathbf{B}^T \mathbf{A}^T, \tag{C.13}$$

$$(\mathbf{A} + \mathbf{B})^T = \mathbf{A}^T + \mathbf{B}^T. \tag{C.14}$$

C.5 Determinant and Matrix Inverse

The **determinant** of a square matrix is defined by Laplace's expansion:

$$\det \mathbf{A} = \sum_{j=1}^{n} a_{ij} \gamma_{ij} \qquad \text{for any } i = 1, 2, \ldots, n, \tag{C.15}$$

where γ_{ij} is called the **cofactor** and

$$\gamma_{ij} = (-1)^{i+j} \det M_{ij}, \tag{C.16}$$

where the scalar $\det M_{ij}$ is called a **minor**. M_{ij} is the same as the matrix $\mathbf{A}$ except that its ith row and jth column have been removed. Note that M_{ij} is always an $(n-1) \times (n-1)$ matrix and that the minors and cofactors are identical except possibly for a sign.

The **adjugate** of a matrix is the transpose of the matrix of its cofactors:

$$\text{adj } \mathbf{A} = [\gamma_{ij}]^T. \tag{C.17}$$

It can be shown that

$$\mathbf{A} \text{ adj } \mathbf{A} = (\det \mathbf{A})\mathbf{I}, \tag{C.18}$$

Identity matrix

where $\mathbf{I}$ is called the **identity matrix**:

$$\mathbf{I} = \begin{bmatrix} 1 & 0 & \cdots & \cdots & 0 \\ 0 & 1 & 0 & \cdots & 0 \\ \vdots & \vdots & \ddots & & \vdots \\ 0 & \cdots & \cdots & 0 & 1 \end{bmatrix},$$

that is, with ones along the diagonal and zeros elsewhere. If $\det \mathbf{A} \neq 0$, then the **inverse** of a matrix $\mathbf{A}$ is defined by

$$\mathbf{A}^{-1} = \frac{\text{adj } \mathbf{A}}{\det \mathbf{A}} \tag{C.19}$$

and has the property that

$$\mathbf{A}\mathbf{A}^{-1} = \mathbf{A}^{-1}\mathbf{A} = \mathbf{I}. \tag{C.20}$$

Note that a matrix has an inverse—that is, it is **nonsingular**—if its determinant is nonzero.

The inverse of the product of two matrices is the product of the inverse of the matrices in reverse order:

$$(\mathbf{AB})^{-1} = \mathbf{B}^{-1}\mathbf{A}^{-1} \tag{C.21}$$

Inversion

and

$$(\mathbf{ABC})^{-1} = \mathbf{C}^{-1}\mathbf{B}^{-1}\mathbf{A}^{-1}. \tag{C.22}$$

C.6 Properties of the Determinant

When dealing with determinants of matrices, the following elementary (row or column) operations are useful:

1. If any row (or column) of $\mathbf{A}$ is multiplied by a scalar α, the resulting matrix $\bar{\mathbf{A}}$ has the determinant

$$\det \bar{\mathbf{A}} = \alpha \det \mathbf{A}. \tag{C.23}$$

Hence

$$\det(\alpha\mathbf{A}) = \alpha^n \det \mathbf{A}. \tag{C.24}$$

2. If any two rows (or columns) of $\mathbf{A}$ are interchanged to obtain $\bar{\mathbf{A}}$, then

$$\det \bar{\mathbf{A}} = -\det \mathbf{A}. \tag{C.25}$$

3. If a multiple of a row (or column) of $\mathbf{A}$ is added to another to obtain $\bar{\mathbf{A}}$, then

$$\det \bar{\mathbf{A}} = \det \mathbf{A}. \tag{C.26}$$

4. It is also easy to show that

$$\det \mathbf{A} = \det \mathbf{A}^T \tag{C.27}$$

and

$$\det \mathbf{AB} = \det \mathbf{A} \det \mathbf{B}. \tag{C.28}$$

Applying Eq. (C.28) to Eq. (C.20), we have that

$$\det \mathbf{A} \det \mathbf{A}^{-1} = 1. \tag{C.29}$$

If $\mathbf{A}$ and $\mathbf{B}$ are square matrices, then the determinant of the block triangular matrix

$$\det \begin{bmatrix} \mathbf{A} & \mathbf{C} \\ \mathbf{0} & \mathbf{B} \end{bmatrix} = \det \mathbf{A} \det \mathbf{B} \tag{C.30}$$

is the product of the determinants of the diagonal blocks. If $\mathbf{A}$ is nonsingular, then

$$\det \begin{bmatrix} \mathbf{A} & \mathbf{B} \\ \mathbf{C} & \mathbf{D} \end{bmatrix} = \det \mathbf{A} \det(\mathbf{D} - \mathbf{C}\mathbf{A}^{-1}\mathbf{B}). \tag{C.31}$$

Using this identity, the transfer function of a scalar system can be written in a compact form:

$$G(s) = \mathbf{H}(s\mathbf{I} - \mathbf{F})^{-1}\mathbf{G} + J = \frac{\det \begin{bmatrix} s\mathbf{I} - \mathbf{F} & \mathbf{G} \\ -\mathbf{H} & J \end{bmatrix}}{\det(s\mathbf{I} - \mathbf{F})}. \tag{C.32}$$

C.7   Inverse of Block Triangular Matrices

If $\mathbf{A}$ and $\mathbf{B}$ are square invertible matrices, then

$$\begin{bmatrix} \mathbf{A} & \mathbf{C} \\ \mathbf{0} & \mathbf{B} \end{bmatrix}^{-1} = \begin{bmatrix} \mathbf{A}^{-1} & -\mathbf{A}^{-1}\mathbf{C}\mathbf{B}^{-1} \\ \mathbf{0} & \mathbf{B}^{-1} \end{bmatrix}. \tag{C.33}$$

C.8   Special Matrices

Diagonal matrix

Some matrices have special structures and are given names. We have already defined the identity matrix, which has a special form. A **diagonal matrix** has (possibly) nonzero elements along the main diagonal and zeros elsewhere:

$$\mathbf{A} = \begin{bmatrix} a_{11} & & & & \mathbf{0} \\ & a_{22} & & & \\ & & a_{33} & & \\ & & & \ddots & \\ \mathbf{0} & & & & a_{nn} \end{bmatrix}. \tag{C.34}$$

Upper triangular matrix

A matrix is said to be **(upper) triangular** if all the elements below the main diagonal are zeros:

$$
\mathbf{A} = \begin{bmatrix} a_{11} & a_{12} & & \cdots & a_{1n} \\ 0 & a_{22} & & & \\ \vdots & & 0 & & \vdots \\ 0 & \vdots & & \ddots & \ddots \\ 0 & 0 & \cdots & 0 & a_{nn} \end{bmatrix}. \tag{C.35}
$$

The determinant of a diagonal or triangular matrix is simply the product of its diagonal elements.

A matrix is said to be in the **(upper) companion form** if it has the structure

$$
\mathbf{A}_c = \begin{bmatrix} -a_1 & -a_2 & & \cdots & -a_n \\ 1 & 0 & & \cdots & 0 \\ 0 & 1 & 0 & \cdots & 0 \\ \vdots & & & \ddots & \vdots \\ 0 & \cdots & & \cdots & 1 & 0 \end{bmatrix}. \tag{C.36}
$$

Note that all the information is contained in the first row. Variants of this form are the lower, left, or right companion matrices. A **Vandermonde matrix** has the following structure:

$$
\mathbf{A} = \begin{bmatrix} 1 & a_1 & a_1^2 & \cdots & a_1^{n-1} \\ 1 & a_2 & a_2^2 & \cdots & a_2^{n-1} \\ \vdots & \vdots & \vdots & & \vdots \\ 1 & a_n & a_n^2 & \cdots & a_n^{n-1} \end{bmatrix}. \tag{C.37}
$$

C.9   Rank

The **rank** of a matrix is the number of its linearly independent rows or columns. If the rank of $\mathbf{A}$ is r, then all $(r+1) \times (r+1)$ submatrices of $\mathbf{A}$ are singular, and there is at least one $r \times r$ submatrix that is nonsingular. It is also true that

$$
\text{row rank of } \mathbf{A} = \text{column rank of } \mathbf{A}. \tag{C.38}
$$

C.10   Characteristic Polynomial

The **characteristic polynomial** of a matrix $\mathbf{A}$ is defined by

$$
a(s) \triangleq \det(s\mathbf{I} - \mathbf{A})
$$
$$
= s^n + a_1 s^{n-1} + \cdots + a_{n-1}s + a_n, \tag{C.39}
$$

where the roots of the polynomial are referred to as **eigenvalues** of $\mathbf{A}$. We can write

$$a(s) = (s - \lambda_1)(s - \lambda_2) \cdots (s - \lambda_n), \tag{C.40}$$

where $\{\lambda_i\}$ are the eigenvalues of $\mathbf{A}$. The characteristic polynomial of a companion matrix [e.g., Eq. (C.36)] is

$$a(s) = \det(s\mathbf{I} - \mathbf{A}_c)$$

$$= s^n + a_1 s^{n-1} + \cdots + a_{n-1}s + a_n. \tag{C.41}$$

C.11 Cayley–Hamilton Theorem

The Cayley–Hamilton theorem states that every square matrix $\mathbf{A}$ satisfies its characteristic polynomial. This means that if $\mathbf{A}$ is an $n \times n$ matrix with characteristic equation $a(s)$, then

$$a(\mathbf{A}) \stackrel{\triangle}{=} \mathbf{A}^n + a_1 \mathbf{A}^{n-1} + \cdots + a_{n-1}\mathbf{A} + a_n \mathbf{I} = 0. \tag{C.42}$$

C.12 Eigenvalues and Eigenvectors

Any scalar λ and nonzero vector $\mathbf{v}$ that satisfy

$$\mathbf{A}\mathbf{v} = \lambda\mathbf{v} \tag{C.43}$$

are referred to as the eigenvalue and the associated **(right) eigenvector** of the matrix $\mathbf{A}$ [because $\mathbf{v}$ appears to the right of $\mathbf{A}$ in Eq. (C.43)]. By rearranging terms in Eq. (C.43) we get

$$(\lambda\mathbf{I} - \mathbf{A})\mathbf{v} = 0. \tag{C.44}$$

Because $\mathbf{v}$ is nonzero, we have

$$\det(\lambda\mathbf{I} - \mathbf{A}) = 0, \tag{C.45}$$

so λ is an eigenvalue of the matrix $\mathbf{A}$ as defined in Eq. (C.43). The normalization of the eigenvectors is arbitrary; that is, if $\mathbf{v}$ is an eigenvector, so is $\alpha\mathbf{v}$. The eigenvectors are usually normalized to have unit length; that is, $\|\mathbf{v}\|^2 = \mathbf{v}^T\mathbf{v} = 1$.

If $\mathbf{w}^T$ is a nonzero row vector such that

$$\mathbf{w}^T\mathbf{A} = \lambda\mathbf{w}^T, \tag{C.46}$$

then $\mathbf{w}$ is called a **left eigenvector** of $\mathbf{A}$ [because $\mathbf{w}^T$ appears to the left of $\mathbf{A}$ in Eq. (C.46)]. Note that we can write

$$\mathbf{A}^T\mathbf{w} = \lambda\mathbf{w} \tag{C.47}$$

so that $\mathbf{w}$ is simply a right eigenvector of $\mathbf{A}^T$.

C.13   Similarity Transformations

Consider the arbitrary nonsingular matrix $\mathbf{T}$ such that

$$\bar{\mathbf{A}} = \mathbf{T}^{-1}\mathbf{A}\mathbf{T}. \tag{C.48}$$

The matrix operation shown in Eq. (C.48) is referred to as a **similarity transformation**. If $\mathbf{A}$ has a full set of eigenvectors, then we can choose $\mathbf{T}$ to be the set of eigenvectors and $\bar{\mathbf{A}}$ will be diagonal.

Consider the set of equations in state-variable form:

$$\dot{\mathbf{x}} = \mathbf{F}\mathbf{x} + \mathbf{G}u. \tag{C.49}$$

If we let

$$\mathbf{T}\xi = \mathbf{x}, \tag{C.50}$$

then Eq. (C.49) becomes

$$\mathbf{T}\dot{\xi} = \mathbf{F}\mathbf{T}\xi + \mathbf{G}u, \tag{C.51}$$

and premultiplying both sides by $\mathbf{T}^{-1}$, we get

$$\dot{\xi} = \mathbf{T}^{-1}\mathbf{F}\mathbf{T}\xi + \mathbf{T}^{-1}\mathbf{G}u$$
$$= \bar{\mathbf{F}}\xi + \bar{\mathbf{G}}u, \tag{C.52}$$

where

$$\bar{\mathbf{F}} = \mathbf{T}^{-1}\mathbf{F}\mathbf{T},$$
$$\bar{\mathbf{G}} = \mathbf{T}^{-1}\mathbf{G}. \tag{C.53}$$

The characteristic polynomial of $\bar{\mathbf{F}}$ is

$$\det(s\mathbf{I} - \bar{\mathbf{F}}) = \det(s\mathbf{I} - \mathbf{T}^{-1}\mathbf{F}\mathbf{T})$$
$$= \det(s\mathbf{T}^{-1}\mathbf{T} - \mathbf{T}^{-1}\mathbf{F}\mathbf{T})$$
$$= \det[\mathbf{T}^{-1}(s\mathbf{I} - \mathbf{F})\mathbf{T}]$$
$$= \det \mathbf{T}^{-1} \det(s\mathbf{I} - \mathbf{F}) \det \mathbf{T}. \tag{C.54}$$

Using Eq. (C.29), Eq. (C.54) becomes

$$\det(s\mathbf{I} - \bar{\mathbf{F}}) = \det(s\mathbf{I} - \mathbf{F}). \tag{C.55}$$

From Eq. (C.55) we can see that $\bar{\mathbf{F}}$ and $\mathbf{F}$ both have the same characteristic polynomial, giving us the important result that a similarity transformation does not change the eigenvalues of a matrix. From Eq. (C.50) a new state made up of a linear combination of old state has the same eigenvalues as the old set.

C.14   Matrix Exponential

Let $\mathbf{A}$ be a square matrix. The **matrix exponential** of $\mathbf{A}$ is defined as the series

$$e^{\mathbf{A}t} = \mathbf{I} + \mathbf{A}t + \frac{1}{2!}\mathbf{A}^2 t^2 + \frac{\mathbf{A}^3 t^3}{3!} + \cdots. \tag{C.56}$$

It can be shown that the series converges. If $\mathbf{A}$ is an $n \times n$ matrix, then $e^{\mathbf{A}t}$ is also an $n \times n$ matrix and can be differentiated:

$$\frac{d}{dt}e^{\mathbf{A}t} = \mathbf{A}e^{\mathbf{A}t}. \tag{C.57}$$

Other properties of the matrix exponential are

$$e^{\mathbf{A}t_1} e^{\mathbf{A}t_2} = e^{\mathbf{A}(t_1 + t_2)} \tag{C.58}$$

and, in general,

$$e^{\mathbf{A}} e^{\mathbf{B}} \neq e^{\mathbf{B}} e^{\mathbf{A}}. \tag{C.59}$$

(In the exceptional case where $\mathbf{A}$ and $\mathbf{B}$ commute—that is, $\mathbf{AB} = \mathbf{BA}$—then $e^{\mathbf{A}} e^{\mathbf{B}} = e^{\mathbf{B}} e^{\mathbf{A}}$).

C.15   Fundamental Subspaces

The **range space** of $\mathbf{A}$, denoted by $\mathcal{R}(\mathbf{A})$ and also called the **column space** of $\mathbf{A}$, is defined by the set of vectors $\mathbf{x}$ where

$$\mathbf{x} = \mathbf{A}\mathbf{y} \tag{C.60}$$

for some vector $\mathbf{y}$. The **null space** of $\mathbf{A}$, denoted by $\mathcal{N}(\mathbf{A})$, is defined by the set of vectors $\mathbf{x}$ such that

$$\mathbf{A}\mathbf{x} = \mathbf{0}. \tag{C.61}$$

If $x \in \mathcal{N}(\mathbf{A})$ and $y \in \mathcal{R}(\mathbf{A}^T)$, then $\mathbf{y}^T \mathbf{x} = 0$; that is, every vector in the null space of $\mathbf{A}$ is **orthogonal** to every vector in the range space of $\mathbf{A}^T$.

C.16 Singular-Value Decomposition

The **singular-value decomposition (SVD)** is one of the most useful tools in linear algebra and has been widely used in control theory during the last three decades. Let $\mathbf{A}$ be an $m \times n$ matrix. Then there always exist matrices $\mathbf{U}, \mathbf{S}$, and $\mathbf{V}$ such that

$$\mathbf{A} = \mathbf{U}\mathbf{S}\mathbf{V}^T. \tag{C.62}$$

Here $\mathbf{U}$ and $\mathbf{V}$ are **orthogonal matrices**; that is,

$$\mathbf{U}\mathbf{U}^T = \mathbf{I}, \mathbf{V}\mathbf{V}^T = \mathbf{I}, \tag{C.63}$$

$\mathbf{S}$ is a **quasidiagonal matrix** with singular values as its diagonal elements; that is,

$$\mathbf{S} = \begin{bmatrix} \mathbf{\Sigma} & 0 \\ 0 & 0 \end{bmatrix}, \tag{C.64}$$

where $\mathbf{\Sigma}$ is a diagonal matrix of nonzero singular values in descending order:

$$\sigma_1 \geq \sigma_2 \geq \cdots \geq \sigma_r > 0. \tag{C.65}$$

The unique diagonal elements of $\mathbf{S}$ are called the **singular values**. The maximum singular value is denoted by $\bar{\sigma}(A)$, and the minimum singular value is denoted by $\underline{\sigma}(\mathbf{A})$. The rank of the matrix is the same as the number of nonzero singular values. The columns of $\mathbf{U}$ and $\mathbf{V}$,

$$\mathbf{U} = [\, u_1 \quad u_2 \quad \ldots \quad u_m \,],$$
$$\mathbf{V} = [\, v_1 \quad v_2 \quad \ldots \quad v_n \,], \tag{C.66}$$

are called the left and right **singular vectors**, respectively. SVD provides complete information about the fundamental subspaces associated with a matrix:

$$\mathcal{N}(\mathbf{A}) = \mathrm{span}[\, v_{r+1} \quad v_{r+2} \quad \ldots \quad v_n \,]$$
$$\mathcal{R}(\mathbf{A}) = \mathrm{span}[\, u_1 \quad u_2 \quad \ldots \quad u_r \,]$$
$$\mathcal{R}(\mathbf{A}^T) = \mathrm{span}[\, v_1 \quad v_2 \quad \ldots \quad v_r \,]$$
$$\mathcal{N}(\mathbf{A}^T) = \mathrm{span}[\, u_{r+1} \quad u_{r+2} \quad \ldots \quad u_m \,]. \tag{C.67}$$

where $\mathcal{N}$ denotes the null space and $\mathcal{R}$ denotes the range space respectively. The **norm** of the matrix $\mathbf{A}$, denoted by $\|\mathbf{A}\|_2$, is given by

$$\|\mathbf{A}\|_2 = \bar{\sigma}(\mathbf{A}). \tag{C.68}$$

If $\mathbf{A}$ is a function of ω, then the infinity norm of $\mathbf{A}$, $\|\mathbf{A}\|_\infty$, is given by

$$\|\mathbf{A}(j\omega)\|_\infty = \max_{\omega} \bar{\sigma}(\mathbf{A}). \tag{C.69}$$

C.17   Positive Definite Matrices

A matrix $\mathbf{A}$ is said to be **positive semidefinite** if

$$\mathbf{x}^T \mathbf{A} \mathbf{x} \geq 0 \qquad \text{for all } \mathbf{x}. \qquad (C.70)$$

The matrix is said to be **positive definite** if equality holds in Eq. (C.70) only for $\mathbf{x} = 0$. A symmetric matrix is positive definite if and only if all of its eigenvalues are positive. It is positive semidefinite if and only if all of its eigenvalues are nonnegative.

An alternate method for determining positive definiteness is to test the minors of the matrix. A matrix is positive definite if all the leading principal minors are positive, and it is positive semidefinite if they are all non-negative.

C.18   Matrix Identity

If $\mathbf{A}$ is $n \times m$ matrix and $\mathbf{B}$ is $m \times n$ matrix then

$$\det[\mathbf{I}_n - \mathbf{AB}] = \det[\mathbf{I}_m - \mathbf{BA}] \qquad (C.71)$$

where $\mathbf{I}_n$ and $\mathbf{I}_m$ are identity matrices of size n and m, respectively.

D Controllability and Observability

Controllability and observability are important structural properties of dynamic system. First identified and studied by Kalman (1960) and later by Kalman et al. (1962), these properties have continued to be examined during the last four decades. We will discuss only a few of the known results for linear constants systems with one input and one output. In the text we discuss these concepts in connection with control law and estimator designs. For example, in Section 7.2 we suggest that if the square matrix given by

$$\mathcal{C} = [\mathbf{G} \ \mathbf{FG} \ \mathbf{F}^2 \ \mathbf{G} \ \ldots \ \mathbf{F}^{n-1}\mathbf{G}] \tag{D.1}$$

is nonsingular, then by transformation of the state we can convert the given description into control canonical form. We can then construct a control law that will give the closed-loop system an arbitrary characteristic equation.

D.1 Controllability

We begin our formal discussion of controllability with the first of four definitions:

Definition I

> The system $(\mathbf{F}, \mathbf{G})$ is **controllable** if for any given nth-order polynomial $\alpha_c(s)$ there exists a (unique) control law $u = -\mathbf{Kx}$ such that the characteristic polynomial of $\mathbf{F} - \mathbf{GK}$ is $\alpha_c(s)$.

From the results of Ackermann's formula (see Appendix E), we have the following mathematical test for controllability: $(\mathbf{F}, \mathbf{G})$ is a controllable pair if and only if the rank of $\mathcal{C}$ is n. Definition I based on pole placement is a frequency-domain concept. Controllability can be equivalently defined in the time domain.

Definition II

> The system $(\mathbf{F}, \mathbf{G})$ is **controllable** if there exists a (piecewise continuous) control signal $u(t)$ that will take the state of the system from any initial state $\mathbf{x}_0$ to any desired final state $\mathbf{x}_f$ in a finite time interval.

We will now show that the system is controllable by this definition if and only if $\mathcal{C}$ is full-rank. We first assume that the system is controllable but that

$$\text{rank}[\mathbf{G}\ \mathbf{FG}\ \mathbf{F}^2\mathbf{G}\ \ldots\ \mathbf{F}^{n-1}\mathbf{G}] < n. \tag{D.2}$$

We can then find a vector $\mathbf{v}$ such that

$$\mathbf{v}[\mathbf{G}\ \mathbf{FG}\ \mathbf{F}^2\mathbf{G}\ \ldots\ \mathbf{F}^{n-1}\mathbf{G}] = 0, \tag{D.3}$$

or

$$\mathbf{v}\mathbf{G} = \mathbf{v}\mathbf{FG} = \mathbf{v}\mathbf{F}^2\mathbf{G} = \cdots = \mathbf{v}\mathbf{F}^{n-1}\mathbf{G} = 0. \tag{D.4}$$

The Cayley–Hamilton theorem states that $\mathbf{F}$ satisfies its own characteristic equation, namely,

$$-\mathbf{F}^n = a_1\mathbf{F}^{n-1} + a_2\mathbf{F}^{n-2} + \cdots + a_n\mathbf{I}. \tag{D.5}$$

Therefore,

$$-\mathbf{v}\mathbf{F}^n\mathbf{G} = a_1\mathbf{v}\mathbf{F}^{n-1}\mathbf{G} + a_2\mathbf{v}\mathbf{F}^{n-2}\mathbf{G} + \cdots + a_n\mathbf{v}\mathbf{G} = 0. \tag{D.6}$$

By induction, $\mathbf{v}\mathbf{F}^{n+k}\mathbf{G} = 0$ for $k = 0, 1, 2, \ldots$, or $\mathbf{v}\mathbf{F}^m\ \mathbf{G} = 0$ for $m = 0, 1, 2, \ldots$, and thus

$$\mathbf{v}e^{\mathbf{F}t}\mathbf{G} = \mathbf{v}\left(\mathbf{I} + \mathbf{F}t + \frac{1}{2!}\mathbf{F}^2t^2 + \cdots\right)\mathbf{G} = 0 \tag{D.7}$$

for all t. However, the zero initial-condition response ($\mathbf{x}_0 = \mathbf{0}$) is

$$\mathbf{x}(t) = \int_0^t \mathbf{v}e^{\mathbf{F}(t-\tau)}\mathbf{G}u(\tau)\,d\tau$$

$$= e^{\mathbf{F}t}\int_0^t e^{-\mathbf{F}t}\mathbf{G}u(\tau)\,d\tau. \tag{D.8}$$

Using Eq. (D.7), Eq. (D.8) becomes

$$\mathbf{v}\mathbf{x}(t) = \int_0^t \mathbf{v}e^{\mathbf{F}(t-\tau)}\mathbf{G}u(\tau)\,d\tau = 0 \tag{D.9}$$

for all $u(t)$ and $t > 0$. This implies that all points reachable from the origin are orthogonal to $\mathbf{v}$. This restricts the reachable space and therefore contradicts the second definition of controllability. Thus if $\mathcal{C}$ is singular, $(\mathbf{F}, \mathbf{G})$ is not controllable by Definition II.

Next we assume that $\mathcal{C}$ is full-rank but $(\mathbf{F}, \mathbf{G})$ is uncontrollable by Definition II. This means that there exists a nonzero vector $\mathbf{v}$ such that

$$\mathbf{v}\int_0^{t_f} e^{\mathbf{F}(t_f-\tau)}\mathbf{G}u(\tau)d\tau = 0, \tag{D.10}$$

because the whole state space is not reachable. But Eq. (D.10) implies that

$$\mathbf{v}e^{\mathbf{F}(t_f-\tau)}\mathbf{G} = 0, \qquad 0 \le \tau \le t_f. \tag{D.11}$$

If we set $\tau = t_f$, we see that $\mathbf{v}\mathbf{G} = 0$. Also, differentiating Eq. (D.11) and letting $\tau = t_f$ gives $\mathbf{v}\mathbf{F}\mathbf{G} = 0$. Continuing this process, we find that

$$\mathbf{v}\mathbf{G} = \mathbf{v}\mathbf{F}\mathbf{G} = \mathbf{v}\mathbf{F}^2\mathbf{G} = \cdots = \mathbf{v}\mathbf{F}^{n-1}\mathbf{G} = 0, \tag{D.12}$$

which contradicts the assumption that $\mathcal{C}$ is full-rank.

We have now shown that the system is controllable by Definition II if and only if the rank of $\mathcal{C}$ is n, exactly the same condition we found for pole assignment.

Our third definition comes closest to the structural character of controllability:

Definition III

> The system $(\mathbf{F}, \mathbf{G})$ is **controllable** if every mode of $\mathbf{F}$ is connected to the control input.

Because of the generality of the modal structure of systems, we will only treat the case of systems for which $\mathbf{F}$ can be transformed to diagonal form. (The double-integration plant does *not* qualify.) Suppose we have a diagonal matrix $\mathbf{F}_d$ and its corresponding input matrix $\mathbf{G}_d$ with elements g_i. The structure of such a system is shown in Fig. D.1. By definition, for a controllable system the input must be connected to each mode so that the g_i are all nonzero. However, this is not enough if the poles (λ_i) are not distinct. Suppose, for instance, that $\lambda_1 = \lambda_2$. The first two state equations are then

$$\dot{x}_{1d} = \lambda_1 x_{1d} + g_1 u,$$

$$\dot{x}_{2d} = \lambda_1 x_{2d} + g_2 u. \tag{D.13}$$

Figure D.1

Block diagram of a system with a diagonal matrix

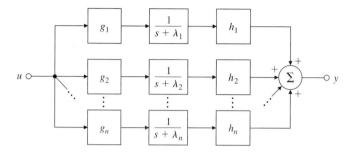

If we define a new state, $\boldsymbol{\xi} = g_2 x_{1d} - g_1 x_{2d}$, the equation for ξ is

$$\dot{\xi} = g_2 \dot{x}_{1d} - g_1 \dot{x}_{2d} = g_2 \lambda_1 x_{1d} + g_2 g_1 u - g_1 \lambda_1 x_{2d} - g_1 g_2 u = \lambda_1 \xi, \qquad (D.14)$$

which does not include the control u; hence $\boldsymbol{\xi}$ is not controllable. The point is that if any two poles are equal *in a diagonal* $\mathbf{F}_d$ *system with only one input*, we effectively have a hidden mode that is not connected to the control, and the system is not controllable [Fig. D.2(a)]. This is because the two state variables move together exactly, so we cannot *independently* control x_{1d} and x_{2d}. There-fore, even in such a simple case, we have two conditions for controllability:

1. All eigenvalues of $\mathbf{F}_d$ are distinct.
2. No element of $\mathbf{G}_d$ is zero.

Now let us consider the controllability matrix of this diagonal system. By direct computation,

$$\mathcal{C} = \begin{bmatrix} g_1 & g_1 \lambda_1 & \cdots & g_1 \lambda_1^{n-1} \\ g_2 & g_2 \lambda_2 & \cdots & \vdots \\ \vdots & \vdots & \cdots & \vdots \\ g_n & g_n \lambda_n & \cdots & g_n \lambda_n^{n-1} \end{bmatrix}$$

$$= \begin{bmatrix} g_1 & & & \mathbf{0} \\ & g_2 & & \\ & & \ddots & \\ \mathbf{0} & & & g_n \end{bmatrix} \begin{bmatrix} 1 & \lambda_1 & \lambda_1^2 & \cdots & \lambda_1^{n-1} \\ 1 & \lambda_2 & \lambda_2^2 & \cdots & \vdots \\ \vdots & \vdots & \vdots & \cdots & \vdots \\ 1 & \lambda_n & \lambda_n^2 & \cdots & \lambda_n^{n-1} \end{bmatrix}. \qquad (D.15)$$

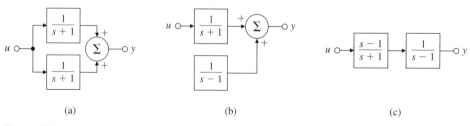

(a)   (b)   (c)

Figure D.2 Examples of uncontrollable systems

Note that the controllability matrix $\mathcal{C}$ is the product of two matrices and is nonsingular if and only if both of these matrices are invertible. The first matrix has a determinant that is the product of the g_i, and the second matrix (called a Vandermonde matrix) is nonsingular if and only if the λ_i are distinct. Thus Definition III is equivalent to having a nonsingular $\mathcal{C}$ also.

Important to the subject of controllability is the **Popov–Hautus–Rosenbrock (PHR) test** (see Rosenbrock, 1970, and Kailath, 1980), which is an alternate way to test the rank (or determinant) of $\mathcal{C}$. The system $(\mathbf{F}, \mathbf{G})$ is controllable if the system of equations

$$\mathbf{v}^T[s\mathbf{I} - \mathbf{F} \quad \mathbf{G}] = \mathbf{0}^T \tag{D.16}$$

has only the trivial solution $\mathbf{v}^T = \mathbf{0}^T$, that is, if the following **matrix pencil** is full-rank for all s

$$\text{rank}[s\mathbf{I} - \mathbf{F} \quad \mathbf{G}] = n, \tag{D.17}$$

or if there is no nonzero $\mathbf{v}^T$ such that [1]

$$\mathbf{v}^T\mathbf{F} = s\mathbf{v}^T, \tag{D.18}$$

$$\mathbf{v}^T\mathbf{G} = 0. \tag{D.19}$$

This test is equivalent to the rank-of-$\mathcal{C}$ test. It is easy to show that if such a vector $\mathbf{v}$ exists, then $\mathcal{C}$ is singular. For, if a nonzero $\mathbf{v}$ exists such that $\mathbf{v}^T\mathbf{G} = 0$, then by Eqs. (D.18) and (D.19) we have

$$\mathbf{v}^T\mathbf{FG} = s\mathbf{v}^T\mathbf{G} = 0. \tag{D.20}$$

Then, multiplying by $\mathbf{FG}$, we find that

$$\mathbf{v}^T\mathbf{F}^2\mathbf{G} = s\mathbf{v}^T\mathbf{FG} = 0, \tag{D.21}$$

and so on. Thus we determine that $\mathbf{v}^T\mathcal{C} = \mathbf{0}^T$ has a nontrivial solution, that $\mathcal{C}$ is singular, and that the system is not controllable. To show that a nontrivial $\mathbf{v}^T$ exists if $\mathcal{C}$ is singular requires more development, which we will not give here (see Kailath, 1980).

We have given two pictures of uncontrollability. Either a mode is physically disconnected from the input [Fig. D.2(b)], or else two parallel subsystems have identical characteristic roots [Fig. D.2(a)]. The control engineer should be aware of the existence of a third simple situation, illustrated in Fig. D.2(c) namely, a **pole-zero cancellation.** Here the problem is that the mode at $s = 1$ appears to be connected to the input but is masked by the zero at $s = 1$ in the preceding subsystem; the result is an uncontrollable system.

---

[1] $\mathbf{v}^T$ is a left eigenvector of $\mathbf{F}$.

This can be confirmed in several ways. First let us look at the controllability matrix. The system matrices are

$$\mathbf{F} = \begin{bmatrix} -1 & 0 \\ 1 & 1 \end{bmatrix}, \qquad \mathbf{G} = \begin{bmatrix} -2 \\ 1 \end{bmatrix},$$

so the controllability matrix is

$$\mathcal{C} = [\, \mathbf{G} \quad \mathbf{FG} \,] = \begin{bmatrix} -2 & 2 \\ 1 & -1 \end{bmatrix}, \tag{D.22}$$

which is clearly singular. The controllability matrix may be computed using the ctrb command in MATLAB [cc] = ctrb(F,G). If we compute the transfer function from u to x_2, we find

$$\mathbf{H}(s) = \frac{s-1}{s+1}\left(\frac{1}{s-1} \right) = \frac{1}{s+1}. \tag{D.23}$$

Because the natural mode at $s = 1$ disappears from the input–output description, it is not connected to the input. Finally, if we consider the **PHR** test,

$$[\, s\mathbf{I} - \mathbf{F} \quad \mathbf{G} \,] = \begin{bmatrix} s+1 & 0 & -2 \\ -1 & s-1 & 1 \end{bmatrix}, \tag{D.24}$$

and let $s = 1$, then we must test the rank of

$$\begin{bmatrix} 2 & 0 & -2 \\ -1 & 0 & 1 \end{bmatrix},$$

which is clearly less than 2. This result means, again, that the system is uncontrollable.

Definition IV

The asymptotically stable system $(\mathbf{F}, \mathbf{G})$ is **controllable** if the controllability Gramian, the square symmetric matrix $\mathcal{C}_g$, given by the solution to the following Lyapunov equation

$$\mathbf{F}\mathcal{C}_g + \mathcal{C}_g\mathbf{F}^T + \mathbf{GG}^T = \mathbf{0}, \tag{D.25}$$

is nonsingular. The controllability **Gramian** is also the solution to the following integral equation

$$\mathcal{C}_g = \int_0^\infty e^{\tau\mathbf{F}}\mathbf{GG}^T e^{\tau\mathbf{F}^T}\, d\tau. \tag{D.26}$$

One physical interpretation of the controllability Gramian is that if the input to the system is white gaussian noise, then $\mathcal{C}_g$ is the covariance of the state. The controllability Gramian (for an asymptotically stable system) can be computed with the following command in MATLAB, [cg] = gram(F,G).

In conclusion, the four definitions for controllability—pole assignment (Definition I), state reachability (Definition II), mode coupling to the input (Definition III), and controllability Gramian (Definition IV)—are equivalent. The tests for any of these four properties are found in terms of the rank of the controllability or controllability Gramian matrices or the rank of the **matrix pencil** $[\mathbf{sI} - \mathbf{F}\ \mathbf{G}]$. If $\mathcal{C}$ is nonsingular, then we can assign the closed-loop poles arbitrarily by state feedback, we can move the state to any point in the state space in a finite time, and every mode is connected to the control input.[2] We have shown the latter for diagonal $\mathbf{F}$ only, but the result is true in general.

D.2   Observability

So far we have discussed only controllability. The concept of observability is parallel to that of controllability, and all of the results we have discussed thus far may be transformed to statements about observability by invoking the property of duality, as discussed in Section 7.5.1. The observability definitions analogous to those for controllability are as follows:

1. *Definition I.* The system $(\mathbf{F}, \mathbf{H})$ is **observable** if, for any nth-order polynomial $\alpha_e(s)$, there exists an estimator gain $\mathbf{L}$ such that the characteristic equation of the state estimator error is $\alpha_e(s)$.

2. *Definition II.* The system $(\mathbf{F}, \mathbf{H})$ is **observable** if, for any $\mathbf{x(0)}$, there is a finite time τ such that $\mathbf{x(0)}$ can be determined (uniquely) from $u(t)$ and $y(t)$ for $0 \leq t \leq \tau$.

3. *Definition III.* The system $(\mathbf{F}, \mathbf{H})$ is **observable** if every dynamic mode in $\mathbf{F}$ is connected to the output through $\mathbf{H.}$

4. *Definition IV.* The asymptotically stable system $(\mathbf{F}, \mathbf{H})$ is **observable** if the observability Gramian is nonsingular.

As we saw in the discussion for controllability, mathematical tests can be developed for observability. The system is observable if the observability matrix

$$\mathcal{O} = \begin{bmatrix} \mathbf{H} \\ \mathbf{HF} \\ \vdots \\ \mathbf{HF}^{n-1} \end{bmatrix} \tag{D.27}$$

is nonsingular. If we take the transpose of $\mathcal{O}$ and let $\mathbf{H}^T = \mathbf{G}$ and $\mathbf{F}^T = \mathbf{F}$, then we find the controllability matrix of $(\mathbf{F}, \mathbf{G})$, another manifestation of duality. The observability matrix, $\mathcal{O}$, may be computed using the obsv command in

---

[2] We have shown the latter for diagonal $\mathbf{F}$ only, but the result is true in general.

MATLAB [oo] = obsv(F,H). The system $(\mathbf{F}, \mathbf{H})$ is observable if the following **matrix pencil** is full-rank for all s:

$$\text{rank} \begin{bmatrix} s\mathbf{I} - \mathbf{F} \\ \mathbf{H} \end{bmatrix} = n. \tag{D.28}$$

The observability Gramian, $\mathcal{O}_g$, which is a symmetric matrix, and the solution to the integral equation

$$\mathcal{O}_g = \int_0^\infty e^{\tau \mathbf{F}^T} \mathbf{H}^T \mathbf{H} e^{\tau \mathbf{F}} d\tau, \tag{D.29}$$

as well as the Lyapunov equation

$$\mathbf{F}^T \mathcal{O}_g + \mathcal{O}_g \mathbf{F} + \mathbf{H}^T \mathbf{H} = \mathbf{0}, \tag{D.30}$$

can also be computed (for an asymptotically stable system) using the gram command in MATLAB [og] = gram(F',H'). The observability Gramian has an interpretation as the "information matrix" in the context of estimation.

E Ackermann's Formula for Pole Placement

Given the plant and state-variable equation

$$\dot{\mathbf{x}} = \mathbf{Fx} + \mathbf{G}u, \tag{E.1}$$

our objective is to find a state feedback control law

$$u = -\mathbf{Kx} \tag{E.2}$$

such that the closed-loop characteristic polynomial is

$$\alpha_c(s) = \det(s\mathbf{I} - \mathbf{F} + \mathbf{GK}). \tag{E.3}$$

First we have to select $\alpha_c(s)$, which determines where the poles are to be shifted; then we have to find $\mathbf{K}$ such that Eq. (E.3) will be satisfied. Our technique is based on transforming the plant equation into control canonical form.

We begin by considering the effect of an arbitrary nonsingular transformation of the state,

$$\mathbf{x} = \mathbf{T}\bar{\mathbf{x}}, \tag{E.4}$$

where $\bar{\mathbf{x}}$ is the new transformed state. The equations of motion in the new coordinates, from Eq. (E.4), are

$$\dot{\mathbf{x}} = \mathbf{T}\dot{\bar{\mathbf{x}}} = \mathbf{Fx} + \mathbf{G}u = \mathbf{FT}\bar{\mathbf{x}} + \mathbf{G}u, \tag{E.5}$$

$$\dot{\bar{\mathbf{x}}} = \mathbf{T}^{-1}\mathbf{FT}\bar{\mathbf{x}} + \mathbf{T}^{-1}\mathbf{G}u = \bar{\mathbf{F}}\bar{\mathbf{x}} + \bar{\mathbf{G}}u. \tag{E.6}$$

Now the controllability matrix for the original state,

$$\mathcal{C}_x = [\mathbf{G} \quad \mathbf{FG} \quad \mathbf{F}^2\mathbf{G} \quad \cdots \quad \mathbf{F}^{n-1}\mathbf{G}], \tag{E.7}$$

provides a useful transformation matrix. We can also define the controllability matrix for the transformed state:

$$\mathcal{C}_{\bar{x}} = [\bar{\mathbf{G}} \quad \bar{\mathbf{F}}\bar{\mathbf{G}} \quad \bar{\mathbf{F}}^2\bar{\mathbf{G}} \quad \cdots \quad \bar{\mathbf{F}}^{n-1}\bar{\mathbf{G}}], \tag{E.8}$$

The two controllability matrices are related by

$$\mathcal{C}_{\bar{x}} = [\mathbf{T}^{-1}\mathbf{G} \quad \mathbf{T}^{-1}\mathbf{F}\mathbf{T}\mathbf{T}^{-1}\mathbf{G} \quad \cdots] = \mathbf{T}^{-1}\mathcal{C}_x \tag{E.9}$$

and the transformation matrix

$$\mathbf{T} = \mathcal{C}_x \mathcal{C}_{\bar{x}}^{-1}. \tag{E.10}$$

From Eqs. (E.9) and (E.10) we can draw some important conclusions. From Eq. (E.9) we see that if $\mathcal{C}_x$ is nonsingular, then for any nonsingular $\mathbf{T}$, $\mathcal{C}_{\bar{x}}$ is also nonsingular. This means that a similarity transformation on the state does not change the controllability properties of a system. We can look at this in another way. Suppose we would like to find a transformation to take the system $(\mathbf{F}, \mathbf{G})$ into control canonical form. As we shall shortly see, $\mathcal{C}_{\bar{x}}$ in that case is *always* nonsingular. From Eq. (E.9) we see that a nonsingular $\mathbf{T}$ will always exist if and only if $\mathcal{C}_x$ is nonsingular. We conclude the following:

We can always transform $(\mathbf{F}, \mathbf{G})$ into control canonical form if and only if the system is controllable.

Let us take a closer look at control canonical form and treat the third-order case, although the results are true for any nth-order case:

$$\bar{\mathbf{F}} = \mathbf{F}_c = \begin{bmatrix} -a_1 & -a_2 & -a_3 \\ 1 & 0 & 0 \\ 0 & 1 & 0 \end{bmatrix}, \qquad \bar{\mathbf{G}} = \mathbf{G}_c = \begin{bmatrix} 1 \\ 0 \\ 0 \end{bmatrix}. \tag{E.11}$$

The controllability matrix, by direct computation, is

$$\mathcal{C}_{\bar{x}} = \mathcal{C}_c = \begin{bmatrix} 1 & -a_1 & a_1^2 - a_2 \\ 0 & 1 & -a_1 \\ 0 & 0 & 1 \end{bmatrix}. \tag{E.12}$$

Because this matrix is upper triangular with ones along the diagonal, it is always invertible. Also note that the last row of $\mathcal{C}_{\bar{x}}$ is the unit vector with all zeros, except for the last element, which is unity. We shall use this fact in the following.

As we pointed out in Section 7.3, the design of a control law for the state $\bar{x}$ is trivial if the equations of motion happen to be in control canonical form. The characteristic equation is

$$s^3 + a_1 s^2 + a_2 s + a_3 = 0, \tag{E.13}$$

and the characteristic equation for the closed-loop system comes from

$$\mathbf{F}_{cl} = \mathbf{F}_c - \mathbf{G}_c \mathbf{K}_c \tag{E.14}$$

and has the coefficients

$$s^3 + (a_1 + K_{c1})s^2 + (a_2 + K_{c2})s + (a_3 + K_{c3}) = 0. \tag{E.15}$$

To obtain the desired closed-loop pole locations, we must make the coefficients of s in Eq. (E.15) match those in

$$\alpha_c(s) = s^3 + \alpha_1 s^2 + \alpha_2 s + \alpha_3, \tag{E.16}$$

so

$$a_1 + K_{c1} = \alpha_1, \qquad a_2 + K_{c2} = \alpha_2, \qquad a_3 + K_{c3} = \alpha_3, \tag{E.17}$$

or, in vector form,

$$\mathbf{a} + \mathbf{K}_c = \boldsymbol{\alpha}, \tag{E.18}$$

where $\mathbf{a}$ and $\boldsymbol{\alpha}$ are row vectors containing the coefficients of the characteristic polynomials of the open-loop and closed-loop systems, respectively.

We need now to find a relationship between these polynomial coefficients and the matrix $\mathbf{F}$. The requirement is achieved by the Cayley–Hamilton theorem, which states that a matrix satisfies its own characteristic polynomial. For $\mathbf{F_c}$ this means

$$\mathbf{F}_c^n + a_1 \mathbf{F}_c^{n-1} + a_2 \mathbf{F}_c^{n-2} + \cdots + a_n \mathbf{I} = \mathbf{0}. \tag{E.19}$$

Now suppose we form the polynomial $\alpha_c(\mathbf{F})$, which is the *closed-loop* characteristic polynomial with the matrix $\mathbf{F}$ substituted for the complex variable s:

$$\alpha_c(\mathbf{F}_c) = \mathbf{F}_c^n + \alpha_1 \mathbf{F}_c^{n-1} + \alpha_2 \mathbf{F}_c^{n-2} + \cdots + \alpha_n \mathbf{I}. \tag{E.20}$$

If we solve Eq. (E.19) for $\mathbf{F}_c^n$ and substitute in Eq. (E.20), we find that

$$\alpha_c(\mathbf{F}_c) = (-a_1 + \alpha_1)\mathbf{F}_c^{n-1} + (-a_2 + \alpha_2)\mathbf{F}_c^{n-2} + \cdots + (-\alpha_n + \alpha_n)\mathbf{I}. \tag{E.21}$$

But, because $\mathbf{F}_c$ has such a special structure, we observe that if we multiply it by the transpose of the nth unit vector, $\mathbf{e}_n^T = [\,0 \quad \cdots \quad 0 \quad 1\,]$, we get

$$\mathbf{e}_n^T \mathbf{F}_c = [\,0 \quad \cdots \quad 0 \quad 1 \quad 0\,] = \mathbf{e}_{n-1}^T, \tag{E.22}$$

as we can see from Eq. (E.11). If we multiply this vector again by $\mathbf{F}_c$,

$$(\mathbf{e}_n^T \mathbf{F}_c)\mathbf{F}_c = [\, 0 \quad \cdots \quad 0 \quad 1 \quad 0\,]\mathbf{F}_c$$

$$= [\, 0 \quad \cdots \quad 0 \quad 1 \quad 0 \quad 0\,] = \mathbf{e}_{n-2}^T, \tag{E.23}$$

and continue the process, successive unit vectors are generated until

$$\mathbf{e}_n^T \mathbf{F}_c^{n-1} = [\, 1 \quad 0 \quad \cdots \quad 0\,] = \mathbf{e}_1^T. \tag{E.24}$$

Therefore, if we multiply Eq. (E.21) by $\mathbf{e}_n^T$, we find

$$\mathbf{e}_n^T \alpha_c(\mathbf{F}_c) = (-a_1 + \alpha_1)\mathbf{e}_1^T + (-a_2 + \alpha_2)\mathbf{e}_2^T + \cdots + (-a_n + \alpha_n)\mathbf{e}_n^T$$

$$= [\, K_{c1} \quad K_{c2} \quad \cdots \quad K_{cn}\,] = \mathbf{K}_c, \tag{E.25}$$

where we use Eq. (E.18) which relates $\mathbf{K}_c$ to $\mathbf{a}$ and $\boldsymbol{\alpha}$.

We now have a compact expression for the gains of the system in control canonical form as represented in Eq. (E.25). However, we still need the expression for $\mathbf{K}$, the gain on the original state. If $u = -\mathbf{K}_c\bar{\mathbf{x}}$, then $u = -\mathbf{K}_c\mathbf{T}^{-1}\mathbf{x}$, so that

$$\mathbf{K} = \mathbf{K}_c\mathbf{T}^{-1} = \mathbf{e}_n^T \alpha_c(\mathbf{F}_c)\mathbf{T}^{-1}$$

$$= \mathbf{e}_n^T \alpha_c(\mathbf{T}^{-1}\mathbf{F}\mathbf{T})\mathbf{T}^{-1}$$

$$= \mathbf{e}_n^T \mathbf{T}^{-1}\alpha_c(\mathbf{F}). \tag{E.26}$$

In the last step of Eq. (E.26) we used the fact that $(\mathbf{T}^{-1}\mathbf{F}\mathbf{T})^k = \mathbf{T}^{-1}\mathbf{F}^k\mathbf{T}$ and that α_c is a polynomial, that is, a sum of the powers of $\mathbf{F}_c$. From Eq. (E.9) we see that

$$\mathbf{T}^{-1} = \mathcal{C}_c \mathcal{C}_{\mathbf{x}}^{-1}. \tag{E.27}$$

With this substitution, Eq. (E.26) becomes

$$\mathbf{K} = \mathbf{e}_n^T \mathcal{C}_c \mathcal{C}_{\mathbf{x}}^{-1}\alpha_c(\mathbf{F}). \tag{E.28}$$

Now we use the observation made earlier for Eq. (E.12) that the last row of $\mathcal{C}_c$, which is $\mathbf{e}_n^T \mathcal{C}_c$, is again $\mathbf{e}_n^T$. We finally obtain Ackermann's formula:

Ackermann's formula

$$\mathbf{K} = \mathbf{e}_n^T \mathcal{C}_{\mathbf{x}}^{-1}\alpha_c(\mathbf{F}). \tag{E.29}$$

We note again that forming the explicit inverse of $\mathcal{C}_{\mathbf{x}}$ is not advisable for numerical accuracy. Thus we need to solve $\mathbf{b}^T$ such that

$$\mathbf{e}_n^T \mathcal{C}_{\mathbf{x}}^{-1} = \mathbf{b}^T. \tag{E.30}$$

We solve the linear set of equations

$$\mathbf{b}^T \mathcal{C}_{\mathbf{x}} = \mathbf{e}_n^T \tag{E.31}$$

and then compute

$$\mathbf{K} = \mathbf{b}^T \alpha_c(\mathbf{F}). \tag{E.32}$$

Ackermann's formula, Eq. (E.29), even though elegant, is not recommended for systems with a large number of state variables. Even if it is used, Eqs. (E.31) and (E.32) are recommended for better numerical accuracy.

F MATLAB Commands

MATLAB Function (.m file)	Description	Page (s)
acker	Ackermann's formula for pole-placement	522, 530, 544, 547, 681
axis	Control axis scaling	399, 401
bode	Bode frequency response	102, 369, 384, 456, 572
c2d	Continuous to discrete conversion	250, 423, 566, 581, 661
canon	State-space canonical forms	507
conv	Polynomial multiplication	111, 214, 534, 536, 822
ctrb	Controllability matrix	854
damp	Damping and natural frequency	674, 675
eig	Eigenvalues and eigenvectors	504, 507, 679
feedback	Feedback connection of two systems	128, 324, 674
gram	Controllability/observability Gramian	854, 856
impulse	Impulse response	138, 144, 154
initial	Initial condition response	544, 551
inv	Matrix inverse	506, 508
loglog	Log-log plot	102, 369, 384, 572
logspace	Logarithmically-spaced frequency points	102
lqe	Linear Quadratic Estimator design	572, 789
lqr	Linear Quadratic Regulator design	537, 538, 788
lsim	Linear system simulation	122, 171
margin	Gain and phase margins	427, 472, 572
max	Largest component	457
nichols	Nichol's chart	445
nyquist	Nyquist plot	395, 399, 401
obsv	Observability matrix	856
ode23	Solution to non-stiff differential equations	172
ones	Array of ones	122, 171

MATLAB Function (.m file)	Description	Page (s)
pade	Pade approximation for time delay	332
parallel	Parallel connection of two systems	127
place	Pole-placement	524, 547, 562, 568, 595
plot	Plot function	34, 119, 122, 215
poly	Form polynomial from its roots	115
printsys	Print system in transfer function format	118
pzmap	Pole-zero map	137
residue	Residues in partial fraction expansion	111, 115, 822
rlocfind	Find root locus gain	304, 472
rlocus	Root locus	276, 472, 534, 536
rltool	Interactive root locus tool	350
roots	Roots of a polynomial	162, 296, 514, 524
semilogx	Semi-log plot	102, 369, 572
semilogy	Semi-log plot	456
series	Series connection of two systems	127, 572, 674
sqrt	Square root	595
ss2tf	State-space to transfer function	118, 121, 510, 513
ss2zp	State-space to pole-zero conversion	121, 135
ss	Conversion to state-space	44, 122, 507, 512, 515
ssdata	Create a state-space model	681
step	Step response	34, 44, 119, 162, 214, 682
tf2ss	Transfer function to state-space	120, 497
tf2zp	Transfer function to pole-zero conversion	121, 135
tf	Conversion to transfer function	34, 102, 127, 138, 154, 214, 534
tzero	Transmission zeros	512, 514, 680
zeros	Array of zeros	122

G

Solutions to Review Questions

CHAPTER 1

1. What are the main components of a feedback control system?
   The process, the actuator, the sensor, and the controller.

2. What is the purpose of the sensor?
   To measure the output variable and, usually, to convert it to an electrical voltage.

3. Give three important properties of a good sensor.
   A good sensor is linear (the output is proportional to the input signal) over a large range of amplitudes and a large range of frequencies at its input, has low noise, is unbiased, is easy to calibrate, and has low cost. The relative values of these properties vary with the particular application.

4. What is the purpose of the actuator?
   The actuator takes an input, usually electrical, and converts it to a signal such as a force or torque that causes the process output to move or change over the required range.

5. Give three important properties of a good actuator.

A good actuator has fast response and adequate power, energy, speed, torque, and so on, to be able to cause the process output to meet the design specifications and is efficient, light weight, small, cheap, and so on. As with sensors, the relative value of these properties varies with the application.

6. What is the purpose of the compensator? Give the input(s) and output(s) of the compensator.

The compensator is to take the sensor output (the input to the compensator) and compute the control signal (the output of the compensator) to be sent to the actuator.

7. What physical variable(s) of a process can be directly measured by a Hall effect sensor?

A Hall effect device measures the strength of a magnetic field and can be most easily configured to measure relative positions of two bodies or relative angles.

8. What physical variable is measured by a tachometer?

A tachometer measures speed of rotation or angular velocity.

9. Describe three different techniques for measuring temperature.

*In each case indicated below, it is important to realize that the devices mentioned need to be **calibrated** and often corrected for nonlinearity in order to give a reliable, accurate measure of temperature.*

(a) *An early technique still used in many home thermostats is based on the **bimetallic strip** composed of two strips of different metals that expand with different coefficients with temperature. As a result, the strip bends with temperature and the resulting motion can be used as a measure of temperature. This principle was introduced in the eighteenth century to maintain a constant length to a clock pendulum for precision time keeping.*

(b) *A technique related to the bimetallic strip is based on the fact that metals with different work functions placed in contact will produce a voltage that is proportional to temperature. Such a device is called a **thermocouple** and is the basis of a standard laboratory technique for measuring temperature.*

(c) *A number of materials have electrical resistance that is dependent in a monotonic way on temperature and a resistance bridge can be used with one of these to indicate temperature. Such devices are called **thermistors**.*

(d) *For high temperatures, it is well known that the color of the radiation due to heat depends on temperature. A piece of iron placed in a fire will glow orange, and then red and will finally become white hot at high temperatures. An instrument for measuring the frequency of the radiation and thus the temperature is a **pyrometer**.*

(e) *In ceramic kilns, cones of different materials that melt at different and known temperatures are placed near the products in the kiln to indicate when the design temperature has been reached. The potter watches until the cone of importance begins to sag and then knows that the products should be removed. These give a quantized measure of temperature.*

10. Why do most sensors have an electrical output, regardless of the physical nature of the variable being measured?
Electrical signals are the most easily manipulated and therefore most controllers are electrical devices, either analog or digital. To provide the signal input to such a device, the sensor needs to produce an electrical output.

CHAPTER 2

1. What is a "free-body" diagram?
To write the equations of motion of a system of connected bodies, it is useful to draw each body in turn with the influence of all other bodies represented by forces and torques on the body in question. A drawing of the collection of such isolated bodies is called a "free-body diagram."

2. What are the two forms of Newton's law?
Translational motion is described by $F = ma$. Rotational motion is described by $M = I\alpha$.

3. Why is it convenient to write equations of motion in the state-variable form?
It provides a standard way to describe the differential equations for any dynamic system so that computer-aided analysis can be carried out more conveniently. It is also more convenient to analyze linear systems in terms of the standard system description matrices.

4. For a structural process to be controlled such as a robot arm, what is the meaning of "collocated control"? "Noncollocated control"?
When the actuator and the sensor are located on the same rigid body, the control is said to be "collocated." When they are on different bodies that are connected by springs, the control is "noncollocated."

5. When, why, and by whom was the device named an "operational amplifier"?
In a paper in 1947, Ragazzini, Randall, and Russell named the high-gain, wide-bandwidth amplifier used in feedback to realize operational calculus 'operations' the operational amplifier.

6. What is the major benefit of having zero input current to an operational amplifier?
With zero input current the amplifier does not load the input circuit and thus the transfer function of the device is not dependent on the amplifier characteristics. Also, the analysis of the circuit is simplified in this case.

7. State Kirchhoff's current law.
The algebraic sum of all currents entering a junction or circuit is zero.

8. State Kirchhoff's voltage law.
The algebraic sum of voltages around a closed path in an electric circuit is zero.

9. Why is it important to have a small value for the armature resistance, R_a, of an electric motor?

The armature resistance causes power loss when the armature current flows and thus reduces the efficiency of the motor.

10. What are the definition and units of the electric constant of a motor?

A rotating motor produces a voltage (called the back emf) in its armature proportional to the rotational speed. The electric constant K_e is the ratio of this voltage to the speed so that $e = K_e\dot{\theta}$. The units are volt-sec/radian.

11. What are the definition and units of the torque constant of an electric motor?

When current, i_a, flows in the armature of an electrical motor, a torque, τ, is produced that is proportional to the current. The torque constant, K_t, is the constant of proportionality so that $\tau = K_t i_a$. The units are Newton-meters/amp.

12. Give the relationships for (a) heat flow across a substance and (b) heat storage in a substance.

(a) *Heat flow is proportional to the temperature difference divided by the thermal resistance, that is,*

$$q = \frac{1}{R}(T_1 - T_2).$$

(b) *The differential equation describing the heat storage is*

$$\dot{T} = \frac{1}{C}q,$$

where C is the thermal capacity of the material.

13. Name and give the equations for the three relationships governing fluid flow.

$$\text{Continuity:} \quad \dot{m} = w_{in} - w_{out}.$$

$$\text{Force equilibrium:} \quad f = pA.$$

$$\text{Resistance:} \quad w = \frac{1}{R}(p_1 - p_2)^{1/\alpha}.$$

14. Why do we approximate a physical model of the plant (which is *always* nonlinear) with a linear model?

Analysis and design of linear models is vastly simpler than with nonlinear models. Furthermore, it has been shown (by Lyapunov) that if the linear approximation is stable, then there is at least some region of stability for the nonlinear model.

CHAPTER 3

1. What is the definition of "transfer function"?

The Laplace transform of the output of a linear, time-invariant system, $Y(s)$, is proportional to the transform of its input, $U(s)$. The function of proportionality is the transfer function, $F(s)$, so that $Y(s) = F(s)U(s)$. It is assumed that all initial conditions are zero.

2. What are the properties of systems whose responses can be described by transfer functions?

The system must be both linear (superposition applies) and time-invariant (the parameters do not vary with time).

3. What is the Laplace transform of $f(t - \lambda)1(t - \lambda)$ if the transform of $f(t)$ is $F(s)$?

$$\mathcal{L}\{f(t - \lambda)1(t - \lambda)\} = e^{-s\lambda}F(s).$$

4. State the Final Value Theorem (FVT).

If all the poles of $sF(s)$ are in the left half-plane, then the final value of $f(t)$ is given by $\lim_{t \to \infty} f(t) = \lim_{s \to 0} sF(s)$.

5. What is the most common use of the FVT in control?

A standard test of a control system is the step response, and the FVT is used to determine the steady-state error to such an input.

6. Given a second-order transfer function with damping ratio ζ and natural frequency ω_n, what is the estimate of the step response rise time? What is the estimate of the percent overshoot in the step response? What is the estimate of the settling time?

These are given by $t_r \cong 1.8/\omega_n$; M_p is set by the damping ratio (see the curve in Figure 3.28) and $t_s \cong 4.6/\sigma$.

7. What is the major effect of an extra zero in the left half-plane on the second-order step response?

Such a zero causes additional overshoot; and the closer the zero is to the imaginary axis, the higher the overshoot. If the zero is more than 6 times the real part of the complex poles, the effect is negligible.

8. What is the most noticeable effect of a zero in the right half-plane on the step response of the second-order system?

*Such a zero often causes an initial **undershoot** of the response.*

9. What is the main effect of an extra real pole on the second-order step response?

A pole slows down the response and makes the rise time longer. The closer the pole is to the imaginary axis, the more pronounced is the effect. If the pole is more than 6 times the real part of the complex poles, the effect is negligible.

10. Why is stability an important consideration in control system design?

Almost any useful dynamic system must be stable to perform its function. Feedback around a system that is normally stable can actually introduce instability so control designers must be able to assure the stability of their designs.

11. What is the main use of Routh's criterion?
With this method, we can find (symbolically) the range of a parameter such as the loop gain for which the system will be stable.

12. Under what conditions might it be important to know how to estimate a transfer function from experimental data?
In many cases, the equations of motion are either extremely complex or not known at all. Chemical processes such as a paper-making machine are often of this kind. In these cases, if one wishes to build a good control, it is very useful to be able to take transient data or steady-state frequency response data and to estimate a transfer function from these.

CHAPTER 4

1. Give three advantages of feedback in control.
   (a) *Feedback can reduce the steady-state error in response to disturbances.*
   (b) *Feedback can reduce the steady-state error in tracking a reference.*
   (c) *Feedback can reduce the sensitivity of a transfer function to parameter changes.*
   (d) *Feedback can stabilize an unstable process.*

2. Give two disadvantages of feedback in control.
   (a) *Feedback requires a sensor that can be very expensive and may introduce additional noise.*
   (b) *Feedback systems are often more difficult to design and operate than open-loop systems.*

3. What is the main objective of introducing integral control?
Integral control will make the error to a constant input go to zero. It removes the effects of process noise bias. It cannot remove the effects of sensor bias.

4. What is the major objective of adding derivative control?
Derivative control typically makes the system better damped and more stable.

5. Why might a designer wish to put the derivative term in the feedback rather than in the error path?
When a reference input might include sudden changes, including it in the derivative action might cause unnecessary large controls.

▲ 6. What is integrator windup?
If the plant actuator output signal saturates, then it may take a long time for the error to be brought back to zero from an initial upset and during this time the integrator output may grow or 'wind up' much more than it would if the system were linear. Special 'anti-windup' circuits are designed to prevent windup.

▲   **7.** Why is an anti-windup circuit important?

When a control includes integral action and is subject to saturation, large inputs can cause large over shoots and slow recovering unless an anti-windup circuit is included.

▲   **8.** Using the nonlinear saturation function having gain 1 and limits ± 1, sketch the block diagram of saturation for an actuator which has gain 7 and limits ± 20.

If the output of the actuator is u_{out} and its input is u_{in}, the control is given by

$$u_{out} = 20 sat \left(\frac{7u_{in}}{20} \right).$$

9. A temperature control system is found to have zero error to a constant tracking input and an error of $0.5°$C to a tracking input that is linear in time, rising at the rate of $40°$C/sec. What is the system type of this control system and what is the relevant error constant $[K_p$ or K_v or etc.$]$?

The system is type 1 and the K_v is the ratio of input rate to error or $K_v = 40/.5 = 80/$ sec.

10. What are the units of K_p, K_v, and K_a?

K_p *is dimensionless, K_v is sec^{-1}, and K_a is sec^{-2}.*

11. What is the definition of system type with respect to reference inputs?

With only a polynomial of degree k reference input (no disturbances), the type is the largest value of k for which the error is a constant.

12. What is the definition of system type with respect to disturbance inputs?

With only a polynomial of degree k disturbance input (no reference), the type is the largest value of k for which the error is a constant.

13. Why does system type depend on where the external signal enters the system?

Because the error depends on where the input enters, so does the value of type.

14. Give two reasons to use a digital controller rather than an analog controller.

(a) *The control law is easier to change if the controller is digital.*

(b) *A digital controller can perform logic and other nonlinear operations much easier than an analog controller.*

(c) *The hardware of a digital controller can be fixed in the design before the details of the actual control design is finished.*

15. Give two disadvantages to using a digital controller.

(a) *The bandwidth of a digital controller is limited by the possible sample frequency.*

(b) *The digital controller introduces noise by the quantization process.*

16. Give the substitution in the discrete operator z for the Laplace operator s if the approximation to the integral in Eq. (4.108) is taken to be the rectangle of height $e(kT_s)$ and base T_s.

$$s = \frac{z - 1}{T_s}.$$

CHAPTER 5

1. Give two definitions for the root locus.

(a) *The root locus is the locus of points in the s-plane where the equation $a(s) + Kb(s) = 0$ has a solution.*

(b) *The root locus is the locus of points in the s-plane where the angle of $G(s) = b(s)/a(s)$ is $180°$.*

2. Define the negative root locus.

The negative root locus is the locus of points where the equation $a(s) - Kb(s) = 0$ has a solution or where the angle of $G(s) = b(s)/a(s)$ is $0°$.

3. Where are the sections of the (positive) root locus on the real axis?

Segments of the real axis to the left of an odd number of zeros and poles are on the root locus.

4. What are the angles of departure from two coincident poles at $s = -a$ on the real axis. Assume there are no poles or zeros to the right of $-a$.

The loci depart at $\pm90°$.

5. What are the angles of departure from *three* coincident poles at $s = -a$ on the real axis. Assume there are no poles or zeros to the right of $-a$.

The loci depart at $\pm60°$ and $180°$.

6. What is the requirement on the location of a lead compensation zero that will cause the locus to pass through a desired root location, r_o?

The angle from the lead zero to r_o must be such that the angle of the compensated transfer function at r_o is $180°$.

7. What is the value of the compensator gain that will cause a closed-loop pole to be at r_o?

If the compensated open-loop transfer function is KDG, then $K = 1/|D(r_o)G(r_o)|$.

8. What is the principal effect of a lead compensation on a root locus?

The lead compensation generally causes the locus to bend toward the left half plane, moving the dominant roots to a place of higher damping.

9. What is the principal effect of a lag compensation on a root locus in the vicinity of the dominant closed-loop roots?

The lag compensation is normally placed so near the origin that it has negligible effect on the root locus in the vicinity of the dominant closed-loop roots.

10. What is the principal effect of a lag compensation on the steady-state error to a polynomial reference input?

A lag compensation normally raises the gain at $s = 0$ and thus increases the velocity constant of a type 1 system and lowers the error to polynomial inputs.

11. Why is the angle of departure from a pole near the imaginary axis especially important?

If the locus starts toward the right half plane, then feedback will make the system less stable. On the other hand if the locus departs toward the left half-plane, then feedback is going to make the system more stable.

12. Define a conditionally stable system.

*A system that becomes **unstable** as gain is reduced is considered to be conditionally stable. That is, its stability is conditioned on having an operating compensator with at least a minimum value of gain.*

13. Show, with a root locus argument, that a system having three poles at the origin *must* be conditionally stable.

With three poles at the origin, the angles of departure insure that two poles leave the origin at $180°$, $\pm 60°$, or, if there are poles on the real axis in the right-half plane, they may leave at $0°$, $\pm 120°$ which is to say that at least one pole begins by moving into the right half-plane. As gain is reduced from the operating level at least one root must pass into the right half-plane for gain low enough and therefore the system must be conditionally stable.

CHAPTER 6

1. Why did Bode suggest plotting the magnitude of a frequency response on log–log coordinates?

In log–log coordinates, the plot for a rational transfer function can be well-guided by linear asymptotes and thus easily plotted and visualized.

2. Define a decibel.

If a power ratio is P_1/P_2, then the measure in decibels is $10\log(P_1/P_2)$. Because power is proportional to voltage squared, and a transfer function would give a ratio of voltages, then the gain of a transfer function $G(j\omega)$ in decibels is $G_{db} = 20\log|G(j\omega)|$.

3. What is the transfer function magnitude if the gain is listed as 14 db?

$14 = 20\log M$, therefore $M = 5.01$.

4. Define gain crossover.

The gain crossover, ω_c, is the value of frequency where the magnitude gain is 1 (or 0 db).

5. Define phase crossover.

The phase crossover, ω_{cp}, is the value of the frequency where the phase crosses $-180°$.

6. Define phase margin, PM.

The phase margin, PM, is a measure of how far in phase the Nyquist plot is from instability. In the typical case, if the phase of the system at gain crossover is ϕ, then phase margin is $180° + \phi$. For example, if $\phi = -150°$ then the phase margin is $30°$.

7. Define gain margin, GM.

The gain margin is a measure of how far the system is from instability by changes in gain alone. If the gain at phase crossover, where the system phase is $180°$, is $|G(j\omega_{cp})|$, then the gain margin is $GM * |G(j\omega_{cp})| = 1.0$ or $GM = 1/|G(j\omega_{cp})|$.

8. What Bode plot characteristic is the best indicator of the closed-loop step response overshoot?

The phase margin is related to the equivalent closed loop damping ratio approximately by $\zeta_{eq} = PM/100$. As we saw in Chapter 3, the step response overshoot is monotonically related to the damping ratio.

9. What Bode plot characteristic is the best indicator of the closed-loop step response rise time?

The rise time is measured by the closed-loop natural frequency which in turn is adequately approximated by the gain crossover. Thus the best indicator of rise time is ω_{cg}.

▲   **10.** A certain control system is required to follow sinusoids that may be any frequency in the range $0 \le \omega_\ell \le 450$ rad/sec and have amplitudes up to 5 units with (sinusoidal) steady-state error to be never more than 0.01. Sketch (or describe) the corresponding performance function $W_1(\omega)$.

The magnitude of W_1 is given by the ratio $|R|/e_b = 5/.01 = 500$. The performance function would then have the value 500 for frequencies up to 450 rad/sec. The Bode magnitude plot would be required to be above this curve for these frequencies.

11. What is the principal effect of a lead compensation on Bode plot performance measures?

The lead compensation usually is used to raise the phase margin at a desired gain crossover frequency.

12. What is the principal effect of a lag compensation on Bode plot performance measures?

The lag compensation is usually used to raise the low frequency gain to reduce the steady-state error to polynomial or low-frequency sinusoidal inputs. It can also be used to lower the crossover frequency, ω_{cg}, where a more favorable phase exists.

13. How do you find the K_v of a type 1 system from its Bode plot?

The K_v is determined by the low-frequency asymptote, which has a slope of -1 for a type 1 system and is given by the expression K_v/ω. The value of the constant may be found either from the frequency where the asymptote reaches 1.0 (or 0 db) or else as the value of the asymptote at the frequency of $\omega = 1$.

14. Why do we need to know beforehand the number of open-loop unstable poles in order to tell stability from the Nyquist plot?

The Nyquist plot encirclements counts the difference in the number of zeros and the number of poles in the right-hand plane of $1 + KDG$. In order to know the number of zeros of this function (which are closed-loop poles and thus unstable poles of the closed-loop), we must know the number of unstable open-loop poles for the plot.

15. What is the main advantage in control design of counting the encirclements of $-1/K$ of $D(j\omega)G(j\omega)$ rather than encirclements of -1 of $KD(j\omega)G(j\omega)$?

If we plot DG alone, then the stability depends on the encirclements of $-1/K$. The designer can thus easily look at the entire range of real K and determine the best value of gain for the design without having to make any more plots.

16. Define a conditionally stable feedback system. How can you identify one on a Bode plot?

A conditionally stable feedback system becomes unstable as gain is reduced. If the low-frequency phase drops below $-180°$ and then there is a reduction of gain until gain crossover occurs where there is no phase margin, then the system is almost surely unstable. A look at the Nyquist plot is necessary to be certain. This condition can also be seen easily from a root locus; the locus will have segments in the RHP for low values of gain.

CHAPTER 7

The following questions are based on a system in state variable form with matrices $\mathbf{F}, \mathbf{G}, \mathbf{H}$, and J, input u, output y, and state $\mathbf{x}$.

1. Give an expression for the transfer function of this system.

$$G(s) = \mathbf{H}(s\mathbf{I} - \mathbf{F})^{-1}\mathbf{G} + J.$$

2. Give two expressions for the poles of the transfer function of the system.
   (a) $\mathsf{p} = \mathsf{eig}(\mathsf{F})$.
   (b) $p = roots\ of\ \det[s\mathbf{I} - \mathbf{F}] = a(s) = 0.$

3. Give an expression for the zeros of the system transfer function.

$$z = roots\ of\ \det \begin{bmatrix} s\mathbf{I} - \mathbf{F} & -\mathbf{G} \\ \mathbf{H} & J \end{bmatrix} = b(s) = 0$$

4. Under what condition will the state of the system be controllable?
   (a) *If the pair $(\mathbf{F,G})$ are controllable, that is, if the matrix*

   $$\mathcal{C} = [\,\mathbf{G} \quad \mathbf{FG} \quad \cdots \quad \mathbf{F}^{n-1}\mathbf{G}\,]$$

   is full rank.
   (b) *If the system can be put into control canonical form.*

5. Under what conditions will the system be observable from the output y?
   (a) *If the matrices* $(\mathbf{F}, \mathbf{H})$ *are observable, that is, if the matrix*

$$
\mathcal{O} =
\begin{bmatrix}
\mathbf{H} \\
\mathbf{HF} \\
\vdots \\
\mathbf{HF}^{(n-1)}
\end{bmatrix}
$$

   has full rank.
   (b) *If the system can be put into observable canonical form.*

6. Give an expression for the *closed-loop* poles if state feedback of the form $u = -\mathbf{Kx}$ is used.
   (a) $\mathsf{p}_c = \mathsf{eig}(\ \mathsf{F} - \mathsf{G} * \mathsf{K})$.
   (b) $p_c = $ *roots of* $\det(s\mathbf{I} - \mathbf{F} + \mathbf{GK}) = \alpha_c(s) = 0$.

7. Under what conditions can the feedback matrix, $\mathbf{K}$, be selected so that the roots of $\alpha_c(s)$ are placed arbitrarily?
   If the system is controllable.

8. What is the advantage of using the LQR or symmetrical root locus in designing the feedback matrix $\mathbf{K}$?
   With LQR, the closed-loop system will be more robust to parameter changes, and the designer has some control over the control effort used by the closed-loop system.

9. What is the main reason for using an estimator in feedback control?
   When the state is not available (usually because it is too expensive or impractical to put sensors on each state variable), then an estimator using only the output, y, *can give an estimate that can be used in place of the actual state.*

10. If the estimator gain, $\mathbf{L}$, is used, give an expression for the closed-loop poles due to the estimator.
    (a) $\mathsf{p}_e = \mathsf{eig}(\mathsf{F} - \mathsf{L} * \mathsf{H})$.
    (b) $p_e = $ *roots of* $\det(s\mathbf{I} - \mathbf{F} + \mathbf{LH}) = \alpha_e(s) = 0$.

11. Under what conditions can the estimator gain, $\mathbf{L}$, be selected so that the roots of $\alpha_e(s) = 0$ are placed arbitrarily?
    If the system is observable.

12. If the reference input is arranged so that the input to the estimator is identical to the input to the process, what will be the overall closed-loop transfer function?

$$
\mathcal{T}(s) = N\frac{b(s)}{\alpha_c(s)}.
$$

13. If the reference input is introduced in such a way as to permit the zeros to be assigned as the roots of $\gamma(s)$, what will the overall closed-loop transfer function be?

$$
\mathcal{T}(s) = N\frac{\gamma(s)b(s)}{\alpha_e(s)\alpha_c(s)},
$$

usually $\gamma(s) = \alpha_e(s)$.

14. What are the three standard techniques for introducing integral control in the state feedback design method?

   (a) *By augmenting the process state to include an integrator state variable.*

   (b) *By the internal model approach.*

   (c) *By using the extended estimator approach.*

CHAPTER 8

1. What is the Nyquist rate? What are its characteristics?
The Nyquist rate is half the sample rate, or $= \omega_s/2$. Above this rate, no frequencies can be represented by a sampled signal.

2. Describe the emulation process.
The controller for a system is designed as if the controller will be analog. The resulting controller is then approximated by a digital equivalent.

3. Describe how to arrive at a $D(z)$ if the sample rate is $30 \times \omega_{BW}$.
Use of the emulation process typically yields satisfactory results. But after using emulation, check the result using a simulation that includes the effect of sampling or else perform an exact discrete linear analysis. It is best to use a simulation that includes all known sampling effects and system delays.

4. Describe how to arrive at a $D(z)$ if the sample rate is $5 \times \omega_{BW}$.
Start by using emulation but include an approximation of the effect of the delay in the plant model when carrying out the analog design. Then check the result via an exact discrete analysis by converting the plant to its discrete equivalent and combining that with the discrete controller. If performance is degraded from that desired, modify the discrete controller using discrete design methods. Finish by using a simulation that includes all known sampling effects and system delays.

5. For a system with a 1-rad/sec bandwidth, describe the consequences of various sample rates.
An absolute minimum sample rate is 2 rad/sec (or 0.32 Hz and $T = 3$ sec). From 2 rad/sec to 10 or 20 rad/sec, the control will be jerky with noticeable steps in the control and the design needs to be done very carefully. Between 20 and 30 rad/sec, the magnitude of the control steps become progressively smaller and design using emulation works reasonably well. Above 30 rad/sec, the control steps are hardly noticeable and emulation can be used with confidence.

6. How can the effects of quantization in an A/D converter be studied?
The roundoff error from the quantization can be considered to be a random input to the system and can be studied by propagating that random error through the system. There may also be an offset error from the quantization which is studied by multiplying the quantization level by the DC gain from the A/D converter to the system output.

7. How can the effects of parameter quantization in the controller be minimized?

(a) *Using a large number of bits to represent the parameters,*

(b) *implementing a high-order controller in parallel low-order blocks, or*

(c) *using floating point arithmetic.*

8. Give two advantages for selecting a digital processor rather than analog circuitry to implement a controller.

(a) *The physical layout of a digital controller can be done before the final design is complete, often resulting in completing the hardware implementation in much less time than required to get an analog controller specified and constructed.*

(b) *A digital processor is more flexible in making design changes because it is easier to re-program software than it is to rewire and/or add op-amps to a printed circuit board.*

(c) *A digital processor can much more easily include nonlinear terms and logic decision steps in the overall controller design to permit adaptive control or gain-scheduling, for example.*

(d) *Many models of the same basic controller can be accommodated by simply using different PROMS with the same hardware design. For example, an automobile manufacturer might have one engine controller hardware design for its entire product line, but have a different PROM for each engine/vehicle combination.*

(e) *Digital controllers are less sensitive to temperature variations than are analog controllers.*

9. Give two disadvantages of selecting a digital processor rather than analog circuitry to implement a controller.

(a) *The finite sampling rate of the A/D and D/A converters and the finite compute speed of the processor limit the bandwidth of the controller to about 1/10 of the sample frequency.*

(b) *The finite accuracy or bit length of the converters introduce extra noise or offsets into the control loop if using low-end controllers.*

(c) *Cost. For simple controllers, a digital implementation will typically be more expensive than an analog implementation.*

CHAPTER 9

1. Why is a collocated actuator and sensor arrangement for a lightly damped structure such as a robot arm easier to design than a noncollocated setup?
In the collocated case, the process naturally has zeros near the lightly damped poles which keep the root locus in the left half-plane.

2. Why should the control engineer be involved in the design of the process to be controlled?

In many cases, the characteristics and locations of the actuators and sensors can have a major impact on the complexity and difficulty in design of the controller. If the needs of control are included in the process design, the final systems are often more effective (better closed-loop performance) and less expensive.

3. Give examples of an actuator and a sensor for the following control problems:

(a) Attitude control of a geosynchronous communication satellite.
Actuators: Cold gas-jet thrusters, momentum wheels, magnetic torquers (coils, torque rod), plasma thruster.
Sensors: Earth sensor (roll, pitch), digital integrated rate assembly (DIRA) gyro (for rates), star tracker.

(b) Pitch control of a Boeing 747 airliner.
Actuators: Elevator.
Sensors: Pitch rate and/or pitch angle is measured using a gyro or a ring-laser gyro.

(c) Track-following control of a CD player.
Actuators: DC motor to move the (dual stage sledge) arm mechanism, magnetic coils (two) for focusing on tracks.
Sensors: Array of photodiodes.

(d) Fuel/air ratio control of a spark-ignited automobile engine.
Actuators: Fuel injection.
Sensors: Zirconium oxide sensor.

(e) Position control for an arm of a robot used to paint automobiles.
Actuators: Hydraulic actuators or electric motors.
Sensors: Encoders to measure arm rotations, pressure sensors, and force sensors.

(f) Heading control of a ship.
Actuators: Rudder.
Sensors: Gyrocompass.

(g) Attitude control of a helicopter.
Actuators: Moving swash plate (either via direct link or servo) rotates main blade angle of attack.
Sensors: Same as aircraft (pitot tube, accelerometers, rate gyros).

References

Abramovitch, D., and G. F. Franklin, A Brief History of Disk Drive Control, to appear in the *IEEE Control System Magazine, 2002*.

Ackermann, J., Der Entwurf Linearer Regelungssysteme im Zustandsraum, *Regelungstech. Prozess-Datenverarb.*, vol. 7, pp. 297–300, 1972.

Airy, G. B., On the Regulator of the Clock-work for Effecting Uniform Movement of Equatorials, *Mem. R. Astron. Soc.*, vol. 11, pp. 249–267, 1840.

Anderson, B. D. O., and J. B. Moore, *Optimal Control: Linear Quadratic Methods*. Englewood Cliffs, NJ: Prentice-Hall, 1990.

Anderson, E., et al., *LAPACK User's Guide*, 3rd ed., Philadelphia: SIAM, 1999.

Ashley, H., *Engineering Analysis of Flight Vehicles*. Reading, MA: Addison-Wesley, 1974.

Åström, K. J., Frequency Domain Properties of Otto Smith Regulators, *Int. J. Control*, vol. 26, no. 2, pp. 307–314, 1977.

Åström, K. J., *Introduction to Stochastic Control Theory*. New York: Academic Press, 1970.

Åström, K. J., and T. Hägglund, *PID Controllers: Theory, Design, and Tuning*, 2nd ed., Research Triangle Park, NC, ISA: 1995.

Athans, M., A Tutorial on the LQG/LTR Method, in *Proc. Am. Control Conf.*, pp. 1289–1296, June 1986.

Bellman, R., *Dynamic Programming*, Princeton, NJ: Princeton University Press, 1957.

Bellman, R., and R. Kalaba, Eds., *Mathematical Trends in Control Theory*. New York: Dover, 1964.

Blakelock, J. H., *Automatic Control of Aircraft and Missiles*, 2nd ed., New York: John Wiley, 1991.

Bode, H. W., Feedback: The History of an Idea, in *Conf. on Circuits and Systems*, New York, 1960; reprinted in Bellman and Kalaba (1964).

Bode, H.W., *Network Analysis and Feedback Amplifier Design*. New York: Van Nostrand, 1945.

Boyd, S. P., and C. H. Barratt, *Linear Controller Design: Limits of Performance*. Englewood Cliffs, NJ: Prentice-Hall, 1991.

Brown, J. W., and R. V. Churchill, *Complex Variables and Applications*. 6th ed., New York: McGraw-Hill, 1996.

Bryson, A. E., Jr., and W. F. Denham, A Steepest-Ascent Method for Solving Optimum Programming Problems, *J. Appl. Mech.*, June 1962.

Bryson, A. E., Jr. and Y. C. Ho, *Applied Optimal Control*. Waltham, MA: Blaisdell, 1969.

Bryson, A. E., Jr. *Control of Spacecraft and Aircraft*. Princeton, NJ: Princeton University Press, 1994.

Buckley, P. S., *Techniques of Process Control*. New York: John Wiley & Sons, 1964.

Butterworth, S., On the Theory of Filter Amplifiers, *Wireless Eng.*, vol. 7, pp. 536–541, 1930.

Callender, A., D. R. Hartree, and A. Porter, Time Lag in a Control System, *Philos. Trans. R. Soc. London A*, vol. 235, pp. 415–444, 1936.

Campbell, D. P., *Process Dynamics*. New York: John Wiley & Sons, 1958.

Campbell, G. A., and R. N. Foster, *Fourier Integrals for Practical Applications*. New York: Van Nostrand, 1948.

Cannon, R. H., Jr., *Dynamics of Physical Systems*. New York: McGraw-Hill, 1967.

Churchill, R. V., *Operational Mathematics*, 3rd ed. New York: McGraw-Hill, 1972.

Clark, R. N., *Introduction to Automatic Control Systems*. New York: John Wiley & Sons, 1962.

Craig, J. J., *Introduction to Robotics: Mechanics and Control*. 2nd ed., Reading, MA: Addison-Wesley, 1989.

de Roover, D., A. Emami-Naeini, J. L. Ebert, Model-Based Control of Fast-Ramp RTP Systems, *6th International Conference on Advanced Thermal Processing of Semiconductors*, pp. 177–186, Kyoto, Japan, September 1998.

Dorato, P., *Analytic Feedback System Design: An Interpolation Approach*. Pacific Grove, CA, Brooks/Cole, 2000.

Dorf, R. C., and R. Bishop, *Modern Control Systems*, 9th ed., Upper Saddle River, NJ: Prentice-Hall, 2001.

Doyle, J. C., B. A. Francis, and A. Tannenbaum, *Feedback Control Theory*. New York: Macmillan, 1992.

Doyle, J. C., and G. Stein, Multivariable Feedback Design: Concepts for a Classical/Modern Synthesis, *IEEE Trans. Autom. Control*, vol. AC–26, no. 1, pp. 4–16, February 1981.

Doyle, J. C., Guaranteed Margins for LQG Regulators, *IEEE Trans. Automat. Control*, AC–23, pp.756–757, 1978.

Doyle, J. C., and G. Stein, Robustness with Observers, *IEEE Trans. on Automat. Control*, pp. 607–611, vol. AC–24, August 1979.

Ebert, J. L., A. Emami-Naeini, H. Aling, and R. L. Kosut, Thermal Modeling of Rapid Thermal Processing Systems, in *3rd International Rapid Thermal Processing Conference*, pp. 343–355, Amsterdam, August 1995a.

Ebert, J. L., A. Emami-Naeini, and R. L. Kosut, Thermal Modeling and Control of Rapid Thermal Processing Systems, *Proc. 34th IEEE Conf. Dec. Control*, pp. 1304–1309, December 1995b.

Elgerd, O. I., *Electric Energy Systems Theory*. New York: McGraw-Hill, 1982.

Elgerd, O.I., and W. C. Stephens, Effect of Closed-loop Transfer Function Pole and Zero Locations on the Transient Response of Linear Control Systems, *Trans. Am. Inst. Electr. Eng. Part 1*, vol. 42, pp. 121–127, 1959.

Emami-Naeini, A., and G. F. Franklin, Zero Assignment in the Multivariable Robust Servomechanism, in *Proc. IEEE Conf. Dec. Control*, December 1982, pp. 891–893.

Emami-Naeini, A., and P. Van Dooren, Computation of Zeros of Linear Multivariable Systems, *Automatica*, vol. 18, no. 4, pp. 415–430, 1982a.

Emami-Naeini, A., and P. Van Dooren, On Computation of Transmission Zeros and Transfer Functions, in *Proc. IEEE Conf. Dec. Control*, pp. 51–55, December 1982b.

Etkin, B., and L. D. Reid, *Dynamics of Flight: Stability and Control*. 3rd ed., New York: John Wiley & Sons, 1996.

Evans, W. R., Graphical Analysis of Control Systems, *Trans. Am. Inst. Electr. Eng.*, vol. 67, pp. 547–551, 1948.

Francis, B. A., A Course in H_∞ Control Theory, in *Lecture Notes in Control and Information Sciences*, vol. 88. Berlin: Springer-Verlag, 1987.

Francis, B. A., and W. M. Wonham, The Internal Model Principle of Control Theory, *Automatica*, vol. 12, pp. 457–465, 1976.

Francis, J. G. F., The QR Transformation: A Unitary Analogue to the LR Transformation, Parts I and II, *Comput. J.*, vol. 4, pp. 265–272, 332–345, 1961.

Franklin, G. F., and A. Emami-Naeini, A New Formulation of the Multivariable Robust Servomechanism Problem, Internal report, Information Systems Laboratory, Stanford University, Stanford, CA, July 1983.

Franklin, G. F., and A. Emami-Naeini, Robust Servomechanism Design Applied to Control of Reel-to-reel Digital Tape Transports, in *Proc. Asilomar Conf.*, pp. 108–113, Asilomar, CA, 1981.

Franklin, G. F., J. D. Powell, and M. L. Workman, *Digital Control of Dynamic Systems*, 3rd ed. Reading, MA: Addison-Wesley, 1998.

Freudenberg, J. S., and D. P. Looze, Right Half Plane Zeros and Design Trade-offs in Feedback Systems, *IEEE Trans. Automat. Control*, vol. AC–30, pp. 555–561, June 1985.

Fuller, A. T., The Early Development of Control Theory, *J. Dyn. Syst. Meas. Control*, vol. 98, pp. 109–118, 224–235, 1976.

Gantmacher, F. R., *The Theory of Matrices*, vols. I and II. New York: Chelsea Publishing, 1959.

Gardner, M. F., and J. L. Barnes, *Transients in Linear Systems*. New York: John Wiley & Sons, 1942.

Goodwin, G. C., S. F. Graebe, and M. E. Salgado, *Control System Design,* Upper Saddle River, NJ: Prentice-Hall, 2001.

Gopinath, B., On the Control of Linear Multiple Input–Output Systems, *Bell Syst. Tech. J.*, vol. 50, pp. 1063–1081, March 1971.

Graham, D. and R. C. Lathrop, The Synthesis of Optimum Response: Criteria and Standard Forms, *Trans. Am. Inst. Electr. Eng.*, vol. 72, pt. 1, pp. 273–288, 1953.

Gunckel, T. L., III and G. F. Franklin, A General Solution for Linear Sampled Data Control, *J. Basic Eng.*, vol. 85-D, pp. 197–201, 1963.

Gyugyi, P., Y. Cho, G. F. Franklin, and T. Kailath, Control of Rapid Thermal Processing: A System Theoretic Approach, in *Proc. IFAC World Congress*, 1993.

Halliday, D., R. Resnick, and J. Walker, *Fundamentals of Physics*, 6th ed. New York: John Wiley & Sons, 2001.

Hang, C. C., K. J. Åström, and W. K. Ho, Refinements of the Ziegler-Nichols Tuning Formula, Rep. CI–90–1, National University of Singapore, Dept. of Electrical Engineering, April 1990.

Hanselman, D. C., and B. C. Littlefield, *Mastering MATLAB 6*. Upper Saddle River, NJ: Prentice-Hall, 2001.

Heffley, R. K., and W. F. Jewell, *Aircraft Handling Qualities,* Tech. Rep. 1004–1, System Technology, Inc., Hawthorne, CA, May 1972.

Higham, D. S., and N. J. Higham, *MATLAB Guide*. Philadelphia, PA: SIAM, 2000.

Ho, M.-T., A. Datta, and S. P. Bhattacharyya, An Elementary Derivation of the Routh-Hurwitz Criterion, *IEEE Trans. Automat. Control*, pp. 405–409, 1998.

Huang, J-J., and D. B. DeBra, Automatic Tuning of Smith-Predictor Design Using Optimal Parameter Mismatch, in *Proc. IEEE Conf. Dec. Control*, pp. 3307–3312, December 2000.

Hubbard, M., Jr., and J. D. Powell, Closed-Loop Control of Internal Combustion Engine Exhaust Emissions, SUDAAR No. 473, Deptartment of Aero/Astro, Stanford University, Stanford, CA, February 1974.

Hurewicz, W., Filters and Servo Systems with Pulsed Data, in James et al. (1947) Chapter 5, *Theory of Servomechanisms*, Radiation Lab. Series, 25. New York: McGraw-Hill, 1947.

James, H. M., N. B. Nichols, and R. S. Phillips, *Theory of Servomechanisms*, Radiation Lab. Series, 25. New York: McGraw-Hill, 1947.

Johnson, R. C., Jr., A. S. Foss, G. F. Franklin, R. V. Monopoli, and G. Stein, Toward Development of a Practical Benchmark Example for Adaptive Control, *IEEE Control Syst. Mag.*, vol. 1, no. 4, pp. 25–28, December 1981.

Kailath, T., *Linear Systems*, Englewood Cliffs, NJ: Prentice-Hall, 1980.

Kalman, R. E., A New Approach to Linear Filtering and Prediction Problems, *J. Basic Eng.*, vol. 85, pp. 34–45, 1960.

Kalman, R. E., On the General Theory of Control Systems, in *Proc. 1st Int. Congr. Autom. Control*, Moscow, pp. 481–493, 1960.

Kalman, R. E., and J. E. Bertram: Control System Analysis and Design via the Second Method of Lyapunov. II. Discrete Systems, *J. Basic Eng.*, vol. 82, pp. 394–400, 1960.

Kalman, R. E., Y. C. Ho, and K. S. Narendra, Controllability of Linear Dynamical Systems, in *Contributions to Differential Equations*, vol. 1. New York: John Wiley & Sons, 1962.

Kane, T. R., and D. A. Levinson, *Dynamics; Theory and Application*, New York: McGraw-Hill, 1985.

Kautsky, J., N. K. Nichols, and P. Van Dooren, Robust Pole Assignment in Linear State Feedback, *Int. J. Control*, vol. 41, no. 5, pp. 1129–1155, 1985.

Khalil, H. K., *Nonlinear Systems*, 2nd ed., Englewood Cliffs, NJ: Prentice-Hall, 1996.

Kochenburger, R. J., A Frequency Response Method for Analyzing and Synthesizing Contractor Servomechanisms, *Trans. Am. Inst. Electr. Eng.*, vol. 69, pp. 270–283, 1950.

Kuo, B. C., Ed., *Proc. Symp. Incremental Motion Control Systems and Devices, Pt. 1: Step Motors and Controls*. Champaign-Urbana: University of Illinois, 1972.

Kuo, B. C., Ed., *Incremental Motion Control*, vol. 2: *Step Motors and Control Systems*. Champaign, IL: SRL Publishing, 1980.

Kuo, B. C., *Automatic Control Systems*, 7th ed., New York: John Wiley & Sons, 1995.

Lanchester, F. W., *Aerodometics*. London: Archibald Constable, 1908.

LaSalle, L. P., and S. Lefschetz, *Stability by Lyapunov's Direct Method*. New York: Academic Press, 1961.

Laub, A. J., A. Linnemann, and M. Wette, Algorithms and Software for Pole Assignment by State Feedback, in *Proc. 2nd Symp. CACSD,* March 1985.

León de la Barra, B. A., A. Emami-Naeini, and E. R. Chinchón, Linear Multivariable Servomechanisms Revisited: System Type and Accuracy Trade-offs, *Automatica*, vol. 34, no. 11, pp. 1449–1452, 1998.

Leonard, N. E., and W. S. Levine, *Using MATLAB to Analyze and Design Control System,* 2nd ed., Reading, MA: Addison-Wesley, 1995.

Lyapunov, A. M., "Problème général de la stabilité du mouvement," *Ann. Fac. Sci. Univ. Toulouse Sci. Math. Sci. Phys.*, vol. 9, pp. 203–474, 1907; original paper published in 1893 in *Commun. Soc. Math. Kharkow*, 1893; reprinted as vol. 17 in *Annals of Math Studies.* Princeton, NJ: Princeton University Press, 1949.

Ljüng, L., *System Identification: Theory for the User*, 2nd ed., Englewood Cliffs, NJ: Prentice-Hall, 1999.

Ljüng, L., and T. Söderström, *Theory and Practice of Recursive Identification.* Cambridge, MA: MIT Press, 1983.

Luenberger, D. G., Observing the State of a Linear System, *IEEE Trans. Mil. Electron.*, vol. MIL–8, pp. 74–80, 1964.

Maciejowski, J. M., *Multivariable Feedback Design.* Reading, MA: Addison-Wesley, 1989.

Marsden, J. E., and M. J. Hoffman, *Basic Complex Analysis*, 3rd ed., Freeman, 1998.

Mason, S. J., Feedback Theory: Some Properties of Signal Flow Graphs, *Proc. IRE*, vol. 41, pp. 1144–1156, 1953.

Mason, S. J., Feedback Theory: Further Properties of Signal Flow Graphs, *Proc. IRE*, vol. 44, pp. 920–926, 1956.

Maxwell, J. C., On Governors, *Proc. R. Soc. London*, vol. 16, pp. 270–283, 1868.

Mayr, O., *The Origins of Feedback Control.* Cambridge, MA: MIT Press, 1970.

McRuer, D. T., I. Askenas, and D. Graham, *Aircraft Dynamics and Automatic Control.* Princeton, NJ: Princeton University Press, 1973.

Mees, A. I., *Dynamics of Feedback Systems.* New York: John Wiley & Sons, 1981.

Messner, W. C., and D. M. Tilburry, *Control Tutorials for MATLAB and Simulink: A Web-Based Approach*, Reading, MA: Addison-Wesley, 1999.

Minimis, G. S., and C. C. Paige, An Algorithm for Pole Assignment of Time Invariant Systems, *Int. J. Control*, vol. 35, no. 2, pp. 341–354, 1982.

Misra, P., and R. V. Patel, Numerical Algorithms for Eigenvalue Assignment by Constant and Dynamic Output Feedback, *IEEE Trans. Autom. Control*, vol. AC–34, no. 6, pp. 577–588, 1989.

Moler, C., and C. van Loan, Nineteen Dubious Ways to Compute the Exponential of a Matrix, *SIAM Rev.*, vol. 20, no. 4, pp. 801–836, 1978.

Norman, S. A., Wafer Temperature Control in Rapid Thermal Processing, Ph.D. dissertation, Stanford University, 1992.

Nyquist, H., Regeneration Theory, *Bell Syst. Technol. J.*, vol. 11, pp. 126–147, 1932.

Ogata, K., *Modern Control Engineering*, 3rd ed., Englewood Cliffs, NJ: Prentice-Hall, 1997.

Oswald, R. K., Design of a Disk File Head Positioning Servo, *IBM Journal of Research and Development*, vol. 18, pp. 506–512, November, 1974.

Panagopoulous, H., K. Åström, and T. Hagglund, Design of PID Controllers Based on Non-Convex Optimization, *Proceedings of American Control Conference*, pp. 3858–3862, San Diego, CA, June 1999.

Perkins, W. R., P. V. Kokotovic, T. Boureret, and J. L. Schiano, Sensitivity Function Methods in Control System Education, *IFAC Conf. on Control Educ.*, June 1991.

Pontryagin, L. S., V. G. Boltyanskii, R. V. Gamkrelidze, and E. F. Mishchenko, *The Mathematical Theory of Optimal Processes*. New York: John Wiley & Sons, 1962.

Ragazzini, J. R., R. H. Randall, and F. A. Russell, Analysis of Problems in Dynamics by Electronic Circuits, *Proc. IRE*, vol. 35, no. 5, pp. 442–452, May 1947.

Ragazzini, J. R., and G. F. Franklin, *Sampled-Data Control Systems*, New York: McGraw-Hill, 1958.

Reliance Motion Control Corp., *DC Motor Speed Controls Servo Systems* 5th ed. Eden Prairie, MN: Reliance Motion Control Corp., 1980.

Rosenbrock, H. H., *State Space and Multivariable Theory*. New York: John Wiley & Sons, 1970.

Routh, E. J., *Dynamics of a System of Rigid Bodies*. London: Macmillan, 1905.

Saberi, A., B. M. Chen, P. Sannuti, *Loop Transfer Recovery: Analysis and Design*. Berlin: Springer-Verlag, 1993.

Safonov, M. G., A. J. Laub, and G. Hartmann, Feedback Properties of Multivariable Systems: The Role and Use of Return Difference Matrix, *IEEE Trans. Autom. Control*, vol. AC–26, pp. 47–65, 1981.

Sastry, S. S., *Nonlinear Systems; Analysis, Stability, and Control*. Berlin: Springer-Verlag, 1999.

Schmitz, E., Robotic Arm Control, Ph.D. thesis, Deptartment of Aero/Astro, Stanford University, Stanford, CA, 1985.

Shampine, L. F., and M. K. Gordon, *Computer Solution of Ordinary Differential Equations: The Initial Value Problem*. New York: W. H. Freeman, 1975.

Sinha, N. K., and B. Kuszta, *Modeling and Identification of Dynamic Systems*. New York: Van Nostrand, 1983.

Smith, O. J. M., *Feedback Control Systems*. New York: McGraw-Hill, 1958.

Smith, R. J., *Electronics: Circuits and Devices*, 4th ed. New York: John Wiley & Sons, 1987.

Stein, G., and M. Athans, The LQG/LTR Procedure for Multivariable Feedback Control Design, *IEEE Trans. Automat Control*, vol. AC–32, pp. 105–114, February 1987.

Strang, G., *Linear Algebra and Its Applications*, 3rd ed. New York: Harcourt Brace, 1988.

Swift, J., *On Poetry: A Rhapsody*, 1973, in *Familiar Quotations*, 15th ed., J. Bartlett, Ed. Boston: Little Brown, 1980.

Sze, S. M., Ed., *VLSI Technology*, 2nd ed., New York: McGraw-Hill, 1988.

Thomson, W. T., and M. D. Dahleh, *Theory of Vibration with Applications*, 5th ed., Englewood Cliffs, NJ: Prentice-Hall, 1998.

Trankle, T. L., Development of WMEC Tampa Maneuvering Model from Sea Trial Data, Rep. MA-RD–760–87201. Palo Alto, CA: Systems Control Technology, March 1987.

Truxal, J. G., *Control System Synthesis*. New York: McGraw-Hill, 1955.

van der Linden, G., J. L. Ebert, A. Emami-Naeini, and R. L. Kosut, RTP Robust Control Design: Part II: Controller Synthesis, in *4th International Rapid Thermal Processing Conference*, pp. 263–271, September 1996.

Van der Pol, B., and H. Bremmer, *Operational Calculus*. New York: Cambridge University Press, 1955.

Van Dooren, P., A. Emami-Naeini, and L. Silverman, Stable Extraction of the Kronecker Structure of Pencils, in *Proc. IEEE Conf. Dec. Control*, San Diego, pp. 521–524, December 1978.

Vidyasagar, M., *Nonlinear Systems Analysis*, 2nd ed., Englewood Cliffs, NJ: Prentice-Hall, 1993.

Wiener, N., *The Extrapolation, Interpolation and Smoothing of Stationary Time Series*. New York: John Wiley, 1949.

Wiener, N., Generalized Harmonic Analysis, *Acta Math.*, vol. 55, pp. 117, 1930.

Woodson, H. H., and J. R. Melcher, *Electromechanical Dynamics*, Part I: *Discrete Systems*. New York: John Wiley & Sons, 1968.

Workman, M. L., Adaptive Proximate Time-Optimal Servomechanisms, Ph.D. dissertation, Stanford University, 1987.

Zames, G., On the Input–Output Stability of Time-varying Nonlinear Feedback Systems—Part I: Conditions Derived Using Concepts of Loop Gain, Conicity and Positivity, *IEEE Trans. Autom. Control*, vol. AC–11, pp. 465–476, 1966.

Zames, G., On the Input-output Stability of Time-varying Nonlinear Feedback Systems—Part II: Conditions Involving Circles in the Frequency Plane and Sector Nonlinearities, *IEEE Trans. Autom. Control*, vol. AC–11, pp. 228–238, 1966.

Zhou, K. and J. C. Doyle, *Essentials of Robust Control*. Englewood Cliffs, NJ: Prentice-Hall, 1998.

Ziegler, J. G., and N. B. Nichols, Optimum Settings for Automatic Controllers, *Trans. ASME*, vol. 64, pp. 759–768, 1942.

Ziegler, J. G., and N. B. Nichols, Process Lags in Automatic Control Circuits, *Trans. ASME*, vol. 65, no. 5, pp. 433–444, July 1943.

Index

Commonly Used MATLAB Commands

MATLAB Function (.m file)	Description	Page (s)
acker	Ackermann's formula for pole-placement	522, 530, 544, 547, 681
bode	Bode frequency response	102, 369, 384, 456, 572
c2d	Continuous to discrete conversion	250, 423, 566, 581, 661
canon	State-space canonical forms	507
conv	Polynomial multiplication	111, 214, 534, 536, 822
damp	Damping and natural frequency	674, 675
eig	Eigenvalues and eigenvectors	504, 507, 679
feedback	Feedback connection of two systems	128, 324, 674
gram	Controllability/observability Gramian	854, 856
impulse	Impulse response	138, 144, 154
initial	Initial condition response	544, 551
inv	Matrix inverse	506, 508
logspace	Logarithmically-spaced frequency points	102
lqe	Linear Quadratic Estimator design	572, 789
lqr	Linear Quadratic Regulator design	537, 538, 788
lsim	Linear system simulation	122, 171
margin	Gain and phase margins	427, 472, 572
nyquist	Nyquist plot	395, 399, 401
ode23	Solution to non-stiff differential equations	172
pade	Pade approximation for time delay	332
parallel	Parallel connection of two systems	127
place	Pole-placement	524, 547, 562, 568, 595
plot	Plot function	34, 119, 122, 215
poly	Form polynomial from its roots	115
pzmap	Pole-zero map	137
residue	Residues in partial fraction expansion	111, 115, 822
rlocfind	Find root locus gain	304, 472
rlocus	Root locus	276, 472, 534, 536
rltool	Interactive root locus tool	350
roots	Roots of a polynomial	162, 296, 514, 524
series	Series connection of two systems	127, 572, 674
ss2tf	State-space to transfer function	118, 121, 510, 513
ss2zp	State-space to pole-zero conversion	121, 135
ss	Conversion to state-space	44, 122, 507, 512, 515
ssdata	Create a state-space model	681
step	Step response	34, 44, 119, 162, 214, 682
tf2ss	Transfer function to state-space	120, 497
tf2zp	Transfer function to pole-zero conversion	121, 135
tf	Conversion to transfer function	34, 102, 127, 138, 154, 214, 534
tzero	Transmission zeros	512, 514, 680